Encyclopedia of VIROLOGY

THIRD EDITION

EDITORS-IN-CHIEF

Dr BRIAN W J MAHY
and
Dr MARC H V VAN REGENMORTEL

AMSTERDAM • BOSTON • HEIDELBERG • LONDON • NEW YORK • OXFORD
PARIS • SAN DIEGO • SAN FRANCISCO • SINGAPORE • SYDNEY • TOKYO
Academic Press is an imprint of Elsevier

ACADEMIC PRESS

Academic Press is an imprint of Elsevier
Linacre House, Jordan Hill, Oxford, OX2 8DP, UK
525 B Street, Suite 1900, San Diego, CA 92101-4495, USA

Copyright © 2008 Elsevier Inc. All rights reserved

The following articles are US government works in the public domain and are not subject to copyright:
Bovine Viral Diarrhea Virus, Coxsackieviruses, Prions of Yeast and Fungi, Human Respiratory Syncytial Virus, Fish Rhabdoviruses, Varicella-Zoster Virus: General Features, Viruses and Bioterrorism, Bean Common Mosaic Virus and Bean Common Mosaic Necrosis Virus, Metaviruses, Crimean-Congo Hemorrhagic Fever Virus and Other Nairoviruses, AIDS: Global Epidemiology, Papaya Ringspot Virus, Transcriptional Regulation in Bacteriophage.

Nepovirus, Canadian Crown Copyright 2008

No part of this publication may be reproduced, stored in a retrieval system or transmitted in any form or by any means electronic, mechanical, photocopying, recording or otherwise without the prior written permission of the publisher

Permissions may be sought directly from Elsevier's Science & Technology Rights Department in Oxford, UK: phone (+44) (0) 1865 843830; fax (+44) (0) 1865 853333; email: permissions@elsevier.com. Alternatively you can submit your request online by visiting the Elsevier web site at (http://elsevier.com/locate/permission), and selecting *Obtaining permission to use Elsevier material*

Notice
No responsibility is assumed by the publisher for any injury and/or damage to persons or property as a matter of products liability, negligence or otherwise, or from any use or operation of any methods, products, instructions or ideas contained in the material herein. Because of rapid advances in the medical sciences, in particular, independent verification of diagnoses and drug dosages should be made

British Library Cataloguing in Publication Data
A catalogue record for this book is available from the British Library

Library of Congress Catalog Number: 200892260

ISBN: 978-0-12-373935-3

For information on all Elsevier publications
visit our website at books.elsevier.com

PRINTED AND BOUND IN SLOVENIA
08 09 10 11   10 9 8 7 6 5 4 3 2 1

Working together to grow
libraries in developing countries

www.elsevier.com | www.bookaid.org | www.sabre.org

ELSEVIER   BOOK AID International   Sabre Foundation

Encyclopedia of VIROLOGY
THIRD EDITION

EDITORS-IN-CHIEF

Brian W J Mahy MA PhD ScD DSc
Senior Scientific Advisor,
Division of Emerging Infections and Surveillance Services,
Centers for Disease Control and Prevention,
Atlanta GA, USA

Marc H V Van Regenmortel PhD
Emeritus Director at the CNRS,
French National Center for Scientific Research,
Biotechnology School of the University of Strasbourg,
Illkirch, France

ASSOCIATE EDITORS

Dennis H Bamford, Ph.D.
Department of Biological and Environmental Sciences
and Institute of Biotechnology, Biocenter 2,
P.O. Box 56 (Viikinkaari 5),
00014 University of Helsinki,
Finland

Charles Calisher, B.S., M.S., Ph.D.
Arthropod-borne and Infectious Diseases Laboratory
Department of Microbiology, Immunology and Pathology
College of Veterinary Medicine and Biomedical Sciences
Colorado State University
Fort Collins
CO 80523
USA

Andrew J Davison, M.A., Ph.D.
MRC Virology Unit
Institute of Virology
University of Glasgow
Church Street
Glasgow G11 5JR
UK

Claude Fauquet
ILTAB/Donald Danforth Plant Science Center
975 North Warson Road
St. Louis, MO 63132

Said Ghabrial, B.S., M.S., Ph.D.
Plant Pathology Department
University of Kentucky
201F Plant Science Building
1405 Veterans Drive
Lexington
KY 4050546-0312
USA

Eric Hunter, B.Sc., Ph.D.
Department of Pathology and Laboratory Medicine, and
Emory Vaccine Center
Emory University
954 Gatewood Road NE
Atlanta Georgia 30329
USA

Robert A. Lamb, Ph.D., Sc.D.
Department of Biochemistry,
Molecular Biology and Cell Biology
Howard Hughes Medical Institute
Northwestern University
2205 Tech Dr.
Evanston
IL 60208-3500
USA

Olivier Le Gall
IPV, UMR GDPP, IBVM,
INRA Bordeaux-Aquitaine, BP 81,
F-33883 Villenave d'Ornon Cedex
FRANCE

Vincent Racaniello, Ph.D.
Department of Microbiology
Columbia University
New York, NY 10032
USA

David A. Theilmann, Ph.D., B.Sc., M.Sc
Pacific Agri-Food Research Centre
Agriculture and Agri-Food Canada
Box 5000, 4200 Highway 97
Summerland
BC V0H 1Z0
Canada

H. Josef Vetten, Ph.D.
Julius Kuehn Institute, Federal Research Centre for
Cultivated Plants (JKI)
Messeweg 11-12
38104 Braunschweig
Germany

Peter J Walker, B.Sc., Ph.D.
CSIRO Livestock Industries
Australian Animal Health Laboratory (AAHL)
Private Bag 24
Geelong
VIC 3220
Australia

PREFACE

This third edition of the *Encyclopedia of Virology* is being published nine years after the second edition, a period which has seen enormous growth both in our understanding of virology and in our recognition of the viruses themselves, many of which were unknown when the second edition was prepared. Considering viruses affecting human hosts alone, the worldwide epidemic of severe acute respiratory syndrome (SARS), caused by a previously unknown coronavirus, led to the discovery of other human coronaviruses such as HKU1 and NL63. As many as seven chapters are devoted to the AIDS epidemic and to human immunodeficiency viruses. In addition, the development of new molecular technologies led to the discovery of viruses with no obvious disease associations, such as torque-teno virus (one of the most ubiquitous viruses in the human population), human bocavirus, human metapneumovirus, and three new human polyomaviruses.

Other new developments of importance to human virology have included the introduction of a virulent strain of West Nile virus from Israel to North America in 1999. Since that time the virus has become established in mosquito, bird and horse populations throughout the USA, the Caribbean and Mexico as well as the southern regions of Canada.

As in the two previous editions, we have tried to include information about all known species of virus infecting bacteria, fungi, invertebrates, plants and vertebrates, as well as descriptions of related topics in virology such as antiviral drug development, cell- and antibody-mediated immunity, vaccine development, electron microscopy and molecular methods for virus characterization and identification. Many chapters are devoted to the considerable economic importance of virus diseases of cereals, legumes, vegetable crops, fruit trees and ornamentals, and new approaches to control these diseases are reviewed.

General issues such as the origin, evolution and phylogeny of viruses are also discussed as well as the history of the different groups of viruses.

To cover all these subjects and new developments, we have had to increase the size of the Encyclopedia from three to five volumes.

Throughout this work we have relied upon the 8th Report of the International Committee on Taxonomy of Viruses published in 2005, which lists more than 6000 viruses classified into some 2000 virus species distributed among more than 390 different genera and families. In recent years the criteria for placing viruses in different taxa have shifted away from traditional serological methods and increasingly rely upon molecular techniques, particularly the nucleotide sequence of the virus genome. This has changed many of the previous groupings of viruses, and is reflected in this third edition.

Needless to say, a work of this magnitude has involved many expert scientists, who have given generously of their time to bring it to fruition. We extend our grateful thanks to all contributors and associate editors for their excellent and timely contributions.

Brian W J Mahy
Marc H V van Regenmortel

HOW TO USE THE ENCYCLOPEDIA

Structure of the Encyclopedia

The major topics discussed in detail in the text are presented in alphabetical order (see the Alphabetical Contents list which appears in all five volumes).

Finding Specific Information

Information on specific viruses, virus diseases and other matters can be located by consulting the General Index at the end of Volume 5.

Taxonomic Groups of Viruses

For locating detailed information on the major taxonomic groups of viruses, namely virus genera, families and orders, the Taxonomic Index in Volume 5 (page...) should be consulted.

Further Reading sections

The articles do not feature bibliographic citations within the body of the article text itself. The articles are intended to be a first introduction to the topic, or a 'refresher', readable from beginning to end without referring the reader outside of the encyclopedia itself. Bibliographic references to external literature are grouped at the end of each article in a Further Reading section, containing review articles, 'seminal' primary articles and book chapters. These point users to the next level of information for any given topic.

Cross referencing between articles

The "See also" section at the end of each article directs the reader to other entries on related topics. For example. The entry *Lassa, Junin, Machupo and Guanarito Viruses* includes the following cross-references:

See also: Lymphocytic Choriomeningitis Virus: General Features.

CONTRIBUTORS

S T Abedon
The Ohio State University, Mansfield, OH, USA

G P Accotto
Istituto di Virologia Vegetale CNR, Torino, Italy

H-W Ackermann
Laval University, Quebec, QC, Canada

G Adam
Universität Hamburg, Hamburg, Germany

M J Adams
Rothamsted Research, Harpenden, UK

C Adams
University of Duisburg–Essen, Essen, Germany

E Adderson
St. Jude Children's Research Hospital, Memphis, TN, USA

S Adhya
National Institutes of Health, Bethesda, MD, USA

C L Afonso
Southeast Poultry Research Laboratory, Athens, GA, USA

P Ahlquist
University of Wisconsin – Madison, Madison, WI, USA

G M Air
University of Oklahoma Health Sciences Center, Oklahoma City, OK, USA

D J Alcendor
Johns Hopkins School of Medicine, Baltimore, MD, USA

J W Almond
sanofi pasteur, Lyon, France

I Amin
National Institute for Biotechnology and Genetic Engineering, Faisalabad, Pakistan

J Angel
Pontificia Universidad Javeriana, Bogota, Republic of Colombia

C Apetrei
Tulane National Primate Research Center, Covington, LA, USA

B M Arif
Great Lakes Forestry Centre, Sault Ste. Marie, ON, Canada

H Attoui
Faculté de Médecine de Marseilles, Etablissement Français Du Sang, Marseilles, France

H Attoui
Université de la Méditerranée, Marseille, France

H Attoui
Institute for Animal Health, Pirbright, UK

L Aurelian
University of Maryland School of Medicine, Baltimore, MD, USA

L A Babiuk
University of Alberta, Edmonton, AB, Canada

S Babiuk
National Centre for Foreign Animal Disease, Winnipeg, MB, Canada

A G Bader
The Scripps Research Institute, La Jolla, CA, USA

S C Baker
Loyola University of Chicago, Maywood, IL, USA

T S Baker
University of California, San Diego, La Jolla, CA, USA

J K H Bamford
University of Jyväskylä, Jyväskylä, Finland

Y Bao
National Institutes of Health, Bethesda, MD, USA

M Bar-Joseph
The Volcani Center, Bet Dagan, Israel

H Barker
Scottish Crop Research Institute, Dundee, UK

A D T Barrett
University of Texas Medical Branch, Galveston, TX, USA

J W Barrett
The University of Western Ontario, London, ON, Canada

T Barrett
Institute for Animal Health, Pirbright, UK

R Bartenschlager
University of Heidelberg, Heidelberg, Germany

N W Bartlett
Imperial College London, London, UK

S Basak
University of California, San Diego, CA, USA

C F Basler
Mount Sinai School of Medicine, New York, NY, USA

T Basta
Institut Pasteur, Paris, France

D Baxby
University of Liverpool, Liverpool, UK

P Beard
Imperial College London, London, UK

M N Becker
University of Florida, Gainesville, FL, USA

J J Becnel
Agriculture Research Service, Gainesville, FL, USA

K L Beemon
Johns Hopkins University, Baltimore, MD, USA

E D Belay
Centers for Disease Control and Prevention, Atlanta, GA, USA

M Benkő
Veterinary Medical Research Institute, Hungarian Academy of Sciences, Budapest, Hungary

M Bennett
University of Liverpool, Liverpool, UK

M Bergoin
Université Montpellier II, Montpellier, France

H U Bernard
University of California, Irvine, Irvine, CA, USA

K I Berns
University of Florida College of Medicine, Gainesville, FL, USA

P Biagini
Etablissement Français du Sang Alpes-Méditerranée, Marseilles, France

P D Bieniasz
Aaron Diamond AIDS Research Center, The Rockefeller University, New York, NY, USA

Y Bigot
University of Tours, Tours, France

C Billinis
University of Thessaly, Karditsa, Greece

R F Bishop
Murdoch Childrens Research Institute Royal Children's Hospital, Melbourne, VIC, Australia

B A Blacklaws
University of Cambridge, Cambridge, UK

C D Blair
Colorado State University, Fort Collins, CO, USA

S Blanc
INRA–CIRAD–AgroM, Montpellier, France

R Blawid
Institute of Plant Diseases and Plant Protection, Hannover, Germany

G W Blissard
Boyce Thompson Institute at Cornell University, Ithaca, NY, USA

S Blomqvist
National Public Health Institute (KTL), Helsinki, Finland

J F Bol
Leiden University, Leiden, The Netherlands

J-R Bonami
CNRS, Montpellier, France

L Bos
Wageningen University and Research Centre (WUR), Wageningen, The Netherlands

H R Bose Jr.
University of Texas at Austin, Austin, TX, USA

H Bourhy
Institut Pasteur, Paris, France

P R Bowser
Cornell University, Ithaca, NY, USA

D B Boyle
CSIRO Livestock Industries, Geelong, VIC, Australia

C Bragard
Université Catholique de Louvain, Leuven, Belgium

J N Bragg
University of California, Berkeley, Berkeley, CA, USA

R W Briddon
National Institute for Biotechnology and Genetic Engineering, Faisalabad, Pakistan

M A Brinton
Georgia State University, Atlanta, GA, USA

P Britton
Institute for Animal Health, Compton, UK

J K Brown
The University of Arizona, Tucson, AZ, USA

K S Brown
University of Manitoba, Winnipeg, MB, Canada

J Bruenn
State University of New York, Buffalo, NY, USA

C P D Brussaard
Royal Netherlands Institute for Sea Research, Texel, The Netherlands

J J Bugert
Wales College of Medicine, Heath Park, Cardiff, UK

J J Bujarski
Northern Illinois University, DeKalb, IL, USA and Polish Academy of Sciences, Poznan, Poland

R M Buller
Saint Louis University School of Medicine, St. Louis, MO, USA

J P Burand
University of Massachusetts at Amherst, Amherst, MA, USA

J Burgyan
Agricultural Biotechnology Center, Godollo, Hungary

F J Burt
University of the Free State, Bloemfontein, South Africa

S J Butcher
University of Helsinki, Helsinki, Finland

J S Butel
Baylor College of Medicine, Houston, TX, USA

M I Butler
University of Otago, Dunedin, New Zealand

S Bühler
University of Heidelberg, Heidelberg, Germany

P Caciagli
Istituto di Virologia Vegetale – CNR, Turin, Italy

C H Calisher
Colorado State University, Fort Collins, CO, USA

T Candresse
UMR GDPP, Centre INRA de Bordeaux, Villenave d'Ornon, France

A J Cann
University of Leicester, Leicester, UK

C Caranta
INRA, Montfavet, France

G Carlile
CSIRO Livestock Industries, Geelong, VIC, Australia

J P Carr
University of Cambridge, Cambridge, UK

R Carrion, Jr.
Southwest Foundation for Biomedical Research, San Antonio, TX, USA

J W Casey
Cornell University, Ithaca, NY, USA

R N Casey
Cornell University, Ithaca, NY, USA

S Casjens
University of Utah School of Medicine, Salt Lake City, UT, USA

R Cattaneo
Mayo Clinic College of Medicine, Rochester, MN, USA

D Cavanagh
Institute for Animal Health, Compton, UK

A Chahroudi
University of Pennsylvania School of Medicine, Philadelphia, PA, USA

S Chakraborty
Jawaharlal Nehru University, New Delhi, India

T J Chambers
Saint Louis University School of Medicine, St. Louis, MO, USA

Y Chang
University of Pittsburgh Cancer Institute, Pittsburgh, PA, USA

J T Chang
Baylor College of Medicine, Houston, TX, USA

D Chapman
Institute for Animal Health, Pirbright, UK

D Chattopadhyay
University of Calcutta, Kolkata, India

M Chen
University of Arizona, Tucson, AZ, USA

J E Cherwa
University of Arizona, Tucson, AZ, USA

V G Chinchar
University of Mississippi Medical Center, Jackson, MS, USA

A V Chintakuntlawar
University of Oklahoma Health Sciences Center, Oklahoma City, OK, USA

W Chiu
Baylor College of Medicine, Houston, TX, USA

J Chodosh
University of Oklahoma Health Sciences Center, Oklahoma City, OK, USA

I-R Choi
International Rice Research Institute, Los Baños, The Philippines

P D Christian
National Institute of Biological Standards and Control, South Mimms, UK

M G Ciufolini
Istituto Superiore di Sanità, Rome, Italy

P Clarke
University of Colorado Health Sciences, Denver, CO, USA

J-M Claverie
Université de la Méditerranée, Marseille, France

J R Clayton
Johns Hopkins University Schools of Public Health and Medicine, Baltimore, MD, USA

R J Clem
Kansas State University, Manhattan, KS, USA

C J Clements
The Macfarlane Burnet Institute for Medical Research and Public Health Ltd., Melbourne, VIC, Australia

L L Coffey,
Institut Pasteur, Paris, France

J I Cohen
National Institutes of Health, Bethesda, MD, USA

J Collinge
University College London, London, UK

P L Collins
National Institute of Allergy and Infectious Diseases, Bethesda, MD, USA

A Collins
University of Wisconsin School of Medicine and Public Health, Madison, WI, USA

D Contamine
Université Versailles St-Quentin, CNRS, Versailles, France

K M Coombs
University of Manitoba, Winnipeg, MB, Canada

J A Cowley
CSIRO Livestock Industries, Brisbane, QLD, Australia

J K Craigo
University of Pittsburgh School of Medicine, Pittsburgh, PA, USA

M St. J Crane
CSIRO Livestock Industries, Geelong, VIC, Australia

J E Crowe, Jr.
Vanderbilt University Medical Center, Nashville, TN, USA

H Czosnek
The Hebrew University of Jerusalem, Rehovot, Israel

T Dalmay
University of East Anglia, Norwich, UK

B H Dannevig
National Veterinary Institute, Oslo, Norway

C J D'Arcy
University of Illinois at Urbana-Champaign, Urbana, IL, USA

A J Davison
MRC Virology Unit, Glasgow, UK

W O Dawson
University of Florida, Lake Alfred, FL, USA

L A Day
The Public Health Research Institute, Newark, NJ, USA

J C de la Torre
The Scripps Research Institute, La Jolla, CA, USA

X de Lamballerie
Faculté de Médecine de Marseille, Marseilles, France

M de Vega
Universidad Autónoma, Madrid, Spain

P Delfosse
Centre de Recherche Public-Gabriel Lippmann, Belvaux, Luxembourg

B Delmas
INRA, Jouy-en-Josas, France

M Deng
University of California, Berkeley, CA, USA

J DeRisi
University of California, San Francisco, San Francisco, CA, USA

C Desbiez
Institut National de la Recherche Agronomique (INRA), Station de Pathologie Végétale, Montfavet, France

R C Desrosiers
New England Primate Research Center, Southborough, MA, USA

A K Dhar
Advanced BioNutrition Corp, Columbia, MD, USA

R G Dietzgen
The University of Queensland, St. Lucia, QLD, Australia

S P Dinesh-Kumar
Yale University, New Haven, CT, USA

L K Dixon
Institute for Animal Health, Pirbright, UK

C Dogimont
INRA, Montfavet, France

A Domanska
University of Helsinki, Helsinki, Finland

L L Domier
USDA–ARS, Urbana, IL, USA

L L Domier
USDA-ARS, Urbana-Champaign, IL, USA

A Dotzauer
University of Bremen, Bremen, Germany

T W Dreher
Oregon State University, Corvallis, OR, USA

S Dreschers
University of Duisburg–Essen, Essen, Germany

R L Duda
University of Pittsburgh, Pittsburgh, PA, USA

J P Dudley
The University of Texas at Austin, Austin, TX, USA

W P Duprex
The Queen's University of Belfast, Belfast, UK

R E Dutch
University of Kentucky, Lexington, KY, USA

B M Dutia
University of Edinburgh, Edinburgh, UK

M L Dyall-Smith
The University of Melbourne, Parkville, VIC, Australia

J East
University of Texas Medical Branch – Galveston, Galveston, TX, USA

A J Easton
University of Warwick, Coventry, UK

K C Eastwell
Washington State University – IAREC, Prosser, WA, USA

B T Eaton
Australian Animal Health Laboratory, Geelong, VIC, Australia

H Edskes
National Institutes of Health, Bethesda, MD, USA

B Ehlers
Robert Koch-Institut, Berlin, Germany

R M Elliott
University of St. Andrews, St. Andrews, UK

A Engel
National Institutes of Health, Bethesda, MD, USA and
D Kryndushkin
National Institutes of Health, Bethesda, MD, USA

J Engelmann
INRES, University of Bonn, Bonn, Germany

L Enjuanes
CNB, CSIC, Madrid, Spain

A Ensser
Virologisches Institut, Universitätsklinikum, Erlangen, Germany

M Erlandson
Agriculture & Agri-Food Canada, Saskatoon, SK, Canada

K J Ertel
University of California, Irvine, CA, USA

R Esteban
Instituto de Microbiología Bioquímica CSIC/University de Salamanca, Salamanca, Spain

R Esteban
Instituto de Microbiología Bioquímica CSIC/University of Salamanca, Salamanca, Spain

J L Van Etten
University of Nebraska–Lincoln, Lincoln, NE, USA

D J Evans
University of Warwick, Coventry, UK

Ø Evensen
Norwegian School of Veterinary Science, Oslo, Norway

D Falzarano
University of Manitoba, Winnipeg, MB, Canada

B A Fane
University of Arizona, Tucson, AZ, USA

R-X. Fang
Chinese Academy of Sciences, Beijing, People's Republic of China

D Fargette
IRD, Montpellier, France

A Fath-Goodin
University of Kentucky, Lexington, KY, USA

C M Fauquet
Danforth Plant Science Center, St. Louis, MO, USA

B A Federici
University of California, Riverside, CA, USA

H Feldmann
National Microbiology Laboratory, Public Health Agency of Canada, Winnipeg, MB, Canada

H Feldmann
Public Health Agency of Canada, Winnipeg, MB, Canada

F Fenner
Australian National University, Canberra, ACT, Australia

S A Ferreira
University of Hawaii at Manoa, Honolulu, HI, USA

H J Field
University of Cambridge, Cambridge, UK

K Fischer
University of California, San Francisco, San Francisco, CA, USA

J A Fishman
Massachusetts General Hospital, Boston, MA, USA

B Fleckenstein
University of Erlangen – Nürnberg, Erlangen, Germany

R Flores
Instituto de Biología Molecular y Celular de Plantas (UPV-CSIC), Valencia, Spain

T R Flotte
University of Florida College of Medicine, Gainesville, FL, USA

P Forterre
Institut Pasteur, Paris, France

M A Franco
Pontificia Universidad Javeriana, Bogota, Republic of Colombia

T K Frey
Georgia State University, Atlanta, GA, USA

M Fuchs
Cornell University, Geneva, NY, USA

S Fuentes
International Potato Center (CIP), Lima, Peru

T Fujimura
Instituto de Microbiología Bioquímica CSIC/University of Salamanca, Salamanca, Spain

R S Fujinami
University of Utah School of Medicine, Salt Lake City, UT, USA

T Fukuhara
Tokyo University of Agriculture and Technology, Fuchu, Japan

D Gallitelli
Università degli Studi and Istituto di Virologia Vegetale del CNR, Bari, Italy

F García-Arenal
Universidad Politécnica de Madrid, Madrid, Spain

J A García
Centro Nacional de Biotecnología (CNB), CSIC, Madrid, Spain

R A Garrett
Copenhagen University, Copenhagen, Denmark

S Gaumer
Université Versailles St-Quentin, CNRS, Versailles, France

R J Geijskes
Queensland University of Technology, Brisbane, QLD, Australia

T W Geisbert
National Emerging Infectious Diseases Laboratories, Boston, MA, USA

E Gellermann
Hannover Medical School, Hannover, Germany

A Gessain
Pasteur Institute, CNRS URA 3015, Paris, France

S A Ghabrial
University of Kentucky, Lexington, KY, USA

W Gibson
Johns Hopkins University School of Medicine, Baltimore, MD, USA

M Glasa
Slovak Academy of Sciences, Bratislava, Slovakia

Y Gleba
Icon Genetics GmbH, Weinbergweg, Germany

U A Gompels
University of London, London, UK

D Gonsalves
USDA, Pacific Basin Agricultural Research Center, Hilo, HI, USA

M M Goodin
University of Kentucky, Lexington, KY, USA

T J D Goodwin
University of Otago, Dunedin, New Zealand

A E Gorbalenya
Leiden University Medical Center, Leiden, The Netherlands

E A Gould
University of Reading, Reading, UK

A Grakoui
Emory University School of Medicine, Atlanta, GA, USA

M-A Grandbastien
INRA, Versailles, France

R Grassmann
University of Erlangen – Nürnberg, Erlangen, Germany

M Gravell
National Institutes of Health, Bethesda, MD, USA

M V Graves
University of Massachusetts–Lowell, Lowell, MA, USA

K Y Green
National Institutes of Health, Bethesda, MD, USA

H B Greenberg
Stanford University School of Medicine and Veterans Affairs Palo Alto Health Care System, Palo Alto, CA, USA

B M Greenberg
Johns Hopkins School of Medicine, Baltimore, MD, USA

I Greiser-Wilke
School of Veterinary Medicine, Hanover, Germany

D E Griffin
Johns Hopkins Bloomberg School of Public Health, Baltimore, MD, USA

T S Gritsun
University of Reading, Reading, UK

R J de Groot
Utrecht University, Utrecht, The Netherlands

A J Gubala
CSIRO Livestock Industries, Geelong, VIC, Australia

D J Gubler
John A. Burns School of Medicine, Honolulu, HI, USA

A-L Haenni
Institut Jacques Monod, Paris, France

D Haig
Nottingham University, Nottingham, UK

F J Haines
Oxford Brookes University, Oxford, UK

J Hamacher
INRES, University of Bonn, Bonn, Germany

J Hammond
USDA-ARS, Beltsville, MD, USA

R M Harding
Queensland University of Technology, Brisbane, QLD, Australia

J M Hardwick
Johns Hopkins University Schools of Public Health and Medicine, Baltimore, MD, USA

D Hariri
INRA – Département Santé des Plantes et Environnement, Versailles, France

B Harrach
Veterinary Medical Research Institute, Budapest, Hungary

P A Harries
Samuel Roberts Noble Foundation, Inc., Ardmore, OK, USA

L E Harrington
University of Alabama at Birmingham, Birmingham, AL, USA

T J Harrison
University College London, London, UK

T Hatziioannou
Aaron Diamond AIDS Research Center, The Rockefeller University, New York, NY, USA

J Hay
The State University of New York, Buffalo, NY, USA

G S Hayward
Johns Hopkins School of Medicine, Baltimore, MD, USA

E Hébrard
IRD, Montpellier, France

R W Hendrix
University of Pittsburgh, Pittsburgh, PA, USA

L E Hensley
USAMRIID, Fort Detrick, MD, USA

M de las Heras
University of Glasgow Veterinary School, Glasgow, UK

S Hertzler
University of Illinois at Chicago, Chicago, IL, USA

F van Heuverswyn
University of Montpellier 1, Montpellier, France

J Hilliard
Georgia State University, Atlanta, GA, USA

B I Hillman
Rutgers University, New Brunswick, NJ, USA

S Hilton
University of Warwick, Warwick, UK

D M Hinton
National Institutes of Health, Bethesda, MD, USA

A Hinz
UMR 5233 UJF-EMBL-CNRS, Grenoble, France

A E Hoet
The Ohio State University, Columbus, OH, USA

S A Hogenhout
The John Innes Centre, Norwich, UK

T Hohn
Basel university, Institute of Botany, Basel, Switzerland

J S Hong
Seoul Women's University, Seoul, South Korea

M C Horzinek
Utrecht University, Utrecht, The Netherlands

T Hovi
National Public Health Institute (KTL), Helsinki, Finland

A M Huger
Institute for Biological Control, Darmstadt, Germany

L E Hughes
University of St. Andrews, St. Andrews, UK

R Hull
John Innes Centre, Colney, UK

E Hunter
Emory University Vaccine Center, Atlanta, GA, USA

A D Hyatt
Australian Animal Health Laboratory, Geelong, VIC, Australia

T Hyypiä
University of Turku, Turku, Finland

T Iwanami
National Institute of Fruit Tree Science, Tsukuba, Japan

A O Jackson
University of California, Berkeley, CA, USA

P Jardine
University of Minnesota, Minneapolis, MN, USA

J A Jehle
DLR Rheinpfalz, Neustadt, Germany

A R Jilbert
Institute of Medical and Veterinary Science, Adelaide, SA, Australia

P John
Indian Agricultural Research Institute, New Delhi, India

J E Johnson
The Scripps Research Institute, La Jolla, CA, USA

R T Johnson
Johns Hopkins School of Medicine, Baltimore, MD, USA

W E Johnson
New England Primate Research Center, Southborough, MA, USA

S L Johnston
Imperial College London, London, UK

A T Jones
Scottish Crop Research Institute, Dundee, UK

R Jordan
USDA-ARS, Beltsville, MD, USA

Y Kapustin
National Institutes of Health, Bethesda, MD, USA

P Karayiannis
Imperial College London, London, UK

P Kazmierczak
University of California, Davis, CA, USA

K M Keene
Colorado State University, Fort Collins, CO, USA

C Kerlan
Institut National de la Recherche Agronomique (INRA), Le Rheu, France

K Khalili
Temple University School of Medicine, Philadelphia, PA, USA

P H Kilmarx
Centers for Disease Control and Prevention, Atlanta, GA, USA

L A King
Oxford Brookes University, Oxford, UK

P D Kirkland
Elizabeth Macarthur Agricultural Institute, Menangle, NSW, Australia

C D Kirkwood
Murdoch Childrens Research Institute Royal Children's Hospital, Melbourne, VIC, Australia

R P Kitching
Canadian Food Inspection Agency, Winnipeg, MB, Canada

P J Klasse
Cornell University, New York, NY, USA

N R Klatt
University of Pennsylvania School of Medicine, Philadelphia, PA, USA

R G Kleespies
Institute for Biological Control, Darmstadt, Germany

D F Klessig
Boyce Thompson Institute for Plant Research, Ithaca, NY, USA

W B Klimstra
Louisiana State University Health Sciences Center at Shreveport, Shreveport, LA, USA

V Klimyuk
Icon Genetics GmbH, Weinbergweg, Germany

N Knowles
Institute for Animal Health, Pirbright, UK

R Koenig
Biologische Bundesanstalt für Land- und Forstwirtschaft, Brunswick, Germany

R Koenig
Institut für Pflanzenvirologie, Mikrobiologie und biologische Sicherheit, Brunswick, Germany

G Konaté
INERA, Ouagadougou, Burkina Faso

C N Kotton
Massachusetts General Hospital, Boston, MA, USA

L D Kramer
Wadsworth Center, New York State Department of Health, Albany, NY, USA

P J Krell
University of Guelph, Guelph, ON, Canada

J Kreuze
International Potato Center (CIP), Lima, Peru

M J Kuehnert
Centers for Disease Control and Prevention, Atlanta, GA, USA

R J Kuhn
Purdue University, West Lafayette, IN, USA

G Kurath
Western Fisheries Research Center, Seattle, WA, USA

I Kusters
sanofi pasteur, Lyon, France

I V Kuzmin
Centers for Disease Control and Prevention, Atlanta, GA, USA

M E Laird
New England Primate Research Center, Southborough, MA, USA

R A Lamb
Howard Hughes Medical Institute at Northwestern University, Evanston, IL, USA

P F Lambert
University of Wisconsin School of Medicine and Public Health, Madison, WI, USA

A S Lang
Memorial University of Newfoundland, St. John's, NL, Canada

H D Lapierre
INRA – Département Santé des Plantes et Environnement, Versailles, France

G Lawrence
The Children's Hospital at Westmead, Westmead, NSW, Australia and
University of Sydney, Westmead, NSW, Australia

H Lecoq
Institut National de la Recherche Agronomique (INRA), Station de Pathologie Végétale, Montfavet, France

B Y Lee
Seoul Women's University, Seoul, South Korea

E J Lefkowitz
University of Alabama at Birmingham, Birmingham, AL, USA

J P Legg
International Institute of Tropical Agriculture, Dar es Salaam, Tanzania,
UK and
Natural Resources Institute, Chatham Maritime, UK

P Leinikki
National Public Health Institute, Helsinki, Finland

J Lenard
University of Medicine and Dentistry of New Jersey (UMDNJ), Piscataway, NJ, USA

J C Leong
University of Hawaii at Manoa, Honolulu, HI, USA

K N Leppard
University of Warwick, Coventry, UK

A Lescoute
Université Louis Pasteur, Strasbourg, France

D-E Lesemann
Biologische Bundesanstalt für Land- und Forstwirtschaft, Brunswick, Germany

J-H Leu
National Taiwan University, Taipei, Republic of China

H L Levin
National Institutes of Health, Bethesda, MD, USA

D J Lewandowski
The Ohio State University, Columbus, OH, USA

H-S Lim
University of California, Berkeley, Berkeley, CA, USA

M D A Lindsay
Western Australian Department of Health, Mount Claremont, WA, Australia

R Ling
University of Warwick, Coventry, UK

M L Linial
Fred Hutchinson Cancer Research Center, Seattle, WA, USA

D C Liotta
Emory University, Atlanta, GA, USA

W Ian Lipkin
Columbia University, New York, NY, USA

H L Lipton
University of Illinois at Chicago, Chicago, IL, USA

A S Liss
University of Texas at Austin, Austin, TX, USA

J J López-Moya
Instituto de Biología Molecular de Barcelona (IBMB), CSIC, Barcelona, Spain

G Loebenstein
Agricultural Research Organization, Bet Dagan, Israel

C-F Lo
National Taiwan University, Taipei, Republic of China

S A Lommel
North Carolina State University, Raleigh, NC, USA

G P Lomonossoff
John Innes Centre, Norwich, UK

M Luo
University of Alabama at Birmingham, Birmingham, AL, USA

S A MacFarlane
Scottish Crop Research Institute, Dundee, UK

J S Mackenzie
Curtin University of Technology, Shenton Park, WA, Australia

R Mahieux
Pasteur Institute, CNRS URA 3015, Paris, France

B W J Mahy
Centers for Disease Control and Prevention, Atlanta, GA, USA

E Maiss
Institute of Plant Diseases and Plant Protection, Hannover, Germany

E O Major
National Institutes of Health, Bethesda, MD, USA

V G Malathi
Indian Agricultural Research Institute, New Delhi, India

A Mankertz
Robert Koch-Institut, Berlin, Germany

S Mansoor
National Institute for Biotechnology and Genetic Engineering, Faisalabad, Pakistan

A A Marfin
Centers for Disease Control and Prevention, Atlanta, GA, USA

S Marillonnet
Icon Genetics GmbH, Weinbergweg, Germany

G P Martelli
Università degli Studi and Istituto di Virologia vegetale CNR, Bari, Italy

M Marthas
University of California, Davis, Davis, CA, USA

D P Martin
University of Cape Town, Cape Town, South Africa

P A Marx
Tulane University, Covington, LA, USA

W S Mason
Fox Chase Cancer Center, Philadelphia, PA, USA

T D Mastro
Centers for Disease Control and Prevention, Atlanta, GA, USA

A A McBride
National Institutes of Health, Bethesda, MD, USA

L McCann
National Institutes of Health, Bethesda, MD, USA

M McChesney
University of California, Davis, Davis, CA, USA

J B McCormick
University of Texas, School of Public Health, Brownsville, TX, USA

G McFadden
University of Florida, Gainesville, FL, USA

G McFadden
The University of Western Ontario, London, ON, Canada

D B McGavern
The Scripps Research Institute, La Jolla, CA, USA

A L McNees
Baylor College of Medicine, Houston, TX, USA

M Meier
Tallinn University of Technology, Tallinn, Estonia

P S Mellor
Institute for Animal Health, Woking, UK

X J Meng
Virginia Polytechnic Institute and State University, Blacksburg, VA, USA

A A Mercer
University of Otago, Dunedin, New Zealand

P P C Mertens
Institute for Animal Health, Woking, UK

T C Mettenleiter
Friedrich-Loeffler-Institut, Greifswald-Insel Riems, Germany

H Meyer
Bundeswehr Institute of Microbiology, Munich, Germany

R F Meyer
Centers for Disease Control and Prevention, Atlanta, GA, USA

P de Micco
Etablissement Français du Sang Alpes-Méditerranée, Marseilles, France

B R Miller
Centers for Disease Control and Prevention (CDC), Fort Collins, CO, USA

C J Miller
University of California, Davis, Davis, CA, USA

R G Milne
Istituto di Virologia Vegetale CNR, Torino, Italy

P D Minor
NIBSC, Potters Bar, UK

S Mjaaland
Norwegian School of Veterinary Science, Oslo, Norway

E S Mocarski
Emory University School of Medicine, Atlanta, GA, USA

E S Mocarski, Jr.
Emory University School of Medicine, Emory, GA, USA

V Moennig
School of Veterinary Medicine, Hanover, Germany

P Moffett
Boyce Thompson Institute for Plant Research, Ithaca, NY, USA

T P Monath
Kleiner Perkins Caufield and Byers, Menlo Park, CA, USA

R C Montelaro
University of Pittsburgh School of Medicine, Pittsburgh, PA, USA

P S Moore
University of Pittsburgh Cancer Institute, Pittsburgh, PA, USA

F J Morales
International Center for Tropical Agriculture, Cali, Colombia

H Moriyama
Tokyo University of Agriculture and Technology, Fuchu, Japan

T J Morris
University of Nebraska, Lincoln, NE, USA

S A Morse
Centers for Disease Control and Prevention, Atlanta, GA, USA

L Moser
University of Wisconsin – Madison, Madison, WI, USA

B Moury
INRA – Station de Pathologie Végétale, Montfavet, France

J W Moyer
North Carolina State University, Raleigh, NC, USA

R W Moyer
University of Florida, Gainesville, FL, USA

E Muller
CIRAD/UMR BGPI, Montpellier, France

F A Murphy
University of Texas Medical Branch, Galveston, TX, USA

A Müllbacher
Australian National University, Canberra, ACT, Australia

K Nagasaki
Fisheries Research Agency, Hiroshima, Japan

T Nakayashiki
National Institutes of Health, Bethesda, MD, USA

A A Nash
University of Edinburgh, Edinburgh, UK

N Nathanson
University of Pennsylvania, Philadelphia, PA, USA

C K Navaratnarajah
Purdue University, West Lafayette, IN, USA

M S Nawaz-ul-Rehman
Danforth Plant Science Center, St. Louis, MO, USA

J C Neil
University of Glasgow, Glasgow, UK

R S Nelson
Samuel Roberts Noble Foundation, Inc., Ardmore, OK, USA

P Nettleton
Moredun Research Institute, Edinburgh, UK

A W Neuman
Emory University, Atlanta, GA, USA

A R Neurath
Virotech, New York, NY, USA

M L Nibert
Harvard Medical School, Boston, MA, USA

L Nicoletti
Istituto Superiore di Sanità, Rome, Italy

N Noah
London School of Hygiene and Tropical Medicine, London, UK

D L Nuss
University of Maryland Biotechnology Institute, Rockville, MD, USA

M S Oberste
Centers for Disease Control and Prevention, Atlanta, GA, USA

W A O'Brien
University of Texas Medical Branch – Galveston, Galveston, TX, USA

D J O'Callaghan
Louisiana State University Health Sciences Center, Shreveport, LA, USA

W F Ochoa
University of California, San Diego, La Jolla, CA, USA

M R Odom
University of Alabama at Birmingham, Birmingham, AL, USA

M M van Oers
Wageningen University, Wageningen, The Netherlands

M B A Oldstone
The Scripps Research Institute, La Jolla, CA, USA

G Olinger
USAMRIID, Fort Detrick, MD, USA

K E Olson
Colorado State University, Fort Collins, CO, USA

A Olspert
Tallinn University of Technology, Tallinn, Estonia

G Orth
Institut Pasteur, Paris, France

J E Osorio
University of Wisconsin, Madison, WI, USA

N Osterrieder
Cornell University, Ithaca, NY, USA

S A Overman
University of Missouri – Kansas City, Kansas City, MO, USA

R A Owens
Beltsville Agricultural Research Center, Beltsville, MD, USA

M S Padmanabhan
Yale University, New Haven, CT, USA

S Paessler
University of Texas Medical Branch, Galveston, TX, USA

P Palese
Mount Sinai School of Medicine, New York, NY, USA

M A Pallansch
Centers for Disease Control and Prevention, Atlanta, GA, USA

M Palmarini
University of Glasgow Veterinary School, Glasgow, UK

P Palukaitis
Scottish Crop Research Institute, Invergowrie, Dundee, UK

I Pandrea
Tulane National Primate Research Center, Covington, LA, USA

O Papadopoulos
Aristotle University, Thessaloniki, Greece

H R Pappu
Washington State University, Pullman, WA, USA

S Parker
Saint Louis University School of Medicine, St. Louis, MO, USA

C R Parrish
Cornell University, Ithaca, NY, USA

R F Pass
University of Alabama School of Medicine, Birmingham, AL, USA

J L Patterson
Southwest Foundation for Biomedical Research, San Antonio, TX, USA

T A Paul
Cornell University, Ithaca, NY, USA

A E Peaston
The Jackson Laboratory, Bar Harbor, ME, USA

M Peeters
University of Montpellier 1, Montpellier, France

J S M Peiris
The University of Hong Kong, Hong Kong, People's Republic of China

P J Peters
Centers for Disease Control and Prevention, Atlanta, GA, USA

M Pfeffer
Bundeswehr Institute of Microbiology, Munich, Germany

H Pfister
University of Köln, Cologne, Germany

O Planz
Federal Research Institute for Animal Health, Tuebingen, Gemany

L L M Poon
The University of Hong Kong, Hong Kong, People's Republic of China

M M Poranen
University of Helsinki, Helsinki, Finland

K Porter
The University of Melbourne, Parkville, VIC, Australia

A Portner
St. Jude Children's Research Hospital, Memphis, TN, USA

R D Possee
NERC Institute of Virology and Environmental Microbiology, Oxford, UK

R T M Poulter
University of Otago, Dunedin, New Zealand

A M Powers
Centers for Disease Control and Prevention, Fort Collins, CO, USA

D Prangishvili
Institut Pasteur, Paris, France

C M Preston
Medical Research Council Virology Unit, Glasgow, UK

S L Quackenbush
Colorado State University, Fort Collins, CO, USA

F Qu
University of Nebraska, Lincoln, NE, USA

B C Ramirez
CNRS, Paris, France

A Rapose
University of Texas Medical Branch – Galveston, Galveston, TX, USA

D V R Reddy
Hyderabad, India

A J Redwood
The University of Western Australia, Crawley, WA, Australia

M Regner
Australian National University, Canberra, ACT, Australia

W K Reisen
University of California, Davis, CA, USA

T Renault
IFREMER, La Tremblade, France

P A Revill
Victorian Infectious Diseases Reference Laboratory, Melbourne, VIC, Australia

A Rezaian
University of Adelaide, Adelaide, SA, Australia

J F Ridpath
USDA, Ames, IA, USA

B K Rima
The Queen's University of Belfast, Belfast, UK

E Rimstad
Norwegian School of Veterinary Science, Oslo, Norway

F J Rixon
MRC Virology Unit, Glasgow, UK

Y-T Ro
Konkuk University, Seoul, South Korea

C M Robinson
University of Oklahoma Health Sciences Center, Oklahoma City, OK, USA

G F Rohrmann
Oregon State University, Corvallis, OR, USA

M Roivainen
National Public Health Institute (KTL), Helsinki, Finland

L Roux
University of Geneva Medical School, Geneva, Switzerland

J Rovnak
Colorado State University, Fort Collins, CO, USA

D J Rowlands
University of Leeds, Leeds, UK

P Roy
London School of Hygiene and Tropical Medicine, London, UK

L Rubino
Istituto di Virologia Vegetale del CNR, Bari, Italy

R W H Ruigrok
CNRS, Grenoble, France

C E Rupprecht
Centers for Disease Control and Prevention, Atlanta, GA, USA

R J Russell
University of St. Andrews, St. Andrews, UK

B E Russ
The University of Melbourne, Parkville, VIC, Australia

W T Ruyechan
The State University of New York, Buffalo, NY, USA

E Ryabov
University of Warwick, Warwick, UK

M D Ryan
University of St. Andrews, St. Andrews, UK

E P Rybicki
University of Cape Town, Cape Town, South Africa

K D Ryman
Louisiana State University Health Sciences Center at Shreveport, Shreveport, LA, USA

K D Ryman
Louisiana State University Health Sciences Center, Shreveport, LA, USA

K H Ryu
Seoul Women's University, Seoul, South Korea

M Safak
Temple University School of Medicine, Philadelphia, PA, USA

M Salas
Universidad Autónoma, Madrid, Spain

S K Samal
University of Maryland, College Park, MD, USA

J T Sample
The Pennsylvania State University College of Medicine, Hershey, PA, USA

C E Sample
The Pennsylvania State University College of Medicine, Hershey, PA, USA

R M Sandri-Goldin
University of California, Irvine, Irvine, CA, USA

H Sanfaçon
Pacific Agri-Food Research Centre, Summerland, BC, Canada

R Sanjuán
Instituto de Biología Molecular y Cellular de Plantas, CSIC-UPV, Valencia, Spain

N Santi
Norwegian School of Veterinary Science, Oslo, Norway

C Sarmiento
Tallinn University of Technology, Tallinn, Estonia

T Sasaya
National Agricultural Research Center, Ibaraki, Japan

Q J Sattentau
University of Oxford, Oxford, UK

C Savolainen-Kopra
National Public Health Institute (KTL), Helsinki, Finland

B Schaffhausen
Tufts University School of Medicine, Boston, MA, USA

K Scheets
Oklahoma State University, Stillwater, OK, USA

M J Schmitt
University of the Saarland, Saarbrücken, Germany

A Schneemann
The Scripps Research Institute, La Jolla, CA, USA

G Schoehn
CNRS, Grenoble, France

J E Schoelz
University of Missouri, Columbia, MO, USA

L B Schonberger
Centers for Disease Control and Prevention, Atlanta, GA, USA

U Schubert
Klinikum der Universität Erlangen-Nürnberg, Erlangen, Germany

D A Schultz
Johns Hopkins University School of Medicine, Baltimore, MD, USA

S Schultz-Cherry
University of Wisconsin – Madison, Madison, WI, USA

T F Schulz
Hannover Medical School, Hannover, Germany

P D Scotti
Waiatarua, New Zealand

B L Semler
University of California, Irvine, CA, USA

J M Sharp
Veterinary Laboratories Agency, Penicuik, UK

M L Shaw
Mount Sinai School of Medicine, New York, NY, USA

G R Shellam
The University of Western Australia,
Crawley, WA, Australia

D N Shepherd
University of Cape Town, Cape Town, South Africa

N C Sheppard
University of Oxford, Oxford, UK

F Shewmaker
National Institutes of Health, Bethesda, MD, USA

P A Signoret
Montpellier SupAgro, Montpellier, France

A Silaghi
University of Manitoba, Winnipeg, MB, Canada

G Silvestri
University of Pennsylvania, Philadelphia, PA, USA

T L Sit
North Carolina State University, Raleigh, NC, USA

N Sittidilokratna
Centex Shrimp and Center for Genetic Engineering and
Biotechnology, Bangkok, Thailand

M A Skinner
Imperial College London, London, UK

D W Smith
PathWest Laboratory Medicine WA, Nedlands, WA,
Australia

G L Smith
Imperial College London, London, UK

L M Smith
The University of Western Australia,
Crawley, WA, Australia

E J Snijder
Leiden University Medical Center, Leiden, The
Netherlands

M Sova
University of Texas Medical Branch – Galveston,
Galveston, TX, USA

J A Speir
The Scripps Research Institute, La Jolla, CA, USA

T E Spencer
Texas A&M University, College Station, TX, USA

P Sreenivasulu
Sri Venkateswara University, Tirupati, India

J Stanley
John Innes Centre, Colney, UK

K M Stedman
Portland State University, Portland, OR, USA

D Stephan
Institute of Plant Diseases and Plant Protection,
Hannover, Germany

C C M M Stijger
Wageningen University and Research Centre, Naaldwijk,
The Netherlands

L Stitz
Federal Research Institute for Animal Health, Tuebingen,
Gemany

P G Stockley
University of Leeds, Leeds, UK

M R Strand
University of Georgia, Athens, GA, USA

M J Studdert
The University of Melbourne, Parkville, VIC, Australia

C A Suttle
University of British Columbia, Vancouver, BC,
Canada

N Suzuki
Okayama University, Okayama, Japan

J Y Suzuki
USDA, Pacific Basin Agricultural Research Center, Hilo,
HI, USA

R Swanepoel
National Institute for Communicable Diseases,
Sandringham, South Africa

S J Symes
The University of Melbourne, Parkville, VIC, Australia

G Szittya
Agricultural Biotechnology Center, Godollo, Hungary

M Taliansky
Scottish Crop Research Institute, Dundee, UK

P Tattersall
Yale University Medical School, New Haven, CT, USA

T Tatusova
National Institutes of Health, Bethesda, MD, USA

S Tavantzis
University of Maine, Orono, ME, USA

J M Taylor
Fox Chase Cancer Center, Philadelphia, PA, USA

D A Theilmann
Agriculture and Agri-Food Canada, Summerland, BC,
Canada

F C Thomas Allnutt
National Science Foundation, Arlington, VA, USA

G J Thomas Jr.
University of Missouri – Kansas City, Kansas City, MO, USA

J E Thomas
Department of Primary Industries and Fisheries, Indooroopilly, QLD, Australia

H C Thomas
Imperial College London, London, UK

A N Thorburn
The University of Melbourne, Parkville, VIC, Australia

P Tijssen
Université du Québec, Laval, QC, Canada

S A Tolin
Virginia Polytechnic Institute and State University, Blacksburg, VA, USA

L Torrance
Scottish Crop Research Institute, Invergowrie, UK

S Trapp
Cornell University, Ithaca, NY, USA

S Tripathi
USDA, Pacific Basin Agricultural Research Center, Hilo, HI, USA

E Truve
Tallinn University of Technology, Tallinn, Estonia

J-M Tsai
National Taiwan University, Taipei, Republic of China

M Tsompana
North Carolina State University, Raleigh, NC, USA

R Tuma
University of Helsinki, Helsinki, Finland

A S Turnell
The University of Birmingham, Birmingham, UK

K L Tyler
University of Colorado Health Sciences, Denver, CO, USA

A Uchiyama
Cornell University, Ithaca, NY, USA

C Upton
University of Victoria, Victoria, BC, Canada

A Urisman
University of California, San Francisco, San Francisco, CA, USA

J K Uyemoto
University of California, Davis, CA, USA

A Vaheri
University of Helsinki, Helsinki, Finland

R Vainionpää
University of Turku, Turku, Finland

A M Vaira
Istituto di Virologia Vegetale, CNR, Turin, Italy

N K Van Alfen
University of California, Davis, CA, USA

R A A Van der Vlugt
Wageningen University and Research Centre, Wageningen, The Netherlands

M H V Van Regenmortel
CNRS, Illkirch, France

P A Venter
The Scripps Research Institute, La Jolla, CA, USA

J Verchot-Lubicz
Oklahoma State University, Stillwater, OK, USA

R A Vere Hodge
Vere Hodge Antivirals Ltd., Reigate, UK

H J Vetten
Federal Research Centre for Agriculture and Forestry (BBA), Brunswick, Germany

L P Villarreal
University of California, Irvine, Irvine, CA, USA

J M Vlak
Wageningen University, Wageningen, The Netherlands

P K Vogt
The Scripps Research Institute, La Jolla, CA, USA

L E Volkman
University of California, Berkeley, Berkeley, CA, USA

J Votteler
Klinikum der Universität Erlangen-Nürnberg, Erlangen, Germany

D F Voytas
Iowa State University, Ames, IA, USA

J D F Wadsworth
University College London, London, UK

E K Wagner
University of California, Irvine, Irvine, CA, USA

P J Walker
CSIRO Australian Animal Health Laboratory, Geelong, VIC, Australia

A L Wang
University of California, San Francisco, CA, USA

X Wang
University of Wisconsin – Madison, Madison, WI, USA

C C Wang
University of California, San Francisco, CA, USA

L-F Wang
Australian Animal Health Laboratory, Geelong, VIC, Australia

R Warrier
Purdue University, West Lafayette, IN, USA

S C Weaver
University of Texas Medical Branch, Galveston, TX, USA

B A Webb
University of Kentucky, Lexington, KY, USA

F Weber
University of Freiburg, Freiburg, Germany

R P Weir
Berrimah Research Farm, Darwin, NT, Australia

R A Weisberg
National Institutes of Health, Bethesda, MD, USA

W Weissenhorn
UMR 5233 UJF-EMBL-CNRS, Grenoble, France

R M Welsh
University of Massachusetts Medical School, Worcester, MA, USA

J T West
University of Oklahoma Health Sciences Center, Oklahoma City, OK, USA

E Westhof
Université Louis Pasteur, Strasbourg, France

S P J Whelan
Harvard Medical School, Boston, MA, USA

R L White
Texas A&M University, College Station, TX, USA

C A Whitehouse
United States Army Medical Research Institute of Infectious Diseases, Frederick, MD, USA

R B Wickner
National Institutes of Health, Bethesda, MD, USA

R G Will
Western General Hospital, Edinburgh, UK

T Williams
Instituto de Ecología A.C., Xalapa, Mexico

K Willoughby
Moredun Research Institute, Edinburgh, UK

S Winter
Deutsche Sammlung für Mikroorganismen und Zellkulturen, Brunswick, Germany

J Winton
Western Fisheries Research Center, Seattle, WA, USA

J K Yamamoto
University of Florida, Gainesville, FL, USA

M Yoshida
University of Tokyo, Chiba, Japan

N Yoshikawa
Iwate University, Ueda, Japan

L S Young
University of Birmingham, Birmingham, UK

R F Young, III
Texas A&M University, College Station, TX, USA

T M Yuill
University of Wisconsin, Madison, WI, USA

A J Zajac
University of Alabama at Birmingham, Birmingham, AL, USA

S K Zavriev
Shemyakin and Ovchinnikov Institute of Bioorganic Chemistry, Russian Academy of Sciences, Moscow, Russia

J Ziebuhr
The Queen's University of Belfast, Belfast, UK

E I Zuniga
The Scripps Research Institute, La Jolla, CA, USA

CONTENTS

Editors-in-Chief	v
Associate Editors	vii
Preface	ix
How to Use the Encyclopedia	xi
Contributors	xiii

VOLUME 1

A

Adenoviruses: General Features	*B Harrach*	1
Adenoviruses: Malignant Transformation and Oncology	*A S Turnell*	9
Adenoviruses: Molecular Biology	*K N Leppard*	17
Adenoviruses: Pathogenesis	*M Benkő*	24
African Cassava Mosaic Disease	*J P Legg*	30
African Horse Sickness Viruses	*P S Mellor and P P C Mertens*	37
African Swine Fever Virus	*L K Dixon and D Chapman*	43
AIDS: Disease Manifestation	*A Rapose, J East, M Sova and W A O'Brien*	51
AIDS: Global Epidemiology	*P J Peters, P H Kilmarx and T D Mastro*	58
AIDS: Vaccine Development	*N C Sheppard and Q J Sattentau*	69
Akabane Virus	*P S Mellor and P D Kirkland*	76
Alfalfa Mosaic Virus	*J F Bol*	81
Algal Viruses	*K Nagasaki and C P D Brussaard*	87
Allexivirus	*S K Zavriev*	96
Alphacryptovirus and *Betacryptovirus*	*R Blawid, D Stephan and E Maiss*	98
Anellovirus	*P Biagini and P de Micco*	104
Animal Rhabdoviruses	*H Bourhy, A J Gubala, R P Weir and D B Boyle*	111
Antigen Presentation	*E I Zuniga, D B McGavern and M B A Oldstone*	121
Antigenic Variation	*G M Air and J T West*	127
Antigenicity and Immunogenicity of Viral Proteins	*M H V Van Regenmortel*	137

Antiviral Agents *H J Field and R A Vere Hodge*	142
Apoptosis and Virus Infection *J R Clayton and J M Hardwick*	154
Aquareoviruses *M St J Crane and G Carlile*	163
Arboviruses *B R Miller*	170
Arteriviruses *M A Brinton and E J Snijder*	176
Ascoviruses *B A Federici and Y Bigot*	186
Assembly of Viruses: Enveloped Particles *C K Navaratnarajah, R Warrier and R J Kuhn*	193
Assembly of Viruses: Nonenveloped Particles *M Luo*	200
Astroviruses *L Moser and S Schultz-Cherry*	204

B

Baculoviruses: Molecular Biology of Granuloviruses *S Hilton*	211
Baculoviruses: Molecular Biology of Mosquito Baculoviruses *J J Becnel and C L Afonso*	219
Baculoviruses: Molecular Biology of Sawfly Baculoviruses *B M Arif*	225
Baculoviruses: Apoptosis Inhibitors *R J Clem*	231
Baculoviruses: Expression Vector *F J Haines, R D Possee and L A King*	237
Baculoviruses: General Features *P J Krell*	247
Baculoviruses: Molecular Biology of Nucleopolyhedroviruses *D A Theilmann and G W Blissard*	254
Baculoviruses: Pathogenesis *L E Volkman*	265
Banana Bunchy Top Virus *J E Thomas*	272
Barley Yellow Dwarf Viruses *L L Domier*	279
Barnaviruses *P A Revill*	286
Bean Common Mosaic Virus and Bean Common Mosaic Necrosis Virus *R Jordan and J Hammond*	288
Bean Golden Mosaic Virus *F J Morales*	295
Beet Curly Top Virus *J Stanley*	301
Benyvirus *R Koenig*	308
Beta ssDNA Satellites *R W Briddon and S Mansoor*	314
Birnaviruses *B Delmas*	321
Bluetongue Viruses *P Roy*	328
Border Disease Virus *P Nettleton and K Willoughby*	335
Bornaviruses *L Stitz, O Planz and W Ian Lipkin*	341
Bovine and Feline Immunodeficiency Viruses *J K Yamamoto*	347
Bovine Ephemeral Fever Virus *P J Walker*	354
Bovine Herpesviruses *M J Studdert*	362
Bovine Spongiform Encephalopathy *R G Will*	368
Bovine Viral Diarrhea Virus *J F Ridpath*	374
Brome Mosaic Virus *X Wang and P Ahlquist*	381
Bromoviruses *J J Bujarski*	386

Bunyaviruses: General Features *R M Elliott*	390
Bunyaviruses: Unassigned *C H Calisher*	399

C

Cacao Swollen Shoot Virus *E Muller*	403
Caliciviruses *M J Studdert and S J Symes*	410
Capillovirus, Foveavirus, Trichovirus, Vitivirus *N Yoshikawa*	419
Capripoxviruses *R P Kitching*	427
Capsid Assembly: Bacterial Virus Structure and Assembly *S Casjens*	432
Cardioviruses *C Billinis and O Papadopoulos*	440
Carlavirus *K H Ryu and B Y Lee*	448
Carmovirus *F Qu and T J Morris*	453
Caulimoviruses: General Features *J E Schoelz*	457
Caulimoviruses: Molecular Biology *T Hohn*	464
Central Nervous System Viral Diseases *R T Johnson and B M Greenberg*	469
Cereal Viruses: Maize/Corn *P A Signoret*	475
Cereal Viruses: Rice *F Morales*	482
Cereal Viruses: Wheat and Barley *H D Lapierre and D Hariri*	490
Chandipura Virus *S Basak and D Chattopadhyay*	497
Chrysoviruses *S A Ghabrial*	503
Circoviruses *A Mankertz*	513
Citrus Tristeza Virus *M Bar-Joseph and W O Dawson*	520
Classical Swine Fever Virus *V Moennig and I Greiser-Wilke*	525
Coltiviruses *H Attoui and X de Lamballerie*	533
Common Cold Viruses *S Dreschers and C Adams*	541
Coronaviruses: General Features *D Cavanagh and P Britton*	549
Coronaviruses: Molecular Biology *S C Baker*	554
Cotton Leaf Curl Disease *S Mansoor, I Amin and R W Briddon*	563
Cowpea Mosaic Virus *G P Lomonossoff*	569
Cowpox Virus *M Bennett, G L Smith and D Baxby*	574
Coxsackieviruses *M S Oberste and M A Pallansch*	580
Crenarchaeal Viruses: Morphotypes and Genomes *D Prangishvili, T Basta and R A Garrett*	587
Crimean–Congo Hemorrhagic Fever Virus and Other Nairoviruses *C A Whitehouse*	596
Cryo-Electron Microscopy *W Chiu, J T Chang and F J Rixon*	603
Cucumber Mosaic Virus *F García-Arenal and P Palukaitis*	614
Cytokines and Chemokines *D E Griffin*	620
Cytomegaloviruses: Murine and Other Nonprimate Cytomegaloviruses *A J Redwood, L M Smith and G R Shellam*	624
Cytomegaloviruses: Simian Cytomegaloviruses *D J Alcendor and G S Hayward*	634

VOLUME 2

D

Defective-Interfering Viruses	L Roux	1
Dengue Viruses	D J Gubler	5
Diagnostic Techniques: Microarrays	K Fischer, A Urisman and J DeRisi	14
Diagnostic Techniques: Plant Viruses	R Koenig, D-E Lesemann, G Adam and S Winter	18
Diagnostic Techniques: Serological and Molecular Approaches	R Vainionpää and P Leinikki	29
Dicistroviruses	P D Christian and P D Scotti	37
Disease Surveillance	N Noah	44
DNA Vaccines	S Babiuk and L A Babiuk	51

E

Ebolavirus	K S Brown, A Silaghi and H Feldmann	57
Echoviruses	T Hyypiä	65
Ecology of Viruses Infecting Bacteria	S T Abedon	71
Electron Microscopy of Viruses	G Schoehn and R W H Ruigrok	78
Emerging and Reemerging Virus Diseases of Plants	G P Martelli and D Gallitelli	86
Emerging and Reemerging Virus Diseases of Vertebrates	B W J Mahy	93
Emerging Geminiviruses	C M Fauquet and M S Nawaz-ul-Rehman	97
Endogenous Retroviruses	W E Johnson	105
Endornavirus	T Fukuhara and H Moriyama	109
Enteric Viruses	R F Bishop and C D Kirkwood	116
Enteroviruses of Animals	L E Hughes and M D Ryan	123
Enteroviruses: Human Enteroviruses Numbered 68 and Beyond	T Hovi, S Blomqvist, C Savolainen-Kopra and M Roivainen	130
Entomopoxviruses	M N Becker and R W Moyer	136
Epidemiology of Human and Animal Viral Diseases	F A Murphy	140
Epstein–Barr Virus: General Features	L S Young	148
Epstein–Barr Virus: Molecular Biology	J T Sample and C E Sample	157
Equine Infectious Anemia Virus	J K Craigo and R C Montelaro	167
Evolution of Viruses	L P Villarreal	174

F

Feline Leukemia and Sarcoma Viruses	J C Neil	185
Filamentous ssDNA Bacterial Viruses	S A Overman and G J Thomas Jr.	190
Filoviruses	G Olinger, T W Geisbert and L E Hensley	198
Fish and Amphibian Herpesviruses	A J Davison	205
Fish Retroviruses	T A Paul, R N Casey, P R Bowser, J W Casey, J Rovnak and S L Quackenbush	212

Fish Rhabdoviruses G Kurath and J Winton	221
Fish Viruses J C Leong	227
Flaviviruses of Veterinary Importance R Swanepoel and F J Burt	234
Flaviviruses: General Features T J Chambers	241
Flexiviruses M J Adams	253
Foamy Viruses M L Linial	259
Foot and Mouth Disease Viruses D J Rowlands	265
Fowlpox Virus and Other Avipoxviruses M A Skinner	274
Fungal Viruses S A Ghabrial and N Suzuki	284
Furovirus R Koenig	291
Fuselloviruses of Archaea K M Stedman	296

G

Gene Therapy: Use of Viruses as Vectors K I Berns and T R Flotte	301
Genome Packaging in Bacterial Viruses P Jardine	306
Giardiaviruses A L Wang and C C Wang	312

H

Hantaviruses A Vaheri	317
Henipaviruses B T Eaton and L-F Wang	321
Hepadnaviruses of Birds A R Jilbert and W S Mason	327
Hepadnaviruses: General Features T J Harrison	335
Hepatitis A Virus A Dotzauer	343
Hepatitis B Virus: General Features P Karayiannis and H C Thomas	350
Hepatitis B Virus: Molecular Biology T J Harrison	360
Hepatitis C Virus R Bartenschlager and S Bühler	367
Hepatitis Delta Virus J M Taylor	375
Hepatitis E Virus X J Meng	377
Herpes Simplex Viruses: General Features L Aurelian	383
Herpes Simplex Viruses: Molecular Biology E K Wagner and R M Sandri-Goldin	397
Herpesviruses of Birds S Trapp and N Osterrieder	405
Herpesviruses of Horses D J O'Callaghan and N Osterrieder	411
Herpesviruses: Discovery B Ehlers	420
Herpesviruses: General Features A J Davison	430
Herpesviruses: Latency C M Preston	436
History of Virology: Bacteriophages H-W Ackermann	442
History of Virology: Plant Viruses R Hull	450
History of Virology: Vertebrate Viruses F J Fenner	455

Hordeivirus J N Bragg, H-S Lim and A O Jackson	459
Host Resistance to Retroviruses T Hatziioannou and P D Bieniasz	467
Human Cytomegalovirus: General Features E S Mocarski Jr. and R F Pass	474
Human Cytomegalovirus: Molecular Biology W Gibson	485
Human Eye Infections J Chodosh, A V Chintakuntlawar and C M Robinson	491
Human Herpesviruses 6 and 7 U A Gompels	498
Human Immunodeficiency Viruses: Antiretroviral Agents A W Neuman and D C Liotta	505
Human Immunodeficiency Viruses: Molecular Biology J Votteler and U Schubert	517
Human Immunodeficiency Viruses: Origin F van Heuverswyn and M Peeters	525
Human Immunodeficiency Viruses: Pathogenesis N R Klatt, A Chahroudi and G Silvestri	534
Human Respiratory Syncytial Virus P L Collins	542
Human Respiratory Viruses J E Crowe Jr.	551
Human T-Cell Leukemia Viruses: General Features M Yoshida	558
Human T-Cell Leukemia Viruses: Human Disease R Mahieux and A Gessain	564
Hypovirulence N K Van Alfen and P Kazmierczak	574
Hypoviruses D L Nuss	580

VOLUME 3

I

Icosahedral dsDNA Bacterial Viruses with an Internal Membrane J K H Bamford and S J Butcher	1
Icosahedral Enveloped dsRNA Bacterial Viruses R Tuma	6
Icosahedral ssDNA Bacterial Viruses B A Fane, M Chen, J E Cherwa and A Uchiyama	13
Icosahedral ssRNA Bacterial Viruses P G Stockley	21
Icosahedral Tailed dsDNA Bacterial Viruses R L Duda	30
Idaeovirus A T Jones and H Barker	37
Iflavirus M M van Oers	42
Ilarvirus K C Eastwell	46
Immune Response to Viruses: Antibody-Mediated Immunity A R Neurath	56
Immune Response to Viruses: Cell-Mediated Immunity A J Zajac and L E Harrington	70
Immunopathology M B A Oldstone and R S Fujinami	78
Infectious Pancreatic Necrosis Virus Ø Evensen and N Santi	83
Infectious Salmon Anemia Virus B H Dannevig, S Mjaaland and E Rimstad	89
Influenza R A Lamb	95
Innate Immunity: Defeating C F Basler	104
Innate Immunity: Introduction F Weber	111
Inoviruses L A Day	117
Insect Pest Control by Viruses M Erlandson	125
Insect Reoviruses P P C Mertens and H Attoui	133
Insect Viruses: Nonoccluded J P Burand	144

Interfering RNAs	K E Olson, K M Keene and C D Blair	148
Iridoviruses of Vertebrates	A D Hyatt and V G Chinchar	155
Iridoviruses of Invertebrates	T Williams and A D Hyatt	161
Iridoviruses: General Features	V G Chinchar and A D Hyatt	167

J

Jaagsiekte Sheep Retrovirus	J M Sharp, M de las Heras, T E Spencer and M Palmarini	175
Japanese Encephalitis Virus	A D T Barrett	182

K

Kaposi's Sarcoma-Associated Herpesvirus: General Features	Y Chang and P S Moore	189
Kaposi's Sarcoma-Associated Herpesvirus: Molecular Biology	E Gellermann and T F Schulz	195

L

Lassa, Junin, Machupo and Guanarito Viruses	J B McCormick	203
Legume Viruses	L Bos	212
Leishmaniaviruses	R Carrion Jr, Y-T Ro and J L Patterson	220
Leporipoviruses and Suipoxviruses	G McFadden	225
Luteoviruses	L L Domier and C J D'Arcy	231
Lymphocytic Choriomeningitis Virus: General Features	R M Welsh	238
Lymphocytic Choriomeningitis Virus: Molecular Biology	J C de la Torre	243
Lysis of the Host by Bacteriophage	R F Young III and R L White	248

M

Machlomovirus	K Scheets	259
Maize Streak Virus	D P Martin, D N Shepherd and E P Rybicki	263
Marburg Virus	D Falzarano and H Feldmann	272
Marnaviruses	A S Lang and C A Suttle	280
Measles Virus	R Cattaneo and M McChesney	285
Membrane Fusion	A Hinz and W Weissenhorn	292
Metaviruses	H L Levin	301
Mimivirus	J-M Claverie	311
Molluscum Contagiosum Virus	J J Bugert	319
Mononegavirales	A J Easton and R Ling	324
Mouse Mammary Tumor Virus	J P Dudley	334
Mousepox and Rabbitpox Viruses	M Regner, F Fenner and A Müllbacher	342
Movement of Viruses in Plants	P A Harries and R S Nelson	348

Mumps Virus	B K Rima and W P Duprex	356
Mungbean Yellow Mosaic Viruses	V G Malathi and P John	364
Murine Gammaherpesvirus 68	A A Nash and B M Dutia	372
Mycoreoviruses	B I Hillman	378

N

Nanoviruses	H J Vetten	385
Narnaviruses	R Esteban and T Fujimura	392
Nature of Viruses	M H V Van Regenmortel	398
Necrovirus	L Rubino and G P Martelli	403
Nepovirus	H Sanfaçon	405
Neutralization of Infectivity	P J Klasse	413
Nidovirales	L Enjuanes, A E Gorbalenya, R J de Groot, J A Cowley, J Ziebuhr and E J Snijder	419
Nodaviruses	P A Venter and A Schneemann	430
Noroviruses and Sapoviruses	K Y Green	438

O

Ophiovirus	A M Vaira and R G Milne	447
Orbiviruses	P P C Mertens, H Attoui and P S Mellor	454
Organ Transplantation, Risks	C N Kotton, M J Kuehnert and J A Fishman	466
Origin of Viruses	P Forterre	472
Orthobunyaviruses	C H Calisher	479
Orthomyxoviruses: Molecular Biology	M L Shaw and P Palese	483
Orthomyxoviruses: Structure of Antigens	R J Russell	489
Oryctes Rhinoceros Virus	J M Vlak, A M Huger, J A Jehle and R G Kleespies	495
Ourmiavirus	G P Accotto and R G Milne	500

VOLUME 4

P

Papaya Ringspot Virus	D Gonsalves, J Y Suzuki, S Tripathi and S A Ferreira	1
Papillomaviruses: General Features of Human Viruses	G Orth	8
Papillomaviruses: Molecular Biology of Human Viruses	P F Lambert and A Collins	18
Papillomaviruses of Animals	A A McBride	26
Papillomaviruses: General Features	H U Bernard	34
Paramyxoviruses of Animals	S K Samal	40
Parainfluenza Viruses of Humans	E Adderson and A Portner	47
Paramyxoviruses	R E Dutch	52
Parapoxviruses	D Haig and A A Mercer	57

Partitiviruses of Fungi *S Tavantzis*	63
Partitiviruses: General Features *S A Ghabrial, W F Ochoa, T S Baker and M L Nibert*	68
Parvoviruses of Arthropods *M Bergoin and P Tijssen*	76
Parvoviruses of Vertebrates *C R Parrish*	85
Parvoviruses: General Features *P Tattersall*	90
Pecluvirus *D V R Reddy, C Bragard, P Sreenivasulu and P Delfosse*	97
Pepino Mosaic Virus *R A A Van der Vlugt and C C M M Stijger*	103
Persistent and Latent Viral Infection *E S Mocarski and A Grakoui*	108
Phycodnaviruses *J L Van Etten and M V Graves*	116
Phylogeny of Viruses *A E Gorbalenya*	125
Picornaviruses: Molecular Biology *B L Semler and K J Ertel*	129
Plant Antiviral Defense: Gene Silencing Pathway *G Szittya, T Dalmay and J Burgyan*	141
Plant Reoviruses *R J Geijskes and R M Harding*	149
Plant Resistance to Viruses: Engineered Resistance *M Fuchs*	156
Plant Resistance to Viruses: Geminiviruses *J K Brown*	164
Plant Resistance to Viruses: Natural Resistance Associated with Dominant Genes *P Moffett and D F Klessig*	170
Plant Resistance to Viruses: Natural Resistance Associated with Recessive Genes *C Caranta and C Dogimont*	177
Plant Rhabdoviruses *A O Jackson, R G Dietzgen, R-X Fang, M M Goodin, S A Hogenhout, M Deng and J N Bragg*	187
Plant Virus Diseases: Economic Aspects *G Loebenstein*	197
Plant Virus Diseases: Fruit Trees and Grapevine *G P Martelli and J K Uyemoto*	201
Plant Virus Diseases: Ornamental Plants *J Engelmann and J Hamacher*	207
Plant Virus Vectors (Gene Expression Systems) *Y Gleba, S Marillonnet and V Klimyuk*	229
Plum Pox Virus *M Glasa and T Candresse*	238
Poliomyelitis *P D Minor*	243
Polydnaviruses: Abrogation of Invertebrate Immune Systems *M R Strand*	250
Polydnaviruses: General Features *A Fath-Goodin and B A Webb*	257
Polyomaviruses of Humans *M Safak and K Khalili*	262
Polyomaviruses of Mice *B Schaffhausen*	271
Polyomaviruses *M Gravell and E O Major*	277
Pomovirus *L Torrance*	283
Potato Virus Y *C Kerlan and B Moury*	288
Potato Viruses *C Kerlan*	302
Potexvirus *K H Ryu and J S Hong*	310
Potyviruses *J J López-Moya and J A García*	314
Poxviruses *G L Smith, P Beard and M A Skinner*	325
Prions of Vertebrates *J D F Wadsworth and J Collinge*	331
Prions of Yeast and Fungi *R B Wickner, H Edkes, T Nakayashiki, F Shewmaker, L McCann, A Engel and D Kryndushkin*	338

Pseudorabies Virus *T C Mettenleiter*	342
Pseudoviruses *D F Voytas*	352

Q

Quasispecies *R Sanjuán*	359

R

Rabies Virus *I V Kuzmin and C E Rupprecht*	367
Recombination *J J Bujarski*	374
Reoviruses: General Features *P Clarke and K L Tyler*	382
Reoviruses: Molecular Biology *K M Coombs*	390
Replication of Bacterial Viruses *M Salas and M de Vega*	399
Replication of Viruses *A J Cann*	406
Reticuloendotheliosis Viruses *A S Liss and H R Bose Jr.*	412
Retrotransposons of Fungi *T J D Goodwin, M I Butler and R T M Poulter*	419
Retrotransposons of Plants *M-A Grandbastien*	428
Retrotransposons of Vertebrates *A E Peaston*	436
Retroviral Oncogenes *P K Vogt and A G Bader*	445
Retroviruses of Insects *G F Rohrmann*	451
Retroviruses of Birds *K L Beemon*	455
Retroviruses: General Features *E Hunter*	459
Rhinoviruses *N W Bartlett and S L Johnston*	467
Ribozymes *E Westhof and A Lescoute*	475
Rice Tungro Disease *R Hull*	481
Rice Yellow Mottle Virus *E Hébrard and D Fargette*	485
Rift Valley Fever and Other Phleboviruses *L Nicoletti and M G Ciufolini*	490
Rinderpest and Distemper Viruses *T Barrett*	497
Rotaviruses *J Angel, M A Franco and H B Greenberg*	507
Rubella Virus *T K Frey*	514

S

Sadwavirus *T Iwanami*	523
Satellite Nucleic Acids and Viruses *P Palukaitis, A Rezaian and F García-Arenal*	526
Seadornaviruses *H Attoui and P P C Mertens*	535
Sequiviruses *I-R Choi*	546
Severe Acute Respiratory Syndrome (SARS) *J S M Peiris and L L M Poon*	552
Shellfish Viruses *T Renault*	560
Shrimp Viruses *J-R Bonami*	567

Sigma Rhabdoviruses	*D Contamine and S Gaumer*	576
Simian Alphaherpesviruses	*J Hilliard*	581
Simian Gammaherpesviruses	*A Ensser*	585
Simian Immunodeficiency Virus: Animal Models of Disease	*C J Miller and M Marthas*	594
Simian Immunodeficiency Virus: General Features	*M E Laird and R C Desrosiers*	603
Simian Immunodeficiency Virus: Natural Infection	*I Pandrea, G Silvestri and C Apetrei*	611
Simian Retrovirus D	*P A Marx*	623
Simian Virus 40	*A L McNees and J S Butel*	630
Smallpox and Monkeypox Viruses	*S Parker, D A Schultz, H Meyer and R M Buller*	639
Sobemovirus	*M Meier, A Olspert, C Sarmiento and E Truve*	644
St. Louis Encephalitis	*W K Reisen*	652
Sweetpotato Viruses	*J Kreuze and S Fuentes*	659

VOLUME 5

T

Taura Syndrome Virus	*A K Dhar and F C T Allnutt*	1
Taxonomy, Classification and Nomenclature of Viruses	*C M Fauquet*	9
Tenuivirus	*B C Ramirez*	24
Tetraviruses	*J A Speir and J E Johnson*	27
Theiler's Virus	*H L Lipton, S Hertzler and N Knowles*	37
Tick-Borne Encephalitis Viruses	*T S Gritsun and E A Gould*	45
Tobacco Mosaic Virus	*M H V Van Regenmortel*	54
Tobacco Viruses	*S A Tolin*	60
Tobamovirus	*D J Lewandowski*	68
Tobravirus	*S A MacFarlane*	72
Togaviruses Causing Encephalitis	*S Paessler and M Pfeffer*	76
Togaviruses Causing Rash and Fever	*D W Smith, J S Mackenzie and M D A Lindsay*	83
Togaviruses Not Associated with Human Disease	*L L Coffey,*	91
Togaviruses: Alphaviruses	*A M Powers*	96
Togaviruses: Equine Encephalitic Viruses	*D E Griffin*	101
Togaviruses: General Features	*S C Weaver, W B Klimstra and K D Ryman*	107
Togaviruses: Molecular Biology	*K D Ryman, W B Klimstra and S C Weaver*	116
Tomato Leaf Curl Viruses from India	*S Chakraborty*	124
Tomato Spotted Wilt Virus	*H R Pappu*	133
Tomato Yellow Leaf Curl Virus	*H Czosnek*	138
Tombusviruses	*S A Lommel and T L Sit*	145
Torovirus	*A E Hoet and M C Horzinek*	151
Tospovirus	*M Tsompana and J W Moyer*	157
Totiviruses	*S A Ghabrial*	163

Transcriptional Regulation in Bacteriophage	R A Weisberg, D M Hinton and S Adhya	174
Transmissible Spongiform Encephalopathies	E D Belay and L B Schonberger	186
Tumor Viruses: Human	R Grassmann, B Fleckenstein and H Pfister	193
Tymoviruses	A-L Haenni and T W Dreher	199

U

Umbravirus	M Taliansky and E Ryabov	209
Ustilago Maydis Viruses	J Bruenn	214

V

Vaccine Production in Plants	E P Rybicki	221
Vaccine Safety	C J Clements and G Lawrence	226
Vaccine Strategies	I Kusters and J W Almond	235
Vaccinia Virus	G L Smith	243
Varicella-Zoster Virus: General Features	J I Cohen	250
Varicella-Zoster Virus: Molecular Biology	W T Ruyechan and J Hay	256
Varicosavirus	T Sasaya	263
Vector Transmission of Animal Viruses	W K Reisen	268
Vector Transmission of Plant Viruses	S Blanc	274
Vegetable Viruses	P Caciagli	282
Vesicular Stomatitis Virus	S P J Whelan	291
Viral Killer Toxins	M J Schmitt	299
Viral Membranes	J Lenard	308
Viral Pathogenesis	N Nathanson	314
Viral Receptors	D J Evans	319
Viral Suppressors of Gene Silencing	J Verchot-Lubicz and J P Carr	325
Viroids	R Flores and R A Owens	332
Virus Classification by Pairwise Sequence Comparison (PASC)	Y Bao, Y Kapustin and T Tatusova	342
Virus Databases	E J Lefkowitz, M R Odom and C Upton	348
Virus Entry to Bacterial Cells	M M Poranen and A Domanska	365
Virus Evolution: Bacterial Viruses	R W Hendrix	370
Virus-Induced Gene Silencing (VIGS)	M S Padmanabhan and S P Dinesh-Kumar	375
Virus Particle Structure: Nonenveloped Viruses	J A Speir and J E Johnson	380
Virus Particle Structure: Principles	J E Johnson and J A Speir	393
Virus Species	M H V van Regenmortel	401
Viruses and Bioterrorism	R F Meyer and S A Morse	406
Viruses Infecting Euryarchaea	K Porter, B E Russ, A N Thorburn and M L Dyall-Smith	411
Visna-Maedi Viruses	B A Blacklaws	423

W

Watermelon Mosaic Virus and Zucchini Yellow Mosaic Virus *H Lecoq and C Desbiez* 433
West Nile Virus *L D Kramer* 440
White Spot Syndrome Virus *J-H Leu, J-M Tsai and C-F Lo* 450

Y

Yatapoxviruses *J W Barrett and G McFadden* 461
Yeast L-A Virus *R B Wickner, T Fujimura and R Esteban* 465
Yellow Fever Virus *A A Marfin and T P Monath* 469
Yellow Head Virus *P J Walker and N Sittidilokratna* 476

Z

Zoonoses *J E Osorio and T M Yuill* 485

Taxonomic Index 497
Subject Index 499

Icosahedral dsDNA Bacterial Viruses with an Internal Membrane

J K H Bamford, University of Jyväskylä, Jyväskylä, Finland
S J Butcher, University of Helsinki, Helsinki, Finland

© 2008 Elsevier Ltd. All rights reserved.

Glossary

DNA packaging Energy-requiring process where the empty virus capsid is filled by the virus genome.
Protein-primed DNA replication Duplication of a linear DNA genome utilizing a protein covalently linked to the DNA terminus to initiate the reaction.
Triangulation number T number (T), a parameter for icosahedral capsids that describes the geometrical arrangement of the protein subunits.

Introduction

Bacterial viruses with an internal membrane all have double-stranded DNA (dsDNA) genomes. They are classified into two families, the *Tectiviridae* and the *Corticoviridae*. The former family consists of several viruses, whereas the latter has only one representative. The type species of the *Tectiviridae* is PRD1, which infects a wide variety of Gram-negative bacteria. Its host range is limited to bacteria that contain a conjugative antibiotic resistance plasmid, since it utilizes the plasmid-encoded cell surface DNA-transfer complex as a receptor. The type species of the family *Corticoviridae* is PM2.

The family *Tectiviridae* can be divided into two groups: viruses infecting Gram-negative bacteria and viruses infecting Gram-positive bacteria. The group infecting Gram-negative bacteria contains six extremely similar phages (PRD1, PR3, PR4, PR5, PR772, and L17) with a linear dsDNA genome. Their sequence similarity is between 91.9% and 99.8%, which is surprising since they have been isolated from different parts of the world. The other group infects Gram-positive bacteria (different *Bacillus* species). The length of the genomes (about 15 kbp) and the order of the genes is conserved in all the tectiviruses, but there is no sequence similarity between the two groups. The genomes of the tectiviruses encode about 35 proteins (**Table 1**).

The corticovirus PM2 was isolated in 1968 from seawater off the coast of Chile. The host is a marine bacterium, *Pseudoalteromonas espejiana*. PM2 has a negatively supercoiled circular dsDNA of about 10 kbp. It encodes about 17 proteins (**Table 2**). The combination of a membrane and a supercoiled DNA genome makes PM2 a unique virus. The entry and the DNA-packaging mechanisms are likely to differ from those of other viruses.

Virion Structure and Properties

Overall Structure

The virion of PRD1 is an icosahedrally symmetric particle approximately 65 nm in diameter (**Figure 1**). It is composed of about 70% protein, 15% lipid, and 15% DNA and has a mass of about 66 MDa. The structure of the virion has been studied extensively using electron microscopy and X-ray crystallography (**Figure 2**). X-ray crystallography results have indicated the roles of four proteins in controlling virus assembly. There are 240 hexagonally shaped trimers of the major capsid protein, P3 (**Figure 2(b)**), occupying the surface of the capsid on a pseudo $T = 25$ lattice, an arrangement that is also found in the human adenovirus capsid. In PRD1, a dimer of the linear glue protein P30 extends from one vertex to the next, cementing the P3 facets together (**Figure 2(c)**). At the vertex, pentamers of P31 interlock with P3 and the transmembrane protein P16.

The majority of the PRD1 vertices have two proteins attached to them: one is a trimer of P5 (**Figure 3(a)**) attached by the N-terminus, the other is a monomer of P2 (**Figure 3(b)**), the receptor-binding protein. Both P5 and P2 are elongated molecules (P5 is 17 nm long, P2 is 15.5 nm long). P2 is a club-shaped molecule with a pseudo β-propeller head and a long tail formed from extended β-sheet. The head is proposed to be the site of

Table 1 PRD1 genes, corresponding proteins, and protein functions

Gene	Protein	Mass (kDa)	Description[a]
I	P1	63.3	DNA polymerase (N)
II	P2	63.7	Receptor binding (S)
III	P3	43.1	Major capsid protein (C)
V	P5	34.2	Trimeric spike protein (S)
VI	P6	17.6	Minor capsid protein, DNA packaging (C, P)
VII	P7	27.1	DNA delivery, transglycosylase (L, M)
VIII	P8	29.5	Genome terminal protein (N)
IX	P9	25.8	Minor capsid protein, DNA packaging ATPase (C, P)
X	P10	20.6	Assembly (A, N)
XI	P11	22.2	DNA delivery (M)
XII	P12	16.6	ssDNA binding protein (N)
XIV	P14	15.0	DNA delivery (M)
XV	P15	17.3	Muramidase (L)
XVI	P16	12.6	Infectivity (M)
XVII	P17	9.5	Assembly (A, N)
XVIII	P18	9.8	DNA delivery (M)
XIX	P19	10.5	ssDNA binding protein (N)
XX	P20	4.7	DNA packaging (M, P)
XXII	P22	5.5	DNA packaging (M,P)
XXX	P30	9.0	Minor capsid protein (C)
XXXI	P31	13.7	Pentameric base of spike (S)
XXXII	P32	5.4	DNA delivery (M)
XXXIII	P33	7.5	Assembly (A, N)
XXXIV	P34	6.7	(M)
XXXV	P35	12.8	Holin (L)

[a](N) nonstructural early protein; (M) integral membrane protein; (S) spike complex protein; (A) assembly protein; (P) packaging protein; (C) capsid protein; (L) lysis protein.

Table 2 PM2 genes, corresponding proteins, and protein functions

Gene	Protein	Mass (kDa)	Description
I	P1	37.5	Spike protein
II	P2	30.2	Major capsid protein
III	P3	10.8	Membrane protein
IV	P4	4.4	Membrane protein
V	P5	17.9	Membrane protein
VI	P6	14.3	Membrane protein
VII	P7	3.6	Membrane protein
VIII	P8	7.3	Membrane protein
IX	P9	24.7	Potential ATPase
X	P10	29.0	Membrane protein
XII	P12	73.4	Replication initiation protein
XIII	P13	7.2	Transcription factor
XIV	P14	11.0	Transcription factor
XV	P15	18.1	Transcription factor
XVI	P16	10.3	Transcription factor
XVII	P17	6.0	Lysis
XVIII	P18	5.7	Lysis

receptor binding, lying distal to the virus. A specific vertex is used for packaging the phage DNA.

The overall size and structure of the phage Bam35 is very similar to that of PRD1. However, the exact counterparts of many of the PRD1 structural proteins have not yet been clearly identified. One of the major differences between PRD1 and Bam35 is the presence of a large transmembrane protein complex in Bam35 that modulates the curvature of the membrane under the capsid facets. Also, the host cell recognizing spike complex is likely to be different.

The corticovirus PM2 particle is icosahedral and measures 77 nm in diameter from spike to spike. The capsid is approximately 60 nm in diameter. Like the tectiviruses, it does not have a tail. The mass of the virion is $\sim 4.5 \times 10^7$ Da and it is composed of protein (72%), lipid (14%), and DNA (14%). The capsid is composed of 200 trimers of the major capsid protein arranged on a pseudo $T = 21$ lattice. Pentameric receptor-binding spikes protrude from the vertices.

Membrane and DNA

In PRD1, about half of the virion proteins are associated with the membrane (**Table 1**). The lipid headgroups are predominantly phosphatidylethanolamine (53%) and phosphatidylglycerol (43%) with 4% of cardiolipin. The membrane is well ordered, following the icosahedral outline of the capsid. Many interactions occur between the membrane and both the capsid and the underlying DNA. The average separation of the concentric layers of DNA is approximately 2.5 nm, similar to that found in other bacteriophages and animal viruses. Removal of the capsid and spike proteins by heat or guanidinium hydrochloride treatment results in aggregation of the membrane vesicle.

In PM2, the membrane, lying underneath the capsid, follows the shape of the capsid as in PRD1, and there are many interactions mediated by additional minor proteins. The lipid composition is approximately 64% phosphatidylglycerol, 27% phosphatidylethanolamine, and 8% neutral lipids and a small amount of acyl phosphatidylglycerol. Release of the capsid and spike proteins from the virion by freeze–thawing or by chelation of calcium ions with ethylene glycol tetraacetic acid (EGTA) results in a soluble vesicle called the lipid core.

Life Cycle

PRD1 is a lytic phage that exploits the transcription functions of its host. The host cell is selected by specific recognition of a plasmid-encoded receptor on the cell surface. PRD1 belongs to the class of broad-host-range, donor-specific phages, which infect cells only when an IncP-, IncN-, or IncW-type multiple-drug resistance conjugative plasmid is present. The primary function of the receptor is in bacterial conjugation. Among the hosts are several opportunistic human pathogens such as *Escherichia coli*, *Salmonella enterica*, and *Pseudomonas aeruginosa*. After adsorption, the phage genome is injected into the cell cytosol, leaving the capsid outside. After the production of the phage components, both virus- and host-encoded factors assist in particle assembly. Host cell lysis releases some 500 progeny viruses. Bam35 is a temperate phage, either growing lytically like PRD1, or existing in a dormant state within the host. In contrast to PRD1, the host range of Bam35, and its relatives GIL01 and GIL16, is limited to one species, *Bacillus thuringiensis*. AP50 infects *Bacillus anthracis*, and a defect phage replicating as a linear plasmid (pBClin15) has been described for *Bacillus cereus*. The corticovirus PM2 is lytic. Of all the dsDNA phages with an internal membrane, the life cycle of PRD1 is best understood and is described below.

Receptor Recognition

A single phage structural protein, P2, is responsible for PRD1 attachment to its host. Each of the PRD1 receptor recognition vertices is a metastable structure and possibly capable of DNA release. The injection vertex is likely to be determined by P2 binding to the receptor. The association of P2 with the receptor activates, possibly by P2 removal, the injection process. This leads to irreversible binding. Both empty and DNA-containing particles are

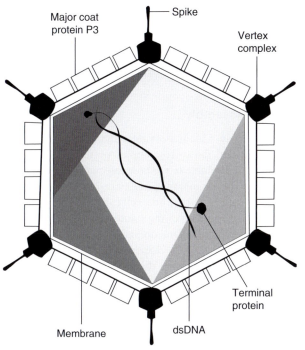

Figure 1 Schematic presentation of PRD1 virion.

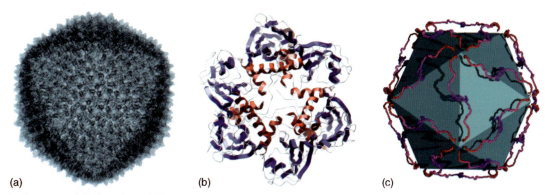

Figure 2 Structure of the icosahedral PRD1 capsid based on X-ray crystallography. (a) Electron density of the PRD1 virion at 4 Å resolution. (b) The major coat protein P3, top view. P3 has a double β-barrel fold, resulting in a hexameric shape of the trimer. (c) The structure of the virion revealed the location of the cementing protein P30 at the twofold positions. The icosahedron is representing the viral membrane. Dimers of P30 are lying underneath the capsid stabilizing the virion structure.

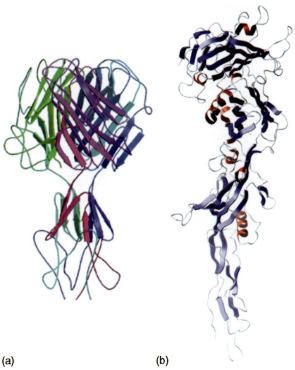

Figure 3 Structure of the vertex proteins of PRD1 based on X-ray crystallography. (a) The trimeric spike protein P5. (b) The monomeric receptor-binding protein P2. (a) Reproduced from Merckel MC, Huiskonen JT, Bamford DH, Goldman A, and Tuma R (2005) The structure of the bacteriophage PRD1 spike sheds light on the evolution of viral capsid architecture. *Molecular Cell* 18: 161–170, with permission from Elsevier.

bound equally tightly to cells, indicating that DNA injection is not a prerequisite for this tight interaction.

DNA Entry

Isolation and analysis of PRD1 mutants have resulted in the identification of eight, phage-specific, structural proteins essential for infectivity. In addition to the spike-complex proteins (P2, P5, and P31) needed for adsorption, protein P11 starts the DNA delivery process and membrane proteins P14, P18, and P32 are involved in later stages of the DNA delivery. Mutant particles missing protein P7 (a lytic transglycosylase) are infectious but the DNA entry process is delayed.

The PRD1 membrane can undergo a structural transformation from a spherical vesicle to a tubular form. A similar process has been described for Bam35 and AP50, thus occurring in all members of the *Tectiviridae*. This tube formation might be required for DNA translocation.

Genome Replication

The genome of PRD1 is a linear double-stranded DNA molecule with proteins covalently attached to both 5′ termini and having 110 bp terminal repeats. PRD1 replicates its DNA by a protein-primed replication mechanism similarly to other viruses with linear dsDNA genomes, including adenovirus and the φ29-type phages. PRD1 DNA replication starts with the formation of a covalent bond between the genome terminal protein, P8, and the 5′ terminal nucleotide, dGMP, in a reaction catalyzed by the phage DNA polymerase, P1. The minimal origin of replication resides in the 20 first-terminal base pairs of both genome ends, and the fourth base from the 3′ end of the template directs, by base complementation, the linking of deoxyribonucleoside monophosphate (dNMP) to the terminal protein. The 3′ end DNA sequence is maintained by sliding back of the polymerase complex.

After initiation, elongation of the initiation complex by the same DNA polymerase takes place resulting in the formation of full-length daughter DNA molecules. Two phage-encoded single-stranded DNA binding proteins, P12 and P19, are involved in replication *in vivo*.

Particle Assembly

Approximately 15 min post infection, the major capsid protein P3 and the spike-complex proteins P2, P5, and P31 are found soluble in the host cell cytosol, whereas the phage-encoded membrane proteins (e.g., P7, P11, P14, and P18) are addressed to the host cell cytoplasmic membrane (CM). Correct folding of the soluble proteins and assembly of a number of viral membrane proteins are dependent on the host GroEL/ES chaperonins. Upon assembly, a virus-specific patch from the host CM is translocated into the forming procapsid using the membrane-bound scaffolding protein P10. In addition, two small phage-encoded proteins are implicated in the assembly process: P17 and P33.

Correct assembly results in an empty capsid enclosing a membrane rich in phage-specific proteins. The linear double-stranded DNA genome is packaged into the prohead by the packaging ATPase P9. Unlike packaging ATPases of most other icosahedral dsDNA bacteriophages, P9 is part of the mature virus structure. It resides at a single vertex that also contains proteins P6, P20, and P22. P6 is a soluble protein needed for efficient DNA packaging and the latter two are integral membrane proteins connecting the portal structure to the viral membrane.

Cell Lysis

At the end of the infection cycle, the newly synthesized progeny virions are released via host cell lysis. Two genes, *XV* and *XXXV*, are involved in this step, which means that a two-component, holin–endolysin system operates in phage PRD1. The product of gene *XV*, protein P15, is a soluble β-1,4-N-acetylmuramidase that degrades the peptidoglycan of the Gram-negative cell causing host cell lysis. The PRD1 particle carries another muramidase, protein P7, which has a lytic transglycosylase activity assisting in genome entry. The presence of two lytic activities probably reflects the broad host range of PRD1.

In addition to lytic enzymes, bacteriophages quite often encode helper protein factors (holins) that facilitate the access of lytic enzymes to the susceptible bond in the cell wall and control the timing of lysis. The PRD1 holin is protein P35.

Genomes and Genomics

The length of the PRD1 genome is 14 927 bp and the guanine–cytosine (GC) content is 48.1%. It has 110 bp long inverted repeat sequences at the ends, which are 100% identical. The genomes of the other Gram-negative bacteria-infecting tectiviruses are very close to that of PRD1, varying between 14 935 and 14 954 bp. They are remarkably similar in nucleotide sequence, the overall identity being 91.9–99.8%. The Bam35 genome is 14 935 bp long with a GC content of 39.7%. The inverted repeat for Bam35 is 74 bp long. These Bam35 inverted terminal repeats (ITRs) have 81% identity. PRD1 and Bam35 do not share much sequence similarity. The discovery of the almost invariant genomes for the two *Tectiviridae* groups contrasts sharply with the situation in the tailed bacteriophages. The nucleotide identity is 100% between Bam35 and GIL01 and 83% between GIL01 and GIL16.

PRD1 genome is organized into two early and three late operons (**Figure 4**). Bam35 operons have not been mapped, but the gene order and the length of the genes are similar to those of PRD1 (**Figure 4**). Although there is no overall sequence similarity between PRD1 and Bam35 genomes, Bam35 genes for DNA polymerase, packaging ATPase, and lytic enzyme can be recognized by corresponding conserved amino acid sequence motifs in the databases. The coat protein gene can be identified by comparing the N-terminal amino acid sequence of the major virion protein to the DNA sequence. The only genes of Bam35, which seem to be in different positions in the genome compared to PRD1, are those responsible for the host cell recognition (the spike protein).

The circular PM2 genome is 10 079 bp long with a GC content of 42.2%. It contains 21 putative genes, of which 17 have so far been shown to be functional. Promoter mapping by primer extension has revealed three operons, which are expressed in a timely fashion during infection. The first early operon is highly similar to the maintenance region of the *Pseudoalteromonas* plasmid pAS28. The second early operon contains genes for DNA replication and regulation of late phage functions. The PM2 genome replicates via a rolling-circle mechanism. Protein P12 has conserved sequence motifs common to superfamily I replication initiation proteins. This superfamily consists of the A proteins of certain bacteriophages, such as ϕX174 and G4, and the initiation proteins of cyanobacterial and archaeal plasmids. The function of the late operon is activated by two phage-encoded transcription factors, P13 and P14. P14 has sequence similarity to the TFIIS-type general eukaryotic transcription factors most closely resembling those of the archaeal organisms *Thermococcus celer* and *Sulfolobus acidocaldaricus*.

The structural proteins encoded by the late genes are, similar to the corresponding proteins of tectiviruses, either membrane associated or soluble. Based on the conserved amino acid sequence motifs deduced from the nucleotide sequence, one of the structural virion proteins

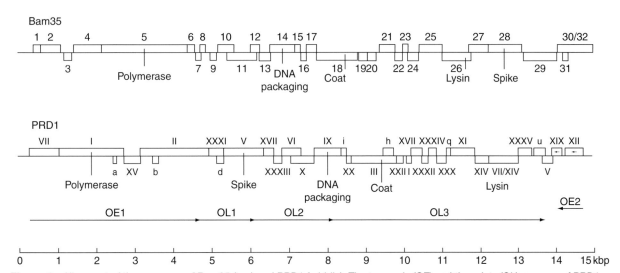

Figure 4 Alignment of the genomes of Bam35 (top) and PRD1 (middle). The two early (OE) and three late (OL) operons of PRD1 are shown at the bottom. In Bam35, the open reading frames (ORFs) are marked by Arabic numerals from left to right. PRD1 ORFs have a Roman numeral if they have been shown to encode protein, otherwise they are marked with a lowercase letter.

is a putative packaging ATPase. The packaging process of the circular supercoiled PM2 DNA is not understood at the moment.

See also: Capsid Assembly: Bacterial Virus Structure and Assembly; Genome Packaging in Bacterial Viruses; Lysis of the Host by Bacteriophage; Replication of Bacterial Viruses; Transcriptional Regulation in Bacteriophage; Virus Evolution: Bacterial Viruses.

Further Reading

Abrescia NGA, Cockburn JJB, Grimes JM, et al. (2004) Insights into assembly from structural analysis of bacteriophage PRD1. *Nature* 432: 68–74.

Bamford DH (2005) Family *Tectiviridae*. In: Fauquet CM, Mayo MA, Maniloff J, Desselberger U, and Ball LA (eds.) *Virus Taxonomy. Eighth Report of the International Committee on Taxonomy of Viruses*, pp. 81–85. San Diego, CA: Elsevier Academic Press.

Bamford DH and Bamford JKH (2006) Lipid-containing bacteriophage PM2, the type-organism of *Corticoviridae*. In: Calendar R (ed.) *The Bacteriophages*, pp. 171–174. New York: Oxford University Press.

Bamford JKH (2005) Family *Corticoviridae*. In: Fauquet CM, Mayo MA, Maniloff J, Desselberger U, and Ball LA (eds.) *Virus Taxonomy: Eighth Report of the International Committee on Taxonomy of Viruses*, pp. 87–90. San Diego, CA: Elsevier Academic Press.

Benson SD, Bamford JKH, Bamford DH, and Burnett RM (1999) Viral evolution revealed by bacteriophage PRD1 and human adenovirus coat protein structures. *Cell* 98: 825–833.

Benson SD, Bamford JKH, Bamford DH, and Burnett RM (2004) Does common architecture reveal a viral lineage spanning all three domains of life? *Molecular Cell* 16: 673–685.

Cockburn JJB, Abrescia NGA, Grimes JM, et al. (2004) Membrane structure and interactions with protein and DNA in bacteriophage PRD1. *Nature* 432: 122–125.

Grahn AM, Butcher SJ, Bamford JKH, and Bamford DH (2006) PRD1-dissecting the genome, structure and entry. In: Calendar R (ed.) *The Bacteriophages*, pp. 161–170. New York: Oxford University Press.

Grahn AM, Daugelavicius R, and Bamford DH (2002) Sequential model of phage PRD1 DNA delivery: Active involvement of the viral membrane. *Molecular Microbiology* 46: 1199–1209.

Merckel MC, Huiskonen JT, Bamford DH, Goldman A, and Tuma R (2005) The structure of the bacteriophage PRD1 spike sheds light on the evolution of viral capsid architecture. *Molecular Cell* 18: 161–170.

Ravantti JJ, Gaidelyte A, Bamford DH, and Bamford JKH (2003) Comparative analysis of bacterial viruses Bam35, infecting a Gram-positive host, and PRD1, infecting Gram-negative hosts, demonstrates a viral lineage. *Virology* 313: 401–414.

Salas M (1991) Protein-priming of DNA replication. *Annual Review of Biochemistry* 60: 39–71.

Icosahedral Enveloped dsRNA Bacterial Viruses

R Tuma, University of Helsinki, Helsinki, Finland

© 2008 Elsevier Ltd. All rights reserved.

Glossary

Carrier state A partial or complete viral genome is maintained inside the host cell as a stable episome without lysis or infection.

Genomic precursor ssRNA destined for packaging.

Genomic segment A piece of genomic dsRNA.

Packaging A process of genomic precursor acquisition.

Pac site A specific signal at 5' end of a genomic precursor which targets it for selective packaging.

Polymerase complex Viral assembly which contains the polymerase and is able to replicate and transcribe the viral genome. Usually it also contains the viral genomic dsRNA.

Procapsid Empty capsid assembly which is capable of acquiring the genomic precursors.

Reverse genetics RNA virus genome manipulation and rescue from cDNA clones.

Viral core dsRNA containing inner part of the virus which is capable of transcription.

Introduction – The *Cystoviridae* Family

Members of the family *Cystoviridae* are lipid-containing bacteriophages with segmented dsRNA genomes. The first recognized member, the bacteriophage φ6, was isolated in 1973. Several other members were discovered in the late 1990s. All known cystoviruses were isolated from the leaves of various plants and consequently infect primarily plant-pathogenic bacteria (pseudomonads). All share a similar architecture with a proteinaceous viral core hosting the three genomic segments (L, M, and S, approx. 6.5, 4, and 3 kbp in size, respectively). The core is enclosed in a lipid envelope which contains the host cell attachment proteins. The viral core contains an RNA-dependent RNA polymerase (RdRP) which plays a central role in RNA metabolism. Despite infecting prokaryotic hosts these phages exhibit structural and functional features that parallel those of the family *Reoviridae*. Because reverse genetics had been developed early on for φ6, the system has been a model for studying the assembly and replication of other dsRNA viruses.

An *in vitro* assembly of infectious nucleocapsids from purified constituents has been achieved for bacteriophages φ6 and φ8. Similarly, RNA packaging and replication have been extensively studied *in vitro*. Together with the reverse genetics system the *in vitro* methods have been instrumental in delineating the virus replication mechanisms. The host entry mechanism mimics that of viruses infecting eukaryotic cells and involves both membrane fusion and an endocytotic-like event. Structural and extensive biochemical characterization of individual proteins and subviral assemblies made the cystoviral system a paradigm for studying mechanisms of molecular machines in atomic details. The self-assembly and replication system shows promise in biotechnology and nanotechnology applications.

Classification and Host Range

Based on their sequences and host range five out of the recently discovered cystoviruses (φ7, φ9, φ10, φ11, φ14) are close relatives of φ6 while the remaining three (φ8, φ12, and φ13) are only distantly related to the φ6 group and among themselves. φ6 and the related phages attach to the host cell via a specific type IV pilus while the members of the second group utilize rough lipopolysaccharides (LPSs). This limits the host range to *Pseudomonas syringae* (*Pseudomonas pseudoalcaligenis* for certain mutants) for the former group while phages belonging to the latter can also infect rough LPS of other Gram-negative bacteria without forming plaques. Only φ8 form plaques on a heptose-less strain of *Salmonella typhimurium*. A common laboratory host of φ6 is *P. syringae* strain HB10Y. The φ8 phage is usually propagated on *P. syringae* strain LM2509.

Virion Structure and Properties

Physico-Chemical Properties

Virion molecular weight (MW, in dalton units, Da) $\sim 9.9 \times 10^7$ Da, sedimentation coefficient S_{20w} $\sim 405S$, density $1.24\,\mathrm{g\,cm^{-3}}$ (sucrose), nucleocapsid MW $\sim 4.0 \times 10^7$ Da. Composition: 70% (w/w) protein, 14% dsRNA, 16% phospholipids. RNA packaging density $350\,\mathrm{g\,cm^{-3}}$.

Architecture of the Polymerase Complex and the Nucleocapsid

Figure 1 depicts schematically the architecture and localization of the proteins for the φ6 virion. The innermost structure is the polymerase complex (PC) which is composed of 120 copies of major structural protein P1 (MW 85 kDa), 12 copies of RdRP P2 (75 kDa), 12 hexamers of packaging motor P4 (subunit MW 35 kD), and 60 copies of assembly factor P7 (17 kDa). Structural details of the dodecahedral P1 skeleton are depicted in **Figure 2**. The $T=1$ lattice is composed of 60 asymmetric P1 dimers as seen for other dsRNA viruses. However, the arrangement of dimers within the asymmetric unit is different from that seen in the reovirus or bluetongue virus cores. P2 monomers are attached to the inner surface of the PC

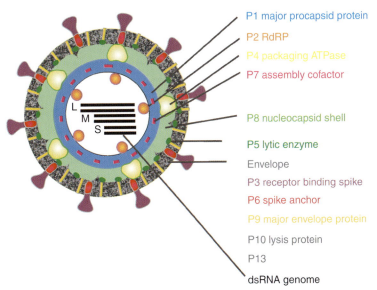

Figure 1 Schematics of φ6 virion architecture. Function of individual proteins: P1-major structural protein of PC, forms the dodecahedral skeleton of the procapsid; P2-RdRP; P3-spike protein, host cell attachment; P4-packaging ATPase, ssRNA translocation; P5-lytic enzyme (lysozyme); P6-integral membrane protein, P3 anchoring and membrane fusion; P7-assembly, packaging and replication cofactor, PC stabilization; P8-nucleocapsid coat protein; P9-membrane protein; P10-host cell lysis, perhaps holin; P11-membrane protein (most likely nonessential).

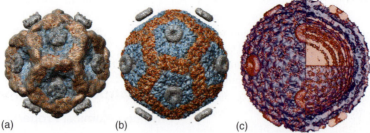

Figure 2 Structure of the polymerase complex and the nucleocapsid of φ6 as revealed by electron cryo-microscopy. (a) Modeled empty PC (void of RNA, precursor for packaging) which is also called the procapsid. (b) The $T=1$ (triangulation number) shell of the expanded, mature, PC which contains dsRNA genome and is transcription competent. (c) The $T=13$ P8 shell which makes the outer surface of NC together with the protruding P4 hexamers. P4 is gray, the two P1 monomers in the asymmetric dimer are colored red and blue, respectively. P8 shell is rendered purple. Reproduced from Huiskonen JT, de Haas F, Bubeck D, Bamford DH, Fuller SD, and Butcher SJ (2006) Structure of the bacteriophage φ6 nucleocapsid suggests a mechanism for sequential RNA packaging. *Structure* 14: 1039–1048, with permission from Elsevier.

shell, possibly at the fivefold vertices. P4 hexamers form turret-like protrusions on the outer surface of the fivefold vertices. Location of the P7 within the PC is not known.

PC undergoes a series of large conformational changes (expansions) during RNA packaging and genome replication. The empty dodecahedral PC (procapsid) expands into a more round, icosahedral dsRNA-filled PC, which is sometimes called the viral core. The expansion consists of motions of whole P1 subunits around hinges that run roughly parallel with the twofold edges of the dodecahedral skeleton (**Figure 2**). The expansion increases the inner volume by a factor of 2.4 and makes room for the dsRNA genome. The mature PC is subsequently enclosed in an icosahedral shell ($T=13$) made of 200 P8 (subunit MW 16 kDa) trimers (**Figure 2**). The resulting structure is called the nucleocapsid (NC). The packaged dsRNA genome is arranged in concentric shells (layers) with average spacing 31 Å. This is larger than the 24–26 Å spacing which has been observed for other dsDNA and dsRNA viruses and reflects the lower packaging density.

The structures of P1 (PC) and P8 (NC) shells are related to the structure of dsRNA containing cores isolated from viruses belonging to the family *Reoviridae*. However, the empty packaging competent precursors have not been detected nor has a conformational change akin to the PC expansion been observed for the latter viruses.

Structure of Individual Proteins

P1

Current structural model of φ6 P1 monomer is based on high-resolution (7.5 Å) electron cryo-microscopy (**Figure 3**). The structure is composed of α-helices that are packed together to form a flat sheet. The few remaining densities have been tentatively assigned to β-sheets. The exact tracing of the polypeptide chain was not possible at this resolution; however, the two conformers in the asymmetric unit were clearly resolved.

P2

Several structures of φ6 P2 alone and in complex with various substrates were solved by X-ray diffraction. P2 is a compact, globular, monomer with α/β fold (**Figure 3**) with a domain arrangement resembling a hand (i.e., palm, fingers, and thumb domains). The fold is similar to those of other viral polymerases, for example, hepatitis C virus and HIV reverse transcriptase. RNA substrate binds in a template channel and the putative dsRNA exit site is partially occluded by the C-terminal domain in the crystal structure (**Figure 3**).

P4

The structures of φ12 P4 hexamer in apo and nucleotide-bound forms have been solved by X-ray diffraction (**Figure 3**). The hexamer has a dome-like structure with a central channel through which RNA passes during packaging. Six equivalent ATP binding sites are located between the subunits at the hexamer periphery. The subunit structure encompasses a catalytic core which is conserved among many ATPases, including Rec-A and hexameric helicases. P4 hexamer interacts with the procapsid dodecahedral framework predominantly via its C-terminal facet.

P8

The structure of the φ6 P8 trimer was derived by fragmentation of the cryo-EM density map at 7.5 Å resolution. The P8 subunit is highly α-helical. The trimer is flat (only 25 Å thick, 75 Å effective diameter) and each subunit is composed of the central, four-helix bundle core and a peripheral four helix bundle (**Figure 3**).

Membrane Envelope

Lipids are derived from the host cytoplasmic membrane but the viral envelope is enriched in phosphatidylglycerol and contains less phosphatidylethanolamine presumably due to high viral membrane curvature and charge repulsion.

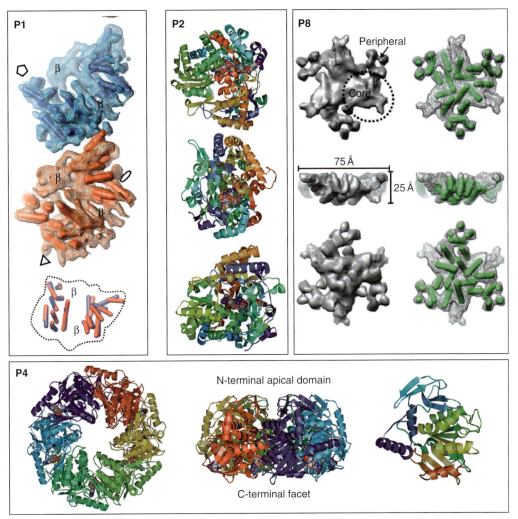

Figure 3 Structure of individual cystoviral proteins. (P1) φ6 P1 structure was derived from cryo-EM. The electron densities of the two conformers found in the asymmetric unit (see **Figure 2**) are shown in blue and red, respectively. Bottom panel shows the tentative secondary structure assignment to the electron density. Reproduced (P2) Ribbon diagram of P2 polymerase structure from phage φ6 in complex with RNA (spacefill representation). Based on PDB id 1UVI, Salgado *et al.* (2004) *Structure* 12: 307–316. Top view shows the C-terminal domain (red) blocking the RNA exit site. Middle panel displays view down the RNA template channel. Bottom panel exhibits view of the template RNA through the substrate channel. Ribbon coloring from N-terminus (blue) to C-terminus (red). (P4) P4 hexamer, the packaging motor (ATPase) structure solved by X-ray diffraction. Based on PDB id 1W44, Mancini EJ, Kainov DE, Grimes JM, Tuma R, Bamford DH, and Stuart DI (2004) Atomic snapshots on an RNA packaging motor reveal conformational changes linking ATP hydrolysis to RNA translocation. *Cell* 118: 743–755. Left panel: top view down the central channel, subunits in different colors. Middle: Side view. Right: One subunit colored from N-terminus (blue) to C-terminus (red). (P8) P8 structure derived from cryo-EM. Reproduced from Huiskonen JT, de Haas F, Bubeck D, Bamford DH, Fuller SD, and Butcher SJ (2006) Structure of the bacteriophage φ6 nucleocapsid suggests a mechanism for sequential RNA packaging. *Structure* 14: 1039–1048, with permission from Elsevier.

Lipids constitute about 40% of membrane mass while the remaining 60% are membrane proteins. The membrane is loosely associated with the nucleocapsid and the membrane-associated proteins lack icosahedral symmetry.

Genome Organization and Sequence Similarity

The φ6 genome (13385 bp) is organized into three segments (**Figure 4**): L (6374 bp in φ6) codes for the procapsid structural proteins, M (4063 bp in φ6) codes for the spike and membrane proteins, and S (2948 bp in φ6) codes for nucleocapsid and membrane proteins. The (coding) plus strand of each segment contains a short conserved sequence at the 5′-end. This sequence is followed by a packaging signal that is unique to each segment (*pac* site, about 200 nt). The dsRNA genome remains sequestered inside the PC throughout the viral life cycle and only the plus strands are extruded into the cytoplasm where they serve as messenger RNAs or packaging precursors.

Two genes, 12 and 14, code for the two nonstructural proteins identified thus far. While the former is essential for membrane morphogenesis, the latter is dispensable under laboratory conditions.

Genomic maps of the other cystoviruses are similar to φ6 with several exceptions. The viruses which attach directly to rough LPS (φ8, 12, 13) possess several separate genes for the receptor binding spike structure (3a, 3b, and 3c in φ12 and φ13). In addition, gene 7 is positioned after gene 1 in the φ8 L-segment and its expected place is occupied by ORF H. The functions of the nonessential ORFs F, G, and H are not known.

Identity at the amino acid sequence level is high among φ6, 7, 9, and 10, while φ8, 12, and 13 exhibit only limited sequence similarity to φ6 and among themselves. The limited similarity is restricted to the conserved motifs of the polymerase and the packaging ATPase. Sequence similarities for the structural proteins P1 and P7 are generally statistically insignificant for the latter group of phages despite conservation of the procapsid structure. However, high sequence identity detected in restricted regions (e.g., lysis proteins of φ12 and φ6, identical 3′-end of M segments in φ6 and φ13) indicate significant genetic interaction (e.g., recombination) among cystoviruses.

Replication Cycle

The virus life cycle is schematically illustrated in **Figure 5**. The steps were elucidated for φ6 but many features are likely to be similar for the other cystoviruses.

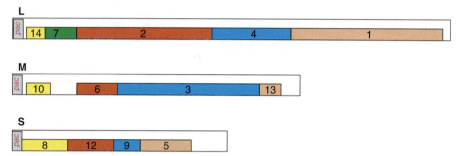

Figure 4 The phi6 genome map showing the locations of genes (gene numbers) and the packaging signal sequences (*pac*).

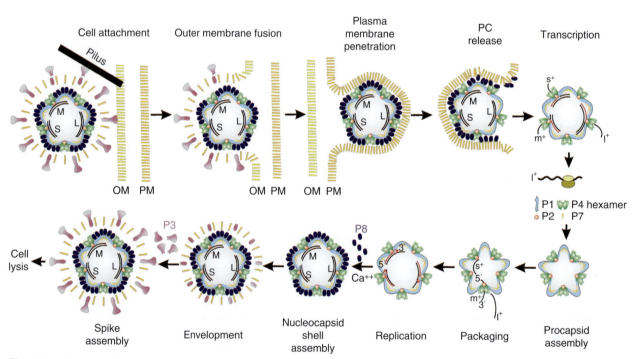

Figure 5 Life cycle diagram of φ6. See the text for details.

Host Cell Attachment and Outer Membrane Fusion

The φ6 and related phages attach to the pilus using the P3 spike protein. The pilus contracts and brings the virion into contact with the outer cell membrane (OM). The φ8, 12, 13 phages utilize a spike composed of proteins P3a, P3b, and P3c (missing in φ8) to attach directly to the rough LPS. The viral envelope then fuses with the OM using the virion fusogenic protein P6. Then the virion-associated lysozyme (P5) degrades the peptidoglycan layer to allow direct contact of the nucleocapsid with the plasma membrane (PM).

Plasma Membrane Penetration

The φ6 nucleocapsid is first enclosed in an endocytic-like vesicle which is brought into the cytoplasm. The NC coat protein P8 disassembles and disrupts the membrane vesicle in order to release the PC into the cytoplasm. The mechanism of the P8-assisted release is still elusive.

Transcription

The depolymerization of the P8 layer activates the PC to perform transcription of the dsRNA genome which produces the plus-strand copies of the three segments (s^+, m^+, l^+). These copies are made inside the PC by the P2 polymerase using a semiconservative mechanism. Several polymerase molecules may engage each segment simultaneously. The resulting RNAs are extruded via P4 hexamers into the cytoplasm where they serve as mRNA for the synthesis of phage proteins and as ssRNA precursors for packaging. Transcription of the three segments is regulated, s^+ and m^+ being produced in a several-fold excess over l^+. The regulation is due to specific interactions between the 3'-end of the minus strand and the polymerase.

PC assembly

The major structural protein P1 co-assembles with P2, P4, and P7 into empty PC (procapsid). The assembly is nucleated by a P1–P4 (P1–P2 in φ8 phage) interaction and intermediates are further stabilized by protein P7 (a 'glue' protein).

Packaging

The unique *pac* sites at the 5'-end of RNA plus strands are specifically recognized in a sequential fashion by the assembled P1 framework. This initial and selective binding brings the 5'-end to the vicinity of the packaging ATPase (P4) hexamer. The P4 ring opens and topologically encloses the RNA inside the central channel. The presence of RNA stimulates the ATPase activity. P4 translocates the ssRNA in the 5'–3' direction into the procapsid at the expense of ATP hydrolysis. The sequential packaging order is accomplished by increasing the PC affinity for m^+ (l^+) *pac* sequence after packaging of s^+ (m^+). This switching is mediated by the expansion which is in turn driven by the increased amount of the packaged RNA (cf. the headfull packaging of dsDNA bacteriophages). The *in vivo* packaging fidelity is high. However, specific *in vitro* packaging conditions may result in acquisition of multiple s^+ segments or of m^+ segment in the absence of prior packaging of s^+. A heterologous RNA can be packaged by φ8 PCs and the sequential dependence of *in vitro* packaging is weak for this phage.

Replication and NC Maturation

During φ8 or after φ6 packaging, the polymerase performs replication (minus-strand synthesis) of ssRNA into the genomic dsRNA. P2 polymerase interacts with a specific sequence at the 3'-end of each plus strand to initiate the self-primed reaction and makes a full copy of each packaged template. This reaction doubles the amount of RNA inside the PC and yields the mature icosahedral PC. The mature PC can be coated with a layer of P8 trimers in a calcium-dependent reaction. This yields the nucleocapsid. Alternatively, PC may remain in the transcription mode and produce more plus-strand RNA. In φ8 virus the P8 protein is directly associated with the membrane and consequently the PC becomes directly enveloped. The mature φ8 PC is also infectious to host cell spheroplasts.

Membrane Acquisition and Host Cell Lysis

NC is enveloped with a lipid bilayer, which is derived from the host cell plasma membrane. It contains only the viral membrane proteins. The virion-associated lytic enzyme P5 is also incorporated at this stage. Structural protein P9 and the nonstructural protein P12 are essential and sufficient for membrane envelopment in φ6. These two proteins alone can produce lipid vesicles inside the host cell suggesting that envelopment takes place within the cytoplasm. The envelope is subsequently decorated with P3 receptor binding spike which is anchored by the integral membrane protein P6. Virions are released by host cell lysis which is assisted by phage-encoded lytic proteins P5 and P10.

Recombination

Recombination in the *Cystoviridae* is mostly heterologous requiring only about three identical nucleotides. The mechanism of recombination is template switching. The frequency of recombination is increased by truncating the 3′-end of one of the packaged plus strands. Such truncation may be a result of cellular nuclease cleavage and yields a poor substrate for replication. Hence, recombination may serve as an effective way of correcting the nuclease damage. In addition, homologous recombination, requiring an identical sequence of about 600 nucleotides, has been demonstrated for the φ8 phage. Another mode of recombination is the exchange of whole genomic segments as also seen for other segmented genome viruses.

Genetic Tools and Applications

Reverse Genetics

A reverse genetics system allows systematic testing of the effects of engineered mutations on specific steps in the phage life cycle. Thus, it greatly facilitates delineation of the viral packaging and replication mechanisms. There are two ways to introduce a foreign gene into the dsRNA genome. (1) By constructing a recombinant ssRNA transcript containing the foreign sequences and a 5′-end *pac* site. This substrate is then packaged into procapsids, replicated and matured into NC *in vitro*. The NCs infect host cell spheroplasts to produce the engineered virus progeny. (2) A more efficient method for reverse genetics is based on electroporation of cDNA plasmids, containing the three genomic segments, into cells expressing either SP6 or T7 RNA polymerase. The foreign sequence may be introduced between the 5′-end *pac* sites and the conserved 3′-end replication sequences of each segment.

In Vitro Assembly, Packaging, and Replication System

φ6 procapsid can be efficiently produced in *Escherichia coli* by simultaneously expressing proteins P1, P2, P4, and P7 from a plasmid containing a cDNA copy of the L segment. These procapsids are capable of packaging and replicating the three genomic precursors *in vitro*. Even under optimal conditions only about 5–10% of these procapsids are successfully packaged *in vitro*. Despite the low efficiency the *in vitro* packaging system has been instrumental in delineating the sequential packaging rules and the replication specificity. Incomplete procapsids (containing proteins P1P4, P1P2P4, P1P4P7, P1P2P7) can be produced in *E. coli* and tested for packaging and polymerase activity or used for assembly of specifically labeled procapsids.

Procapsids can be assembled *in vitro* from purified proteins. The *in vitro* assembly system allowed the identification of the minimum assembly requirements and the nucleation mechanism. The *in vitro* system also allows incorporation of chemically labeled or modified subunits to facilitate biophysical and structural studies.

Carrier State

Bacteriophages φ6 and φ8 can establish a carrier state during which the viral genome (or portion of it) replicates as a stable episome inside the host cell cytoplasm. The stable episome constitutes the phage dsRNA inside polymerase complexes that assemble in the cytoplasm. Consequently, the L segment cDNA coding for the PC protein is required to establish the state. The carrier state is maintained by a selective resistance marker engineered into one of the phage genomic segments (usually M or S).

The carrier state is established by electroporating ColE1 suicide plasmids containing the cDNAs of the phage segments into a pseudomonas host that expresses T7 RNA polymerase. These plasmids cannot replicate in pseudomonads and are used to introduce the phage ssRNA into the cytoplasm.

A carrier state may be used to probe RNA packaging or to select new phage mutants. Given the high number of mutations introduced during the phage RNA replication cycle, the carrier state constitutes a good vehicle for targeted evolution of proteins or RNA with novel functions or a way to produce large quantities of siRNA.

Polymerase Applications

φ6 RdRP has the unique capacity to perform self-primed, primer-independent, replication and transcription of RNA substrates. This feature can be used, for example, to obtain complete copies of viral RNA genomes (amplification), or for primer-independent RNA sequencing. An engineered version of RdRP is commercially available.

See also: Replication of Bacterial Viruses; Virus Entry to Bacterial Cells.

Further Reading

Bamford DH, Ojala PM, Frilander M, Walin L, and Bamford JKH (1995) Isolation, purification, and function of assembly intermediates and subviral particles of bacteriophages PRD1 and φ6. In: Adolph KW (ed.) *Methods in Molecular Genetics, Vol. 6: Microbial Gene Techniques*, pp. 455–474. San Diego: Academic Press.

Butcher SJ, Grimes JM, Makeyev EV, Bamford DH, and Stuart DI (2001) A mechanism for initiating RNA-dependent RNA polymerization. *Nature* 410: 235–240.

Huiskonen JT, de Haas F, Bubeck D, Bamford DH, Fuller SD, and Butcher SJ (2006) Structure of the bacteriophage φ6 nucleocapsid suggests a mechanism for sequential RNA packaging. *Structure* 14: 1039–1048.

Kainov DE, Tuma R, and Mancini EJ (2006) Hexameric molecular motors: P4 packaging ATPase unravels the mechanism. *Cellular and Molecular Life Sciences* 63: 1095–1105.

Makeyev EV and Bamford DH (2000) The polymerase subunit of a dsRNA virus plays a central role in the regulation of viral RNA metabolism. *EMBO Journal* 19: 6275–6284.

Makeyev EV and Bamford DH (2004) Evolutionary potential of an RNA virus. *Journal of Virology* 78: 2114–2120.

Mancini EJ, Kainov DE, Grimes JM, Tuma R, Bamford DH, and Stuart DI (2004) Atomic snapshots on an RNA packaging motor reveal conformational changes linking ATP hydrolysis to RNA translocation. *Cell* 118: 743–755.

Mindich L (1999) Precise packaging of the three genomic segments of the double-stranded-RNA bacteriophage φ6. *Microbiology and Molecular Biology Reviews* 63: 149–160.

Mindich L (1999) Reverse genetics of the dsRNA bacteriophage φ6. *Advances in Virus Research* 53: 341–353.

Mindich L (2005) Phages with segmented double-stranded RNA genomes. In: Calendar R (ed.) *The Bacteriophages*, 2nd edn., pp. 197–207. New York: Oxford University Press.

Poranen MM, Paatero AO, Tuma R, and Bamford DH (2001) Self-assembly of a viral molecular machine from purified protein and RNA constituents. *Moleculer Cell* 7: 845–854.

Poranen MM, Tuma R, and Bamford DH (2005) Assembly of double-stranded RNA bacteriophages. *Advances in Virus Research* 64: 15–43.

Sun Y, Qiao X, and Mindich L (2004) Construction of carrier state viruses with partial genomes of the segmented dsRNA bacteriophages. *Virology* 319: 274–279.

Icosahedral ssDNA Bacterial Viruses

B A Fane, M Chen, and J E Cherwa, University of Arizona, Tucson, AZ, USA
A Uchiyama, Cornell University, Ithaca, NY, USA

© 2008 Elsevier Ltd. All rights reserved.

Glossary

Procapsid A viral assembly intermediate containing a full complement of scaffolding proteins but devoid of genome.

Scaffolding protein A protein that directs the assembly of the virus, found in the procapsid assembly intermediate but not in the mature virion.

S Sedimentation constant in Svedberg units, measured by analytical ultracentrifugation.

T = 1 icosahedron A geometric shape consisting of 20 faces and 12 vertices. A T = 1 virion is composed of 60 viral coat proteins.

History

Bacteriophage øX174, the most well known virus of the family *Microviridae* (micro: Greek for small) was isolated in the 1920s by Sertic and Bulgakov. From the onset, this phage appeared very different from the other bacteriophages isolated during this extensive period of discovery. It was unusually tiny, readily passing through the smallest of ultrafilters. This biophysical characteristic defined its 'race': race X (Roman numeral ten) and the isolate was placed in a vial labeled #174, from which the phage derived its name, race X phage in test tube #174. And there it sat, relatively unperturbed, for decades. The first electron micrographs revealed small, vague isometric particles, vastly different from the tailed morphologies, which came to represent bacteriophages in general. Robert Sinsheimer unraveled the odd nature of the genome in 1959 and øX174 became the first recognized single-stranded (ss) DNA phage. While genetic maps of other phages consisted of orderly linear gene progressions, the øX174 map was most peculiar with genes located within genes. When Sanger and colleagues sequenced the genome, the first one ever sequenced, the complex arrangement of overlapping reading frames was confirmed, so beguiling that many suspected an extraterrestrial origin. As the *New York Times* reported the theory, an advanced race engineered øX174 and disseminated it into the cosmos where it would "persist until the evolution of intelligent life and finally of investigators interested in the genetics of phage." Although attempts were actually made to decipher the hypothesized hidden message, they all failed. However, rumors persist that the code has been broken, it reads, "Behold this marvelous little thing."

This small virus would continue to have a large impact on molecular biology. Arthur Kornberg and colleagues used it to elucidate the molecular mechanism of prokaryotic DNA replication. Masaki Hayashi and colleagues defined the øX174 assembly pathway and were the first to demonstrate that viral DNA packaging could be achieved *in vitro*. As the genome sequence initially defied imagination in 1978, the atomic structure of the viral procapsid, the first such structure solved, beguiled structural virologists, with its unusual external scaffolding protein lattice. And when events were at their most bizarre, the skeletons in the family *Microviridae* closet emerged, the gokushoviruses (Japanese for very small). These viruses escaped detection for decades, hiding out in their obligate intracellular parasitic bacterial hosts.

Virion Morphology and Genome Content

All members of the family *Microviridae* have ssDNA circular genomes of positive polarity and a T = 1 icosahedral capsid. The bulk of the capsid is composed of 60 copies of a major capsid protein (**Figure 1** and **Table 1**), which exhibits the common β-barrel motif found in most icosahedral virion capsid proteins. In addition to the major capsid protein, microviruses also contain 60 copies each of the major spike protein G and DNA binding protein J. The fourth structural protein, the DNA pilot protein, is buried within vertex channels running through the

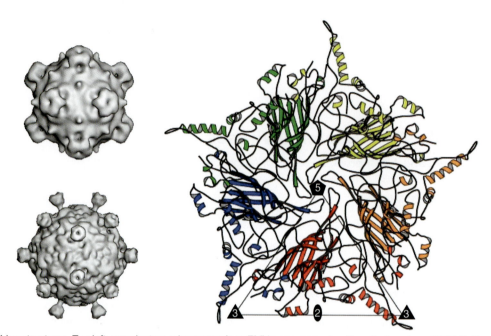

Figure 1 Virion structures. Top left, cryoelectron microscopy (cryoEM) image reconstruction of microvirus øX174. Bottom left, cryoEM of gokushovirus SpV4. Right, the arrangement of five øX174 coat proteins around a fivefold axis of symmetry. Numbers indicate axis of symmetry and the number of coat proteins that would be interacting at that axis.

Table 1 Microviridae gene products

Microvirus protein	Gokushovirus protein	Function
A	VP4	Stage II and stage III DNA replication.
A*	UD[a]	An unessential protein for viral propagation. It may play a role in the inhibition of host cell DNA replication and superinfection exclusion.
B	VP3	Internal scaffolding protein, required for procapsid morphogenesis and the assembly of early morphogenetic intermediates. Sixty copies present in the procapsid.
C	VP5?[b]	Facilitates the switch from stage II to stage III DNA replication. Required for stage III DNA synthesis.
D	NP[c]	External scaffolding protein, required for procapsid morphogenesis. 240 copies present in the procapsid.
E	UD	Host cell lysis.
F	VP1	Major coat protein. Sixty copies present in the virion and procapsid.
G	NP	Major spike protein. Sixty copies present in the virion and procapsid.
H	VP2?	DNA pilot protein needed for DNA injection, also called the minor spike protein. Twelve copies present in the procapsid and virion.
J	VP8?	DNA binding protein, needed for DNA packaging, 60 copies present in the virion.
K	UD	An unessential protein for viral propagation. It may play a role optimizing burst sizes in various hosts.

[a]UD, undetermined.
[b]indicates a hypothesized function based on bioinformatic data.
[c]NP, not present in gokushoviruses.

fivefold axes of symmetry. Particles lacking this protein extrude DNA from these locations. Protein H stoichiometry varies among the microviruses, in which there are three major clades represented by bacteriophages G4, α3, and øX174. In øX174, there are 10–12 H proteins per virion, while in bacteriophage α3 only 4–6 copies are required for viability.

Bacteriophage øX174 typifies the microvirus morphology (**Figure 1**), in which 70 Å-diameter G protein pentamers adorn the fivefold axes of symmetry, rising 30 Å from the surface of the 250 Å-diameter capsid. The gokushoviruses lack major spike proteins: hence, the fivefold axes of symmetry are not decorated. The cryoelectron microscopy image reconstruction of gokushovirus SpV4 reveals mushroom-shaped protrusions at the threefold axes of symmetry, which rise 54 Å above the surface of the 270 Å-diameter capsids (**Figure 1**). These threefold related structures appear to be composed of amino acid sequences of three interacting coat proteins. The gokushovirus VP2 is also a structural protein. Although its function has not been experimentally defined, bioinformatic approaches suggest it is analogous to the microvirus DNA pilot protein H. These proteins share a predicted N-terminal transmembrane and a central coiled-coil domain. VP8 may be analogous to the DNA binding protein J, both highly basic short peptides. However, it has not yet been detected in a gokushovirus virion. This may be due to the small size and the difficulties associated with propagating large quantities of gokushoviruses for biochemical characterization.

The genetic maps of microvirus øX174 and gokushovirus Chp2 are depicted in **Figure 2**. Gokushoviral genomes encode neither external scaffolding D nor major spike G proteins. The absence of these genes accounts for the smaller genomes. The external scaffolding protein has at least three known functions in øX174 morphogenesis. It mediates procapsid assembly, stabilizes the assembled procapsid at the two- and threefold axes of symmetry, and directs the placement of the major spike protein G pentamers into the fivefold related capsid craters. These functions are either not required or performed by different proteins in the gokushoviruses. In the gokushoviruses, procapsid assembly is most likely mediated by the internal scaffolding protein VP3, as is seen with the øX174 internal scaffolding protein B. Despite the small compact nature of microvirus genomes, two proteins A* and K with obscure or unknown functions are very strongly conserved. Three additional small ORFs are found in the H-A intercistronic region of the bacteriophage α3-like proteins. Other proteins are involved in DNA replication and lysis. These will be discussed in the context of their functions during viral replication.

Host Cell Recognition, Attachment, and Penetration

Bacteriophage øX174 attachment requires lipopolysaccharides (LPSs) containing specific terminal glucose and galactose moieties. Most microvirus research has been conducted with øX174. Six coat protein residues constitute this Ca^{2+}-dependent glucose binding site. Due to the icosahedral symmetry of the capsids, each virion has 60 glucose binding sites. Attachment is a reversible reaction. Although the molecular basis of the following irreversible penetration step remains obscure, it most likely involves the spike proteins G and H. Host range mutations map to these two proteins and to coat protein residues directly surrounding

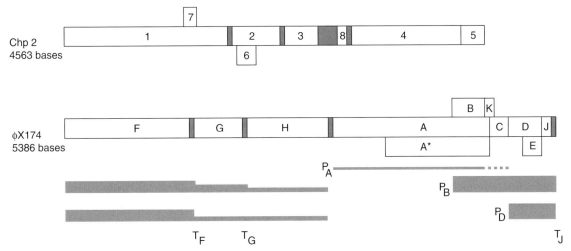

Figure 2 Genetic maps. Top, gokushovirus Chp2. Middle, microvirus øX174. Bottom, transcripts found in øX174 infected cells. The promoters and transcription terminators are indicated on the linear map of øX174. Line thickness indicates the relative abundance of the transcripts. The gene A transcript is very unstable; the terminator for this transcript is unknown. For protein functions, see **Table 1**.

the fivefold related spike complexes. Both spike proteins bind LPS of øX174 sensitive cells, but the specific details of these interactions are unknown. The location of the host range mutations argues for the existence of a second host cell penetration receptor. An analogy can be made to the large tailed bacteriophages that 'walk' along the surface of the cell, via tailspike interactions, until they find a receptor for penetration. But instead of walking, øX174 would merrily roll along.

In electron micrographs, the majority of eclipsed øX174 particles are imbedded at points of adhesion between the cell wall and inner membrane, suggesting the location of the hypothesized second receptor and indicating that the genome may be ejected directly into the cytoplasm. Penetration requires the DNA pilot protein H, which enters the cell along with the genome. Both genetic and preliminary structural data indicate that conformational changes in the capsid accompany or facilitate DNA ejection. Second-site suppressors of cold-sensitive DNA pilot proteins map to two places in the major spike protein. One set of suppressing residues lines the channel that passes through each fivefold vertex. The second set alters the interface between β-strands B and I of the G protein β-barrel, suggesting that conformational switches within this interface may mediate the channel opening.

Due to their recent discovery, the details of gokushovirus attachment and penetration have yet to be defined. The results of bioinformatic approaches combined with host range studies with the chlamydiaphages suggest that the threefold related protrusion may govern host cell specific attachment. However, there is no direct experimental evidence to support this hypothesis. Unlike microviruses, gokushoviruses appear to utilize a host cell protein, as opposed to LPS, as a receptor molecule. The exact location of the hypothesized DNA pilot protein, VP2, at the three- or fivefold axes of symmetry, is unknown as is the conduit through with the genome is ejected.

DNA Replication

The studies of øX174 DNA replication are of historical significance. Positive polarity ssDNA replication strategies are complex, occurring in three separate stages, which are described below (**Figure 3**). Kornberg and colleagues reconstituted the first two stages *in vitro* while Hayashi and colleagues reconstituted the third stage of DNA replication, which includes concurrent packaging of the single-stranded genome. Collectively, these studies established the first defined viral genome replication process on the biochemical level.

Single-stranded viral DNA is delivered into the infected cell. Stage I DNA replication involves the conversion of the ss genome into a covalently closed, double-stranded (ds), circular molecule, called replicative form one (RFI) DNA. Since purified single-stranded microvirus DNA produces progeny when transfected, host cell proteins are both necessary and sufficient for stage I replication *in vivo*. A stem–loop structure in the FG intercistronic region serves as the host cell and protein recognition site, which initiates the assembly of the primosome. After primosome assembly, the complex migrates along the ssDNA in a $5'\rightarrow 3'$ direction synthesizing the requisite RNA primers for DNA replication. Addition of the holoenzyme leads to chain elongation. The 13 host cell proteins required for this stage of synthesis are the same proteins involved in replicating the host cell chromosome.

During stage II DNA synthesis, RF1 DNA is amplified. In addition to the host cell proteins required for stage I replication, stage II replication is dependent on the viral A protein and the host cell *rep* protein, which functions as a helicase. The viral A protein binds to the origin of replication and nicks it to initiate $(+)$ strand synthesis, which will occur via a rolling circle mechanism. After nicking, protein A forms a covalent ester bond with the DNA, forming a relaxed circular molecule called RFII. The host cell *rep* protein unwinds the helix and the host ssDNA binding protein (ssb) stabilizes the separated strands. After one round of rolling circle synthesis, protein A cuts the newly generated origin and acts as a ligase, generating a covalently closed circular molecule. Minus strand synthesis is mechanistically similar to stage I DNA synthesis.

Stage III DNA synthesis involves the concurrent synthesis and packaging of the ssDNA genome. Procapsids and viral protein C are required for this reaction along with all the proteins involved in the previous stages of replication with the exception of the ssb. Thus, a single-stranded genome is not synthesized unless there is a procapsid in which to package it. The competition between ssb and protein C for binding ssDNA most likely signals the commencement of stage III synthesis. Proteins A, C, and *rep* form a complex on the dsDNA that docks to procapsids, presumably in a groove that spans one of the twofold axes of symmetry. Procapsid binding does not occur in the absence of protein C. Mechanistically, stage III DNA synthesis is similar to the stage II $(+)$ rolling circle synthesis. As the new $(+)$ strand is synthesized, it is translocated into the procapsid. After one round of synthesis, protein A, which is covalently attached to the origin of replication, cuts the newly synthesized origin and acts as a ligase to generate a covalently closed circular molecule.

Gene Expression

Since microviruses contain single-stranded genomes of positive polarity, stage I DNA synthesis, which generates the negative strand, must occur before transcription.

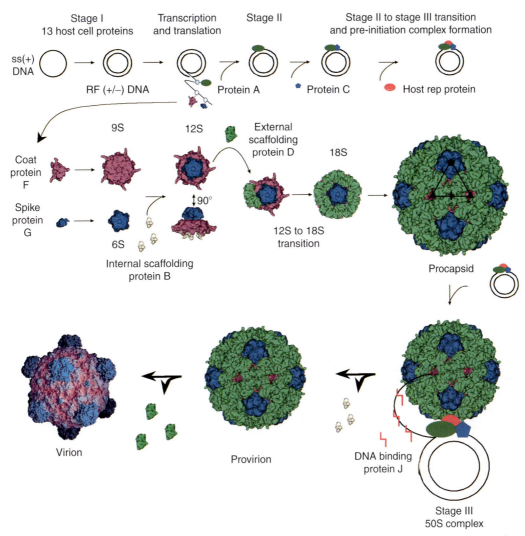

Figure 3 Morphogenesis and DNA replication of ϕX174.

Unlike large bacteriophages, gene expression in the members of the family *Microviridae* is neither temporal nor mediated by *trans*-acting mechanisms. Thus, the timing and relative production of viral proteins is entirely dependent on *cis*-acting regulation signals: promoters, transcription terminators, mRNA stability sequences, and ribosome binding sites. Promoters are found upstream of genes A, B, and D and terminators are found after genes J, F, G, and H (**Figure 2**). Since terminators are not 100% efficient, a wide variety of transcripts are produced. Yet there is a rough correlation between the abundance of a gene transcript and the amount of the encoded protein required for the viral life cycle. For example, gene D transcripts, which encode the external scaffolding protein, are the most abundant in the cell and the requirement of protein D is the greatest for progeny production. Similarly, there are more gene F, J, and G transcripts than transcripts of gene H. The relative stoichiometry of these structural proteins are 5:5:5:1, respectively. Protein expression is also affected by mRNA stability. Each mRNA species decays with a characteristic rate. Transcripts of gene A decay very rapidly, ensuring that this nonstructural protein is not overexpressed. Finally, regulation can also be achieved on the translational level. Despite gene E's location within gene D, the most abundant transcript, few E proteins, which mediate cell lysis, are translated due to an extremely ineffective ribosome binding site.

Morphogenesis

Despite extensive searches, a dependence on host molecular chaperones, such as groEL, and groES, has never been documented. Thus, chaperone independence is another factor that distinguishes the microviruses from dsDNA bacteriophages. The first virally encoded microvirus assembly intermediates are the 9S and 6S particles, respective pentamers of the major coat F and spike G proteins (**Figure 3**). After 9S pentamer formation, five internal scaffolding proteins bind to the 9S underside, which will become the capsid's internal surface, forming the 9S* intermediate. 9S* particles then associate with spike protein pentamers to produce 12S assembly intermediates. The internal scaffolding protein also facilitates the incorporation of the DNA pilot protein H. However, the exact time protein H enters the assembly pathways remains somewhat obscure. The addition of 20 external scaffolding proteins results in the construction of the 18S particle.

The external scaffolding protein has been the focus of many research endeavors due to its inherent ability to achieve different structures. The four D subunits (D1, D2, D3, and D4) per coat protein (**Figure 4**), are arranged as two similar, but not identical, asymmetric dimers (D1D2 and D3D4). Each subunit makes a unique set of contacts with the underlying coat protein, the spike protein, and neighboring D protein subunits. Accordingly, the structure of each subunit is unique. The atomic structure of the assembly naive D protein dimer has also been determined. The subunits within that dimer, DA and DB, appear poised to achieve the four structures found in the procapsid. DA has a structure somewhat between D1 and D3, while DB has a structure midway between D2 and D4.

External scaffolding proteins are rare, only observed in parasitic satellite virus systems such as bacteriophage P4, in which the satellite virus encodes an external scaffolding protein that forces the helper's virus capsid to form a smaller capsid. Thus, microviruses are the only nonsatellite viruses known to encode an external scaffolding protein. This protein performs many of the functions typically associated with internal scaffolding proteins in one-scaffolding-protein systems, the organization of assembly precursors into a procapsid, and the stabilization of that structure. However, its function is physically and temporally dependent on the internal scaffolding protein, which can be eliminated from the pathway

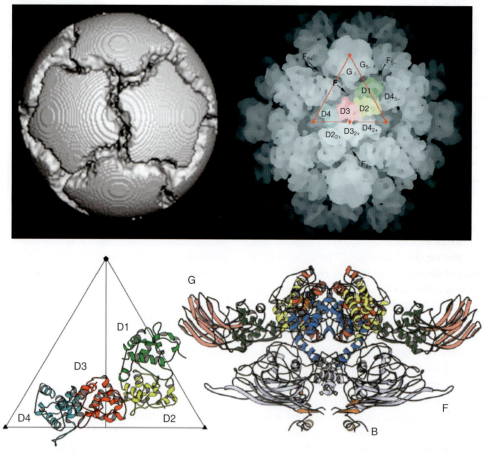

Figure 4 CryoEM and atomic renderings of the ɸX174 procapsid. Top left, rendering of the procapsid with the removal of the external scaffolding protein. Note that the coat protein pentamers do not contact each other. Top right, isosurface tracing of the ɸX174 procapsid, the four structurally unique external scaffolding proteins are depicted as D1–D4. Bottom left, the Cα backbones of the four external scaffolding protein subunits. Bottom right, cross-section of the procapsid along a twofold axis of symmetry, showing the positions of the internal B and external D scaffolding proteins relative to the coat F and spike G proteins.

(see the section titled 'Evolution and evolutionary studies' below). Procapsid morphogenesis is completed with the association of the 12 18S particles. The structure is almost exclusively held together via external scaffolding protein contacts across twofold axes of symmetry. Remarkably, there appear to be no associations between coat protein pentamers (**Figure 4**).

Gokushovirus genomes do not encode an external scaffolding protein. The details of the early assembly pathway remain to be elucidated. A gokushovirus particle containing VP3 in addition to the structural coat and DNA pilot proteins has been isolated. Unlike virions, these particles are devoid of DNA. This particle most likely represents a procapsid and indicates that VP3 functions as an internal scaffolding protein.

DNA Packaging and the DNA Binding Protein

Genome biosynthesis and packaging are concurrent processes in øX174. The pre-initiation complex, consisting of the host cell *rep*, viral A and C proteins, associates with the procapsid forming the 50S complex. As described above, the viral A protein binds the origin of replication in replicative form DNA. The results of genetic studies indicate that the pre-initiation docking site resides along a twofold axis of symmetry. The DNA binding protein J enters the procapsid during packaging and is absolutely required for genome encapsidation. Once in the procapsid, the C-terminus of the protein, which is very hydrophobic and aromatic, competes with the internal scaffolding protein for binding to a cleft in the viral coat protein. This competition results in the extrusion of the internal scaffolding protein during the packaging reaction.

Microvirus J proteins are extremely basic proteins that bind the genome via electrostatic interactions. The genome also interacts with a small cluster of basic amino acid residues in the capsid. Unlike large dsDNA bacteriophages, the øX174 genome does not exist as a dense core in the capsid. Instead, the DNA is tethered to the capsid's inner surface. In the atomic model, the protein forms an S-shaped polypeptide chain devoid of secondary structure. The C-terminus is tightly associated with the cleft located near the center of the coat protein. Each of the 60 proteins traces a path toward the fivefold axis of symmetry, crosses over to the adjacent capsid protein, and veers toward the C-terminus of the adjacent J protein. Thus, the DNA binding protein guides the incoming genome into a somewhat ordered conformation and a portion of the genome is ordered in the X-ray structure. The biophysical characterization of fully packaged particles with foreign genome-length DNA or øX174 genomes with mutant DNA binding proteins suggest that protein–DNA interactions influence the final stages of morphogenesis. Morphogenesis terminates with the provirion to virion transition: the dissociation of the external scaffolding protein and an 8.5 Å radial collapse of capsid pentamers around the genome. The tethered genome constrains the spatial orientation and secondary structure of the remaining nucleotides. Therefore, altering the tether or the base composition of the packaged nucleic acid may affect the magnitude or the integrity of the collapse. However, the role of DNA–capsid interactions in øX174 is not as dramatic as those seen in other viral systems in which abrogating genome–capsid interactions lead to severely aberrant particles.

Lysis

Unlike large double-stranded bacteriophages, microviruses do not have the genetic capacity to encode two-component endolysin and holin lysis systems. Instead, they have evolved a small protein, lysis protein E that inhibits peptidoglycan biosynthesis. Thus, lysis is dependent on host cell division, during which the cell becomes sensitive to osmotic pressure. The results of several elegant genetic studies elucidated protein E function. By selecting for lysis resistant cells, Ryland Young and colleagues first uncovered the *slyD* (*s*ensitivity to *l*ysis) gene, which encodes a peptidyl-prolyl cis-transferase-isomerase, or PPIase. However, it is unlikely that the *slyD* gene product is the E protein target. Considering the function of PPIase's in protein folding and the E protein's five prolyl bonds, it seemed more likely that the E protein was a substrate for the host cell enzyme. In fact, gene E mutants, *Epos*-plates on *slyD*, can be readily isolated. The *Epos* proteins were used in a second round of selection with *slyD* cells, surviving colonies containing mutations in the *mraY* gene, which encodes tanslocase I. This enzyme catalyzes the formation of the first lipid-linked intermediate in cell wall biosynthesis and most likely the target of protein E.

Evolution and Evolutionary Studies

Due to the small genomes and the ability to interpret amino acid substitutions within the context of the virion and procapsid atomic structures, the microviruses have become one of the leading systems for molecular evolution analyses. In studies pioneered by Jim Bull, Holly Wichman, and colleagues, viruses are placed under selective conditions, and grown for numerous generations in a chemostat. At various time intervals, individual genomes are sequenced, allowing the appearance and disappearance of beneficial mutations to be monitored. Of course, the mutations obtained differ according to the selection conditions. While these studies identify beneficial changes in both structural and nonstructural proteins, many mutations are in genetic regulatory sequences, which most likely optimize the relative level of viral proteins synthesized under the experimental conditions.

The results of an exhaustive search for new *Escherichia coli* microviruses reveal that the 47 known species can be divided into three separate clades, typified by bacteriophages øX174, G4, and α3. Although there is evidence for some horizontal gene transfer between the species, the extent is considerably lower than that observed for dsDNA viruses. The one gene that seems to be the most recent acquisition, at least in its present form, is the external scaffolding protein gene, which appears to have originated in the øX174 clade and spread to the other two.

The evolutionary relationship between the gokusho and microviruses remains somewhat obscure. There is a deep evolutionary rift between the two groups, with no known intermediate species. The rift appears to be a function of the biology of their hosts, not the hosts' evolutionary relationship. The gokushoviruses have been isolated from obligated intracellular parasitic bacteria or mollicutes. For example, the *Bdellovibrio* host for the gokushovirus øMH2K is a proteobacterium, like other microvirus hosts, but øMH2K is closely related to the phages of chlamidia. One of the primary differences between the two groups is the external scaffolding protein.

While the existence of two scaffolding proteins is unique in any system, it is particularly peculiar for the $T=1$ microviruses. No other $T=1$ virus requires a scaffolding protein, let alone two. An accumulation of both genetic and structural data suggests that the external scaffolding protein may be more essential for morphogenesis. The internal scaffolding protein, on the other hand, may be better viewed as en efficiency protein, facilitating several morphogenetic reactions, but not absolutely essential for any one in particular. This hypothesis has recently been tested by the isolation of a sextuple mutant strain of øX174 that no longer requires the internal scaffolding protein. Although mutations in structural and external scaffolding proteins did arise, two mutations reside in the external scaffolding gene promoters, leading to the overexpression of the mutant external scaffolding protein. These three mutations appear to have a kinetic effect on virion assembly, indicating that one function of the internal scaffolding is to lower the critical concentration of the external scaffolding protein needed to nucleate the assembly. The morphogenesis of wild-type øX174 is extremely rapid, progeny virions can be detected as quickly as 5 min post-infection, which may be critical for a small phage without the genome content to encode superinfection exclusion functions. At 5 min postinfection, most other phages are just concluding early gene expression. Rapid morphogenesis may be a consequence of having two scaffolding proteins, allowing microviruses to compete with the larger and vastly more prevalent dsDNA phages. In this evolutionary model, the external scaffolding protein is a recent acquisition. Those phages that did not acquire the gene, the gokushoviruses, persisted by finding a niche free of competition, obligate intracellular parasitic hosts like chlamidia.

Acknowledgments

The authors thank Drs. M. G. Rossmann and Timothy Baker and colleagues for assistance with figures and the support of the National Science Foundation.

See also: Plant Virus Vectors (Gene Expression Systems); Transcriptional Regulation in Bacteriophage.

Further Reading

Bernhardt TG, Struck DK, and Young R (2001) The lysis protein E of phi X174 is a specific inhibitor of the MraY-catalyzed step in peptidoglycan synthesis. *Journal of Biological Chemistry* 276(9): 6093–6097.

Brentlinger KL, Hafenstein S, Novak CR, et al. (2002) *Microviridae*, a family divided: Isolation, characterization, and genome sequence of phiMH2K, a bacteriophage of the obligate intracellular parasitic bacterium Bdellovibrio bacteriovorus. *Journal of Bacteriology* 184(4): 1089–1094.

Bull JJ, Badgett MR, Wichman HA, et al. (1997) Exceptional convergent evolution in a virus. *Genetics* 147(4): 1497–507.

Chipman PR, Agbandje-McKenna M, Renaudin J, Baker TS, and McKenna R (1998) Structural analysis of the Spiroplasma virus, SpV4: Implications for evolutionary variation to obtain host diversity among the Microviridae. *Structure* 6(2): 135–145.

Dokland T, McKenna R, Ilag LL, et al. (1997) Structure of a viral procapsid with molecular scaffolding. *Nature* 389(6648): 308–313.

Fane BA, Brentlinger KL, Burch AD, et al. (2006) øX174 et al. In: Calendar (ed.) *The Bacteriophages*, pp. 129–146. London: Oxford University Press.

Fane BA and Prevelige PE, Jr. (2003) Mechanism of scaffolding-assisted viral assembly. *Advances in Protein Chemistry* 64: 259–299.

Hafenstein S and Fane BA (2002) phi X174 genome-capsid interactions influence the biophysical properties of the virion: Evidence for a scaffolding-like function for the genome during the final stages of morphogenesis. *Journal of Virology* 76(11): 5350–5356.

Hayashi M, Aoyama A, Richardson DL, and Hayashi NM (1988) Biology of the bacteriophage øX174. In: Calendar R (ed.) *The Bacteriophages*, vol. 2, pp. 1–71. New York: Plenum.

Liu BL, Everson JS, Fane B, et al. (2000) Molecular characterization of a bacteriophage (Chp2) from Chlamydia psittaci. *Journal of Virology* 74(8): 3464–3469.

McKenna R, Xia D, Willingmann P, et al. (1992) Atomic structure of single-stranded DNA bacteriophage phi X174 and its functional implications. *Nature* 355(6356): 137–143.

Morais MC, Fisher M, Kanamaru S, et al. (2004) Conformational switching by the scaffolding protein D directs the assembly of bacteriophage phiX174. *Molecular Cell* 15(6): 991–997.

Novak CR and Fane BA (2004) The functions of the N terminus of the phiX174 internal scaffolding protein, a protein encoded in an overlapping reading frame in a two scaffolding protein system. *Journal of Molecular Biology* 335(1): 383–390.

Rokyta DR, Burch CL, Caudle SB, and Wichman HA (2006) Horizontal gene transfer and the evolution of microvirid coliphage genomes. *Journal of Bacteriology* 188(3): 1134–1142.

Uchiyama A and Fane BA (2005) Identification of an interacting coat-external scaffolding protein domain required for both the initiation of phiX174 procapsid morphogenesis and the completion of DNA packaging. *Journal of Virology* 79(11): 6751–6756.

Icosahedral ssRNA Bacterial Viruses

P G Stockley, University of Leeds, Leeds, UK

© 2008 Elsevier Ltd. All rights reserved.

Glossary

Pfu Plaque-forming unit; that is, the number of phage particles per host cell required to generate productive infections. Theoretically, this value could be 1, but in reality only very few of the phages that attach to the edge of a pilus will get access to the inside of a cell, so values in the 100s are not uncommon.

Polycistronic RNA The term cistron was coined a long time ago to mean any section of an mRNA that gets translated. The RNA phage genomes are 'polycistronic', because they encode multiple protein products.

Quasi-equivalent conformers These are different conformations of the identical polypeptide sequences that form at the differing symmetry-related positions of viral capsids.

Translational repression Refers to the process whereby a protein product interacts with an mRNA molecule, thus preventing its translation.

Triangulation number, T Defined mathematically as $T = (h^2 + hk + k^2)$, where h and k are any pair of positive integers, including zero. For instance, giving rise to triangulation numbers of 1 for $h,k = 1,0$, and 3 for $h,k = 1,1$. These h and k values can be determined by analysis of real capsids viruses by analyzing them in terms of networks of equilateral triangles that cover the surface of a sphere.

Background

Not all *Escherichia coli* cells are the same. Some of them produce a protein tube at the surface known as a pilus. It turns out that the information to build this structure is encoded by a separate genetic element, the F factor not carried by all cells. The RNA phages were initially identified because of their ability to infect *E. coli* carrying this factor. The pili encoded by F are used by the cells to swap DNA with other bacteria that lack the factor, resulting in a primitive form of sexual genetic transfer known as conjugation and the use of the term 'male' for bacteria that carry the F factor. Thus bacteria are able to swap useful genes, for example, for antibiotic resistance, rapidly within their populations. All processes in biology can be seen as opportunities ripe for exploitation, and the F pilus system, so useful to the bacteria in sharing useful genetic characteristics, presents an opening for a group of bacterial viruses to gain entry into F$^+$ cells. Tim Loeb working in Norton Zinder's laboratory at the Rockefeller University in New York was the first to realize that there might be male-specific bacterial viruses, bacteriophages. Together they isolated and began to characterize such phages from an initial sample of raw sewage. One intensively studied version known as MS2 is believed to have been named because it was male-specific factor 2, although there are unconfirmed stories that the MS also stands for metropolitan sewer.

Large numbers of similar phages were rapidly isolated worldwide after the initial discovery, and we now know that such phages are extremely common in the environment with up to 10^7 Pfu ml^{-1} in sewage. Both human and animal hosts appear to harbor *E. coli* populations able to support such phages, which are hence termed coliphage. In addition, virtually identical phages have been found that are specific to *Acinetobacter* (AP205) and other Gram-negative bacteria. In such phages, the infection process still occurs via pili but these are not the same as the sex pili of *E. coli*. For instance, phage PRR1 infects hosts via pili encoded by incompatibility type P plasmids which are widely dispersed, allowing the phage to infect a range of different bacterial cells.

All these bacteriophages turn out to have very closely related structural and genetic features. They form spherical capsids of ~250 Å diameter that enclose a single-stranded RNA (ssRNA) genome of ~3500–4300 nt in length. The ssRNA is the positive-sense version of the genome and acts as a messenger RNA (mRNA) while in infected cells. On the basis of immunological cross-reactivity and genome organization, the phages have been classified into two genera and these have been subdivided in turn into two groups. All these phages are members of the *Leviviridae* or the *Alloleviridae*, the latter being distinguished by having slightly longer genomes and because of the presence of read-through products from the coat protein gene. Subgroups I and II belong to the former genus and typical members are MS2 and GA, respectively, whereas group III and IV phages belong to the latter genus with typical members being Qβ and SP, respectively.

These very simple pathogens have been used since their discovery to understand many basic processes of molecular biology. They have been used to probe the structure of the bacterial pilus, to dissect the function of the protein synthesis pathway, to understand RNA replication, and to identify the initiation and termination signals on prokaryotic mRNAs. These studies led to the discovery of novel forms of genetic regulation along a polycistronic RNA, including translational repression.

Phage RNA was the first nucleic acid to be replicated in a test tube and was used as a model for the development of nucleic acid sequencing technology. Detailed insights into viral assembly processes and the molecular basis of sequence-specific RNA–protein recognition have also emerged. Research applications of these phages and their biology are still very active (see below).

Capsid Structure and Life Cycle

At the time of their discovery, the three-dimensional structures of viruses were not known but it was realized that the limited coding capacity of viral nucleic acids was likely to place a constraint on the number and size of gene products that could be dedicated to building a protective container, or capsid. This in turn implied that capsids are likely to be composed of multiple copies of one or very few 'coat' protein subunits necessarily resulting in highly symmetric structures. The two most 'efficient' designs for such structures, in terms of allowing multiple copies of the smallest protein units to enclose the maximum volume, and hence encompass the largest genome, are the helix and the icosahedron. Simple spherical virus capsids are therefore based on an icosahedral surface lattice (**Figure 1**).

A mathematically perfect icosahedron would have 60 facets in it so the stoichiometry for such viruses would be

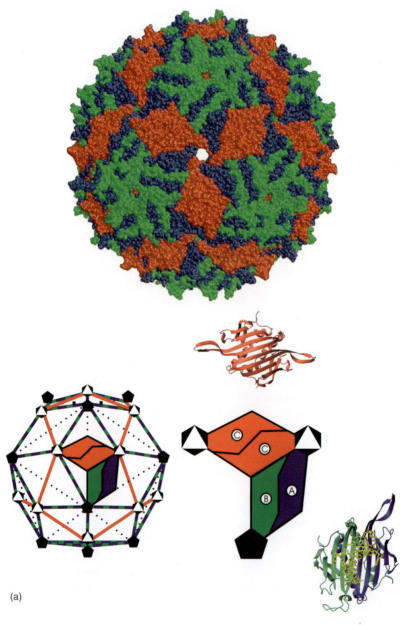

(a)

Figure 1 Continued

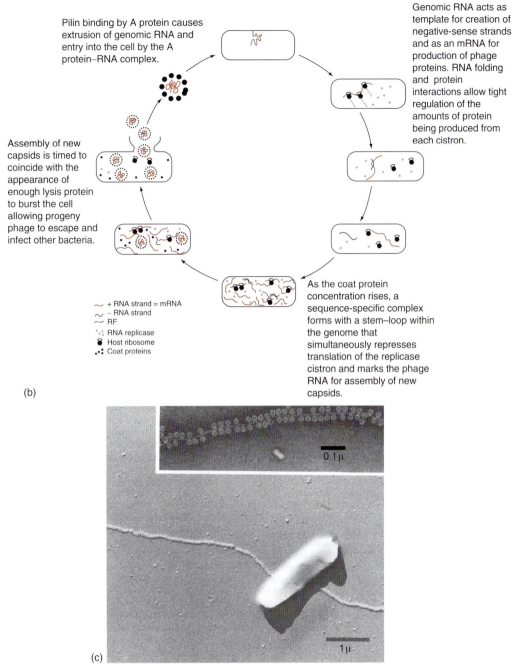

Figure 1 Capsid structure and life cycle. (a) Shows a space-filling representation of one face of the MS2 bacteriophage particle as determined using X-ray crystallography viewed along the icosahedral threefold axis. The differing quasi-equivalent protein conformers are colored red, green, and blue. There are 60 of each type accounting for the 180 coat protein copies per $T=3$ shell. Note, the position of the A protein in the shell is not observed from the X-ray structure for technical reasons. Below is a cartoon of an icosahedron with the threefold (open triangles) and fivefold (solid pentagons) symmetry axes highlighted. The MS2 coat protein subunits form noncovalent dimers in solution and in the final capsid, and the way in which they pack around the axes is illustrated on the right both as a cartoon and as detailed three-dimensional representations of the differing quasi-equivalent conformers. Proteins are shown as ribbon diagrams representing their polypeptide backbones. The A/B dimer is shown bound to the translational repression RNA stem–loop (yellow); see also **Figure 3**. (b) Shows a schematic of the phage life cycle, including the attachment process, entry of the genomic RNA, replication, self-assembly of progeny phage particles, and subsequent lysis of the host. (c) Shows the process of infection with phage particles adhered along the length of a bacterial pilus. The magnification is ~16000× and reveals the relative sizes of the host and its virus particles. (b) Reproduced from Watson JD, Hapkins NH, Roberts JW, Steltz JA, and Winer AM (1993) *Molecular Biology of the Gene*, *Vol. I*: *General Principle*, 5th edn. New Delhi: Pearson Education, Inc., with permission. (c) Reproduced from Paranchych W (1975) Attachment, ejection and penetration stages of the RNA phage infectious process. In: Zinder ND (ed.) *RNA Phages*, p. 89. New York: Cold Spring Harbor Laboratory Press, with permission from Cold Spring Harbor Laboratory Press (NY).

60 coat proteins per shell. However, from many early studies, it was clear that viruses often had many more than 60 subunits in their capsids, although in cases where detailed biochemistry was possible it appeared that such viruses contained simple multiples of the expected value, for example, 180, 240, etc. Don Caspar and Aaron Klug offered an explanation for such protein stoichiometries via a theory they called quasi-equivalence. They argued that viral coat proteins are three-dimensional objects that are partially flexible due to the chemical properties of their constituent polypeptide backbones and amino acid side chains. It would therefore be possible to conserve inter-subunit bonding if the precise molecular interactions at any position in the capsid were flexible enough to accommodate small changes from ideality. This theory leads immediately to predictions of 'allowed' stoichiometries for capsids with subunit numbers of 180, 240, etc., derived from what is called a subtriangulation of the sphere and defined by the triangulation number, T. This value is allowed to vary from 1, 3, 4, etc., yielding the observed stoichiometries of 60, 180, 240, etc. subunits. The implication of this idea is that the coat proteins within the capsids with more than 60 subunits must be able to adopt slightly different conformations depending on their relative locations in the shell. These are known as quasi-equivalent conformers and this idea has subsequently been confirmed by X-ray structure determination of many viruses. The RNA phages have 180 coat protein subunits per capsid and are therefore examples of $T = 3$ shells.

As well as the major coat protein, it is known that the capsids of RNA phages contain a minor protein that is essential for infectivity known as the maturation (or A) protein. In some of the RNA phages, another minor protein is associated with the capsids due to the production of a small percentage of read-through versions of the coat protein. The function of these minor components is largely unknown. Mature infectious capsids must contain at least one copy of the maturation protein as well as the icosahedral shell composed of the major coat protein. Infection is initiated by the interaction of the A protein with the edge of a bacterial pilus, confirming that part of this protein must project on the outer surface of the virion. This interaction triggers a release of the phage RNA from inside the phage shell and proteolytic cleavage of the A protein into two major fragments. The RNA and the A protein domains then enter the cell leaving the remainder of the capsid, that is, the major coat protein, on the outside. This implies that at least one fragment of the A protein must form a tight interaction with the phage RNA. Indeed, careful measurement of the kinetics of uptake of both phage RNA and the two fragments of the A protein suggest that they enter the cell as a single complex. Two RNA-binding sites for the A protein along the genome, one toward the 5' end and the other close to the 3' end, have been identified. It is tempting to speculate that pilin attachment leads to conformational change in the A protein, that is made irreversible by the proteolytic cleavage, and this in turn results in expulsion of the phage genomic RNA from its capsid. The precise mechanism of how the A protein–RNA complex is taken up into the cell has not been worked out in any detail. It is assumed that these complexes can transit into the cell like DNA but this does not seem to have been firmly established.

Whatever the mechanism of cell entry, once inside the cell the phage RNA then serves as an mRNA (see below). Expression of phage proteins results in replication of the phage RNA via the formation of an RNA-dependent RNA polymerase consisting of one phage-encoded replicase subunit, the host proteins S1, and elongation factors Tu and Ts, and an additional host factor. This polymerase creates negative strand versions of the genome that in turn serve as templates for the production of progeny positive-sense genomes. In an exquisite example of timing controlled by macromolecular interactions, this replication phase of infection is reined in by a translational repression event. Translation of the replicase subunit becomes blocked by interaction of a coat protein dimer with a stem–loop operator site of just 19 nt in length that includes the start codon for that cistron.

The resultant coat protein–genomic RNA complexes then serve as the assembly initiation complexes for formation of progeny phage particles. At some point, the A protein must become a part of this complex. The final act of infection is lysis of the host cell by the action of a lysis protein leading to release of up to 10 000 progeny particles. During this process, the phage has dramatically differing needs for each of its gene products: replicase, coat protein, maturation protein, and lysis protein. Each protein also needs to act at different times. Remarkably, both the amounts and timing of the appearance of each of these products is very tightly regulated, even though each cistron is present once on every copy of the genomic RNA. The details of that regulation are described in the following sections.

Control of Phage Gene Expression

When the phage-encoded protein expression is analyzed (**Figure 2**), it can be seen that the first protein to be produced after the RNA gains entry to the cell is the coat protein, and its expression remains high throughout an infection cycle. There seems little regulation of this expression other than the sequence and stability of the initiator hairpin structure that binds the ribosome. There are also some data that suggest that upstream flanking sequences can act positively to promote translation, perhaps by forming structures that favor binding of the ribosome to this site.

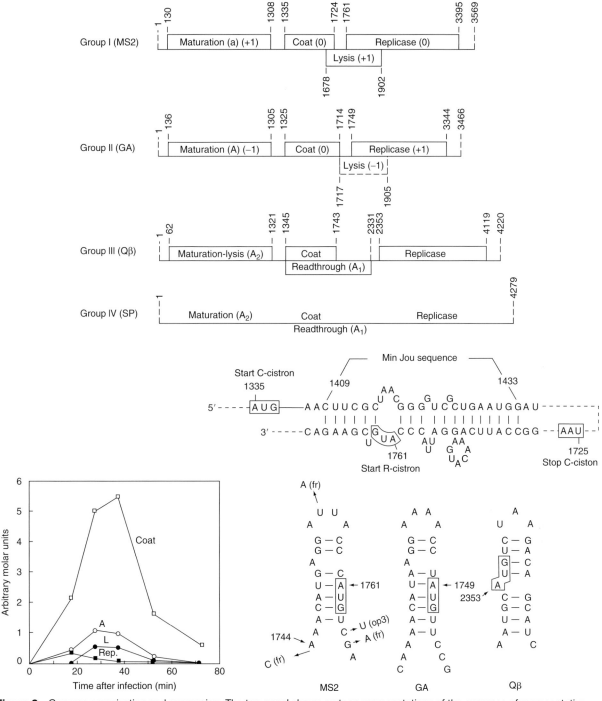

Figure 2 Genome organization and expression. The top panel shows cartoon representations of the genomes of representative members of each of the four subgroups of the RNA bacteriophages. The locations of the encoded proteins are also shown. The middle panel left shows the amounts of each of the encoded proteins produced during an infection cycle. Although each gene is present in equal amounts along each copy of every genome, there is a dramatic difference in the amounts of each protein in the cell, reflecting the requirements for different numbers of these protein products for replication, self-assembly of progeny phage, and lysis of the host. The implication is that the expression of these protein products is tightly controlled. The middle panel right shows one type of regulation that allows control of the amount of each protein product produced. At various times in the phage life cycle, the stem–loop structures shown at the bottom of the figure are sequestered by long range Watson–Crick base-pairing interactions to a segment of the coat protein gene (illustrated here for MS2). This section of the genome contains the start codon for the phage-encoded replicase subunit, so that the tertiary interaction prevents expression of replicase until the coat protein gene begins to be translated. As a ribosome transits this base-paired region, it frees up the downstream section allowing the ribosome-binding site and replicase translation to occur. At a later stage, the re-folded stem–loop is recognized by the coat protein – a coat protein dimer binds to the site and prevents further translational initiation, an example of translational repression (**Figure 3**). (a–c) Reproduced from Van Duin J (1988) Single-stranded RNA bacteriophages. In: Calendar R (ed.) *The Bacteriophages*, vol. 1, pp. 117–167. New York: Plenum, with permission.

The maturation protein is obviously needed in much smaller amounts and its appearance is slower and it is produced in much lower amounts than coat protein. It is believed that a number of regulatory features lead to this poor expression level. The start codon for the maturation gene is GUG instead of the more usual and more efficient AUG. In addition, it has been proposed that the ribosome-binding site in this cistron can be sequestered in internal base pairing with another section of the maturation protein message and, as a result, maturation protein is only expressed on nascent replicating genomic RNAs, that is, at a point when the downstream base pairing partners are yet to be replicated.

The replicase subunit is initially produced with similar kinetics and in similar amounts to the maturation protein but by 20 min after infection its levels are reduced. In fact, the phage makes more maturation protein than is strictly necessary, but the replicase is made at a level of ~5 copies/genome and its levels are always tightly controlled. This regulation is achieved in two distinct ways. Initially, the ribosome start site on the replicase subunit is sequestered in another long-range base-pairing interaction, this time with a section of the coat protein gene. Translation of the coat protein gene sequesters this complementary sequence, known as the Min Jou sequence after its discoverer, and thus frees the region around the start of the replicase gene so that it can fold to form a ribosome-binding site.

However, replication is further controlled via a translational repression complex that forms with the phage coat protein. A stem–loop operator hairpin (**Figure 3**) can form that is bound sequence-specifically by a dimer of the coat protein in a concentration-dependent fashion. This operator encompasses the start codon for replicase and formation of this repression complex effectively switches off replicase translation. The complex also acts to trigger self-assembly of the phage coat protein shell around the phage RNA. Regulated expression is obviously tightly controlled because of the importance of timing the switch in the viral life cycle from replication to assembly of progeny phage. There may also be an additional reason to ensure that large quantities of replicase are not produced because of its role(s) in creating the active replication complex.

The phage-encoded replicase subunit is the RNA-dependent RNA polymerase responsible for copying the phage genome. However, it does not function as a separate subunit. Instead, it forms complexes with at least three cellular proteins normally involved in other functions; these are the ribosomal protein S1, which functions during the loading and initiation of mRNAs on the ribosome, and translational elongation factors EF-Tu and EF-Ts. These species form a heterotypic tetramer that appears to be the active replication enzyme for positive-sense strands of the genome, that is, in the production of the negative-sense strand. The opposite reaction, that is, the generation of the positive-sense strand from the negative strand appears not to need the S1 subunit. An explanation for this altered stoichiometry in the replication machinery has been proposed based on the fact that the positive-strand must also serve as a template for peptide synthesis. This would lead to situations in which ribosomes and replication complexes, which would be moving in opposite directions on the RNA, would collide. It is believed that S1 functions in these cases by competing with the ribosome for loading onto the RNA, thus preventing replication on actively translating RNAs. The replication complexes also appear to interact with another host protein, which appears to be hexameric, but its role(s) is unclear.

A final reason for tight regulation of the expression of replicase is that the subunit or its complexes with cellular proteins are toxic to the cell. This would interrupt the phage life cycle before progeny phages are assembled. Cell death is however the ultimate goal of the infection, thus releasing progeny phage particles, but this is actively controlled by a phage-encoded lysis protein. This cistron is read from the +1 reading frame initiating within the coat protein gene. As can be seen from **Figure 2**, the lysis protein is the last to be produced, consistent with its final function in the phage life cycle. The reason for the delayed production of this gene product appears again to be a consequence of translational control. In this case, the ribosome binding site allowing initiation of the lysis protein translation appears inaccessible until the ribosome translating the coat protein completes the translation in its defined termination site. Experiments where the coat protein gene stop codons are moved further 3′ prevent lysis expression, implying that correctly regulated expression must be due to a defined RNA folding event similar to those controlling replicase translational repression. In principle, variations of the details of this mechanism may partially account for the frequency of read-through products in some of the coat proteins in some phages.

The Translational Repression/Assembly Initiation Complex

As discussed above, the expression of the phage-encoded replicase subunit is controlled by a translational repression event that occurs when a coat protein dimer binds to a 19 nt stem–loop structure (**Figure 3**), thus sequestering the AUG start codon from replicase. In a set of seminal experiments, Janette Carey and Olke Ulhenbeck were able to show that all the sequence information for the specific RNA–coat protein interaction was contained within this short fragment. Furthermore, they established rapid assays for the interaction *in vitro*, allowing extensive investigation of the effects of changes to nucleotide base sequence and the chemical functional groups on the RNA that contribute to this interaction.

Icosahedral ssRNA Bacterial Viruses

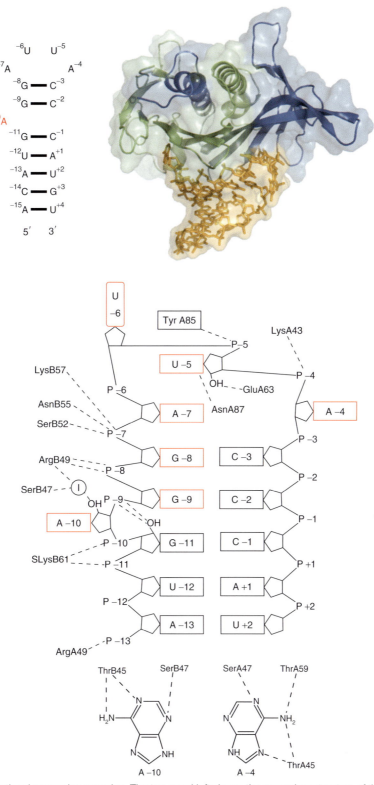

Figure 3 The MS2 translational repression complex. The top panel left shows the secondary structure of the translational operator stem–loop and the number of the bases relative to the replicase start codon. To the right is a detailed view of the X-ray structure of the interaction of the operator stem–loop with a coat protein dimer (shown as the A/B quasi-equivalent conformers). The polypeptide (blue and green) and RNA (gold) are shown as ribbon and stick models, respectively, within a transparent representation of the molecular surfaces. The bottom panel shows the intermolecular contacts that occur within the complex. The detailed contacts at A-4 and A-10 are shown below for clearity. The dotted lines indicate hydrogen bond interactions.

In addition, it has proved possible to determine X-ray crystal structures for the RNA stem–loop and over 30 of its variants in complex with the coat protein shell (**Figures 3** and **1**).

The detailed intermolecular contacts revealed and confirmed by these techniques have made this complex one of the most intensively studied examples of an RNA–protein complex. It has also been possible to determine structures for variants of the coat protein mutant at the RNA-binding site. The results of these studies suggest that the interaction is remarkably robust to simple changes outside a few key residues, principally the loop adenine at position −4 (the bases are numbered with respect to the first nucleotide of the replicase start codon). The half-life of the RNA–coat protein complex *in vitro* and presumably *in vivo* is very short (<1 min), implying that translational repression is short-lived. However, this is a simplification because the RNA–protein complex created then serves as a site of assembly initiation for formation of the intact capsid (see below). The repressive function of the coat protein reconfirms the mechanism of infection described above. If significant numbers of coat protein subunits were to enter the cell with the genomic RNA, the replicase gene would never be expressed. RNA variants that have higher affinity for coat protein can be identified *in vitro* but these would be lethal *in vivo*.

When cells are co-infected with MS2 and Qβ phages, the progeny phages that result consist of MS2 and Qβ RNAs inside their respective coat protein shells, that is, there is no evidence that the proteins can mispackage the competing genome. In part, this is because of the significant structural differences between the translational operator sites (**Figure 2**), but it also implies that there is direct coupling in the RNA–protein interaction leading to translational repression and assembly of progeny phage particles. It is believed that the RNA–coat protein complex that forms initially with a single coat protein dimer marks the phage RNA for assembly by addition of further coat protein subunits. All that is required to avoid misincorporation is that the coat proteins from different phages are unable to co-assemble and that seems to be the case even though there is a great deal of structural similarity between them.

The precise mechanism by which the phage capsid is generated is still unclear. When and how is the maturation protein incorporated into the structure, for instance? What we do know is that the stem–loop interaction appears to cause a conformational change in the coat protein subunits that bind, possibly between differing quasi-equivalent conformers, and provides a binding site for the stepwise addition of other coat protein dimers. These must somehow adjust their precise conformations in a structural context in order to produce a capsid of the correct size and symmetry. Recent evidence suggests that the folded genomic RNA may play a role in this process.

The major structural difference between quasi-equivalent conformers in a capsid occurs in a short loop of polypeptide connecting the F and G β-strands in each subunit. These FG loops have identical amino acid sequences in all subunits, but around the icosahedral fivefold symmetry axes they fold back toward the body of the subunit helping to create space to pack the subunits at this position. In contrast, at the icosahedral threefold axes, FG loops from two distinct quasi-equivalent subunits alternate in extended conformations, allowing six such loops to be packaged at these points. We now know that binding of the RNA translational operator causes a conformation change to occur at the FG loop, and it is tempting to assume that this interaction sets up the initial asymmetry to define a capsid structure via self-assembly. Precisely how the RNA is folded to fit into this structure and whether it folds during or before the assembly reaction is still a matter of debate.

Applications of RNA Phages – Three-Hybrid Assay

RNA phages have become model viruses for the development of a large number of novel scientific techniques. These are remarkably varied and include the controlled release of these phages into the wider environment as markers for the passage of viruses and of the ability of water treatment systems to remove viral particulates from drinking water. One such test was even performed on the International Space Station.

A more medically orientated set of applications has also been developed. For instance, it is known that viruses present defined epitopes to the immune system of their hosts. If those epitopes are common and unchanging, the immune response that occurs after an initial infection will render the host immune to further infection by that particular virus. Many viruses however constantly alter their outer protein surfaces presenting differing epitopes to the immune system, thus escaping immune neutralization. However, modern structural immunology techniques can identify conserved functional domains/sequences within these viral proteins that cannot be altered without also causing the virus to lose the ability to replicate or assemble. Raising immune responses against such refined constant epitopes leads to potent vaccines against a range of human and animal pathogens. In fact, work in this area has shown that the immune system is particularly well evolved to recognize repeated copies of particular epitopes, such as those that appear on the surfaces of pathogenic viruses. Therefore, researchers are busy trying to turn the tables on viruses by presenting conserved epitopes from pathogenic viruses as repeated arrays on the surfaces of harmless virus particles, such as the RNA phages. Not all such applications are directed at preventing viral infections. One artificial construct based on the phage Qβ presents nicotine as an artificially linked epitope and it is hoped that this virus-like particle will allow 'vaccination' of smokers who otherwise have failed to break their addictions to tobacco.

Our detailed knowledge about the coat protein–RNA operator interaction that regulates replicase translation and triggers self-assembly of progeny phage particles has also been adapted to allow encapsidation of non-phage molecules ranging from large protein toxins, to peptides and DNA oligonucleotides. These potentially therapeutic species can then be directed to specific cells by decoration of the external phage capsid surface with ligands for cell-surface receptor molecules. Once complexed at the cell surface, receptor-mediated endocytosis results in internalization of the encapsidated drug into endosomes where the pH is slightly acidic, conditions that favor disassembly of phage shells, and hence release of the 'drug' entities into the cell. Encapsidation of the plant toxin ricin A chain, that normally needs an additional B chain to cross into cells, allows this highly toxic protein to be delivered into the interior of cells carrying defined external signals resulting in reagents with effective lethal doses in the picomolar range. Controls lacking the targeting ligand are essentially nontoxic, suggesting that encapsidation reduces the nonspecific uptake of ricin A chain known from other studies to create side effects. Such artificially created virus systems have additional advantages, since they can be produced carrying a range of different therapeutic cargoes that can all be targeted to the same cell via attachment of the identical targeting signals. The result is the possibility of refined multidrug therapy without the problems associated with each component having differing pharmacokinetic properties.

An important and very intensively used additional application of the translational repression complex is in the three-hybrid assay system. Macromolecular interactions *in vivo* are central to understanding the integrated functions of all gene products. However, there are relatively few techniques that can easily be used to detect such interactions. One screen, known as the two-hybrid assay, was developed to detect protein–protein interactions using a simple genetic phenotypic screen. It works by creating two fusion proteins. The first carries the information to encode a DNA-binding protein domain fused to a 'bait' protein whose cellular partners are sought. The second fusion protein carries a series of potential partners for the bait domain translationally fused to a transcriptional activation domain. Both fusion constructs are expressed in yeast that carries a binding site for the DNA-binding domain in close proximity to a reporter gene inferring resistance to some chemical challenge, for example, an antibiotic or essential metabolite. Only in those yeast cells where bait and target interact will the transcriptional activation domain drive production of the

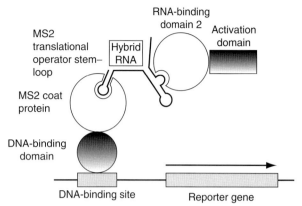

Figure 4 Three-hybrid assay. A schematic view of the components of the three-hybrid assay for detecting RNA–protein interaction *in vivo*. A DNA-binding domain is fused to the MS2 coat protein (RNA-binding domain 1). A second potential RNA-binding domain (#2) is fused to a transcriptional activation domain. Hybrid RNAs carrying the MS2 translational operator stem–loop are also expressed in yeast cells carrying the DNA-binding target adjacent to a reporter gene. Reproduced from SenGupta DJ, Zhang B, Kraemer B, Pochart P, Fields S, and Wickens M (1996) A three-hybrid system to detect RNA-protein interactions *in vivo*. *Proceedings of the National Academy of Sciences, USA* 93(16): 8496–8501, Copyright (1996) National Academy of Sciences, USA.

essential reporter gene. All other cells will not grow. Recovery of the live cells and sequencing of the constructs they carry thus reveals potential partners of the protein–protein interaction *in vivo*, although there is known to be a rate of false positives.

In the three-hybrid variant of this approach, the protein–protein interaction is replaced by a protein–RNA interaction. This is achieved by fusing the DNA-binding domain to the MS2 coat protein, thus producing a tethered protein that will bind to RNAs carrying the translational operator site. The translational operator sequence is then embedded in transcripts that include other potential RNA-binding domains. The transcriptional activation domain is then expressed as fusions with potential binding partners for specific RNA sites (**Figure 4**). The end result is the same as for a two-hybrid assay. Only those yeast cells in which the hybrid RNA is bound by a fusion partner will survive and the RNA sequences and or binding partners can then be identified by sequencing.

See also: Replication of Bacterial Viruses; Virus Particle Structure: Nonenveloped Viruses; Virus Particle Structure: Principles.

Further Reading

Caspar DLD and Klug A (1962) Physical principles in the construction of regular viruses. *Cold Spring Harbor Symposia on Quantitative Biology* 24: 1–24.

Paranchych W (1975) Attachment, ejection and penetration stages of the RNA phage infectious process. In: Zinder ND (ed.) *RNA Phages*, pp. 85–112. New York: Cold Spring Harbor Laboratory Press.

SenGupta DJ, Zhang B, Kraemer B, Pochart P, Fields S, and Wickens M (1996) A three-hybrid system to detect RNA–protein interactions in vivo. *Proceedings of the National Academy of Sciences, USA* 93: 8496–8501.

Stockley PG and Stonehouse NJ (1999) Virus assembly and morphogenesis. In: Russo VEA, Cove DJ, Edgar LG, Jaenisch R, and Salamini F (eds.) *Development: Genetics, Epigenetics, and Environmental Regulation*, pp. 3–20. Berlin: Springer.

Stockley PG, Stonehouse NJ, and Valegård K (1994) Molecular mechanism of RNA phage morphogenesis. *International Journal of Biochemistry* 26: 1249–1260.

Valegård K, Murray JB, Stockley PG, Stonehouse NJ, and Liljas L (1994) Crystal structure of a bacteriophage–RNA coat protein-operator complex. *Nature* 371: 623–626.

Valegård K, Murray JB, Stonehouse NJ, van den Worm S, Stockley PG, and Liljas L (1997) The three dimensional structures of two complexes between recombinant MS2 capsids and RNA operator fragments reveal sequence specific protein–RNA interactions. *Journal of Molecular Biology* 270: 724–738.

Van Duin J (1988) Single-stranded RNA bacteriophages. In: Calendar R (ed.) *The Bacteriophages*, pp. 117–167. New York: Plenum.

Watson JD, Hapkins NH, Roberts JW, Steltz JA, and Winer AM (1993) *Molecular Biology of the Gene, Vol. I: General Principle*, 5th edn. New Delhi: Pearson Education, Inc.

Witherell GW, Gott JM, and Uhlenbeck OC (1991) Specific interaction between RNA phage coat proteins and RNA. *Progress in Nucleic Acid Research and Molecular Biology* 40: 185–220.

Wu M, Sherwin T, Brown WL, and Stockley PG (2005) Delivery of antisense oligonucleotides to leukaemia cells by RNA bacteriophage capsids. Nanomedicine: Nanotechnology. *Biology and Medicine* 1: 67–76.

Icosahedral Tailed dsDNA Bacterial Viruses

R L Duda, University of Pittsburgh, Pittsburgh, PA, USA

© 2008 Elsevier Ltd. All rights reserved.

Glossary

Contractile tail The tail type that defines the *Myoviridae*, composed of a baseplate that contacts the host surface, an outer protein cylinder or sheath, and an inner tail tube or core. When a host is bound the sheath contracts, driving the tube into the cell to deliver the genome to the host.

DNA packaging The energy-driven process by which a procapsid is filled by a phage chromosome to become a mature capsid.

Head or capsid The protein container for the *Caudovirales* phage genome, usually icosahedral in shape, that protects it from the environment until it is delivered to a new host.

Horizontal gene exchange Transfer of genetic material between different species.

Portal A grommet-like ring, composed of 12 copies of the portal protein, through which DNA is packaged and to which the tail is attached.

Prohead or procapsid An immature capsid that has a spherical shape and contains no DNA.

T-number The triangulation number (*T*) is a formal descriptor for complex icosahedral structures which describes the geometric arrangement of subunits for a viral capsid. The number of capsid protein subunits is $\sim 60 \times T$.

Tail The part of the virion responsible for attachment to the host and genome injection. This is attached to one vertex of the capsid, at the portal.

Tail fiber An elongated protein fiber attached to a phage tail that helps recognize a host by binding to the host by the tip of the fiber.

Virion The entire viral particle composed of a capsid containing the viral genome, tail, tail fibers, and other appendages.

Introduction

The icosahedral, tailed double-stranded DNA (dsDNA) bacterial viruses or bacteriophages of the order *Caudovirales* are among the most widely recognized icons of modern molecular biology. Anyone who has participated in an advanced biology, genetics, or molecular biology class is likely to have been exposed to images of the familiar lunar-lander-shaped bacteriophage T4 virion (family *Myoviridae*) with its elongated icosahedral capsid and machine-like contractile tail (**Figure 1(a)**) or the plain icosahedral capsid and elegant, gently curving tail of bacteriophage λ (family *Siphoviridae*; **Figure 1(b)**). Both T4 and λ are bacteriophages that infect the enteric bacterial host *Escherichia coli* and have been studied extensively. Bacteriophages were originally discovered in the early twentieth century (1915–17) by English microbiologist Frederick W. Twort and French-Canadian microbiologist Felix d'Herelle. Although there was much early interest in bacteriophages because of their potential for use in treating bacterial diseases (phage therapy), research on

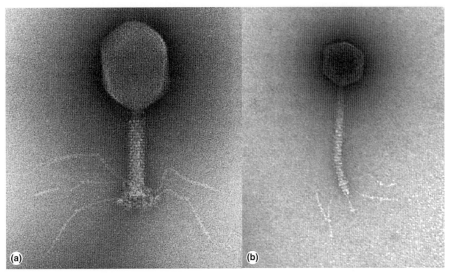

Figure 1 Electron micrographs of bacteriophage T4 and bacteriophage λ (variant Ur-λ). (a) Negatively stained electron micrograph of bacteriophage T4, the prototypical member of the *Myoviridae*. The T4 tail is 100 nm long and the T4 head has a width of 85 nm. Many of the features labeled in **Figure 2** are visible in the micrograph, including the baseplate, tail sheath, whiskers, collar, and long tail fibers. (b) Negatively stained electron micrograph of bacteriophage λ. The micrograph is of an original isolate of λ, called Ur-λ, that has long side tail fibers. Most common laboratory strains of λ carry a mutation in the side tail fiber gene that eliminates the fibers from the particle. The length of the lambda tail is ~150 nm (not including the protruding central fiber) and the lambda head is ~63 nm in diameter.

bacteriophages for disease therapy was largely abandoned after the discovery of antibiotics. Bacteriophage research was revived in the 1940s by Max Delbrück and colleagues (the well-known Phage Group that often met at the Cold Spring Harbor Laboratories on Long Island, NY), and the focused phage research started by this group of scientists led to many fundamental biological discoveries during the birth of molecular biology. The modern-day emergence of multiply antibiotic-resistant strains of human pathogens has rekindled an interest in developing phage therapy as a means of treating human diseases.

The Structure of Tailed dsDNA Bacteriophages

The invention and commercialization of the electron microscope led to the first images of phages as sperm-like particles with a head and tail using a technique called metal shadowing, in which images were formed by heavy metal atoms such as uranium that were evaporated in a vacuum and allowed to strike a dried specimen at an angle. The later introduction of negative stains (salts of heavy metals that dry as a thin layer without forming crystals and in which small particles such as phages could be embedded) for electron microscopy resulted in images that were far richer in detail than earlier methods and revealed the complexity and variety of phage morphology. **Figure 1** shows electron micrographs of phages T4 and λ made using the negative stain technique. Electron microscopy has remained a major tool in the study of bacteriophages and, in fact, the current taxonomic system for phages relies heavily on phage morphology as determined by electron microscopy as a major discriminating factor. **Table 1** shows a taxonomic table for the order *Caudovirales* and some of their characteristics. Members of the *Myoviridae* have contractile tails and include bacteriophages T4 (**Figures 1(a) and 2**) and P1. Members of the *Siphoviridae* have long noncontractile tails and include phages λ (**Figure 1(b)**) and T5. Members of the *Podoviridae* have shorter stubby or stumpy tails and include phages P22 and T7. The utility of morphological classification as a basis for phage taxonomy has more recently come into question as new insights into phage evolution have revealed that horizontal exchange of genes is widespread among phages, and further that some structural genes of bacteriophages and viruses that infect organisms from other domains of life appear to share common ancestors.

Tailed Bacteriophage Structure and Function

Capsids

In the tailed dsDNA bacteriophages, the capsid is the container for the phage genome that protects it from the environment until it is delivered to a new host by the phage tail, the organelle of attachment and genome

Table 1 Tailed dsDNA bacteriophages: Order *Caudovirales*

Family	Genus	Defining example	Capsid T-number	Genome size (bp)
Myoviridae (phages with contractile tails)	T4-like viruses	Enterobacteria phage T4	$T=13; Q=20$	168 903
	P1-like viruses	Enterobacteria phage P1	$T=13?$	93 601
	P2-like viruses	Enterobacteria phage P2	$T=7$	33 593
	Mu-like viruses	Enterobacteria phage Mu	$T=7$	36 717
	SP01-like viruses	Bacillus phage SP01	$T=16$	~132 500
	ϕ-H-like viruses	Halobacterium virus ϕH	$T=7?$	~57 000
Siphoviridae (phages with thin noncontractile tails)	λ-like viruses	Enterobacteria phage λ	$T=7$	48 502
	T1-like viruses	Enterobacteria phage T1	$T=7?$	48 836
	T5-like viruses	Enterobacteria phage T5	$T=13$	121 750
	c2-like viruses	Lactococcus phage c2	$T=4?; Q=7?$	22 172
	L5-like viruses	Mycobacterium phage L5	$T=7$	52 297
	ψM1-like viruses	Methanobacterium ψM1	Unknown	~23 246
Podoviridae (phages with short stubby tails)	T7-like viruses	Enterobacteria phage T7	$T=7$	39 937
	ϕ-29-like viruses	Bacillus phage ϕ29	$T=3; Q=5$	19 366
	P22-like viruses	Enterobacteria phage P22	$T=7$	41 724

A ? denotes unknown or unconfirmed values.

injection. In many other types of viruses, the capsid is taken up by cells with the genome still inside and the virus particle uncoats or disassembles within the new host to initiate viral replication. The dsDNA bacteriophages' capsids do not have to disassemble in this manner to initiate infection, so they can be constructed to be highly stable – resistant to a wide variety of chemical and physical assaults from the environment – as they travel from one host to another. The dsDNA genome in bacteriophage capsids is packed to a very high density ($\sim 0.5\,g\,ml^{-1}$) which results in high pressure, so the capsid shell has to be strong enough to withstand the high internal pressure from DNA. Most members of the *Caudovirales* have symmetric icosahedral capsids, but some have capsids with an elongated icosahedral shape, like T4 (**Figures 1(a)** and **2**) that have icosahedral ends and an elongated tubular middle section.

Tails and Tail Fibers

Phage tails function to identify and bind to the correct host and to deliver the phage genome into the new host to initiate replication. Phage tails have a large variety of morphological forms with many parts and appendages, such as those labeled on the diagram of T4 in **Figure 2** and visible in the electron micrograph in **Figure 1**. We know little about the functions of some tail parts, but others are understood in considerable detail. For example, in bacteriophage T4, both the long tail fibers and the short tail fibers (which are folded under the baseplate until after the long tail fibers are bound to the host; see **Figure 2**) recognize specific host receptors. The long tail fibers of phage T4 are the primary determinants of host range, and mutations that change the T4 host range cause alterations in the fibers near their distal tips. The whiskers that are attached to the top of the T4 tail have multiple functions.

The whiskers act as assembly jigs for adding the long tail fibers to the virion, and they also act as environmental sensors that sequester the long tail fibers under unfavorable conditions and thus prevent attachment to a host.

The long tail fibers of T4 have an analog in bacteriophage λ, but in the ordinary laboratory strains of λ these fibers are not made because of a mutation in the side tail fiber gene. The electron micrograph in **Figure 1(b)** shows the long side tail fibers present on Ur-λ, a primordial λ that lacks this mutation. These long side tail fibers of Ur-λ speed up the adsorption of this phage to its host, but are not required for infection, because the primary interaction of λ with its host is mediated by the central tail fiber (a trimer of the tail protein gpJ) protruding from the end of the lambda tail. The central tail fiber of λ binds to a host maltose transport protein (called *lam*B or *mal*B) in the outer membrane of the host. Binding of phage λ to its receptor under suitable conditions acts as a signal that triggers the injection of the phage DNA genome into the host to initiate replication.

The structure and functions of many parts of the contractile tails of *Myoviridae* have been revealed in depth, especially for the phage T4 tail. After the T4 long tail fibers attach to the host, the short tail fibers unfold from the baseplate and bind to a second receptor on the host surface. The baseplate remains attached to the cell and undergoes a dramatic reorganization in which it changes from a hexagon to a star shape, causing the sheath to contract and drive the internal tail tube into the cell. As the nail-shaped tip of the inner tail tube passes through the outer membrane, tail lysozyme molecules are released to create a small hole in the cell wall peptidoglycan layer. These combined actions allow the tip of the tail tube to reach the inner membrane, where a channel is created that allows the dsDNA genome to enter the cell and initiate replication.

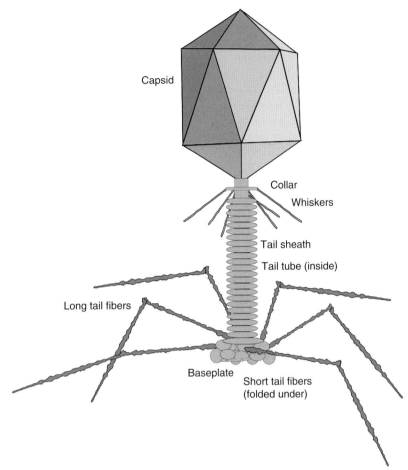

Figure 2 Schematic drawing of bacteriophage T4, with some of the structural components identified.

Temperate versus Virulent Bacteriophages

Virulent bacteriophages, once they have infected a host, are committed to go through their entire growth cycle including replication of the bacteriophage chromosome, production of progeny viral particles, and the lysis of the host cell. Temperate phages, on the other hand, are able to grow lytically like their virulent counterparts, but they are also capable of going into a dormant state within the host. This dormant state is called lysogeny, and often involves integration of the phage chromosome into the host chromosome, as occurs with phage λ, or conversion of the phage chromosome into an autonomously replicating plasmid. The seemingly normal host containing a dormant phage chromosome is called a lysogen. The dormant phage chromosome is maintained in a mostly silent state, until some change in the condition of the host (often related to stress, such as damage to the host DNA) acts as a signal to induce the phage to begin a lytic growth cycle, all the way through progeny release. The temperate phage in the lysogenic state makes a protein factor or repressor protein that can repress (nearly) all phage functions, except for the synthesis of the repressor. If the repressor is subsequently destroyed, as it may be after DNA damage occurs within the cell, the phage initiates a normal growth cycle, usually excising the phage chromosome from the host chromosome as an early step. How a temperate phage decides between a virulent and lysogenic mode of infection is a complex process that appears to depend on the conditions within the host cell at the time of infection. In a simplified version of the process, it is a balance between the production of sufficient repressor protein to shut down phage development and the antagonistic action of other factors that prevents repressor synthesis and function.

Injection

DNA Injection

As described above, all *Caudovirales* bacteriophage particles have tails consisting of highly specialized parts with specific roles to mediate attachment and infection. The fundamental goal of attachment is the introduction of the viral DNA into the host to begin production of progeny virions. The detailed mechanism of DNA entry varies widely from phage to phage. Although this process is called DNA injection, the phage particles do not act like syringes, as the name implies. In some cases it appears that chromosome entry is largely passive, a result of the DNA being very densely packed into the virion under very high DNA pressure. The energy stored by packaging the chromosome under pressure is used to drive the DNA out of the capsid into the host; this appears to be true for many phages, including phage T4 and phage λ. Other viruses such as phage T7 (family *Podoviridae*) and phage T5 (family *Siphoviridae*), have more elaborate mechanisms of DNA injection with an initial stage that appears to be largely passive, followed by a pause after only a fraction of the chromosome has been injected. Such paused DNA injection resumes only after proteins (encoded by the genes injected in the initial stage) act to restart chromosome entry or to pull the rest of the chromosome into the host in a controlled manner.

Protein Injection

Many of these viruses inject both DNA and proteins into their hosts. The function of a few of these injected proteins is known, but in many cases remains mysterious. For example, phage T4 injects multiple copies of several proteins into the host. These include a protein that modifies the host RNA polymerase and other proteins of unknown function. Surprisingly, most of phage T4's injected proteins are not absolutely required for infection, and mutants that lack them are able to grow on many ordinary laboratory hosts. Phage T7 injects several large proteins that possibly play important roles in the injection process, for example, by forming channels for the passage of DNA and other proteins into the host. The injected proteins of phage T7 are essential for phage growth.

Host Interactions and Regulation of Gene Expression

Classes of Genes

A regulated program of gene expression begins soon after the virus has injected its DNA. The expression of phage genes can be divided by function, such as host takeover, replication, virion assembly, or lysis, but more often they are sorted by their temporal order of expression. In most cases gene expression is divided into early genes and late genes, although for some phages there may be a cascade of gene expression modes that occur sequentially. In a cascading series, early phage gene products control the expression of a set of middle gene products, which in turn control the expression of late genes.

Early Genes

Phage early genes are usually transcribed by the normal host RNA polymerase, while the expression of late genes is dependent on the expression of the early genes. The earliest genes expressed might be ones that produce proteins to counteract the host's antibacteriophage defensive systems, as occurs in phage T7. Early genes of temperate phages are those required to make the decision to become dormant or to grow normally and carry out the first steps of these processes, including, for example, the repressor gene and integration genes, or DNA replication and recombination genes. Early genes of virulent bacteriophages, such as phage T4, include genes for genome replication and genes that shut down unneeded host functions, such as host transcription and host protein synthesis, and may include other genes required to recycle host resources, for example, by specifically degrading the host chromosome to intermediates that can be reused during the replication of the phage genome.

Late Genes

The proteins encoded by the late genes include the structural proteins required for the assembly of the bacteriophage virion, as well as those needed for packaging the genome into the virion and for cell lysis. Late gene expression is often dependent on early (or middle) gene expression because the late genes have transcriptional promoters that are not recognized by the normal host polymerase. The early genes may direct the synthesis of transcription factors that either (1) change the specificity of the host RNA polymerase or (2) modify the polymerase in other ways, in order to express the whole set of late genes at the appropriate time with high efficiency. Alternatively, one of the early genes may encode for an entirely new RNA polymerase with a new specificity that recognizes late gene promoters. Expression of the structural proteins to high levels allows the production of a large number of progeny phages.

Virion Assembly

Assembly Pathways

The assembly of *Caudovirales* virions is a highly ordered process in which separate subassemblies of the virion are built and then joined to form the mature virion. The assembly of capsids, tails, and tail fibers each follows an independent assembly pathway in which individual protein components are added, usually sequentially, to a growing structure until it is complete.

Tail Assembly

The major components of a *Caudovirales* tail are a baseplate or tail tip, major tail proteins (one for a noncontractile tail, and two for contractile tails (an inner tube protein and an outer sheath protein)), and termination or capping proteins to stabilize the completed tail and connect it to the head. Tail assembly begins with the formation of an initiator complex, which may be either a complete tail tip or a complete baseplate that is assembled via a separate pathway. The initiator complex includes a template that specifies the length of the tail, the tape measure protein complex in a compact form. The major tail proteins bind to the initiator complex and assemble into a tube around the tape measure protein until the full length of the tape measure protein is enclosed and the termination proteins can bind and stabilize the tail.

Capsid Assembly

The capsids of these viruses are built as precursor procapsids, into which the phage genome is later packed. Procapsids are initially assembled from several types of components: a portal complex (which will connect the capsid to the tail), a major capsid protein, a scaffolding protein (which may be a separate protein or may be a disposable part of the major capsid protein), and decoration or stabilization proteins (which add to and stabilize the capsid only after the genome is packed inside). The number of capsid protein subunits needed for assembly is equal to $\sim 60 \times T$ subunits, where T is the triangulation number listed in **Table 1** (or, for phages with elongated capsids, $\sim 30 \times (T + Q)$ subunits, where T and Q are specified in **Table 1**). So for a $T = 7$ virus, such as λ, ~ 420 copies are needed and more copies are needed for larger phages. Capsid assembly begins with the completion of the assembly initiator complex, usually the portal. The major capsid protein together with hundreds of copies of the scaffolding protein co-assembles onto the initiator to produce a complete shell with the scaffolding protein on the inside. After assembly is complete, the scaffolding protein is expelled intact or digested by a special protease that is also incorporated into the procapsid. Procapsids appear spherical and usually have a smaller diameter than mature capsids. When DNA is packaged into the procapsid, the capsid usually expands and changes shape, transforming into the typical angular, icosahedral shape of mature capsids. The decoration proteins add to the outside of the capsid after it expands and help to stabilize the structure. Once DNA is packaged, the capsid is made ready to join to a tail by the addition of proteins that bind to the portal.

Lysis

Lysis of an infected host requires two phage-encoded protein factors, an endolysin or lysozyme, and a second protein called a holin. The endolysin is a soluble enzyme with the capacity to break the bonds holding the host cell wall together. The endolysin molecules accumulate within the host cytoplasm during the late stages of infection, but are unable to attack the cell wall because they are sequestered within the cytoplasmic membrane. Holin proteins allow the endolysin to get across the cytoplasmic membrane by forming holes in the inner membrane. Holins are synthesized and inserted into the membrane in a form that does not form holes at first. As the bacteriophage infection proceeds, the holins accumulate in the membrane until a predetermined time at which they suddenly and catastrophically induce membrane breakdown. The membrane holes produced by holins allow the endolysin molecules to attack the bacterial cell wall, causing rapid cell lysis and releasing the progeny virus from the cell.

Genomes and Genomics

Chromosome Diversity and Replication

The chromosomes that are packaged into the capsids of the *Caudovirales* are linear dsDNA. *Caudovirales* virions always contain an entire genome or slightly more than an entire genome to ensure that a complete genome is packaged into each and that every particle can initiate an infection. To achieve this, a site-specific mechanism may be used, in which DNA packaging begins and ends at defined sites that are recognized by the packaging machinery to create exactly unit-genome-length DNA chromosomes. Alternatively, the amount of virion DNA may be regulated by a head-full packaging mechanism which fills a capsid that has the capacity to hold slightly more than one genome's worth of DNA. The head-full-packaged chromosomes have the same sequence at each end and are said to be terminally redundant.

After injection, the phage genomes often rearrange to form a circular chromosome that is the primary replicative form of the genome of many dsDNA phages; however, many other phages replicate without forming such DNA circles. Chromosome replication often takes place bidirectionally from a single origin, but more complex replication schemes and multiple replication origins have also been observed. Late in infection, most phages switch to a mode of replication that produces DNA concatamers, or long strings of genome copies joined end to end, either by a rolling circle mode of replication from circular chromosomes, or by recombination between multiple overlapping genome copies. Such DNA concatamers, whether linear or branched, are the forms of the genome that are the usual substrate for DNA packaging for both head-full and site-specific mechanisms. When it does occur, circularization of the phage chromosome takes place via one of the two mechanisms. The first is by the annealing of complementary single-stranded DNA sticky ends left by the packaging enzymes (in the cases where the packaged chromosomes have defined endpoints). The second is by a recombination-like mechanism that joins the complementary regions of the overlapping, terminally redundant chromosome ends to form circular DNA molecules. Circular chromosomes are the most common form for bacterial chromosomes, but some bacteria also have linear chromosomes, and a small subset of members of this order also forms linear chromosomes that replicate as linear plasmids when in their lysogenic form.

Diversity in Genome Size and Organization

The genomes of these viruses range in size from $\sim$20 000 bp in phage ϕ29, to $\sim$170 000 bp in phage T4, and up to $\sim$500 000 bp in other known examples. The smaller phages have proportionally fewer genes than the large phages, but are nonetheless functional and successful phages. Given this wide range of sizes and diversity of these viruses, it is not surprising that there is not a common conserved genome structure in the members of the order *Caudovirales*. Within a genus of phages there is often a common genetic structure, but within and across families there is often little resemblance in genome organization. There are notable exceptions to these generalizations; for example, the genus 'P22-like viruses' and the genus 'λ-like viruses' have rather closely related genomes.

Common Themes in Genome Structure

Despite the differences mentioned, there are many common general features in *Caudovirales* genomes. The genes for the structural proteins, such as those for capsids, tails, or tail fibers, tend to be found clustered together, and within these clusters, the genes that encode proteins that physically interact with each other also tend to be grouped together. For example, the capsid protein genes for the portal, the scaffolding protein, and the major capsid protein usually occur together and in the order listed. Sets of late genes are often grouped together in clusters and, in some cases, the entire set of structural genes are grouped together in clusters and transcribed together from a single promoter, as is the case for phage λ.

Horizontal Exchange of Genes Is Widespread

A large number of complete bacteriophage genomes have been fully sequenced, and hundreds of these have been deposited in GenBank and other databases. In addition, there are also many temperate phages residing in the genomes of bacterial chromosomes – some defective (or cryptic), and some complete and able to form a viable phage. Analyses of the sequences of these genomes have shown that many of the genes have highly similar counterparts in other genomes and in many cases the similarity can be inferred to be truly homologous. An important conclusion of these analyses is that there is extensive horizontal exchange of genes between phages that are not within the same species or genus or order. It appears that genetic exchange between phages by both homologous and nonhomologous recombination mechanisms is widespread and that many phages are mosaic combinations of genes found elsewhere. Many of the homologous arise genes within the *Caudovirales*, but some are from outside this order. In some cases, large modules of genes in a pair of phages are closely related (e.g., the head genes of phage HK97 (*Siphoviridae*) and phage SfV (*Myoviridae*)) despite belonging to different taxonomic groups. In other cases, a single phage tail fiber encoding gene in one phage may contain several distinct regions of homology with several other different phages.

Common Ancestry

The exact nature of the evolutionary relationships among phages and how it relates to phage taxonomy is a controversial subject, but at the level of individual protein-coding genes, it is fairly certain that proteins with a high degree of sequence similarity in analogous proteins almost certainly share a common evolutionary ancestor. The power and utility of bioinformatics to tease out weak (sequence) similarities between distantly related proteins provides the evolutionary scientist powerful tools to detect relationships that escape casual examination. However, the reliance on sequence similarity matches to detect homologous relationships breaks down when homologous proteins have diverged so far that sequence similarity is undetectable. A notable case is that of the major capsid proteins of bacteriophages and other viruses. The three-dimensional structure and protein fold of the major capsid protein of phage HK97 was determined to high resolution by X-ray crystallography. Sensitive bioinformatic techniques were able to detect weak sequence similarities between the HK97 capsid protein sequence and the capsid proteins of a large number of other members of the *Caudovirales*, suggesting that the HK97 capsid protein fold is quite common. Subsequently, the determination of the protein folds of phages T4 (*Myoviridae*), P22, ϕ29, and ϵ15 (*Podoviridae*) by structural methods has shown that all of these phages also have the same protein fold as HK97, despite the lack of any detectable sequence similarity.

See also: Capsid Assembly: Bacterial Virus Structure and Assembly; Genome Packaging in Bacterial Viruses; History of Virology: Bacteriophages; Replication of Bacterial Viruses; Virus Evolution: Bacterial Viruses.

Further Reading

Cairns J, Stent GS, and Watson JD (eds.) (1992) *Phage and the Origins of Molecular Biology*, expanded edition. Cold Spring Harbor, New York: Cold Spring Harbor Laboratory Press.

Calendar R (ed.) (2006) *The Bacteriophages*, 2nd edn. Oxford: Oxford University Press.
Caspar DL and Klug A (1962) Physical principles in the construction of regular viruses. *Cold Spring Harbor Symposium on Quantitative Biology* 27: 1–24.
Hendrix RW (2003) Bacteriophage genomics. *Current Opinion in Microbiology* 6: 506–511.
Hendrix RW and Casjens S (1988) Control mechanisms in dsDNA bacteriophage assembly. In: Calendar R (ed.) *The Bacteriophages*. New York: Plenum.
Hendrix RW and Duda RL (1998) Bacteriophage HK97 head assembly: A protein ballet. *Advances in Virus Research* 50: 235–288.
Hendrix RW, Roberts JW, Stahl FW, and Weisberg RA (1983) *Lambda II*. Cold Spring Harbor, New York: Cold Spring Harbor Laboratory Press.
Karam JD, Drake JW, Kreuzer KN, et al. (eds.) (1994) *Molecular Biology of Bacteriophage T4*. Washington, DC: ASM Press.
Katsura I (1990) Mechanism of length determination in bacteriophage lambda tails. *Advances in Biophysics* 26: 1–18.
Leiman PG, Kanamaru S, Mesyanzhinov VV, Arisaka F, and Rossmann MG (2003) Structure and morphogenesis of bacteriophage T4. *Cellular and Molecular Life Sciences* 60: 2356–2370.

Idaeovirus

A T Jones and H Barker, Scottish Crop Research Institute, Dundee, UK

© 2008 Elsevier Ltd. All rights reserved.

Introduction

The genus *Idaeovirus* contains a single virus species, *Raspberry bushy dwarf virus*, named after the disease (characterized by stunting and proliferation (bushiness) of canes) with which it was first associated in red raspberry (*Rubus idaeus*). This mechanically transmissible virus was detected consistently in such diseased plants. Later, however, it was shown that raspberry bushy dwarf virus (RBDV) on its own in young raspberry plants did not induce obvious disease but, in mixed infection with black raspberry necrosis virus (BRNV), an aphid-borne virus, it intensified the degeneration in vigor caused by infection with BRNV alone. RBDV therefore does play a causal part in the total 'bushy dwarf' syndrome.

RBDV is serologically indistinguishable from loganberry degeneration virus and synonymous with raspberry yellows virus and is probably the same as raspberry line pattern virus reported from Poland.

Studies on the molecular biology of RBDV have shown it to possess a novel combination of properties and it has been assigned, as the sole known member, to the genus *Idaeovirus*. Though RBDV has a bipartite genome, its closest affinities are with viruses that have tripartite genomes in the family *Bromoviridae*.

Biological Properties

Geographical Distribution

RBDV probably occurs wherever red raspberry (*R. idaeus, R. strigosus*), black raspberry (*R. occidentalis*), or blackberry (*R. fruticosus*), is grown. The virus has been reported from Australasia, China, Eastern and Western Europe, North and South America, South Africa, and the former Soviet Union. However, at least three variants of RBDV have been characterized and some are more restricted in their geographical distribution.

Common or Scottish (S) isolates

The Scottish type isolate (S) is representative of isolates that have been identified in almost every country where raspberry is grown and are therefore sometimes referred to as isolates of the 'common strain'. They produce characteristic symptoms in herbaceous test plants and, under suitable environmental conditions, in sensitive raspberry cultivars. They are distinguished by a restricted *Rubus* host range, and many red raspberry cultivars are highly resistant, probably immune to infection. S isolates also fail to infect some hybrid berry cultivars.

Resistance-breaking (RB) isolates

These isolates are largely indistinguishable biologically from S isolates except that they are able to infect red raspberry cultivars and hybrid berry cultivars immune to S isolates. Indeed, they infect by grafting nearly all *Rubus* species and cultivars tested. RB isolates seem restricted in their distribution to Central and Eastern Europe, Russia and Siberia, and, probably through importation of infected material, to isolated areas in England and some parts of Western Europe.

Black raspberry (B) isolates

Two isolates of RBDV from the American black raspberry (*R. occidentalis*), which were serologically closely related to S isolates, were distinguishable from them by spur formation in double-diffusion serological tests in agarose gel. This is the only evidence of serological variability in field isolates of RBDV. B isolates are more difficult to maintain in culture in herbaceous plants and their particles are more difficult to purify in quantity than those of S isolates. They are also difficult to detect in *R. occidentalis* extracts by inoculation to *Chenopodium quinoa* test plants but are readily detected by enzyme-linked immunosorbent assay (ELISA), suggesting that they may have a low specific infectivity. Only B isolates were found in naturally infected *R. occidentalis*. However, more data are needed to determine whether B isolates are restricted to *R. occidentalis* and whether S isolates can occur in *R. occidentalis* in nature. B isolates have been found in nature only in North America, the sole region where *R. occidentalis* is grown extensively.

Natural and Experimental Transmission

In red raspberry and black raspberry, RBDV is readily transmitted to progeny seedlings via either the pollen or the ovule of infected plants and transmission is greatest when both parents are infected. Transmission in association with pollen is confirmed in field and glasshouse studies. Healthy plants can be infected after the first flowering season when planted close to virus sources.

Experimentally, RBDV is readily transmitted from *Rubus* by grafting to infectible *Rubus* species and cultivars. However, graft-inoculated plants usually need to go through a dormant season before the virus becomes fully systemic. RBDV is also transmissible from infected *Rubus* to herbaceous test plants by mechanical inoculation of sap extracts in aqueous solutions of either 2% nicotine or 1% polyethylene glycol (PEG) (mol. wt. 6000 Da). Mechanical transmission from RBDV-infected herbaceous plants to some infectible red raspberry cultivars is also possible by the same procedure, but with great difficulty.

Disease Symptoms and Effects

Naturally infected plants

In nature, RBDV occurs in wild and cultivated *Rubus* species and cultivars. When infecting alone, RBDV is often symptomless in *Rubus* plants, especially in Britain. However, in sensitive raspberry cultivars and under ill-defined environmental conditions, it induces symptoms of yellows disease. This disease is characterized by an initial yellowing of the main or minor veins of the lower leaves or leaflets that progressively extends to other veins and ultimately to the interveinal lamina, sometimes affecting the whole leaf (**Figure 1**). Some raspberry cultivars may also develop chlorotic/yellow rings or line patterns in leaves.

RBDV infection also causes 'crumbly fruit' in some raspberry cultivars. This condition, arising from the abortion of some drupelets and the uneven development of others, frequently results in an abnormally shaped fruit, which, on picking, may disintegrate into individual drupelets or drupelet clusters (**Figure 2**). The same disease syndrome can be induced by other unrelated causes, such as poor pollination conditions, infection with other viruses, or genetic aberrations.

Figure 1 Leaves of RBDV-infected red raspberry showing, from upper left through to lower right, the progressive development of yellows disease symptoms. Reproduced with the permission of the Scottish Crop Research Institute (SCRI).

Figure 2 Symptoms of crumbly fruit of red raspberry caused by RBDV infection. Reproduced with the permission of the Scottish Crop Research Institute (SCRI).

Additional symptoms have been attributed to RB isolates. In field studies in England, an association was reported between infection with RBDV-RB and premature defoliation of fruiting canes, decreased vigor, leaf curling, necrosis and death of laterals, and increased winter kill in some red raspberry cultivars that are immune to S isolates of this virus. Although this evidence is only circumstantial, it suggests that these severe symptoms are caused by infection with RBDV-RB.

Experimentally infected plants

RBDV has been transmitted by graft inoculation from *Rubus* to perennial plants in other genera. It symptomlessly infected *Fragaria vesca* and *Prunus mahaleb* seedlings and induced pronounced chlorotic veinbanding and line patterns in *Cydonia oblonga* cv. C7-1. RBDV has a moderately wide host range in herbaceous test plants, but infects most hosts symptomlessly. By mechanical inoculation of RBDV-infective sap, over 50 plant species from 12 dicotyledonous families were infected.

Particle Properties

Properties of Virus Particles

RBDV particles are best purified from leaf tissue of *C. quinoa*. Purified virus particles are about 33 nm in diameter, and appear quasi-isometric (**Figure 3**) because they tend to collapse and deform on the electron microscope grid. They appear more spherical in shape when aldehyde-fixed. Particle preparations have A260/A280 of 1.62, suggesting that RBDV particles contain about 24% RNA. The particles are unstable and readily disrupt in the presence of 0.01% sodium dodecyl sulfate, indicating that they are stabilized by protein–RNA linkages.

The particles sediment as a rather broad band in sucrose density gradients, with a sedimentation coefficient of 115 S in 0.05 M citrate buffer, pH 6. Formaldehyde-fixed virus particles, centrifuged to equilibrium at 20 °C in solutions of CsCl or RbBr, form a band at a density of $1.37\,\mathrm{g\,ml}^{-1}$. Very rarely, preparations also contain particles $c.$ 15 nm in diameter that sediment at $c.$ 34 S and are serologically identical to the 33-nm-diameter particles and may be re-aggregates of viral coat protein.

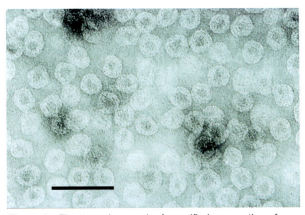

Figure 3 Electron micrograph of a purified preparation of RBDV particles stained in uranyl formate. Scale = 100 nm. Reproduced with the permission of the Scottish Crop Research Institute (SCRI).

Composition of Virus Particles

Purified preparations of RBDV particles typically yield a single major protein estimated by polyacrylamide gel electrophoresis to have an M_r of $c.$ 30 kDa. However, some preparations, even highly purified ones, contain minor amounts of smaller proteins, presumably of plant origin.

The particles contain three species of single-stranded RNA (ssRNA) with M_r about 2.0×10^6 (RNA-1), 0.8×10^6 (RNA-2), and 0.3×10^6 (RNA-3). The three RNA species are sometimes present in approximately equimolar amounts but in some preparations and with some isolates, RNA-3 (or RNA material of about the same size) is often present in much greater molar concentration than RNA-1 and RNA-2. The way in which the RNA molecules are packaged in the particles is unknown, but it is unlikely that all three RNA species are contained within the same particle.

Molecular Biology

Nucleotide Sequences

The complete nucleotide sequences have been determined for the three RNA species present in particles of isolates S and RB. **Figure 4** is a diagram of the genome structure of RBDV deduced from these sequences. RNA-3 is 946 nt long and contains a single ORF, which encodes the 30 kDa coat protein. This predicted size corresponds to that of protein obtained from purified virus particles and that of the main polypeptide made by *in vitro* translation of RNA-3.

RNA-2 is 2231 nt long and its 3′ half contains the entire sequence of RNA-3; RNA-3 is thus a subgenomic mRNA. In addition, RNA-2 contains, in its 5′-half, a second ORF which encodes a 39 kDa protein. Polypeptides obtained by

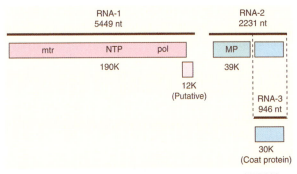

Figure 4 Diagram showing the genome structure of RBDV based on the RNA sequences. RNA molecules are represented by solid lines and the putative positions of the products of the ORFs are indicated as boxes. The positions of domains in the polyproteins are shown as: MP, residues shared among putative transport proteins; mtr, methyltransferase; NTP, NTP-binding; pol, RNA-dependent RNA polymerase. The dotted lines indicate that RNA-3 is derived from the 3′ terminal portion of RNA-2. Reproduced with the permission of the Scottish Crop Research Institute (SCRI).

in vitro translation of RNA-2 appeared to be larger than this: translation in wheat germ extracts produced a 46 kDa protein and translation in reticulocyte lysate produced a 42 kDa protein.

RNA-1 is 5449 nt long and contains one large ORF which encodes a 190 kDa protein. This protein corresponds in size to the largest polypeptide made by *in vitro* translation. A small ORF is also present in a different translation frame and overlapping the ORF for the 190 kDa protein (**Figure 4**). This small ORF encodes a 12 kDa polypeptide. All RBDV isolates sequenced to date contain this ORF in this 3' region. No corresponding subgenomic RNA was detected in RNA extracted from RBDV-infected plants, and infectious RNA transcripts of a 12 kDa deletion mutant of RBDV were still able to infect herbaceous host plants indicating that the 12 kDa protein is not essential for infectivity. Intriguingly, the position of this gene corresponds to that of the 2b gene in the tripartite genome of cucumber mosaic virus (CMV) that encodes a 10 kDa polypeptide and overlaps the C-terminal 69 codons of ORF 2a that encodes the RNA polymerase protein. In their respective RNA molecules both the CMV 2b and putative RBDV 12 kDa proteins are encoded in the +1 reading frame of the major ORF. The CMV 2b protein is important in both long-distance systemic movement and the suppression of post-transcriptional gene silencing but it is also not necessary for infectivity. However, experiments failed to demonstrate the activity of the 12 kDa RBDV protein in suppressing post-translational gene silencing.

There are similarities among the 5' or 3' terminal noncoding sequences of each of the three RBDV RNA species. At the 5' end of RBDV RNA, the first six to eight nucleotides are similar in all three RNA species, the consensus being UAUUU. At the 3' end, RNA-3 is identical to RNA-2 (being derived from it), and RNA-1 and RNA-2 share sequence and structural similarities. The 3'-terminal 18 nucleotides are identical in RNA-1 and RNA-2, and the terminal 71 nucleotides of each RNA can be folded into similar stem–loop structures.

The noncoding region between the two ORFs in RNA-2 contains the start of RNA-3 and presumably also contains sequences responsible for the initiation and promotion of its transcription. Like the comparable region of bicistronic RNAs of other viruses, this part of RBDV RNA-2 is rich in A and U residues. No double-stranded RNA (dsRNA) was found in infected tissue which could correspond to a double-stranded form of RNA-3, and RNA-3 is therefore probably transcribed from a minus-sense copy of RNA-2.

Virus Protein Sequences

The 30 kDa protein (coat protein)

The coat protein of RBDV shows no striking similarities with the coat proteins of other viruses. In comparisons made by using CLUSTALV, RBDV coat protein had a sequence identity of between 10% (alfalfa mosaic virus, AMV) and 19% (tobacco streak virus, TSV) to coat proteins of viruses in the family *Bromoviridae*. An alternative to this sequence-based alignment program is to use secondary structure predictions to produce an alignment. The RBDV coat protein structure predicted by such an alignment was of a long N-terminal sequence containing two α-helices and followed by a region containing eight β-sheets. If correct, this arrangement would resemble that of the coat proteins of many viruses that have isometric particles. The N-terminal parts of the coat proteins of RBDV and of viruses in the family *Bromoviridae* are relatively rich in basic amino acids which are thought to be internal and to interact with the encapsidated RNA.

The 39 kDa protein

RNA-2 of RBDV corresponds in size and gene content to RNA-3 of viruses in the family *Bromoviridae*. The ORF for the 39 kDa protein therefore corresponds in position to the ORF for the putative movement protein of, for example, AMV. A small sequence identity was detected in multiple alignment tests between the 39 kDa protein and the putative or actual movement proteins of several viruses. In further alignment tests of this sort, Mushegian and Koonin have detected a match between part of the 39 kDa protein sequence and a motif conserved in the movement proteins of some other viruses, such as red clover necrotic mosaic virus and soil-borne wheat mosaic virus. These proteins were proposed to belong to a '30 kDa superfamily' of movement proteins.

The 190 kDa protein

The sequence of the 190 kDa protein has similarities with replicase proteins of viruses in the Sindbis-like supergroup. There was good correspondence between parts of the 190 kDa protein and parts of the 183 kDa protein of tobacco mosaic virus but the greatest similarity was with a hypothetical composite protein consisting of the 90 kDa protein of AMV (encoded by RNA-2) attached to the C-terminus of the AMV 125 kDa protein (which is encoded by RNA-1). The three regions of greatest similarity correspond to the locations in the AMV proteins of the methyltransferase, NTP-binding, and RNA-dependent RNA polymerase domains. The first two domains occur in the 125 kDa protein of AMV and the polymerase domain occurs in the 90 kDa protein of AMV. RBDV RNA-1 therefore corresponds in size and gene functions to RNA-1 plus RNA-2 of AMV and related viruses. Comparisons of the three domains of several viruses by using CLUSTALV suggested that RBDV is marginally more similar to viruses in the family *Bromoviridae* than to other viruses. The maximum similarity was to AMV.

Detection and Control

Detection in Plants

In *Rubus*, RBDV may induce yellowing, line patterns, or crumbly fruit, but such symptoms are unreliable for diagnosis because they may be due to infection with other disease agents or to other causes. Furthermore, RBDV often infects plants symptomlessly. Detection and diagnosis therefore depend on bioassays and/or serological tests.

RBDV can be reliably detected by one or other of the various forms of ELISA and this is the preferred detection method for B isolates, which seem to have a low specific infectivity. Other studied isolates are readily detected by mechanical transmission to *C. quinoa* test plants. However, identification of RBDV in symptom-bearing herbaceous test plants is dependent on serological tests. As the distribution of RBDV in individual *Rubus* plants can be erratic, leaves/leaflets for mechanical transmission or ELISA should be taken and pooled from at least three different nodes.

Polyclonal antisera raised to S isolates have a moderate titre of *c.* 1/512 in gel double diffusion tests. RBDV antigen in purified virus preparations or in infective sap of *C. quinoa* reacts with such antisera in microprecipitin tests and in agarose gel double-diffusion serological tests. As already mentioned, spur formation in this latter test can be used to distinguish S isolates from B isolates. In Canada, monoclonal antibodies to RBDV were able to distinguish three different epitopes of a Canadian isolate of RBDV. However, these monoclonal antibodies reacted in double antibody sandwich ELISA (DAS-ELISA) with all tested isolates from red raspberry, including B, S, and RB isolates.

Therapy

Unlike many other viruses infecting *Rubus*, RBDV is not readily eradicated from *Rubus* plants by heat treatment alone. However, heat treatment for several weeks at 36 °C followed by propagation from shoot tips or apical meristems eliminated the virus from some plants. Using only tissue culture of meristem tips, it was necessary to subculture up to three times to eliminate the virus.

Control in Crops

Because RBDV is transmitted in association with infected pollen, the only methods of controlling it in crops are to grow resistant cultivars, or to plant healthy stock of infectible cultivars and grow them in isolation from possible sources of infection. Many red raspberry cultivars are resistant, possibly immune, to S isolates (the most widespread isolates worldwide). This resistance is conferred by the presence of a single dominant gene, *Bu*. However, almost all *Bu*-containing cultivars are infectible by grafting with RB isolates. Resistance to graft inoculation with RB isolates was detected in only very few red raspberry cultivars. Studies on this form of resistance indicated that its mode of inheritance was complex but seemed to depend on the presence of gene *Bu*, together with a second resistance component whose inheritance was probably multigenic. In field studies, some raspberry cultivars, including some that do not contain gene *Bu*, are more resistant than others to natural infection with RB isolates. This might suggest that inherent resistance to field infection with RB isolates may exist in some cultivars that are infectible by grafting.

Relationships with Other Viruses

RBDV resembles viruses in the genus *Ilarvirus*, family *Bromoviridae* in having easily deformable particles that are transmitted in association with pollen. RNA-2 resembles RNA-3 of ilarviruses in the arrangement and sizes of its encoded gene products, the generation of a 3' terminal subgenomic RNA (sgRNA) and in the structured nature of the 3' ends of the molecules. Nevertheless, RBDV is serologically unrelated to all recognized ilarviruses tested. Also, RBDV differs from ilarviruses in the sedimentation behavior of its particles, in the number and sizes of its RNA molecules, and in having a bipartite genome. Taken together, these properties distinguish RBDV from all other well-characterized viruses and justify its classification as the sole member of the genus *Idaeovirus*. The genus has not been assigned to a family, but, because of its molecular properties, the genus *Idaeovirus* may be tentatively classified in the family *Bromoviridae*, albeit as an atypical member of a family whose four recognized genera contain viruses with tripartite genomes. The sequence of the translation product of RBDV RNA-1 resembles in different parts, sequences in the translation products of viruses in the family *Bromoviridae* and to a lesser extent the sequences of the helicase + polymerase protein ($M_r \sim 183 \times 10^3$) of tobamoviruses. Idaeoviruses, therefore, belong to the 'Sindbis-like' supergroup.

See also: Alfalfa Mosaic Virus; Brome Mosaic Virus; Bromoviruses; Cucumber Mosaic Virus; Ourmiavirus; Phycodnaviruses.

Further Reading

Jones AT (2005) Genus *Idaeovirus*. In: In: Fauquet CM, Mayo MA, Maniloff J, Desselberger U, and Ball LA (eds.) *Virus Taxonomy: Eighth Report of the International Committee on Taxonomy of Viruses*, pp. 1063–1065. New York: Academic Press.

Jones AT and Mayo MA (1998) Raspberry bushy dwarf virus. In: Description of Plant Viruses No. 360, 6pp. Association of Applied Biologists. http://www.dpvweb.net/dpv/showdpv.php?dpvno=360 (accessed January 2008).

Jones AT, Mayo MA, and Murant AF (1996) Raspberry bushy dwarf idaeovirus. In: In: Harrison BD and Murant AF (eds.) *The Plant Viruses, Vol. 5: Polyhedral Virions and Bipartite Genomes*, pp. 283–301. New York: Plenum.

Iflavirus

M M van Oers, Wageningen University, Wageningen, The Netherlands

© 2008 Elsevier Ltd. All rights reserved.

Glossary

Internal ribosome entry site (IRES) Complex secondary RNA structure that allows translation initiation independent of a 5′ cap (m^{7}GpppN) structure.

UTR Untranslated region in an mRNA molecule on either the 5′ or 3′ side of an open reading frame.

Introduction

Iflaviruses have been described so far in lepidopteran and hymenopteran insects as well as in bee parasitic mites. In insects these viruses can cause mortality or developmental malformations. Sacbrood virus (SBV), for instance, lethally affects honeybee larvae (*Apis mellifera*) and the first descriptions of sacbrood disease date back as early as 1913. SBV was initially characterized in 1964. Other examples of iflaviruses causing disease in honeybees are deformed wing virus (DWV), which causes morphological abnormalities, and the related Kakugo virus (KV), the causative agent of aggressiveness in infected bees. Bee parasitic Varroa mites (*Varroa destructor*) function as vectors for these (and other) bee viruses, but DWV and the related Varroa destructor virus 1 (VDV-1) can also replicate in these mites. Infectious flacherie virus (IFV) infects the silkworm *Bombyx mori* and is a major cause of cocoon loss in silk production in Japan. In 1998 IFV was the first iflavirus for which the complete genome sequence was determined. Iflaviruses were previously referred to as insect picorna-like viruses together with viruses currently classified as *Dicistroviridae*.

Characteristics and Classification

Iflaviruses are characterized by icosahedral particles, which are nonenveloped and have a size of approximately 30 nm (**Figure 1**). The virions contain a single copy of a single-stranded, linear RNA molecule of positive polarity. The RNA genome is nonsegmented and encodes one large polyprotein, which is cleaved post-translationally into structural (capsid) and nonstructural proteins.

Infections flacherie virus is the type species of the genus *Iflavirus* which is not yet assigned to a virus family. Other species with an official status in this genus are *Sacbrood virus* and *Perina nuda virus*. Tentative species in this genus according to the Universal Virus Database (ICTVdBAse, version 4) are Deformed wing virus, Kakugo virus, Varroa destructor virus 1, and Ectropis obliqua picorna-like virus. The recently discovered Venturia canescens small RNA-containing virus (VcSRV) also has typical iflavirus features.

In the past, the name insect picorna-like viruses has been used to indicate insect viruses that share major features with vertebrate picornaviruses but are now classified either in the genus *Iflavirus* or in the family *Dicistroviridae*. Since the vernacular name 'picorna-like viruses' does not refer to a definite taxonomic unit, it is better avoided. In the light of this, the species *Ectropis obliqua picorna-like virus* will probably change in the future to *Ectropis obliqua virus*, similar to the classified *Perina nuda virus* which was known as *Perina nuda picorna-like virus* in older literature. However, to prevent confusion with other viruses isolated from these insect species, such as *Perina nuda nuclear polyhedrovirus*, the addition of iflavirus in the species name needs to be considered (i.e., *Perina nuda iflavirus*).

Viral Capsid Proteins

Three major capsid proteins have been detected for iflaviruses (VP1–VP3), ranging in size between 28 and 35 kDa. Although structural analyses have not been performed for iflavirus capsids, it is anticipated that heterotrimers of these three capsid proteins serve as the building blocks of the viral capsids in a similar fashion as for mammalian picornaviruses and cricket paralysis virus (CrPV; *Dicistroviridae*).

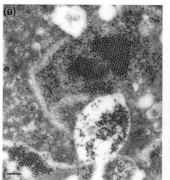
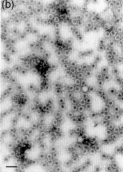

Figure 1 Virus particles of VOV-1. (a) Electron microscope photograph of a thin section of mite tissue with VDV-1 particles. (b) Purified VDV-1 particles. Scale: 150 nm (a), 50 nm (b). Photographs courtesy of Dick Peters.

A smaller less prominent structural protein is described as VP4 (4–12 kDa). The position of VP4 in the iflavirus capsids is unresolved and cannot be extrapolated from related viruses due to the variation in VP4 conformations observed in the virion structures. The various names used in the literature to refer to the structural proteins of iflaviruses are confusing. Here the nomenclature used in the *Eighth Report of the International Committee on Taxonomy of Viruses* will be followed where the capsid proteins appear in the order VP2, VP4, VP3, and VP1 from N to C in the polyprotein (**Figure 2**). The final maturation step for the iflavirus virion is proposed to be the cleavage between VP3 and VP4, thereby making the virus infectious. In picornaviruses, which have a different order of VP proteins in the polyprotein (VP4, VP2, VP3, and VP1 from N to C), this final cleavage occurs between VP4 and VP2.

The viral entry mechanism into cells is unknown, but most likely involves the release of the VP4 capsid protein as the first step after host cell recognition, similar to the situation in dicistroviruses and mammalian picornaviruses.

Genome Organization

As of February 2007, a total of seven complete iflavirus genome sequences are available (IFV, SBV, Perina nuda virus (PnV), Ectropis obliqua picorna-like virus (EoPV), DWV, KV, and VDV-1). The main characteristics of these genomes are summarized in **Table 1**. The genome size of these viruses varies between 8832 and 10 152 nt, excluding a 3′ polyA tail.

The RNA genome is nonsegmented and is translated into one large polyprotein, as shown by *in vitro* translation experiments. The polyprotein is cleaved into functional proteins by a virus-encoded protease. The information for the capsid proteins is located in the 5′ part of the genome and is followed downstream by a cluster encoding the nonstructural proteins (**Figure 2**).

For PnV and VDV-1 nucleotide sequence information was combined with N-terminal protein sequencing data for capsid proteins, allowing the demarcation of the protease cleavage sites and therefore the borders of the capsid proteins. In the iflavirus polyprotein, the structural proteins are preceded by a leader protein (L) of unknown function. This iflavirus L protein is not homologous to the leader protein of mammalian picornaviruses, which has protease activity.

The nonstructural proteins appear in the polyprotein in the order helicase (NTP-binding protein), protease, and RNA-dependent RNA polymerase (RdRp), as is the case in all members of a proposed picornavirus-like superfamily (see below). The RdRp harbors a canonical RGD motif typical for RNA polymerases of positive-sense single-stranded RNA (ssRNA) viruses.

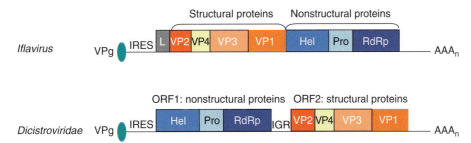

Figure 2 Genome organization of viruses in the genus *Iflavirus* compared to *Dicistroviridae*. Iflavirus genomes encode a single polyprotein with the structural proteins VP1–VP4 located in the N-terminal part and the nonstructural proteins (helicase, protease, and RNA-dependent RNA polymerase (RdRp)) in the C-terminal part. A function for the N-terminal L polypeptide has not been assigned yet. The presence of an internal ribosomal entry site (IRES) element in the 5′ untranslated region (UTR) has been confirmed for several iflaviruses. Dicistroviruses have a bicistronic genome, encoding two polyproteins separated by an intergenic region (IGR) that serves as IRES for the second open reading frame (ORF). ORF1 encodes the nonstructural proteins, ORF2 the capsid proteins.

Table 1 Iflavirus genome data[a]

Virus name	Genome size	ORF size	5′ UTR length	3′ UTR length	Accession number
IFV	9650	9255	156	239	AB000906
SBV	8832	8574	178	80	AF092924
PnV	9476	8958	473	45	AF323747
EoPV	9394	8961	390	43	AY365064
DWV	10 166	8682	1139	345	AY292384
KV	10 152	8682	1156	313	AB070959
VDV-1	10 112	8682	1117	313	AY251269

[a]All sizes in nt.

Comparative surveys of variation in iflavirus genomes have been performed for both DWV and VDV-1 in mites collected from beehives in distinct European locations. This study showed similarities of 96–99% between the various isolates, indicating that iflavirus genomes are not a fixed entity but subject to genetic variation in the field.

Relation to Other Taxa

Iflavirus genomes share several features with other viruses with positive-sense RNA genomes, for which a picornavirus-like superfamily has been proposed containing the families *Comoviridae*, *Picornaviridae*, *Potyviridae*, and *Sequiviridae*. The overall genome organization of iflaviruses strongly resembles that of mammalian picornaviruses, and iflaviruses have therefore been described as entomogenous picornaviruses. In these two groups of viruses the RNA genome encodes one large open reading frame (ORF) with the structural proteins followed by the nonstructural ones, although the order of the capsid proteins in the polyprotein and the number of proteases differ. The position of VP4 in iflaviruses is between VP2 and VP3, whereas in the *Picornaviridae* VP4 is the N-terminal capsid protein. The order of the capsid proteins in the polyprotein is similar in iflaviruses and dicistroviruses (**Figure 2**). Picornaviruses encode a 2C protease that is not found in iflavirus and dicistrovirus genomes.

In phylogenetic trees iflaviruses form a clade separated from dicistroviruses: their distinct genome structures justify their classification in different taxa (**Figure 3**). The major genomic difference between iflaviruses and dicistroviruses is that the latter encode separate polyproteins for nonstructural and structural proteins (hence have bicistronic genomes), while iflavirus genomes are monocistronic and encode a single polyprotein. In order to translate these ORFs into polyproteins dicistroviruses have both a 5′ and an intergenic internal ribosomal entry site (IRES) element (**Figure 2**). These two IRES elements vary from each other in terms of sequence, structure, and mechanism of action.

Another insect virus related to iflaviruses is kelp fly virus (KFV), which encodes one large polyprotein with an inhibitor of apoptosis (IAP) preceding the capsid and nonstructural protein segments. The encoded IAP protein has homology with baculovirus IAP proteins. Whether this *iap* gene is functional and plays a crucial role in the infection process is not known. Phylogenetic analysis using the RdRp sequence placed KFV in a clade separated from iflaviruses and dicistroviruses (**Figure 3**).

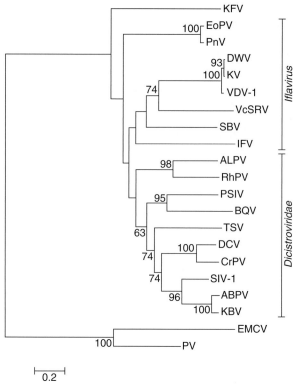

Figure 3 Phylogenetic relationship between viruses in the genus *Iflavirus* and the family *Dicistroviridae* based on RdRp sequences. Encephalomyocarditis virus (EMCV [ABC25550]) and poliovirus (PV [CAA24465]) were used as outgroups. Phylogenetic and molecular evolutionary analyses were conducted using MEGA version 3.1. Amino acid sequences were aligned using the Clustal W module and phylogenetic relationships and evolutionary distances were calculated using the neighbor-joining method. Bootstrap analysis was performed with 100 replicates. The distance marker indicates 0.2 substitutions per amino acid. The RdRp sequences were extracted form GenBank: acute bee paralysis virus (ABPV) [AAN63803], aphid lethal paralysis virus (ALPV) [NP_733845], black queen cell virus (BQV) [NP_620564], CrPV [NP_647481], DWV [AAP49283], drosophila C virus (DCV) [NP_044945], EoPV [NP_919029], IFV [NP_620559], KV [YP_015696], Kashmir bee virus (KBV) [NP_851403], KFV [YP_415507], PnV [NP_277061], plautia stali intestine virus (PSIV) NP_620555], rhipalopsiphum padi virus (RhPV) [NP_046155], SBV [AAL79021], solenopsis invicta virus 1 (SIV-1) [YP_164440], tauro syndrome virus (TSV) [ABB17263], VDV-1 [AY251269], and VcSRV [AY534885].

Tissue Tropism, Pathology, and Transmission

The primary infection sites of iflaviruses may vary. For IFV, the goblet cells of the midgut epithelium are the primary targets, suggesting virus uptake via food contamination. IFV infection finally leads to a flacherie phenotype, where excessive amounts of body fluids accumulate under the skin. This flacherie is also observed in other lepidopterans infected by iflaviruses.

Bee iflaviruses may spread by food contamination, although they also exploit other infection routes. An important site of infection for DWV and VDV-1 (and probably also KV) in honeybees are the feeding sites of *Varroa* mite nymphs and female adults on bee larvae. These mites can serve as vectors for the transmission of bee viruses. In addition, their constant feeding makes bees more susceptible to pathogens in general, due to exhaustion and immunomodulation. Direct transmission from bee to bee without the intervention of *Varroa* mites cannot be ruled out, but the incidence of infection is much lower in colonies which are not infested by these mites.

DWV and VDV-1 have been detected in head, thorax, and abdomen of bees. DWV infection may result in the development of bees with underdeveloped, malformed wings. KV, which varies only slightly from DWV in sequence, exhibits a strong neurotropism and is correlated with aggressive behavior of infected bees. Iflaviruses may also persist in insect populations in a latent symptomless state.

Replication

Replication of iflaviruses occurs in the cytoplasm where crystalline aggregates of progeny virus can be detected (**Figure 1(a)**). Iodination of purified IFV RNA showed that it was covalently linked to a protein of 11.5 kDa. This so-called VPg protein was removed upon translation of IFV RNA in a rabbit reticulocyte lysate, as is also the case for VPg protein of mammalian picornaviruses. It is anticipated that the 5′ ends of both the genomic RNA and the complementary strand RNA are covalently linked to a terminal VPg, implying that the iflavirus VPg protein also serves as a primer for RNA synthesis in iflaviruses, similar to the situation in the *Picornaviridae*. The location of VPg in the polyprotein is unresolved.

The presence of negative-strand iflavirus RNA is a hallmark to demonstrate replication of these positive-strand RNA viruses in insects and/or mites. For DWV and VDV-1, for example, negative-strand RNA and hence replication has been demonstrated in honeybees, as well as in bee-parasitic *Varroa* mites.

Translation

The sequenced iflaviruses can be grouped based on the length of their 5′ untranslated region (UTR) (**Table 1**). SBV and IFV have very short 5′ UTRs (156–178 nt) with hardly any predicted secondary structure. PnV and EoPV have longer 5′ UTRs (390–473 nt), which are homologous to each other and for which a mostly nonbranched secondary structure has been predicted. DWV, KV, and VDV-1 have much longer 5′ UTRs, which also have mutual homology and for which complex-branched secondary structures have been predicted. The existence of covariation in these homologous sequences, where a mutation at one side is accompanied by a second mutation to preserve secondary structures, is a measure for the likelihood of these predicted secondary structures. A three-branched or clover-leaf structure is shared between the DWV/VDV-1/KV group of iflaviruses and the entero/rhino IRES type of mammalian picornaviruses.

The presence of an IRES in the 5′ UTR, which would allow translation of the polyprotein in a cap-independent manner, has been shown for both VDV-1 and EoPV, using bicistronic plasmid constructs. For EoPV this was combined with mutational analysis to determine the functional domains in the IRES element. The EoPV core IRES was located in the third and fourth stem loop (out of six). Due to the high degree of similarity in sequence and predicted secondary structure between EoPV and PnV on the one hand and between VDV-1 and DWV/KV on the other, IRES elements are also predicted for these viruses. The presence of a functional IRES element in the very short unstructured IFV and SBV 5′ UTR sequences seems unlikely and is probably unnecessary but this awaits further analysis.

The interaction of host translation initiation factors with these IRES elements has not been studied yet. Nor is it known how iflavirus infection affects the translational machinery of the host.

Detection Methods

Iflaviruses can be detected and identified by serological methods such as enzyme-linked immunosorbent assay (ELISA) and Western blot analysis. However, this does not always result in easily interpretable data as virus preparations may contain mixtures of iflaviruses, leading to the production of nonspecific antisera. For instance, DWV and VDV-1 have been shown to coexist in a variety of European beehives. A more specific and therefore reliable method is reverse transcriptase-polymerase chain reaction (RT-PCR) with specific primers based on unique genome sequences. However, relying on single PCR primer pairs for detection is not a reliable method since iflavirus genomes are prone to variation.

RT-PCR also allows the discrimination between positive- and negative-strand RNA molecules by choosing a plus- or minus-strand-specific primer for the reverse transcription step. This makes it possible to establish if a virus is replicating in a certain insect species. Virus titers can be determined by quantitative PCR (QT-PCR), for instance to compare susceptibility of various developmental stages or to determine transmission rates.

See also: Dicistroviruses; Picornaviruses: Molecular Biology; Potyviruses; Sequiviruses.

Further Reading

Bowen-Walker PL, Martin SJ, and Gunn A (1999) The transmission of deformed wing virus between honeybees (*Apis mellifera* L.) by the ectoparasitic mite *Varroa jacobsoni* Oud. *Journal of Invertebrate Pathology* 73: 101–106.

Ghosh RC, Ball BV, Willcocks MM, and Carter MJ (1999) The nucleotide sequence of sacbrood virus of the honey bee: An insect picorna-like virus. *Journal of General Virology* 80: 1541–1549.

Hartley CJ, Greenwood DR, Gilbert RJC, *et al.* (2005) Kelp fly virus: A novel group of insect picorna-like viruses as defined by genome sequence analysis and a distinctive virion structure. *Journal of Virology* 79: 13385–13398.

Isawa H, Asano S, Sahara K, Iizuka T, and Bando H (1998) Analysis of genetic information of an insect picorna-like virus, infectious flacherie virus of silkworm: Evidence for evolutionary relationships among insect, mammalian, and plant picorna(-like) viruses. *Archives of Virology* 143: 127–143.

Lu J, Zhang J, Wang X, Jiang H, Liu C, and Hu Y (2006) *In vitro* and *in vivo* identification of structural and sequence elements in the 5′ untranslated region of Ectropis obliqua picorna-like virus required for internal initiation. *Journal of General Virology* 87: 3667–3677.

Ongus JR, Peters D, Bonmatin JM, Bengsch E, Vlak JM, and van Oers MM (2004) Complete sequence of a picorna-like virus of the genus *Iflavirus* replicating in the mite *Varroa destructor*. *Journal of General Virology* 85: 3747–3755.

Ongus JR, Roode EC, Pleij CW, Vlak JM, and van Oers MM (2006) The 5′ non-translated region of Varroa destructor virus 1 (genus *Iflavirus*): Structure prediction and IRES activity in *Lymantria dispar* cells. *Journal of General Virology* 87: 3397–3407.

Reineke A and Asgari S (2005) Presence of a novel small RNA-containing virus in a laboratory culture of the endoparasitic wasp *Venturia canescens* (Hymenoptera: Ichneumonidae). *Journal of Insect Physiology* 51: 127–135.

Rossmann MG and Tao Y (1999) Structural insight into insect viruses. *Nature Structural Biology* 6: 717–719.

Tate J, Liljas L, Scotti P, Christian P, Lin T, and Johnson JE (1999) The crystal structure of cricket paralysis virus: The first view of a new virus family. *Nature Structural Biology* 6: 765–774.

Wang X, Zhang J, Lu J, Yi F, Liu C, and Hu Y (2004) Sequence analysis and genomic organization of a new insect picorna-like virus Ectropis obliqua picorna-like virus, isolated from *Ectropis obliqua*. *Journal of General Virology* 85: 1145–1151.

Wu CY, Lo CF, Huang CJ, Yu HT, and Wang CH (2002) The complete genome sequence of Perina nuda picorna-like virus, an insect-infecting RNA virus with a genome organization similar to that of the mammalian picornaviruses. *Virology* 294: 312–323.

Yang X and Cox-Foster DL (2005) Impact of an ectoparasite on the immunity and pathology of an invertebrate: Evidence for host immunosuppression and viral amplification. *Proceedings of the National Academy of Sciences, USA* 102: 7470–7475.

Yue C and Genersch E (2005) RT-PCR analysis of deformed wing virus in honeybees (*Apis mellifera*) and mites (*Varroa destructor*). *Journal of General Virology* 86: 3419–3424.

Ilarvirus

K C Eastwell, Washington State University – IAREC, Prosser, WA, USA

© 2008 Elsevier Ltd. All rights reserved.

Glossary

Cross-protection A virus-infected plant often exhibits resistance to infection by strains of the same or closely related viruses. The degree of cross-protection has been used as a measure of the relatedness of viruses.

Dicistronic An RNA that encodes two gene products, often with a noncoding region separating them.

Recovery A phenomenon in which a virus-infected plant displays severe symptoms after initial infection, but then symptoms decline; this is often associated with a decline in virus concentration.

Subgenomic RNA A segment of RNA generated from a genomic RNA molecule through an internal promoter sequence. Since eukaryotic systems only efficiently translate the open reading frame located at the 5′ terminus of mRNA, this permits efficient translation of downstream open reading frames.

Suppressor of virus-induced gene silencing A molecule produced through virus infection that mitigates the natural plant defense system.

Virus-induced gene silencing A natural plant defense system that uses small fragments of double-stranded RNA to direct the RNA-degrading or silencing enzymes to a target RNA molecule. Double-stranded RNA results from the synthesis of complementary minus- and plus-sense RNA sequences during virus replication.

Taxonomy

The genus *Ilarvirus* was officially recognized in 1975. It is currently placed within the family *Bromoviridae* that also includes the genera *Alfamovirus*, *Bromovirus*, *Cucumovirus*, and *Oleavirus*. The quasi-isometric nature of ilarvirus particles and their relative instability in the absence of antioxidants are reflected in the sigla which originates from the descriptive phrase: isometric labile ringspot viruses. Other distinguishing features of the genus are the distribution of the genome over three plus-sense RNA segments and the absolute requirement for the coat protein to activate the RNA genome to initiate infection. These traits are shared with the genus *Alfamovirus*.

Experimental host ranges of ilarviruses are quite broad, although many have a relatively narrow range of natural hosts consisting of woody plants. Woody hosts may require several years after initial infection for the virus to uniformly infect the plant. The ringspot symptom for which the genus is named is a common symptom of many virus–host combinations. Infections by ilarviruses frequently result in a shock phase during which severe chlorotic or necrotic rings form, followed by a period of recovery with greatly reduced symptoms. This recovery phenomenon has now been described for many virus groups. No inclusion bodies have been described in tissue infected with members of the genus *Ilarvirus*.

Members of the genus *Ilarvirus* are currently divided into six subgroups (**Table 1**). Historically, this separation was based solely on serological reactivity, but it is generally supported by comparison of the amino acid sequences of structural proteins (**Figure 1**). The exception is the coat protein sequence of *Humulus japonicus* latent virus (HJLV) of subgroup 3 which is most similar to that of prune dwarf virus (PDV) of subgroup 4 rather than other members of subgroup 3. *Alfalfa mosaic virus* (AMV) is the monotypic member of the *Alfamovirus* genus; the sequence of the coat protein encoded by AMV also segregates with HJLV and PDV in this analysis. Analyses of amino acid sequences of the nonstructural proteins encoded by all three RNA segments are consistent with current subgroup designations.

In addition to the currently recognized members of the genus *Ilarvirus*, many synonyms have appeared in the literature, most of which refer to specific strains of virus species. These are listed in **Table 1**. The taxonomy of ilarvirus isolates that infect hop plants (*Humulus lupulus* L.) is still evolving. Viruses designated as hop virus C, NRSV-intermediate, and NRSV-HP-2 are related serologically to prunus necrotic ringspot virus (PNRSV) and to apple mosaic virus (ApMV) while hop virus A, apple mosaic virus-hop, and prunus necrotic ringspot virus-HP-1 are serologically related to ApMV only. Analyses of coat protein sequences of viruses obtained from hop plants of diverse geographic origins indicate that both serotypes are strains of ApMV.

Particle Structure and Genome Organization

The virions of ilarviruses are generally quasi-isometric (**Figure 2**), and the size of particle can vary within a species in relation to RNA content; particles range in diameter from 20 to 35 nm. Some ilarviruses also produce a small proportion of bacilliform particles. Virus particles contain subunits of a single structural protein (coat protein). The molecular mass of the coat protein subunits in the range of 19–30 kDa, with the majority of members having coat protein subunits within the rather narrow range of 24–26 kDa. The entire genome of ilarviruses is distributed over three RNA molecules; each is encapsidated separately. However, preparations of virions contain a fourth RNA species that is 3′-co-terminal with RNA3; this subgenomic RNA encodes the coat protein. Recently, a fifth RNA species corresponding to the 3′-untranslated region (UTR) of RNAs 1–3 was found to be encapsidated in virions of PNRSV. This is believed to represent the chance encapsidation of partially degraded RNA that is frequently found within purified virus particles. All of the major RNA species bear a 5′-7-methyl-G cap structure.

The relationship between the genera *Alfamovirus* and *Ilarvirus* is particularly close. The unique dependence on coat protein for infectivity of viral RNA is shared by members of both genera. Moreover, the coat proteins of AMV and some ilarviruses can reciprocally activate infectivity of RNA of the other. This phenomenon has led some investigators to suggest that the genera *Ilarvirus* and *Alfamovirus* should be combined. In as much as these viruses also share several features in common with other genera of the family *Bromoviridae*, the two genera remain separate. Studies of AMV have contributed much of the basic information from which the structure and replication of ilarviruses were determined and will be included here in subsequent comparisons of ilarviruses.

The type member of the genus *Ilarvirus* is *Tobacco streak virus* (TSV). The total genome of the WC isolate of TSV consists of 8622 nucleotides distributed over three RNA segments of 3491, 2926, and 2205 nucleotides (**Figure 3**). The largest segment of RNA, RNA1, contains a single open reading frame (ORF1) encoding a protein of 123.4 kDa. This protein is the replicase and contains the methyltranferase and helicase motifs. RNA2 contains ORF2a encoding a protein of 91.5 kDa. This protein is the RNA-dependent RNA-polymerase based on the presence of conserved RNA-binding motifs. RNA2 of TSV and other members of subgroups 1 and 2 contain a second potential overlapping ORF, ORF2b; AMV and ilarviruses of the remaining subgroups apparently lack this second ORF on RNA2. The function of the ORF2b product with a molecular mass of 22.4 kDa has not been determined. The dicistronic organization of RNA2 detected in ilarvirus subgroups 1 and 2 is similar to that of members of the genus *Cucumovirus*; the putative product of ORF2b of cucumoviruses has been proposed to be a key element in suppressing virus-induced gene silencing. The sequence of *Fragaria chiloensis* latent virus (FCiLV) of ilarvirus subgroup 6 is unique in that its RNA2 has the potential to encode a second small protein located toward the 5′ end of RNA2 (nucleotide positions 350–856 of RNA2). However, the putative product of this ORF bears no significant similarity to protein 2b of the cucumoviruses, nor other proteins thought to be encoded by ilarviruses.

Table 1 Virus members of the genus *Ilarvirus*

Virus (abbreviation) Synonyms	GenBank accession numbers used for comparison		
	RNA1	RNA2	RNA3
Subgroup 1			
Tobacco streak virus (TSV)	NC_003844	NC_003842	NC_003845
Annulus orae virus			
Asparagus stunt virus			
Bean red node virus			
Black raspberry latent virus			
Datura quercina virus			
Nicotiana virus 8			
Nicotiana virus vulnerans			
Sunflower necrosis virus			
Tractus orae virus			
Parietaria mottle virus (PMoV)	NC_005848	NC_005849	NC_005854
Tentative members			
Blackberry chlorotic ringspot virus (BCRSV)	DQ091193	DQ091194	DQ091195
Grapevine angular mosaic virus	—	—	—
Strawberry necrotic shock virus (SNSV)	—	AY743591	AY363228
Subgroup 2			
Asparagus virus 2 (AV-2)	—	—	X86352
Asparagus latent virus			
Asparagus virus C			
Citrus leaf rugose virus (CiLRV)	NC_003548	NC_003547	NC_003546
Citrus crinkly leaf virus			
Citrus variegation virus (CVV)	—	—	U17389
Citrus psorosis virus complex (infectious variegation component)			
Elm mottle virus (EMoV)	NC_003569	NP_619575	NC_003570
Hydrangea mosaic virus (HdMV)			AF172965
Lilac streak mosaic virus			
Lilac white mosaic virus			
Spinach latent virus (SpLV)	NC_003808	NC_003809	NC_003810
GE 36 virus			
Tulare apple mosaic virus (TAMV)	NC_003833	NC_003834	NC_003835
Subgroup 3			
Apple mosaic virus (ApMV)	NC_003464	NC_003465	NC_003480
Birch line pattern virus			S78319[a]
Birch ringspot virus			AAL84586[b]
Dutch plum line pattern virus			AAN01243[c]
Hop virus A			
Horse chestnut yellow mosaic virus			
Mild apple mosaic virus			
Mountain ash variegation virus			
Plum (Dutch) line pattern virus			
Severe apple mosaic virus			
Blueberry shock virus (BlShV)	—	—	—
Humulus japonicus latent virus (HJLV)	NC_006064	NC_006065	NC_006066
Humulus japonicus virus			
Prunus necrotic ringspot virus (PNRSV)	NC_004362	NC_004363	NC_004364
Cherry rugose mosaic virus			CAB37309[d]
Currant (red) necrotic ringspot virus			
Danish plum line pattern virus			
European plum line pattern virus			
Hop virus B			
Hop virus C			
North American plum line pattern virus			
Peach ringspot virus			
Plum (Danish) line pattern virus			
Plum (European) line pattern virus			
Plum (North American) line pattern virus			
Plum line pattern virus			
Prunus ringspot virus			

Continued

Table 1 Continued

Virus (abbreviation)	GenBank accession numbers used for comparison		
Synonyms	RNA1	RNA2	RNA3
Red currant necrotic ringspot virus			
Rose chlorotic mottle virus			
Rose line pattern virus			
Rose mosaic virus			
Rose vein banding virus			
Rose yellow mosaic virus			
Rose yellow vein mosaic virus			
Sour cherry necrotic ringspot virus			
Subgroup 4			
Prune dwarf virus (PDV)	PDU57648	AF277662	L28145
Cherry chlorotic ringspot virus			
Chlorogenus cerasae virus			
Peach stunt virus			
Prunus chlorotic necrotic ringspot virus			
Sour cherry yellows virus			
Subgroup 5			
American plum line pattern virus (APLPV)	NC_003451	NC_003452	NC_003453
Peach line pattern virosis virus			
Plum (American) line pattern virus			
Plum line pattern virus			
Prunus virus 10			
Subgroup 6			
Fragaria chiloensis latent virus (FCiLV)	NC_006566	NC_006567	NC_006568
Lilac ring mottle virus (LiRMoV)	—	—	U17391

[a]ApMV-G strain, CP sequence only.
[b]ApMV-hop strain, CP sequence only.
[c]ApMV-Fuji strain, CP sequence only.
[d]PNRSV-MRY1 isolate, CP sequence only.

RNA3 of ilarviruses is dicistronic with two potential ORFs. ORF3a encodes a protein of 31.7 kDa with cell-to-cell movement function (MP) and ORF3b encodes the structural coat protein (CP) with a molecular mass of 26.2 kDa. RNA3 of FCiLV contains a third potentially functional ORF, whose putative product does not have any sequence similarity to proteins expressed by other ilarviruses. It is unknown if this ORF is expressed *in vivo*.

Virus Replication

Members of this genus require an 'activated' RNA genome to initiate infection, that is, viral RNA must form a complex with the coat protein to be infectious. This is a distinguishing feature of ilarvirus and alfamovirus replication that differentiates them from other genera in the family *Bromoviridae*. Activation of the genome can be accomplished either through the co-inoculation of viral RNA with coat protein, or by the inclusion of subgenomic RNA4 that is then translated to produce the activating coat protein *in vivo*. This activation can be accomplished through heterologous complex formation, that is, the coat protein from one ilarvirus can bind to and potentiate the replication of RNA from several other but not all members of the genus *Ilarvirus*. Furthermore, the coat protein of AMV can activate the genome of several ilarviruses and vice versa. More recently, it has been demonstrated that RNA3 is able to fulfil the role of either coat protein or RNA4, but with much reduced efficiency. The presence of RNA4 or coat protein increases the efficiency of infection up to 1000-fold relative to the RNA activation achieved by RNA3. This is the result of very inefficient expression of the coat protein gene directly from the dicistronic RNA3.

The mechanism by which genome activation occurs is the result of the specific interaction of the coat protein with sequences located at the 3' terminus of viral RNA. All ilarviruses for which sequences are known possess sequences at the 3' terminus of genomic RNAs that are predicted to assume secondary structures that play a role in regulating the relative efficiency of the RNA species for transcription versus translation. Indeed, in the presence of coat protein, transcription of viral RNA is inhibited and translation is favored. The 3'-UTR of each RNA segment contains a series of conserved (A/U) (U/A/G)GC motifs that occur as single-stranded regions at the base of stem-loop structures that are recognized by and bound to coat protein molecules (**Figures 4** and **5**). The sequence and

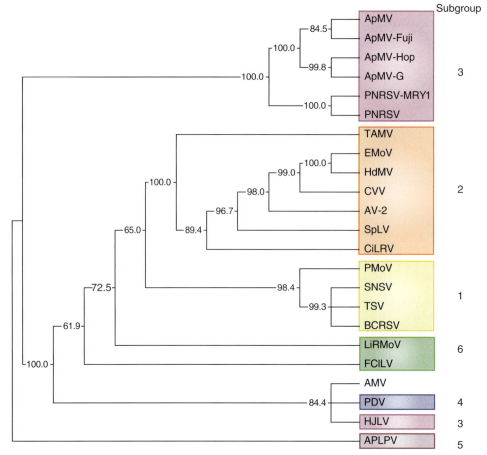

Figure 1 Cladogram illustrating the sequence relationships of ilarviruses based on the deduced amino acid sequence of the coat protein encoded by ORF3b of RNA3. Numbers at branch points indicate the percentage of bootstrap replicates in which that branch occurs. Subgroup designations are derived primarily from serological as well as molecular data and are indicated to the right of the shaded boxes. The sequence of the alfamovirus alfalfa mosaic virus is included in the analysis. See **Table 1** for abbreviations.

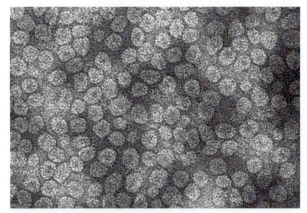

Figure 2 Electron micrograph of negative-stained particles of prune dwarf virus. In addition to quasi-isometric particles 20 and 23 nm in diameter, preparations from some ilarviruses including prune dwarf virus also contain bacilliform particles 19 × 33 nm and 19 × 38 nm.

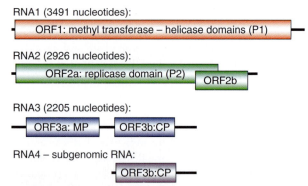

Figure 3 Genome organization of tobacco streak virus (TSV). The positive-sense RNA genome of the WC isolate of TSV is distributed between three segments. ORF2b is present only in genus *Ilarvirus* subgroups 1 and 2. ORF3a encodes a protein (MP) associated with virus movement and ORF3b encodes that coat protein (CP). The subgenomic RNA4 is transcribed from an internal promoter in the intercistronic region of minus-sense RNA3; RNA4 is 3′ co-terminal with RNA3. ORF2b is present only in genus *Ilarvirus* subgroups 1 and 2.

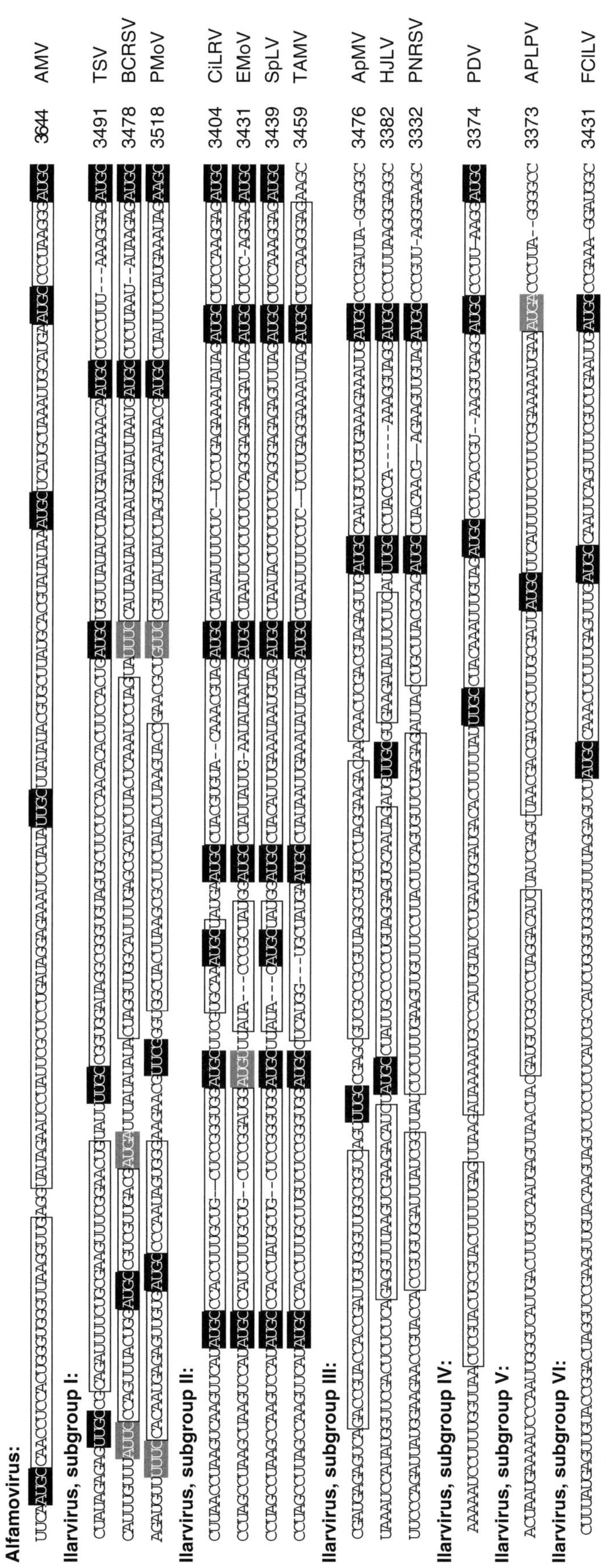

Figure 4 The 3′-untranslated region of RNA1 of ilarviruses contains a repeated (A/U) (U/A/G)GC sequence. The repeated single-stranded regions are indicated by black highlighting. Corrupted copies of (A/U) (U/A/G)GC motifs are highlighted in gray. Potential stem–loop structures are indicated by boxes.

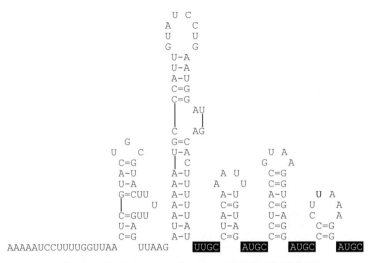

Figure 5 The 3′-untranslated region of RNA1 of ilarviruses contains a repeated (A/U) (U/A/G)GC sequence. These conserved regions are single-stranded motifs (black highlighting) that appear between base-paired stem-loops and are critical for coat protein binding and potentially for replicase binding. The secondary structure of the 3′ terminus of prune dwarf virus RNA1 illustrates the potential stem-loops configuration with intervening conserved single-stranded motifs.

the size of the stem–loop structures vary considerably when all of the ilaviruses are considered. However, within each subgroup of ilarvirus, the overall length and secondary structures of the binding region(s) are well conserved. In the case of TSV, the (A/U) (U/A/G)GC motif is repeated five times flanking four potential stem–loop structures within the 260 nucleotides located at the 3′ terminus of RNA1. This region may represent two coat protein binding sites as has been demonstrated for AMV. Ilarviruses belonging to the subgroups 4, 5, and 6 possess shorter discernable recognition sequences and may contain single coat protein binding sites. Deletion analyses of AMV RNA4 have shown that the first and third 3′-most stem loops are critical for coat protein binding and hence translation, whereas the other stem loops play a role in enhancing efficient transcription but are not strictly required for that function. There is an absolute requirement for the single-stranded (A/U) (U/A/G)GC motifs in virus function.

The role of the RNA–coat protein complex was elucidated through studies with recombinant viruses. The 3′-UTR and its ability to form a complex with coat protein subunits is functionally analogous to a poly-A tail; a poly-A tail of 40–80 residues can experimentally compensate for the absence of activating coat protein molecules. The formation of a RNA–coat protein complex or the addition of a poly-A tail both act to stimulate the translation of viral RNA resulting in the synthesis of proteins needed for virus replication. In eukaryotes, translation efficiency of mRNA is enhanced through the formation of a loop formed by the interaction of a poly-A binding protein (PABP) with poly-A, and then subsequent binding of this complex to the 5′-cap structure of RNA in combination with other cellular initiation factors. It has been suggested that the complex formed between viral RNA and coat protein mimics the complex formed by the PABP and the poly-A tail. This proposal is further strengthened by the observation that the RNA–coat protein complex binds elongation factors associated with cellular mRNA translation. Based on data from AMV, it appears that the binding of coat protein to 3′-terminal sequences of viral RNA not only facilitates translation, but also effectively blocks transcription of the RNA.

The essential complex formation between the coat protein and RNA has been examined in detail to determine critical features of the coat protein. Limited digests indicate that the N-terminus of the coat protein is primarily responsible for RNA binding. Inspection of the amino acid sequences at the N-terminus of the coat protein reveals a high percentage of basic amino acid residues that are consistent with a nucleic acid binding function. More detailed RNA protection assays and competitive binding studies suggest that a consensus sequence represented by (Q/K/R) (P/N)TXRS(R/Q) (Q/N/S) (W/F/Y)A is involved in coat protein binding to the 3′-UTR of AMV and ilarvirus genomic RNA (**Figure 6**). Although this functional domain is not strictly conserved in all ilarviruses, a core motif centered on the critical arginine residue is generally well preserved in all ilarviruses and AMV, suggesting a key role in virus function. Despite significant variation in primary sequence, the striking similarity in protein characteristics across all ilarviruses and AMV may account for the reciprocity in the ability of coat protein from several ilarviruses and AMV to activate the virus genome in heterologous combinations.

In addition to the functional domain of the coat protein described above, additional features of the coat protein may contribute to RNA binding. The critical basic arginine residue and its flanking residues are

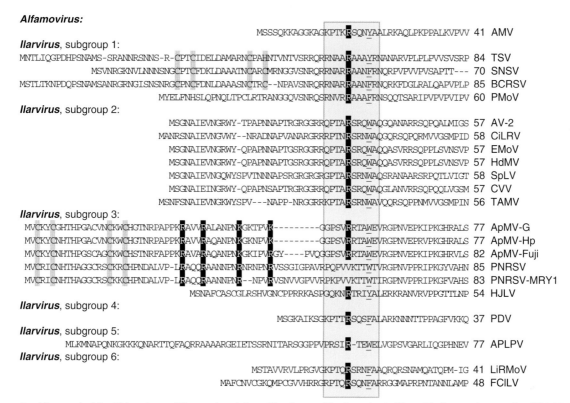

Figure 6 Alignment of the N-terminus of the coat proteins of ilarviruses. The box shaded in red indicates the putative RNA binding region with the consensus sequence (Q/K/R) (P/N)TXRS(R/Q) (Q/N/S) (W/F/Y)A and the central arginine residue highlighted in black. It has been demonstrated that in prunus necrotic ringspot virus, four arginine residues upstream of the conserved binding site are essential for RNA binding; these are highlighted with black, as are comparable residues in the apple mosaic virus coat protein sequence. Members of subgroups 1 and 3 also contain a putative zinc finger that may be involved in RNA binding; the cysteine/histidine residues potentially involved in chelating zinc cations are highlighted in gray. See **Table 1** for abbreviations.

preceded by a putative zinc finger in the coat proteins of members of subgroups 1 and 3, but not in other subgroups of ilarviruses, nor in AMV (**Figure 6**). Zinc fingers are typically involved in nucleic acid binding. Cysteine and/or hisitidine residues are critical elements of the motif represented by $CN_2CN_{10}(C/H)N_2(C/H)$; the cysteine and histidine residues are the functional amino acids capable of chelating zinc cations. The participation or requirement of this putative zinc finger in virus replication is uncertain, particularly since it is not present in all ilarviruses.

The RNA-binding region of the coat protein that is conserved in other ilarviruses appears to be absent or at least highly diverged in PNRSV. In particular, an arginine expected to be critical for RNA binding is absent in the coat protein of PNRSV in this central conserved position, although a basic lysine residue is located nearby. Moreover, electrophoretic shift assay suggests that other basic residues are involved in RNA binding. Four arginine residues of PNRSV coat protein have been experimentally determined to be involved in RNA binding and are located towards the N-terminus of the site of the conserved motif present in other ilarviruses (**Figure 6**). A similar distribution of four basic residues is observed in ApMV in addition to the conserved (Q/K/R) (P/N) TXRS(R/Q) (Q/N/S) (W/F/Y)A motif.

RNA folding predictions suggest that the 3′-UTR of some ilarviruses can fold into complex tRNA-like structures or pseudoknots that do not accommodate coat protein binding. By analogy to information obtained from AMV, the pseudoknot conformation favors transcription and hence synthesis of viral RNAs. Mutational analysis indicates that stem–loop structures located just upstream of the coat protein binding site as well as the ability to form a tRNA-like pseudoknot are required for transcription and synthesis of minus-sense RNA. Thus, the relative abundance of RNA in the pseudoknot conformation relative to that bound to coat protein provides a feedback mechanism to coordinate phases of virus replication. Translation to form nonstructural proteins is favored in the presence of coat protein, while transcription as part of the replication process is favored in the absence of the coat protein.

In addition to a requirement for regulating the use of positive-sense RNA for translation versus transcription, is the requirement for coordinating the rates of synthesis of positive-sense versus minus-sense RNA molecules. The virus replication cycle requires the switch from minus-sense RNA synthesis to the repetitive transcription into positive-sense progeny viral RNA. The genomic features responsible for this transition are not well understood. However, recent experiments with transformed plants suggest that coat protein is not involved in this process. Secondary structural features of the minus-sense RNAs are believed to regulate the synthesis of plus-sense RNA.

The subgenomic RNA encoding the coat protein, RNA4, is transcribed from a minus-sense copy of RNA3. In brome mosaic virus, the intercistronic region between the ORFs of RNA3 contains a subgenomic promoter region consisting of sequences enriched for A/U flanking the core promoter and a poly-U tract that enhances transcription. However, the intercistronic regions of AMV and the ilarviruses are considerably shorter than that of brome mosaic virus. Moreover, although still slightly enriched in U residues, the poly-U tract and strong bias for U are absent. The transcriptional start site and a strong promoter for the synthesis of positive-sense subgenomic RNA4 occurs immediately 3' of the position corresponding to the termination codon of ORF3a. The subgenomic promoter of AMV has been characterized; a critical component of the promoter region is a three-nucleotide loop on a base-paired stem that bears a 'bulge' at the fifth nucleotide from the loop along the base-paired stem. The stem sequence below the bulge is less critical for promoter activity. Potential stem–loop structures are evident in the intercistronic regions between ORF3a and ORF3b of ilarviruses and these may fulfill the same critical role as those in AMV and brome mosaic virus.

Based on studies of transgenic plants conducted with AMV, replication of RNA1 and RNA2 are coordinated in that a mutation leading to changes in the encoded protein of either RNA product prevents the replication of both. Moreover, this requirement is for a *cis*-encoded function since transgenic expression of the wild-type protein does not rescue the mutated RNA. In contrast, the replication of RNA3 can be supported by RNA1 and RNA2 gene products expressed in *trans*. This is a critical distinction highlighting the independence of RNA3 replication from that of RNA1 and RNA2. During the natural infection cycle, binding of the coat protein in the inoculum to the 3' terminus of RNA1 and RNA2 is sufficient to facilitate their translation into a replication complex in the cytoplasm of the infected cell. This is accomplished by the interaction of the coat protein–RNA complex with host initiation factors. However, replication of viral RNA can only be initiated once the coat protein is dissociated from the 3' of the RNA molecules, allowing the 3' terminus to assume the tRNA-like pseudoknot conformation. Regulation of this dissociation process is unknown. However, it has been speculated that the association of the replication complex with membrane structures could displace the coat protein, or proteolysis of the coat protein could result in its removal from RNA. Alternatively, the coat protein could be displaced by binding of the replicase to stem–loop structures upstream of the coat protein binding sites. Replication then proceeds through synthesis of minus-sense RNA and the replication of progeny plus-sense RNA segments, including the synthesis of the plus-sense subgenomic RNA encoding the coat protein. It has been demonstrated that both the coat protein and the movement protein are required for systemic infection. It appears that the infective agent moves through modified plasmodesmata as a protein–RNA complex rather than as intact virions.

Epidemiology and Control

One or more modes of transmission have been demonstrated for each member of the genus *Ilarvirus*. All ilarviruses can be transmitted by mechanical inoculation to experimental hosts, although some with difficulty. Ilarviruses tend to have a wide experimental host range, but many have woody plants as their primary natural hosts. The role of mechanical transmission between woody plants is likely small. However, succulent hosts such as hop and asparagus plants may be more prone to field spread through mechanical transmission associated with horticultural practices.

The dissemination of viruses through vegetative propagation is one of the most critical factors affecting the distribution of ilarviruses that infect perennial crops. Budding and grafting are important means of transmission of ilarviruses that infect woody plants such as fruit trees and roses. These plants are routinely budded onto rootstock to allow rapid expansion of a particular clone into hundreds or thousands of plants. Other hosts such as hop or asparagus plants can be vegetatively propagated by stem cuttings and/or crown division. In both cases, there is high likelihood that virus will be present in the progeny plants. Chance grafting of root systems is also an important means of transmission of some viruses of perennial crops.

Many ilarviruses are transmitted through seed when the seed-bearing plant is infected. Transmission of pollen-borne virus occurs to seedlings with variable frequency, and transmission of virus from pollen to the pollinated plant has also been demonstrated for PNRSV, PDV, and TSV. Seed and pollen transmission poses a particularly high risk in fruit trees since seedling rootstocks are frequently used in production and virus infection can originate from the rootstock, the scion, or both. The most effective means of control is through the use of virus-tested material during the establishment of plantings. This approach is implemented through programs where plants are propagated from virus-tested plants and grown under conditions that minimize re-infection by viruses, including the pollen-borne ilarviruses. Many industries rely on informal virus control programs whereas others have developed official certification programs. The use of spatial separation of the propagation source material from potentially infected host plants is a basic strategy to minimize the introduction of virus through pollen. The risk associated with seed transmission is further reduced by screening samples of seed lots for the presence of virus. Where seed is produced from perennial plants such as *Prunus* species and asparagus, routine virus testing and rouging of mother plantings is implemented to limit the distribution of virus through seed.

The association of virus transmission with pollen raises concern about the role of bees in virus epidemiology. In the case of PNRSV and PDV, it has been shown that the virus is detectable by serological assays on honeybees emerging from hives weeks after exposure to flowers with virus-laden pollen. Pollen from hives is no longer viable at this stage, but the virus is infectious. The ability of this virus to infect trees then depends on whether the virus is transmitted to the tree by fertilization or by some other physical process. This issue remains largely unresolved. Holding bees in hives for several days before allowing them to enter a new orchard has been suggested as one means of minimizing this risk.

It has been demonstrated that TSV is transmitted to natural hosts in a process known to be facilitated by thrips. The thrips species demonstrated to transmit TSV include *Frankliniella occidentalis* (Pergande), *Frankliniella schultzei* (Trybom), *Microcephalothrips abdominalis* (Crawford), *Thrips tabaci* (Lindeman), and *Thrips parvispinus* (Karny). The association of thrips with pollen transmission of TSV is one of the major factors leading to the increased importance of this virus in vegetable production in many areas of the world. The thrips facilitate transmission by mechanical abrasion of the leaf cells, allowing pollen-associated virus particles to enter cells and initiate infection. Pollen-to-leaf transmission has been demonstrated, but leaf-to-leaf transmission has not. There have been many attempts to confirm the role of thrips in the transmission of ilarviruses other than TSV. Other ilarviruses have been transmitted from pollen to the leaves of experimental hosts in a thrips-mediated manner, but transmission to natural hosts by thrips has not been convincingly demonstrated. Unlike AMV, there is no evidence of aphid transmission of any members of the genus *Ilarvirus*.

Most ilarviruses have a limited natural host range. This feature in combination with a narrow temporal window for transmission imposed by the short flowering period of many hosts means that plant-to-plant spread of ilarviruses is relatively slow. TSV is an exception where a wide natural host range sustains sources of infection throughout the year. This may explain the growing impact of TSV on production of vegetables and legumes such as soybean. Agronomic strategies such as removing perennial weeds that would otherwise provide a green bridge for the thrips vectors and the virus has helped reduce the carryover of virus inoculum in a limited number of situations. Since rainfall abates thrips populations, in some areas, sowing seed so that germination will occur during periods of increased rain also helps reduce crop losses caused by TSV. As with the ilarviruses that infect fruit trees and asparagus, the use of certified, virus-tested seed is critical for effective virus control.

Rugose mosaic is a serious debilitating disease of sweet cherry (*Prunus avium* L.) that is caused by a severe strain of PNRSV. The long latent period between infection and visible expression of disease symptoms complicates efficient control strategies. In this case, trees can be inoculated prophylactically with less severe strains of PNRSV. The inoculated plant is then resistant to subsequent infection by the severe virus strain. This phenomenon is termed cross-protection. In combination with accurate diagnosis and prompt inoculum removal, cross-protection is an effective tool in controlling rugose mosaic disease.

See also: Alfalfa Mosaic Virus; Bromoviruses; Plant Antiviral Defense: Gene Silencing Pathway; Plant Virus Diseases: Fruit Trees and Grapevine; Replication of Viruses; Vector Transmission of Plant Viruses; Vegetable Viruses; Viral Suppressors of Gene Silencing.

Further Reading

Ansel-McKinney P and Gehrke L (1998) RNA determinants of a specific RNA-coat protein peptide interaction in alfalfa mosaic virus: Conservation of homologous features in ilarvirus RNAs. *Journal of Molecular Biology* 278: 767–785.

Ansel-McKinney P, Scott SW, Swanson M, Ge X, and Gehrke L (1996) A plant viral coat protein RNA binding consensus sequence contains a crucial arginine. *EMBO Journal* 15: 5077–5084.

Aparicio F, Vilar M, Perez-Payá E, and Pallás V (2003) The coat protein of *Prunus* necrotic ringspot virus specifically binds to and regulates the conformation of its genomic RNA. *Virology* 313: 213–223.

Bol JF (2005) Replication of Alfamo- and Ilarviruses: Role of the coat protein. *Annual Review of Phytopathology* 43: 39–42.

Garrett RG, Cooper JA, and Smith PR (1985) Virus epidemiology and control. In: Francki RIB (ed.) *The Plant Viruses, Vol 1:Polyhedral Virions with Tripartite Genomes,* ch. 9, pp. 269–297. New York: Plenum.

Greber RS, Teakle DS, and Mink GI (1992) Thrips-facilitated transmission of prune dwarf and *Prunus* necrotic ringspot viruses from cherry pollen to cucumber. *Plant Disease* 76: 1039–1041.

Haasnoot PCJ, Brederode FTh, Olsthoorn RCL, and Bol JF (2000) A conserved hairpin structure in *Alfamovirus* and *Bromovirus* subgenomic promoters is required for efficient RNA synthesis *in vitro*. *RNA* 6: 708–716.

Li WX and Ding SW (2001) Viral suppressors of RNA silencing. *Current Opinion in Biotechnology* 12: 150–154.

Mink GI (1983) The possible role of honeybees in long-distance spread of *Prunus* necrotic ringspot virus from California into Washington sweet cherry orchards. In: Plumb RT and Thresh JM (eds.) *Plant Virus Epidemiology: The Spread and Control of Insect-Borne Viruses,* pp. 85–91. Oxford: Blackwell Scientific Publications.

Roossinck MJ, Bujarski MJ, Ding SW, et al. (2005) *Bromoviridae*. In: Fauquet CM, Mayo MA, Maniloff J, Desselberger U, and Ball LA (eds.) *Virus Taxonomy: Eighth Report of the International Committee on Taxonomy of Viruses*, pp. 1049–1058. San Diego, CA: Elsevier Academic Press.

Scott SW, Zimmerman MT, and Ge X (2003) Viruses in subgroup 2 of the genus *Ilarvirus* share both serological relationships and characteristics at the molecular level. *Archives of Virology* 148: 2063–2075.

Swanson MM, Ansel-McKinney P, Houser-Scott F, Yusibov V, Loesch-Fries LS, and Gehrke L (1998) Viral coat protein peptides with limited sequence homology bind similar domains of alfalfa mosaic virus and tobacco streak virus RNAs. *Journal of Virology* 72: 3227–3234.

Xin H-W, Ji L-H, Scott SW, Symons RH, and Ding S-W (1998) Ilarviruses encode a cucumovirus-like 2b gene that is absent in other genera within the *Bromoviridae*. *Journal of Virology* 72: 6956–6959.

Immune Response to Viruses: Antibody-Mediated Immunity

A R Neurath, Virotech, New York, NY, USA

© 2008 Elsevier Ltd. All rights reserved.

Milestones in History

The primeval notion about immunity to disease arose in Greece about 25 centuries ago from the observation that those who recovered from an apparently contagious disease became resistant to a subsequent similar sickness. The earliest known attempts to intentionally transfer immunity to an infectious disease occurred in China in the tenth century. It involved exposing uninfected people to material from lesions caused by smallpox. This not always successful practice was introduced in the seventeenth century to the Ottoman Empire and subsequently to England and its colonies in North America.

This approach was revolutionized by replacing material from human lesions by that derived from cowpox lesions first in 1774 in England by a farmer, Benjamin Jesty, who used it on his family, and 22 years later by Edward Jenner. Thereafter, the widely used vaccination has been considered as the first example of immunization with a life-attenuated virus. Nevertheless, both the cause of the disease and the success of vaccination remained unexplained.

An understanding started to emerge about a century later when (1) Koch and Pasteur set forth the germ theory of infectious diseases, and Pasteur developed attenuated vaccines against anthrax and rabies; and (2) Von Behring and Kitasato demonstrated that immunity could be transferred by a soluble serum component(s). Several decades later, these components were shown to be antibody immunoglobulins (Heidelberger and Kabat) whose fundamental structure was elucidated by Porter and Edelman in the late 1950s, and further characterized by numerous X-ray crystallography and sequencing studies.

However, several findings indicated that antibodies are not the only mediators of specific immunity. Immune responses corresponding to delayed hypersensitivity and allograft rejections appeared unrelated to the presence of serum antibodies. Immunity could also be transferred by cells from immunized to naive animals (Landsteiner and Chase). It was shown by Gowans and colleagues in 1962 that lymphocytes are essential for immune responses.

Antibodies to very many distinct antigens can be produced upon immunization. Therefore, it was presumed that the diverse antibodies were not preformed but would be generated on demand following antigenic stimuli. In theory, there were two possibilities: that an antigen either directs or selects for the formation of a specific antibody. Further studies fully supported the clonal selection theory (Burnet, Jerne, and Talmage). The repertoire of diverse antibodies undergoing a process of maturation leading to high-affinity antibodies is generated by somatic rearrangements and hypermutation of immunoglobulin genes (Tonegawa).

Although it became evident that both immunoglobulins and cells are essential for immunity, the details of the process whereby antigens and viruses elicit immune responses were contributed to by results of studies in transplantation biology and notably the discovery of the major histocompatibility complex (MHC; HLA molecules on human cells) (Benacerraf, Dausset, and Snell). Their role in presentation of antigen fragments and interactions of specific antibody producing B-cells with T-cells, as well as the function of cytokines, chemokines, and adhesion molecules will be discussed in later sections of this article.

Immunoglobulins, lymphocytes, MHC molecules, and antigen receptors are all components of adaptive (acquired) immunity. Components of innate immunity, responding rapidly to an invading pathogen, play a key role in initiating and orchestrating the adaptive response (Janeway).

The major milestones in the history of immunology and their time lines are summarized in **Table 1**.

Table 1 History of immunology timeline

Year	Event	Person
1798	Smallpox vaccination	Edward Jenner
1876	Validation of germ theory of disease by discovering bacterial basis of anthrax	Robert Koch
1879	Chicken cholera vaccine development	Louis Pasteur
1890	Discovery of diphtheria 'antitoxin' in blood	Emil Von Behring and Shibasaburo Kitasato
1882	Isolation of the tubercle bacillus	Robert Koch
1883	Delayed type hypersensitivity	Robert Koch
1884	Phagocytosis; cellular theory of immunity	Elie Metchnikoff
1891	Proposal that antibodies are responsible for immunity	Paul Ehrlich
1891	Passive immunity	Emil Roux
1894	Complement and antibody activity in bacteriolysis	Jules Bordet
1900	A, B, and O blood groups	Karl Landsteiner
1901	Cutaneous allergic reaction	Maurice Arthus
1903	Opsonization by antibody	Almroth Wright and Stewart Douglas
1907	Discipline of immunochemistry founded	Svante Arrhenius
1910	Viral immunology theory	Peyton Rous
1917	Haptens discovered	Karl Landsteiner
1921	Cutaneous reactions	Carl Prausnitz and Heinz Kustner
1924	Reticuloendothelial system	Ludwig Aschoff
1939	Discovery that antibodies are gamma globulins	Elvin Kabat
1942	Adjuvants	Jules Freund and Katherine McDermott
1942	Cellular transfer of sensitivity in guinea pigs (anaphylaxis)	Karl Landsteiner and Merill Chase
1944	Immunological hypothesis of allograft rejection	Peter Medawar
1948	Demonstration of antibody production in plasma B-cells	Astrid Fagraeus
1949	Distinguishing self vs. nonself and its role in maintaining immunological unresponsiveness (tolerance) to self	Macfarlane Burnet and Frank Fenner
1952	Discovery of agammaglobulinemia (antibody immunodeficiency)	Ogden Bruton
1953	Immunological tolerance hypothesis	Rupert Billingham, Leslie Brent, Peter Medawar, and Milan Hasek
1955–59	Clonal selection theory	Niels Jerne, David Talmage, and Macfarlane Burnet
1957	Discovery of interferon	Alick Isaacs and Jean Lindenmann
1958	Identification of first autoantibody and first recognition of autoimmune disease	Henry Kunkel
1959–62	Elucidation of antibody structure	Rodney Porter and Gerald Edelman
1959	Lymphocytes as the cellular units of clonal selection; discovery of lymphoid circulation	James Gowans
1961	Discovery of thymus involvement in cellular immunity	Jacques Miller
1968	Recognition of B- and T-cells in immunodeficiencies	Robert Good
1968	Distinction of bone marrow- and thymus-derived lymphocyte populations; discovery or T- and B-cell collaboration	Jacques Miller and Graham Mitchell
1965	Demonstration of allelic exclusion in B-cells	Benvenuto Pernis
1968–70	Elaboration of two-signal model of lymphocyte activation	Peter Bretscher and Melvin Cohn
1970	Discovery of membrane immunoglobulins	Benvenuto Pernis and Martin Raff
1971	Recognition of hypervariable regions in Ig chains	Elvin Kabat
1972	Elucidation of the major histocompatibility complex	Baruj Benacerraf, Jean Dausset, and George Snell
1974	Discovery of MHC restriction	Rolf Zinkernagel and Peter Doherty
1974	HLA-B27 predisposes to an autoimmune disease	Derek Brewerton
1975	Monoclonal antibodies used in genetic analysis	Georges Kohler and Cesar Milstein
1976	First demonstration of cross-priming	Michael Bevan
1978	Direct evidence for somatic rearrangement in immunoglobulin genes	Susumu Tonegawa
1978	Recognition that dendritic cells are distinctive and highly potent antigen-presenting cells	Ralph Steinman
1979	Discovery of leukocyte adhesion molecules and their role in lymphocyte trafficking	Eugene Butcher
1984–87	Identification of genes for the T-cell antigen receptor	Mark Davis, Leroy Hood, Stephen Hedrick, and Gerry Siu
1987	Crystal structure of MHC peptide solved	Pam Bjorkman, Jack Strominger, and Don Wiley
1989	Emerging field of innate immunity, infectious nonself model of immune recognition ('stranger hypothesis')	Charlie Janeway

Continued

Table 1 Continued

1994	Danger hypothesis of immune responsiveness	Polly Matzinger
1986	Discovery of T-helper subsets	Tim Mossmann and Bob Coffman
1989	Discovery of first chemokines	Edward Leonard, Teizo Yoshimura, and Marco Baggiolini
1991	Discovery of the first costimulatory pathway (CD28/B7) for T-cell activation	Kevin Urdahl and Mark Jenkins
1992	Cloning of CD40 ligand and recognition of its role in T-cell-dependent B-cell activation	Armitage, Spriggs, Lederman, Chess, Noelle, Aruffo, et al.
1996–7	Discovery of the role of Toll and Toll-like receptors in immunity	R. Medzhitov, CA. Janeway, Jr., and J. Hoffmann

Reprinted with permission from Steven Greenberg, MD, from http://www.columbia.edu/itc/hs/medical/pathophys/immunology/readings/ConciseHistoryImmunology.pdf.

Immunoglobulin Phylogeny

The adaptive immune system, based on clonally diverse repertoires of antigen recognition molecules, arose in jawed vertebrates (gnasthostomes) about 500 million years ago. Homologs of human immunoglobulins, T-cell antigen receptors (TCRs), MHC I and MHC II molecules, and recombination activating genes (RAGs) have been identified in all classes of gnasthostomes. This has provided evolutionary advantages allowing the recognition of potentially lethal pathogens, including viruses, and the initiation of protective responses against them (**Figure 1**), and represented an add-on to a preexisting innate immune system. For immunoglobulins, somatic variation was further expanded through class switching, gene conversion, and somatic hypermutation. Jawless vertebrates (agnathans) were shown to also have an adaptive immune system based on recombinatorial assembly of genetic units different from those of gnasthostomes, to generate a diverse repertoire of lymphocytes, each with distinct receptors (**Figure 2**). The assembly relies on highly variable leucine-rich repeat (LRR) protein modular units.

Most gnasthotomes generate a large part of their immunoglobulin diversity in ways similar to those in humans. However, exceptions have occurred. Two of these, relevant to passive immunotherapy and development of diagnostics using the corresponding immunoglobulins, are mentioned here. (1) Only three classes of H-chain genes, corresponding to IgM, IgY, and IgA, exist in chicken. Class switching occurs from IgM to either IgY or IgA. IgY is considered to be the ancestor of 'present-day' IgG and IgE. Unlike human IgG, IgY does not bind to cellular Fc receptors and to proteins A and G, respectively, and does not activate the complement system. (2) H-chain antibodies devoid of light chains occur in the Camelidae species (camels, dromedaries, and llamas). Single-domain fragments from such antibodies react with specific antigens with high affinity constants (nanomolar range). The binding involves longer than average third variable loops of the antibody molecules. These unique antibodies and their antigen-specific fragments are expected to become important for biotechnological and medical applications, including intrabodies (intracellular antibodies).

Innate Immunity

The innate immune system provides an early defense against pathogens and alerts the adaptive immune system when initial invasion by a pathogen has occurred. The innate system predates evolutionarily the adaptive system and relies on antimicrobial peptides and on germline-encoded pattern-recognition receptors (PRRs). Antimicrobial peptides can damage enveloped viruses and interfere with processes involved in fusion of viruses with target cell membranes. PRRs recognize microbial (viral) components defined as pathogen-associated molecular patterns (PAMPs). Different PRRs are specific for distinct PAMPs, have distinct expression patterns, activate different signaling pathways, thereby eliciting distinct responses against invading pathogens. PRRs can be subdivided into two classes: Toll-like receptors (TLRs) and non-Toll-like innate immune proteins.

Toll was originally identified as a differentiation protein in *Drosophila*, and later (1996) shown to play a role in defense against fungal infection. Searching the entire DNA sequence database for similarities with the coding sequence for *Toll* resulted in identification of Toll-like sequences, and ultimately in discovery of at least 10 functional TLRs in humans. TLRs are type 1 integral membrane glycoproteins the extracellular domain of which contains several LRR motifs and a cytoplasmic signaling domain homologous to that of the interleukin-1 receptor (IL-1R). TLRs 2, 3, 4, 7, 8, and 9 recognize PAMPs characteristic for viruses. Viral DNAs rich in CpG motifs are recognized by TLR9, leading to activation of pro-inflammatory cytokines and type 1 interferon (IFN) secretion. TLR7 and TLR8, expressed within the endosomal membrane, are specific for viral single-stranded RNA (ssRNA). Double-stranded RNA (dsRNA), generated

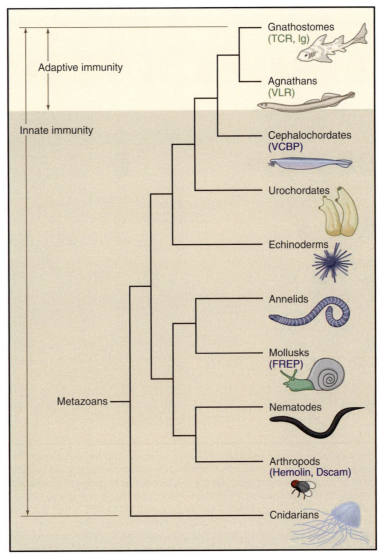

Figure 1 Phylogenetic tree indicating theoretical evolutionary relationships of metazoans and the emergence of adaptive immunity in conjunction with innate immunity. Families of immune molecules, other than Toll-like receptors (TLRs), are indicated in blue: chitin-binding domain-containing proteins (VCBPs), fibrinogen-related proteins (FREPs), hemolin and Down's syndrome cell adhesion molecule (Dscam). The recombinatorial based immune receptors are indicated in green: T-cell receptors (TCRs), immunoglobulins (Ig's), and variable lymphocyte receptors (VLRs). Reprinted from Cooper MD and Alder N (2006) The evolution of adaptive immune systems. *Cell* 124: 815–822, with permission from Elsevier.

during viral infection as an intermediate for ssRNA viruses or during transcription of viral DNA, is recognized by TLR3. TLR3 is expressed in dendritic cells (DCs) and in a variety of epithelial cells, including airway, uterine, corneal, vaginal, cervical, biliary, and intestinal cells. Cervical mucosal epithelial cells also express functional TLR9, suggesting that TLR3 and TLR9 provide an antiviral environment for the lower female reproductive tract. Some viral envelope glycoproteins are recognized by TLR2 and TLR4, each expressed at the cell surface, leading to production of pro-inflammatory cytokines. In general, the engagement of TLRs by microbial PAMPs triggers signaling cascades leading to the induction of genes involved in antimicrobial host defenses. This includes the maturation and migration of DCs from sites where infection occurs to lymphoid organs where DCs can initiate antigen-specific immune responses. Triggering distinct TLRs elicits different cytokine profiles and different immune responses. Engagement of TL3 and TRL 4, respectively, upregulates polymeric immunoglobulin receptor expression on cells. Thus, bridges between innate and adaptive immune responses are established.

TLRs are expressed either at the cell surface or in lysosomal/endosomal membranes. Therefore, they would

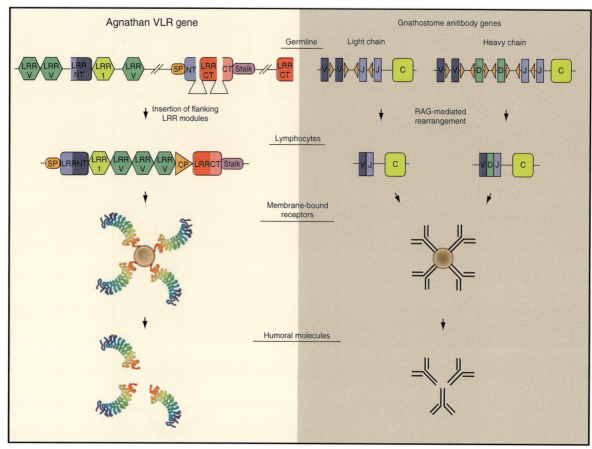

Figure 2 Two recombinatorial systems used for generating diverse antigen receptors in vertebrates. The figure compares the assembly of leucine-rich repeat (LRR) modular genetic units in agnathan lymphocytes to generate variable lymphocyte receptor (VLR) genes vs. the rearrangement of Ig gene segments in gnathostome B-lymphocytes to generate diverse antibody genes. Variable 24-amino-acid LRR (LRRV, green); N-terminal capping LRR (LRRNT, blue); variable 18-amino-acid LRR (LRR1, yellow); signal peptide (SP, orange); first six amino acids of LRRNT (NT, light blue); C-terminal capping LRR (LRRCT, red); last nine amino acids of LRRCT (CT, pink); variable 13 amino acid connecting peptide (CP, orange); and invariant VLR stalk (Stalk, purple). The small orange triangles adjacent to the representative V (blue), D (green), and J (light blue) gene segments represent recombination signal sequences (RSSs). The C (yellow) indicates the immunoglobulin constant region. Reproduced from Cooper MD and Alder N (2006) The evolution of adaptive immune systems. *Cell* 124: 815–822, with permission from Elsevier.

not recognize pathogens that succeeded in invading the cytosolic compartment. These pathogens are detected by a variety of cytoplasmic PRRs. They include retinoic acid inducible protein (RIG-1) with a helicase domain recognizing viral dsRNA, and a related protein, MDA5. Other proteins which may be involved in innate antiviral immunity include a triggering receptor on myeloid cells (TREM-1), myeloid C-type lectins and siglecs, recognizing sialic acid. One of the elements of both innate and adaptive immunity is the complement system. Some viruses or virus-infected cells can directly activate the complement cascade in the absence of antiviral antibodies.

Many viruses are endowed with properties subverting innate immune responses. Vaccinia virus produces proteins suppressing TLR- and IL-1R-induced signaling cascades. Paramyxoviruses produce proteins which associate with MDA5, and thus inhibit dsRNA-induced activation processes. Adenoviruses avoid immune surveillance by TRL9. A nonstructural protein of hepatitis C virus blocks signaling by RIG-1 and MDA5. Marburg and Ebola viruses, members of the family *Filoviridae*, elicit direct activation of TREM-1 on neutrophils. This can lead to vigorous inflammatory responses contributing to fatal hemorrhagic fevers in infected humans. Thus, some viruses have developed strategies to overcome either innate or adaptive (see next section) immune responses.

Adaptive Immunity

Adaptive immunity is a complex anticipatory system triggered by exposure to antigens, including viruses. Its hallmarks are selectivity, diversification, specificity, and memory. The principal effector molecules of the system

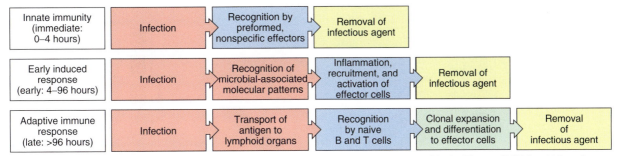

Figure 3 Three phases of a response to an initial infection. Copyright 2005 from Janeway CA, Jr., Travers P, Walport M, and Shlomchik MJ (2005) *Immunobiology. The Immune System in Health and Disease*, 6th edn, figure 2.1, p. 37. New York: Garland Science. Reproduced by permission of Garland Science/Taylor & Francis LLC.

are antigen-binding receptors (immunoglobulin (Ig) and T-cell receptors (TCR)). The following simplified overview (see **Figure 3**) will be limited to Ig's and humoral immune responses.

Differentiation of B-cells into antibody-generating plasma cells occurs through distinct pathways. Two pathways lead to rapid IgM and IgA antibody production by B-1 cells against T-cell-independent antigens. These antibodies have low antigen-binding constants, and immunological memory does not evolve. The third pathway involves clonal selection. The B-cells synthesize IgM and IgD low affinity antibodies which are expressed as antigen receptors on the surface of cells. Each B-cell produces a different receptor recognizing a distinct epitope. Encounter with the appropriate epitope on an antigen elicits cell division, and generation of selected clones with identical receptor specificities. The clones further differentiate into specific antibody-secreting cells. The remarkable diversity of antibodies is attributable to the fact that genes coding for Ig variable regions are inherited as sets of gene fragments, each encoding a portion of the variable region of a particular Ig polypeptide chain. The fragments are joined together to form a complete gene in individual lymphocytes. The joining process involves addition of DNA sequences to the ends of fragments to be joined, thus increasing diversity. Further diversity arises from the assembly of each Ig protein from pairs of H- and L-chains, each B-cell producing only one kind of either chain (**Figure 4**). In addition, the assembled genes for Ig's mutate rapidly when B-cells are activated by binding an antigen. These hypermutations lead to new receptor variants representing the process of affinity maturation of an immune response. Moreover, B-cells and their progeny can produce an additional variation by altering the constant part of the H chain due to gene rearrangements (=Ig isotype and subclass switching) (**Figures 5–7**). These antibodies have identical paratopes but distinct effector functions (complement activation, phagocytosis, transcytosis, etc.). As a consequence of the aforementioned processes, there are B-lymphocytes of at least 100 million distinct specificities in every human individual at any given time.

The majority of Ig is produced in mucosa-associated tissues, predominantly in the intestine, rather than in the bone marrow, spleen, and lymph nodes, in the form of IgA. This harmonizes with the fact that mucosal surfaces represent the predominant sites for entry of pathogens, including viruses. Coincidentally, it has been demonstrated that the gut is the major site for HIV-1 replication and depletion of CD4+ cells. There are two subclasses of IgA, IgA1 and IgA2, which occur in monomeric, dimeric, tetrameric, and polymeric (pIgA) forms. A distinguishing feature of secretory IgA is its association with another glycoprotein, the secretory component (SC) (**Figure 8**). SC is also the extracellular portion (which can be generated by proteolytic cleavage) of an integral epithelial cell membrane protein, the polymeric Ig receptor (pIgR) mediating transcytosis of pIgA and IgM. The latter process allows neutralization of pathogens within intracellular vesicular compartments.

An essential aspect of adaptive immunity is its ability to recall past encounters with a pathogen for decades or even an entire lifetime. This fundamental feature is the foundation of successful vaccination. Germinal center-derived memory B-cells have the following attributes: antigen specificity; hypermutated Ig variable gene segments; and ability to bestow immunological memory following their adoptive transfer to immunologically naïve recipients. The CD27 surface antigen is a marker for memory B-cells.

The multifaceted performance of B-cells is the result of a thoroughly orchestrated ensemble involving CD4+ helper T-cells, DCs, MHC class II antigens, cell-differentiation antigens (CDs), cytokines, etc. In summary, in addition to occupancy of the Ig receptor, B-cells must interact with antigen-specific T-cells. The T-cells, through specific TCRs, recognize peptide fragments generated from the antigen internalized by the B-cell, and displayed on the surface of the B-cell as a peptide–MHC class II complex.

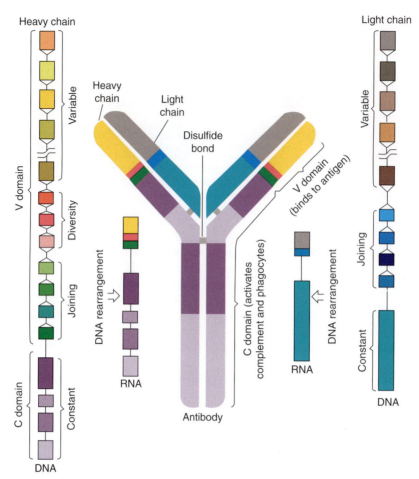

Figure 4 Antibody molecule consisting of pairs of H- and L-chains each encoded by genes assembled from different DNA segments. The segments rearrange to generate genes for chains that are distinct in each B-cell. The joining is variable so that the gene segments can encode the estimated 100 million specific antibodies each human is capable of producing. Reprinted from Janeway CA, Jr. (1993) How the immune system recognizes invaders. *Scientific American* (Sep.): 73–79, with permission of Scientific American and Ian Worpole (artist).

Helper T-cells stimulate the B-cell following binding of the CD40 ligand on the T-cell to CD40 on the B-cell; the interaction of tumor necrosis factor (TNF)–TNF receptor family ligand pairs; and the release of specific cytokines. Further details of these interactions are shown in **Figures 9–11**.

The intricacies of all these tightly coordinated events seem to minimize the possibility that 'rationally designed' synthetic vaccines will be able to successfully recreate the specificity of virus neutralization B-cell epitopes or neotopes.

Biological Functions of Antiviral Antibodies

The surface of viruses is represented by a mosaic cluster of protein or glycoprotein subunits. The subunits correspond to a single or two or more species. The pattern of repetitiveness is usually a key factor responsible for the efficiency of early and rapid B-cell responses, potent IgM antibody production, and efficient downstream antibody class switching. However, the immune response is not restricted to antigenic sites (epitopes and neotopes) on the surface of viruses. Virus particles, following initial infection or provided as a vaccine, also separate into constituent parts. Consequently, unassembled surface subunits (their epitopes and cryptotopes) and internal virus components become exposed to the immune system, ultimately resulting in the production of antibodies having multiple specificities. Only some of these are directed against intact viruses, and may have virus-neutralizing properties. The formation of antibodies with distinct specificities may not be simultaneous but rather sequential. Especially in case of some not directly cytopathic viruses (hepatitis B and C, lymphocytic choriomeningitis,

	Immunoglobulin								
	IgG1	IgG2	IgG3	IgG4	IgM	IgA1	IgA2	IgD	IgE
Heavy chain	γ_1	γ_2	γ_3	γ_4	μ	α_1	α_2	δ	ε
Molecular weight (kDa)	146	146	165	146	970	160	160	184	188
Serum level (mean adult mg ml^{-1})	9	3	1	0.5	1.5	3.0	0.5	0.03	5×10^{-5}
Half-life in serum (days)	21	20	7	21	10	6	6	3	2
Classical pathway of complement activation	++	+	+++	−	+++	−	−	−	−
Alternative pathway of complement activation	−	−	−	−	−	+	−	−	−
Placental transfer	+++	+	++	−/+	−	−	−	−	−
Binding to macrophage and phagocyte Fc receptors	+	−	+	−/+	−	+	+	−	+
High-affinity binding to mast cells and baseophils	−	−	−	−	−	−	−	−	+++
Reactivity with staphylococcal protein A	+	+	−/+	+	−	−	−	−	−

Figure 5 Properties of human immunoglobulin isotypes. The molecular mass of IgM corresponds to that of a pentamer. IgE is associated with immediate hypersensitivity. When attached to mast cells, it has a much higher half-life than in plasma. Copyright 2005 and permission as shown in legend for **Figure 3** (source = figure 4.17).

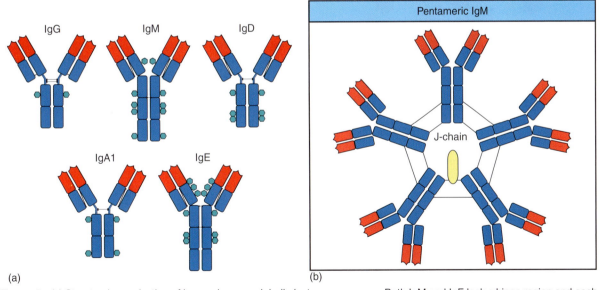

Figure 6 (a) Structural organization of human immunoglobulin isotype monomers. Both IgM and IgE lack a hinge region and each contains an additional heavy chain domain. Disulfide bonds linking the chains are indicated by black lines. N-linked glycans are shown as turquoise hexagons. (b) Pentameric IgM is associated with an additional polypeptide, the J-chain. The monomers are cross-linked by disulfide bonds to each other and to the J-chain. Copyright 2005 and permission as shown in legend for **Figure 3** (source figures 4.18 and 4.23).

Functional activity	IgM	IgD	IgG1	IgG2	IgG3	IgG4	IgA	IgE
Neutralization	+	−	++	++	++	++	++	−
Opsonization	+	−	+++	*	++	+	+	−
Sensitization for killing by NK cells	−	−	++	−	++	−	−	−
Sensitization of mast cells	−	−	+	−	+	−	−	+++
Activates complement system	+++	−	++	+	+++	−	+	−

Distribution	IgM	IgD	IgG1	IgG2	IgG3	IgG4	IgA	IgE
Transport across epithelium	+	−	−	−	−	−	+++ (dimer)	−
Transport across placenta	−	−	+++	+	++	+/−	−	−
Diffusion into extravascular sites	+/−	−	+++	+++	+++	+++	++ (monomer)	+
Mean serum level (mg ml^{-1})	1.5	0.04	9	3	1	0.5	2.1	3×10^{-5}

Figure 7 Functions and distribution of human immunoglobulin isotypes. Copyright 2005 and permission as shown in legend for **Figure 3** (source figure 9.19).

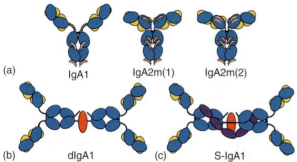

Figure 8 Structural organization of different molecular forms of human IgA. Heavy chains are shown in mid-blue, light chains in yellow, J-chain in red, and the secretory component (SC) in navy blue. (a) Monomeric mIgA. N-linked glycans are shown in orange, and O-linked glycans by small green circles. (b) Dimeric IgA1. (c) Dimeric secretory IgA1 (S-IgA1). For clarity, glycans are not shown for (b) and (c). Reprinted from Woof JM and Mestecky J (2005) Mucosal immunoglobulins. *Immunological Reviews* 206: 64–82, figure 1, with permission from Blackwell Publishing.

and HIV-1, prone to elicit a chronic carrier state), virus-neutralizing antibodies (VNAbs) appear with a delay after non-neutralizing antibodies. The latter may function as a 'decoy' if they are targeted like VNAb to virus surface epitopes.

Virus-Neutralizing and Protective Antibodies

VNAbs are crucial for protection against reinfection by a virus the VNAbs are specific for. Protection by efficacious vaccines correlates closely with *in vitro* determined VNAb titers of sera from immunized individuals. Protection by passive immunization relies on VNAb recognizing neutralization epitopes (or neotopes) on the virus surface. Coating of virus particles by antibodies is necessary but not always sufficient for virus neutralization. The effectiveness of virus neutralization correlates with the rate of antibody binding to critical epitopes and is augmented by slow dissociation of the formed antigen–antibody complexes. These kinetic parameters can be determined experimentally.

Spatial adaptive complementarity, electrostatic interactions, hydrogen bonds, and van der Waals forces contribute to the binding. Experimentally determined virus neutralization can depend on the target cells used. Virus neutralization is a multihit process and is successful when the number of unencumbered viral molecules, essential for initiation of the virus replicative cycle, is brought below a minimum threshold level. The mechanism of neutralization depends on processes obligatory for reproduction of a particular virus, and may involve the following steps: attachment to cell receptors; post-attachment events, internalization (endocytosis); fusion with cell

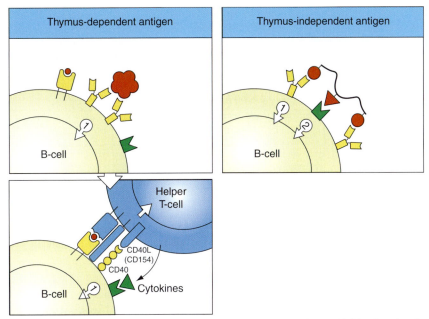

Figure 9 T-helper-cell-dependent initiation of the humoral immune response (two left panels). The first signal required for B-cell activation is delivered by binding of antigen (virus) (large red particle) to Ig cell receptors corresponding to monomeric IgM. Internalization and degradation of the antigen, and complex formation of the resulting peptide(s) (small red circle) with MHC class II molecules on the B-cell, allow the second signal to be delivered, that is, by the interaction between CD40 on the B-cell and the CD40 ligand (=CD154) on the CD4+ helper T-cell, and the engagement of the T-cell receptor (TCR) with the peptide-MHC class II complex on the B-cell. The activation is promoted by binding of cytokines to their specific receptors (see **Figure 10**). For comparison (right panel), in case of T-helper-cell-independent antigens, the second signal can be delivered by the antigen itself, either through binding of a part of the antigen to a receptor of the innate immune system (e.g., TLRs; green), or by extensive cross-linking of the membrane IgM by a polymeric antigen. Copyright 2005 and permission as shown in legend for **Figure 3** (source figure 9.2).

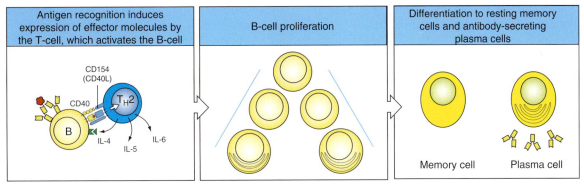

Figure 10 Antigen recognition induces the expression of B-cell stimulatory interleukins IL-4, IL-5, and IL-6 (and/or others) by the T-cell, driving the proliferation and differentiation of B-cells into antibody-secreting plasma cells. Activated B-cells can alternatively become memory B-cells. Copyright 2005 and permission as shown in legend for **Figure 3** (source figure 9.3).

membranes or endosomal vesicles; uncoating and/or intracellular localization; and enzymatic activities (e.g., transcription). Direct occupancy by VNAb of cell receptor binding sites on the virus surface might not be obligatory for neutralization. Steric hindrance or induction of deleterious conformational changes may be sufficiently effective. A unique feature of several anti-HIV-1 human monoclonal VNAb, having distinct specificities, is a very long finger-like third complementarity determining region of the immunoglobulin heavy chain allowing access to a recessed critical site on HIV-1 gp120. Such feature is rare in the human immunoglobulin repertoire but is common in the Ig's of Camelidae.

The mechanism and magnitude of VNAb neutralizing effects are influenced by their immunoglobulin isotype and subtype which affect interactions with complement,

Role of cytokines in regulating Ig isotype expression							
Cytokines	IgM	IgG3	IgG1	IgG2b	IgG2a	IgE	IgA
IL-4	Inhibits	Inhibits	Induces		Inhibits	Induces	
IL-5							Augments production
IFN-γ	Inhibits	Induces	Inhibits		Induces	Inhibits	
TGF-β	Inhibits	Inhibits		Induces			Induces

Figure 11 Role of cytokines in regulating Ig isotype switching. The individual cytokines either induce (violet) or inhibit (red) the production of particular Ig isotypes. IFN, interferon; TGF, transforming growth factor. Copyright 2005 and permission as shown in legend for **Figure 3** (source figure 9.7).

Fc receptors, and transcytosis through mucosal epithelia. Non-neutralizing virus-surface-binding antibodies sometimes enhance the effectiveness of VNAb, limit viral spread in the early phases of infection, and contribute to its suppression through antibody-mediated cellular cytotoxicity (ADCC). Antibodies directed to epitopes (or neotopes) on distinct surface proteins may act synergistically in virus neutralization.

The principle that VNAbs specific for virus surface components provide protection against disease is not absolute. The flavivirus nonstructural protein NS1 elicits a protective immune response against yellow fever, dengue, and tick-borne encephalitis viruses. The paratope binding site containing F(ab′)2 fragments are ineffective. Thus the immunoglobulin Fc portion is obligatory for the protective effect.

Antibody-Dependent Enhancement of Viral Diseases

Some viruses make use of antiviral antibodies to gain entry into target cells thus widening cell receptor usage to initiate infection. The infectious virus–antibody complexes rely upon the Fc portion of IgG antibodies to gain entry into monocytes/macrophages and granulocyte through Fc (FcR) or complement receptors (CR) expressed on these cells. The characteristic feature of the viruses is their propensity to establish persistent infections and their antigenic diversity. Antibody-dependent enhancement (ADE) has been demonstrated to occur *in vitro* with members of the families *Bunyaviridae*, *Coronaviridae*, *Flaviviridae*, *Orthomyxoviridae*, *Paramyxoviridae*, *Retroviridae*, *Rhabdoviridae*, and *Togaviridae*. A link between *in vitro* ADE and clinical manifestations cannot be always established. A relationship between ADE and disease exacerbation has been observed for dengue, measles, yellow fever, and respiratory syncytial viruses (RSVs). ADE may occur in children infected at a time when the level of transferred maternal antiviral antibodies declines to insufficient levels. ADE could represent an obstacle for development of vaccines, as has been the case for anti-RSV vaccines. A vaccine consisting of formaldehyde-treated measles virus hemagglutinin (ineffective to elicit antibodies to the virus fusion protein) induced antibodies causing ADE and led to aggravated atypical disease following infection with measles virus.

FcR- and CR-independent ADE was shown to occur following binding to HIV-1 of antibodies eliciting conformational changes in the gp120 envelope glycoprotein, allowing direct virus binding to cellular co-receptors while bypassing the primary binding to the primary CD4 cell receptor.

Immune System Evasion by Viruses

Persistence in an infected host and repeated reactivation of many viruses rely on several specific evasion strategies of adaptive immunity. Thus common protective responses are redirected or altered to the advantage of the infectious agent. This includes antiviral antibody responses and involves (1) specific paratope–epitope interactions (Fab fragments) and (2) effector mechanisms mediated by the Fc portion of antibodies.

The first mechanism is provided by genetically determined amino acid replacements leading to changes of virus epitopes (or neotopes) involved in virus neutralization. The sites of these escape mutations are usually on the viral surface, result in structural changes in antibody/virus contact sites, and lead to much less favorable kinetic parameters for antibody binding or completely abrogate binding. Presentation of new glycan chains on enveloped viruses or elimination of these chains due to mutations of N-glycosylation sites may cause substantial epitope alterations. The rate of escape mutation appearance is promoted by error-prone replication of the viral genome. Antibodies with new specificities must be produced to bring the mutant viruses under control. The process is repeated, and if not successful, persistent infection is established. A similar process leads to evasion from T-cell-mediated protective responses.

Secondly, several viruses bypass clearance processes facilitated by the Fc portion of bound antibodies by encoding and expressing Fc receptor analogs. Subversion of the complement cascade provides another way how to block clearance of cell-free virus and infected cells.

Additional scenarios for escaping immune surveillance include: interference with MHC class I restricted antigen presentation involving also inhibition of MHC class I cell surface expression or synthesis of viral MHC class I homologs; blocking MHC class II restricted antigen presentation; downregulation of cellular CD4 or its degradation; interference with cytokine effector functions; and other strategies.

Immunoglobulins for Passive Immunization against Human Viruses

Transfer of immunity from an immune donor to an unprotected recipient by serum is one of the early landmarks in the history of immunology (Von Behring). The active serum components have later been identified as immunoglobulins. The half-life of immunoglobulin isotypes in serum is 6–21 days (**Figure 5**). Consequently, administered antivirus immunoglobulins can provide only short term prophylactic and therapeutic benefits, respectively, unless they are administered repeatedly or incorporated into a slow-release medical device. The most common applications are pre-exposure (travel, protection against community-wide infection(s), medical professionals, immunosuppressed individuals, combination with live vaccines to minimize their potential side effects) and post-exposure prophylaxis (passive immunization may provide immediate protection while the benefits from vaccination are delayed). Local mucosal applications of immunoglobulins 'as needed' appear promising against perinatal and sexual transmission, respectively, of several viruses (e.g., herpesviruses and HIV-1).

The immunoglobulins are isolated from serum of individuals pre-screened for high levels of antibodies against a particular virus or from vaccinated individuals. The immunoglobulins are purified, treated to remove or inactivate infectious agents which might be present in the pooled serum source, notwithstanding rigorous screening of the individual sera entering the pool. The products are further processed depending on their intended intramuscular or intravenous applications. All these immunoglobulins are polyclonal with respect to the 'indicated' virus and contain other antibodies originally present in the pooled sera. Oral administration of antibodies produced in bovine colostrums or chicken yolk has been recently suggested.

Alternatively, monoclonal antibodies (mAbs) specific for epitopes, known to elicit virus-neutralizing and protective immune responses, are used. They are prepared using hybridoma technologies, immortalized human peripheral B-cells, transgenic mice and bacteriophage expression libraries. If derived from animal species, the mAbs are 'humanized' using recombinant DNA techniques by replacing amino acid sequences outside the antigen-binding sites with sequences corresponding to human immunoglobulins. By an *in vitro* directed evolution process allowing manipulation of antigen-binding kinetics, mAb variants having much faster antigen association rates and much slower dissociation rates can be produced. Such antibodies have a much improved capacity to neutralize the target virus and may have a great clinical potential. Their production in high yield in plants offers an economically advantageous approach applicable to both IgG and secretory IgA antibodies.

While polyclonal antibodies from human sera provide immunological diversity, mAbs are highly specific for a single virus epitope. Potential alterations of such epitopes may generate virus neutralization escape mutants and decrease or eliminate the effectiveness of mAb prophylactics/therapeutics. Polyclonal antibody preparations prepared from pooled sera are not uniform and vary with the source of serum pools.

This problem can be overcome by the development of human recombinant antigen-specific polyclonal antibodies by a novel Sympress technology (Symphogen, Lyngby, Denmark).

Immunoglobulins, either already in clinical use or in development, are directed against one of the following viruses: hepatitis A and B; cytomegalovirus; rabies; respiratory syncytial virus; smallpox; vaccinia; varicella zoster; measles; mumps; rubella; parvovirus B19; Epstein–Barr virus; herpes simplex; tick-borne encephalitis; poliovirus; Hantavirus; West Nile virus; rotavirus; poliovirus; HIV-1; Ebola virus; and severe acute respiratory syndrome-associated coronavirus.

Unlike vaccines, passive immunization can rapidly deliver protective levels of antibodies directly to susceptible mucosal sites where many virus infections are initiated. Secretory IgA because of its polyvalency and relative stability may have advantages over IgG for passive immunization at these sites.

Antibodies function also as immunomodulators which can bridge innate and acquired, and cellular and humoral immune responses, respectively. Infected host cells can be targeted by linking anticellular toxins to antiviral antibodies or by bispecific antibodies in which one Fab fragment of the antibody is virus specific and the other one recognizes a host cell component or receptor.

Research on antibody-mediated immunity and understanding of antibody-based prophylaxis and therapies for virus diseases have provided a foundation for research on and development of antivirus vaccines.

Vaccines against Human Viral Diseases

Vaccination is the most successful medical intervention against viral diseases. Vaccines prevent or moderate illnesses caused by virus infection in an individual and prevent or diminish virus transmission to other susceptible persons, thus contributing to herd immunity. This effect is expected to be long term, depends on establishment of immunological memory at both the B- and T-cell levels, and may require consecutive or repeated vaccinations. The effectiveness of vaccination might be diminished or compromised for viruses occurring in the form of simultaneous quasi-species or undergoing time-dependent changes of antigenic properties (antigenic drift and antigenic shift). However, in some cases, vaccination predisposes to aggravated disease elicited by infection with a virus identical or related to that used for vaccination, that is, antibody enhancement of virus infection occurs (dengue and respiratory syncytial viruses and HIV-1).

The following categories of vaccines can be distinguished (**Table 2**): (1) live attenuated; (2) whole virus (inactivated); (3) glycoprotein subunit vaccines; and (4) protein vaccines based on recombinant DNA technologies. One vaccine in category (3), hepatitis B surface antigen (HBsAg), is derived from plasma of hepatitis B virus carriers. It is remarkable that vaccines in categories (3) and (4) correspond to multi-subunit self-assembled particles having antigenic specificities closely similar or identical to those expressed on the surface of virus particles. On the other hand, individual virus protein or glycoprotein subunit molecules or their peptide fragment have been less suitable candidates for vaccine development because of insufficient similarities with intact viruses.

Vaccine formulations require the incorporation of an immunological adjuvant to enhance their immunogenicity. Adjuvants are designed to optimize antigen delivery and presentation, enhance the maturation of antigen-presenting dendritic cells, and induce immunomodulatory cytokines.

Currently, most vaccines are administered parenterally using syringes with needles. The procedure is disliked by many, and is questionable for mass vaccination programs in developing countries. Therefore, efforts are being made to produce vaccines which can be delivered onto mucosal surfaces, that is, mostly orally or nasally. In addition, high-workload needle-free injection devices are being developed. At this juncture, the bifurcated needle, developed by Benjamin Rubin over 40 years ago, must be mentioned. It proved to be essential for the successful campaign to eradicate smallpox worldwide.

Veterinary Vaccines

The development and use of veterinary vaccines has the following aims: cost-effective prevention and control of virus diseases in animals; induce herd immunity; improve animal welfare and food production for human consumption; decrease the usage of veterinary drugs, thereby minimizing their environmental impact and food contamination; and decrease the incidence of zoonoses (e.g., infections by avian influenza, rabies, West Nile, Rift Valley fever viruses, respectively).

Research, development, and production of some veterinary vaccines have been on the forefront of the general field of vaccinology. The foot-and-mouth disease virus vaccine was the first one produced at an industrial scale (Frenkel method). A vaccinia-rabies virus G protein recombinant vaccine was among the first biotechnology-based vaccines licensed. The world's first DNA vaccine (against West Nile virus in horses) was approved by the US Department of Agriculture in July 2005. DIVA (Differentiating Infected from Vaccinated Animals; also termed marker) veterinary recombinant vaccines and companion diagnostic tests have been developed. They can be applied to programs to control and eradicate virus infections. These are examples to be considered for the development of human vaccines. The latter will require rigorous evaluations for safety and efficacy which are more difficult to obtain than in veterinary settings. A list of licensed veterinary vaccines is shown in **Table 3**.

Table 2 Past and present vaccines against human viral diseases

Live attenuated	Killed whole virus	Glycoprotein subunit	Genetically engineered
Influenza (nasal)	Influenza	Influenza	Papillomavirus
Rabies	Rabies	Hepatitis B	Hepatitis B
Poliovirus	Poliovirus		
Yellow fever	Japanese encephalitis		
Measles	Tick-borne encephalitis		
Mumps	Hepatitis A		
Rubella			
Adenovirus			
Varicella zoster			
Rotavirus			
Smallpox (vaccinia)			

Table 3 Current veterinary vaccines

Species	Live	Killed	Recombinant	DNA
Avian	Encephalomyelitis		Encephalomyelitis (fowl pox vector)	
	Influenza		Influenza (fowl pox vector)	
	Pneumovirus	Paramyxovirus		
		Polyomavirus		
		Reovirus		
	Bursal disease	Bursal disease	Bursal disease (Marek's disease vector)	
	Marek's disease		Marek's disease (Marek's disease vector)	
	Fowl pox		Fowl pox	
	Newcastle disease	Newcastle disease	Newcastle disease (fowl pox vector)	
		Bronchitis		
	Anemia			
	Laryngotracheitis		Laryngotracheitis (fowl pox vector)	
	Duck enteritis			
	Duck hepatitis			
	Canary pox			
Feline	Calicivirus	Calicivirus immunodeficiency virus		
		Leukemia virus	Leukemia virus (canarypox vector)	
	Infectious peritonitis			
	Rhinotracheitis	Rhinotracheitis		
	Panleukopenia	Panleukopenia		
		Rabies	Rabies (canary pox vector)	
Canine	Adenovirus 2			
	Parvovirus	Parvovirus		
	Coronavirus	Coronavirus		
	Parainfluenza		Parainfluena (canary pox vector)	
	Canine distemper		Canine distemper (canary pox vector)	
	Measles			
	Hepatitis			
	Rabies	Rabies		
Sheep and goat	Bluetongue			
	Ovine ecthyma			
	Poxviruses			
		Louping ill		
Equine	Influenza	Influenza	Influenza (canary pox vector)	
	Rhinopneumonitis	Rhinopneumonitis		
		Rotavirus		
	Arteritis			
	African horse sickness	African horse sickness		
		Encephalomyelitis		
	West Nile virus	West Nile virus	West Nile virus (canary pox vector)	West Nile virus
	Flavivirus chimera			
Bovine	Respiratory syncytial virus	Respiratory syncytial virus		
	Rhinotracheitis	Rhinotracheitis		
	Diarrhea	Diarrhea		
	Bronchitis	Bronchitis		
	Parainfluenza 3			
	Rotavirus	Rotavirus		
	Coronavirus	Coronavirus		
	Herpes 1	Herpes 1		
	Foot-and-mouth disease	Foot-and-mouth disease		
	Rinderpest			

Continued

Table 3 Continued

Species	Live	Killed	Recombinant	DNA
Porcine	Pseudorabies Enterovirus Parvovirus Rotavirus Transmissible gastroenteritis Reproductive and respiratory syndrome Hog cholera	Pseudorabies Influenza Circovirus Rotavirus Transmissible gastroenteritis Reproductive and respiratory Syndrome Hog cholera		

Several of the described vaccines are being administered as combination vaccines.

See also: AIDS: Vaccine Development; Antigen Presentation; Antigenic Variation; Antigenicity and Immunogenicity of Viral Proteins; Cytokines and Chemokines; Diagnostic Techniques: Serological and Molecular approaches; DNA Vaccines; Immune Response to viruses: Cell-Mediated Immunity; Neutralization of Infectivity; Vaccine Production in Plants; Vaccine Strategies.

Further Reading

Ahmed R (ed.) (2006) Immunological memory. *Immunological Reviews* 211, 5–337.

Burton DR (ed.) (2001) Antibodies in viral infection. *Current Topics in Microbiology and Immunology* 260, 1–300.

Casadevall A, Dadachova E, and Pirofsky LA (2004) Passive antibody therapy for infectious diseases. *Nature Reviews Microbiology* 2: 695–703.

Cooper MD and Alder N (2006) The evolution of adaptive immune systems. *Cell* 124: 815–822.

Frank SA (2002) *Immunology and Evolution of Infectious Disease.* Princeton: Princeton University Press.

Hangartner L, Zinkernagel RM, and Hengartner H (2006) Antiviral antibody responses: The two extremes of a wide spectrum. *Nature Reviews Immunology* 6: 231–243.

Janeway CA, Jr. (1993) How the immune system recognizes invaders. *Scientific American* (Sep.): 73–79.

Janeway CA, Jr., Travers P, Walport M, and Shlomchik MJ (2005) *Immunobiology. The Immune System in Health and Disease,* 6th edn. New York: Garland Science.

Levine MP, Kaper JB, Rappuoli R, Liu M, and Good MF (eds.) (2004) *New Generation Vaccines* 3rd edn. New York: Dekker.

O'Neill LA (2005) Immunity's early-warning system. *Scientific American* (Jan.): 38–45.

Paul WE (ed.) (2003) *Fundamental Immunology* 5th edn. Philadelphia: Lippincott Williams and Wilkins.

Plotkin SA and Orenstein WA (2004) *Vaccines,* 4th edn. Philadelphia: Saunders–Elsevier.

Pulendran B and Ahmed R (2006) Translating innate immunity into immunological memory: Implications for vaccine development. *Cell* 124: 849–863.

Van Regenmortel MHV (2002) Reductionism and the search for structure–function relationships in antibody molecules. *Journal of Molecular Recognition* 15: 240–247.

Weissman IL and Cooper MD (1993) How the immune system develops. *Scientific American* (Sep.): 65–71.

Woof JM and Mestecky J (2005) Mucosal immunoglobulins. *Immunological Reviews* 206: 64–82.

Immune Response to Viruses: Cell-Mediated Immunity

A J Zajac and L E Harrington, University of Alabama at Birmingham, Birmingham, AL, USA

© 2008 Elsevier Ltd. All rights reserved.

Glossary

Antigen presentation The process by which proteins are degraded into peptides that are loaded onto MHC molecules and these complexes are targeted to the cell surface.

Central-memory T cell A population of memory T cells which primarily reside in secondary lymphoid organs; characterized by the expression of CD62L and CCR7.

Chemokines Chemotactic cytokines which stimulate the migration of cells.

Cytokines Secreted proteins that regulate cellular actions by signaling via specific receptors.

Cytotoxic T lymphocyte (CTL) T cells which can kill virus-infected cells upon activation.

Effector T cell Cells capable of immediate functional activity resulting in pathogen removal.

Effector-memory T cell A population of memory T cells poised for immediate effector

function that primarily resides outside the lymphoid organs.
Immune homeostasis Maintenance of lymphocyte populations at steady-state levels.
Immunodominance The hierarchy of T-cell responses to the array of individual epitopes which are presented during any given viral infection.
Immunological memory The ability of the host to mount rapid recall responses upon re-exposure to the inducing antigen.
Immunopathology Tissue damage that results from the actions of the host's immune response.
Major histocompatibility complex (MHC) A cluster of genes involved in immune recognition and regulation; MHC class I molecules couple with β2 microglobulin to present peptides to CD8 T cells; MHC class II molecules present peptides to CD4 T cells.
T-cell exhaustion The progressive loss of antiviral T-cell functions which can culminate in the complete deletion of specific T-cell populations during chronic viral infections.
T-cell receptor (TCR) Heterodimeric receptor expressed by T cells that binds specific peptide–MHC complexes.
T helper 1 (Th1) cell Effector CD4 T-cell subset characterized by the production of IFN-γ; associated with immune responses to intracellular bacteria and viruses.
T helper 2 (Th2) cell Effector CD4 T-cell subset characterized by the secretion of IL-4, IL-5, and IL-13; important for helminth infections; linked to allergies and asthma.

General Overview

Cell-mediated immune responses play a critical role in combating viral infections. They are comprised of T-cell responses, which fundamentally differ from antibody (humoral) responses in the way they bring about infection control. The cardinal trait of cell-mediated responses is that the physical presence of reactive T cells is required for immunity, whereas humoral responses are conferred by the presence of soluble antibodies. T cells, together with B cells, form the adaptive immune response to viral infections. The hallmarks of adaptive immunity include antigen specificity and memory. These features allow T cells to elaborate responses which specifically target the numerous viruses which may infect the host. The ability to establish long-lived immunological memory provides a unique mechanism to better protect the host during subsequent viral exposures.

Due to their importance in controlling pathogens, cell-mediated immune responses are widely studied. Significantly, much of our understanding of cell-mediated immunity, including the fundamental concepts of major histocompatibility complex (MHC) restriction, tolerance, T-cell diversity, and immunological memory, has been determined by analyzing immune responses to viruses. Cell-mediated immune responses are dynamic, diverse, and display a broad range of phenotypic and functional properties.

T-Cell Recognition

T cells differ from antibodies (humoral responses) in the way they recognize viral antigens. Antibodies are capable of binding to intact viral proteins, including structural components of viral particles and also viral proteins present at the surface of infected cells. By being able to bind to conformationally complex structures, antiviral antibodies have the unique ability to neutralize the infectivity of viral particles present in the circulation or at mucosal surfaces, a function that cannot be performed by T cells. T cells cannot recognize intact viral proteins and therefore play no role in directly neutralizing whole viral particles. Instead, T cells recognize short peptide fragments presented at the cell surface in a noncovalent association with MHC molecules (see **Figure 1**). Thus, T-cell recognition is referred to as being MHC-restricted, since an individual T cell will only bind strongly to one particular MHC molecule, and is also peptide specific, as a T cell will predominately only recognize one specific short antigenic peptide.

T-Cell Receptors

CD4 and CD8 T cells express a unique surface receptor, the T-cell receptor (TCR) that determines the MHC restriction and peptide specificity of an individual T cell. In the vast majority of T cells, this is a heterodimeric receptor comprised of the TCR α- and β-chains. A smaller population of T cells (~5% of circulating T cells in humans) express an alternative form of the TCR, formed by the noncovalent association of TCR γ- and δ-chains; however, the roles of γδ T cells in controlling viral infections are not well defined. Each T cell expresses only one unique version of the TCR whose precise sequence and structure represents the end result of a series of gene rearrangements. This recombinatorial process, which occurs during T-cell ontogeny in the thymus, generates a massive repertoire of T cells with tremendous diversity between their individual TCR sequences. Estimates of the size of the T-cell repertoire suggest that 2.5×10^7 different TCRs are detectable in human blood. This large ensemble of

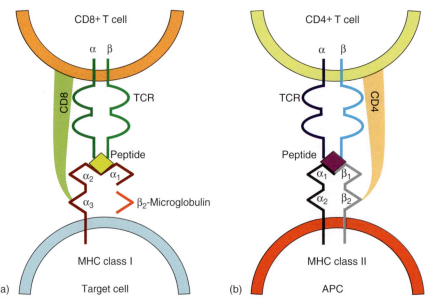

Figure 1 Similarities and differences between CD4 and CD8 T-cell recognition. Both CD4 and CD8 T cells express unique T-cell receptors at the cell surface which determine their antigen specificity and MHC restriction. (a) CD8 T cells recognize MHC class I molecules together with a non-covalently associated antigenic peptide, typically of 8–10 amino acids in length. MHC class I complexes are widely expressed and usually present endogenously synthesized antigens, including peptides derived from the degradation of viral proteins. (b) CD4 T cells recognize peptide antigens presented by MHC class II molecules. These antigenic peptides are typically 13–17 amino acids in length. These peptides are usually derived from extracellular antigens which have been endocytosed into professional antigen-presenting cells where they are proteolytically processed and re-presented at the cell surface bound to MHC class II molecules. MHC class II complexes have a much more limited tissue distribution than MHC class I molecules and are primarily expressed by macrophages, dendritic cells, and B cells.

individual T-cell clones is collectively capable of recognizing and responding to the vast array of antigens, including virally encoded antigens, which may be encountered during the lifespan of the host.

Antigen Processing and Presentation

CD8 T cells recognize MHC class I complexes (**Figure 1**). MHC class I molecules are expressed on virtually all cell types and usually present peptides derived from endogenously synthesized proteins. Antigen processing occurs continuously as newly synthesized proteins become degraded into peptide fragments by proteasomes. These fragments, typically of 8–10 amino acids in length, enter the endoplasmic reticulum and, if they have sufficient binding affinity, associate with MHC class I heavy chains together with the nonpolymorphic protein β2-microglobulin. These assembled MHC peptide complexes are then transported to the cell surface. This process allows MHC class I molecules to sample peptide fragments derived from proteins which are produced within the cell, including normal cellular proteins as well as virally encoded proteins, and present them for inspection by CD8 T cells. This ongoing immunological surveillance allows CD8 T cells to detect, respond to, and remove host cells which express 'non-self' viral proteins.

Although the endogenous pathway of antigen presentation provides a valuable mechanism for revealing the presence of virally infected cells to CD8 T cells, an alternative 'cross-presentation' pathway also operates. During cross-presentation, viral particles (or other antigens) are endocytosed by professional antigen-presenting cells and then undergo proteolytic degradation. The resulting peptide fragments can then bind to MHC class I molecules, and traffic to the cell surface. This enables antigen-presenting cells, such as dendritic cells, to display virally derived peptides to CD8 T cells even if the antigen-presenting cell itself is not capable of supporting virus replication.

CD4 T cells differ from CD8 T cells as they recognize peptides presented by MHC class II rather than MHC class I complexes (**Figure 1**). Unlike MHC class I complexes, which are ubiquitously expressed, MHC class II molecules are only presented by certain cell types such as dendritic cells, macrophages, and B cells. This limits CD4 T-cell recognition to professional antigen-presenting cells, since it is these specialized cells that have the capacity to display peptide–MHC class II complexes to CD4 T cells. Due to structural differences in the peptide-binding groove of MHC class I and MHC class II complexes, the viral peptides that are presented to CD4 T cells are typically longer (~13–17 amino acids in length) compared to those displayed by MHC class I molecules. In addition, whereas MHC class I complexes primarily present endogenously synthesized antigens, MHC class II molecules usually present antigens derived from extracellular sources. These exogenous antigens can include viral particles and also remnants of virally infected cells. Following uptake by professional antigen-presenting cells, these antigens are degraded into peptide fragments within acidified endosomes. Alternatively, if the antigen-presenting cell is actively infected with the virus, then intracellular vesicles containing viral proteins can serve as a source of peptides for associating with MHC class II complexes. Once at the cell surface these presented antigens can be detected by, and activate, CD4 T cells which express TCRs that are capable of specifically recognizing the peptide–MHC class II combination.

Immunodominance

Individual viruses encode multiple potential T-cell epitopes; therefore T-cell responses elicited during viral infections are not monoclonal or monospecific. Instead, oligoclonal subsets of cells are induced and, although each individual T cell is responsive to only one particular peptide–MHC combination, the overall pool of cells is sufficiently diverse to ensure that numerous epitopes can be detected. The kinetics, magnitudes, phenotypic and functional traits, as well as the stability of T responses to each individual virally encoded epitope can differ. Consequently, an ordered hierarchy can emerge as certain epitopes elicit more abundant, or immunodominant responses, whereas others are less prevalent and give rise to subdominant responses (**Figure 2**).

Our understanding of the precise determinants of immunodominance is incomplete; however, the hierarchy of T-cell responses is likely to be shaped by many factors. The magnitude of responses to individual viral epitopes is influenced by properties of the host's T cells including precursor frequencies and the avidity of the T cells for the presented viral antigen, as well as viral related factors including the ability of viral peptides to bind MHC complexes, the abundance of presented antigen, the kinetics of viral protein synthesis, and the types of cells which present the viral antigens. Changes in the patterns of immunodominance have been reported most notably during the course of persistent viral infections, as well as following secondary exposures to viruses which have been previously controlled (**Figure 2**). In the case of lymphocytic choriomeningitis virus (LCMV) infection of C57BL/6 mice, the NP396 epitope is co-dominant following well-controlled acute infections, and CD8 T cells specific for this epitope respond most vigorously during secondary exposures to LCMV. By contrast, during persistent LCMV infections, responses to this usually dominant epitope can become completely undetectable. During Epstein–Barr virus (EBV) infections shifting patterns of immunodominance are observed as responses to immediate early and early viral proteins are initially detected, but as viral latency becomes established responses to lytic cycle proteins decline and responses to latent viral proteins predominate. These observations demonstrate that not all antiviral T cells respond equally and suggest that certain specificities of T cells may be more effective at combating particular viral infections.

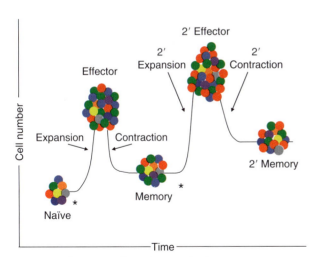

Figure 2 Changes in T-cell immunodominance can occur during primary and secondary immune responses following acute viral infections. Viruses elicit T-cell responses to a range of individual viral epitopes. These responses are not necessarily equal and in the example depicted the T cells specific for the 'green' epitope are immunodominant following the primary infection. This pattern of immunodominance is maintained during the memory phase but secondary exposure to the virus results in an anamnestic recall response during which the 'red' epitope-specific T cells predominate. Asterisk indicates when exposure to the virus occurred.

Induction of Cell-Mediated Immunity during Viral Infections

As a virus infection becomes established in the host, a series of molecular and cellular signals are initiated which activate cell-mediated immune responses. These signals include the production of interferons, other cytokines, and inflammatory mediators, in addition to the mobilization of local dendritic cells. Dendritic cells are thought to provide a critical cellular link for priming naive CD4 and CD8 T cells (**Figure 3**). It has been proposed that these cells are especially prone to infection by viruses which facilitate their role as cellular sensors for signaling the occurrence of an infection. Even if dendritic cells are not permissive to active infection with particular viruses, they can also present antigens through cross-priming to CD8 T cells, as well as to CD4 T cells via the classical exogenous antigen processing pathway.

The primary activation events which induce cell-mediated immunity predominately occur in secondary lymphoid organs including regional lymph nodes and the

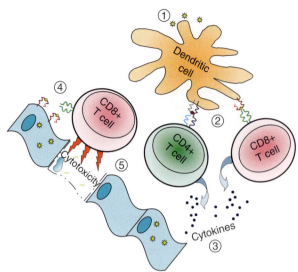

Figure 3 The induction and function of cell-mediated immune responses. (1) The presentation of viral antigens by professional antigen-presenting cells is a critical step in initiating cell-mediated immune responses. (2) Recognition of cognate peptide–MHC complexes by CD4 and CD8 T cells stimulates their proliferation and differentiation into effector cells. (3) Antigen-activated effector CD4 and CD8 T cells can express cytokines such as IL-2, TNF-α, and IFN-γ. These cytokines have important roles in coordinating the antiviral immune response and can also have direct antiviral effects. (4) Once activated antiviral T cells disperse into tissues where can they respond locally, at the sites of viral infection. (5) If viral peptide–MHC complexes are recognized, then effector CD8 T cells can elaborate cytotoxic effector functions which kill the infected cells. CD4 T cells can also become cytotoxic during certain infections; however, their impact is less promiscuous as they recognize the MHC class II complexes which are only expressed by professional antigen-presenting cells.

spleen. During the early stages of viral infections dendritic cells residing at the initial sites of infection take up viral antigens, become activated, and migrate to regional lymph nodes. Within the lymph nodes these dendritic cells encounter naïve T cells which are circulating through these organs as part of their normal immunosurveillance protocol. Engagement of TCRs on the naive T cells with viral–peptide MHC complexes presented by the dendritic cells results in sequestration of the T cells and launches the antiviral T-cell response (**Figure 2**). The ensuing proliferation and differentiation of virus-specific T cells also occur in conjunction with inflammatory mediators such as interferons and other danger signals. Many parameters, including the duration and strength of antigenic stimulation, co-stimulatory interactions, the presence of cytokines, and the provision of CD4 T cells help guide the developing response. These early events play a critical role in driving the generation of both the effector T cells as well as the subsequent establishment of the memory T cell pool.

CD8 T Cells

One of the most impressive aspects of CD8 T-cell responses is the massive proliferation of these cells which occurs during the initial phase of many viral infections (**Figures 2** and **4**). Experimental studies of acute LCMV infection of mice have demonstrated that antiviral CD8 T cells can increase over 10 000-fold during the first week of infection; over 50%, and perhaps even more, of the host's CD8 T cells are LCMV-specific at the peak of the response! Marked expansions of virus-specific CD8 T cells are a common feature of many virus infections including influenza, vaccinia virus, EBV, yellow fever virus, and early following human immunodeficiency virus (HIV) infection. During this expansion phase the patterns of gene expression change promoting the synthesis of cytokines and cytotoxic effector molecules as well as alterations in surface molecules including cytokine receptors and adhesion molecules. This results in an expanded pool of virus-specific effector cells with functional attributes necessary to control the infection. The ensemble of virus-specific effector cells which emerge during the acute phase of the infection is remarkably heterogeneous and comprises of subsets which differ in their epitope specificity, clonal abundance, effector potential, expression of adhesion molecules and cytokine receptors, and ultimate fate. Although the initial activation of T-cell responses occurs in secondary lymphoid organs, the effector cells become dispersed throughout the host. In this way the T cells are available locally, at the sites of infection, where they operate to eliminate the host of virally infected cells.

CD8 T cells are potent antiviral effector cells due to their ability to produce both inflammatory mediators as well as cytotoxic effector molecules (**Figure 3**). CD8 T cells are commonly referred to as cytotoxic T lymphocytes (CTLs), which emphasizes their ability to kill virally infected target cells. These killing functions are triggered as the effector T cell become activated following engagement with a virally infected target cell displaying an appropriate peptide–MHC complex. The subsequent release of perforin and granzyme molecules by the T cells ensures the swift destruction of the infected cell. Ideally, this targeted removal of the infected cell occurs before progeny virus is released. Alternative Fas and TNF-dependent cytotoxic mechanisms have been reported but their *in vivo* significance in killing virus infected cells is not well defined. In addition to their direct killing functions, CD8 T cells also produce a range of cytokines and chemokines. In the laboratory the production of these effector molecules is often used to detect the presence of antiviral T cells. Most importantly, within the infected host the production of these soluble mediators, such as IFN-γ and TNF-α, can also help clear viral infections without causing death

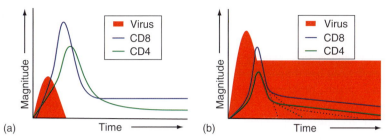

Figure 4 Successful and unsuccessful T-cell responses during acute and chronic viral infections. (a) During acute viral infections massive T-cell responses can be induced which play a principal role in clearing the infection. Following the resolution of the infection, the responding T-cell pool is downregulated but a long-lived pool of memory T cells becomes established which helps protect against subsequent viral exposures. (b) During chronic viral infections T-cell responses are elicited but a variety of phenotypic and functional defects manifest as these responses succumb to exhaustion. A gradation of exhausted phenotypes is often observed, ranging from an inability to produce effector cytokines to the complete deletion of virus-specific T cells.

of infected cells. This cytokine-mediated purging of infected cells has been most convincingly shown during viral hepatitis.

CD4 T Cells

CD4 T cells are traditionally known as helper T cells because of their ability to provide help to B cells and CD8 T cells, resulting in antibody production, class switching, cytotoxic T cell activity, and memory development. In addition to assisting cells of the adaptive immune system, CD4 T cells produce an array of cytokines and chemokines that stimulate cells of the innate immune system, such as macrophages and neutrophils, to traffic to the sites of infection and elaborate their effector activities. It has also been demonstrated that CD4 T cells are directly capable of antiviral functions, through the production of IFN-γ and, in some circumstances, by inducing lysis of virally infected cells (**Figure 3**). Thus, CD4 T cells are critical constituents of the cell-mediated immune response to viral infections; however, it should be emphasized that innate immunity as well as humoral immune responses, which are helped by CD4 T cells, are key components of the host overall antiviral response.

Like CD8 T cells, naive CD4 T cells circulate through secondary lymphoid organs in a relatively quiescent state. Following recognition of antigen in the context of MHC class II, a cascade of signaling events is initiated within the CD4 T cell which results in activation, proliferation, and differentiation into an effector CD4 T cell (**Figure 3**). Classically effector CD4 T cells have been divided into two polarized subsets based on their cytokine production profile. T helper 1 (Th1) cells primarily produce IFN-γ and are critical for the immune responses to various viral infections, as well as infections with intracellular bacteria. This subclass of effector cells is typically associated with antiviral cell-mediated immunity. Conversely, T helper 2 (Th2) cells predominantly secrete the cytokines IL-4, IL-5, and IL-13, assist with the eradication of helminth infections, and have historically been linked with the production of antibodies and humoral immune responses.

In recent years, the definition of CD4 T-cell subsets has expanded beyond Th1 and Th2 cells, with the importance of unique populations of regulatory CD4 T cells and also IL-17 producing 'Th17' cells becoming evident. Regulatory cells are pivotal for preventing autoimmunity by suppressing the activation of autoreactive T cells. Regulatory T cells are typically characterized by the expression of the transcription factor Foxp3 and by the production of the suppressive cytokines IL-10 and TGF-β. Relatively little is known regarding the significance of these CD4 T-cell populations during viral infections; however it has been proposed that regulatory T cells are both beneficial to the host, by limiting immunopathology, and detrimental, by dampening effector functions.

T-Cell Memory

Ideally, the primary immune response overwhelms the infection and results in the complete eradication of the virus from the host. If the infection is successfully resolved then the expanded pool of effector T cells does not remain constitutively activated. Instead, a downregulation phase ensues during which typically the majority (<90%) of the virus-specific T cells present at the peak of the immune response die by apoptosis. The remaining 5–10% of T cells survive the contraction phase and constitute a long-lived pool of memory T cells (**Figures 2** and **4**). In this way, a beneficial memory of past infections is established as, by comparison with naïve hosts, an increased number of virus-specific T cells are maintained which are tuned to rapidly respond if they re-encounter infected cells. The population of memory T cells which emerges following the resolution of the infection and restoration of homeostasis is not uniform, as phenotypic and functional diversity is apparent even within subsets of memory T cells which recognize the same viral epitope. This is well illustrated by the classification of memory T cells into broad categories termed effector- and central-memory T cells. These subsets have been defined based upon their anatomical location, functional quality, proliferative potential, and expression of surface molecules.

Since T cells neither recognize nor neutralize the infectivity of viral particles they do not confer sterilizing immunity and cannot prevent secondary infections. Nevertheless, as the host cells become infected, preexisting memory T cells which developed following prior viral exposures can mount robust recall responses. These anamnestic responses are characteristically more rapidly induced, greater in magnitude, and possibly more functionally competent than primary T-cell responses. Such pronounced secondary responses help protect the host by contributing to the swift control of the infection thereby reducing the morbidity and mortality. Memory T-cell responses are not the only components of secondary immune responses as these cells act in conjunction with antiviral antibodies to protect the host during viral reexposures.

Analysis of both clinical specimens and experimental animal models has demonstrated that acute viral infections can induce very long-lived memory T-cell responses (**Figure 4**). Studies using experimental mice have demonstrated that memory CD8 T cells reactive against various infections such as LCMV, vaccinia virus, and influenza are maintained at remarkably stable levels for over 2 years following infection. By contrast, CD4 T-cell responses are not as consistent and have been reported to gradually decay over time. Although natural exposures to viruses lead to the formation of immunological memory, these advantageous responses are also induced following vaccinations. Vaccines are successful in protecting the host against subsequent infections because of their ability to promote long-lived memory responses. In humans the longevity of T-cell responses has been investigated in detail following smallpox vaccination. Smallpox-specific T-cell responses are detectable in individuals who received a single dose of the vaccine 75 years previously! Notably, the findings suggested that the responses do decline slowly with predicted half-lives of 8–15 years.

Persistent Viral Infections

Although T-cell responses can be highly effective at controlling acute infections and contribute to protective secondary responses, persistent viral infections do arise and are often associated with the development of phenotypically and functionally inferior responses (**Figure 4**). These types of infections include many viral pathogens which are of significant public health importance such as HIV and hepatitis C virus (HCV). A common feature of these infections is that T-cell responses are initially induced but qualitative and quantitative defects become apparent as the generation of robust sets of effector cells, as well as the progression of memory T-cell development are subverted. By comparison with successful T-cell responses elaborated during acute viral infections, a spectrum of phenotypic and functional defects have been detected during persistent infections. The production of cytokines including IL-2, TNF-α, and IFN-γ, as well as cytotoxic effector molecules such as perforin may be diminished or abolished, and decreased proliferative potential has also been observed. The severe loss of effector activity as well as the physical deletion of antiviral T cells which can occur during persistent infections has been termed exhaustion (**Figure 4**).

The parameters which contribute to T-cell exhaustion are not fully understood. Comparative analysis of T-cell responses to viral infections which result in different levels of antigenic exposure, such as influenza, cytomegalovirus, EBV, HCV, and HIV, indicate that antiviral T cells may adopt different preferred phenotypic and functional set points. Experimental studies suggest that many factors including, but not limited to, viral targeting and destruction of dendritic cells, the production of immunosuppressive cytokines such as IL-10, the depletion of CD4 T cell subsets, and the induction of weak neutralizing antibody responses can all contribute inferior cell-mediated immune responses. Changing viral loads may also impact the functional quality of the T-cell response. During acute HCV infection antiviral CD8 T cells transiently lose the ability to produce IFN-γ, but recover from this 'stunned' state as the viral loads are brought under control. Importantly, this suggests that under certain conditions the exhaustion of virus-specific T cells may be prevented or even reversed. Several reports have now demonstrated that during persistent LCMV, HIV, and HCV infections, antiviral T cells can express the inhibitory receptor PD-1. Antibody-based therapeutic treatments to block this receptor have been shown to promote proliferation of previously exhausted T cells and restore their functional activities. This is a promising experimental observation; however, the jury is still out on whether this approach will be a beneficial treatment for persistent infections of humans.

Immunopathology

Since viruses are obligate intracellular pathogens they must infect permissive host cells in order to replicate. Infected cells die as a direct result of the virus' lytic replication cycle or are killed as a consequence of the actions of the antiviral immune response. Although immune-mediated destruction of virally infected cells is necessary to contain the infection, it can also result in immunopathology, which represents collateral damage to the host caused by the actions of the immune response. A classical example of immunopathology occurs following intracranial infection of adult mice with LCMV. Mice infected by this route succumb to a characteristic lethal disease and expire approximately 1 week following infection. Death can be prevented by immunosuppression of the mice, which has shown that the disease is a consequence of a vigorous CD8 T-cell response rather than due to the infection *per se*. HBV-associated viral hepatitis is another instance where anti-viral CD8 T-cell responses, which are attempting to clear the infection, cause liver damage in the infected individual.

Most viral infections are associated with the development of an IFN-γ-producing Th1-CD4 T cell response. In various animal models, the absence of CD4 T cell help during viral infection results in impaired clearance of the infectious agent. However, not all CD4 T-cell responses are beneficial as the induction of inappropriate types of CD4 T-cell responses can be deleterious to the host, due to immunopathology. In the 1960s, a group of young children were administered a formalin-inactivated vaccine for respiratory syncytial (RS) virus and following exposure to live RS virus, these children exhibited enhanced infection rates and immunopathology linked to increased frequencies of eosinophils and neutrophils within the airways. Animal studies have indicated that the vaccine was associated with a Th2-biased virus-specific immune response (increased levels of IL-4 and IL-13, as well as eosinophil recruitment to the lungs) that upon live infection displayed many of the signatures of immunopathology which manifested in these vaccinated children. Supporting experiments suggest that immunization to promote the Th1 responses or ablation of Th2 responses prevents the development of these pathological effects following live viral infection.

Immune Evasion

Arguably, one of best indications of the importance of cell-mediated immunity in controlling viral infections is the observation that many viruses have evolved strategies to evade the actions of the host immune response. There is, however, no one universal evasion mechanism; instead, viruses have adopted various preferred approaches to escape antiviral T-cell responses. A common strategy is to change the amino acid sequence of T-cell epitopes or nearby flanking residues that impede the recognition or processing of the antigenic peptide. Many viruses rely on error-prone polymerases in order to replicate, which favor the incorporation of mutations in progeny viral genomes. The resulting variant viruses will have a selective advantage if the amino acid substitution abolishes the ability of the epitope to associate with MHC molecules, negatively impacts recognition of the epitope by T cells, or prevents antigen processing.

In addition to mutating epitope sequences, many viruses encode specific molecules which function to interfere with the antiviral immune response. Both MHC class I and class II antigen-presenting pathways are targeted by several viral proteins. These immune evasion molecules block antigen presentation in various ways, ranging from preventing the transport of antigenic peptides into the endoplasmic reticulum to inhibiting the egress of peptide-loaded MHC complexes. The end result of these inhibitory strategies is to impair immunological surveillance. Although there is much anecdotal evidence that interference with antigen processing diminishes cell-mediated immune responses, experimental studies using murine cytomegalovirus (MCMV) question this notion. Infection of mice with mutants which lack several viral genes known to block antigen processing did not effect the ability of the host to elaborate an anti-MCMV CD8 T-cell response.

See also: Antigen Presentation; Antigenic Variation; Antigenicity and Immunogenicity of Viral Proteins; Cytokines and Chemokines; Immune Response to viruses: Antibody-Mediated Immunity; Immunopathology; Innate Immunity: Defeating; Innate Immunity: Introduction; Persistent and Latent Viral Infection; Vaccine Strategies; Viral Pathogenesis.

Further Reading

Alcami A and Koszinowski UH (2000) Viral mechanisms of immune evasion. *Immunology Today* 9: 447–445.

Doherty PC and Christensen JP (2000) Accessing complexity: The dynamics of virus-specific T cell responses. *Annual Review of Immunology* 18: 561–592.

Ertl HC (2003) Viral immunology. In: Paul WE (ed.) *Fundamental Immunology*, 5th edn., pp. 1021–1227. Philadelphia, PA: Lippincott Williams and Wilkins.

Finlay BB and McFadden G (2006) Anti-immunology: Evasion of the host immune system by bacterial and viral pathogens. *Cell* 124: 767–782.

Frelinger JA (2006) *Immunodominance – The Choice of the Immune System*. Weinheim, Germany: Wiley-VCH.

Klenerman P and Hill A (2005) T cell and viral persistence: Lessons from diverse infections. *Nature Immunology* 6: 873–879.

Oldstone MBA (2006) Viral persistence: Parameters, mechanisms and future predictions. *Virology* 344: 111–118.

Seder RA and Ahmed R (2003) Similarities and differences in CD4+ and CD8+ effector and memory T cell generation. *Nature Immunology* 4: 835–842.

Yewdell JW and Bennink JR (1999) Immunodominance in major histocompatibilty complex class I-restricted T lymphocyte responses. *Annual Review of Immunology* 17: 51–88.

Zinkernagel RM and Doherty PC (1997) The discovery of MHC restriction. *Immunology Today* 18: 14–17.

Immunopathology

M B A Oldstone, The Scripps Research Institute, La Jolla, CA, USA
R S Fujinami, University of Utah School of Medicine, Salt Lake City, UT, USA

© 2008 Elsevier Ltd. All rights reserved.

Glossary

Adaptive immune system Comprises T cells, B cells, and antibodies. Elements of the innate immune system are required for initiation of the adaptive immune response. The adaptive immune system recognizes specific antigenic epitopes. Immunopathology is mainly mediated by the adaptive immune response to viral infection.

Antibody Antibodies are secreted by plasma cells and can neutralize viruses, bind to the surface of infected cells, activate the complement system, and form immune complexes. Antibodies are an important part of the adaptive immune response.

B cell Each B cell has a specific type of antibody molecule on its surface (B cell receptor). Viral proteins are recognized by B cell receptors. B cells mature into plasma cells that secrete large amounts of antibodies.

Complement A protein cascade system mainly found in serum. The cascade can be activated by immune complexes of viral proteins and antibodies. The protein cascade generates chemotactic peptides for phagocytic cells, proteins that can coat viral particles, and lyse viral membranes.

Cytokines Proteins that can be produced by various cells within the body. These proteins are produced in response to infection. Particularly, immune cells but other cell types have cytokine receptors. Cytokines mediate inflammation and can modulate the immune response through cytokine receptors.

Immune complex A complex between antibody molecules and antigens. These complexes can activate the complement system through the Fc receptor on the antibody molecules and can lodge in glomeruli and small vessels causing immunopathology.

Immunopathology Disease resulting from the immune response to infection causing tissue injury and damage.

Innate immune system Comprising phagocytic cells (macrophages and neutrophils), complement system, natural killer (NK) cells, and various cytokines that can act within minutes to hours after infection. Recognition is to patterns of microbial structures, not specific epitopes.

Major histocompatibility complex (MHC) A region encoding highly polymorphic proteins that are involved in immune recognition of T cells, some cytokines and proteins of the complement cascade. This region is coded for by genes on human chromosome 6 and on mouse chromosome 17.

Natural killer (NK) cells A population of lymphocytes that can kill virus-infected cells. NK cells can secrete large amounts of the antiviral cytokine IFN-γ. IFN-γ can also activate macrophages. NK cells are part of the innate immune system.

T cell T cells are mature in the thymus. These lymphocytes recognize viral peptides in the context of MHC molecules through the T-cell receptor on the surface of T cells. T cells can kill virus-infected cells and release potent antiviral cytokines such as interferons.

Introduction

When a virus infects a vertebrate host, an immune response is rapidly generated against the virus. This immune response will often determine the survival or death of the host in acute infections or the establishment of a persistent infection. Tissue damage or pathology can result due to the antiviral immune response. This is different from direct viral effects where virus infection of cells can lead to tissue damage and apoptosis. A vigorous antiviral immune response is responsible for the immunopathology associated with acute and persistent/chronic virus infections. There is a necessary balance between viral clearance by the immune system and pathology induced by the immune system in attempting to clear the virus (**Figure 1**).

One important component of the adaptive immune system is the lymphocyte. Due to specialized receptors on the surface of lymphocytes (T cell receptor (TCR), B cell receptor (BCR)/antibody molecule), these cells are able to recognize or discriminate between different antigenic peptides derived from microbes or self-proteins. Lymphocytes can be divided into several broad populations. These cells can act directly to lyse or kill virus-infected cells (discussed below) or produce antiviral substances such as cytokines (interferons (IFNs), tumor necrosis factor (TNF)) that have antiviral activity but also can disturb the host's cell function or by themselves cause cellular and tissue injury. Lymphocytes in collaboration with cells of the innate immune system can also kill virus-infected cells or secrete antibodies with antiviral neutralizing capacity (discussed below).

One population of lymphocytes are the B cells which are bursa or bone marrow-derived lymphocytes. These lymphocytes express BCRs on their surfaces called immunoglobulins or antibodies (**Figure 2**). The antibodies are secreted from differentiated B cells called plasma cells. Different antibodies can bind to specific viral proteins or antigens; but in general, one specific antibody is originally derived from one B cell and can recognize a unique epitope. Antibodies can neutralize viruses by attaching to the surface of the virion, thus inhibiting the attachment, penetration, or uncoating phases of viral replication (**Figure 2**, arrow 1). Antibodies can also bind to virions in the blood (viremia) or to viral proteins (such as hepatitis B virus surface antigen), and form antigen–antibody complexes (immune complexes) (**Figure 2**, arrow 2). If the complexes are of the appropriate size, glomerulonephritis can occur due to immune complex deposition (discussed below). Immune complexes are also engulfed by Fc receptor bearing cells of the innate immune system found in the liver and spleen. However, since immune complexes are often infectious, by this means, Fc or complement expressing cells can also be infected. However, this process aids in clearance of virus from the blood, thereby limiting the viremia. Infected cells can be killed by antibodies specific for viral proteins expressed on the surface of virus-infected cells. This requires the presence of complement proteins. Complement proteins binding to the Fc portion of the immunoglobulin (**Figure 2**, arrow 3) activate the complement cascade leading to the formation of the complement membrane attack complex on the cell membrane and eventually cell lysis. Antibodies via the Fc portion of the immunoglobulin molecule can bind to Fc receptor positive cells (macrophages and NK cells – two cell types of the innate immune system) (**Figure 2**, arrows 4 and 5). The Fab portion of the immunoglobulin (antigen recognition site) gives specificity for the killing of the virus-infected cells. Therefore, antibodies can participate in clearing/neutralizing circulating free virus, as well as

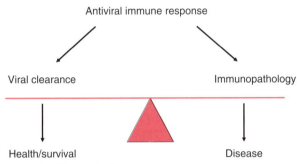

Figure 1 The antiviral immune response is a balance between viral clearance and immunopathology. One direction leads to survival of the host, whereas the other can lead to disease.

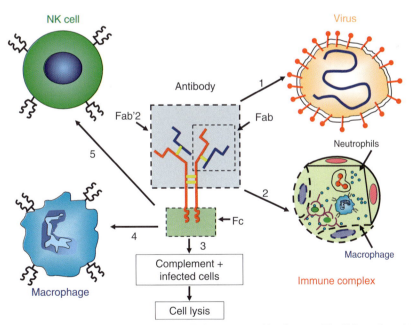

Figure 2 Antiviral antibodies can play an important role in viral clearance and/or disease. The Fab portion of the antibody molecule can bind and neutralize the virus (arrow 1). Antibodies binding to viral proteins or virions can form immune complexes that can lead to immune complex disease (arrow 2). Antibody binding to virus-infected cells through the Fc portion of the antibody molecule can activate the complement system leading to lysis of the infected cell (arrow 3). The Fc portion of the antibody molecule can also bind to Fc receptors in macrophages and NK cells. The antibody-coated macrophages and NK cells can kill infected cells via antibody-dependent cell-mediated cytotoxicity (arrows 4 and 5).

killing virus-infected cells with the help of complement and cells of the innate immune system. However, the number of antibody molecules required to kill a cell is usually in excess of 1×10^6, which makes this an inefficient process.

T cells represent another population of lymphocytes that participates in viral clearance and immunopathology. T cells can be further divided into two subpopulations, $CD4^+$ and $CD8^+$ T cells (**Figure 3**). T cells with the surface molecule CD4 are called T-helper cells (Th) due to their ability to secrete certain cytokines that 'help' B cells to differentiate into antibody producing plasma cells and 'help' the subpopulation of T cells expressing CD8 molecules to acquire cytotoxic or killer cell activity. $CD4^+$ T cells are 'restricted' by major histocompatibility complex (MHC) class II molecules. The MHC class II molecule presents viral peptides as a complex that is recognized by TCRs on $CD4^+$ T cells. The $CD4^+$ T cells can be further subdivided, depending on the types of cytokines they secrete, into $CD4^+$ Th1 T cells that are involved in delayed-type hypersensitivity immune responses and $CD4^+$ Th2 T cells that help B cells. When activated $CD4^+$ Th1 T cells secrete IFN-γ that has antiviral activity and can also in turn activate macrophages. These activated macrophages, cells of the innate immune system, can kill virus-infected cells indirectly by releasing reactive oxygen species and toxic cytokines, such as TNF or lymphotoxin (LT) that lyse cells by what is known as bystander killing. The $CD4^+$ T-cell population is also involved in the maintenance of antiviral T-cell memory. T-cell memory is key for immune individuals to respond to subsequent infections with the same or similar viruses. The $CD4^+$ T-cell population also contains a subset of T cells that also express a surface protein, CD25, and an internal protein called FoxP3. These T cells have the ability to modulate the adaptive immune response, and therefore are known as T-regulatory cells.

The $CD8^+$ T-cell population is intimately involved in viral clearance and immunopathology. $CD8^+$ T cells kill virus-infected cells by two mechanisms and are very efficient usually requiring recognition of ten or fewer viral peptide molecules in association with the MHC class I complex (see below). The first mechanism is by release of perforin molecules that form pores in the target cell's membrane similar to the complement membrane attack complex. The second mechanism involves the interaction of FasL-Fas on the T cell and the infected cell, respectively. Fas is a member of the TNF receptor family. Fas binding to its ligand leads to activation of caspase 8, a death signal in the infected cells that results in apoptosis. The cytotoxic T-lymphocyte (CTL) killing of virus infected cells occurs when infected cells are killed in a MHC class I restricted manner (TCR on $CD8^+$ T cells recognizing viral peptide complexed to class I molecules) prior to the maturation of infectious virus thus reducing the virus' ability to replicate and disseminate within the body. This supplements the action of antiviral antibodies that can neutralize infectious virus. Memory $CD8^+$ T cells form during the antiviral immune response and are long-lived T cells that can protect animals in the event of subsequent infections with the same virus.

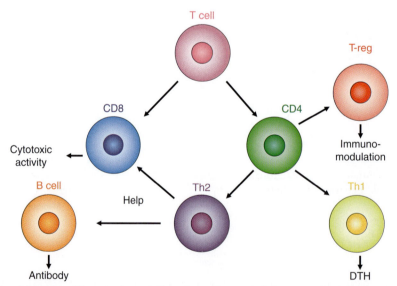

Figure 3 T cells can be divided into different subpopulations depending on their function. $CD8^+$ T cells have the ability to kill virus-infected cells and are restricted by MHC class I molecules. $CD4^+$ T cells can be further subdivided. Some $CD4^+$ T cells have the ability to provide 'help' to other T cells and B cells. Others can act as regulatory cells modulating the function of other T cells. $CD4^+$ T cells can also be directly involved in inflammatory responses through the production of various pro-inflammatory cytokines.

Thus B cells, $CD4^+$ and $CD8^+$ T cells, in conjunction with complement and cells (NK, macrophage, neutrophil) of the innate immune system, are responsible for viral clearance. These are also the cellular participants that are involved in immunopathology associated with virus infections.

Immunopathology

Antiviral immune responses are responsible for the immunopathology associated with acute and persistent/chronic virus infection. Immunopathology is determined by several factors. These involve contributions from both the host and the virus. Host factors include the age of the individual, genetic makeup of the host (MHC and non-MHC genes), route and/or site of infection, and whether the immune system is intact or immunocompromised. Characteristics of the virus consist of the strain or type and the size of the inoculum/dose.

Age of the host can play an important role in determining whether an individual survives, becomes immune, or acquires a persistent infection. For example, infection with lymphocytic choriomeningitis virus (LCMV), an arenavirus, has different outcomes depending on the age of the host. The natural host of the virus is the mouse. Its genome is comprised of two segments of RNA and is ambisense. The two RNA segments are encased by a nucleocapsid protein forming the nucleocapsid that is surrounded by a lipid envelope embedded with two viral glycoproteins. For the most part LCMV is noncytolytic and does not induce cell death after infection. Neonatal infection, infection of newborn mice with LCMV, or of adult mice with immunosuppressive variant(s) results in a persistent infection where many tissues are infected and mice survive. At this early age, the mouse's immune system is not fully developed and the mouse is not able to mount a functional CTL response. It was first proposed by Burnet that mice persistently infected with LCMV were immunologically tolerant to the virus (the host's immune system is not able to recognize the virus and mount an immune response). Accordingly, this purported immune tolerance to LCMV was why the virus was able to persist. However, seminal studies by Oldstone and colleagues demonstrated that mice persistently infected with LCMV were not tolerant. These mice produced antivirus antibodies and routinely developed glomerulonephritis. The inflammatory lesions in the kidneys were due to LCMV–anti-LCMV immune complex deposition in the glomeruli activating the complement cascade and recruiting of cells of the innate immune system leading to disease (to be discussed below) (**Figure 2**, arrow 2). Similar findings followed for congenital murine retrovirus infections.

In contrast, if mice are infected as adults with a moderate dose of LCMV (nonimmunosuppressive variant) by an extraneural route of infection (outside the central nervous system (CNS)) little or no immunopathologic disease is observed. Anti-LCMV CTLs are induced that are able to clear virus-infected cells. A portion of these LCMV-specific T cells go on to become memory T cells. Subsequent intracerebral (ic) infection of an adult immunocompetent immune mouse results in far less inflammation and immunopathology in the CNS and the animal survives. Interestingly, after ic LCMV infection of adult nonimmune mice, virus is able to replicate in certain epithelial cells. After about 5–7 days post infection the infected mouse mounts a robust anti-LCMV CTL immune response which is responsible for inflammation in the choroid plexus, ependyma, and leptomeninges causing choriomeningitis. Death is due to the action of the anti-LCMV CTL immune response altering the blood–brain barrier and initiating brain edema. This infection illustrates several of the factors involved in immunopathology. The age of the host is critical, since in very young animals the immune systems are not fully developed. These animals are immunocompromised, that is, unable to mount an effective CTL response allowing a noncytolytic virus-like LCMV to persist. Other studies have shown the importance of CTL in initiating immunopathology and death. Adult mice whose T cells are ablated or inhibited and then infected with LCMV via the ic route of infection develop a persistent infection and do not die of the choriomeningitis. Adoptive transfer of immune spleen cells, immune T cells, or LCMV-specific $CD8^+$ T cells into LCMV carrier mice leads to clearance of viral infection. Thus, this experiment complements studies of acute infection and substantiates the role of the antiviral T cells in clearance of virus and termination of infection.

Mice that are infected as newborns or earlier with LCMV develop a persistent infection, but are not without immunopathology. As mice infected at birth or earlier develop into adults, they produce anti-LCMV antibodies that can bind to viral antigens (present due to the persistent infection) forming immune complexes. Depending on the size of the immune complexes and numbers formed, the complexes deposit in the glomeruli of the kidneys due to the filtering action within the glomeruli. As a consequence of immune complex deposition in the glomeruli, the complement cascade is initiated releasing chemotactic peptides that recruit and activate polymorphonuclear leukocytes (PMNs) and macrophages into the glomeruli. These activated cells release inflammatory cytokines, such as TNF and interleukin (IL)-1, and reactive oxygen intermediates, as well as lysosomal enzymes that induce inflammation and immunopathology resulting in glomerulonephritis. The extent of glomerulonephritis induced by persistent LCMV infection varies and depends both on the strain of mouse (genetic influences discussed below) and strain of virus.

The genetic composition of the host can play an important role in whether immunopathology occurs. For example, infection with Theiler's murine encephalomyelitis virus (TMEV), a picornavirus related to the cardioviruses, has different outcomes depending on the strain of the host. As with LCMV the mouse is the natural host. Because of the availability of inbred mice and mapping of their molecules, particularly MHC and background genes involved in innate immunity, the mouse has been an exquisite model to measure not only host genetics but influence of viral genetics as well. TMEV is a positive stranded RNA virus. The genome is approximately 8000 bases in length. The icosohedral capsid is comprised of four viral proteins. The virion is not encapsidated by a membrane. TMEV can be divided into two groups depending on neurovirulence. The GDVII subgroup is comprised of viruses that are highly neurovirulent. Members of the TO subgroup are less neurovirulent. Ic infection with the DA strain of TMEV (TO subgroup) into adult C57BL/6 mice leads to an antiviral $CD8^+$ T-cell response that clears the virus from the CNS and other tissues of the body. There is limited inflammation in the CNS but mice survive and are immune to subsequent infections with TMEV.

In contrast, ic infection of SJL/J mice with the DA strain of TMEV induces an antiviral $CD8^+$ T-cell response which is not able to clear the virus. Therefore, DA virus is able to establish a persistent infection in the CNS. Immunopathology results as a consequence of a chronic anti-TMEV immune response attempting to clear virus-infected cells in the CNS. Infected SJL/J mice eventually develop large areas of inflammation and demyelination in the spinal cord with a spastic paralysis. Genetic differences in the MHC complex between the two strains of mice are implicated in whether one mouse strain is able to clear the virus and the other mouse strain is not able to clear the virus resulting in disease. In addition, innate immune responses are also genetically determined. Therefore, the early inflammation in response to infection is encoded within the genome and varies from individual to individual.

The initial site of infection is important. A peripheral infection with LCMV in an adult mouse generates a robust anti-LCMV CTL response and immunity. Since the virus is not present in the CNS or only a few cells are infected, the CTL response clears the virus and little pathology ensues. Infection by the ic route also induces a vigorous CTL response. However, a significant number of meningial cells are infected with the virus. CTLs recognize virus-infected cells in the meninges lysing the cells via a perforin-mediated mechanism. This contributes to the death of the mouse. Therefore, the site or route of infection and, which cells are infected by the virus, are important factors.

The strain of the virus can play a major role in immunopathology, as shown by the following two examples. As mentioned previously TMEV can be divided into two groups depending on neurovirulence. The GDVII subgroup comprises viruses that are highly neurovirulent. An intravenous (IV) or ic infection of mice with the GDVII virus will result in the virus killing the mouse in 7–10 days. The death is due to the virus' ability to infect large numbers of neurons and directly kill these cells. In contrast, infection of mice (SJL/J) with viruses from the TO subgroup with 100 000–1 000 000 times more virus leads to mild acute disease with TMEV establishing a persistent infection in the CNS. The chronic anti-TMEV immune response results in extensive demyelination and inflammation. The pathology is caused by the antiviral immune response comprising antibodies capable of neutralizing the virus, $CD4^+$ and $CD8^+$ T cells recognizing virus-infected cells and either directly killing the infected cells, as in the case of antiviral $CD8^+$ T cells or TMEV antibody and complement, or killing the infected cells via bystander mechanisms which are mediated by $CD4^+$ T cells and the effector/killer cells are macrophages (as in delayed-type hypersensitivity responses).

CNS infection of adult mice with LCMV leads to the generation of antiviral $CD8^+$ T cells and infected mice die about 7–10 days later as the $CD8^+$ T-cell response develops. As mentioned above there is extensive infiltration of T cells and mononuclear cells in the meninges, alterations in the blood–brain barrier, and edema. In contrast, infection of adult mice with a variant of LCMV, clone 13, that differs in two amino acids from wild-type Armstrong LCMV, induces an immunosuppression due to infection of dendritic cells which are then not able to arm and expand T- and B-viral-specific cells. The variant LCMV is able to persist in adult mice. These persistently infected mice also develop immune complexes in their kidneys. Different strains of virus, such as clone 13, can initiate different types of disease leading to immunopathology.

The amount of virus the host first encounters can set the stage for later disease. For example, high doses of certain strains of LCMV when given by the IV route results in dissemination to the various lymphoid tissues. Virus can replicate to high titers at these sites constantly stimulating the antiviral T cells leading to 'CTL exhaustion'. This is likely due to both the continual stimulation of the TCR by antigen presenting cells leading to the apoptosis of LCMV-specific CTLs and to infection of dendritic cells. With the depletion or failure to generate virus-specific CTLs, LCMV is able to establish a persistent infection. This is in contrast when low or moderate amounts of LCMV are used to infect mice. The infected mice are able to mount an effective CTL response, virus is cleared, and animals are immune.

In summary, immunopathology results from an imbalance between the immune system's ability to clear the virus and resulting tissue damage due to the antiviral immune response. These can take the form of antibody and complement, $CD4^+$ T cells and $CD8^+$ CTLs causing tissue damage in the process of eliminating virus and virus-infected cells.

See also: Antigen Presentation; Immune Response to viruses: Antibody-Mediated Immunity; Immune Response to viruses: Cell-Mediated Immunity.

Further Reading

Borrow P and Oldstone MBA (1997) Lymphocytic choriomeningitis virus. In: Nathanson N, Ahmed R, Gonzalez-Scarano F, et al. (eds.) Viral Pathogenesis, pp. 593–627. Philadelphia: Lippincott-Raven Publishers.

Buchmeier MJ and Oldstone MBA (1978) Virus-induced immune complex disease: Identification of specific viral antigens and antibodies deposited in complexes during chronic lymphocytic choriomeningitis virus infection. Journal of Immunology 120: 1297–1304.

Henke A, Huber S, Stelzner A, and Whitton JL (1995) The role of $CD8^+$ T lymphocytes in coxsackievirus B3-induced myocarditis. Journal of Virology 69: 6720–6728.

Oldstone MBA (1982) Immunopathology of persistent viral infections. Hospital Practice (Office Education) 17: 61–72.

Oldstone MBA and Dixon FJ (1969) Pathogenesis of chronic disease associated with persistent lymphocytic choriomeningitis viral infection. I. Relationship of antibody production to disease in neonatally infected mice. Journal of Experimental Medicine 129: 483–505.

Selin LK, Cornberg M, Brehm MA, et al. (2004) CD8 memory T cells: Cross-reactivity and heterologous immunity. Seminars in Immunology 16: 335–347.

Tsunoda I and Fujinami RS (1996) Two models for multiple sclerosis: Experimental allergic encephalomyelitis and Theiler's murine encephalomyelitis virus. Journal of Neuropathology and Experimental Neurology 55: 673–686.

Infectious Pancreatic Necrosis Virus

Ø Evensen and N Santi, Norwegian School of Veterinary Science, Oslo, Norway

© 2008 Elsevier Ltd. All rights reserved.

Glossary

Fingerlings A life-cycle stage when young salmonids are one finger in length.
Fry Newly spawned fish that have fully absorbed their yolk sac.
Milt The sperm from the male fish.
Parr A young salmonid with parr marks (dark blotches) on the sides, before migration to the sea.
Smolts Salmonid fish going through smoltification, physiological changes that will allow fish to change from life in freshwater to life in the sea. The smolt state follows the parr state.

Introduction

Infectious pancreatic necrosis (IPN) was originally recognized as an acute and highly contagious disease of juvenile salmonids which typically occurs at the time of commencement of feeding of fry. The clinical manifestation of the disease has, however, changed over the last two decades. IPN outbreaks are still observed during the first few weeks after commencement of feeding of fry and also later in the freshwater phase, typically in 10–20 g fish. The most remarkable change in the clinical manifestation of the disease, observed since the mid-1980s, is the number of outbreaks in post-smolts after seawater transfer, and this has emerged as a significant problem, particularly in salmon farming. The disease pattern is seen in many of the major Atlantic salmon-producing countries such as Chile, UK, and Norway. The mortality rates in the freshwater stage vary considerably from negligible to almost 100%, while disease outbreaks in seawater typically result in 10–20% cumulative mortality, and can reach 70% in individual sea cages. This variation in mortality has been ascribed to factors related to the host species (e.g., the age and/or genetic resistance of fish) and environmental stressors. Virus characteristics such as variations between virus serotypes and strains and viral infection loads can also play an important role in determining mortality rates.

Virus Classification

Infectious pancreatic necrosis virus is the type species of the genus *Aquabirnavirus* within the family *Birnaviridae*. The disease caused by infectious pancreatic necrosis virus (IPNV) in salmonid fish is characterized by necrosis of the exocrine pancreas, which is the origin of the name of both the disease and the type species. However, aquabirnaviruses have been isolated from a wide variety of freshwater and marine fish species, as well as from marine invertebrates worldwide, and are associated with a wide range of host pathologies. Aquatic birnaviruses are an antigenically diverse group of virus, and somewhat difficult to classify.

Serological classification based on neutralizing antibodies against surface epitopes separates isolates into serogroups A and B. Serogroup A contains nine serotypes (A1–A9) comprising most of the isolates, whereas serogroup B consists of only one serotype. Due to a lack of standardization of the methods, serotyping has been supplemented with, and partly replaced by, genetic classification. Phylogenetic analysis based on the VP2 coding region of genomic segment A suggests that aquabirnaviruses may be clustered into six genogroups that correspond to the previously established serotypes. Genogrouping based on 310 basepairs (bp) at the VP2/NS junction region has identified seven genogroups, the additional genogroup comprising Japanese aquabirnavirus isolates (**Figure 1**). More distantly related is the blotched snakehead virus which is currently an unassigned member of the family *Birnaviridae* and it has been proposed to be classified in a new genus.

There are no agreed species demarcation criteria for the genus *Aquabirnavirus*, and this has led to difficulty in nomenclature. Distantly related strains may cause similar pathologies, whereas closely related strains may range from avirulent to very virulent. Closely related strains are found in both farmed fish and marine species in the same geographical region. Therefore, the species concept does not apply very well to members of the genus *Aquabirnavirus*. Another aspect contributing to this is the intrinsic variability of RNA viruses such as the aquabirnaviruses, is the error-prone process of nucleic acid replication.

Genomic Organization and Virus Structure

The IPNV virion is 60 nm in diameter and a single-shelled, icosahedral structure with no envelope (**Figure 2**). The virus genome consists of two segments of double-stranded (ds) RNA. For serotype Jasper, segment A comprises 3097 bp and the smaller segment B comprises 2784 bp.

Genome segment A (**Figure 3**) contains a large open reading frame (ORF) encoding a 106 kDa polyprotein (NH_2-VP2-protease (VP4)-VP3-COOH), which is co-translationally cleaved to generate pVP2 (the precursor of VP2), VP4, and VP3. The processing of the large polyprotein is controlled by VP4, which is a viral protease with a serine/lysine catalytic dyad. The cleavage sites have been identified at the pVP2–VP4 and VP4–VP3 junctions, and the structure of VP4 has recently been elucidated by X-ray crystallography. The pVP2 precursor is processed further into the mature VP2 which is the major capsid protein and, in addition, three small peptides are generated that can be detected in virus particles. VP2 spontaneously self-assembles into particles with a diameter of about 25 nm, suggesting that late maturation is to avoid premature assembly during particle morphogenesis. Although studies have indicated that some elements of VP3 may be exposed on the surface of the virion, serotype-specific and neutralizing antibodies are almost exclusively directed against continuous and discontinuous epitopes of VP2. VP3 is an internal capsid protein associated with the viral genome. By applying the yeast two-hybrid system in combination with co-immunoprecipitation, VP3 has recently been shown to bind to VP1 and to self-associate strongly. In addition, VP3 has been shown to bind specifically to dsRNA in a sequence-independent manner. The binding of VP3 and VP1 is not dependent on the presence of dsRNA.

Genome segment A also contains a small ORF that proceeds and partly overlaps the large ORF encoding the 106 kDa polyprotein. The smaller ORF encodes a highly basic, arginine-rich 17 kDa polypeptide that has been detected in infected cells but has not, to date, been conclusively demonstrated to be present in purified virus particles. This apparently nonstructural protein, designated VP5, is encoded by most but not all IPNV strains that have been examined. In an Asian IPNV strain, an anti-apoptotic effect has been demonstrated for VP5 and this may aid viral replication in the initial phase of infection. However, the amino acid sequence of VP5 is not well conserved and the anti-apoptotic effect does not occur in all IPNV strains encoding the polypeptide.

Segment B encodes a 90 kDa protein (VP1) which functions as an RNA-dependent RNA polymerase (RdRp). This enzyme directs the transcription of nonpolyadenylated mRNA from genome segments A and B. Within the virus particle, VP1 occurs as both a free polypeptide and covalently linked to the 5' terminus of each genome segment.

Antigenic Structure and Neutralization Sites of the Virus

The VP2 is the type-specific antigen in which neutralizing sites reside within the hypervariable region. Several attempts have been made to identify the neutralization domains of VP2 and the findings reported are summarized in **Table 1**. The region comprising amino acid residues 183–335 has been found by several authors to contain neutralizing epitopes.

Virus Propagation and Cytopathic Effects

IPNV replicates in the cytoplasm and a single cycle of replication takes 16–20 h. IPNV infects a wide range of cultured fish cells, including CHSE-214, RTG-2, and BF-2 cells, all of which are routinely used for laboratory propagation of the virus. IPNV replicates well at temperatures ranging from 15 to 22 °C, resulting in a characteristic cytopathic effect. Available information on attachment, penetration, and uncoating is limited. Purified virus

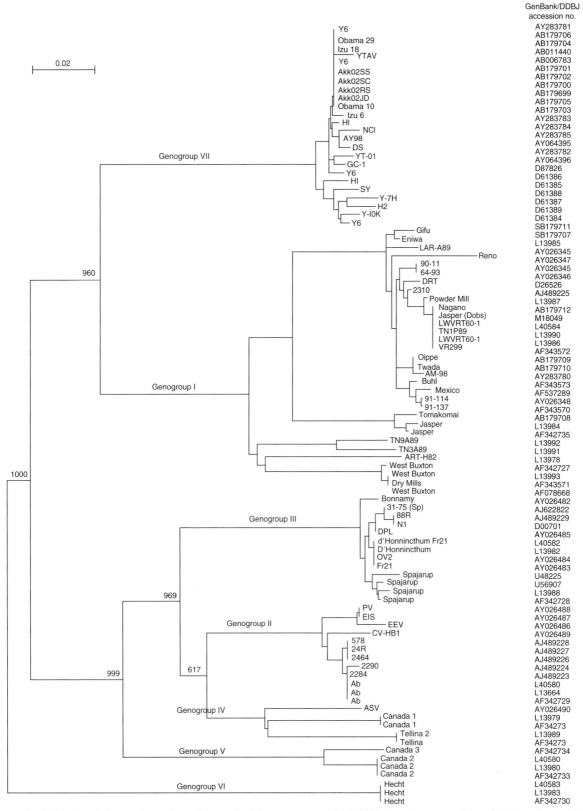

Figure 1 Molecular phylogenetic tree based on nucleotide sequences of the VP2/NS junction among 93 worldwide isolates of IPNV and other aquabirnaviruses, and corresponding GenBank/DDBJ accession numbers. Bootstrap values form 100 replicates are shown at major nodes. Scale: 0.02 replacement nucleotides per site. Reproduced from Nishizawa T, Kinoshita S, and Yoshimizu M (2005) An approach for genogrouping of Japanese isolates of aquabirnaviruses in a new genogroup, VII, based on the VP2/NS junction region. *Journal of General Virology* 86: 1973–1978, with permission from Society for General Microbiology.

attaches to CHSE-214 cells by specific and nonspecific binding to cell membrane components. After binding, IPNV can be observed in endosomes after 20–30 min, suggesting that viral entry occurs by endocytosis. Entry does not appear to be dependent on the acidic pH of the endosomes, and the virion RdRp (VP1) is active without proteolytic treatment of the virion, indicating that uncoating may not be a precondition for virus replication. VP1 can be guanylylated *in vitro* whereupon it becomes a primer for *in vitro* RNA synthesis. Unlike other viral VPg polypeptides, VP1 self-guanylylates *in vitro* in a template-independent manner, and the guanylylation site has been mapped to serine 163. Segment A mRNA is synthesized in larger amounts (2–3 times) than segment B, reflecting the relative abundance of the viral proteins. Assembly of virions takes place in the cytoplasm of infected cells, and virus release occurs via cell lysis.

Apoptosis is induced in CHSE-214 cells following IPNV infection and the so-called 'McKnight' cells associated with sloughing of mucosal cells in the intestines of juvenile salmonids suffering from IPN resemble apoptotic cells morpologically. Similar apoptotic bodies are also found in the liver of Atlantic salmon post-smolts suffering from IPN (**Figure 4**).

Acute Infections

IPNV causes acute disease in juvenile salmonids and is the major pathogen of concern in brook trout (*Salvelinus fontinalis*), rainbow trout (*Oncorhyncus mykiss*), and Atlantic salmon (*Salmo salar* L.) hatcheries. Outbreaks also occur in fingerlings and parr later in the freshwater phase, as well as in post-smolts after transfer to seawater.

Salmonid fish suffering acute IPNV infection display darkened skin, abnormal behavior such as whirling or swimming on the side, and a visibly distended abdomen. Whitish threads of debris are often seen hanging from the anal orifice. At necropsy, common findings are ascites and hemorrhages in perivisceral adipose tissue, and the liver is often pale. Histopathologically, IPN is characterized by acute multifocal necrosis of the exocrine pancreas. In cases of more protracted disease, evidence of a cellular immune response can be observed. In addition to pancreatic lesions, multifocal hepatic necrosis and acute catarrhal enteritis with necrosis and sloughing of the intestinal epithelium have been reported. Viral antigen can be detected within lesions by immunohistochemistry or indirect fluorescent antibody tests.

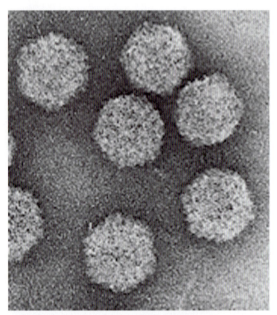

Figure 2 Negative staining of IPNV serotype Sp grown on Chinook salmon embryonic (CHSE-214) cells and purified by density gradient centrifugation.

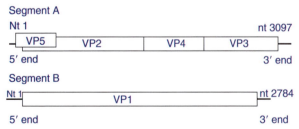

Figure 3 Schematic presentation of the genomic organization of IPNV depicting the proteins encoded by genome segments A and B.

Persistent Infections

A high proportion of fish surviving an IPN outbreak become persistently infected without apparent signs of disease. These subclinically infected individuals are of

Table 1 Summary of studies addressing the neutralizing part of IPNV VP2 protein

Author(s)	Year	Reference	VP2 region
Mason C	1992	PhD Thesis, Oregon State University	207–314
Liao and Dobos	1995	Virology 209: 684–687	200–350 (243–357) 83
Heppell *et al.*	1995	Virology 214: 40–49	183–335
Tarrab *et al.*	1995	Journal of General Virology 76: 551–558	183–337
Frost *et al.*	1995	Journal of General Virology 76: 1165–1172	204–330
Blake *et al.*	2001	Diseases of the Aquatic Organisms 45: 89–102	243–335

major importance as sources of horizontal transmission of the virus. In the carrier state, there is no direct negative impact on infected individuals, although there are some indications of suppressed immune response. However, recent studies have shown that there are differences in the ability of different strains of the Sp serotype of IPNV to establish persistent infections in Atlantic salmon fry. Carrier fish shed virus in the feces but titers fluctuate over time and increase during periods of stress. In addition, persistently infected brood fish can transfer IPNV to their progeny.

Virus can be isolated from peripheral leukocytes a short time after experimental infection but not during chronic stages. The virus is located in head kidney leukocytes in fish that recover from an acute infection and most likely in macrophages or macrophage-like cells of the head kidney in persistently infected fish. It has also been shown that IPNV multiplies in adherent leucocytes isolated from carriers, although a lytic infection is not seen and, concordantly, virus is not found in the supernatant medium from cultures of isolated head kidney macrophages. Similarly, IPNV is able to induce persistent, nonlytic infections in cell cultures in which a high proportion of the cells are infected but produce relatively low quantities of virus. Different strains of IPNV appear to vary in their ability to induce persistence in cell culture. Infected cells are resistant to superinfection with either homologous or heterologous strains or serotypes, and cultures can be cleared of IPNV infection by several passages in the presence of virus neutralizing antiserum. The underlying mechanisms of the establishment and maintenance of IPNV persistence *in vivo* and *in vitro* are largely unknown.

Molecular Determinants of Virulence

The mortality rate during an IPN outbreak can vary and this has been ascribed to factors related both to the host and to the virus. Viral factors influencing mortality rates are the dose of infection as well as virulence variations between serotypes and strains. IPNV strains of different serotypes are highly heterogenous and the Sp serotype is considered to be more virulent than the Ab serotype. Within the Sp serotype, different genotypes exhibit different virulence characteristics which have been studied by the use of reverse genetics.

In vitro transcribed positive-sense cRNAs of IPNV are infectious when transfected together into permissive cell cultures, allowing the recovery of genetically modified isolates. The reverse genetics technique is a valuable tool for studying the IPNV life cycle and the role of individual viral proteins in the pathogenesis of infection. Experimental studies using such recombinant IPNV strains have demonstrated that VP5 expression is not required for efficient replication *in vitro* or *in vivo*, or for virulence, or persistence of the virus. Reassortant viruses containing segment A from a high-virulence strain and segment B from a low-virulence strain, prepared by either reverse genetics or co-infection techniques, have also been used to demonstrate that segment B has no direct influence on IPNV virulence. For IPNV serotype Sp, it has been shown that certain amino acid sequence motifs in VP2 strongly correlate with virulence. Residues 217 and 221 appear to be critical determinants of virulence, and genetic signatures involving these amino acids are strongly associated with avirulent, moderately and highly virulent strains, as shown in **Table 2**. Furthermore, serial passage of the Sp serotype in cell culture leads to the attenuation of virulence as a result of an A→T mutation at amino acid residue 221.

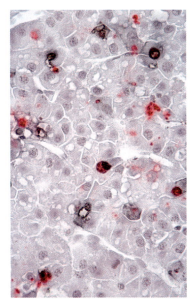

Figure 4 Liver tissue from IPNV-infected Atlantic salmon, fixed in 10% buffered formalin and embedded in paraffin wax. The tissue section was double-stained for IPNV antigen using a rabbit immune serum to IPNV Sp (red color) and for apoptotic cells based on the TUNEL method (black coloration). Note double-staining in some of the hepatocytes.

Table 2 Point mutations in VP2 result in a marked reduction to almost complete loss of virulence whereby 217 and 221 are key determinants

Amino acid residues of VP2			In vivo virulence characteristics
217	221	247	
T	A	T	Highly virulent
T	T	T	Avirulent
P	A	A	Moderately virulent
P	T	A	Avirulent

Substitution of threonine for proline at residue 217 (T217P) reduces the virulence of the strain, producing a mortality rate approx. 50 % lower than the high-virulent variant (TAT). Similarly, single amino acid substitutions of alanine for threonine at residue 221 render highly virulent strains (TAT) almost avirulent. Passage in culture of high virulent strains is associated with motif changes in position 221 (A to T) for TAT ant PAA variants, each of which results in avirulence.

Table 3 Summary of neutralizing domain, virulence determinants and the cell culture adaptation characteristics of IPN virus

Characteristics	VP2 amino acid residues
Immunogenic domain	183–337
Virulence determinants	217, 221, 247 (Sp serotype)
Cell culture adaptation	221 (Sp serotype)
Antibody neutralization	Not known

Serotype-specific findings have been indicated.

In summary, the neutralizing domain, the virulence determinants, and the cell culture adaptation with subsequent loss of *in vivo* virulence seem to be located in the VP2 hypervariable region for IPNV, although there is a limited number of serotypes and strains of the IPN virus that have been characterized for their virulence signatures and cell culture adaptation traits (**Table 3**).

Epidemiology

It has long been known that survivors of an IPN outbreak become carriers and surviving fish continue to shed virus without any apparent clinical symptoms, and subclinical infection is important for horizontal transmission of virus. The efficiency by which the virus is spread in a population of susceptible individuals varies between strains and there are indications that this might correlate with virulence characteristics whereby highly virulent variants are more efficient at spreading. Transmission of the virus to progeny has been shown to occur vertically via eggs in brook trout and rainbow trout. The virus is possibly present inside the eggs since there are indications that surface disinfection of the eggs is not fully effective in preventing the introduction of IPNV to hatcheries. One possible route of intra-ovum introduction is via the milt, by adsorption of virus to the sperm head.

Surveys for the prevalence of IPNV have concluded that most North Sea salmon farms harbor IPNV carrier fish, making selection for IPNV-free brood fish difficult (prevalence in Scotland is 80%). Once introduced to a hatchery, the virus is difficult to eliminate, as it is stable in water and relatively resistant to many disinfectants. It is also still not clear whether seawater outbreaks of IPN are the result of persistent infection obtained during the freshwater period or exposure from a marine source.

Control of IPN through Management

Good management includes the environment of fish to minimize stress and to reduce the likelihood of transmission of the pathogen. This is particularly important at the time of first feeding of fry, when smolt are transferred to sea and during the first few months after sea water transfer.

Transfers of fish between farms are considered as one of the highest risks in the transmission of IPNV. Another factor is the movement of eyed eggs and fry between hatcheries, common in Norway and other salmon-producing countries.

Vertical transmission has not been conclusively demonstrated in Atlantic salmon but there is evidence from other salmonid species to indicate that vertical transmission is a strategy utilized by IPNV.

Disease Control through Vaccination

There are few scientific publications addressing or documenting the effect of vaccination against IPN and most of what has been published has focused on the immune responses elicited by vaccination. Vaccination with an *Escherichia coli*-expressed VP2 recombinant protein or insect baculovirus-expressed subunit vaccine based on all structural proteins of IPNV induces specific antibody responses in Atlantic salmon. For *E. coli*-expressed proteins, it has also been shown that vaccination will confer higher secondary antibody titers in immunized fish compared to nonvaccinated controls. Similar observations have been made for inactivated vaccines based on whole virus. Reliable challenge models for IPN in Atlantic salmon post-smolts have only recently been established and this partly accounts for the lack of scientific data on protective immunity. It has been shown that partial protection can be induced in Atlantic salmon immunized with a baculovirus/insect larvae-expressed subunit vaccine. However, the level of protection was below that required under field conditions in a commercial setting.

DNA vaccination against IPNV has also been attempted in Atlantic salmon. Fish were immunized with plasmids encoding the entire polyprotein of segment A and shorter variants of the protein, and subsequently challenged with virulent IPNV strains. A reasonable protection was attained with constructs expressing the entire segment A but it is difficult to draw a firm conclusion as to the potential that lies in the use of DNA vaccines for protection against IPNV infection since the number of published studies is few.

Despite the sparse knowledge of the immunity to IPNV, the first commercial, recombinant subunit (VP2) vaccine was introduced in Norway in 1995 and has come into use throughout the industry. By 2 years after introduction of the vaccine, 68% of the fish transferred to sea were vaccinated and the national coverage had increased to 85% in 2002. Vaccines based on killed, whole-virus are also currently available in Norway, the UK, and Chile. The extent to which the use of the vaccine has influenced the prevalence of IPNV has not been ascertained.

Disease Control through Breeding

There are few published papers related to the genetic resistance to IPN in Atlantic salmon or other fish species. In Norway, IPNV challenge experiments in Atlantic salmon fry (around 0.20 g size) have shown 15–30 % of the variation in mortality between family groups can be explained by genetic factors (heritability). Therefore, there is a potential for improving resistance to IPNV through breeding programs where brood fish are selected on the basis of off-spring performance.

Another approach is to select brood fish based on survival following disease outbreaks, but the latter strategy presents a risk of bringing the infection over to offspring. One option is to expose pedigreed populations of fish to the field challenge, with some of the unchallenged fill-siblings remaining in biosecure conditions for breeding of unchallenged brood-stock.

See also: Necrovirus.

Further Reading

Blake S, Ma J-Y, Caporale DA, Jairath S, and Nicholson BL (2001) Phylogenetic relationships of aquatic birnaviruses based on deduced amino acid sequences of genome segment A cDNA. *Diseases of Aquatic Organisms* 45: 89–102.

Coulibaly F, Chevalier C, Gutsche I, et al. (2005) The birnavirus crystal structure reveals structural relationships among icosahedral viruses. *Cell* 120: 761–772.

Dobos P (1995) The molecular biology of infectious pancreatic necrosis virus (IPNV). *Annual Review of Fish Diseases* 5: 25–54.

Duncan R, Nagy E, Krell PJ, and Dobos P (1987) Synthesis of the infectious pancreatic necrosis virus polyprotein, detection of a virus-encoded protease, and fine structure mapping of genome segment A coding regions. *Journal of Virology* 61: 3655–3664.

Feldman AR, Lee J, Delmas B, and Paetzel M (2006) Crystal structure of a novel viral protease with a serine/lysine catalytic dyad mechanism. *Journal of Molecular Biology* 358: 1378–1389.

Frost P, Håvarstein LS, Lygren B, Ståhl S, Endresen C, and Christie KE (1995) Mapping of neutralization epitopes on infectious pancreatic necrosis viruses. *Journal of General Virology* 76: 1165–1172.

Heppell J, Tarrab E, Lecomte J, Berthianme L, and Arella M (1995) Strain variability and localization of important epitopes on the major structural protein (VP2) of infectious pancreatic necrosis virus. *Virology* 214: 40–49.

Hill BJ and Way K (1995) Serological classification of infectious pancreatic necrosis (IPN) virus and other aquatic birnaviruses. *Annual Review of Fish Diseases* 5: 55–77.

Hong JR and Wu JL (2002) Induction of apoptotic death in cells via bad gene expression by infectious pancreatic necrosis virus infection. *Cell Death and Differentiation* 9: 113–124.

Liao L and Dobos P (1995) Mapping of a serotype specific epitope of the major capsid protein VP2 of infectious pancreatic necrosis virus. *Virology* 209: 684–687.

Nishizawa T, Kinoshita S, and Yoshimizu M (2005) An approach for genogrouping of Japanese isolates of aquabirnaviruses in a new genogroup, VII, based on the VP2/NS junction region. *Journal of General Virology* 86: 1973–1978.

Pedersen T, Skjesol A, and Jorgensen JB (2007) VP3, a structural protein of the infectious pancreas necrosis virus, interacts with the RNA-dependent RNA polymerase VP1 and with double-stranded RNA. *Journal of Virology* 81: 6652–6663.

Rodriguez Saint-Jean S, Borrego JJ, and Perez-Prieto SI (2003) Infectious pancreatic necrosis virus: Biology, pathogenesis, and diagnostic methods. *Advances in Virus Research* 62: 113–165.

Santi N, Sandtro A, Sindre H, et al. (2005) Infectious pancreatic necrosis virus induces apoptosis *in vitro* and *in vivo* independent of VP5. *Virology* 342: 13–25.

Santi N, Song H, Vakharia VN, and Evensen O (2005) Infectious pancreatic necrosis virus VP5 is dispensable for virulence and persistence. *Journal of Virology* 79: 9206–9216.

Shivappa RB, McAllister PE, Edwards GH, Santi N, Evensen O, and Vakharia VN (2005) Development of a subunit vaccine for infectious pancreatic necrosis virus using a baculovirus insect/larvae system. *Developments in Biological Standardization* 121: 165–174.

Song H, Santi N, Evensen O, and Vakharia VN (2005) Molecular determinants of infectious pancreatic necrosis virus virulence and cell culture adaptation. *Journal of Virology* 79: 10289–10299.

Tarrab E, Berthianme L, Grothé S, O'Connor-McCourt M, Heppell J, and Lecomte J (1995) Evidence of a major neutralizable conformational epitope region on VP2 of infectious pancreatic necrosis virus. *Journal of General Virology* 76: 551–558.

Xu HT, Si WD, and Dobos P (2004) Mapping the site of guanylylation on VP1, the protein primer for infectious pancreatic necrosis virus RNA synthesis. *Virology* 322: 199–210.

Infectious Salmon Anemia Virus

B H Dannevig, National Veterinary Institute, Oslo, Norway
S Mjaaland and E Rimstad, Norwegian School of Veterinary Science, Oslo, Norway

© 2008 Elsevier Ltd. All rights reserved.

Glossary

Ascites Accumulation of serous fluid in the peritoneal cavity.
Coprophagy Eating of feces.
Exophthalmia Protrusion of the eyeballs.
Fry Young offsprings, presmolt stage.
Hemagglutinin-esterase A surface glycoprotein responsible for both receptor-binding (hemagglutinin) and receptor-destroying (esterase) activities.
Pathognomonic Pathological changes typical for a specific disease.

Smolt The stage in the life cycle of young anadromous salmonids when the fish is physiologically adapted to seawater, that is, farmed salmonids can be moved from freshwater to seawater.
Well boats Boats that are used for transportation of live fish.

Introduction

Infectious salmon anemia virus (ISAV) is the causative agent of infectious salmon anemia (ISA), a disease of farmed Atlantic salmon (*Salmo salar*). ISA primarily affects fish held in or exposed to seawater. The disease appears as a systemic condition characterized by severe anemia and hemorrhages in several organs. Mortality during an outbreak of ISA varies significantly. Daily mortality in affected net pens ranges from 0.2% to 1%, but may increase during an outbreak and the cumulative mortality may exceed 90% in severe cases.

ISA was recognized in 1984 in Norway, and was soon identified as a contagious viral disease. The disease increased in prevalence and showed a peak in 1990. In the following years, the incidence of ISA was greatly reduced by the implementation of legislatory measures or husbandry practices based on general hygiene. These included mandatory health control in hatcheries and health certification for fish, restrictions on transportation of live fish, regulations of disinfection of wastewater from fish slaughterhouses and of water supplies to hatcheries. Approximately 10 years after the first recognition of ISA, the causative virus, ISAV, was isolated in cell culture using the SHK-1 cell line established from Atlantic salmon head kidney. The subsequent investigations showed that the morphological, physiochemical, and genetic properties of ISAV are consistent with classification in the *Orthomyxoviridae*.

Virus Properties and Classification

ISAV is a pleiomorphic, enveloped virus, 100–130 nm in diameter, with 10–12 nm surface projections (**Figure 1**). The virus hemagglutinates erythrocytes of several fish species and it has receptor-destroying and membrane fusion activities. Endothelial and leukocytic cells are the main target cells and the virus replicates by budding from the cell membrane. ISAV has two main surface glycoproteins: the hemagglutinin-esterase (HE) responsible for the receptor-binding and receptor-destroying activities, and a fusion protein (F).

The buoyant density of virus particles in sucrose or cesium chloride is $1.18 \, \text{g ml}^{-1}$. The virus is stable at

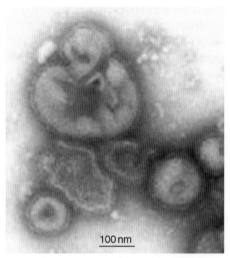

Figure 1 Electron micrograph of negative-stained ISAV particles purified from infected cell culture medium. Photo by Ellen Namork, The Norwegian Institute of Public Health, Oslo, Norway.

pH 5.7–9.0. The virus replicates in the salmon cell lines SHK-1, TO, and ASK, with a replication optimum of 10–15 °C. Some strains also replicate in the CHSE-214 cell line. There is no replication in SHK-1 cells at 25 °C or higher and, even at 20 °C, the yield of virus in SHK-1 cells is only 1% of the yield at 15 °C.

The 14.3 kb genome consists of eight single-stranded RNA segments of negative polarity in the size range of 1.0–2.4 kb. The amino acid identity between the ISAV proteins and those of other orthomyxoviruses is low (13–25%), and the organization of ISAV genes is unique. ISAV is therefore classified as the type species of the genus *Isavirus* within the family *Orthomyxoviridae*.

Geographic Distribution and Host Range

After the first recorded outbreak in 1984, ISA was considered to be a uniquely Norwegian disease. In 1996, a disease in Canada (New Brunswick) designated as hemorrhagic kidney syndrome was verified as ISA. The disease has thereafter been reported in Scotland (1998), the Faroe Islands (2000), and in the USA (Maine, 2001).

While natural outbreaks of ISA have only been described in Atlantic salmon, the virus may survive and replicate in other salmonid fish under experimental conditions. ISAV has been detected in wild Atlantic salmon and brown trout (*Salmo trutta*). Outbreaks of ISA in farmed Atlantic salmon have occurred mainly during the seawater stage, but indications of disease outbreaks in the freshwater stage have been reported. Disease and transmission are readily induced experimentally in Atlantic salmon kept in either freshwater or seawater. Wild Atlantic salmon are susceptible to ISA and show the same clinical signs as farmed fish when experimentally infected. The virus has been isolated from apparently healthy rainbow trout (*Oncorhynchus mykiss*) in Ireland (2002). ISAV has also been reported to have been isolated from Coho salmon (*Oncorhynchus kitsutch*) in Chile, but this observation needs to be confirmed. Subclinically infected feral salmonids (Atlantic salmon, brown trout, and sea brown trout) have been identified in Scotland and Norway by reverse transcriptase-polymerase chain reaction (RT-PCR). However, reports of RT-PCR-positive samples from marine, nonsalmonid fish (Atlantic cod (*Gadus morhua*) and pollock (*Pollachius virens*)) need to be corroborated as samples were collected from the vicinity of cages holding ISA-diseased Atlantic salmon.

ISAV replication has been demonstrated in experimentally infected brown trout and in rainbow trout, but disease and mortality have only been produced in rainbow trout by experimental infection. ISAV has also been detected by RT-PCR in experimentally infected Arctic charr (*Salvelinus alpinus*) but neither virus replication nor clinical signs of disease have been demonstrated. ISAV could be reisolated from Pacific salmon species (*O. mykiss, Oncorhynchus keta, O. kitsutch, Oncorhynchus tshawytscha*) injected intraperitoneally with various Canadian and Norwegian virus isolates but no mortality was observed. Pacific salmon are therefore considered more resistant to ISA than Atlantic salmon but should not be ignored as potential virus carriers. Among marine fish, ISAV is able to propagate in herring (*Clupea harengus*) after bath challenge, but attempts to induce infection in pollock have failed. There are no indications that ISAV can infect mussel (*Mytilus edulis*) or scallops (*Pecten maximus*).

Transmission, Vectors and Reservoir Hosts

Epidemiological and experimental studies have shown that ISA may spread by water-borne transmission. The risk of ISA is closely linked to geographical proximity to farms with ISA outbreaks or to slaughterhouses and processing plants. The spread of the disease over long distances may be caused by transportation of infected fish by well boats.

The virus may be shed into the water by various routes such as skin, mucus, feces, and urine. The most likely route of virus entry is through the gills and skin lesions but transmission by coprophagy has also been proposed. ISAV may retain infectivity for long periods outside the host. No significant loss in virus titer was observed after incubation of virus supernatants for 14 days at 4 °C and 10 days at 15 °C, and infectivity of tissue preparations is retained for at least 48 h at 0 °C, 24 h at 10 °C, and 12 h at 15 °C. A 3-$\log_{10}$ reduction in virus titer after 4 months storage in sterile seawater at 4 °C has been observed, but ISAV survival time in natural seawater may be shorter.

There are indications that ISAV may be transmitted vertically. ISAV has been detected by real-time RT-PCR in fertilized eggs from ISA-diseased brood fish. However, it is not known if this represents infective virus as recovery of virus by isolation in cell culture was not performed. Eggs, fry, and juveniles from ISAV-infected parents have been demonstrated ISAV-positive by real-time RT-PCR. On the other hand, ISAV could not be detected in progeny from healthy but virus-positive brood fish. Furthermore, ISAV has been detected by real-time RT-PCR in Atlantic salmon parr and smolt sampled from hatcheries, that is, from the freshwater stage. However, there are no verified field observations that can confirm vertical transmission of ISA disease. Nevertheless, the detection of virus in progeny from infected brood fish and the few reported outbreaks in the freshwater stage indicate that this route of transmission cannot be excluded.

The sea louse (*Lepeophtheirus salmonis*) has been suggested as a possible vector for ISAV, but it is not clear if this is by passive transfer or active virus replication. Reservoir hosts of ISAV have not been identified, but the virus replicates in sea trout, which are abundant in Norway in fiords and coastal areas in the vicinity of the fish farms. However, the possible role of sea trout as a reservoir host for ISAV can only be speculated at this time.

Genetics

Nucleotide sequences of all eight ISAV genome segments have been described. The genome encodes at least 10 proteins (**Table 1**). Segments 1, 2, and 4 encode the viral polymerase subunits PB1, PB2, and PA, respectively. Segment 3 encodes the 68 kDa nucleoprotein (NP). Segments 5 and 6 encode the two major surface glycoproteins: the 50 kDa fusion (F) protein, and the 42 kDa HE responsible for receptor-binding and receptor-destroying activities. The two smallest ISAV genomic segments each contain two overlapping reading frames (ORFs). Two mRNAs are transcribed from genome segment 7 – one is collinear with the viral RNA (vRNA) and the other is spliced in an arrangement similar to the two smallest gene segments of influenza A virus. The unspliced mRNA of segment 7 encodes a nonstructural protein that has been suggested to interfere with the interferon type 1 response of the cell. The protein encoded by the spliced mRNA has not yet been characterized. No splicing of transcripts from segment 8 ORFs has been detected, indicating that ISAV uses a bicistronic coding strategy for this genomic segment. The smaller ORF1 encodes a 22 kDa matrix protein, while the collinear mRNA transcript from the larger ORF2 encodes an RNA binding structural protein of about 26 kDa with putative interferon antagonistic properties. ISAV gene segments and the respective encoded proteins are summarized in **Figure 2**.

Table 1 Genome segments and encoded proteins of ISA virus

Segment (kb)	Encoded protein	ORF (bp)	Protein (kDa)
1 (2.3)	Polymerase, PB2	2127	80.5[a]
2 (2.3)	Polymerase, PB1	2169	79.5[a]
3 (2.2)	Nucleoprotein, NP	1851	66–74[b]
4 (2.0)	Polymerase, PA	1840	65.3
5 (1.7)	Fusion, F	1332	53
6 (1.5)	Hemagglutinin-esterase, HE	1176[c]	38–46[b]
7 (1.3)	Two open reading frames		
	• ORF1 = nonstructural (NS)	903	34
	• ORF2 = spliced protein	369	17.5
8 (1.0)	Two open reading frames		
	• Matrix, M	588	22–24
	• as RNA-binding protein	726	26

[a]Estimations based on amino acid sequence.
[b]The estimated molecular masses of some of the proteins differ slightly in the literature, probably due to differences in experimental conditions. For HE, this could also be due to differences in glycosylation as well as in the highly polymorphic region (HPR).
[c]Length of ORF for the reference ISAV isolate Glesvaer/2/90. The length varies between isolates due to the HPR characterized by deletions/gaps. The proposed ancestral ORF sequence is estimated to 1236 bp.

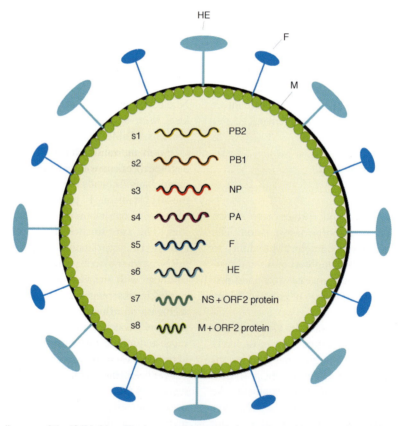

Figure 2 Schematic diagram of the ISAV virion. The two smallest genomic segments (segments 7 and 8) each contain two open reading frames (ORFs). The proteins encoded by each ORF is yet to be characterized. s, Genomic segment; PB2/PB1/PA, polymerase subunits; NP, nucleoprotein; F, fusion protein; HE, hemagglutinin-esterase; NS, nonstructural protein; M, matrix protein. Adapted from Ida Skaar.

The 3′- and 5′-terminal structure of vRNAs, and the ISAV transcription strategy, resemble those of influenza viruses. As for all members of the *Orthomyxoviridae*, each ISAV genome segment contains partially self-complementary termini that are essential for the replication process. However, compared to the 12–13 nucleotides conserved in the influenza A virus, only 8–9 nucleotides are conserved in ISAV, most likely reflecting the lower replication temperature optimum for this virus. As for the influenza viruses, the terminal 21–24 nucleotides are predicted to form self-complementary panhandle structures important for transcriptional regulation of viral RNA. ISAV also exerts cap-stealing in the host cell's nucleus, transferring 8–18 nucleotide 5′-cap structures from host mRNAs to the viral mRNAs and affecting the normal transcription and protein synthesis in the cell. ISAV mRNAs are polyadenylated, a process that is regulated from a signal 13–14 nucleotides downstream the 5′-terminus of the vRNAs. Altogether, the ISAV replication strategy is similar to those of the other orthomyxoviruses.

ISAV isolates vary in virulence, as observed by differences in disease development and clinical signs in field outbreaks and in experimental trials using genetically standardized fish. However, analysis of important virulence factors is still in its infancy. The ISAV genome is highly conserved, both over time and geographical distance, with 98–99% overall nucleotide similarity between Norwegian isolates. However, small, but relevant, differences between isolates both within and between defined geographical areas do exist, and analysis of the 5′-terminal end of the HE gene has demonstrated the presence of two major groups of isolates, one European and one North American. The European group may be further divided into three major subgroups.

The surface glycoprotein HE gene displays the highest sequence variation (94–97%) which is mainly concentrated in a small highly polymorphic region (HPR) of the HE protein that is predicted to lie immediately outside the viral envelope. Variation in this region is characterized by the presence of gaps rather than single-nucleotide substitutions. The polymorphism has been suggested to arise from differential deletions of a full-length precursor gene (HPR0) as a consequence of strong functional selection, possibly related to a recent or ongoing crossing of a species barrier. All HPRs described to date can be derived from such a full-length precursor sequence. Furthermore, the presence of a long HPR0 gene has been detected by RT-PCR in wild Atlantic salmon. The variation in HPR is most likely an important virulence factor as virus isolated from diseased fish always contains a deletion in this region.

Other genes are also most certainly important for virulence as isolates with identical HPR induce large differences in mortality in genetically standardized (major histocompatibility complex (MHC)-compatible) fish. Candidate genes include genome segment 5 encoding the surface glycoprotein with fusion activity and genome segment 7 and 8 encoding the putative IFN I antagonists in alternative reading frames. The synthetic IFN I inducer poly I:C induces no or only minor protection against ISAV infection in Atlantic salmon or cell culture, demonstrating the efficiency of the ISAV IFN I antagonistic properties. In orthomyxoviruses, reassortment of gene segments occurs frequently and is a major contributor to the evolution of these viruses and the emergence of new virulent strains. Alignments and phylogenetic studies of full-length sequenced ISAV isolates provide evidence for both genome segment reassortment and recombination. A 30 bp insert found in close proximity to the putative cleavage site of the fusion protein in several unrelated isolates combined with extensive sequence internal homology in this region suggests the presence of a recombinational hot-spot in this gene.

Pathogenesis

The mortality rate during an ISA outbreak may vary significantly. Variation in the seriousness in disease outbreaks may be influenced by the environmental and management parameters, genetics of the fish, and the virus strain.

In affected populations of farmed fish, individuals may harbor the virus for weeks or months before the development of disease. Prior to an outbreak, a slightly increased mortality over a period of 1–3 weeks is often seen. Outbreaks are often restricted initially to one or two net pens and up to 12 months can pass before clinical ISA spreads to neighboring pens in a farm. The signs exhibited by infected fish range from none to severe. The disease may appear throughout the year. The course of the disease may vary from an acute form with rapid development and high mortality, to a chronic form in which a slow increase in mortality is observed over several months, but several immediate forms may exist.

Experimentally, ISA can be induced in Atlantic salmon following intraperitoneal injection with tissue preparations or purified virus, or through infected cohabitants. The incubation period is usually 10–20 days. Studies indicate that the major portal of ISAV entry is the gills but oral entry cannot be excluded. Viral spread within the body is detectable from 5 days post-infection and cumulates at approximately 15 days. This is followed by a temporary decrease in viral load, reaching a minimum at around 25 days post-infection, followed by a second rise that continues to the terminal stage of ISA.

ISAV is found budding from endothelial cells that seem to be the main target cells. Their ubiquitous presence enables a systemic infection. By *in situ* hybridization, the most prominent ISAV-specific signals are detected in the endothelial cells of the heart. The presence of infectious virus has also been reported in leukocytes. ISAV specifically binds to glycoproteins containing 4-O-acetylated sialic acids. The viral esterase is specific for this sialic acid, indicating that it may be a receptor determinant for ISAV. ISAV infects cells via the endocytic pathway and fusion between virus and cell membrane takes place in the acidic environment of endosomes.

Clinical Features and Pathology

Fish infected with ISAV show a range of pathological changes, from none to severe. None of the described lesions is considered pathognomonic. Diseased fish appear lethargic and, in terminal stages, often sink to the bottom of the cage. The disease appears as a systemic condition characterized by severe anemia (hematocrit values below 10) and hemorrhages in several organs. The most prominent external signs are pale gills, exophthalmia, distended abdomen, and petechia in the eye chamber; skin hemorrhages in the abdomen, and scale edema. The major gross postmortem findings are circulatory disturbances in several organs caused by endothelial injury in peripheral blood vessels. These pathological manifestations of ISA are mainly recognized in the liver, kidney, gut, and gills, but not all of these organs are affected to the same extent during a disease outbreak. In some cases, the pathological manifestations are more clearly seen in one single organ than in others.

Abundant ascitic fluid is often present. A fibrinous layer may cover the liver capsule and the liver may be partly or diffusely dark red. The spleen and kidney may appear dark and swollen and the intestinal wall congested and dark. Petechial hemorrhages are seen on the surface of several organs as well as the adipose tissue and skeletal muscle. The major histopathological findings have been observed in the liver, kidneys, and intestine. The lesions result in extensive congestion of the liver with dilated sinusoids and, in later stages, the appearance of blood-filled spaces. Multifocal to confluent hemorrhages and/or necrosis of hepatocytes at some distance from large vessels in the liver are often seen. In the kidney, interstitial hemorrhage with tubular necrosis in the hemorrhagic areas, and accumulation of erythrocytes in the glomeruli may occur. Accumulation of erythrocytes in blood vessels of the intestinal lamina propria and hemorrhage into the lamina propria may be observed. In the spleen, accumulation of erythrocytes and distention of stroma may be found, and numerous erythrocytes may be present in central venous sinus and lamellar capillaries of the gills.

In ISA outbreaks on the American East Coast in the 1990s, histopathological changes were prominent in the kidneys. The major pathological findings included renal interstitial hemorrhage with tubular necrosis and casts, branchial lamellar and filamental congestion, and congestion of the intestine and pyloric cecae. The disease therefore initially became known as hemorrhagic kidney syndrome (HKS). The gross appearance of the fish was somewhat similar to that reported for ISA in Norway. However, liver congestion was a rare finding in fish from outbreaks in New Brunswick. HKS was later confirmed to be ISA by RT-PCR and cell culture isolation of the virus. Similar kidney and intestinal manifestations of the disease has now also been reported in Norway.

Diagnosis

Diagnosis of ISA was initially based on macroscopic, histological, and hematological findings. A dark liver was considered as a typical finding and the presence of multifocal, hemorrhagic liver necrosis with a 'zonal' appearance and hematocrit values below 10 confirmed the diagnosis. However, following the isolation of ISAV in SHK-1 cells, a number of methods for detection of virus in tissue samples were established.

Detection of ISAV antigens in kidney imprints from Atlantic salmon exhibiting clinical signs with the use of an indirect fluorescent antibody technique (IFAT) has been an important method for verification of ISA. The production of an anti-ISAV HE monoclonal antibody to be used as the primary antibody in the IFAT was an important step in the development of specific diagnostic tools. Polyclonal antibodies suitable for immunohistochemistry have also been developed, and this method has recently been implemented in the routine diagnostics of ISA.

ISAV was first isolated in the salmonid cell line SHK-1, but other salmonid cell lines such as CHSE-214, ASK, and TO also support viral propagation. Development of cytopathic effect (CPE) may be observed in infected cell cultures, but the extent of CPE may vary dependent on cell type and virus strain. The isolated virus may be identified using IFAT on infected cell cultures. Strain variations with respect to cell susceptibility have been observed. The Atlantic salmon cell lines SHK-1, TO, and ASK do not support growth of all ISAV variants as virus cannot be recovered from some RT-PCR-positive samples. The CHSE-214 cell line supports growth of some ISAV isolates, but others, such as the European variants of ISAV, do not replicate in this cell line at all. This limits the utility of the CHSE-14 cell line for virus isolation. Currently available fish cell lines appear to be either not sensitive enough or not permissive for all ISAV strains.

RT-PCR techniques have been widely in use since the RT-PCR for ISAV was described in 1997, and several different primer sets have been found to be suitable for detection of ISAV. Real-time RT-PCR has also been established for ISAV, allowing a further increase in specificity and sensitivity.

Immune Response, Prevention and Control

Due to lack of standardized molecular and experimental tools in fish, detailed knowledge of immune responses to viral agents is limited. This represents a major problem for the development of effective vaccines against fish viral diseases. For ISA, the ability to induce a strong lymphocyte proliferative response correlates with survival and virus clearance, while induction of a humoral response is less protective, as shown in experimental trials using MHC-compatible fish.

Currently available vaccines against ISA contain inactivated whole virus grown in cell culture added to mineral oil adjuvants. Vaccines are approved by the United States Department of Agriculture (USDA) and the Canadian Food Inspection Agency (CFIA) and are commercially available in Canada and the USA. Vaccination has also been used in the Faroe Islands. The use of vaccines against ISA in the rest of Europe is subject to EU/European Fair Trade Association approval. The present policy is that control of ISA should normally be based on a nonvaccination strategy. However, due to the severe disease problems in the Faroe Islands, exceptions to these rules have been made. The efficacy of ISAV vaccines has been evaluated using a vaccination-challenge model in the laboratory. Data on the evaluation of vaccines in field situations have so far not been presented. Currently available vaccines do not result in total clearance of virus in immunized fish.

The incidence of ISA may be greatly reduced by geneal husbandry practices and control of the movement of fish, mandatory health controls, and transport and slaughterhouse regulations. Specific measures including health restrictions on affected, suspected, and neighboring farms, enforced sanitary slaughtering, segregation of different generations of fish, as well as disinfection of offal and wastewater from fish slaughterhouses and fish-processing plants, also contribute to reduction in the incidence of the disease.

See also: Orthomyxoviruses: Molecular Biology; Orthomyxoviruses: Structure of antigens.

Further Reading

Falk K, Aspehaug V, Vlasak R, and Endresen C (2004) Identification and characterization of viral structural proteins of infectious salmon anemia virus. *Journal of Virology* 78: 3063–3071.

Falk K, Namork E, Rimstad E, Mjaaland S, and Dannevig BH (1997) Characterization of infectious salmon anaemia virus, an orthomyxo-like virus isolated from Atlantic salmon (*Salmo salar* L.). *Journal of Virology* 71: 9016–9023.

Koren CWR and Nylund A (1997) Morphology and morphogenesis of infectious salmon anaemia virus replicating in the endothelium of Atlantic salmon *Salmo salar*. *Diseases of Aquatic Organisms* 29: 99–109.

Krossoy B, Devold M, Sanders L, et al. (2001) Cloning and identification of the infectious salmon anaemia virus haemagglutinin. *Journal of General Virology* 82: 1757–1765.

Mjaaland S, Hungnes O, Teig A, et al. (2002) Polymorphism in the infectious salmon anemia virus hemagglutinin gene: Importance and possible implications for evolution and ecology of infectious salmon anemia disease. *Virology* 302: 379–391.

Mjaaland S, Rimstad E, and Cunningham C (2002) Molecular diagnosis of infectious salmon anaemia. In: Cunningham CO (ed.) *Molecular Diagnosis of Salmonid Diseases*, pp. 1–22. Boston: Kluwer Academic.

Mjaaland S, Rimstad E, Falk K, and Dannevig BH (1997) Genomic characterization of the virus causing infectious salmon anaemia in Atlantic salmon (*Salmo salar* L.): An orthomyxo-like virus in a teleost. *Journal of Virology* 71: 7681–7686.

Rimstad E and Mjaaland S (2002) Infectious salmon anaemia virus. An orthomyxovirus causing an emerging infection in Atlantic salmon. *Acta Pathologica Microbiologica et Immunologica Scandinavica* 110: 273–282.

Influenza

R A Lamb, Howard Hughes Medical Institute at Northwestern University, Evanston, IL, USA

© 2008 Elsevier Ltd. All rights reserved.

Glossary

Antigenic drift The gradual accumulation of amino acid changes in the surface glycoproteins of influenza viruses.

Antigenic shift The sudden appearance of antigenically novel surface glycoproteins in influenza A viruses.

Pandemic The rapid spread of an infection involving many countries.

Reassortment The exchange of genome RNA segments among influenza A viruses.

Recombinant virus A virus generated through reverse genetics techniques.

> **Reverse genetics** Techniques which allow the introduction of specific mutations into the genome of an RNA virus.
> **Subtype** Classification of influenza A virus strains according to the antigenicity of their HA and NA proteins.

Introduction: The Disease

The disease influenza is caused by influenza virus (family *Orthomyxoviridae*). Symptoms include fever, headache, cough, nasal congestion, sneezing, and whole-body aches. Influenza virus remains an important viral pathogen of significant medical importance causing mortality statistics comparable to human immunodeficiency virus (HIV) and producing considerable morbidity in the population. There were about 22 000 deaths from acquired immune deficiency syndrome (AIDS) in the US in 1993 compared with an estimated 20 000 influenza-associated deaths in the US in each of the epidemic years between 1972–73 and 2005–06. Influenza epidemics continue to infect large numbers of people worldwide, despite the availability of inactivated vaccines derived from the current circulating strains, because of frequent natural variation of the hemagglutinin (HA) and neuraminidase (NA) envelope proteins of the virus. This variation allows the virus to escape neutralization by preexisting circulating antibody in the blood stream, present as a result of either previous natural infection or immunization.

Taxonomy and Nomenclature

The *Orthomyxoviridae* comprises five genera: *influenzavirus A*; *influenzavirus B*; *influenzavirus C*; *thogotovirus* which are tick-borne viruses; and *isavirus* which includes infectious salmon anemia virus. Influenza A viruses are further classified into subtypes based on the antigenic properties of their surface spike glycoproteins, the HA, and NA. There are 16 known HA subtypes (H1–H16) and nine NA subtypes (N1–N9). Both influenza A and B virus strains cause disease, but influenza A virus strains, in most epidemic years, are usually more widespread. All human pandemics have been caused by influenza A virus strains. Retrospective seroepidemiology indicates that the 1889–91 pandemic was caused by an H3-like virus. The 1918–19 pandemic was caused by an H1N1 virus, the 1957 pandemic by an H2N2 virus, and the 1967–68 pandemic by an H3N2 virus. In 1977, there was a reintroduction in the human population of a co-circulating H1N1 virus, and since 2001 the reassortant virus H1N2 has been isolated. Since 1997 a highly pathogenic avian virus (H5N1) has infected millions of domestic fowl, and as of 12 March 2007 is known to have infected 278 humans resulting in 168 deaths.

Transmission and Tissue Tropism

Influenza viruses of humans and other mammals are spread by aerosols, including sneezing. The virus replicates in the cells of the upper and lower respiratory tract, reaching a peak at 2–3 days after infection, and the virus is usually cleared in 7 days. Children experiencing their first infection can shed virus up to 13 days.

Influenza viruses of birds are usually spread by fecal contamination of water but can also be spread by aerosols (e.g., high pathogenicity H5N1). Avian influenza viruses replicate in both the respiratory tract and the lower intestinal tract, hence the shedding of high concentrations of virus into feces.

Virus Isolation and Propagation

The first influenza virus was isolated from pigs in 1930 and the first human virus in 1933 (influenza A virus). Influenza B virus was isolated in 1940 and influenza C virus in 1946. Influenza viruses can either be propagated in tissue culture cells (particularly Madin–Darby canine kidney cells) or in the allantoic cavity of embryonated chicken eggs.

The Virus – Structure, Genome, and Proteins

The *Orthomyxoviridae* are enveloped viruses of 150–200 nm diameter. In the electron microscope, the shape of the viruses ranges from roughly spherical to pleomorphic, and filamentous viruses over a micron in length are observed from some hosts (**Figure 1**). The lipid envelope of influenza viruses is derived from the plasma membrane of the host cell in which the virus is grown. Influenza A and B virions are morphologically indistinguishable, whereas influenza C virions can be distinguished from the other genus as the glycoprotein spike is organized into orderly hexagonal arrays.

The influenza A, B, and C viruses can be distinguished on the basis of antigenic differences between their nucleocapsid (NP) and matrix (M) proteins. Influenza A viruses are further divided into subtypes based on the antigenic nature of their HA and NA glycoproteins. Other important characteristics that distinguish influenza A, B, and C viruses are given as follows.

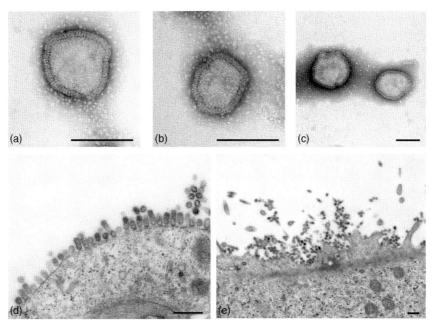

Figure 1 Electron micrographs of purified influenza virus virions (A/Udorn/72) (a–c) and A/WSN/33 virions budding from the surface of infected MDCK cells (d, e). Scale = 100 nm (a–c), 500 nm (d, e). Data provided by Dr. George Leser, Northwestern University, Evanston, IL, USA.

1. Influenza A viruses infect naturally a wide variety of avian species, humans, and several other mammalian species including swine and horses. Influenza B virus appears to naturally infect only humans, but influenza C virus has been isolated mainly from humans and also from swine in China.
2. The HA and NA of influenza A viruses exhibit much greater amino-acid-sequence variability than their counterparts in the influenza B viruses. Influenza C virus has only a single multifunctional glycoprotein, HA-esterase-fusion protein (HEF).
3. Although influenza A, B, and C viruses possess similar proteins, each virus type has distinct mechanisms for encoding proteins.
4. Influenza A and B viruses each contain eight distinct RNA segments whereas influenza C viruses contain seven RNA segments.

The most striking feature of the influenza A virion is a layer of about 500 spikes radiating outward (10–14 nm) from the lipid envelope. These spikes are of two types: rod-shaped spikes of HA and mushroom-shaped spikes of NA. The ratio of HA to NA varies but is usually 4–5 to 1. Influenza A, B, and C viruses also encode other integral membrane proteins, the M2 (influenza A virus), NB and BM2 (influenza B virus), and CM2 (influenza C virus) proteins, respectively. Biochemical evidence indicates that these proteins are only present in a few copies in virions. The viral matrix protein (M1) is thought to underlie the lipid bilayer and to associate with the cytoplasmic tails of the glycoproteins and the ribonucleoprotein (RNP) core of the virus.

Inside the virus, observable by thin sectioning of virus or by disrupting particles, are the RNP structures which can be separated into different size classes and contain eight different segments of single-stranded RNA. The RNPs have the appearance of flexible rods. The RNP strands often exhibit loops on one end and a periodicity of alternating major and minor grooves, suggesting that the structure is formed by a strand that is folded back on itself and then coiled on itself to form a type of twin-stranded helix. The RNPs consist of four protein species and RNA. NP is the predominant protein subunit of the nucleocapsid and coats the RNA, approximately 20 nucleotides per NP subunit. Associated with the RNPs is the RNA-dependent RNA polymerase complex consisting of the three P (polymerase) proteins – PB1, PB2, and PA – which are only present at 30–60 copies per virion. The NS2/NEP protein is present at 130–200 molecules per virion (**Figure 2**).

Influenza A and B viruses each contain eight segments of single-stranded RNA (for influenza A virus 2341 nucleotides to 890 nucleotides chain length), and influenza C viruses contain seven segments of single-stranded RNA (influenza C viruses lack an NA gene). The gene assignment for influenza A virus is as follows: RNA segment 1 codes for PB2, 2 for PB1 and in some strains also PB1-F2, 3 for PA, 4 for HA, 5 for NP, 6 for NA, 7 for M1 and M2, and 8 for NS1 and NS2/NEP. PB1, PB2, and PA form the RNA-dependent RNA polymerase and together with the NP protein form the RNPs. HA has receptor (cell surface expressed sialic acid (SA)) binding activity and HA through multivalent interactions attaches the virus to the surface of

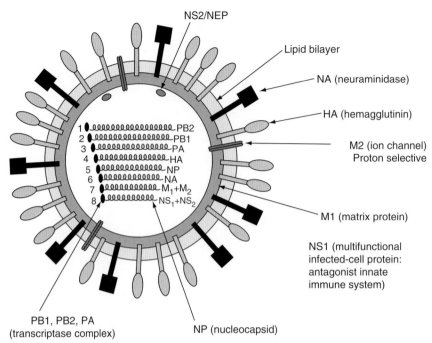

Figure 2 A schematic diagram of the structure of the influenza A virus particle. Three types of integral membrane protein – hemagglutinin (HA), neuraminidase (NA), and small amounts of the M2 ion channel protein – are inserted through the lipid bilayer of the viral membrane. The virion matrix protein M1 is thought to underlie the lipid bilayer and also to interact with the helical RNPs. The NS2/NEP protein associates with the M1 protein and is required for export of the RNPs out of the nucleus. Within the envelope are eight segments of single-stranded genome RNA (ranging from 2341 to 890 nucleotides) contained in the RNP. Associated with the RNPs are small amounts of the transcriptase complex, consisting of the proteins PB1, PB2, and PA. The coding assignment of the eight RNA segments are also illustrated. RNA segments 7 and 8 each code for more than one protein (M1 and M2, and NS1 and NS2/NEP, respectively). NS1 is found only in infected cells and is not thought to be a structural component of the virus. NS1 is a multifunctional protein and its functions include being an antagonist of the innate immune system. Adapted from Lamb RA and Krug RM (2001) *Orthomyxoviridae*: The viruses and their replication. In: Knipe DM and Howley PM (eds.) *Fields Virology*, 4th edn., pp. 1487–1531. Philadelphia, PA: Lippincott Williams and Wilkins.

a cell. HA also mediates the entry of the virus into cells by causing virus–cell membrane fusion. NA has neuraminidase activity that cleaves SA from complex carbohydrate molecules. Although seemingly an opposing activity to HA-binding SA, HA binding and NA activity work at different pHs and at different times in the virus life cycle. Thus, NA activity is the receptor-destroying activity necessary for virus release from cells. The viral matrix protein (M1) is thought to underlie the lipid bilayer and to associate with the cytoplasmic tails of the glycoprotein and the RNP core of the virus. The NS2 protein forms an association with the M1 protein, which is thought to be an essential interaction in the virus life cycle for export of the RNP complex from the nucleus. The NS1 protein is a multifunctional protein. It mediates the block of transport of host cell mRNAs from the nucleus, it interacts with phosphatidylinositol-3-kinase blocking phosphorylation of the downstream effector molecule Akt, and NS1 has major roles in defeating the innate immune system and limiting the production of interferons. PB1-F2 is thought to be involved in promoting apoptosis of host cells. The influenza A virus M2 protein and the influenza B virus BM2 protein are proton-selective ion channels that cause acidification of the interior of the virus particle during uncoating of the virion in endosomes. M2 protein is the target of the antiviral drug amantadine. The function of the influenza B virus NB protein is not known. NB is not essential for replication of the virus in tissue culture cells but it appears to confer a growth advantage in infections of mice. The influenza C virus CM2 protein has been suggested to be the counterpart of M2 and BM2 and to have ion channel activity, but this has not been shown rigorously (**Figure 3**).

Evolution of Influenza Virus

The finding that all known HA and NA subtypes are maintained in avian species together with phylogenic analysis of genome nucleotide sequences led to the hypothesis that all mammalian influenza A viruses are derived from the avian influenza virus pool. The evolutionary rates for avian virus both at the nucleotide and amino acid level are significantly lower than for human viruses, a finding which suggests that the influenza viruses

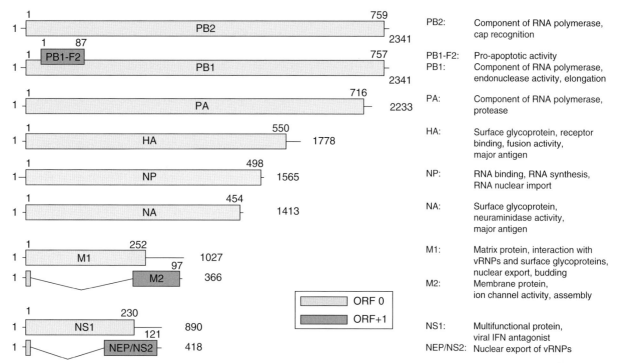

Figure 3 Genome structure of influenza A/Puerto Rico/8/34 virus. RNA segments (in nucleotides) shown in positive sense and their encoded proteins (in amino acids). The lines at the 5′ and 3′ termini represent the noncoding regions. The PB1 segment contains a second ORF in the +1 frame resulting in the PB1-F2 protein. The M2 and NEP/NS2 proteins are encoded by spliced mRNAs (the introns are indicated by the V-shaped lines). Adapted from Palese P and Shaw ML (2006) *Orthomyxoviridae*: The viruses and their replication. In: Knipe DM and Howley PM (eds.) *Fields Virology*, 5th edn., pp. 1647–1689. Philadelphia, PA: Lippincott Williams and Wilkins.

of aquatic birds are in evolutionary stasis and thus are optimally adapted to their hosts. Thus, although mutations occur, they do not lead to amino acid changes. In contrast, mammalian and domestic poultry viruses continue to accumulate amino acid changes. Among human influenza viruses, the genes accumulate mutations at different rates despite the RNA polymerase having a constant mutation rate, for example, HA is evolving faster than the genes for the internal proteins, probably because of selection pressure. The H3 HA gene has evolved in a single lineage since 1968 when it was first introduced into humans with a mutation rte of 4×10^{-3} substitutions per nucleotide per year and 5×10^{-3} substitutions per residues per year in HA1. Most of these amino acid changes occur in the antigenic epitopes (sites) and it is for this reason that the influenza virus vaccine has to be reformulated each year. In contrast to influenza A virus, type B and C viruses appear to be at or near an evolutionary equilibrium in humans.

Influenza Virus Genetics

Reassortment

Reassortment is the switching of viral RNA (gene) segments in cells infected with two different influenza viruses. The term recombination is often used incorrectly for the process of reassortment. Although reassortment occurs for influenza A, B, and C viruses, it does not occur among the different types. The 1957 and 1968 pandemic viruses have been determined to be reassortments between HA and NA (1957) and HA and PB1 (1967) of avian virus origin into a human virus genetic background, and the H5N1 viruses circulating between 1997 and 2007 arose from multiple reassortment events among avian influenza viruses.

Recombination

Recombination is a rare event caused by the virion-associated RNA polymerase switching templates. Recombination has led to the insertion of nonviral nucleotide sequences into the HA gene and to the generation of defective-interfering (DI) RNAs.

Reverse Genetics

Several systems are available for the generation of influenza viruses from cloned DNA. The process is highly efficient and novel viruses, including reconstruction of 1918 influenza virus, have been generated. The reverse genetic systems for influenza virus have permitted a detailed structure–function analysis of the viral proteins and their role in the virus life cycle, pathogenesis, and host range.

Influenza Pandemics

Pandemics are outbreaks that occur in very large geographic areas, usually involving more than one continent, and affecting large percentages of the population in a short period of time.

The Pandemic of 1918–19 – Spanish Influenza

The 1918–19 pandemic killed more than 25 million people worldwide and reduced life expectancy in the US by 10 years. Approximately one-third of the US population became sick and the mortality rate is estimated to be greater than 2.5%. The mortality pattern was different from that observed for other influenza virus pandemics. Usually the highest death rates are found in very young children and in the elderly. However, for 1918, a large number of deaths occurred in young adults (15–35 years old). Analysis of the reconstructed 1918 virus indicates that this virus is highly pathogenic for mice and nonhuman primates and the available data suggest that the 1918 virus causes a 'cytokine storm' leading to inflammatory cell infiltration and hemorrhage.

The Pandemics of 1957 (Asian Influenza) and 1968 (Hong Kong Influenza)

In 1957, a newly identified H2N2 influenza virus that is a reassortment between avian and human genes (see the section titled 'Reassortment') was identified as the causative agent of influenza. Although the ensuing pandemic was not extraordinarily pathogenic, the increased mortality (70 000 deaths in the US and over 1 million worldwide) is attributed to the lack of preexisting immunity for HA and NA among humans. In 1968 the H2N2 Asian influenza virus was completely replaced by an H3N2 virus and this virus was also a reassortment between avian and human viruses. This virus was moderate in its pathogenicity (in the US, 33 800 excess mortality) but the attack rate (40%) was highest in 10–14 year olds. Probably, preexisting antibodies to the N2NA moderated the disease in older humans.

The Re-Emergence of H1N1 Viruses in 1977 – Russian Influenza

In May 1977, an outbreak of influenza occurred on the Russian–Chinese border that spread rapidly in the former Soviet Union and China, and by 1978 reached the US. The virus was identified as an H1N1 virus that was very closely related to H1N1 viruses that circulated in the 1950s. It was suggested at the time, and it is still the prevailing view, that an accidental release of virus from a laboratory started the pandemic. In 2007, this H1N1 virus is still co-circulating with the H3N2 virus.

The H5N1 'Bird Flu' Outbreak

In Hong Kong, in May 1997 a 3-year-old boy died of influenza that was identified as being caused by an H5N1 virus that was entirely of avian origin. This was the first known occurrence of transmission of an avian virus to humans with a fatal outcome. In the fall of 1997, 17 additional cases were reported with five fatal outcomes. However, importantly, there was no evidence of human-to-human transmission. Culling of all poultry in Hong Kong's live bird markets is attributed to preventing further human infections. Nonetheless, the potential for avian influenza viruses transmitting to humans was not fully realized until these cases occurred. The observed mortality rate of 33% is atypical for influenza virus infections and this formed the basis for the worldwide concern that there may be a looming pandemic of grave consequences.

In July 2003 a new outbreak of H5N1 virus started in Thailand, Vietnam, and Indonesia. Since then the virus has spread over much of Asia, reached southeast Europe, and Africa, particularly Nigeria. The virus had led to the depopulation or death through disease of over 100 million poultry. These more recent H5N1 viruses have caused severe disease in limited numbers of humans. As of 12 March 2007, there have been 278 known human infections resulting in 168 deaths. These more recent H5N1 viruses cause systemic infections in humans, with virus being recovered from the stool and cerebrospinal fluid in addition to respiratory organs. Again, to date, there is no conclusive evidence for extensive human-to-human spread of disease.

Influenza Virus in Humans

Antigenic Drift

Antigenic drift is caused by point mutations and is defined as the minor gradual antigenic changes in the HA or NA protein. Influenza A virus drift variants result from the positive selection of spontaneously arising mutants by neutralizing antibodies, that is, antibody escape mutants. Mutations on the human virus HA or NA amino acid sequence occur at a frequency of less than 1% per year. Nonetheless, antigenic drift variants can cause epidemics and often prevail for 2–5 years before being replaced by a different variant.

HA is the major antigenic spike glycoprotein and from the atomic structure of prefusion HA and mapping of both natural virus HA variants and laboratory-derived monoclonal antibody escape mutants, it was possible to determine that HA possessed five antigenic sites (epitopes). These antigenic sites are all located in HA1 at or near the top of the molecule and mostly are found in protein loops. Similarly, antigenic drift has been found for NA and the sites of antigenic drift mapped to specifc regions of the NA atomic structure.

Antigenic Shift

Antigenic shift involves major antigenic changes and it occurs through the introduction, in a new human virus, of immunologically distinct HA and/or NA molecules. Antigenic shift leads to high infection rates in an immunologically naïve population and is the cause of influenza pandemics. As discussed above (reassortment) antigenic shift occurs through mixing of human and avian genes.

Influenza Virus in Animals

The natural host (reservoir) for influenza A viruses are aquatic birds, but influenza A viruses infect a wide variety of animals in addition to humans. These include land birds and poultry, swine, horses, dogs, cats, whales, and seals. In the hosts, apart from humans and poultry, influenza viruses usually cause asymptomatic infections and the viruses are in evolutionary stasis. Even H5N1 viruses are in evolutionary stasis in aquatic birds (with notable exceptions). However, on introduction into land-based poultry or mammalian species, they evolve rapidly.

In aquatic birds, influenza viruses replicate in the epithelial cells of the intestinal tract and these birds shed influenza viruses in high concentration in feces. Influenza viruses have been isolated routinely from lakes or ponds where migratory birds have congregated. In contrast, human influenza viruses replicate in the upper respiratory tract of ducks, but not in their intestinal tract. In part, this is thought to be due to the difference in linkage of receptor molecules used by avian and human viruses (SA linked to galactose via $\alpha 2,3$ vs. $\alpha 2,6$ linkages) and corresponding amino acids located in the receptor-binding sites on the tip of the HA molecule.

Avian influenza viruses are classified as highly pathogenic (HPAI) or low pathogenicity (LPAI) viruses. LPAI viruses cause mild respiratory disease, whereas HPAI viruses cause extensive mortality. Although multiple genes are involved in virulence, high pathogenicity is always associated with the HA molecule containing a series of basic residues at the cleavage site such that the HA molecules are cleaved to the two chains HA1 and HA2 intracellularly in the *trans*-Golgi network by the endogenous enzyme furin. In contrast, low pathogenicity viruses contain an HA molecule that contains a single basic residue in the HA cleavage site and HA is cleaved/activated by extracellular trypsin-like enzymes.

There have been many outbreaks of HPAI since 1955 and these have always involved the H5 or H7 subtypes. Although LPAI viruses had been thought to be innocuous, it is now clear that HPAI arises from LPAI of the same H5 and H7 subtypes.

Several studies have indicated a role for pigs in the emergence of pandemic influenza. This includes the finding that pigs can be naturally or experimentally infected with avian influenza viruses and that the pig trachea contains both avian ($\alpha 2,3$-linked SA) and human-type receptors ($\alpha 2,6$-linked SA). Replication of avian viruses in pigs leads to variants that prefer human-type receptors. Taken together, these finding suggest that pigs can host genetically diverse viruses and lead to the notion that pigs may be the 'mixing vessel', that is, pigs can be infected simultaneously with avian and human viruses which permits the generation of reassortant viruses that could cause pandemic influenza.

H7N7 and H3N8 viruses have historically been associated with equine influenza. The horse-racing industry has promoted the use of both killed and live-attenuated vaccines for these equine-tropic viruses.

In the laboratory, although not a natural host, mice are used extensively as an animal model. Except for mouse adapted strains (e.g., A/WSN/33 and A/PR/8/34) that have been passaged multiple times through mouse brains, most human viruses do not cause mouse lethality, but they do replicate in mice and cause transient weight loss. Interestingly, recently isolated H5N1 viruses are lethal for mice without prior adaptation. Most inbred strains of mice lack the interferon-induced Mx gene and thus although mice may be a convenient small animal model it is not a perfect model.

Ferrets are readily infected by influenza A or B viruses and they develop a febrile rhinitis. The long trachae of the ferret is thought to resemble the human trachea and pathological changes of bronchitis and pneumonia resemble those seen in humans. The drawback of ferrets is that they are much more difficult to handle than mice.

Both Old and New World primates can be infected with influenza viruses. Recently, cynomolgus macaques have been used as a model system for H5N1 viruses and found to develop acute respiratory distress syndrome with fever, analogous to the disease observed in H5N1-infected humans.

Molecular Determinants of Host Range Restriction and Pathogenesis

Three proteins, HA, PB2, and NS1, have been identified as major determinants in host range restriction and pathogenicity of influenza viruses.

The HA Protein

Cleavage of HA
HA is synthesized as a precursor polypeptide chain that for HA to be active in mediating virus-cell fusion, the precursor HA0 has to be cleaved at a specific site – the cleavage site – in two chains HA1 and HA2. The atomic structure of HA shows that HA1 contains the receptor-binding site and the sites of the antigenic epitopes and

HA2 is the domain of the protein-mediating membrane fusion. The cleavage site of HPAI H5 and H7 viruses contains mutiple basic residues and are recognized by proteases resident in the *trans*-Golgi network such as furin or PC6 and thus can be cleaved in multiple different organs. In contrast, HA of LPAI and human viruses contain a single arginine residue in the cleavage site that has to be cleaved by extracelullar trypsin-like proteases and is cleaved in only a few organs.

Receptor specificity

The HA of human viruses bind preferentially to SA that is linked to the penultimate galactose (Gal) residue by an $\alpha 2,6$-linkage, whereas most avian and equine viruses have a higher binding affinity for SA linked to galactose via an $\alpha 2,3$-linkage. Nonciliated epithelial cells express SA$\alpha 2,6$Gal oligosaccharides and are predominantly infected by human influenza virus. It has also been found that whereas SA$\alpha 2,6$Gal oligosaccharides are dominant on epithelial cells in nasal mucosa, trachea, and bronchi, SA$\alpha 2,3$Gal oligosaccharides are found on nonciliated bronchiolar cells at the junction between the respiratory bronchiole and alveolus and also on type cells lining the alveolar wall.

Receptor specificity is determined by the nature of the amino acids that form the receptor-binding pocket. The residues at positions 226 and 228 play an important role in receptor specificity for the H2 and H3 HAs and receptor specificity can be switched by mutations of these amino acid residues. For H1 HAs receptor specificity is determined in large part by the nature of the residues found at position 190 (H3 residue numbering).

The NS1 Protein

In addition to affecting the transport of host mRNAs out of the nucleus, NS1 functions as an interferon (INF) antagonist. Probably, all viruses need a system to defeat INF action in INF-competent hosts. The NS1 protein targets both IFN β production and the activation of INF-induced genes that are required for establishing the antiviral state. Although there are 200–300 IFN-stimulated genes, the biological properties of the encoded proteins are known only for a few, for example, dsRNA-activated protein kinase (PKR), Mx proteins, $2',5'$ oligoadenylate synthetase and RNAase L. NS1 interferes with PKR and RNase L.

NS1 protein from different influenza viruses appears to have different abilities to counteract the IFN system and thus this affects pathogenicity of the virus. Viruses that defeat the IFN system well are more pathogenic than those that only cause partial inactivation of the IFN system. Several findings suggest that the NS1 protein of highly pathogenic viruses may cause the cytokine imbalance and that has been observed in patients infected with H5N1 virus or macaques infected with 1918 H1N1 virus.

The PB2 Protein

The PB2 protein is part of the RNA-dependent RNA polymerase complex. PB2 binds to ^{7}MeGpppG cap structures on host mRNAs; the cap structure and 10–13 nucleotides of the host mRNA are cannibalized by the influenza virus RNA polymerase to act as primers for mRNA transcription. The PB2 protein and the residue found at position 627 (glutamic acid in LPAI and lysine in HPAI) have emerged as important determinants of virulence. How the mechanism by which these residue changes affect the polymerase activity is not fully elucidated. However, it appears that viruses with lysine at position 627 grow faster than those with glutamic acid at this position.

The NA Protein

HA of the A/WSN/33 human influenza virus contains a single arginine residue at the HA cleavage site; yet, for plaque formation in tissue culture unlike other human influenza virus strains, WSN does not require trypsin for multiple rounds of virus replication and plaque formation. However, the ability of A/WSN/33 to form plaques without trypsin requires the A/WSN/NA gene. It was determined that the WSN NA protein has lost a carbohydrate chain as compared to other N1NA proteins and that the lack of the carbohydrate chain permits the WSN NA to bind plasminogen, the precursor to the protease plasmin. It is envisaged that the tethered plasmin cleaves HA.

Antivirals

Amantadine and Rimantadine: M2 Ion Channel Blockers

Amantadine, and its methyl derivative rimantadine, were discovered as anti-influenza viral drugs in the 1960s by workers at DuPont. Rimantadine is licensed in the US for both the treatment and prophylaxis of influenza and at one time was being used extensively for prophylaxis in nursing homes. The drugs target the pore (transmembrane domain) of the M2 proton-selective ion channel. A drawback to the use of rimantadine is that drug-resistant variant viruses arise in humans within 4 days of drug treatment. Quite remarkably in 2005–06, the circulating H3N2 viruses were almost all rimantadine resistant despite little use of rimantadine. It has been postulated that the rimantadine-resistance mutation in M2 piggy-backed with another change in HA which conferred viral fitness.

NA Inhibitors

Two neuraminidase inhibitors are licensed in the US for the treatment and prophylaxis of influenza. Zanamivir is a derivative of the transition state intermediate 2-deoxy-2, 3-dihydro-*N*-acetylneuraminic acid and the drug binds in the NA active site and has a K_i of 2×10^{-10}. The drug has to be administered intranasally or inhaled. Oseltamivir is another derivative of the transition state intermediate 2-deoxy-2,3-dihydro-N-acetylneuraminic acid and it contains a lipophilic group, which improves its bioavailability, and the drug can be taken orally. Although when used for treatment of influenza these drugs only reduce the duration of illness by a day or two, this may be highly significant (beneficial) in infections of high pathogenicity viruses.

Resistant variants to the NA inhibitors have been selected both *in vitro* and *in vivo* but at a much lower frequency than for amantadine-resistant variants. In part, this is because the drugs interact with several residues in the active site and each of these residues is important for enzyme activity. Thus, mutation of a residue in the active site leads to a lowered K_i and a lowered enzyme activity. Interestingly, emergence of resistance is first detected by finding mutations in the SA-binding site of HA, which compensates for reduced NA activity; the affinity of HA for its receptor is lowered. Later mutations occur in NA in the framework residues that surround the active site and stabilize its structure.

Vaccines

Inactivated Vaccines

Inactivated influenza A and B virus vaccines are licensed for administration in humans. The vaccine is reformulated each year to include the strains thought most likely to be prevailing. The choice of seed virus is made by the World Health Organization (WHO) and the US Centers for Disease Control (CDC). The vaccine virus is currently grown in embryonated eggs but because many human influenza virus isolates do not grow to high yield in eggs, a reassortment virus is made using the high egg-yielding PR/8/34 genetic backbone and the HA and NA genes of the candidate virus. Reverse genetics procedures speed up the time needed to produce a high-yielding reassortment over the former method of mixed infection and selecting the virus from random plaques. Vaccine is manufactured by harvesting allantoic fluid and concentrating the virus by zonal centrifugation and inactivation of infectivity with formalin or beta-propiolactone. One egg yields one to three doses of vaccine. The vaccine is administered intramuscularly. Vaccination with inactivated virus has been shown to consistently confer resistance to illness (reduced frequency and severity of disease), and to a somewhat lesser extent infection with influenza A and B viruses.

Live Virus Vaccines

A live attenuated vaccine (Flumist) has been licensed. The vaccine virus is based on the genetic backbone of a cold-adapted virus (A/Ann Arbor/6/60) and a reassortment made to incorporate the current HA and NA genes. The cold-adapted virus replicates efficiently in the nasopharynx to induce protective immunity. However, replication is restricted at higher temperatures, including those present in the lower airways and lungs. In clinical studies with matched strains, the live virus vaccine demonstrated 87% efficacy in children and 85% in adults. There is continued interest in the use of a live influenza virus vaccine because infection of the respiratory tract stimulates both systemic and local immunity, and in principle should stimulate cell-mediated immunity. Thus, all components of the human immune response are brought into action. Furthermore, there is the added advantage of acceptance of a nasal spray rather than a needle injection in young children.

See also: Antigenic Variation; Antiviral Agents; Assembly of Viruses: Enveloped Particles; Defective-Interfering Viruses; Diagnostic Techniques: Microarrays; Electron Microscopy of Viruses; Emerging and Reemerging Virus Diseases of Vertebrates; History of Virology: Vertebrate Viruses; Immune Response to viruses: Antibody-Mediated Immunity; Innate Immunity: Defeating; Innate Immunity: Introduction; Membrane Fusion; Orthomyxoviruses: Molecular Biology; Orthomyxoviruses: Structure of antigens; Replication of Viruses; Vaccine Strategies; Viral Membranes; Viral Pathogenesis; Viral Receptors; Viruses and Bioterrorism; Zoonoses.

Further Reading

Fouchier RA, Schneeberger PM, Rozendaal FM, *et al.* (2004) Avian influenza A virus (H7N7) associated with human conjunctivitis and a fatal case of acute respiratory distress syndrome. *Proceedings of the National Academy of Sciences, USA* 101: 1356–1361.

Gamblin SJ, Haire LF, Russell RJ, *et al.* (2004) The structure and receptor binding properties of the 1918 influenza hemagglutinin. *Science* 303: 1838–1842.

Garcia-Sastre A (2004) Identification and characterization of viral antagonists of type 1 interferon in negative-strand RNA viruses. *Current Topics in Microbiology and Immunology* 283: 249–280.

Goto H and Kawaoka Y (1998) A novel mechanism for the acquisition of virulence by a human influenza A virus. *Proceedings of the National Academy of Sciences, USA* 95: 10224–10228.

Guan Y, Poon LL, Cheung CY, *et al.* (2004) H5N1 influenza: A protean pandemic threat. *Proceedings of the National Academy of Sciences, USA* 101: 8156–8161.

Gubareva LV, Kaisr L, and Hayden FG (2000) Influenza virus neuraminidase inhibitors. *Lancet* 355: 827–836.

Kawaoka Y and Webster RG (1988) Sequence requirements for cleavage activation of influenza virus hemagglutinin expressed in mammalian cells. *Proceedings of the National Academy of Sciences, USA* 85: 324–328.

Lamb RA, Holsinger LJ, and Pinto LH (1994) The influenza A virus M_2 ion channel protein and its role in the influenza virus life cycle. In: Wimmer E (ed.) *Receptor-Mediated Virus Entry into Cells*, pp. 303–321. Cold Spring Harbor, NY: Cold Spring Harbor Press.

Lamb RA and Krug RM (2001) Orthomyxoviridae: The viruses and their replication. In: Knipe DM and Howley PM (eds) *Fields Virology*,

4th edn., pp. 1487–1531. Philadelphia, PA: Lippincott Williams and Wilkins.

Mastrosovich MN, Mastrosovich TY, Gray T, *et al.* (2004) Human and avian influenza viruses target different cell types in cultures of human airway epithelium. *Proceedings of the National Academy of Sciences, USA* 101: 4620–4624.

Neumann G, Watanabe T, Ito H, *et al.* (1999) Generation of influenza A viruses entirely from cloned cDNAs. *Proceedings of the National Academy of Sciences, USA* 96: 9345–9350.

Palese P and Shaw ML (2006) *Orthomyxoviridae*: The viruses and their replication. In: Knipe DM and Howley PM (eds.) *Fields Virology*, 5th edn, pp. 1647–1689. Philadelphia, PA: Lippincott Williams and Wilkins.

Skehel JJ and Wiley DC (2000) Receptor binding and membrane fusion in virus entry: The influenza hemagglutinin. *Annual Review of Biochemistry* 69: 531–569.

Tumpey TM, Basler CF, Aguilar PV, *et al.* (2005) Characterization of the reconstructed 1918 Spanish influenza pandemic virus. *Science* 310: 77–80.

Varghese JN, Laver WG, and Colman PM (1983) Structure of the influenza virus glycoprotein antigen neuramindase at 2.9 A resolution. *Nature* 303: 35–40.

Wright PF, Neumann G, and Kawaoka Y (2006) Orthomyxoviruses. In: Knipe DM and Howley PM (eds.) *Fields Virology*, 5th edn, pp. 1691–1740. Philadelphia, PA: Lippincott Williams and Wilkins.

Innate Immunity: Defeating

C F Basler, Mount Sinai School of Medicine, New York, NY, USA

© 2008 Elsevier Ltd. All rights reserved.

Glossary

Complement A system of serum proteins that function to promote the phagocytosis or lysis of pathogens or pathogen-infected cells.

Conventional dendritic cell (cDC) A phagocytic cell type that serves as a sentinel for the immune system and that links innate and adaptive immune responses. Upon encountering an antigen, cDCs can be induced to undergo maturation, such that they produce cytokines, upregulate to their surface co-stimulatory molecules, and promote T-cell activation.

Cytokine A secreted protein that promotes or modulates immune responses.

Interferon α/β (IFN-α/β) A family of structurally related cytokines the expression of which can be induced by virus infection and other stimuli. IFN-α/β proteins bind to the IFN-α/β receptor activating a signaling pathway that modulates the expression of hundreds of genes and induces in cells an antiviral state.

Natural killer (NK) cell A lymphocyte that can kill target cells in a non-antigen-specific manner. Killing by NK cells is tightly regulated by positive and negative signals.

Opsonization The coating of a particle so as to promote its phagocytosis.

Pathogen-associated molecular patterns (PAMPs) Molecular patterns found on pathogens that are recognized as 'foreign' and that trigger innate immune responses.

Pattern recognition receptor (PRR) A host molecule that recognizes PAMPs and signals to stimulate innate immune responses.

Toll-like receptor (TLR) A member of a family of type I transmembrane proteins that serve as pattern-recognition receptors. TLRs signal through association with cytoplasmic adaptor proteins.

Innate Immunity

Immune responses have evolved, in part, to eliminate or to contain infectious agents such as viruses. Innate immunity consists of a variety of relatively nonspecific and short-lived responses typically triggered soon after the appearance of an antigen (such as a virus or other pathogen). Innate immune mechanisms can be contrasted with adaptive immune responses in which responses are antigen specific and are characterized by immunological memory.

Innate immunity can be mediated by proteins, for example, interferons (IFNs), other cytokines, or complement; and by specific cell types such as macrophages, dendritic cells (DCs), and natural killer (NK) cells. These innate responses act rapidly to suppress infection. Given that viruses have co-evolved with their hosts, it is perhaps not surprising that viruses have evolved ways to defeat innate immune responses. For virtually any effector of the innate immune response, an example can be found whereby a virus overcomes or counteracts this host response. The strategies employed by viruses to evade these responses will vary. In general, however, viruses with large genomes, such as are found in many DNA viruses, may produce 'accessory proteins' that specifically carry out immune-evasion functions. In contrast, viruses with small genomes, such as are common among many RNA viruses, often encode multifunctional proteins. These may carry out both immune-evasion functions and functions essential for virus replication. However, regardless of the general strategy employed, suppression of innate immunity is generally critical for viral pathogenesis.

The IFN-α/β System

The IFN-α/β response serves as a major component of the innate immune response to virus infection, can be activated in most cell types, and triggers in cells an 'antiviral state'. It should be noted that there is also an IFN-γ which is structurally distinct from IFN-α/β and which is produced mainly by immune cells. Despite these differences, IFN-γ can activate signaling pathways similar to those activated by IFN-α/β and can also induce in cells an antiviral state. The IFN-α/β system can be activated in most cell types and can be viewed, in simplified terms, in two phases (**Figure 1**). The first is an induction phase in which virus infection or another stimulus activates latent transcription factors. These induce expression of the IFN-α/β genes which encode multiple IFN-α proteins and, in humans, a single IFN-β protein. In the second phase, the secreted, structurally related IFN-α/β proteins bind to the IFN-α/β receptor. This activates a Jak-STAT signaling pathway that induces expression of numerous genes, some of which have antiviral properties. There is only a single IFN-β gene but multiple IFN-α genes, and, in most cell types, it is primarily IFN-β that is produced after initial virus infection. However, in a positive-feedback loop, IFN-α/β

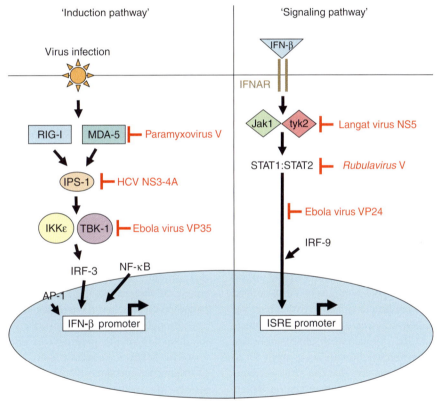

Figure 1 A simplified, schematic diagram of the IFN-α/β system and examples of viral proteins that disrupt the IFN-α/β response. Here the IFN-α/β system is depicted in two phases, each subject to virus intervention. First, the 'induction pathway' is activated by virus infection. Products of the infection activate signaling pathways, either through cytoplasmic sensors of virus infection such as retinoic acid inducible gene I protein (RIG-I) or melanoma differentiation antigen 5 protein (MDA-5) or through select Toll-like receptor (not shown). Signaling through RIG-I or MDA-5 requires IPS-1 and results in activation of the interferon regulatory factor 3 (IRF-3) kinases IKKε and TBK-1. These signaling pathways trigger activation of the transcription factors IRF-3, nuclear factor-kappa B (NF-κB), and AP-1. These cooperate to induce IFN-β gene expression. Examples of viral proteins that target the induction phase include paramyxovirus V proteins that bind and inhibit MDA-5, the hepatitis C virus (HCV) NS3-4A protease that cleaves IPS-1, and the Ebola virus VP35 protein that may act at the level of the IRF-3 kinases. Second, in the 'signaling pathway', secreted IFN-α/β binds to the IFN-α/β receptor, activating a Jak-STAT signaling cascade, resulting in the expression of numerous genes. Examples of viral proteins that target the IFN signaling pathways are the NS5 protein of Langat virus that prevents Jak1 and tyk2 kinase activation, the V proteins of the paramyxovirus genus *Rubulavirus*, which target STAT proteins for proteasome-dependent degradation, and the Ebola virus VP24 protein that prevents nuclear accumulation of activated STAT1. A third aspect of the response is the production of IFN-induced genes that encode antiviral gene products (not shown). Viral inhibitors of specific IFN-induced antiviral effector proteins also exist. Examples of inhibitors described in the text are depicted in red font in the figure.

can prime cells to produce numerous IFN-α types and larger overall amounts of IFN-α/β. The consequence is an amplification of the antiviral response.

Pathogen-Associated Molecular Patterns and Pattern-Recognition Receptors

Innate immune mechanisms, including IFN-α/β responses, may be triggered by diverse stimuli. Therefore, mechanisms have evolved to recognize as foreign structures and molecules that are conserved among groups of pathogens but distinct from host molecules. The pathogen structures and molecules recognized by the host are generically referred to as pathogen-associated molecular patterns (PAMPs), while the host molecules that recognize such structures are referred to as pattern-recognition receptors (PRRs). PRRs important for detection of virus infection include select members of the Toll-like receptor (TLR) family of transmembrane PRRs, including TLR3, TLR4, TLR7, TLR8, and TLR9, and particular cellular RNA helicases including the retinoic acid inducible gene I protein (RIG-I) and the melanoma differentiation antigen 5 protein (MDA-5).

TLRs are expressed primarily on cells, such as DCs and macrophages, which function to trigger innate immune responses and to act as antigen-presenting cells, initiating adaptive immune responses. Several TLRs have the capacity to recognize viral nucleic acids. TLR3 recognizes double-stranded RNA (dsRNA), such as polyI:polyC experimentally added to cells or dsRNA produced during the course of virus infection. TLR7 and TLR8 can be activated by virus-derived single-stranded RNAs (ssRNAs) rich in guanosine or uridine, such as those derived from influenza virus or human immunodeficiency virus (HIV). TLR9, in contrast recognizes dsDNA; and TLR9 signaling can be stimulated by CpG DNA motifs or by virus-derived dsDNA such as that from herpes simplex virus. There are also examples of other, non-nucleic acid, viral products signaling through TLRs, although such observations are often viewed as controversial because it is difficult to completely exclude the possibility that the virus preparations used contain some low level contamination with a bacterial TLR ligand. Examples of viral protein recognition by TLRs include the recognition of respiratory syncytial virus (RSV) fusion (F) protein by TLR4 and measles virus hemagglutinin protein activation of TLR2 signaling.

In general, activation of TLR signaling would seem to be a disadvantage for a virus, as it will promote innate antiviral responses. However, in some cases, viruses may take advantage of the innate response. For example, mouse mammary tumor virus (MMTV) infects B cells early during the course of infection *in vivo*. This virus can also activate TLR4 signaling. Because replication of this retrovirus requires that its host cell be dividing, the activation of B cells via TLR4 may facilitate establishment of virus infection. Similarly, activation of DCs, another early target of MMTV infection, via TLR4 signaling increases DC numbers and increases expression of CD71 the viral receptor on these cells, again promoting the infection process.

Different TLRs have different distributions among cell types, and the consequences of TLR activation will vary, depending both upon the specific TLR in question as well as the cell type on which it is activated. For example, conventional dendritic cells (cDCs) function as sentinels which, in an immature state, constitutively sample their environment for foreign antigen. Upon encountering antigen and activation of PRRs, the cDCs mature and become potent activators of T-cell responses. Of the IFN-α/β-inducing TLRs, human cDCs express TLRs 3 and 4. Their activation results in IFN-β production and the maturation of the cDC, thereby providing a link between the relatively nonspecific innate response to infection and the antigen-specific adaptive immunity of T-cell responses. In contrast, human plasmacytoid DCs (pDCs) express TLRs 7, 8, and 9, and activation of pDCs results in the production of copious amounts of IFN-α.

Many cell types do not express TLRs. However, such cells typically have the capacity to mount an IFN-α/β response to virus infection. In such cells, intracellular PRRs detect and signal in response to virus infection. Two such intracellular PRRs are RIG-I and MDA-5. RIG-I binds to and is activated by RNA molecules possessing 5' triphosphate groups, although activation of signaling may be influence by other properties of the RNAs such as secondary structure or modifications to the RNAs. MDA-5 can be activated by intracellular dsRNA, such as transfected polyI:polyC. These properties appear to fit with data obtained from mice in which either RIG-I or MDA-5 were knocked out. Thus, viruses that produce RNAs with 5' triphosphates during the course of their replication, including influenza viruses, paramyxoviruses, and flaviviruses, trigger IFN-β production through RIG-I. In contrast, RIG-I was not essential for detection of the picornavirus encephalomyocarditis virus in MDA-5 knockout cells, presumably because picornaviruses produce RNAs where the 5' end is occupied by the covalently linked viral protein, VPg. MDA-5, in this case, probably activates IFN-α/β production due to recognition of viral dsRNA. It is notable however that expression from viruses of inhibitors of RIG-I or MDA-5 might influence the outcome of such analyses. For example, multiple paramyxoviruses have been found to encode 'V' proteins (discussed further below) that inhibit MDA-5 function. Thus, it is possible that paramyxovirus infection yields products that activate not only RIG-I but also MDA-5. However, because the presence of the V proteins blocks MDA-5 function, it may appear that paramyxoviruses can only activate RIG-I. The evolutionary pressure to retain an inhibitor of MDA-5 may in fact suggest a role for detection of negative-strand RNA viruses by MDA-5.

Signaling Leading to IFN-α/β Production

The cellular molecules immediately downstream of TLRs differ, depending upon which TLR is engaged. Similarly, RIG-I and MDA-5 share common downstream signaling molecules but these differ to some extent from the TLR-activated pathways.

TLR signaling is mediated by specific downstream 'TIR-domain'-containing adaptor proteins. TLR3 signaling to induce IFN-α/β expression appears to require the adapter TRIF, while TLR4 signaling appears to require TRIF and a second adaptor, TRAM. For the RIG-I and MDA-5 helicases, recognition of the appropriate nucleic acid molecule is thought to induce structural rearrangements in the helicase, exposing an otherwise repressed caspase recruitment domain (CARD, a protein:protein interaction domain), allowing the helicase to interact with downstream signaling molecules. A mitochondria-localized protein IPS-1 (also called MAVS, Cardif, or VISA) lies downstream of RIG-I and MDA-5; it is essential for the signaling from RIG-I and MDA-5. Each of these pathways leads to the activation of the cellular kinases IKKε and TBK-1 which phosphorylate interferon regulatory factor 3 (IRF-3) and also activate nuclear factor-kappa B (NF-κB), transcription factors critical for IFN-β gene expression. In contrast, TLRs 7, 8, and 9 utilize signaling pathways that are dependent upon the adapter MyD88, which involve activation of IRF-7 and which leads to the production of IFN-α.

IRF-3 and IRF-7

In most cell types, IFN-β is the primary form of IFN that is first produced. The IFN-β promoter transcription occurs following the coordinated activation of IRF-3, NF-κB, and the ATF-2/c-Jun form of AP1. The initial production of IFN-β, as well as select IFN-α genes, induces expression of a similar IRF family member, IRF-7. IRF-7, like IRF-3, is activated by phosphorylation in response to virus infection. However, unlike IRF-3, IRF-7 is able to participate in the activation of numerous IFN-α genes, thus providing a mechanism by which IFN-α/β production can be greatly amplified in response to infection. In contrast to this model, in pDCs, IRF-7 appears to be constitutively expressed and this contributes to the copious production of IFN-α by these cells in response to TLR 7, 8, or 9 agonists.

IFN Signaling

IFN-α/β and IFN-γ bind to two distinct receptors but activate similar signaling pathways. For both pathways, ligand binding activates receptor-associated Jak family tyrosine kinases. These undergo auto- and transphosphorylation and phosphorylate the cytoplasmic domains of the receptor subunits. The receptor-associated phosphotyrosine residues then serve as docking sites for the SH2 domains of STAT proteins. The receptor-associated STATs then undergo tyrosine phosphorylation and form homo- or heterodimers via reciprocal SH2 domain–phosphotyrosine interactions. Signaling from the IFN-α/β receptor results predominately in the formation of STAT1:STAT2 heterodimers which additionally interact with IRF-9. IFN-γ signaling results predominately in the formation of STAT1:STAT1 homodimers. Upon dimerization, the STAT1:STAT2 heterodimer or the STAT1:STAT1 homodimer interacts with a specific member of the karyopherin α (also known as importin α) family of nuclear localization signal receptors, karyopherin α1 (importin α5). This interaction with karyopherin α1 mediates the nuclear accumulation of these STAT1-containing complexes. The consequence of the activation and nuclear accumulation of these complexes is the specific transcriptional regulation of numerous genes, some of which have antiviral properties.

IFN-Induced Genes Encoding Antiviral Proteins

Treatment of cells with interferon induces expression of more than 100 genes and induces in cells an 'antiviral state', meaning that the cells become resistant to virus infection and/or replication. Our understanding of the IFN-induced antiviral state is very incomplete. However, the products of several IFN-induced genes have been demonstrated to exert antiviral properties. Examples include the dsRNA-activated protein kinase, PKR, which upon activation, phosphorylates the translation initiation factor eukaryotic initiation factor 2α (eIF2α). This leads to a general inhibition of translation which can suppress virus replication. Similarly, 2′-5′ oligoadenylate synthetases (OASs) are a family of IFN-induced, dsRNA-activated enzymes that catalyze formation of 2′-5′ oligoadenylates. These activate RNase L, an enzyme that destroys mRNAs and cellular RNAs and suppresses translation of proteins. Mx proteins, including Mx1 in mice or MxA in humans, belong to the dynamin family of GTPases. Their expression is IFN-inducible, and members of this family have been shown to inhibit replication of specific virus families. For example, human MxA can inhibit replication of bunyaviruses, orthomyxoviruses, paramyxoviruses, rhabdoviruses, togaviruses, picornaviruses, and hepatitis B virus. At least part of the antiviral effect of these proteins is exerted by their ability to alter the intracellular trafficking of viral proteins; for example, human MxA can sequester orthomyxovirus nucleocapsids in the cytoplasm, preventing entry into the nucleus where virus RNA synthesis takes place.

Viral Evasion of the Host IFN Response

It is likely that most, if not all, viruses successfully maintained in nature have evolved mechanisms to counteract IFN responses.

Examples of the mechanisms by which viruses defeat the IFN system include inactivation of the PRR signaling pathways, inactivation of the IFN-induced signaling pathways, and the targeting of IFN-induced antiviral 'effectors'.

Targeting Components of the PRR Signaling Pathways

Numerous examples now exist of viruses targeting the signaling pathways activated by TLRs, RIG-I, or MDA-5. Mechanisms range from targeting of the PRRs themselves, to the targeting of the transcription factors that these pathways activate, to the targeting of kinases that activate these transcription factors. The RIG-I pathway is targeted by several viruses, including Ebola virus and hepatitis A virus. In poliovirus-infected cells, the cytoplasmic PRR MDA-5 is cleaved in a caspase- and proteosome-dependent manner, presumably due to poliovirus activation of apoptosis. This observation is consistent with the idea that MDA-5 is particularly relevant to picornavirus infection. Hepatitis C virus serves as an example of a virus that can target both a TLR (TLR3) and intracellular PRR (RIG-I) pathways. HCV is an enveloped, positive-strand RNA virus of the flavivirus family associated with persistent liver infections, cirrhosis, and liver cancer. The HCV genome encodes NS3-4A, a noncovalent complex with protease and RNA helicase activity. NS3-4A has been found to inhibit, in a protease-dependent manner, virus-induced activation of IRF-3, which, as noted above, plays a major role in activating IFN-α/β responses. TLR3 signaling requires the function of the adaptor TRIF, and the protease activity of NS3-4A cleaves and inactivates TRIF, to prevent TLR3-induced IFN-α/β expression. NS3-4A was also found to cleave IPS-1, the mitochondria-localized adaptor downstream of both RIG-I and MDA-5, similarly preventing RIG-I-induced IFN-α/β expression.

Targeting Components of the IFN Signaling Pathways

Numerous examples also exist of virus-mediated inhibition of the IFN-activated Jak-STAT signaling cascades. There are many examples among the members of the paramyxovirus family itself. It should be noted that several of these also inhibit IFN-α/β production and might therefore also be classified as proteins that target multiple aspects of the IFN system (see the section titled 'Viruses targeting multiple aspects of the IFN system'). Viruses of the genus *Rubulavirus* of the paramyxovirus family encode V proteins which target STAT proteins for ubiquitin-dependent degradation. However, the specific STAT targeted by a particular V protein will vary. For example, the simian virus 5 (SV5) V protein targets STAT1 while human parainfluenza virus type 2 (HPIV2) encodes a V protein that promotes STAT2 degradation. In these cases, the V proteins appear to direct the STAT protein to a ubiquitin ligase complex that includes the proteins DDB1 and cullin 4A. In contrast, Nipah virus, an emerging zoonotic paramyxovirus that causes highly lethal encephalitis in humans, encodes proteins (P, V, and W) that share a common N-terminal domain. This domain contains sequences that mediate an interaction with STAT1, and for the V protein, also STAT2. These Nipah virus proteins inhibit STAT1 not by targeting it for degradation; rather, the P, V, and W appear to sequester STAT1 preventing the tyrosine phosphorylation that would otherwise activate it.

Upstream and downstream steps in the IFN signaling pathways are also subject to interruption by virus infection. The tyrosine kinases that activate STAT1 and STAT2 typically undergo tyrosine phosphorylation following exposure of cells to IFN. However, in cells infected with Langat virus, a flavivirus of the tick-borne encephalitis virus complex, neither STAT1/2 nor tyk2/Jak1 phosphorylation is seen. This inhibition is mediated by as association of the viral polymerase protein NS5 with both the IFN-α/β receptor and the IFN-γ receptor. In contrast, Ebola virus infection does not prevent STAT1 tyrosine phosphorylation. However, in Ebola virus-infected cells, tyrosine-phosphorylated STAT1 fails to enter the nucleus. The failure of otherwise activated STAT1 to traffic to the nucleus appears to be due to the action of the Ebola virus VP24 protein. Tyrsoine-phosphorylated STAT1 has been reported to enter the nucleus through an interaction with the nuclear localization signal receptor karyopherin alpha 1 (importin $\alpha 5$). VP24 interacts with this specific karyopherin alpha, and in doing so, inhibits STAT1–karyopherin alpha 1 interaction, thus explaining the defect in STAT1 nuclear import.

Targeting of Antiviral Effector Proteins

Among the IFN-induced genes with demonstrated antiviral activity, PKR has been the most heavily studied, and antagonists of PKR function were among the earliest described viral inhibitors of the IFN response. Upon activation by dsRNA, PKR phosphorylates the translation factor eIF2α, a subunit of the translation initiation factor eIF2 that brings methionyl (Met) initiator tRNA to the ribosome. Phosphorylation of eIF2α on serine 51 inhibits a GDP-to-GTP exchange reaction required for the continuing function of eIF2, and therefore this phosphorylation arrests translation at the initiation phase. As a consequence, numerous viruses have devised diverse ways to preserve the translation of their mRNAs in the face of PKR activation. Examples include the adenovirus VA$_I$ RNA and the vaccinia virus E3L protein, which impede PKR activation. By inhibiting PKR function, these molecules not only help preserve translation of viral mRNAs but also facilitate virus replication in the presence of an IFN response. Another notable counterbalance to PKR function is the herpes simplex virus $\gamma_1 34.5$ protein. In this case, the viral protein recruits cellular phosphatase 1 alpha (PP1α) to dephosphorylate eIF-2α. A functional $\gamma_1 34.5$ has been found to be required for HSV-1 neurovirulence, demonstrating its importance *in vivo*, although subsequent studies have identified additional herpes simplex virus proteins, including US11, that also facilitate translation of viral mRNAs.

Viruses Targeting Multiple Aspects of the IFN System

As was noted above, it has become apparent that viruses will often target multiple components of the IFN response. Influenza viruses serve as such an example. The influenza A virus NS1 protein is a multifunctional inhibitor of innate immune responses. Evidence suggests that NS1 directly targets RIG-I, the PRR that detects influenza viruses in most cell types. This has the consequence of preventing signaling that would lead to the activation of IRF-3 and NF-κB and to IFN-α/β production. However, NS1 proteins also can affect, in a more global way, the expression of cellular gene expression. This occurs because NS1 can interfere with host cell pre-mRNA processing and can inhibit the nuclear export of cellular mRNAs. The ability of NS1 to suppress gene expression in this way has the potential to inhibit expression of IFN-α/β genes. In addition, this function may also suppress expression of genes induced by IFN thus preventing induction of an antiviral state. Finally, NS1 is able to inhibit the activation of PKR and therefore can directly target the function of at least one IFN-induced, dsRNA-activated antiviral protein. That the NS1 protein is critical for viral evasion of IFN responses is highlighted by *in vivo* experiments. Mutant influenza viruses either lacking or encoding altered NS1s display attenuated phenotypes in mice able to mount an IFN response. However, this attenuation can be largely reversed in animals lacking a fully competent IFN system, demonstrating that the attenuation in the wild-type mice is directly related to the absence of an IFN-antagonist protein.

Defeating IFN Responses – Consequences for Adaptive Immune Responses

It should be recognized that innate cytokine responses, including IFN-α/β responses, not only serve to suppress virus replication, but also act to promote adaptive immune responses. IFN-α/β can, for example, affect T-cell responses, influencing the ability of CD8+ T-cell populations to expand and generate memory cells, although the magnitude of the IFN-effect depends upon the pathogen administered. Similarly, pathogen-specific effects of IFN-α/β upon clonal expansion of CD4+ T cells have also been described. Connections between the IFN-α/β response and the function of antigen-presenting cells have also been identified. For example, the same stimuli that activate IFN-α/β production can also activate DC maturation, promoting the upregulation of major histocompatibility complex molecules and co-stimulatory molecules, such as CD80 and CD86, and the production of cytokines. DC maturation then promotes activation of T-cell responses. Therefore, the virus-encoded mechanisms that tend to suppress IFN-α/β responses, particularly those that inhibit IFN-α/β production, may serve to suppress adaptive immune responses as well. This hypothesis is supported in part by the observation that expression of the influenza virus NS1 protein, a suppressor of IFN-α/β production, rendered viruses less able to promote human DC maturation and function compared with viruses that lacked the IFN-antagonist NS1 protein.

Other Cellular Antiviral Proteins

Other cellular proteins may also serve as innate defense against virus infection. Prime examples are members of the APOBEC3 family of cytidine deaminases which function as a defense against retrovirus infection. APOBEC3 proteins deaminate cytidines to uridines in single-stranded DNA. Family member APOBEC3G can incorporate into budding HIV-1 particles and, during reverse transcription in the subsequently infected cell, deaminate cytidines. Because this results in introduction of dC to dU changes on minus-strand DNA, the result is the dG to dA hypermutation of the positive, coding-strand of the virus genome. This can result in the introduction of lethal (to the virus) mutations as well as susceptibility to cleavage by uracil DNA glycosylase and apurinic-apyrimidinic (AP) endonuclease enzymes. However, HIV-1 can defeat the antiviral effect of APOBEC3G through the function of its protein, viral infectivity factor (Vif). Vif is required for HIV-1 replication in many cell types. This is due, in part, to the ability of Vif to bind APOBEC3G and target it for ubiquitin-dependent degradation via a ubiquitin ligase complex. The ability of Vif to eliminate APOBEC3G from cells of a particular species can determine its ability to replicate in that species. Thus, HIV-1 Vif can target APOBEC3G in humans and chimpanzees. However, the inability of HIV-1 to target APOBEC3G in African green monkeys or rhesus macaques restricts the ability of HIV-1 to infect these species, thus providing a demonstration of the critical role of Vif in overcoming the innate antiviral effect of APOBEC3G.

Cellular Components of Innate Immunity to Virus Infection

As noted above, cellular components of the innate immune response also exist. Included among such cell types are DCs (discussed above), macrophages, and NK cells. NK cells function to produce cytokines or to lyse cells in a non-antigen-specific manner. Cytokines released by NK cells may exert antiviral effects or regulate other aspects of innate or adaptive immune responses. NK killing of target cells can occur through exocytosis of cytotoxic granules or through engagement of death receptors. A role for NK function in controlling virus replication has been described for several virus types including arenaviruses, paramyxoviruses, HIV, and, most notably, herpes viruses. NK activity is regulated via a series of positive and negative regulatory signals, and it is the balance of these regulatory signals that appears to determine NK cell activation. The best-studied examples of virus evasion of NK responses are those used by cytomegalovirus (CMV). One classic mechanism of NK cell activation is recognition of a cell with low levels of class I major histocompatibility complex (MHC-I) on its surface. Because NK cells express upon their surface receptors that recognize MHC-I and negatively regulate their function, targets lacking MHC-I drive this balance toward NK cell activation. CMV-infected cells typically express low levels of MHC-I upon their surface. This is due, at least in part, to virus-encoded proteins designed to downregulate MHC-I expression, so as to permit evasion of antigen-specific T-cell responses. CMV must therefore also evade NK responses that would otherwise be activated by low MHC-I. One such mechanism is the production by CMV of MHC-I-like proteins. In addition, CMV infection downregulates ligands on the infected cell that would trigger activating signals in NK cells.

Complement

Complement is a group of proteins, both soluble and cell associated, that serve multiple innate immune functions. Antiviral properties of the complement system include its ability to enhance activation (priming) of adaptive immune responses, to lyse membrane-bound viruses and infected cells, and to promote phagocytosis of viruses and infected cells. Three major complement pathways exist and are referred to as the classical, the alternate, and the lectin-binding pathways. Although activated by different means, each pathway can, through a cascade of events, lead to the deposition of the complement protein C3b on a target. This process, termed opsonization, promotes phagocytosis. Deposition of C3b can also lead to the formation of a 'membrane attack complex' which can lyse the coated particle or cell. The classical pathway is activated by binding of a complement protein, C1q, which attaches to the Fc portion of an antibody. Thus, virus or virus-infected cells can be targeted by complement and subject to C3b deposition and subsequent lysis or phagocytosis when they become bound by antibody. Activation of the alternative pathway, in contrast, occurs in the absence of antibody and occurs when an activated form of the complement protein C3 recognizes a (usually) foreign surface. In the lectin pathway, lectins may bind to carbohydrates on antigens such as viruses, triggering a pathway similar to the classical pathway.

Because complement is strongly pro-inflammatory and cytolytic, it must be tightly regulated to protect the host, and some of these host-encoded protections can either be mimicked or co-opted by viruses to evade the antiviral effects of complement. For example, several either encode 'regulators of complement activation (RCA)' proteins or package host-encoded RCA proteins. For example, gamma herpes viruses (γHVs), large DNA viruses that establish lifelong infection, encode RCA homologs. One well-studied example is the murine γHV68 RCA which can inhibit C3 and C4 deposition *in vitro*. Deletion of this γHV68-encoded RCA decreased virulence during acute infection, but virulence was restored by deletion of host C3, demonstrating the *in vivo* importance of the viral RCA as an antagonist of the complement system. An alternate strategy can be employed by HIV-1 which is able to incorporate into virus particles two glycosylphosphatidylinositol (GPI)-linked complement control proteins, CD55 and CD59. These make virus resistant to complement-mediated lysis. Finally, viruses may also produce complement inhibitors with no obvious homology to the host complement system. West Nile virus, for example, produces a secreted glycoprotein NS1. NS1 can bind to cell surfaces but it also accumulates in serum. NS1 binds to and recruits a complement regulatory protein, factor H, and the ability of NS1 to interact with factor H correlates with its ability to impair complement activation and to suppress deposition of C3 and formation of membrane attack complexes on infected cells.

Conclusion

As is illustrated by the selected examples provided above, innate mechanisms targeting virus infection are numerous, and the manner in which viruses overcome such responses are equally diverse. Defeating such mechanisms is also critical for maintenance of viruses in nature and for viral pathogenesis. It should be recognized, however, that it may not be evolutionarily advantageous for a virus to absolutely defeat host immune responses, as this might result in the rapid demise of the host with reduced opportunity for virus amplification and transmission. Therefore, viruses do not necessarily fully block innate immune responses. Rather many viruses will have evolved mechanisms to attenuate innate immunity only to such a degree that their spread to new hosts is facilitated.

See also: Immune Response to viruses: Antibody-Mediated Immunity; Immune Response to viruses: Cell-Mediated Immunity; Innate Immunity: Introduction.

Further Reading

Andoniou CE, Andrews DM, and Degli-Esposti MA (2006) Natural killer cells in viral infection: More than just killers. *Immunological Reviews* 214: 239–250.

Basler CF and Garcia-Sastre A (2002) Viruses and the type I interferon antiviral system: Induction and evasion. *International Reviews of Immunology* 21: 305–337.

Chiu YL and Greene WC (2006) Multifaceted antiviral actions of APOBEC3 cytidine deaminases. *Trends in Immunology* 27: 291–297.

Finlay BB and McFadden G (2006) Anti-immunology: Evasion of the host immune system by bacterial and viral pathogens. *Cell* 124: 767–782.

Gale M, Jr., and Foy EM (2005) Evasion of intracellular host defence by hepatitis C virus. *Nature* 436: 939–945.

Hilleman MR (2004) Strategies and mechanisms for host and pathogen survival in acute and persistent viral infections. *Proceedings of the National Academy of Sciences, USA* 101(supplement 2): 14560–14566.

Horvath CM (2004) Weapons of STAT destruction. Interferon evasion by paramyxovirus V protein. *European Journal of Biochemistry* 271: 4621–4628.

Kawai T and Akira S (2007) Antiviral signaling through pattern recognition receptors. *Journal of Biochemistry (Tokyo)* 141(2): 137–145.

Levy DE and Garcia-Sastre A (2001) The virus battles: IFN induction of the antiviral state and mechanisms of viral evasion. *Cytokine and Growth Factor Reviews* 12: 143–156.

Lopez CB, Moran TM, Schulman JL, and Fernandez-Sesma A (2002) Antiviral immunity and the role of dendritic cells. *International Reviews of Immunology* 21: 339–353.

Mastellos D, Morikis D, Isaacs SN, Holland MC, Strey CW, and Lambris JD (2003) Complement: Structure, functions, evolution, and viral molecular mimicry. *Immunologic Research* 27: 367–386.

Innate Immunity: Introduction

F Weber, University of Freiburg, Freiburg, Germany

© 2008 Elsevier Ltd. All rights reserved.

Glossary

Apoptosis A form of programmed cell death.
Apoptotic bodies Remnants of cells which underwent apoptosis.
Complement system A pathogen-triggered cascade of biochemical reactions involving more than 20 soluble and cell-bound proteins. Complement activation results in opsonization, priming of humoral immune responses, and perforation of membranes.
Cytokines Proteins which mediate cell–cell communication related to pathogen defense. Secreted by immune cells or tissue cells.
Innate immunity Physical and chemical barriers, cells, cytokines, and antiviral proteins which exclude, inhibit, or slow down infection with little specificity and without adaptation or generation of a protective memory.
Interferons (IFNs) Cytokines mediating antiviral activity. Distinguished into type I (IFN-α/β), type II (IFN-γ), and type III (IFN-λ). Type I and type III IFNs directly mediate antiviral activity in responding cells, whereas type II IFN is more immunomodulatory.
Interferon-stimulated response element (ISRE) A promoter element common to all type I IFN-stimulated genes.
Opsonization Tagging of infected cells or pathogens for destruction by phagocytic cells.
Pathogen-associated molecular patterns (PAMPs) Molecular signatures of pathogens used by the innate immune system to distinguish self from non-self. Often highly repetitive patterns.
Pattern recognition receptors (PRRs) Intracellular and extracellular receptors recognizing specific PAMPs.
Phagocytosis Uptake of particles by cells.

Introduction

Viruses attempting to conquer a mammalian body are faced with an array of problems. 'Innate immunity' in a wider sense means all sorts of factors which exclude, inhibit, or slow down infections in a rapid manner but with little specificity and without adaptation or generation of a protective memory. Many of these efficient and not at all primitive defenses are evolutionarily old and can be found in all metazoans. For the sake of brevity, however, the discussion in this article is restricted to mammals as these are the best investigated organisms in that respect. RNA interference, the innate immune system of plants and nonvertebrates, is not covered here.

Mammalian innate immune defenses against virus infections can be divided into several distinct parts such as mechanical and chemical barriers (not further mentioned here), complement system, phagocytic/cytolytic cells of the immune system which act in a nonspecific manner, and cytokines (most prominently the type I interferons).

The Complement System

The complement system (which 'complements' the adaptive immune system in the defense against pathogens) primes the adaptive immune response and is also directly effective against pathogens. Complement activation is achieved by specific receptors recognizing pathogens or immunocomplexes. Three different pathways are being distinguished which are termed the classical pathway (triggered by antigen–antibody complexes), the mannan-binding lectin pathway (triggered by lectin binding of pathogen surfaces), and the alternative pathway (triggered by complement factor C3b-coated pathogen surfaces). They all activate a cascade of reactions involving more than 20 soluble and cell-bound proteins, thus resulting in a rapid and massive response. The complement system is able to (1) tag infected cells and pathogens for destruction by phagocytic cells (opsonization), (2) prime humoral immune responses, and (3) perforate membranes of infected cells by the membrane-attack complex. In response, viruses have evolved effective countermeasures such as incorporation of cellular complement-regulatory proteins into particles or expressing specific inhibitors in infected cells.

Cellular Innate Immunity

Macrophages/monocytes, granulocytes, natural killer cells, and dendritic cells belong to the cellular branch of the innate immune system. Monocytes circulate in the bloodstream for several hours before they differentiate into macrophages. These potent phagocytic cells either continue patrolling or they permanently settle in particular tissues (i.e., the Kupffer cells of the liver), being able to rapidly remove viral particles and apoptotic bodies. Activated macrophages also synthesize inflammatory cytokines such as interferon (IFN)-γ and tumor necrosis factor (TNF)-α, thus triggering an adaptive immune response. Granulocytes are also able to remove viral particles and apoptotic bodies by phagocytosis. They are rapidly attracted to inflammatory sites and enter the tissue by transendothelial migration. Both macrophages and granulocytes cleave the ingested viral proteins into fragments and present them to T lymphocytes.

Natural killer (NK) cells are able to recognize infected cells in an antigen-independent manner and destroy them by their cytotoxic activity. Also, they rapidly produce large amounts of IFN-γ to activate the adaptive immune system. NK cells are regulated by a fine balance between stimulatory and inhibitory receptors. One of their prominent features is their ability to destroy cells which lack MHC I molecules on their surface. As many viruses downregulate MHC expression in order to avoid an adaptive immune response, NK surveillance represents an important early warning and attack system against virus infections.

A key connection between the innate and the adaptive immune system is provided by dendritic cells (DCs). These specialized immune cells sample antigen at the site of infection, activate themselves and the surrounding tissue cell by cytokine synthesis, and then migrate to secondary lymphatic organs in order to mobilize T cells against the presented antigen. The differentiation into efficient antigen-presenting cells (APCs) is achieved by cytokine production which, in turn, is triggered by stimulation of receptors recognizing pathogen-specific molecular patterns (PAMPs). Two main types of DCs are being distinguished: myeloid DCs (mDCs) and plasmacytoid DCs (pDCs). mDCs are an early split-off of the myeloid bone marrow precursors, that is, the stem cells which are also giving birth to macrophages/monocytes and granulocytes, among others. Depending on the location, several subsets of mDCs such as Langerhans cells or interstitial cells are being distinguished. pDCs, which are not segregated into subpopulations, are thought to be derived from lymphatic precursor cells. Both mDCs and pDCs can sense viral infection by several intra- and extracellular PAMP receptors (see below). Depending on the DC type, high levels of interleukins or interferons are being produced which coin the subsequent immune reaction. pDCs are potent producers of the main antiviral cytokines, the type I interferons.

Antiviral Cytokines: The Type I Interferons

Isaacs and Lindenmann discovered in 1957 that cells which had been in contact with virus particles secrete a soluble factor which confers resistance to influenza viruses, a phenomenon called 'interference'. In the subsequent years, it became more and more clear that the so-called type I interferon (IFN-α/β) system is our primary defense mechanism against viral infections. In fact, humans with genetic defects in the IFN signaling pathway have a bad prognosis as they die at an early age of viral diseases which would otherwise pose little problems. Similarly, knockout mice with a defective IFN system quickly succumb to viral pathogens of all sorts although they have an intact adaptive immune system.

In response to virus infection, pDCs are particularly well equipped to synthesize and secrete IFN-α/β, but in principle all nucleated cells are able to do so. In an autocrine and paracrine manner, IFNs trigger a signaling chain leading to the expression of potent antiviral proteins which limit further viral spread. In addition, IFNs initiate, modulate, and enhance the adaptive immune response. The signaling events which culminate in the direct IFN-dependent restriction of virus growth can be divided into three steps, namely (1) transcriptional induction of IFN synthesis, (2) IFN signaling, and (3) antiviral mechanisms.

Interferon Induction

A number of pattern recognition receptors (PRRs) recognize conserved PAMPs of viruses and initiate induction of IFN genes (see **Figure 1**). PRRs can be divided into the extracellular/endosomal toll-like receptors (TLRs) and the intracellular receptors RIG-I, MDA-5, and PKR. The main PAMPs of viruses appear to be nucleic acids, namely double-stranded RNA (dsRNA), single-stranded RNA (ssRNA), and double-stranded DNA (dsDNA).

dsRNA is an almost ubiquitous transcriptional by-product of RNA and DNA viruses. It is recognized by TLR3, the related RNA helicases RIG-I and MDA-5, and the protein kinase PKR. A third dsRNA-binding member of the RIG-I helicase family, LGP2, acts as a negative-feedback inhibitor.

Viruses with a negative-strand ssRNA genome (e.g., influenza virus) are unique in that they do not produce substantial amounts of intracellular dsRNA. Their genomic ssRNA is recognized in the endosome by TLR7 and -8. Interestingly, in the cytoplasm, RIG-I recognizes

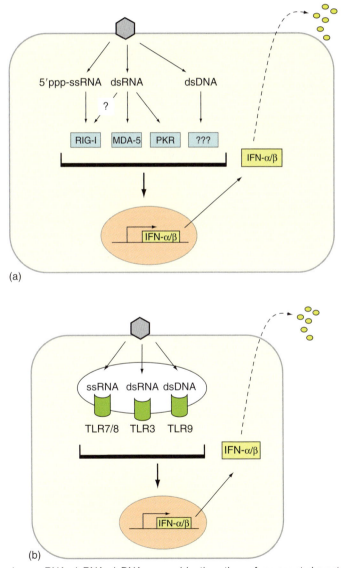

Figure 1 Depending on the virus, ssRNA, dsRNA, dsDNA, or combinations thereof represent characteristic by-products of infection which lead to induction of IFN-α/β genes. (a) These signature molecules are recognized by the intracellular PRRs RIG-I, MDA-5, PKR, and an unknown receptor for viral dsDNA. RIG-I recognizes dsRNA, but was shown to be important for recognition of 5′-triphosphate-containing ssRNA *in vivo* (see text). (b) Intracellular PAMP recognition is mirrored by the endosomal TLR pathways recognizing the same characteristics, except that ssRNAs do not need to be 5′-triphosphorylated.

the influenza virus genome in a 5′-triphosphate-dependent manner. The question how much the well-documented dsRNA-binding and unwinding activity of RIG-I contributes to its 5′-triphosphate-dependent recognition of viral genomes remains to be solved.

The third important PAMP, viral dsDNA, is again recognized both by an endosomal receptor, TLR9, and an unknown intracellular receptor. Thus, for all three nucleic acid-based PAMPs of viruses, there are specific PRRs present both in the endosomal and the intracellular compartment.

Besides nucleic acids, some viral proteins can provoke a TLR response such as the envelope proteins of respiratory syncytial virus and measles virus by activating TLR4 and TLR2, respectively.

All PRR-triggered signaling pathways eventually culminate in a strong activation of type I IFN transcription. The 'classic' intracellular pathway of IFN-β gene expression involves RIG-I and MDA-5 which both contain two N-terminal caspase-recruiting domain (CARD)-like regions and a C-terminal DExD/H box RNA helicase domain (**Figure 2(a)**). RNA binding to the helicase domain

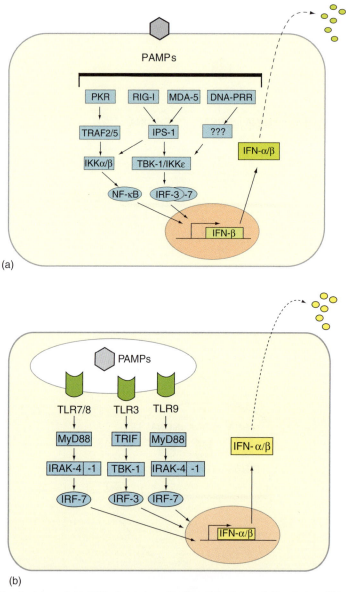

Figure 2 (a) PAMP recognition by intracellular PRRs leads to activation of the transcription factors NF-κB, IRF-3, and AP-1 (not shown). The cooperative action of these factors is required for full activation of the IFN-β promoter. IRF-3 is phosphorylated by the kinases TBK-1 and IKKε which in turn are activated by RIG-I and MDA5 via IPS-1. NF-κB is activated by the PKR pathway as well as by IPS-1. The IFN-induced IRF-7 later enhances IFN gene transcription, but is also essential for immediate early IFN-β transcription. (b) PAMP recognition by endosomal PRRs. IRF-7 is activated by IRAK-1, which in turn is phosphorylated by IRAK-4 in an MyD88-dependent manner. Both TLR7/8 and TLR9 use the MyD88 adaptor, whereas TLR3 activates IRF-3 via TRIF and TBK1.

induces a conformational change which liberates the CARD domain to interact with the signaling partner IPS-1 (also called Cardif, MAVS, or VISA). This adaptor mediates RIG-I and MDA-5 signaling and needs to be located at the mitochondrial membrane. IPS-1 has a CARD-like domain which binds to RIG-I and MDA5 and a C-terminal region which activates the kinases IKKε and TBK-1. These kinases are known to phosphorylate the transcription factor IFN regulatory factor (IRF)-3, a member of the IRF family. Phosphorylated IRF-3 homodimerizes and is transported into the nucleus. In addition, the transcription factor nuclear factor-kappa B (NF-κB) is recruited in a PKR/TRAF- and IPS-1-dependent way. Together, IRF-3 and NF-κB strongly upregulate IFN gene expression. This leads to a 'first wave' of IFN production (IFN-β and IFN-α4 in mice) which triggers the expression of the transcription factor IRF-7. Recent evidence has shown that IRF-7 is a master regulator of IFN gene expression and that IRF-3 seems to cooperate with IRF-7 for full activity. IRF-7 can be activated in the same way as IRF-3 and is responsible for a positive-feedback loop that initiates the synthesis of several IFN-α subtypes as the 'second-wave' IFNs.

mDCs can sense dsRNA by the classic intracellular pathway and, in addition, by TLR3 (**Figure 2(b)**). dsRNA-induced triggering of endosomal TLR3 proceeds via TRIF and TRAF3 which activate the kinase TBK-1, leading to phosphorylation of IRF-3 and, subsequently, to the activation of IFN-β gene expression.

pDCs sense the presence of viral ssRNA or dsDNA by TLR7, TLR8, and TLR9 (**Figure 2(b)**). Upon activation, TLR7, -8, and -9 signal through their adaptor molecule MyD88, the IRAK kinases, and IRF-7 to transcriptionally activate multiple IFN-α genes. In contrast to other cell types, pDCs contain considerable amounts of constitutively expressed IRF-7. IRF-7 is further upregulated in response to IFN and generates a positive-feedback loop for high IFN-α and IFN-β production. Furthermore, TLR7 and TLR9 are retained in the endosomes of pDCs to allow prolonged IFN induction signaling.

Type I IFN Signaling

IFN-β and the multiple IFN-α subspecies activate a common type I IFN receptor (IFNAR) which signals to the nucleus through the so-called JAK–STAT pathway (**Figure 3**). The STAT proteins are latent cytoplasmic transcription factors which become phosphorylated by the Janus kinases JAK-1 and TYK-2. Phosphorylated STAT-1 and STAT-2 recruit a third factor, IRF-9, to form a complex known as IFN-stimulated gene factor 3 (ISGF-3) which translocates to the nucleus and binds to the IFN-stimulated response element (ISRE) in the promoter region of interferon-stimulated genes (ISGs). Specialized proteins serve as negative regulators of the JAK–STAT pathway. The suppressor of cytokine signaling

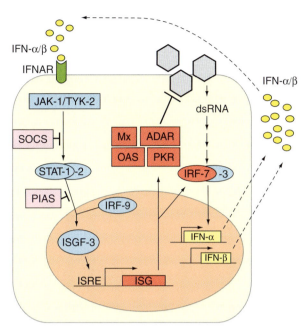

Figure 3 IFN-α and IFN-β bind to the type I IFN receptor (IFNAR) and activate the expression of numerous ISGs via the JAK/STAT pathway. IRF-7 amplifies the IFN response by inducing the expression of several IFN-α subtypes. Mx, ADAR, OAS, and PKR are examples of proteins with antiviral activity. Modified from Haller O, Kochs G, and Weber F (2006) The interferon response circuit: Introduction and suppression by pathogenic viruses. *Virology* 344: 119–130, with permission from Elsevier.

(SOCS) proteins prevent STAT activation whereas protein inhibitor of activated STAT (PIAS) family members function as small ubiquitin-like modifier (SUMO) E3 ligases and inhibit the transcriptional activity of STATs.

Direct Antiviral Effects of Type I IFNs

Type I IFNs activate the expression of several hundred IFN-stimulated genes (ISGs) with multiple functions. To date, five antiviral pathways have been studied in great detail, namely the protein kinase R (PKR), the RNA-specific adenosine deaminase 1 (ADAR 1), the 2–5 OAS/RNaseL system, the product of the ISG56 gene (p56), and the Mx proteins. PKR, ADAR1, and 2–5 OAS are constitutively expressed in normal cells in a latent, inactive form. Basal mRNA levels are upregulated by IFN-α/β and these enzymes need to be activated by viral dsRNA. PKR is a serine-threonine kinase that phosphorylates – among other substrates – the α-subunit of the eukaryotic translation initiation factor eIF2. As a consequence, translation of cellular and viral mRNAs is blocked. PKR also plays a role in virus-induced NF-κB activation, as described above. ADAR 1 catalyzes the deamination of adenosine on target dsRNAs to yield inosine. As a result the secondary structure is destabilized due to a change from an AU base pair to the less stable IU base pair and mutations accumulate within the viral genome. The 2–5 OAS catalyzes the synthesis of short $2'$–$5'$ oligoadenylates that activate the latent endoribonuclease RNaseL. RNaseL degrades both viral and cellular RNAs, leading to viral inhibition. P56 binds the eukaryotic initiation factor 3e (eIF3e) subunit of the eukaryotic translation initiation factor eIF3. It functions as an inhibitor of translation initiation at the level of eIF3 ternary complex formation and is likely to suppress viral RNA translation. Mx proteins belong to the superfamily of dynamin-like large GTPases and have been discovered as mediators of genetic resistance against orthomyxoviruses in mice. They most probably act by enwrapping viral nucleocapsids, thus preventing the viral polymerase from elongation of transcription.

The antiviral profiles of the IFN effectors listed above are distinct but often overlapping. Mx proteins, for example, mainly inhibit segmented negative-strand RNA viruses and also Semliki Forest virus, whereas the 2–5 OAS/RNaseL system appears more important against positive-strand RNA viruses. Moreover, only rarely the presence of one particular IFN effector determines host resistance. Rather, it is the sum of antiviral factors affecting, for example, genome stability, genetic integrity, transcription, and translation that confers the full antiviral power of IFN.

Indirect Antiviral Effects of Type I IFNs

Besides the effector proteins listed above, several ISGs contribute in a more indirect manner to the enhancement of both innate and adaptive immune responses. Virus-sensing (and in part antiviral) PRRs such as TLR3, PKR, RIG-I, and MDA5 are by themselves upregulated in a type-I-IFN-dependent manner. Similarly, IRF-7 and and STAT1, the key factors of type I IFN and ISG transcription, respectively, are ISGs. The strong positive-feedback loop mediated by the upregulation of these PRRs and transcription factors is counterbalanced by several negative regulators such as LGP2, SOCS, and PIAS, which are either ISGs or depend on IFN signaling for their suppressive action.

Type I IFNs can directly enhance clonal expansion and memory formation of $CD8^+$ T cells. Also, IFNs promote NK cell-mediated cytotoxicity and trigger the synthesis of cytokines such as IFN-γ or IL-15 which modulate the adaptive immune response, enhance NK cell proliferation, and support $CD8^+$ T-cell memory. Moreover, by upregulating TLRs, MHCs, and costimulatory molecules, IFNs enable APCs (most prominently DCs) to become competent in presenting viral antigens and stimulating the adaptive immune response.

Good Cop–Bad Cop

Given their massive impact on the cellular gene expression profile, it is quite expected that type I IFNs have not only antiviral, but also antiproliferative and immunomodulatory effects. Treatment with IFNs is an established therapy against several viral and malignant diseases such as hepatitis B, hepatitis C, Kaposi's sarcoma, papillomas, multiple sclerosis, and several leukemias and myelomas. However, the strong and systemic effects of IFNs do not come without a price. Administration of IFN can locally produce inflammation, and systemically cause fever, fatigue, malaise, myalgia, and anemia. It is no coincidence that these latter are 'flu-like' symptoms, since in many acute infections IFNs play a dominant role. The effects of IFN which are desired and beneficial if restricted to the site of first infection can turn into a life-threatening 'cytokine storm' if it becomes systemic. Severe acute respiratory syndrome (SARS) and human infections with H5N1 influenza viruses are examples of such out-of-control innate immune responses. Another 'dark side' aspect is that patients with autoimmune diseases have chronically elevated levels of IFNs, and that IFN therapy can aggravate autoimmune disorders. It is thought that pDCs (and in part B cells) are autostimulated by self-DNA via TLR9 and by small nuclear RNA complexes (snRNPs) via TLR7. Chronic production of IFNs causes maturation of mDCs, which in turn activate autoreactive T and B cells.

Concluding Remarks

The concept of innate immunity certainly comprises more than the IFN system (see above), but type I IFNs represent a central part. These cytokines not only have direct antiviral effects but also orchestrate the first defense reactions and the subsequent adaptive immune response, thus determining the course of infection. The recent findings that basically every virus appears to have evolved one or several countermeasures for controlling the IFN response is testament to its importance. In addition, IFNs are not only antiviral, but also effective tumor suppressors. Tumor cells often eliminate the IFN system during the transformation process. The payoff is an increased susceptibility to infection, an Achilles heel which is exploited by the therapeutic concept of oncolytic viruses. Tumor selectivity of such viruses can be even more increased by using IFN-sensitive mutants. The inability of those mutants to fight the IFN response is complemented by the mutations of the tumor cells, thus allowing virus growth. At the same time, these viruses are unable to infect the IFN-competent body cells.

Recently, it became apparent that there exists a hitherto unnoticed parallel world called the type III IFN system. The cytokines IFN-$\lambda 1$, -$\lambda 2$, and -$\lambda 3$ are induced by virus infection or dsRNA and signal through the JAK/STAT cascade, but use a separate common receptor. They are able to activate antiviral gene expression and have been shown to inhibit replication of several viruses.

Thus, the IFN response has a backup system to enforce the first line of defense against virus infections.

Future studies will have to address the relative contribution of type I and type III IFNs to antiviral protection and the coming years may have even more surprises in stock. The innate immune system may be old, but as long as there are viruses and tumors, it will never come out of fashion.

See also: Immune Response to viruses: Antibody-Mediated Immunity; Innate Immunity: Defeating; Interfering RNAs; Polydnaviruses: Abrogation of Invertebrate Immune Systems.

Further Reading

Akira S and Takeda K (2004) Toll-like receptor signalling. *Nature Reviews Immunology* 4: 499–511.
Ank N, West H, and Paludan SR (2006) IFN-λ: Novel antiviral cytokines. *Journal of Interferon and Cytokine Research* 26: 373–379.
Colonna M, Trinchieri G, and Liu YJ (2004) Plasmacytoid dendritic cells in immunity. *Nature Immunology* 5: 1219–1226.
Diefenbach A and Raulet DH (2003) Innate immune recognition by stimulatory immunoreceptors. *Current Opinion in Immunology* 15: 37–44.
Garcia MA, Gil J, Ventoso I, et al. (2006) Impact of protein kinase PKR in cell biology: From antiviral to antiproliferative action. *Microbiology and Molecular Biology Reviews* 70: 1032–1060.
Garcia-Sastre A and Biron CA (2006) Type 1 interferons and the virus–host relationship: A lesson in detente. *Science* 312: 879–882.
Haller O and Kochs G (2002) Interferon-induced mx proteins: Dynamin-like GTPases with antiviral activity. *Traffic* 3: 710–717.
Haller O, Kochs G, and Weber F (2006) The interferon response circuit: Induction and suppression by pathogenic viruses. *Virology* 344: 119–130.
Hoebe K, Janssen E, and Beutler B (2004) The interface between innate and adaptive immunity. *Nature Immunology* 5: 971–974.
Honda K, Takaoka A, and Taniguchi T (2006) Type I interferon gene induction by the interferon regulatory factor family of transcription factors. *Immunity* 25: 349–360.
Reis e Sousa C (2006) Dendritic cells in a mature age. *Nature Reviews Immunology* 6: 476–483.
van Boxel-Dezaire AH, Rani MR, and Stark GR (2006) Complex modulation of cell type-specific signaling in response to type I interferons. *Immunity* 25: 361–372.
Vilcek J (2006) Fifty years of interferon research: Aiming at a moving target. *Immunity* 25: 343–348.
Volanakis JE (2002) The role of complement in innate and adaptive immunity. *Current Topics in Microbiology and Immunology* 266: 41–56.

Inoviruses

L A Day, The Public Health Research Institute, Newark, NJ, USA

© 2008 Published by Elsevier Ltd.

Glossary

Everted DNA Inside-out DNA helix with phosphates at the center.
Integrative phage A phage capable of having its genome integrated into that of its host.
Productive state Of a bacterium actively exporting progeny filamentous phage without lysis.
Prophage state Of a bacterium with a repressed phage genome inserted in its genome or in one of its plasmids.

Discovery, Ecology, and Host Range

In 1960, T. Loeb reported finding phages that formed plaques on *Escherichia coli* carrying the F transfer factor but not on those without it. His finding was confirmed by workers around the world, and publications in 1963 by H. Hoffmann-Berling and D. A. Marvin, by Loeb and N. Zinder and co-workers, and by H. P. Hofschneider established that such plaque formers, independently isolated, were small spherical RNA phages (fr, f2, and M12), and small filamentous DNA phages (fd, f1, and M13), the latter now collectively referred to as Ff phage (**Figure 1**). The capital F is for the F factor specifying the host, and the small f stands for filamentous. These new DNA viruses had intriguing characteristics. Surprisingly, they did not lyse their hosts in culture yet formed the plaques on bacterial lawns, their filamentous virions contained DNA genomes that were both single stranded and circular, and it was observed, by E. Trenkner, that capsid protein of the infecting phage is incorporated into progeny virions. The fd, f1, and M13 genomes were among the first viral genomes sequenced. They each have either 6407 or 6408 nucleotides encoding the same 11 genes but fewer than 100 nucleotide differences even though the isolates were from different places in the world. This near constancy seemed remarkable, and has become more so in view of the plasticity of the Ff genome made evident in a plethora of subsequent mutagenesis studies.

Now over 50 different species of filamentous DNA phages are known, and their bacterial hosts occupy many habitats. Taxonomically the phages comprise the genus *Inovirus* in the family *Inoviridae*, the terms coming from the Greek word for filament. Infections by many

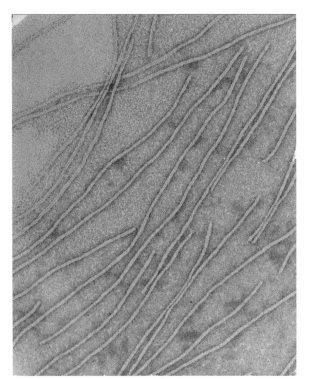

Figure 1 A uranyl acetate negative stain electron micrograph of fd showing one complete virion and the ends of others. One end is more rounded or blunt, and the other is tapered with additional tassel-like mass sometimes perceptible extending from the tip. Ff (fd) particles are 890 nm long under these dry conditions, but atomic force micrographs and cryoelectron micrographs indicate that lengths of fully hydrated phage might be slightly longer.

individual species lead to productive carrier states that extrude virus, sometimes copiously, while host cells continue to divide, yet infections by many other species lead to prophage carrier states with repressed genomes integrated into the bacterial genomes. **Tables 1** and **2** list the amino acid sequences of the major coat proteins for several selected species, as well as the names of known phages and those of some of the bacterial hosts. The protein sequences show, on the one hand, enormous diversity, yet, on the other, extensive homology as, for example, the common sequences for Pf1 and Pf4, and for CUS-1 and CUS-2, as well closely related sequences for CUS-2 (YP01Φ), X, and C2. The homologies reflect extensive horizontal genetic transfer, as well as the effects of evolution on sequences perfoming the same functions. Most of the phage/host systems are in Gram-negative bacteria, such as coliforms, pseudomonads, xanthomonads, vibrios, and many others, but two of the systems are Gram positive. Filamentous phages have been discovered through (1) characterizations of plaque formers on specific bacterial strains, (2) applications of phage concentration procedures to culture supernatants, (3) sequence studies of plasmid populations, (4) screens for virulence factors in pathogenic bacteria, and (5) *in silico* analyses of bacterial genomic sequences. Like other phages, inoviruses are ubiquitous, and their number on Earth must be astronomical.

Host ranges of the Gram-negative systems are determined usually by the multidomain gene 3 protein located at one end of the virion interacting with a specific, type IV pilus and the Tol QRA protein complex. The combined receptor systems are highly efficient, approaching one infection per physical particle in some cases. The pili receptors are encoded chromosomally or on plasmids of different incompatibility groups, that is, phage Ff adsorbs to IncF pili, IKe to IncN pili, Pf3 to IncP pili, etc., and transmission of the plasmids to new bacterial species can transfer phage sensitivity. Some phages can infect, at low efficiency, cells without pili by reaching the tolQRA co-receptors some other way. Some host ranges are quite broad such as that of phage X. Transfections of non-natural hosts with naked phage DNA are sometimes possible, as shown by the transfection of female strains of *E. coli* with DNA from Pf1, a phage of *Ps. aeruginosa*.

Many, but not all, inoviruses form plaques on bacterial lawns even though they do not lyse their hosts in culture. Plaque formation and the size and appearance of plaques depend on relative rates of bacterial growth and phage production. These can be affected by phage mutations and/or the presence of prophages in the host, as well as details of plate conditions.

Morphology

Except for differences in length, filamentous phage virions are morphologically similar to the fd virions shown in **Figure 1**. Most appear to have a tapered end, with additional structures extending from it, and a rounded or blunt end. Diameters of the Gram-negatives are invariably about 7 nm, but a value of 12 nm was reported for the Gram-positive phage B5. The length range for Gram-negative wild types is 700 nm (Pf3) to 3500 nm (Pf4). The length depends upon both the size and the conformation of the packaged genome, and this conformation depends on the chemical nature of the DNA interaction domain (**Tables 1** and **2**). For example, in Ff, the DNA–protein interface is largely electrostatic, and reduction of the number of lysine residues in DNA interaction domain from 4 to 3 leads to stable virions approximately 4/3 longer than wild type, hence a more extended conformation for a DNA of unaltered size. On the other hand, DNA insertions or deletions, *in vitro* or *in vivo*, can lead to virions having unaltered DNA conformations but lengths that are longer or shorter than wild-type. Measurements of lengths have been by conventional electron microscopy (EM), scanning transmission electron microscopy (STEM), cryo-EM, atomic force microscopy, and hydrodynamic methods. In electron micrographs of static particles on grids, some phages appear more flexible than others, and

Table 1 Amino acid sequences of major capsid proteins for several strains

Phage/prophage	Solvent interaction	DNA interaction
KSF-1Φ		AIALGLGWYPWYLEVIVMVAFVSDALTVVVAVAYFMAFAYGFYTGVNAS
Pf3		MQSVITDVTGQLTAVQADITTIGGAIIVLAAVVLGIRWIKAQFF
Pf1, (**Pf4**)		GVIDTSAVESAITDGQGDMKAIGGYIVGALVILAVAGLIYSMLRKA
f237, Vf04K68		EVDITGAINSAVSGGQANVSLVVAGLIGMAALGFGVTMVVGFLRR
ΦSMA9		MVTVIADAFQPPNKEQLAVWAGGPFGLLLFLFVAGRIAGSVATFFDKNR
CTXΦ		ADAGLVTEVTKTLGTSKDTVIALGPLIMGVVGAIVLIVTVIGLIRKAK
VSK		MPLCKKFSLQSTLLVSLLSLLLLVLPSLASTWLSKASLSANALSTKPNR
CUS-1, CUS-2		AEGAASSGVDLSPLTNSIDFSTVLVAIMAVAASLVTLYAGVAGVRWVLRTVKSA
X		AEGDVVGGKGIDLTPLTNSVNFGSVLTGIMAVAGSLIVLYAGSAGVRWILRMVRGA
C2		MGPTAPTDIASLASSVDFSSVGLGILAIAGTVITLYVTWKGAQFVIRQVRGA
Ff (f1, fd)		AEGDDPAKAAFDSLQASATEYIGYAWAMVVVIVGATIGIKLFKKFTSKAS
I2-2		ADDGTSTATSYATEAMNSLKTQATDLIDQTWPVVTSVAVAGLAIRLFKKFSSKAV
B5		MVILEAPSDVGGTVSAAITALGPQITPIIGVAIGVSLIPFAAKWIFRKAKSLVS
Xf		SGGGGVDVGDVVSAIQGAAGPIAAIGGAVLTVMVGIKVYKWVRRAM
Cf1		SGGGGDFDGTAIIGKVTTYTAIGVTILAPSRSVVGRFATRSDRRQVSQSAA

The names of integrative strains are in bold typeface, and potential charges are indicated for acidic (D, E) and basic residues (K, R). The sequences are right justified to compare DNA interaction domains, and are ordered on the basis of the number of basic residues in these domains. KSF-1Φ protein has no basic residues at all, but there are six aromatic residues (F, Y) in the DNA domain. Pf3 protein has only two basic residues, but there are 2.5 phosphate charges per subunit in the virion. Pf1 protein has two basic residues, yet one phosphate per subunit in the virion. VSK has only basic residues, even in the solvent interaction domain. Many sequences have proline (P) residues, often near the middle. Xf is much less α-helical than the others, having a proline in the middle as well as ten glycines (G). The basic residues in Cf1c are farther from its C terminus than is the case for the others. Over 50 inoviruses are known, and genomic sequences are available for most. Listed alphabetically, with integrative strains in bold and Ff phages in italics, they include: **493**, *AE2*, B5(Gram +), CAK-1(Gram +), **Cf1c, Cf16, Cf1tv, CTXΦ, CUS-1, CUS-2**, *δA, Ec9, F12*, **f237**, fs-1, fs-2, FXP, *HR*, I2-2, **If1**, IKe, Ivpf5, **KSF-1Φ, Lf,** *M13*, **MDA, Nf1-A, Nf4-G3**, Pf1, Pf2, Pf3, Pf4, Pf5, PH75, **ΦRSS1, ΦRSM1**, ΦSMA9, Pr64FS, SF, SW1, tf-1, V6, **Vf04K68, VGJΦ**, VSK, X, X2, Xf, Xo, **Vf12, Vf33**, Xv, **Xf1fΦ, Yp01Φ** & **Ypf1** (alternate names for CUS-2), and *ZJ/2*.

C2 and X show distinct waviness. On the other hand, phages fd, Xf, and Pf3 appear to flex only gently by EM and their lengths deduced from rotational and translational diffusion rates indicate rodlike behavior. It is worth noting that filament morphology is dynamic. The lateral and torsional flexing modes can be quite fast so it is best to consider these structures as dynamic entities exhibiting motions on many timescales. A new aspect of inovirus morphology has come from recent liquid crystal studies that indicate overall large length-scale chiralities superimposed on filament flexing for several species. This type of chirality, called coiling, has been shown to reflect diversity in the structural details of the DNA packaging. So far only one virus, Pf1, out of eight studied shows no tendency to coil.

Virion Structure

There is as yet no consensus detailed structure model for any of the virions. Nevertheless, there are many constraints defining the structures of the DNA and capsids of several of them, especially Ff, Pf1, and Pf3. First, none contains lipids or carbohydrates, only DNA and protein. Second, the structures of the packaged genomes and the protein capsids are mutually interdependent. Empty capsids do not exist, and conformations DNA and protein have in the intact virions have not been observed under other conditions. The DNA contents range from 6% to ~15% of the total virion mass. The major coat proteins that form the shell constitute 97–99% of the protein mass, and the small remainder is in proteins at the ends critical for assembly and infection. The virions are put together in complex, multicomponent reactions at the membranes, so they are not self-assembly systems like tobacco mosaic virus (TMV). Also, they are taken apart by multicomponent reactions of a different sort at the membrane of the cells being infected.

The most extensive information is for the Ff group, and the ideas from Ff structure apply to the others as well, although there are remarkable differences in the details. Some of these ideas are presented here in conjunction with the schematic diagram in **Figure 2**. The gene names

Table 2 Bacterial hosts for phages that are in **Table 1** and/or are relevant to the text

B5	*Proprionibacterium freudenreichii (Gm +)*
C2	*Escherichia coli*, IncC
CAK-1	***Clostridium acetobutylicum (Gm +)***
Cf1	*Xanthamonas campestris pv citri*
CTXΦ	***Vibrio cholera***
CUS-1	*Escherichia coli 018:K1:H7*
CUS-2 (Yp01Φ)	*Yersinia pestis*
f237	*Vibrio parahaemolyticus*
Ff (f1, fd, M13)	*Escherichia coli*, IncF
I2–2	*Escherichia coli*, IncI$_2$
KSF-1Φ	***Vibrio cholera***
MDA, Nf1-A	*Neisseria meningitidis*
Nf4-G3	*Neisseria gonorrhea*
Pf1	*Pseudomonas aeruginosa K (PAK)*
Pf3	*Pseudomonas aeruginosa 01 (PA01)*
Pf4	***Pseudomonas aeruginosa 01 (PA01)***
PH75	*Thermus thermophilus*
ΦRSS1, ΦRSM1	***Ralstonia solanacearum***
SW1	*Shewanella piezotolerans* WP3
tf-1	*Escherichia coli*, IncT
Vf04K68	***Vibrio parahaemolyticus***
Vf12, Vf33	*Vibrio parahaemolyticus*
VGJΦ, VSK	***Vibrio cholera***
X	*Escherichia coli*, IncX
Xf	*Xanthamonas oryzea pv oryzea*
ΦSMA9	*Stenotrophomaltophilia*

Integrative phages and their hosts are in bold. Only two systems, B5 and CAK-1, are Gram positive. Genomic sequences of bacteria in many taxonomic families contain filamentous prophage genomes.

for Ff are based on numbered complementation groups assigned in early f1 genetic studies. Here Arabic numbers are used and gene products are labeled gp3, gp6, etc. X-ray crystallographic studies of gp3 have revealed that both its N2 and its N1 domains are accessible for interactions with the pilus and the Tol QRA receptor proteins, respectively. The gp7/9 end has the DNA packaging signal (PS), which has a DNA hairpin with a stem of two helical turns of almost perfect base pairing. Compensating mutations in the proteins and the DNA provided genetic evidence for direct interaction between the DNA of PS and the gp7/9 complex.

In Ff phages the genomes (6407 or 6408 nt) are packaged by approximately 2700 subunits of gp8, corresponding to about 2.4 nt per subunit. This noninteger ratio is the consequence of a meshing of nonequivalent helical symmetries of the DNA and of the capsid to form the structure. As the DNA helix winds within the core of the capsid, it contacts different points on the successive subunits, as indicated in **Figure 2**. Note that while the two DNA strands point in opposite directions the subunits point in the same direction. For these reasons, subunit structures are inherently nonequivalent (polymorphic) at the DNA–protein interface. The nonequivalence, or polymorphism, was proposed to be the source of large length-scale coiling effects on overall morphology

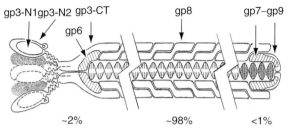

Figure 2 A scale diagram of an end-to-end section through a conjectural model of Ff. The phage diameter is 6 nm, the DNA diameter is 2 nm, and the shell thickness is also 2 nm. The structure has fivefold rotational symmetry. The diagram was drawn as if a blade angled at 144° were passed from end to end. Axial displacements of 3.2 nm align gp8 subunits next to each other, and the azimuthal angle of 144° puts two subunits opposite to each other in this particular diagram. The gp8 subunits are largely α-helical with diameters of 1 nm in the inner and outer portions. The overlaps, which are complex, are merely adumbrated by the gp8 shape, and the bend in the middle is exaggerated. The gp8 C-terminal regions are near the DNA and are the positively charged. The DNA pitch is 2.8 nm and the nucleotide rise 0.28 nm. The DNA helix has stacked bases with H-bonding between bases on opposite strands, but only about 25% can be of the Watson–Crick (W–C) type. The shaded part of the DNA represents a W–C base-paired packaging signal (PS). All five proteins are accessible to the outside as required by results from phage display studies. The gp3 part of the diagram is adapted from Lubkowski J, Hennecke F, Pluckthun A, and Wlodawer A (1998) The structural basis of phage display elucidated by the crystal structure of the N-terminal domains of g3p. *Nature Structural Biology* 5: 140–147. Overall diagram modified from Day LA and Maniloff J (2000) *Inoviridae*. In: van Regenmortel MHV, Fauquet CM, Bishop DHL, *et al.* (eds.) *Virus Taxonomy*: Seventh Report of the International Committee on Taxonomy of Viruses, pp. 267–275. San Diego, CA: Academic Press, with permission from Elsevier.

deduced from the liquid crystal behaviors and micrographs of several phages. Subunit polymorphism is present in filamentous phages in which the DNA and capsid helical symmetries are not the same, and can be subtle or pronounced depending on the details of the chemical nature of the protein–DNA interface, whether more electrostatic or more hydrophobic.

Protein Subunit Conformation

Electronic spectroscopy (ultraviolet absorbance, linear and circular dichroism, and fluorescence), vibrational spectroscopy (infrared, Raman, resonance Raman, and Raman optical activity), and solid-state nuclear magnetic resonance (NMR) have been used to determine the protein conformations of many strains, especially Ff, Pf1, and Pf3. These three are 85% or more α-helical, and the α-helices are tilted about 20° on average with respect to the virion axes. However, there are several exceptions to the α-helix rule for inovirus capsids. **Tables 1** and **2** shows many prolines, either alone or in combination with small glycine residues and sometimes near the middle of the sequence, allowing bends or kinks. Solid-state NMR provides information about such kinks as well as atomic level information on backbone and side chain structure and dynamics, and hence provides essential constraints for full structure determinations.

Capsid Helical Symmetry

X-ray fiber diffraction patterns have established two basic types of capsid helical symmetry. The patterns for Ff, If1, and IKe, all show layer spacings of 32 Å (3.2 nm) and sharp diffraction on the meridian at 16 Å. It was eventually realized that these patterns indicated two pentamers of subunits every 32 Å. This symmetry characterizes class I capsid structures. Each pentamer is interdigitated with the pentamers before it and after it in such a way that DNA interaction domains of ten subunits contact the DNA at any point along the filament. Early radiochemical results indicated five copies of gp3 in fd, and one expects the fivefold rotational symmetry to extend into the two ends of the capsid.

Class II capsid symmetry is characteristic of Pf1, Xf, Pf3, and PH75. These viruses all have X-ray patterns with layer spacings of 75–80 Å and intensity close to the meridian on every fifth layer and certain other layers. These patterns indicate a helix of 5.4 subunits per turn repeating every five turns, but the patterns change with hydration and are difficult to interpret in detail. However, patterns for fibers of Pf1 prepared at temperatures below about 10 °C reliably define a capsid of 5.46 subunits per turn of a one-start helix. Subunits in successive turns interdigitate with those in the preceding and following turns so that the DNA interaction domains of 11 subunits contact the DNA at any point along the filament.

Theoretical studies of the packing of α-helical subunits around DNA have helped clarify why these two packing symmetries are preferred over a few other possible symmetries. However, the molecular mechanisms determining choice symmetry types are not understood.

Virion Stoichiometry and DNA Conformations

The DNA structures in filamentous phages differ dramatically, and three types have been proposed for the DNA helices in Ff, Pf1, and Pf3. The first type is much like classical double-stranded DNA (dsDNA) with H-bonded and stacked bases, although only about 25% of the H-bonding can be of the Waston–Crike type. This type is in Ff, in which the axial distance between nucleotides is 2.8 Å and off-layer X-ray diffraction for a sample held at low pH indicated a pitch of 28 Å. At face value, these data indicate a tenfold DNA helix, in accord with the fivefold rotational aspect of the capsid symmetry. Many spectroscopic measurements, including, a chiral method, established the type of right-handed DNA structure in **Figure 2**. Several other phages in both capsid symmetry classes I and II have been shown to have this type of DNA helix, with pitches near 30 Å and stacked bases facing in with the possibility of ∼25% Watson–Crick (W–C) hydrogen bonding. Stoichiometric ratios lie between 2.0 and 2.5 nt per subunit.

The second type is that proposed for the DNA in the Pf1 virion. The Pf1 virion has 1.0 nt per subunit and the axial distance between bases 6.1 Å. Although some debate remains as to the exact structure, these two constraints and several others from various spectroscopies are satisfied by an everted DNA helix model having paraxial phosphates only 2 Å from the center. The bases are unstacked, strongly tilted, and extended outward into the capsid. The DNA has the 16 Å pitch of the capsid and 2.73 nucleotides per turn in each DNA strand to match exactly the 5.46 subunits per helical turn of the capsid helix. The helical symmetry of the DNA was established by the positions of phosphorus and two nondiester oxygens of the nucleotides as deduced from the Pf1 electron density map, and from theoretical analysis of the positioning forces exerted by electrostatic interactions between the phosphates. Further support is from single-molecule studies of DNA twist, and from Raman and infrared spectra of oriented Pf1. Pf1-like DNA structures are expected in f237 and Vf04K68, as well as in FXP, and certainly in Pf4, based on sequence (**Tables 1** and **2**).

The third type of DNA helix is that in the Pf3 virion. It is proposed to have 16 Å pitch but the axial base separation is only 2.4 Å yet the bases are not stacked. The Pf3 stoichiometry is 2.5 nt per subunit but the Pf3 subunit can provide only two positive charges. The constraints have produced an everted DNA model with phosphates at about 5 Å radius, leaving a hole in the center for metal ions to neutralize the charge. These three types of DNA packing are thus very different. The differences are reflected in their central mass densities which are high for fd, very high for Pf1, but low for Pf3, as shown by STEM analysis. Other types of DNA packing are anticipated. It seems certain that there are critical factors in addition to the sequences in **Tables 1** and **2** that determine DNA structure type but none are yet known.

Genome Organization

Inovirus genomes range from just below 6 kbp to over 12 kbp, with 6–8 kbp required for productive infections. With few exceptions, the genes for productive infections are ordered according to their functions in the life cycle. Two or three DNA replication genes are followed in order by five virion structure genes, then by two virion morphogenesis genes, and finally by DNA sequences for controlling DNA packaging and DNA replication. All five structure proteins are membrane proteins prior to their incorporation into virions, while other gene products function enzymatically as inner membrane proteins and another forms a pore in the outermembrane through which virions are released. On other genomes, the patterns of virion structural genes are invariant, but the relative positions of morphogenesis genes can switch or be absent. For CTXF, the outer membrane pore function is provided by a host secretin and its position on the genome is taken by cholera toxin genes *ctxA* and *ctxB*. In the Gram-positive B5 system, there is no outer membrane, and hence a gene for a pore forming protein is missing completely from the phage genome.

Integrative strains encode integrases and repressors together with operator sequences, and the genes for them are located upstream and downstream of the core genes just described. Their sizes and detailed arrangements vary. Prophage sequences, partial or complete, are found integrated at one or many chromosomal sites. In several systems, the principal site is the bacterial *dif* site and the enzymes involved are the host XerC/D recombinases. In the CTXΦ system, integration at *dif* places one or more phage genomes juxtaposed with a smaller helper phage element called RS1. This arrangement allows complete phage genomes of circular topology to be generated without excision of the integrated prophage.

Life Cycles

Carrier States That Extrude Virions

Inoviruses infect their natural hosts without lysing or otherwise killing them, and the infected cells divide and produce virus indefinitely. Three groups of complex processes occur to produce virions: attachment of the phage and import of its ssDNA, conversion of it to a circular dsDNA for initial viral gene expression followed by viral dsDNA replication leading to progeny ssDNA sequestered in cytoplasm and structure proteins ready in the membrane, and finally virion assembly and export. These processes are delicately regulated to produce the appropriate numbers of the various proteins for continual cell division and copious phage production. The most extensive information is from the Ff/*E. coli* systems.

Phage attachment and DNA penetration

Phage attaches initially to the end of the F pilus through domain N2 of gp3 (**Figure 2**) which facilitates contact of domain N1 of gp3 with Tol A, a protein anchored in the inner membrane that extends into the periplasm. This interaction in turn facilitates interactions between gp3 and membrane proteins Tol Q and Tol R that translocate the ssDNA into the cytoplasm and insert the protein subunits of the incoming phage into the membrane, and these subunits are later found in progeny virions. The Tol QRA proteins are absolutely required for DNA entry. The pili receptors, whether nonconjugative, like a toxin co-regulated pilus receptor for CTXΦ and a mannose-sensitive hemagglutinin pilus receptor for KSF-1 and other vibriophages, or conjugative like the F-pilus, enhance the specificity and efficiency of infection but they are not always essential. The receptors in Gram-positive systems have not been established.

DNA replication and gene expression

The ssDNA is converted to a parental supercoiled dsDNA by cellular enzymes starting with RNA polymerase which creates an RNA primer for the initial complementary strand synthesis. The DNA site for the priming has an unusual hairpin structure with a high affinity for the RNA pol holoenzyme. Priming is followed by complementary strand synthesis by a DNA polymerase, then displacement of the RNA primer and ligation, and finally gyrase action, to produce the parental supercoiled dsDNA. Transcripts are generated from promoters of various strengths and various rho-dependent and -independent terminators. Translation is controlled by overlaps of genes in different frames or alternate starts in the same frame as well as by codon usage. Replication of this dsDNA begins with a nick at a specific, high-symmetry site once a viral endonuclease is expressed from it. Progeny dsDNA circles are produced via ssDNA intermediates in rolling circle replication. The dsDNA circles lead to further mRNA and protein synthesis. Increasing amounts of another replication protein and a ssDNA-binding protein eventually downregulate the nicking activity of the nuclease. Large amounts of the DNA binding protein (gene 5 protein) accumulate to form complexes with nascent plus strands being displaced in the rolling circle replication. About 100 progeny ssDNA plus strands become sequestered within flexible unbranched filamentous structures having stoichiometric ratios of 9 or 10 nt per gp5 dimer. The crystal structure of gp5 has been determined and the structures and properties of gp5/nucleic acid complexes have been extensively studied. Nevertheless, the complex mechanism of gp5 slippage on the lengthening DNA to form such structures, which involves the gp5 folding the nascent single strand back on itself and continually adjusting the fold point, is not established. The mechanism leaves the tight hairpin of the packaging site hairpin exposed at one end and the rest of the loop of ssDNA covered with about 700 gp5 dimers. While this is happening, thousands of copies of the major coat protein and smaller numbers of the four other capsid proteins are accumulating in the inner membrane. The morphogenesis proteins are also inserted in the inner membrane, and about 14 molecules of another morphogenetic protein forms a closed pore in the outer membrane.

Assembly and export

Assembly nucleates with the formation of a complex between gp7 and gp9 (**Figure 2**) and the exposed packaging signal at one end of the gp5/ssDNA complex, anchoring it to the membrane. Concerted nucleoside triphosphate (NTP)-driven interactions move the DNA from the complexes into the membranes where gp8 coat protein displaces gp5. The gp8 are present as specific dimers and higher multimers in the membrane and they might remain with the same neighbors but with altered contacts during assembly and in the virion. Virions are completed by additions of gp6 and gp3 which recognize the reverse turn of the DNA when the whole genome is packaged. Mutants of these two latter proteins lead to long virions containing multiple genomes. Adhesions of the inner-membrane assembly sites and the outer-membrane pores create the complete assembly/extrusion sites.

Virus is produced at about 100 sites per cell, without lysis and without preventing the cell from dividing.

Prophage Carrier States

Integrative filamentous phages which enter prophage states infect the cell as described above by means of attachment to specific type IV pili and DNA penetration via Tol QRA proteins. The circular ssDNA is then converted to circular dsDNA which can replicate autonomously or become integrated. It is conceivable that both processes take place in the same cell, or even that integration is initiated on the ssDNA. Whether integrated or not, the phage genomes are subject to repression by their own repressors, like the master regulator Rst R of CTXΦ, or by cellular repressors like Lex A, and the degree of suppression depends on several factors. Total suppression of the integrated prophage blocks all phage replication and export as well as the production of co-regulated or independently regulated toxins carried on the phage genomes. The regulatory state is shifted when the bacterium infects its plant or animal target, or when it is exposed to UV radiation or to mitomycin C, all of which can induce virion production at low or high levels. The filamentous prophage states differ from classic lysogeny of the lambdoid phages in that lysis does not occur, and the cells are less immune to superinfection by more of the same phage. Although widely used, the classic terms lytic and lysogenic are not exactly correct for these systems.

Filamentous Phages in Bacterial Pathogens

Innumerable bacteria inhibit Earth and most cause no harm to plants or animals. Those that do cause harm are variants carrying virulence factors of various types, principally toxins, and these are present in genetic elements or islands in the chromosomes or plasmids in the otherwise harmless strains. They can be transmitted vertically to progeny or transmitted horizontally to nonpathogenic or other pathogenic strains by way of transposons, plasmids, or prophages or infectious phage particles. Detailed characterizations of virulence factors in *Vibrio cholera* strains, coupled with the prior observation that filamentous phage Cf1 can enter prophage states, set the stage for the discovery that cholera toxin genes are encoded on the genome of an integrative filamentous phage/prophage they named CTXΦ. Transmission of the virulence genes via filamentous phage particles converted nonpathogenic vibrio strains to pathogenic ones. Detailed studies of CTXΦ, have made it the paradigm for filamentous phage-borne virulence factors.

A comprehensive search for principal virulence factors of *E. coli* 018:K1:H7, which is a K1 pathogenic strain responsible for human neonatal meningitis and cystitis in women, led to the discovery of CUS-1, an inducible prophage integrated at the *E. coli dif* site. The 9.5 kbp genome of CUS-1 has five or so open reading frames (ORFs) not in the 6.4 kbp genome of Ff and their functions probably include an integrase and a repressor. Searches by polymerase chain reaction (PCR) of pathogenic and nonpathogenic strains established that prophage CUS-1 is tightly correlated with virulence. A prophage of virtually identical sequence integrates into the *dif* site of *Yersinia pestis* (biovar orientalis), and it was given the name CUS-2. An ORF present in CUS-1 is absent in CUS-2. The presence of CUS-2Φ filamentous particles in supernatants was observed, and PCR analyses of over 50 different *Y. pestis* strains established the wide distribution of CUS-2 in modern pathogenic strains. This same *Yersinia* prophage has also acquired the names Yp01Φ and Ypf1 in separate studies.

A meningococcal disease-associated prophage (MDA phage) of 8 kbp was recently identified through DNA microarray comparisons of the whole genomes of 29 highly invasive *Neisseria meningitis* strains with 20 noninvasive strains. The bacterial isolates were from different worldwide occurrences of disease, and the comparisons showed a single 8 kbp cluster of genes that were in 29/29 of the invasive strains but only in 2/29 of the noninvasive strains. The genes were clearly those of a filamentous phage, and filamentous phage virions were found in the culture supernatants. Statistical analyses pointed unequivocally to the phage as a principal virulence factor. Computer-based genomics of the *Neisseria* strains identified this same gene cluster in *N. memingitidis*, as well as many clusters of filamentous prophage-like genes in the *N. gonorrhea* genome. These were Nf2 through Nf4.

Pf4 is a prophage in the genome of *Pseudomonas aeruginosa*, strain 01 (PA01), that is induced and exported when this bacterium forms biofilms on different substrates. PA01 is a treacherous pathogen that causes infections in burns and nosarcomial infections in hospitals, as well as in the clogged lungs of cystic fibrosis patients. The Pf4 genome of 12.5 kbp contains a 7.5 kbp portion virtually identical to the Pf1 genome so one can project a 3500 nm long virion on the basis of a Pf1-like DNA conformation and a Pf1 length of 2100 nm. Another phage likely to have Pf1-like DNA structure and that is tightly associated with pathogenicity is f237. This virus has been found with high frequency as a prophage in *V. parahaemolyticus* pandemic strains of servar O3:K6 obtained from patients suffering from acute gastroenteritis and septicema caused by shellfish in Japan and North America. Its genome is similar to the CTXΦ genome with a potential virulence factor located in the position of the cholera toxin genes.

Filamentous Phage Display

The now highly developed and expanding technology of displaying libraries of protein structures from which selections are made for desired properties began over 20 years ago when G. P. Smith published his paper on displaying libraries of antigens fused to the gp3 of filamentous phage Ff(f1). The fundamental idea is that the selected property is directly coupled to the gene determining the sequence of the protein with that property. Through thoughtful library design of and efficient selection, it has become routine to obtain peptides or proteins of high specificity and affinity. Thousands of applications of display in many areas of research and biotechnology as well as nanotechnology have brought with them many ingenious solutions to technical problems such as bias in the supposedly random sequences, limitations on the sizes and orientations of the inserts, and their functionalities that can be presented. The work has led to better understanding of filamentous phage structure which in turn has led to new display applications. In addition to Smith's initial unexpected result, it was fully unexpected that all five structural proteins of Ff can be used to display libraries, and that enormous proteins could be fused to properly modified gp8 subunits, and also that even heterodimeric proteins, such as Fab light and heavy chains, can be presented on the gp7/9 end of the virions. These results and many others are due to the great plasticity of the filamentous phage structure. This plasticity itself was unexpected based on the constancy in sequence among the three Ff's mentioned at the beginning of this article and the complexity of the assembly and extrusion processes. The detailed mutagenesis studies of S. S. Sidhu have shown just how plastic the gp8 protein structures can be. Among the countless applications, the display of antibody libraries following the contributions of G. Winter and co-workers and the development of homing peptides following the work of E. Rouslahti seem particularly noteworthy. These types of applications are the subjects of many general and specialized reviews. Display technology using filamentous phage has become routine and powerful, and it has spawned other display technologies based on other phages and on cell surface display in yeast and *E. coli* that extend the possibilities for highly specific diagnostic methods, tissue and tumor specific drugs, therapeutic antibodies, and vaccines. Much of these developments will be based on continued refinements of filamentous phage display systems.

See also: Filamentous ssDNA Bacterial Viruses; History of Virology: Bacteriophages; Icosahedral ssDNA Bacterial Viruses.

Further Reading

Bille E, Zahar JR, Perrin A, et al. (2005) A chromosomally integrated bacteriophage in invasive meningococci. *Journal of Experimental Medicine* 201: 1905–1913.

Bradley DE, Coetzee JN, Bothma T, and Hedges RW (1981) Phage X: A plasmid-dependent, broad host range, filamentous bacterial virus. *Journal of General Microbiology* 126: 389–396.

Campos J, Martinez E, Marrero K, et al. (2003) Novel type of specialized transduction for CTX phi or its satellite phage RS1 mediated by filamentous phage VGJΦ in *Vibrio cholerae*. *Journal of Bacteriology* 185: 7231–7240.

Davis BM and Waldor MK (2005) Virulence-linked bacteriophages of pathogenic vibrios. In: Waldor MK, Freidman DI, and Adhya SL (eds.) *Phages: Their Role in Bacterial Pathogenesis and Biotechnology*, pp. 187–205. Washington, DC: ASM Press.

Day LA and Maniloff J (2000) *Inoviridae*. In: van Regenmortel MHV, Fauquet CM, Bishop DHL, et al. (eds.) *Virus Taxonomy: Seventh Report of the International Committee on Taxonomy of Viruses*, pp. 267–275. San Diego, CA: Academic Press.

Day LA, Marzec CJ, Reisberg SA, and Casadevall A (1988) DNA packing in filamentous bacteriophages. *Annual Review of Biophysics and Biophysical Chemistry* 17: 509–539.

Goldbourt A, Gross BJ, Day LA, and McDermott AE (2007) Filamentous phage studied by magic-angle spinning NMR: Resonance assignment and secondary structure of the coat protein in Pf1. *Journal of the American Chemical Society* 129: 2338–2344.

Gonzalez MD, Lichtensteiger CA, Caughlan R, and Vimr ER (2002) Conserved filamentous prophage in *Escherichia coli* O18:K1:H7 and *Yersinia pestis* biovar orientalis. *Journal of Bacteriology* 184: 6050–6055.

Liu DJ and Day LA (1994) Pf1 virus structure: Helical coat protein and DNA with paraxial phosphates. *Science* 265: 671–674.

Lubkowski J, Hennecke F, Pluckthun A, and Wlodawer A (1998) The structural basis of phage display elucidated by the crystal structure of the N-terminal domains of g3p. *Nature Structural Biology* 5: 140–147.

Model P and Russel M (2006) Filamentous bacteriophage. In: Calendar R (ed.) *The Bacteriophages*, 2nd edn., pp. 375–456. New York: Oxford.

Sidhu SS (2000) Phage display in pharmaceutical biotechnology. *Current Opinion in Biotechnology* 11: 610–616.

Tomar S, Green MM, and Day LA (2007) DNA–protein interactions as the source of large-length-scale chirality evident in the liquid crystal behavior of filamentous bacteriophages. *Journal of the American Chemical Society* 129: 3367–3375.

Tsuboi M, Kubo Y, Ikeda T, Overman SA, Osman O, and Thomas GJ, Jr. (2003) Protein and DNA residue orientations in the filamentous virus Pf1 determined by polarized Raman and polarized FTIR spectroscopy. *Biochemistry* 42: 940–950.

Wang YA, Yu X, Overman S, Tsuboi M, Thomas GJ, Jr., and Egelman EH (2006) The structure of a filamentous bacteriophage. *Journal of Molecular Biology* 361: 209–215.

Welsh LC, Symmons MF, Sturtevant JM, Marvin DA, and Perham RN (1998) Structure of the capsid of Pf3 filamentous phage determined from X-ray fibre diffraction data at 3.1 Å resolution. *Journal of Molecular Biology* 283: 155–177.

Insect Pest Control by Viruses

M Erlandson, Agriculture & Agri-Food Canada, Saskatoon, SK, Canada

© 2008 Elsevier Ltd. All rights reserved.

Glossary

Biological control The use of microbial pathogens to control pest organisms.
Integrated pest management Multifaceted approach in designing insect control strategies.
Optical brighteners Stilbene compounds used to enhance infectivity of biological control agents.
Melanoplus sanguinipes Species of migratory grasshopper.

Introduction

Insects take a tremendous toll on food and fiber production in the agricultural and forestry sectors, with estimates as high as 30% of crop production lost to insect damage in the field and in storage. Insects also act as vectors for many disease organisms infecting plants of agricultural importance and animals and humans. For some 60 years, insect control has relied primarily on the use of broad-spectrum chemical pesticides. This heavy usage has led to the emergence of secondary pest problems due to the loss of arthropod natural enemies and the development of resistance in insect pest populations to chemical insecticides. In addition, negative human health and environmental effects have been noted due to the broad target spectrum of these chemicals and their persistence in the environment. As a result, more environmentally sustainable pest control strategies have been proposed, notably integrated pest management (IPM). IPM incorporates a range of control tactics including cultural methods, judicious use of chemical insecticides, and biological control agents to suppress insect pest populations below levels that cause economic damage.

Biological control agents include microbial pathogens of insects and these are attractive for IPM systems due to their narrow host range. Currently microbial pest control products constitute only about 1.5% of the worldwide pesticide market, and microbial insecticide products are dominated by *Bacillus thuringiensis* (Bt)-based formulations. Much of the developmental research on microbial pathogens of insects has focused on their use in the same fashion as chemical insecticides. However, the potential for microbial pathogens to amplify in hosts and persist in pest populations following application allows the implementation of alternate use strategies. Four potential strategies for the use of microbial pathogens as insect control agents have been suggested: classical biological control, conservation, inoculation, and inundation. Although variable terminology has been used by researchers working on different groups of biological control agents to describe the four strategies, virus examples can be cited for each approach. The suitability of a virus for any of these strategies depends on the basic biology of the virus–host interaction and the specific attributes of the agricultural or forest production system.

Classical biological control involves the introduction of a microbial pathogen from a foreign source into a new ecosystem with an aim to permanent establishment and long-term suppression of an insect pest, most commonly a non-native species. A good example is the European spruce sawfly, *Gilpinia hercyniae*, a major forest pest of European origin, which became established in eastern North America in the 1930s through early 1940s. Following the introduction of a highly specific and efficacious baculovirus originating from northern Europe, the pest population was reduced to below an economically significant level and remains so to the present time.

Conservation aims to conserve or enhance naturally occurring populations of microbial pathogens of insects either by environmental manipulation or changes to existing insect control strategies, including use of chemical insecticides. There are few examples of the implementation of this strategy with microbial pathogens of insects in general and no good virus examples can be cited.

The inoculative release strategy involves the application of a microbial pathogen to an insect pest population at the early stage of an outbreak, based on the expectation of season-long control with the potential for carry-over to multiyear control. This approach has been referred to as augmentation when indigenous microbial pathogens are used and seasonal colonization when non-native species are introduced. There are a number of examples of this approach in forestry, including the baculovirus product TM Biocontrol-1 (Orgyia pseudotsugata multiple nucleopolyhedrovirus (MNPV)) for control of Douglas fir tussock moth. For this and the classical approach to work effectively the virus needs a mechanism for efficient spread through the host population and is often most successful for pest species that exhibit a colonial or gregarious behavior that increases the interaction between individuals.

The strategy of inundation is similar to the use of conventional chemical insecticides in that the microbial insecticide is applied to a pest insect population that has exceeded its economic threshold with the expectation of rapid control. This approach has been used mostly against pests of greenhouse and horticultural crops as, for example, the use of Spodoptera exigua nucleopolyhedrovirus for control of beet armyworm in greenhouse crops in Europe or the codling moth granulovirus (GV) in orchards. To succeed as a microbial insecticide, a virus must be sufficiently infectious and virulent to produce high rates of infection when applied in the field and kill insects in a timely fashion so as to limit economic plant damage or the vectoring potential of the insect pest. In addition, the virus must be relatively cheap to produce, in quantities required for the market, and be in a form that is readily stored and applied. The host specificity must be sufficiently broad to target pests that constitute an economically viable market and yet have enough target specificity to be advantageous in an IPM system. Finally, a virus's target specificity, biology, and ecology need to be well understood in order for government regulatory agencies to be convinced of its environmental and human health safety as a prerequisite for registration as a microbial insecticide.

Viruses are associated with insects in a wide range of ecological relationships: as pathogens, as plant and animal viruses relying on insect vectors for transmission, and as specialized symbionts that allow hymenopteran parasitoids to fully develop in their hosts (e.g., polydnaviruses). The current taxonomy of viruses includes 22 families whose hosts include an insect or other arthropod for at least some members of the group. At least 14 virus families have representatives pathogenic in invertebrates (**Table 1**) and at least some of these families have representative viruses that have been investigated for potential as biological control agents of insects.

In the following sections, those virus families whose members have been investigated most extensively for potential as insect control agents will be described. Research has focused on virus groups with readily observable symptomatology (baculoviruses and iridescent viruses), those producing impressive epizootics in insect populations (baculoviruses), and those easily detected by light microscopy as a consequence of being occluded in relatively large protein crystals (baculoviruses, cypoviruses (CPVs), and entomopoxviruses) within infected host cells. As well, economically important insect orders, notably Lepidoptera, Diptera, Coleoptera, and Orthoptera, have been more thoroughly screened for viruses than other arthropod and insect groups. Although a number of virus groups have been investigated, those in the family *Baculoviridae* show the most promise for development as insect biological control agents.

Table 1 Virus families associated with insects

Family/genus	Nucleic acid	Virion shape	Occlusion body	Related viruses in Vertebrates	Plants
Reoviridae					
Cypovirus	dsRNA, linear 10 segments	Isometric	+	Reovirus (e.g., blue tongue)	Rice dwarf virus Wound tumor virus
Tetraviridae				None	None
Betatetravirus	ssRNA+, 1-linear	Isometric	−		
Omegatetravirus	ssRNA+, 2-linear	Isometric	−		
Dicistroviridae				None	None
Cripavirus	ssRNA+, linear	Isometric	−		
Nodaviridae				Stripe jack virus	None
Alphanodavirus	ssRNA+, 2-linear	Isometric	−		
Picornaviridae	ssRNA+	Isometric	−	Polio, cold virus	Many
Rhabdoviridae	ssRNA−	Bullet-shaped	−	Rabies	
Baculoviridae				None	None
Nucleopolyhedrovirus	dsDNA, circular	Rod	+		
Granulovirus	dsDNA, circular	Rod	+		
Poxviridae				Smallpox	None
Entomopoxvirus	dsDNA, linear	Ovoid	+		
Iridoviridae				African swine fever virus	None
Iridovirus and Chloridovirus	dsDNA	Isometric	−		
Ascoviridae				None	None
Ascovirus	dsDNA, circular	Rod–ovoid	−		
Polydnaviridae				None	None
Bracovirus	dsDNA, m-circular	Rod	−		
Ichnovirus	dsDNA, m-circular	Fusiform	−		
Parvoviridae				Canine distemper virus	None
Densovirus	ssDNA	Isometric	−		

RNA Viruses

Tetraviridae

Tetraviruses are small icosahedral, nonenveloped viruses with linear single-stranded RNA (ssRNA) genomes. Tetraviruses have been isolated only from Lepidoptera. Infection appears to be restricted to the midgut and horizontal transmission by ingestion of infectious virus is likely the major route of infection; however, there is some evidence for vertical transmission. There is much variation in pathogenicity among tetravirus isolates, with symptoms ranging from inapparent to acutely lethal infections. Helicoverpa armigera stunt virus (HaSV) is the most extensively studied virus in this group and may have some potential as a biocontrol agent for pests in the *Helicoverpa* group. The *Helicoverpa* group contains some of the most damaging pests of agricultural crops, including the cotton bollworm, tobacco budworm, and corn earworm. HaSV is highly pathogenic for early instar larvae but has almost no impact on the last two larval stages. HaSV produces rapid cessation of feeding, significant delays in larval growth, and very characteristic stunting or shrinkage of the larval body. With one exception, no tetravirus replicates in cell culture and these viruses are difficult to produce in quantities required for insect control. Recently, HaSV virion assembly was demonstrated in plant protoplasts co-transformed with a plasmid expressing the HaSV capsid gene and plasmids carrying the complete cDNAs of the two HaSV genomic RNAs, suggesting novel methods of tetravirus production are possible.

In a recent field test, HaSV applied against *H. armigera* in sorghum achieved control comparable to that of commercial preparations of Gemstar (a registered strain of the baculovirus Heliothis zea single nucleopolyhedrovirus and standard chemical insecticides. These data along with the possibility of producing infectious HaSV virus particles in transgenic plants suggest that tetraviruses may have potential as viral insecticides or in more novel approaches comparable to transgenic crops expressing *B. thurinigensis* σ-endotoxins.

Reoviridae

Reoviruses are characterized by icosahedral virions with genomes consisting of 10–12 linear dsRNA molecules. The family *Reoviridae* contains a number of genera that replicate in both vertebrate hosts and arthropod vectors such as ticks and mosquitoes, as well as plant viruses that replicate both in plant hosts and arthropod vectors such as leafhoppers. The family also contains a single genus *Cypovirus* whose member viruses exclusively infect insects.

CPVs, like baculoviruses and entomopoxviruses (see below), are occluded within crystalline proteinaceous structures referred to as occlusion bodies (OBs) or polyhedra. The OB provides some stability for the virus particles in the environment. CPVs are transmitted orally and upon ingestion the OB dissolves in the alkaline conditions of the host insect gut, releasing virions to infect gut cells. Typical CPV infections are limited to midgut epithelial cells and disease symptoms are chronic in nature, mimicking those of starvation. These include reduced feeding and larval growth, increased development times, and often malformed adults that have reduced longevity and fecundity.

CPVs have been isolated most frequently from Lepidoptera but a few isolates from Diptera and Hymenoptera have been described. Although CPVs are highly infectious and persist in insect populations, they have received limited attention for development as viral insecticides due to their generally chronic rather than acute symptomology. To date, a single CPV product, Matsukemin, has been registered for use against *Dendrolimus spectabilis*, a pest of pine trees in Japan and China. It is significant that this virus is targeted against a pest of forests where substantial defoliation can be tolerated without economic damage. The other proposed use for CPV is in combination with chemical insecticides for control of insect populations resistant to these specific chemicals. There is evidence that CPV infection of the insect gut is able to suppress chemical insecticide detoxification based on cytochrome P450 pathways.

DNA Viruses

As previously mentioned, some of the most studied insect-specific DNA viruses are those readily detected due to the presence of OB and those producing observable epizootics in insect pests and thus are considered to have good potential as biological control agents of insects. In the following sections those virus families with species showing the best potential will be discussed, including *Parvoviridae*, *Poxviridae*, and *Baculoviridae*. The unassigned species *Oryctes rhinocerus virus* will also be discussed as it is an important example of a virus used successfully in an inoculative strategy.

Parvoviridae: Subfamily Densovirinae

Members of the subfamily *Densovirinae* exclusively infect arthropods. These viruses are small, nonenveloped, icosahedral virions with a single linear 4–6 kbp single-stranded DNA (ssDNA) genome. Densoviruses (DNVs) have been isolated from Lepidoptera, Diptera, Homoptera, Hemiptera, and Orthoptera. DNVs were discovered in laboratory colonies of the wax moth, *Galleria mellonella*, and in insects reared commercially such as wax moth and silkworm. DNVs infect larvae and produce a range of symptoms that include alterations in cuticular pigmentation, progressive paralysis, and death. Some DNVs appear to infect a single host while others have broad experimental host ranges. The virulence of DNVs varies considerably depending on the virus species.

Some of the DNVs infecting economically important pests are quite virulent and host specific, and have been investigated as biocontrol agents. Early studies showed that introductions of *Galleria mellonella* DNV into beehives controlled heavy infestations of wax moth. The potential of several DNVs for the control of the oil palm insect pests *Sibine fusca* and *Casphalia extranea* was tested in spray trials in South America, Africa, and Egypt using homogenates of infected insects. The results were promising but as yet no information is available on the development of commercial products based on these DNVs. The Aedes aegypti densovirus was investigated for mosquito control in the Soviet Union, and a commercial product called Viroden was developed.

Although there is considerable genome sequence homology between DNVs and some of the vertebrate parvoviruses, there is no evidence for DNV-related pathology in mammalian injection trials. However, more extensive safety tests would be required before any DNV could be registered as a viral insecticide under the current regulations in most of the developed nations.

Poxviridae: Subfamily *Entomopoxvirinae*

Entomopoxviruses, subfamily *Entomopoxvirinae*, are typical poxviruses with large brick-shaped enveloped virions and a linear double-stranded DNA (dsDNA) 130–375 kbp genome. These large complex viruses replicate in the cytoplasm of infected cells and mature virions are typically occluded within proteinaceous OBs called spheroids. There are three recognized genera in the *Entomopoxvirinae*, distinguished by virion morphology, genome size, and host range: *Alphaentomopoxvirus* has been isolated exclusively from Coleoptera, *Betaentomopoxvirus* from Lepidoptera and Orthoptera, and *Gammaentomopoxvirus* from Diptera.

Entomopoxviruses have been investigated for their potential use as biological control agents against orthopteran insects from which other insect virus groups, including baculoviruses, have not been described. Melanoplus sanguinipes entomopoxvirus (MSEV) and Oedaleus senegalensis entomopoxvirus are of interest because they infect many of the major grasshopper and locust pest species. MSEV, in particular, has a broad host range among grasshoppers and locusts. This virus is infectious upon ingestion. It initially infects midgut epithelium cells and eventually produces a systemic infection in the host. MSEV infections develop slowly in grasshoppers with mortality occurring only after 3 or more weeks, though other symptoms such as inhibition of pigment production (**Figure 1**), developmental delay, and reduced food consumption occur more quickly. Preliminary field trials in which 1.0×10^{10} MSEV spheroids per hectare, formulated in starch granules or on wheat bran, were applied to rangeland plots produced 25–30% infection levels among grasshoppers within 2 weeks. However, for grasshopper and locust pests, entomopoxviruses are not likely virulent enough to be considered for use as microbial insecticides for crop protection but may have the potential in an inoculative strategy to suppress grasshopper populations over a number of seasons in rangeland habitats where plant damage is not as critical as it is for cereal crops.

The morphological and biochemical similarities of entomopoxviruses to poxviruses infecting mammals have raised potential safety concerns. However, as more genome sequence data have become available, it is clear that entomopoxviruses have a set of collinear core genes

Figure 1 Nymphs of the migratory grasshopper, *Melanoplus sanguinipes*, infected with Melanoplus sanguinipes entomopoxvirus (a) compared to uninfected nymphs (b).

that distinguish them from vertebrate poxviruses. In addition, no serological relationship exists between insect and vertebrate poxviruses, and the evidence for host restriction of entomopoxviruses to insects is strong.

Baculoviridae

The *Baculoviridae* is the most studied group of insect-specific viruses due to their long-recognized potential as biological control agents and their use as eukaryotic expression vector systems. Baculoviruses are characterized by rod-shaped enveloped virions and a circular, dsDNA genome of 80–180 kbp. As with CPVs and entomopoxviruses, virions are contained within OB (0.5–5 μm diameter). The virus is infectious orally and occlusion-derived virions (ODVs) infect midgut epithelial cells after ingestion and dissolution of the OB in the alkaline condition of the midgut. Typically, two virion phenotypes occur in baculovirus infections: ODVs and budded virions (BVs), which spread the infection to tissues throughout the host. Thus the ODVs are responsible for horizontal transmission between insect hosts and BVs for systemic spread of infection within a host.

Baculoviridae contains two genera, *Nucleopolyhedrovirus* and *Granulovirus*, which are distinguished on the basis of OB morphology. GVs have been described only from Lepidoptera whereas NPVs have been isolated from Lepidoptera, Hymenoptera, Diptera, and possibly crustaceans. To date, baculoviruses have been isolated from over 700 species of arthropods.

Baculoviruses are often associated with spectacular epizootics, particularly among lepidopteran and sawfly forest pests. At late stages of baculovirus infection, insects may climb to the top of host plants where they die and are thus quite obvious (**Figure 2**). These infected cadavers serve as a source of inoculum as they degrade and release OB onto the plant canopy or into the soil. Although the virions within OB are quite susceptible to ultraviolet (UV) inactivation, when OBs are protected within intact cadavers or in upper layers of soil, the virus can remain viable for extended periods. In this way, the OBs can serve as a reservoir for infection of subsequent host generations, potentially for several years as the virus is returned to the plant canopy. This pattern of virus cycling is particularly evident for insect populations in more stable ecosystems such as forests.

The obvious nature of baculovirus epizootics led to an early interest in their potential as biological control agents. Baculoviruses have a number of attributes that are potential advantages for a microbial insecticide, particularly in the current context of increased interest in more environmentally sustainable pest control in an IPM system. In general, baculoviruses are quite virulent compared to other insect viruses and kill their hosts in 5–14 days depending on the specific virus–host interaction, virus

Figure 2 Bertha armyworm, *Mamestra configurata*, larva infected with Mamestra configurata nucleopolyhedrovirus.

dose, stage of insect development, food plant, and abiotic factors (notably temperature). Baculoviruses often replicate to high levels, and it is not uncommon for $1.0 \times 10^8 – 1.0 \times 10^9$ OBs to be produced in a single host, resulting in large quantities of inoculum for secondary infection beyond the initial virus application. Baculoviruses are typically quite host-specific, infecting a few closely related species. There are, however, exceptions. For example, Autographa californica multiple nucleopolyhedrovirus and Mamestra brassicae multiple nucleopolyhedrovirus infect upward of 100 species in several lepidopteran families. Narrow host specificity limits the risk of negative nontarget effects and makes baculoviruses attractive for use in IPM systems. The fact that baculoviruses are occluded in OB leads to increased environmental stability and enables baculovirus-based insecticides to be applied using existing spray application technology. Most current strategies for production of baculovirus-based insecticides rely on *in vivo* production in insects, allowing for relatively low technology production systems, which can be advantageous in the developing world. It is also generally considered that the cost of registration of baculoviruses, as low-risk pest control alternatives, may be less than that for registration of chemical insecticides.

However, some of the perceived advantages can also be a liability. For example, the host specificity of a baculovirus may limit its market size to one or a few pest species, and yet significant development and production costs are incurred. In contrast, costs are spread over a broader spectrum of target species in the case of a chemical insecticide. Although relatively virulent, baculoviruses need to be ingested by and replicate in the host before feeding stops and mortality occurs. Thus, the lag time between application and control is significantly longer than for contact chemical insecticides whose knockdown effect occurs very rapidly. Education of agricultural producers and/or pest control companies on strategies for the use of baculovirus insecticides will be essential for successful adoption of this technology. There is also a perceived high cost of production for baculoviruses whether done *in vivo* in cultured insects, largely due to labor costs, or in insect cell culture systems which can be capital intensive and operationally expensive due to media components required for their maintenance. However, one of the major limitations of baculoviruses is their limited field stability in aqueous spray suspensions due to UV inactivation. The half-life of a spray application of unformulated baculovirus may be as little as half a day under certain field conditions, and novel approaches to virus formulation are required to overcome this drawback.

Currently, a number of baculoviruses are registered as insect control products worldwide (**Table 2**). Most of the baculoviruses registered for control of lepidopteran and sawfly forest pests were developed and registered by governmental agencies; for example, Gypcheck and Neocheck-S were produced by the US Forest Services. These viruses have typically been used to treat pest populations before they reach their maximum outbreak phase, in essence creating an epizootic before it would naturally occur in a delayed density-dependent manner. This use pattern is an example of an inoculative release strategy. As many of these forest pest populations are cyclical in nature, the production and stockpiling of baculovirus-based insecticides has not been taken up by commercial ventures but rather by government agencies to be used when required.

A limited number of baculoviruses are used as microbial insecticides or inundative control agents in relatively large scale applications. The Cydia pomonella granulovirus (CpGV) for control the codling moth in orchard

Table 2 Registered and experimental virus-based bioinsecticides

Virus	Product names	Target pest	Production/crop system
Baculoviruses			
Lymantria dispar multiple nucleopolyhedrovirus	Gypcheck	Gypsy moth	Forestry
Orgyia pseudotsugata multiple nucleopolyhedrovirus	TM Biocontrol	Douglas fir tussock moth	Forestry
Neodiprion sertifer nucleopolyhedrovirus	Neocheck-S, Virox	European spruce sawfly	Forestry
Neodiprion lecontei nucleopolyhedrovirus	Lecontivirus	Redheaded pine sawfly	Forestry
Adoxphyes orana granulovirus	Capex 2	Summer fruit tortrix moth	Orchard
Cydia pomonella granulovirus	Madex 3, CYD-X, Granupom, Carposin Virosoft CP-4, Virin-Gyap	Codling moth	Orchard
Autographa californica multiple nucleopolyhedrovirus	VPN 80	Multiple pest targets	Horticulture, Glasshouse and field crops
Anagrapha falcifera multiple nucleopolyhedrovirus		Multiple pest targets	Horticulture, Glasshouse and field Crops
Anticarsia gemmatalis multiple nucleopolyhedrovirus	Polygen, multigen	Velvet bean caterpillar	Soybean
Heliocoverpa armigera single nucleopolyhedrovirus		*Heliothis/Helicoverpa* complex	Cotton
Helicoverpa zea multiple nucleopolyhedrovirus	Gemstar	*Heliothis/Helicoverpa* complex	Cotton, Horticulture
Mamestra brassicae multiple nucleopolyhedrovirus	Mamestrin	Multiple pest target	
Mamestra configurata multiple nucleopolyhedrovirus		Bertha Armyworm (*M. configurata*)	Canola
Spodoptera exigua multiple nucleopolyhedrovirus	Spodex	Beet Armyworm (*S. exigua*)	Horticulture, glasshouse, and field crops
Cypoviruses			
Dendrolimus spectabilis cypovirus 1	Matsukemin	Pine caterpillar	Forestry

crops has been successfully marketed under a number of product names. CpGV is an attractive control alternative for producers both to combat insecticide-resistant populations and where there is pressure to reduce chemical pesticide applications in the orchard industry. There are also a few examples of baculovirus products being used to control lepidopteran pests of field crops. These include: Helicoverpa zea nucleopolyhedrovirus and Helicoverpa armigera single nucleopolyhedrovirus (HaSNPV) for control of members of the *Helicoverpa/Heliothis* pest complex on cotton; Spodoptera exigua multiple nucleopolyhedrovirus for control of beet armyworm, *S. exigua*, in greenhouse ornamental and vegetable crops, as well as field crops; and Anticarsia gemmatalis multiple nucleopolyhedrovirus (AgMNPV) for control of velvetbean caterpillar, *Anticarsia gemmatalis*, in soybean. Although not used extensively, AcMNPV and Anagrapha falcifera multiple nucleopolyhedrovirus (a strain of AcMNPV) have great potential due to their much wider host range. Both HaSNPV and AgMNPV are used extensively. Each year, HaSNPV is applied to more than 100 000 ha of cotton in China. AgMNPV is the most successful example of a baculovirus-based insecticide, being applied annually to more than 2 million ha of soybean in Brazil. This virus is produced from infected field-collected and/or laboratory-reared caterpillars. A number of factors have been cited for the success of AgMNPV as a viral insecticide: (1) it is relatively virulent and gives good control when applied at 1.5×10^{11} OB/ha, a rate 5–10 times lower than that for some other baculovirus products; (2) it is efficiently transmitted within the pest population following application (thus, one AgMNPV application gives control equivalent to that from 1.8 applications of conventional chemical insecticides); (3) soybean can withstand substantial defoliation without yield reductions; (4) formulated AgMNPV costs producers less than chemical insecticides; and (5) an IPM program and education strategy for producers was developed during its introduction in Brazilian agriculture.

Despite the success of AgMNPV as a viral insecticide, the delay between virus application and pest mortality and the rapid inactivation of virus by UV radiation are still considered major drawbacks. The latter problem is being addressed for baculoviruses in general by formulation research. Previous formulation efforts were based on approaches typically used with chemical pesticides in terms of wetting, sticker and emulsifying agents, as well as UV-blocking and absorption agents. Recently, more novel UV protection agents have been investigated, including stilbene-derived optical brighteners. One unexpected outcome of this research was the demonstration of increased infectivity of baculoviruses in the presence of optical brighteners, possibly due to their interaction with the peritrophic matrix, a protective barrier that lines the insect midgut. This has led to increased research efforts to identify other synergistic compounds that increase efficacy when included in a formulation cocktail for baculoviral insecticides.

Significant research effort, based on baculovirus molecular biology and from the development of baculovirus expression vectors, has gone into the genetic manipulation to enhance speed of kill. Two lines of investigation have been pursued: (1) insertion and expression of exogenous genes that impact negatively on normal insect function; and (2) manipulation or deletion of baculovirus genes that affect infectivity or virulence. Exogenous genes that have been expressed in baculoviruses include insect-specific neurotoxin genes from invertebrate sources, and insect genes that encode peptide hormones, enzymes that affect hormones or hormonally regulated processes, and proteases (**Table 3**). Recombinant baculoviruses expressing insect-specific neurotoxins from scorpion and mite sources using baculovirus late and very late gene promoters (*p6.9*, *p10*, and *polyhedrin*) have produced the best results, decreasing the time to kill host insects by 30–50%, thus killing hosts in 48–72 h post-infection. The expression of peptide hormones or enzymes regulating hormone levels, for example juvenile hormone esterase, has met with less success, possibly because of the complex nature of insect endocrine regulation. Significant increases in the speed of kill, up to 50% faster, have been observed for a recombinant AcMNPV virus expressing the flesh fly protease, cathepsin L, an enzyme known to digest proteins associated with the basal lamina of the insect midgut. It is postulated that protease degradation of the midgut basal lamina allows for more rapid spread of virus infection to the rest of the insect.

Table 3 Genetic manipulation of baculoviruses for enhanced efficacy

Invertebrate toxins (insect selective neurotoxins)
  Androctonus australis (North African scorpion) insect toxin 1 (AaIT)
  Leiurus quinquestriatus hebraeus (Israeli scorpion) insect toxin 2 (LqhIT2)
  Bethus eupeus (Central Asian scorpion) insect toxin 1 (BeIT)
  Pyemotes tritici, straw-itch mite insect-selective neurotoxin (tox34)
  Agelenopsis aperta (American funnel web spider) insect–selective neurotoxin (μ-*Aga-IV*)
  Anemonia sulcata (sea anemones) insect-selective neurotoxin (AsII)
Other toxins
  Bacillus thuringiensis σ-endotoxin
Insect hormones
  Diuretic hormone – *Manduca sexta* DH
  Eclosion hormone – *M. sexta* EH
  Prothoracicotropic hormone
  Pheromone biosynthesis activating neuropeptide – *Helicoverpa zea* PBAN
Enzymes
  Juvenile hormone esterase – JHE – from multiple species
  Baculovirus enhancin (metalloprotease) – TnGV and MacoNPV
  Basal lamina degrading enzymes – human gelatinase A, rat stromelysin-1, and flesh fly cathepsin L

Manipulation of baculovirus genes for increased efficacy has for the most part focused on deletion strategies, particularly deletion of the gene for *ecdysteroid UDP-glucosyl transferase* (*egt*). *egt* encodes an enzyme that inactivates the insect hormone ecdysone which controls the timing of the molt. This gene is thought to confer an advantage on baculoviruses because delaying the molt, an energy- and resource-intensive process, leads to prolonged larval development and higher yields of virus. *egt*-deletion viruses kill host insects up to 30% more quickly than the wild-type virus.

Among other baculovirus genes that increase efficacy is *enhancin*, a gene found only in GVs and a few NPVs. *enhancin* encodes a metalloprotease that degrades the peritrophic matrix, and its expression in baculoviruses that do not naturally contain the gene has been demonstrated to lower the dose required for infection of host larvae.

A number of recombinant baculoviruses have been tested in small-scale field trials to evaluate efficacy, persistence, nontarget insect impacts, and safety. The most commonly used approach is a combination of the *egt*-deletion strategy and scorpion toxin gene expression. AcMNPV and HaSNPV recombinants of this type have been field-tested in different crop systems and have demonstrated significant increases in efficacy over wild-type viruses, in some cases giving equivalent crop protection as standard chemical insecticides.

To date, none of the recombinant baculoviruses has been registered or commercialized, but they appear to hold good promise as effective insect control agents. There is some question as to whether wild-type or recombinant baculoviruses will be successfully commercialized using a similar model to that for chemical. Perhaps more innovative strategies for commercially producing and marketing viral insect control agents will need to be devised.

Oryctes Rhinoceros Virus

Oryctes rhinoceros virus (OrV) was previously classified as a nonoccluded baculovirus based on structural similarities including an enveloped, rod-shaped virion and a circular dsDNA genome of approximately 130 kbp. However, there are morphological differences, notably nucleocapsid structure and the lack of an OB. In addition, genomic sequence data indicate that OrV is not related to baculoviruses. OrV was first discovered in the 1960s during a search for biological control agents for adults of the rhinoceros beetle, *Oryctes rhinoceros*, which caused severe damage to coconut palm plantations on South Pacific islands where the insect had been accidentally introduced. Originally discovered in rhinoceros beetle larvae in Malaysia, the subsequent utilization of this virus in an IPM system is credited with saving the palm industry in the South Pacific.

Larvae become infected with OrV by ingestion and produce a systemic infection with significant mortality occurring within 1–4 weeks. Larvae do not cause significant economic damage to palms as they typically feed on decaying vegetation in palm plantations. Thus initial control attempts with OrV targeted the larval stage by applying homogenized infected larvae mixed with sawdust to leaf litter in palm plantations. This strategy significantly reduced the buildup of adult beetle populations and feeding damage on palm shoots. In addition, infected adult beetles disseminated OrV to neighboring plantations. The early stage of infection in adult beetles is restricted to the midgut epithelium; thus, large quantities of OrV are voided in the fecal material, resulting in very efficient horizontal transmission to other adult beetles as well as to larvae. Virus infection of adults also leads to significantly reduced fecundity. Thus, it became common practice to release infected beetles into infested palm plantations as means of introducing the virus into the system. Initial trials conducted in Samoa showed reductions in palm damage of up to 95% in treatment versus control plots. Subsequently, introductions of virus-infected adults were made on other South Pacific islands leading to significantly lower economic damage to palms. This utilization strategy for OrV has led to the long-term control of this important pest with only occasional re-introductions of infected adults being required. This is an example of a very successful inoculative release strategy. The success is largely due to specific knowledge of the virus–pest interaction; the adult beetle produces large quantities of OrV inoculum in fecal material leading to infection of other beetles and the gregarious nature of the pest also contributes to the efficacy by which the virus spreads naturally.

Recently, changes in agriculture practices have led to resurgence in the rhinoceros beetle as an economic pest in Malaysia. This has spurred renewed interest in OrV as a biological control agent and led to additional studies on the ecology and genetics of the virus. OrV strains of differing virulence have been identified in natural populations; thus there are opportunities for selecting the most efficacious strains for particular beetle populations. The ongoing successful application of this virus for beetle control will require more basic knowledge of OrV genetics and full understanding of the environmental factors that determine the virus prevalence and dispersal in the pest population. Finally, it is noteworthy that in the 40 years since the discovery of OrV no similar viruses have been found in any arthropod other than beetles in the family Scarabaeidae.

Several other families of insect-specific DNA viruses have been investigated for their potential as biological control agents but due to their lack of virulence and poorly understood modes of transmission they are not discussed here. These include *Ascoviridae* and *Iridoviridae* (*Iridovirus* and *Chloriridovirus*).

Conclusions

Insect-specific viruses have proven potential for development as biological control agents of insect pests in the context of sustainable IPM systems. Their utilization faces several hurdles that include the lag time between virus application and insect control, virus production technology, and limited field stability, but these issues are being addressed by further research and development. To date, baculoviruses have been the most successfully exploited insect viruses in terms of the number of species registered and commercially developed for use in specific markets. There is potential for other virus groups including CPVs, tetraviruses, and entomopoxviruses to be developed as more detailed information on their genetics and biology is forthcoming. For all these virus groups the possibility of genetic engineering holds promise for increased efficacy. However, it is quite likely that innovative approaches that diverge from the chemical insecticide model may be needed to fully exploit the potential of viruses as insect biological control agents.

See also: Ascoviruses; Baculoviruses: Molecular Biology of Granuloviruses; Baculoviruses: Molecular Biology of Nucleopolyhedroviruses; Baculoviruses: Molecular Biology of Sawfly Baculoviruses; Entomopoxviruses; Oryctes Rhinocerous Virus; Parvoviruses of Arthropods; Tetraviruses.

Further Reading

Bonning BC (2005) Baculoviruses: Biology, biochemistry, and molecular biology. In: Gilbert L, Iatrou K, and Gill S (eds.) *Comprehensive Molecular Insect Science,* vol. 6, pp. 233–270. Amsterdam: Elsevier.

Christian PD, Murray D, Powell R, Hopkinson J, Gibb NN, and Hanzlik TN (2005) Effective control of a field population of *Helicoverpa armigera* by using the small RNA virus Helicoverpa armigera stunt virus (*Tetraviridae*: Omegatetravirus). *Journal of Economic Entomology* 98: 1839–1847.

Copping LG and Menn JJ (2000) Biopesticides: A review of their action, applications and efficacy. *Pest Management Science* 56: 651–676.

Huger AM (2005) The *Oryctes* virus: Its detection, identification, and implementation in biological control of the coconut palm rhinoceros beetle, *Oryctes rhinoceros* (Coleoptera: Scarabaeidae). *Journal of Invertebrate Pathology* 89: 78–84.

Kamita SG, Kang K-D, Hammock BD, and Inceoglu AB (2005) Genetically modified baculoviruses for pest insect control. In: Gilbert L, Iatrou K, and Gill S (eds.) *Comprehensive Molecular Insect Science,* vol. 6, pp. 271–322. Amsterdam: Elsevier.

Lacey LA, Frutos R, Kaya HK, and Vail P (2001) Insect pathogens as biological control agents: Do they have a future? *Biological Control* 21: 230–248.

Moscardi F (1999) Assessment of the application of baculoviruses for control of Lepidoptera. *Annual Review of Entomology* 44: 257–289.

Vail PV, Hostetter DL, and Hoffmann DF (1999) Development of multi-nucleocapsid nucleopolyhedroviruses (MNPVs) infectious to loopers (Lepidoptera: Noctuidae: Plusiinae) as microbial control agents. *Integrated Pest Management Reviews* 4: 231–257.

Insect Reoviruses

P P C Mertens and H Attoui, Institute for Animal Health, Pirbright, UK

© 2008 Elsevier Ltd. All rights reserved.

Glossary

Diptera (also known as true flies) An order of insects that comprises mosquitoes, gnats, midges, sand flies, and other flies. They possess a single pair of wings on the mesothorax (the middle of the three segments of the thorax of an insect) and a pair of halters which are derived from the hind wings present on the metathorax (the posterior of the three segments of the thorax of an insect). The name is derived from the number of wings (*di* means two in Greek and *pteron* refers to the wings).

Hymenoptera This order of insects comprises the sawflies, wasps, bees, and ants. The name refers to the membranous wings of the insects, and is derived from ancient Greek (*humen*: membrane and *pteron*: wings). The hind wings of hymenopterans are connected to their forewings by hooks.

Lepidoptera This order of insects comprises butterflies, skippers, and moths. The name is derived from ancient Greek and refers to the minute scales (*lepidon*) that cover the membranous lanceolate wings (*pteron*) of adults.

Introduction

Insects can become infected by many diverse DNA or RNA viruses, including members of the families *Baculoviridae*, *Poxviridae* (genus *Entomopoxvirus*), *Iridoviridae*, *Polydnaviridae*, *Ascoviridae*, *Birnaviridae*, and *Reoviridae*. The family *Reoviridae* (the reoviruses) includes 15 recognized genera of icosahedral, nonmembranous viruses, with genomes composed of 9–12 segments of linear double-stranded RNA (dsRNA). Many animal and plant reoviruses are transmitted by insect or arthropod vectors that they also infect (genera: *Orbivirus, Fiivirus, Phytoreovirus, Oryzavirus, Seadornavirus,* and *Coltivirus*). However, three established genera of reoviruses (*Cypovirus, Idnoreovirus,* and *Dinovernavirus*) contain viruses that only infect insects.

The genus *Cypovirus* (sigla: *cy*toplamic *po*lyhedrosis *virus*) was recognized by the International Committee for the Taxonomy of Viruses (ICTV) in 1978. Cypovirus (CPV) particles are ∼70 nm in diameter, contain a ten segmented dsRNA genome, and belong to the 'turreted' reoviruses, with surface projections (turrets) situated at each of the 12 'fivefold' vertices. Unlike most other reoviruses, CPVs are single-shelled, with no outer capsid layer. However, they characteristically produce crystalline protein inclusion bodies (polyhedra, usually within the cytoplasm of infected cells), in which virus particles become singly or multiply embedded. More than 230 CPV isolates have been described, with evidence for at least 20 distinct species (based on differences in the migration pattern of their genome segments during electrophoresis–electropherotype and nucleotide sequence divergence). However, very few CPVs have so far been isolated from Africa, and it appears likely that many more species will be recognized in the future.

The genus *Idnoreovirus* (sigla: *i*nsect-*d*erived *non*occluded *reovirus*) was formally recognized by ICTV in 2005. Idnoreoviruses have ten segmented dsRNA genomes, are icosahedral, ∼70 nm in diameter with turreted 'core' particles. However, unlike the CPVs, they also have an outer capsid protein layer and do not produce polyhedra. Idnoreoviruses have been isolated from various flies including the fruit fly, the small fruit fly (drosophila), the olive fly, and housefly. They have also been isolated from wasps.

The recently created genus *Dinovernavirus* (sigla: *d*ouble-stranded *in*sect, *nove*m (nine in Latin, indicating the number of genome segments) *RNA virus*) has initial support from ICTV and includes the first nine-segmented dsRNA virus. Like the CPVs, dinovernaviruses are single-shelled, turreted, and lack an outer capsid layer, but do not produce polyhedra or a 'polyhedrin' protein.

Table 1 summarizes the recognized species within the three genera described above.

Table 1 List of species (and tentative species) within genera *Cypovirus, Idnoreovirus,* and *Dinovernavirus*

Species	Representative isolate	Abbreviation
Genus *Cypovirus*		
Cypovirus 1	Bombyx mori cypovirus 1	BmCPV-1
Cypovirus 2	Aglais urticae cypovirus 2	AuCPV-2
Cypovirus 3	Anaitis plagiata cypovirus 3	ApCPV-3
Cypovirus 4	Actias selene cypovirus 4	AsCPV-4
Cypovirus 5	Euxoa scandens cypovirus 5	EsCPV-5
Cypovirus 6	Aglais urticae cypovirus 6	AuCPV-6
Cypovirus 7	Mamestra brassicae cypovirus 7	MbCPV-7
Cypovirus 8	Abraxas grossulariata cypovirus 8	AgCPV-8
Cypovirus 9	Agrotis segetum cypovirus 9	AsCPV-9
Cypovirus 10	Aporophyla lutulenta cypovirus 10	AlCPV-10
Cypovirus 11	Heliothis armigera cypovirus 11	HaCPV-11
Cypovirus 12	Autographa gamma cypovirus 12	AgCPV-12
Cypovirus 13	Polistes hebraeus cypovirus 13	PhCPV-13
Cypovirus 14	Lymantria dispar cypovirus 14	LdCPV-14
Cypovirus 15	Trichoplusia ni cypovirus 15	TnCPV-15
Cypovirus 16	Choristoneura fumiferana cypovirus 16	CfCPV-16
Tentative cypovirus species	Heliothis armigera cypovirus (B strain)	HaCPV-B
	Plutella xylostella cypovirus	PxCPV
	Maruca vitrata cypovirus (A strain)	MvCPV-A
	Maruca vitrata cypovirus (B strain)	MvCPV-B
Genus *Idnoreovirus*		
Idnoreovirus 1	Diadromus pulchellus idnoreovirus 1	DpIRV-1
Idnoreovirus 2	Hyposoter exiguae idnoreovirus 2	HeIRV-2
Idnoreovirus 3	Musca domestica idnoreovirus 3	MdIRV-3
Idnoreovirus 4	Dacus oleae idnoreovirus 4	DoIRV-4
Idnoreovirus 5	Ceratitis capitata idnoreovirus 5	CcIRV-5
	Drosophila melanogaster idnoreovirus 5	DmIRV-5
Tentative idnoreovirus species	Drosophila S virus	
Genus *Dinovernavirus*		
Dinovernavirus 1	Aedes pseudoscutellaris dinovernavirus 1	ApDNV-1

Historical Overview

CPVs were first recognized by Ishimori in 1934, who observed polyhedra in the midgut cells of diseased silkworm larvae. Cytoplasmic polyhedrosis has subsequently been recognized as an important cause of economic losses in the Japanese sericulture industry. CPVs have a wide host range, infecting many lepidopterous insect species (221 host species have been identified). Recent reports of CPV infections in mosquitoes confirm that they can also infect Diptera. The high levels of sequence divergence between equivalent genome segments of different CPV species, and the large numbers of distinct species, suggest that this is an ancient group. Indeed, CPVs have been detected in biting midges trapped in amber, indicating that they existed almost 100 million years ago.

The first idnoreovirus was isolated from the housefly *Musca domestica* in 1978. Several other idnoreoviruses have been isolated; the most recent (Diadromus pulchellus idnoreovirus 1 (DpIRV-1) has a genome composed of ten dsRNA segments in functional male insects, with an additional segment (derived from genome segment 3) in particles isolated from females or sterile males.

The genus *Dinovernavirus* currently contains a single species (*Aedes pseudoscutellaris dinovernavirus*; the virus Aedes pseudoscutellaris dinovernavirus 1 (ApDNV-1) was isolated from the AP61 cell line established by Varma *et al.* from the mosquito *Aedes pseudoscutellaris*.

Host Range, Diseases, Transmission, and Distribution

Cypoviruses

CPVs have only been isolated from arthropods, including members of over 45 genera of Lepidoptera. However, they have also been isolated from Hymenoptera, including the wasps *Polistes hebraeus* and *Diadromus pulchellus*, and from Diptera, including the mosquitoes *Culex restuans* and *Uranotaenia sapphirina*. One isolate was reported from the freshwater daphnid *Simocephalus expinosus*. Experimental studies indicate that CPVs cannot infect mammals or mammalian cell cultures.

The majority of CPV infections produce chronic disease, often without extensive larval mortality. Consequently, many individuals become adults even though they are heavily diseased. Infected larvae can stop feeding as early as 2 days post infection, producing symptoms of starvation. Their body size and weight are often reduced and diarrhea is common. The duration of the larval stage can also increase by ~1.5 times. Infected pupae are frequently smaller and many diseased adults do not emerge correctly, can be malformed, and flightless. Infected females may exhibit a reduced egg-laying capacity.

CPVs can be transmitted on the surface of eggs, producing high levels of infection in the subsequent generation. However, disinfection of the egg surface destroys the virus, indicating that CPVs are not transovarially transmitted. The minimum infectious dose increases dramatically in later larval instars, although virulence varies significantly with virus strain and host species. Larvae can sometimes recover from CPV infections, possibly because the gut epithelium has considerable regenerative capacity and infected cells are shed at each larval moult.

CPVs are normally transmitted by an oral–fecal route. Ingested polyhedra on contaminated food materials dissolve in the high pH environment of the insect gut, releasing occluded virus particles, which can then infect the cells lining the gut wall. Infection is generally restricted to the columnar epithelial cells of the larval midgut, although goblet cells may also become infected. CPV infection spreads throughout the midgut region (and sometimes infect the fat body), although infection of the entire gut has occasionally been observed in some species. The production of very large numbers of polyhedra gives the gut a characteristically creamy-white appearance. The endoplasmic reticulum of infected cells is progressively degraded, mitochondria enlarged, and the cytoplasm becomes highly vacuolated. In most cases, the nucleus shows few pathological changes. However a CPV strain has been detected that produces polyhedra within the cell nucleus. In the later stages of infection, cellular hypertrophy is common and microvillae are reduced or absent. Polyhedra are released into the gut lumen by cell lysis and are excreted. The gut pH is lowered during infection preventing dissolution of progeny polyhedra in the gut fluid.

Idnoreoviruses

All known members of the genus *Idnoreovirus* infect insects, with host species that include *Diadromus pulchellus* and *Hyposoter exiguae* (wasps: Hymenoptera); *Musca domestica* (housefly); *Dacus oleae* (olive fly); *Ceratitis capitata* (Mediterranean fruit fly); and *Drosophila melanogaster* (small fruit fly) (flies: Diptera).

Unlike the CPVs, idnoreoviruses appear to cause few pathological effects in their hosts, although they may significantly influence the biological properties of individual insects. Drosophila S virus appears to be associated with the 'S' phenotype in *D. simulans* (a maternally inherited morphological trait associated with abnormal bristles). The presence of an additional 3.33 kbp dsRNA segment in DpIRV-1 is related to the sex and ploidy of the host, and may play a role in the biology of the host wasp species, possibly by providing information necessary for larval development.

Dinovernavirus

ApDNV-1, the only representative of this genus, was isolated from persistently infected *A. pseudscutellaris* cells.

Virion Properties, Genome, and Replication

Cypoviruses

Large proteinaceous inclusion bodies are characteristically produced during CPV infections, usually within the cytoplasm of infected insect cells. The crystalline matrix of these 'polyhedra' is primarily composed of a single viral 'polyhedrin' protein that is often encoded by the smallest genome segment (Seg-10). The polyhedrin structure appears to protect the occluded virus particles, possibly filling the role of outer capsid proteins found in other reoviruses. Polyhedra are frequently symmetrical (cubic, icosahedral, or irregular) depending on both the virus strain (polyhedrin sequence) and the host species. The polyhedrin molecules appear to be arranged as a face-centered cubic lattice with center-to-center spacing varying between 4.1 and 7.4 nm. In some virus species, the virus particles are occluded singly within small occlusion bodies, which can also aggregate. However, some CPVs produce large polyhedra containing large numbers of particles, often regularly spaced within the polyhedrin matrix. 'Empty' polyhedra containing no virus particles have also been observed.

CPV particles have a single-layered capsid, composed of a central shell, ~57 nm in diameter (by cryoelectron microscopy), which extends to 71.5 nm when the 12 'turrets' or 'spikes' on the icosahedral fivefold vertices are included. These turrets are hollow, up to 20 nm in length and 15–23 nm wide (conventional electron microscopy and negative staining). The CPV capsid has a central compartment, ~35 nm in diameter, which normally contains the genomic dsRNA.

CPV virions are structurally comparable to the cores of other reoviruses, particularly those from the genera with 'turreted' cores (e.g., *Orthoreovirus*, *Aquareovirus*, and *Oryzavirus*) (**Figures 1(a)** and **2**). CPV particles generally contain five to six distinct proteins, three of which have been identified as the T2 protein (120 copies, equivalent to the VP3(T2) subcore shell protein of bluetongue virus, or orthoreovirus lambda1); 'large protrusion protein' (LPP, 120 copies, comparable to orthoreovirus sigma2); and 'turret protein' (TP, 60 copies, comparable to orthoreovirus lambda2). CPVs also contain transcriptase enzyme complexes attached to the inner surface of the capsid shell at the icosahedral fivefold vertices. Two or three of the CPV structural proteins are usually >100 kDa.

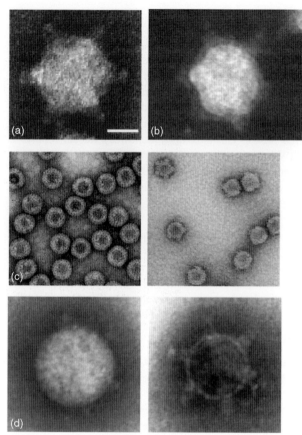

Figure 1 Electron micrographs of CPV, dinovernavirus, and idnoreovirus particles. (a) Contrast electron micrograph of nonoccluded Orgyia pseudosugata cypovirus 5 (OpCPV-5) virion. (b) Contrast electron micrograph of APDNV particles. The bars represent 20 nm. (c) Electron micrographs of purified virus particles (left) and core particles (right) of Hyposoter exiguae idnoreovirus-2 (HeIRV-1) stained with uranyl acetate. (d) Electron micrographs of a virus particle (left) and core particle (right) (stained with sodium phosphotungstate) of purified Dacus oleae idnoreovirus-4 (DoIRV-4). DoIRV-4 virions have small icosahedrally arranged surface projections (probably up to 12 in number), while the cores show much larger 'spikes' or 'turrets', which may lose a portion near to the tip (like those of the CPVs). (a) Courtesy of C. L. Hill. (c) Courtesy of Andrea Makkay and Don Stoltz. (d) Courtesy of Max Bergoin.

The structural proteins of Bombyx mori cypovirus 1 (BmCPV-1) are 148 (VP1), 136 (VP2), 140 (VP3), 120 (VP4), 64 (VP6), and 31 kDa (VP7) (see **Table 2**). Polyhedra also contain the 25–37 kDa 'polyhedrin protein' (28.5 kDa for BmCPV-1) which represents approximately 95% of the polyhedra protein (dry weight). The buoyant density of virions in CsCl is 1.44 g cm^{-3}, c. 1.30 g cm^{-3} for empty particles, and 1.28 g cm^{-3} for polyhedra. Due to the high level of variation between different CPVs, it is unlikely that their homologous proteins can be identified simply by their migration order during polyacrylamide gel electrophoresis (PAGE).

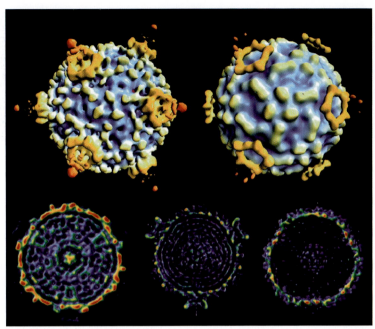

Figure 2 Cryoelectron microscopy reconstruction of a virion of Orgyia pseudosugata cypovirus-5 (OpCPV-5) to 26 Å resolution. Top left: Reconstruction of a nonoccluded virion; top right: reconstruction of an occluded virion. Bottom left: Cross section of a cryoelectron microscopy reconstruction of a fully occluded virion; bottom center: cross section of a full nonoccluded virion; and bottom right: cross section of an empty virion. The cross sections show evidence of dsRNA packaged as distinct layers and suggest localization of the transcriptase complexes at the fivefold axes of symmetry. Courtesy of C. L. Hill.

Table 2 Coding assignments of the CPV type 1 genome segments

Segment	Size (bp)	Protein	Function/location
Seg-1	4190	VP1 (VP1)	Major core (virion)
Seg-2	3854	VP2 (VP2)	RdRp (virion)
Seg-3	3846	(VP3)	(Virion)
Seg-4	3262	VP3 (VP4)	Possible methyltransferase (virion)
Seg-5	2852	NS1* (NS5)	Nonstructural, similar to FMDV 2Apro
Seg-6	1796	VP4 (VP6)	Leucine zipper ATP/GTP binding protein (virion)
Seg-7	1501	NS3, NS4 (VP7)	Nonstructural, with 'structural' cleavage products
Seg-8	1328	VP5 (NSP8)	Unknown (anomalous PAGE migration at 55 kDa)
Seg-9	1186	NS5 (NSP9)	Nonstructural, dsRNA binding
Seg-10	994	Polyhedrin	Polyhedrin protein (Pod)

NS1* (NS5): cleaved into NS2 (NS5a) and NS6 (NS5b).
RdRp, RNA-dependent RNA polymerase.
From information available at: http://www.iah.bbsrc.ac.uk/dsRNA_virus_proteins/BmCPV-1.htm.
Under the 'Protein' heading, an alternative nomenclature for the viral proteins is given in brackets. The nomenclature that is used in the Eighth Report of the ICTV and presented on the website: http://www.iah.bbsrc.ac.uk/dsRNA_virus_proteins/protein-comparison.htm.

CPVs retain infectivity for several weeks at −15, 5, or 25 °C. The virus retains full enzymatic activity (dsRNA-dependent single-stranded RNA (ssRNA) polymerase and capping activity) even after repeated freeze–thawing (up to 60 cycles), even though this disrupts the virus particle into ten active but distinct enzyme/template complexes. Each complex containing one genome segment and a complete transcriptase complex that includes one of the 'spike' structures derived from the fivefold axes of the virion capsid.

Cations have relatively little effect on CPV structure. Heat treatment of virions at 60 °C for 1 h leads to degradation and release of genomic RNA. Virus particles are relatively resistant to treatment with trypsin, chymotrypsin, ribonuclease A, deoxyribonuclease, or phospholipase. Virion enzyme functions also show some resistance to treatment with proteinase K. However, this may reflect retention of enzyme activities despite particle disruption, particularly during the early stages of digestion. CPV particles are resistant to detergents including sodium deoxycholate (0.5–1%), but are disrupted by 0.5–1% sodium dodecyl sulfate (SDS), which releases the genomic dsRNA. Treatment with triton X-100, NP40, or urea also causes disruption of the virus particle structure. One or two fluorocarbon treatments have little effect on virus infectivity, though treatment with ethanol leads to release of RNA from virions. Viruses and polyhedra are readily inactivated by ultraviolet (UV) irradiation which dissociates the dsRNA template from the transcriptase complexes. Polyhedra remain infectious for years at temperatures below 20 °C. Virions can be released from polyhedra by treatment with carbonate buffer at pH greater than 10.5 but are disrupted below pH 5. As in permissive insects' midguts, high pH treatment completely dissolves the polyhedral protein matrix. This process is partly due to increased solubility of polyhedrin at high pH but is also aided by alkaline-activated proteases associated within the polyhedra.

Polyhedra (but not virions) contain significant amounts of adenylate-rich oligonucleotides. In the majority of cases, CPV particles contain ten dsRNA genome segments. However, virus particles containing an 11th small segment have been detected (e.g., Trichoplusia ni cypovirus 15, TnCPV-15). The genome segments of BmCPV-1 vary from 4190 to 994 bp with a total genome size of 24 809 bp. The genome segments of other CPVs have been estimated by electrophoretic comparisons, and have calculated sizes between 5.6 and 0.6 kbp, with a total genome length of 29.2–33.3 kbp.

The size distribution of the genome segments varies widely between different CPVs (e.g., the smallest dsRNA has an estimated size that varies between 0.53 and 1.44 kbp). These size differences formed the initial basis for recognition and classification of distinct species (electropherotypes) of CPVs (which differ significantly in the migration of at least three genome segments during electrophoresis using 1% agarose or 3% SDS-PAGE). The genome segment migration patterns of types 1, 12, and 14 have some overall similarity, suggesting that they are more closely related than some other CPVs (**Figure 3**).

The termini of different genome segments within a single CPV species are often highly conserved but differ from those reported for other species (**Table 3**). Choristoneura fumiferana cypovirus 16 (CfCPV-16) shows high levels of sequence variation when compared to CPV-1, -2, -5, -14, or -15 (**Table 3**) and is therefore considered to be a distinct species, even though it has a similar 5′ end but different 3′ ends to representatives of CPV-5. These data demonstrate that different CPV electropherotypes are likely to have different conserved RNA terminal sequences.

Large size variations in the genome segments of most CPV species (apart from CPV-1, -12, and -14) indicate that the gene assignments for one 'type' or species cannot simply be applied to the other CPVs. Genome segment coding assignment by *in vitro* translation of individual genome segments have been published for isolates of CPV-1 and -2. These data and subsequent sequencing studies indicate that the polyhedrin protein is often encoded by the smallest segment.

CPV uptake appears to be a relatively inefficient process in insect cell cultures, but can be significantly improved by lipofection. Liposomes deliver CPV particles into the cytoplasm where replication occurs. CPVs do not require particle modification to activate transcriptase and capping enzymes. The outer coat proteins

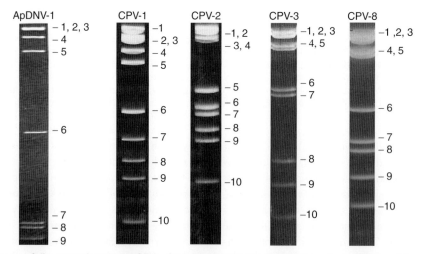

Figure 3 The genomes of dinovernaviruses and CPVs. Genome electropherotypes of members of genera *Dinovernavirus* (ApDNV-1) and *Cypovirus* (CPV-1, -2, -3, and -8). The segment position is indicated to the right of each profile.

Table 3 The conserved terminal sequences of member viruses of genera *Cypovirus*, *Idnoreovirus*, *Dinovernavirus*, and their comparison to those of genera *Fijivirus* and *Oryzavirus* (both are plant viruses)

Virus	(Isolate)	Conserved RNA terminal sequences
Cypovirus 1	(BmCPV-1)	5′-AGUAA............GUUAGCC-3′
	(DpCPV-1)	5′-AGTAA............GUUAGCC-3′
	(LdCPV-1)	5′-AGU$^{A/A}$/$_{G/G}$............G^u/$_c$UAGCC-3′
Cypovirus 2	(LiCPV-2)	5′-AGUUUUA............UAGGUC-3′
Cypovirus 4	(ApCPV-4)	5′-AAUCGACG............GUCGUAUG-3′
	(AaCPV-4)	5′-AAUCGACG............GUCGUAUG-3′
	(AmCPV-4)	5′-AAUCGACG............GUCGUAUG-3′
Cypovirus 5	(OpCPV-5)	5′-AGUU............UUGC-3′
Cypovirus 14	(LdCPV-14)	5′-AGAA............CAGCU-3′
Cypovirus 15	(TnCPV-15)	5′-AUUAAAAA............GC-3′
Cypovirus 16	(CfCPV-16)	5′-AGUUUUU............UUUGUGC-3′
Idnoreovirus 1	DpIRV-1	5′-^{A}/$_{U/G}$CAAUUU............AGUAAAAAAAUn^{A}/$_{G}$G-3′
Dinovernavirus	APDNV	5′-AGUU^{A}/$_U$............^{A}/$_U$AGU-3′
Fijivirus	NLRV	5′-AGU............GUUGUC-3′
	MRCV	5′-AAGUUUUU............GUC-3′
	FDV	5′-AAGUUUUU............GUC-3′
	RBSDV	5′-AAGUUUUU............GUC-3′
Oryzavirus	RRSV	5′-GAUAAA............GUGC-3′

Data concerning the terminal regions of members of the *Reoviridae* are listed at: http://www.iah.bbsrc.ac.uk/dsRNA_virus_proteins/CPV-RNA-Termin.htm.

of other reoviruses need to be removed to activate transcription. CPV polymerase activity can show very pronounced dependence on the presence of S-adenosyl-L-methionine or related compounds, although this varies between different CPV species.

Virus replication and assembly occur within the host cell cytoplasm, although viral RNA synthesis can sometimes occur within the nucleus. Replication is accompanied by the formation of viroplasm (virogenic stroma or 'VIB') within the cytoplasm, containing large amounts of viral proteins and virus particles. The mechanism used to select the individual genome segments for packaging and assembly into progeny particles (exactly one copy of each segment) is unknown. The importance of the conserved terminal regions in this process is indicated by packaging and transcription of a mutant Seg10 of CPV-1 that contained only 121 bp from the 5′ end and 200 bp from the 3′ end. Particles are occluded within polyhedra apparently at the periphery of the VIB, from about 15 h post-infection onward. The polyhedrin protein is produced late in infection and in large excess compared to the other viral proteins. It is unknown how polyhedrin synthesis is regulated.

Idnoreoviruses

Electron microscopy and negative staining of the idnoreovirus particles show that they have no prominent features, are spherical in appearance (icosahedral symmetry), and ~70 nm in diameter. However, they do have a clearly defined outer capsid layer that can be removed to reveal the virus core. Unlike the CPVs, idnoreoviruses do not encode a polyhedrin protein, and the virus particles are 'nonoccluded'. Idnoreovirus core particles have an estimated diameter of ~60 nm, with 12 icoahedrally arranged surface 'turrets' or 'spikes'.

Limited studies of some viruses within the genus indicate that they are resistant to freon (trichlorotrifluoroethane) and CsCl. They may also be resistant to chymotrypsin. Intact virus particles and cores of the prototype isolate Diadromus pulchellus idnoreovirus-1 (DpIRV-1) have densities of 1.370 and 1.385 g ml^{-1}, respectively, while intact virions and empty particles of Dacus oleae idnoreovirus-4 (DoIRV-4) have a density of ~1.38 and ~1.28 g ml^{-1}, respectively (determined by CsCl gradient centrifugation). The outer capsid layer of Hyposoter exiguae idnoreovirus-2 (HeIRV-2) can be disrupted by brief exposure to 0.4% sodium sarcosinate, releasing the virus core.

The genome of most idnoreoviruses consists of ten linear segments of dsRNA that are conventionally identified as 'genome segment 1' to 'genome segment 10' (Seg-1 to Seg-10) in order of reducing molecular weight (and increasing electrophoretic mobility during agarose gel electrophoresis (AGE)). The total genome of DpIRV-1 contains an estimated 25.15 kbp of dsRNA, with individual segments ranging between ~4.8 and ~0.98 kbp, and an electrophoretic migration pattern (by 1% AGE) showing five larger and five smaller segments (a '5/5' pattern). However, the virions of DpIRV-1 may be atypical since they can sometimes also contain an 11th, 3.33-kbp genome segment, the presence of which is related to the sex and ploidy of the individual wasp host. This additional dsRNA (migrating between genome segments 3 and 4) contains sequences similar to and therefore possibly derived from Seg-3 (3.8 kbp).

The genome segments of HeIRV-2 range in size from ~3.9 to ~1.35 kbp, with a '4/6' electrophoretic migration pattern (by PAGE). DoIRV-4 contains an estimated 23.4 kbp, with segments estimated from ~3.8 to ~0.7 kbp and a '5/3/2' electrophoretic migration pattern (by PAGE). Ceratitis capitata idnoreovirus-5 (CcIRV-5) has a '3/3/4' genome segment migration pattern when analyzed by PAGE and has clear similarities to Drosophila melanogaster idnoreovirus-5 (DmIRV-5), suggesting that despite some serological differences, they belong to the same virus species. It is unclear how closely drosophila S virus is related to the other drosophila viruses. It is therefore, currently, classified as a 'tentative species' within the genus.

By analogy with other reoviruses (e.g., the orbiviruses), genome segment migration patterns (by AGE) are likely to be characteristic of each *Idnoreovirus* species. Initial sequencing studies suggest that the 3' termini of DpIRV-1 genome segments are more variable than most other reoviruses with little sign of conservation. However, conserved sequences were detected at the 5' termini (**Table 3**), which are different from those of other reovirus species. No sequence data are currently available for other members of this genus.

Although the proteins of DpIRV-1 have not been extensively characterized, purified virions contain 11 structural proteins with M_r 21–140 kDa (as analyzed by SDS-PAGE). Three of these appeared to be glycosylated (M_r approximately of 21, 15, and 35 kDa). Some of the viral genome segments have been sequenced (as indicated in **Table 4**). The viral proteins are named as VP1–VP10 based on the molecular weight of the genome segment (segment number) from which they are translated.

Table 4 Coding assignments for the DpIRV-1 genome segments

Segment	Size (bp)	Protein	Function/location
Seg-1	4800	VP1	
Seg-2	4230	VP2	
Seg-3	3812	VP3	Contains RdRp motifs
Seg-x	3333		Sequence closely related to Seg-3. Presence related to sex and ploidy of the host
Seg-4	3000	VP4	
Seg-5	2700	VP5	
Seg-6	1750	VP6	
Seg-7	1652	VP7	
Seg-8	1318	VP8	
Seg-9	1240	VP9	
Seg-10	985	VP10	

Seg-x: the supernumerary segment that is linked to the sex and ploidy of the insect host.
RdRp, RNA-dependent RNA polymerase.
More information is available at: http://www.iah.bbsrc.ac.uk/dsRNA_ virus_proteins/Idnoreovirus.htm.

On the basis of their structural and biochemical similarity, it seems likely that many aspects of the genome organization and replication of the idnoreoviruses will show similarities to those of other reoviruses (particularly the other turreted viruses). On this basis it is likely that the virus core will contain transcriptase complexes that synthesize mRNA copies of the individual genome segments. These will be exported and translated to produce the viral proteins within the host cytoplasm. These positive-sense RNAs are also likely to form templates for negative-strand synthesis during progeny virus assembly and maturation. The genome segments that have been characterized represent single genes, with a large ORF, and relatively short noncoding terminal regions.

Dinovernavirus

Transmission electron microscopy of ApDNV (the prototype dinovernavirus) showed that purified virions have a structure similar to core particles of turreted reoviruses (**Figure 1(b)**), with particular similarities to the single-shelled CPVs, with no outer capsid layer. The mean diameter of the virus particle is ~49–50 nm, with a central space (for the viral genome) that is 36–37 nm in diameter. Turrets were clearly visible projecting from the particle surface.

The infectivity of ApDNV-1 particles is destroyed by treatment with 0.1% SDS but unaffected by 1% deoxycholate or repeated treatments with Freon 113 or Vertrel-XF. Freezing at −20 or −80 °C, or heating to 50–60 °C for 30 min abolished ApDNV infectivity, even in presence of 50% fetal bovine serum (FBS). However, infectivity was stable for up to 3 weeks at room temperature and for at least 5 months at 4 °C. ApDNV is stable at pH 6–8, but infectivity was reduced tenfold at pH 4–5 or pH 9–10. The virion morphology (observed by electron microscopy) becomes distorted below pH 5 and is completely disrupted below pH 3.5.

ApDNV persistently infects AP61 cells without any visible cytopathic effects and has been detected in these cells from several different sources. The number of copies of the ApDNV genome in each persistently infected AP61 cell was estimated by quantitative polymerase chain reaction (PCR) as between 1 and 5, although treatment with 2-aminopurine (2-AM) can increase the number to 60–80 genome copies per cell. The virus can also infect and replicate in mosquito cells, including AA23 and C6/36 (from *Aedes albopictus*) cells and A20 cells (from *Aedes aegypti*), but not in AE cells (*A. aegypti*) or Aw-albus (*Aedes w-albus*). ApDNV failed to replicate in mammalian cells or in mice.

Unlike other reoviruses, the genome of ApDNV only contains nine segments of dsRNA, with a 4/1/3 AGE migration profile (**Figure 3**). The termini of ApDNV contain conserved terminal sequences (**Table 3**). The first three nucleotides 'AGU' are conserved between ApDNV, some fijiviruses, and some CPVs; though, unlike the CPVs and fijiviruses, the first and last nucleotides of the ApDNV genome segments are complementary (A and U). The mean G+C content of the ApDNV genome is 34.4%, as compared to 34.8% for fijiviruses, 44.7% for oryzaviruses, 43% for the CPVs, and 63% for idnoreoviruses.

The putative functions of ApDNV proteins VP1–VP9 (based on sequence similarities to other reovirus proteins) are shown in **Table 5**. These comparisons suggest that there is no equivalent to the CPV polyhedrin gene and appear to confirm the status of ApDNV as a distinct and authentic nine-segmented virus.

Antigenic and Genetic Relationships

Cypoviruses

CPVs are currently classified within 16 species that were initially identified by their distinctive dsRNA migration patterns. Cross-hybridization analyses of the dsRNA, serological comparisons of the viral proteins, and comparison of RNA sequences confirmed the validity of this classification and have identified five further tentative species within the genus. However, only relatively few CPV isolates have been characterized, suggesting that there may be many more species that are as yet unidentified.

Virus isolates within a single CPV electropherotype exhibit high levels of antigenic cross-reaction, as well as efficient cross-hybridization of their denatured genomic dsRNA, even under high-stringency conditions. In contrast, there is little or no evidence of serological cross-reaction between viruses from different electropherotypes (representing different CPV species). However, CPV-1 and CPV-12 are exceptions, showing a low but significant level of serological cross-reaction. These viruses also show some overall similarity in their electropherotype pattern and detectable levels of cross-hybridization of their genomic RNA. CPV-14 also shows some similarity in its RNA electropherotype pattern to both CPV-1 and CPV-12, suggesting that it may also show some antigenic relationship and RNA sequence homology with these viruses.

The nomenclature currently used to identify different CPV isolates takes account of both the virus species and the host species from which the virus was originally isolated (e.g., an isolate of CPV-1 from *Bombyx mori* would be identified as BmCPV-1). The sequence data that are available for isolates of CPV-1, -2, -5, -14, -15, and -16 allow a comparison of some genes from these viruses. There is significantly higher conservation in the largest genome segments (possibly as a result of functional constraints) although the level of variation is relatively uniform across the whole genome. Earlier cross-hybridization studies suggested that the level of nucleotide sequence variation is also relatively uniform across the whole CPV genome, possibly reflecting the absence of neutralizing antibodies (and therefore antibody selective pressure) in their insect hosts. Different isolates within a single CPV species usually show very high levels of nucleotide sequence identity. For example, different isolates of CPV-5 show >98% identity in genome Seg 10 (the polyhedrin gene), while different isolates of CPV-1 show 89–98% nucleotide sequence identity in this gene. In contrast, comparisons of unrelated species showed only relatively low levels (20–23%) of sequence identity (**Figure 4**).

The level of amino acid (aa) identity in the putative viral polymerase varied from 92.9% to 99.5% within a single CPV type, and 42.3–43.3% between different types (**Figure 5**); see **Table 6** for abbreviations and accession numbers).

Table 5 Coding assignments for the ApDNV-1 genome segments

Segment	Size (bp)	Protein	Function by similarity
Seg-1	3817	VP1	Possible cell attachment
Seg-2	3752	VP2	RdRp
Seg-3	3732	VP3	T2
Seg-4	3375	VP4	Nonstructural
Seg-5	3227	VP5	Possible capping enzyme
Seg-6	1775	VP6	Possible NTPase
Seg-7	1171	VP7	Nonstructural
Seg-8	1151	VP8	Possible translational regulation
Seg-9	1147	VP9	Viral inclusion bodies

More information is available at: http://www.iah.bbsrc.ac.uk/dsRNA_virus_proteins/Dinovernavirus.htm.

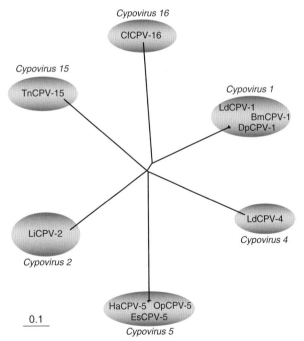

Figure 4 Phylogenetic tree for polyhedrin proteins from isolates belonging to six *Cypovirus* species. The scale bar indicates the number of substitutions per site.

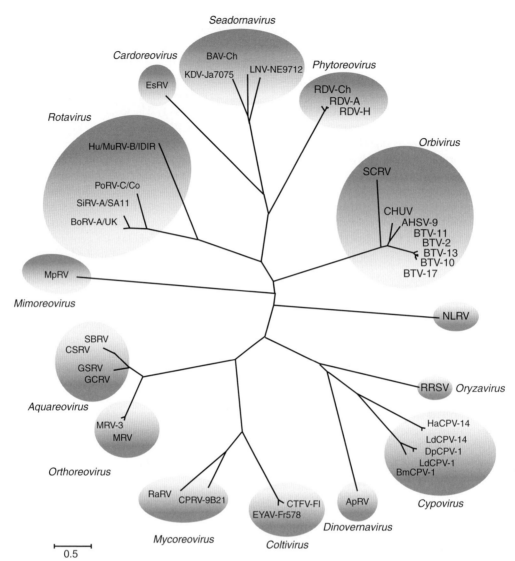

Figure 5 Neighbor-joining phylogenetic tree, built with available polymerase sequences for representative members of 15 recognized genera of family *Reoviridae*. The abbreviations and accession numbers are those provided in **Table 6**. The scale bar indicates the number of substitutions per site.

Idnoreoviruses

Antigenic relations between different idnoreoviruses have not been analyzed. The phylogenetic relationship between different idnoreoviruses is also unknown (only DpRV has been sequenced). However, the sequence of DpRV shows homologies to other reoviruses (aa identities reaching up to 20% with fijiviruses, oryzaviruses, rotaviruses, seadornaviruses, and orbiviruses).

Dinovernaviruses

Sequence comparison of the Aedes pseudoscutellaris sdinovernavirus genome to other reoviruses has shown significant aa identities in each of the viral proteins. Members of the genera *Cypovirus*, *Oryzavirus*, and *Fijivirus*, which are also turreted, showed the highest aa identities (19–31%). Amino acid identity values <30% between RdRps of the various reoviruses can be used to distinguish the members of different genera. The relationship of ApDNV to other reoviruses is illustrated by a phylogenetic tree for the RdRp gene (**Figure 5**). Although these sequence comparisons confirm the validity of ApDNV classification within a separate genus, they also indicate significant relationships with the turreted CPVs, fijiviruses, and oryzaviruses (aa identities of 26%, 23%, and 22%, respectively). Lower levels of aa identity (20%) were also detected with the mycoreoviruses (which are also turreted).

Table 6 Sequences of the RNA-dependent RNA polymerases (RdRps) used in phylogenetic analysis of different reoviruses (**Figure 4**)

Species	Isolate	Abbreviation	Accession number
Genus *Seadornavirus* (12 segments)			
Banna virus	Ch	BAV-Ch	AF168005
Kadipiro virus	Java-7075	KDV-Ja7075	AF133429
Liao ning virus	LNV-NE9712	LNV-NE9712	AY701339
Genus *Coltivirus* (12 segments)			
Colorado tick fever virus	Florio	CTFV-Fl	AF134529
Eyach virus	Fr578	EYAV-Fr578	AF282467
Genus *Orthoreovirus* (10 segments)			
Mammalian orthoreovirus	Lang strain	MRV-1	M24734
	Jones strain	MRV-2	M31057
	Dearing strain	MRV-3	M31058
Genus *Orbivirus* (10 segments)			
African horse sickness virus	Serotype 9	AHSV-9	U94887
Bluetongue virus	Serotype 2	BTV-2	L20508
	Serotype 10	BTV-10	X12819
	Serotype 11	BTV-11	L20445
	Serotype 13	BTV-13	L20446
	Serotype 17	BTV-17	L20447
Palyam virus	Chuzan	CHUV	Baa76549
St. Croix river virus	SCRV	SCRV	AF133431
Genus *Rotavirus* (11 segments)			
Rotavirus A	Bovine strain UK	BoRV-A/UK	X55444
	Simian strain SA11	SiRV-A/SA11	AF015955
Rotavirus B	Human/murine strain IDIR	Hu/MuRV-B/IDIR	M97203
Rotavirus C	Porcine Cowden strain	PoRV-C/Co	M74216
Genus *Aquareovirus* (11 segments)			
Golden shiner reovirus	GSRV	GSRV	AF403399
Grass carp reovirus	GCRV-873	GCRV	AF260511
Chum salmon reovirus	CSRV	CSRV	AF418295
Striped bass reovirus	SBRV	SBRV	AF450318
Genus *Fijivirus* (10 segments)			
Nilaparvata lugens reovirus	Izumo strain	NLRV-Iz	D49693
Genus *Phytoreovirus* (10 segments)			
Rice dwarf virus	Isolate China	RDV-Ch	U73201
	Isolate H	RDV-H	D10222
	Isolate A	RDV-A	D90198
Genus *Oryzavirus* (10 segments)			
Rice ragged stunt virus	Thai strain	RRSV-Th	U66714
Genus *Cypovirus* (10 segments)			
Bombyx mori cytoplasmic polyhedrosis virus 1	BmCPV-1	BmCPV-1	AF323782
Dendrolimus punctatus cytoplasmic polyhedrosis 1	DpCPV-1	DpCPV-1	AAN46860
Lymantria dispar cytoplasmic polyhedrosis 1	LdCPV-1	LdCPV-1	NC_003017
Lymantria dispar cytoplasmic polyhedrosis 14	LdCPV-14	LdCPV-14	AAK73087
Heliothis armigera cypovirus 14	HaCPV-14	HaCPV-14	DQ242048
Genus *Mycoreovirus* (11 or 12 segments)			
Rosellinia anti-rot virus	W370	RaRV	AB102674
Cryphonectria parasitica reovirus	9B21	CPRV	AY277888
Genus *Mimoreovirus* (11 segments)			
Micromonas pusilla reovirus	MpRV	MPRV	DQ126102
Genus *Dinovernavirus* (9 segments)			
Aedes pseudoscutellaris dinovernavirus	ApDNV	ApDNV	DQ087277
Genus *Cardoreovirus* (12 segments)			
Eriocheir sinensis reovirus	Isolate 905	EsRV	AY542965

See also: African Horse Sickness Viruses; Aquareoviruses; Bluetongue Viruses; Coltiviruses; Insect Pest Control by Viruses; Mycoreoviruses; Orbiviruses; Plant Reoviruses; Reoviruses: General Features; Reoviruses: Molecular Biology; Rotaviruses; Seadornaviruses.

Further Reading

Attoui H, Jaafar MF, Belhouchet M, *et al.* (2005) Expansion of family *Reoviridae* to include nine-segmented dsRNA viruses: Isolation and characterization of a new virus designated Aedes pseudoscutellaris reovirus assigned to a proposed genus (Dinovernavirus). *Virology* 343: 212–223.

Mertens PPC, Attoui H, Duncan R, and Dermody TS (2005) *Reoviridae*. In: Fauquet CM, Mayo MA, Maniloff J, Desselberger U, and Ball LA (eds.) *Virus Taxonomy: Eighth Report of the International Committee on Taxonomy of Viruses*, pp. 447–454. San Diego, CA: Elsevier Academic Press.

Mertens PPC, Makkay A, Stoltz D, Duncan R, Bergoin M, and Dermody TS (2005) Idnoreovirus, *Reoviridae*. In: Fauquet CM, Mayo MA, Maniloff J, Desselberger U, and Ball LA (eds.) *Virus Taxonomy: Eighth Report of the International Committee on Taxonomy of Viruses*, pp. 517–521. San Diego, CA: Elsevier Academic Press.

Mertens PPC, Rao S, and Zhou ZH (2005) Cypovirus, *Reoviridae*. In: Fauquet CM, Mayo MA, Maniloff J, Desselberger U, and Ball LA (eds.) *Virus Taxonomy: Eighth Report of the International Committee on Taxonomy of Viruses*, pp. 522–533. San Diego, CA: Elsevier Academic Press.

Payne CC and Mertens PPC (1983) Cytoplasmic polyhedrosis viruses. In: Joklik WK (ed.) *The Reoviridae*, pp. 425–504. New York: Plenum.

Plus N, Gissman L, Veyrunes JC, Pfister H, and Gateff E (1981) Reoviruses of drosophila and ceratitis populations and of drosophila cell lines: A new genus of the *Reoviridae* family. *Annales de Virologie (Institut Pasteur)* 132E: 261–270.

Plus N, Veyrunes JC, and Cavalloro R (1981) Endogenous viruses of *Ceratitis capitata* Wied. 'J.R.C. Ispra strain' and *C. capitata* permanent cell lines. *Annales de Virologie (Institut Pasteur)* 132E: 91–100.

Varma MG, Pudney M, and Leake CJ (1974) Cell lines from larvae of *Aedes* (*Stegomyia*) *malayensis* Colless and *Aedes* (*S*) *pseudoscutellearis* (Theobald) and their infection with some arboviruses. *Transactions of the Royal Society for Tropical Medicine and Hygiene* 68: 374–382.

Insect Viruses: Nonoccluded

J P Burand, University of Massachusetts at Amherst, Amherst, MA, USA

© 2008 Elsevier Ltd. All rights reserved.

Glossary

Agonadal A condition describing insects that have malformed reproductive tissues.

Oryctes rhinoceros The rhinoceros beetle that is an economic insect pest of coconut and palm plantations.

History

The unassigned nonoccluded, insect viruses (NOIVs) are a loosely associated group of rod-shaped, enveloped, DNA-containing, insect viruses that have structural and biological similarities but with possible genetic relatedness. It has been proposed that this group of viruses be classified as a new virus family, the *Nudiviridae*. These viruses were often referred to as nonoccluded baculoviruses since they resemble baculoviruses in their size and shape as well as due to the fact that they contain a double-stranded circular DNA genome. Unlike members of the family *Baculoviridae*, these viruses are not found associated with a protein crystal occlusion body associated with baculoviruses and as such have evolved modes of transmission involving a close association between the virus and the host which often involves the virus becoming latent or persistent.

Only three NOIVs have been studied to a significant extent. The first of these to be identified was the oryctes virus (Or-1V), which was initially isolated from larvae of the rhinoceros beetle, *Oryctes rhinoceros*, in Southeast Asia in 1963 and is currently used throughout the palm-growing tropics of the Asia-Pacific region for the control of this beetle, an economically important pest of oil palm. The successful use of Or-1V for the control of this insect pest is a landmark in classical biological control.

The best-characterized NOIV is Hz-1 virus (Hz-1V), which was first identified in 1978 as a persistent agent in the insect cell line IMC-Hz-1. The IMC-Hz-1 cell line was established from adult ovarian tissues of the corn earworm *Helicoverpa zea*. The first indications that this cell line was persistently infected with a virus came from electron micrographs of these cells in which long, virus-like particles were visible. The virus harbored in these cells was eventually recovered and propagated in a number of lepidopteran cell lines of ovarian origin. Interestingly, not only could the virus undergo productive replication in these cell lines but it could also establish persistent infections in these lines.

The virus Hz-2V, formerly known as gonad-specific virus (GSV), is presently the only known sexually transmitted insect virus (**Figure 1**). This NOIV was first observed in electron micrographs of ovarian tissues of insects from a colony of *H. zea* established at the

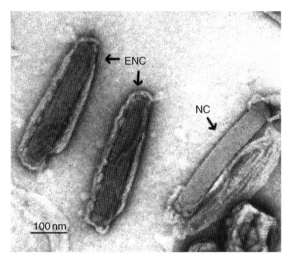

Figure 1 Electron micrograph of Hz-2V virus particle showing enveloped nucleocapsids (ENC) and a nucleocapsid from which the envelope has been partially removed (NC). Reproduced from Burand JP and Lu H (1997) Replication of a gonad-specific insect virus in TN-368 cells in culture. *Journal of Invertebrate Pathology* 70: 88–95, with permission from Elsevier.

USDA-ARS Southern Insect Management Laboratory in Stoneville, MS. A portion of the insects from this colony were sterile and had malformed reproductive tissues, a condition referred to as being agonadal (AG). Some individual females from the Stoneville colony were fertile asymptomatic carriers of the virus and when mated with healthy males gave rise to virus-infected, sterile, AG progeny. AG males from the Stoneville colony were also sterile and could transmit virus to healthy females which also gave rise to AG progeny.

Physical Properties

All three viruses in this group have supercoiled, double-stranded DNA genomes ranging in size from 130 kbp for Or-V1 to approximately 230 kbp for Hz-1V and Hz-2V. The genomes of both Hz-1V and Hz-2V have been sequenced and are 99% identical, indicating that these two viruses are very closely related. Both of these viruses code for over 100 open reading frames (ORFs), several of which share homology with baculovirus genes. This homology with baculovirus genes, including the *per os* infectivity or *pif* genes and inhibitor of apoptosis or *IAP* genes, probably represents domains in these proteins that perform functions which are common in the infection pathway of these viruses in their insect hosts. The absence of the four linked, baculovirus 'central core' genes (*helicase*, *lef-5*, *ac96*, and *38K*) in the Hz-1V and Hz-2V genomes suggests that these two viruses are not members of the family *Baculoviridae*.

The enveloped Or-1V virus particle is 220 nm × 120 nm and is composed of 27 structural proteins ranging in size from 9.5 to 215 kDa with 14 of these proteins residing in the

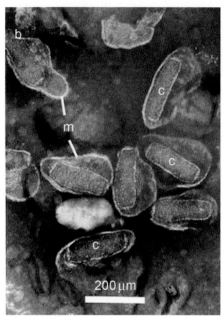

Figure 2 A micrograph showing Or-1V nucleocapsids to permit readers to compare and contrast structure. Reproduced from Huger AM (2005) The *Oryctes* virus: Its detection, identification, and implementation in biological control of the coconut palm rhinoceros beetle, *Oryctes rhinoceros* (Coleoptera: Scarabaeidae). *Journal of Invertebrate Pathology* 89(1): 78–84, with permission from Elsevier.

viral envelope (**Figure 2**). Hz-1V and Hz-2V are longer and narrower particles with enveloped nucleocapsids measuring 414 nm × 80 nm. These virus particles are also complex being comprised of 14–16 proteins of between 190 and 14 kDa, five of which are thought to be components of the viral envelope. The viral envelope of Hz-2V itself is distinctive, consisting of a large number of folds or ridges, each of which is associated with an individual filament located between the envelope and nucleocapsid giving the envelope a 'rope-like' appearance (**Figure 1**).

Replication

The replication of Or-1V has been studied *in vitro* using the cell line DSIR-HA-1179 which was derived from the black beetle, *Heteronychus arator*. The virus apparently enters these cells by pinocytosis with the first signs of viral cytopathic effect (CPE) are visible by 18–24 h post infection (pi), with cell lysis and viral release occurring by 3–4 days pi. Nuclear hypertrophy and migration of nuclear chromatin occur in infected cells by about 7 h pi followed by an accumulation of envelope material in chromatin-free areas of the nucleus. The assembly of the virus begins with the formation of envelopes and nucleocapsid shells in these chromatin-free areas; then the shells are filled with viral DNA which comprises the electron-dense core of the virus. Virus particles enter the cytoplasm between 12 and 36 h pi at which time virus replication peaks followed by the release of virus particles that bud from the cell membrane.

Only eight of the 27 Or-1V structural proteins have been detected in *in vitro* replication studies in which infected, DSIR-HA-1179 cells were pulse-labeled with [^{35}S]-methionine. Two of the structural proteins, p4.6 and p10, were first detected at 4 h pi while the p11.5 protein was first detected at 10 h pi, and six viral proteins, p46.5, p40, p27, p25, p22, and p13, all were first detected at 6 h pi. The inability to detect the synthesis of the remaining virus structural proteins or any additional virus-induced intracellular proteins may be a consequence of Or-1V not actively shutting off cell protein synthesis during replication and/or the method of labeling used in these experiments.

Hz-1V and Hz-2V replication has been examined in the ovarian cell line (TN-368) derived from the cabbage looper, *Trichoplusia ni*. Replication of these two viruses proceeds very rapidly in this cell line culminating in cell lysis by approximately 24 h pi. The first signs of viral replication are rounding of the cells followed by nuclear hypertrophy. As replication proceeds in these cells, membrane vesicles appear in the nucleus, which become filled with electron-dense capsids containing viral DNA and develop into single, mature enveloped nucleocapsids. The replication of Hz-2V in the gypsy moth ovarian cell line Ld652Y proceeds more slowly and more closely resembles *in vivo* replication culminating in the formation of bundles of enveloped virus particles in the cell nucleus.

The Hz-1V replication cycle consists of the sequential expression of virus-specific intracellular proteins and it has been divided into three stages: an early stage, from 0 to 4 h pi, which is prior to the onset of viral DNA replication; an intermediate phase, from 4 to 8 h pi; and a late phase, beyond 8 h pi. The pattern of expression of viral-specific intracellular proteins in cells productively infected with standard virus particles differs from the pattern of protein synthesis in cells infected with defective interfering particles, which has been shown to lead to the establishment of cell lines persistently infected with Hz-1V.

A total of 101 viral-specific transcripts expressed from dispersed regions along the viral genome have been detected during productive Hz-1V replication. These transcripts range in size from 0.8 to over 9.5 kbp in length and follow a cascade of temporal expression similar to that found for virus-specific intracellular protein synthesis. Only three of the 24 viral transcripts detected at 2 h pi are highly expressed and only a single 2.9 kbp transcript is constitutively expressed (CE 2.90) throughout virus replication. At 4 h pi 30 new transcripts are detected with an additional 21 detected at 6 h pi and 16 more transcripts detected at 8 h pi. No new viral transcripts were detected later than 8 h pi. The expression of a single 2.9 kbp viral transcript known as the persistently associated transcript (PAT1) has been found in cells persistently infected with Hz-1V. PAT1 is constitutively expressed in cells persistently infected with the virus and its expression appears to be required for the establishment and maintenance of Hz-1V persistent infections.

Pathology and Transmission

Or-1V infection of rhinoceros beetles occurs *per os* and although the virus was initially discovered in beetle larvae, insects in breeding sites in this stage of development are generally virus free. Transmission of the virus is thought to occur primarily between adults through cooperative nest building and other social activities including mating that are part of the adult behavior of this insect. In both larvae and adults, midgut epithelial cells are the primary site of virus replication from which the infection spreads to other tissues including the fat body. Infected larvae appear translucent during the initial phases of infection as the infected fat body disintegrates. With the progression of viral infection, chalky-white bodies become visible in the midgut epithelium under the insects' integument and the diseased larvae appear shiny, beige, and waxen, dying within 1–4 weeks after infection.

Virus replication in adult beetles results in the proliferation of infected midgut cells to the point where within 1–2 weeks after the onset of infection they fill the midgut lumen, with virus being released from infected cells into the gut and eventually excreted in the feces of infected beetles. This proliferation of cells and the swelling of the midgut is a diagnostic symptom of viral infection in adults. Virus infection in adults is often chronic, lasting several weeks before the insects dies. During this time infected individuals continue to shed virus in their feces as they visit breeding sites, contaminating these sites while they attempt to mate and in this way ultimately contributing to the spread of Or-1V within the beetle population.

Once it was established that infected adults could act as effective vectors and spread the virus to other beetles through the contamination of breeding sites, Or-1V was introduced as a control agent by simply releasing virus-infected beetles into pest populations. Successful releases of infected beetles and the self-spreading of the viral disease resulting in the long-term suppression of the rhinoceros beetle have been demonstrated in areas throughout the South Pacific including Fiji, Tonga, Wallis Island, the Tokelau and Palau Islands, as well as Papua New Guinea and American Samoa.

Hz-1V is not known to cause any pathology *in vivo*; Hz-2V on the other hand shows a very interesting and unique pathology in insects. Hz-2V replication in insects does not result in mortality but rather occurs only in adult reproductive tissues resulting in the malformation of these tissues and the sterility of infected moths. Infected insects appear normal as larvae but emerge as AG adults with abnormal reproductive systems.

The reproductive tissues of Hz-2V-infected male moths are grossly malformed into a 'Y-shaped' structure. In infected males many of the reproductive tissues normally responsible for producing sperm, pheromonostatic peptide (PSP), and the spermatophore are absent or grossly malformed. Interestingly, reproductive tissues essential for the initiation of copulation and transfer of reproductive fluids to female moths during mating appear to be intact, suggesting that AG males may be able to mate with healthy female moths and transfer Hz-2V particles, without fertilizing these individuals or altering their sexual receptivity to further mating with other male moths. Virus-infected female moths lack ovaries as well as other reproductive tissues and have grossly deformed and enlarged common and lateral oviducts. EM observations of these reproductive structures shows a significant amount of virus replication in cells which have proliferated in these tissues. Upon emergence, many Hz-2V-infected females have a clearly visible, virus-filled 'plug' covering the reproductive opening at the end tip of the abdomen (**Figure 3**).

Figure 3 Hz-2V-infected *H. zea* female moth. Arrow indicates presence of virus plug. Reproduced from Burand JP and Lu H (1997) Replication of a gonad-specific insect virus in TN-368 cells in culture. *Journal of Invertebrate Pathology* 70: 88–95, with permission from Elsevier.

Hz-2V-infected female moths produce 5–7 times more of the male-attracting pheromone and as a result attract more males than their healthy counterparts. Healthy male moths attracted to infected females can become contaminated with virus upon coming into contact with the 'virus' plug at the tip of the female's abdomen while attempting to mate. Unlike healthy females that stop calling for mates and are nonreceptive to males after mating, Hz-2V-infected females continue to call and attempt to mate with males, serving as very efficient vectors in facilitating the spread of the virus within the insect population.

Persistence

With Or-1V transmission occurring primarily through fecal contamination of nesting sites and with Hz-2V being sexually transmitted to healthy individuals primarily during matings with infected moths, it is clear that the inability of NOIVs to survive outside their insect host for any extended period of time has led to these viruses evolving modes of transmission which involve a close association between them and their insect host. In the case of Hz-2V the existence of the virus in both a persistent and latent state in the insect host appears to be a part of the normal replication cycle of the virus. In infected insects Hz-2V can be persistent in the early larval stage with virus particles being virtually undetectable. Although viral DNA and some viral gene transcripts can be detected in these individuals, they appear normal and are physically indistinguishable from uninfected insects. For some of infected insects, particularly females, the virus remains persistent and these individuals are fertile and able to pass the virus on transovarially to their progeny, some of which become sterile, AG adults. In other infected insects Hz-2V is latent, becoming activated into productive replication with the onset of the development of adult reproductive tissues and causing the malformation of these tissues. The fate of Hz-2V-infected insects becoming fertile asymptomatic carriers of the virus or sterile AG adults appears to be determined by the amount of virus each individual egg receives from the female parent.

The close association which has evolved between Or-1V and the rhinoceros beetle has proven to be the key to the successful use of this virus as a biological control agent and has resulted in the persistence of this virus in beetle population throughout the Asia-Pacific region. The successful use of this virus in controlling this beetle pest suggests that other NOIVs may also have potential for use in insect pest control programs (**Figure 4**). The ability of Hz-2V to establish persistent infections in *H. zea* insects and be transmitted to the progeny of these insects suggests that this NOIV may

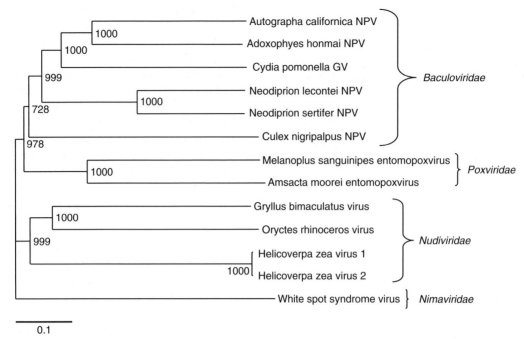

Figure 4 A dendogram showing phylogenetic relationships.

have the potential to serve both as model to study the nature of viral persistence and as a tool for altering the genetic make-up of insects.

See also: Baculoviruses: Molecular Biology of Nucleopolyhedroviruses; Herpesviruses: Latency; Insect Pest Control by Viruses; Oryctes Rhinocerous Virus; Persistent and Latent Viral Infection.

Further Reading

Burand JP (1998) *Nudiviruses*. In: Miller LK and Ball A (eds.) *The Insect Viruses*, 69pp. New York: Plenum.

Burand JP, Kawanishi CY, and Huang Y-S (1986) Persistent baculovirus infections. In: Granados RR and Federici BA (eds.) *The Biology of Baculoviruses,* vol. 1, pp. 159–177. Boca Raton, FL: CRC Press.

Burand JP and LU H (1997) Replication of a gonad-specific insect virus in TN-368 cells in culture. *Journal of Invertebrate Pathology* 70: 88–95.

Burand JP, Rallis CP, and Tan W (2004) Horizontal transmission of Hz-2V by virus infected *Helicoverpa zea* moths. *Journal of Invertebrate Pathology* 85: 128–131.

Hamm JJ, Carpenter JE, and Styer EL (1996) Oviposition day effect on incidence of agonadal progeny of *Helicoverpa zea* (Lepidoptera: Noctuidae) infected with a virus. *Annals of the Entomological Society of America* 89: 266–275.

Huger AM (2005) The *Oryctes* virus: Its detection, identification, and implementation in biological control of the coconut palm rhinoceros beetle, *Oryctes rhinoceros* (Coleoptera: Scarabaeidae). *Journal of Invertebrate Pathology* 89: 78–84.

Huger AM and Kreig A (1991) *Baculoviridae*: Nonoccluded baculoviruses. In: Adams JR and Bonami JR (eds.) *Atlas of Invertebrate Viruses*, pp. 287–319. Boca Raton, FL: CRC Press.

Wang Y, van Oers MM, Crawford AM, Vlak JM, and Jehle JA (2007) Genomic analysis of Oryctes rhinoceros virus reveals genetic relatedness to Heliothis zea virus 1. *Archives of Virology* 152(3): 519–531.

Interfering RNAs

K E Olson, K M Keene, and C D Blair, Colorado State University, Fort Collins, CO, USA

© 2008 Elsevier Ltd. All rights reserved.

Glossary

dsRNA Double-stranded RNA, the trigger for RNAi and intermediate in RNA virus replication.
miRNA MicroRNA; small RNAs, about 21–23 bases in length, that post-transcriptionally regulate the expression of genes by binding to the 3′ untranslated regions (3′ UTR) of specific mRNAs.
RISC RNA-induced silencing complex; a multiprotein complex that brings the guide strand of the siRNA duplex and the cellular mRNA together and cleaves

> the mRNA with associated endonuclease activity. mRNA is then degraded.
> **RNAi** An evolutionarily conserved, sequence-specific antiviral pathway triggered by double-stranded RNA (dsRNA) that leads to degradation of both the dsRNA and mRNA with homologous sequence.
> **siRNA** Small interfering RNAs; 21–25 bp duplexes that provide guides for sequence-specific cleavage of mRNA. siRNAs are considered the hallmark of RNAi.

Introduction

Antiviral innate immune responses in vertebrate hosts restrict viral invasion until humoral and cell-mediated acquired immune responses specifically clear virus infection. In mammals, innate immune responses are induced when infected cells recognize viral components such as nucleic acids through host pattern-recognition receptors. Toll-like receptors are important pattern-recognition sensors that recognize viral components (e.g., double-stranded RNA (dsRNA)) and signal induction of type I interferons and inflammatory cytokines that lead to an antiviral state. Organisms such as insects probably depend entirely on antiviral innate immune responses to overcome viral infection, since acquired immune responses have not been identified in invertebrates. Interferon-like molecules also have not been found in insects, making it unclear how insects cope with viral infections. In 1998, researchers first described an intracellular response in the worm *Caenorhabditis elegans* that could be triggered by dsRNA and efficiently silenced the expression of genes having sequence identity with the dsRNA trigger. This response, termed RNA interference or RNAi, is now known to be an ancient antiviral innate immune response in eukaryotic organisms including worms, plants, insects, and mammals and is a common evolutionary link among these organisms in their fight against virus invasion.

RNA Interference

RNAi, post-transcriptional gene silencing (PTGS), quelling, and sense suppression are terms for related pathways that have been described in different organisms. All are RNAi responses triggered by dsRNA that result in degradation of both dsRNA and mRNA with cognate sequence. These pathways are highly conserved evolutionarily and exist in many organisms including plants, fungi, and animals. RNAi and related pathways have several functions, including regulation of development, silencing and regulation of gene expression, and defense against viruses and transposable elements. In this article, we focus on RNAi as an antiviral, innate immune pathway.

RNAi Mechanism of Action

The RNAi pathway is divided into an initiator phase and an effector phase. The initiator phase consists of the recognition and processing of long dsRNA molecules by the RNaseIII enzyme Dicer into small interfering RNAs (siRNAs) of 21–25 bp. siRNAs, considered the hallmark of the RNAi response, are duplexes with $3'$ overhangs of 2 nt. Each strand has a $5'$-PO_4 and $3'$-OH. In the fruitfly *Drosophila melanogaster* (drosophila), the siRNAs are unwound and incorporated into the RNA-induced silencing complex, or RISC, with the assistance of Dicer-2 and R2D2, to start the effector phase of the pathway. In the effector phase, one strand of the siRNA duplex acts a as guide sequence to target the RISC to complementary mRNAs and determine the cleavage site on the mRNA. The RISC is known to include products of the following genes: *Argonaute2* (*Ago2*), *Vasa intronic gene* (*VIG*), *fragile X mental retardation* (*FXR*), and *Tudor Staphylococcal nuclease* (*Tudor-SN*). Other genes in drosophila encode proteins having RNA helicase activity associated with RNAi and include *spn-E*, *Rm62*, and *armi*. The latter gene products have been implicated with heterochromatin and transposon silencing.

RNAi Components and Function: Dicers, the Sensors of dsRNA and Initiators of siRNA Production

The majority of biochemical information related to Dicer proteins has been elucidated from studies in drosophila, showing that Dicer enzymes are intracellular sensors of dsRNA that initiate RNAi in drosophila cultured cell and embryo lysates. Candidate genes from three families encoding RNase III motifs were expressed in drosophila S2 cells, immunoprecipitated, and tested *in vitro* for their ability to transform long dsRNA molecules (>30 bp) into small RNAs of ~ 21 nt (**Figure 1**). The enzyme capable of producing siRNAs was termed Dicer and contained a helicase domain as well as two RNase III domains. The production of siRNAs by Dicer-2 in drosophila was ATP-dependent and the enzyme was inactive in degrading single stranded RNAs. Dicer depletion by immunoprecipitation from drosophila cell lysates resulted in decreased siRNA production. Dicer is a ~ 200 kDa protein with an N-terminal RNA helicase domain, a PAZ (PIWI/Argonaute/Zwille) domain, a conserved domain of unknown function (DUF283), two C-terminal RNase III domains, and an RNA binding motif. In its active form, Dicer is a homodimer. Dicer is evolutionarily conserved, and is found in drosophila, *C. elegans*, humans, mice, trypanosomes, zebrafish, the fungi *Magnaporthe oryzae* and *Neurospora crassa*, budding yeast (*Schizosaccharomyces pombe*), plants, and many other organisms. The number

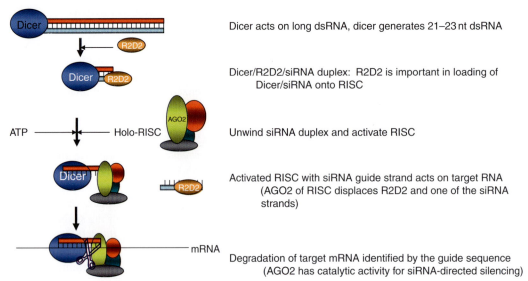

Figure 1 General scheme for RNAi leading to degradation of a specific mRNA.

of *dicer* (*dcr*) genes varies in different organisms. In drosophila there are two *dcr* genes (*dcr1* and *dcr2*); in the plant *Arabidopsis thaliana* there are four, each of which processes siRNAs from different dsRNA sources; however, in the genomes of humans and *C. elegans*, there is only one *dcr* gene.

Dicers and miRNA Biogenesis

The RNAi pathway has two branches leading to production of either siRNA or microRNA (miRNA). The miRNAs *lin-4* and *let-7* were first discovered in *C. elegans* and play a crucial role in development. Subsequently, hundreds of different miRNAs have been found in most eukaryotic organisms and exploring their expression and function is now an important topic of research in a wide range of invertebrates and vertebrates. miRNAs are produced from endogenous pre-miRNA transcripts that form stem–loop precursors. Silencing of genes by the miRNA pathway occurs not by degradation of the mRNA, but rather by translational arrest during protein synthesis. Also, unlike the siRNA pathway, miRNA silencing does not require complete base pairing between the miRNA and the target sequence to be silenced. Distinct RISCs process small RNAs for the effector phase of the two gene-silencing mechanisms. It is not known how the distinction is made between siRNA production and miRNA production with a single Dicer enzyme in *C. elegans*. In humans, interferon and other innate immune responses are induced by long dsRNA, so Dicer-like activity generating siRNAs may not be as crucial. In organisms such as drosophila that encode two Dicer proteins, Dicer-1 is the enzyme that produces miRNAs from endogenous transcripts.

R2D2 Protein and RNAi

R2D2 is a 36 kDa protein with two dsRNA binding domains that bridges the initiator and effector stages of the RNAi pathway. R2D2 co-purifies with Dicer-2 from an siRNA-generating extract of drosophila S2 cells. R2D2 association does not affect the enzymatic activity of Dicer-2, but is required to load the newly formed siRNAs into RISC. Dicer-2 forms a heterodimer complex with R2D2 and the siRNA duplex that appears to detect the thermodynamic asymmetry of the siRNA duplex. The passenger strand of the siRNA duplex is separated from the guide strand, which interacts at its 5' and 3' ends with the PIWI (possessing RNase H-like activity) and PAZ domains of Argonaute-2 (AGO2), respectively. Strand selection (passenger strand vs. guide strand) depends on the thermodynamic stability of the first four nucleotides of the 5' terminus of an siRNA duplex. The siRNA strand whose 5' end has lower base-pairing stability becomes the guide strand, leaving the more stable strand as the passenger strand. R2D2 binds to the thermodynamically more stable end of an siRNA whereas Dicer-2 binds on the opposite end. Release of the degraded passenger strand is believed to be an ATP-dependent reaction. The guide siRNA strand remains associated with RISC and guides AGO2 to the target mRNA containing the complementary sequence. After hybridization, AGO2 cleaves the phosphodiester backbone of the target mRNA. Target RNA cleavage occurs between the 10th and 11th nucleotides of the guide siRNA measured from its 5' phosphate group. The phosphate group of the 5' end of the guide siRNA influences the fidelity of the cleavage position as well as the stability of RISC. Recently it was confirmed that the RNAi pathway of drosophila functions as an antiviral immunity mechanism. Drosophila with null mutations for genes encoding Dicer-2, Argonaute-2, or R2D2 were highly susceptible to RNA viruses such as Flock house virus (FHV) (*Nodaviridae*), Drosophila C virus (*Dicistroviridae*), or Sindbis virus (*Togaviridae*) which in some cases were lethal for the flies.

Argonaute Proteins and RNAi

Another family of genes that has already been discussed in this review as important to RNAi is the *argonaute* gene family. In drosophila, there are five *argonaute* genes. Two of the genes, *argonaute1* (*ago1*) and *argonaute2* (*ago2*), have been implicated in RNAi. Argonaute-2 is an important component of the RISC complex. AGO2 has been termed 'Slicer' and is the only component of human RISC that is required for the degradation of mRNA molecules, and its PIWI domain may contain the endonuclease activity. Drosophila embryo mutants lacking *ago2* are unable to perform siRNA-directed mRNA cleavage, although miRNA-directed cleavage is still possible. The embryos also lack the capacity to load siRNAs into RISC.

Drosophila Argonaute-1 (AGO1) is not involved in siRNA-directed cleavage as is AGO2. AGO1 is believed to function downstream of the production of the siRNAs and not as a component of RISC. These studies also demonstrated that *Drosophila* mutants lacking AGO1 are embryonic lethals, implicating AGO1 in *Drosophila* development. AGO1 is required for miRNA-directed cleavage and dispensable for siRNA-directed cleavage, showing divergent roles for different Argonaute proteins. This functional differentiation of the Argonaute proteins has also been noted in plants.

Argonaute proteins are characterized by the PAZ domain and the PIWI domain. The crystal structures of the PAZ domain of AGO2 from *Drosophila* and the thermophile *Pyrococcus furiosus* were determined and shown to have structural properties similar to proteins that bind single-stranded nucleic acid. The PAZ domain recognizes the 3′ overhangs of the siRNA duplexes. Interestingly, this is the same region that is recognized by certain viral suppressors of RNAi. In addition to its endonuclease motif, the PIWI domain is involved with protein–protein interactions between Argonaute and Dicer and may play a role in siRNA loading onto RISC. Recently, researchers have shown that both human AGO1 and AGO2 proteins reside in intracellular structures known as 'cytoplasmic bodies'. These areas of the cell are believed to be sites of regulation of cellular mRNA turnover. Several other AGO proteins from drosophila are associated with the RNAi pathway. These proteins include Piwi, implicated in transposon silencing, Aubergine, associated with germline gene repression, and AGO3, an argonaute-like protein of unknown function. As already mentioned, a number of other proteins have now been associated with the RNAi pathway, including Vig, Tudor-SN, and Fmr-1, all associated with RISC.

Unique Properties of RNAi in Plants and Animals

In *C. elegans* and plants, an amplification of the RNAi response occurs. In these systems, the original siRNAs act as primers for synthesis of new intracellular dsRNA by the endogenous RNA-dependent RNA polymerase (RdRP). These newly produced dsRNAs are processed by Dicer and generate an additional and potentially more diverse pool of siRNAs. This phenomenon is termed 'transitive RNAi' and allows for the degradation of a full-length mRNA even when the initial trigger sequence represents only a portion of the gene or genome. Amplification of the siRNA signal is bidirectional along the transcript in plants; however, in *C. elegans* the signal can only spread 3′–5′ along the transcript. Transitive RNAi does not occur in species such as *D. melanogaster*, where RdRP genes are absent from the genome. Gene knockdown experiments in organisms lacking transitive RNAi require design of dsRNA for specific disruption of gene expression.

One unique aspect of PTGS in plants is the ability of the silencing signal to spread throughout the organism. The siRNAs, complexed with host proteins, are part of a silencing complex that can move to other tissues in the plant through phloem tissues. The distance that the siRNAs travel is dependent on their exact length and the Dicer enzyme by which they were generated. The exact mechanism for long-distance movement has yet to be elucidated. RNAi in *C. elegans* can spread from the point of induction, especially if the dsRNA is introduced into the intestine either by injection or by feeding.

RNAi has become an important reverse genetics tool for studying gene function. Systematic knockdown of all genes has been accomplished in both drosophila and *C. elegans* with great success, identifying previously unknown functions of genes in various biological pathways and revealing differences in the RNAi pathways of each organism. In drosophila, the siRNA signal appears not to be amplified (transitive RNAi) and does not spread as seen in *C. elegans* and plants. This means that RNAi activity is most likely confined to those cells in which the dsRNA trigger is detected by Dicer. RNA silencing has been used in many other invertebrates, including mosquito disease vectors. Several studies have shown that RNAi can be used to knock down endogenous and exogenous gene expression in mosquitoes that transmit medically important pathogens to animals and humans. Studies of RNAi in mosquitoes have also shown that it is possible to silence RNAi complex genes using RNAi. As an example, the ability to silence the RNAi pathway in a hemocyte cell line of *Anopheles gambiae* was tested by transfecting dsRNA derived from exon sequences of the *A. gambiae dcr1* and *dcr2* and *argonaute 1–5* (*ago1–5*) genes. RNAi in *A. gambiae* cells required expression of Dicer-2, AGO2, and AGO3 proteins. This study also demonstrated that RNAi in the mosquito, as in drosophila, does not spread from the target cells, suggesting that RdRP-mediated transitive amplification is absent in the mosquito.

Viruses and RNAi

Since the discovery of RNA silencing, a number of researchers have hypothesized that RNAi plays an important role in antiviral defense. RNA viruses with positive-sense genomes, in particular, form dsRNA intermediates as they replicate in host cells. Whether this is the general viral trigger of RNAi in infected cells is still controversial, since viruses with negative-strand genomes generate little detectable dsRNA during their replication. Studies to elucidate a mechanism termed pathogen-derived resistance (PDR) in plants were among the first to show that viruses could be targeted by RNA silencing. Transgenic plants were engineered to express a portion of the tobacco etch virus (TEV) coat protein. Upon challenge with TEV, the transgenic plants were found to be resistant to the virus. No overt symptoms were seen, and no virus could be recovered from the leaf tissue. PDR was shown to be virus specific, as the transgenic plants could be infected with a genetically unrelated virus; however, any RNA that was introduced into plant cells that shared homology with the RNA in the transgene was degraded. Plant virologists also observed that the delivery of a fragment of viral RNA to plant cells by a heterologous virus expression vector made the cells resistant to challenge by the first virus, and termed this related phenomenon virus-induced gene silencing. Soon after the discovery that plants used an RNA-mediated defense mechanism against viruses, it was shown that many plant viruses encoded protein suppressors of RNA silencing. This would be expected as the virus and host develop a biological arms race to express countermeasures that limit the ability of one to overcome the other.

Viral Encoded Suppressors of RNAi

Several suppressors have been found that interrupt the RNA-silencing pathway at different steps, indicating that evasion of RNA silencing has evolved more than once. One protein, helper component-proteinase, or HC-Pro, from potyviruses of plants is thought to be the most potent suppressor of RNA silencing found to date. HC-Pro functions at a step that prevents the accumulation of siRNAs by interacting with the RNase III enzyme Dicer. HC-Pro also has the capability to reverse established silencing of a transgene, suggesting that the protein inhibits a mechanism required for the maintenance of silencing. The 19 kDa protein (p19) from tombusviruses acts as a suppressor of PTGS in plants in a different manner from HC-Pro. p19 does not block production of the 21–25 nt siRNAs; rather it binds them via the 2-nt 3′ overhangs. Notably, the protein will bind only double-stranded 21 nt sequences; single-stranded RNAs of the same length are not recognized by p19. In binding the siRNAs, p19 forms a homodimer and sequesters the guide sequences that are required for RISC incorporation and targeting mRNA degradation. Another virus, cucumber mosaic virus, encodes a protein 2b that interferes with the spread of siRNA signal in the plant host, allowing systemic spread of the virus.

In addition to the PTGS suppressors encoded by plant viruses, the insect virus Flock house virus encodes a protein, B2, which can suppress RNAi activity in both drosophila S2 and plant cells. In transgenic plants containing a green fluorescent protein (GFP) gene, in which transient expression of siRNAs targeting GFP mRNA had silenced its expression, the presence of B2 protein reversed silencing of GFP. siRNAs were still detected in the tissues, indicating that the suppression of gene silencing did not occur before the production of siRNAs. It is likely that this protein sequesters the siRNAs from the RNAi machinery, possibly in a manner similar to p19. Finally, two other animal viruses encode suppressors of interferon induction that also have apparent RNAi suppressor activity. These are the influenza virus NS1 protein and the vaccinia virus E3L protein.

Viruses may employ other strategies to evade the RNAi pathway. These strategies include (1) sequestration of the viral dsRNA replicative intermediates in viral cores or in double-membrane structures in the host cell formed during viral replication; (2) viral replication and spread outpacing the RNAi pathway; and (3) replication in tissues that are resistant to RNAi.

miRNAs and Viruses

In the last couple of years, a number of viral-encoded miRNAs have been discovered. The functions of most viral-derived miRNAs are unknown; however, DNA viruses such as polyomaviruses and herpesviruses transcribe miRNAs in infected vertebrate cells that appear to regulate expression of critical viral and host genes during infection. The location of miRNAs within different virus genomes are not conserved, suggesting that miRNAs are likely to be recent acquisitions in viral genomes that help adapt the virus to the host during the virus lifecycle. For instance, a viral miRNA in SV40 virus was recently discovered that regulates the viral T antigen. The SV40 miRNA accumulates in late stages of infection and targets the early T antigen mRNA for degradation, thus reducing its expression. Studies of many families of RNA viruses have failed to identify miRNAs in RNA genomes and this finding is consistent with the prominent role of the cellular DNA-dependent RNA polymerase II in the biogenesis of miRNA precursors.

Use of RNAi for Disease Control

Virus diseases of plants cause extensive economic losses. Control of plant virus diseases has usually been associated with control of insects or nematodes that transmit viruses, or through sanitation and quarantine of infected plants. The transformation of plants with effector genes designed to transcribe inverted-repeat RNA (dsRNAs) that target plant viruses can provide novel virus-resistant varieties. Transgenic plants expressing inverted repeats of viral sequences exhibit varying degrees of resistance to the virus or viruses with genome sequences closely related to the source of the transgene. This resistance is due to PTGS wherein viral mRNA is degraded in the cytoplasm soon after synthesis.

In a similar genetic approach to that described previously in plants, we have genetically modified *Aedes aegypti* mosquitoes to exhibit impaired vector competence for dengue type 2 virus (DENV-2) transmission. DENVs are normally transmitted by *A. aegypti* mosquitoes to humans during epidemic outbreaks of dengue diseases. If a DENV-derived dsRNA trigger is expressed in the cell prior to viral translation and replication, an antiviral state can be induced in the mosquito that blocks virus infection. To do this, mosquitoes were genetically modified to express an inverted-repeat (IR) RNA derived from the premembrane protein coding region of the DENV-2 RNA. The IR RNA formed a 560 bp dsRNA in infected midgut epithelial cells of the mosquitoes to induce the RNAi pathway. A transgenic family, Carb77, was selected that expressed IR RNA in the midgut after a blood meal. Carb77 mosquitoes ingesting an artificial blood meal with 10^7 pfu ml^{-1} of DENV-2 exhibited marked reduction of viral envelope antigen in midguts and salivary glands after infection. Transmission of virus by the Carb77 line was significantly diminished when compared to control mosquitoes. As evidence that the resistance was RNAi mediated, DENV-2-derived siRNAs were readily detected in RNA extracts from midguts following ingestion of a blood meal with no virus. In addition, loss of the resistance phenotype was observed when the RNAi pathway was interrupted by injecting *ago2* dsRNA 2 days prior to induction of IR-RNA transgene, confirming that DENV-2 resistance was caused by an RNAi response.

Targeting of replicating animal viruses using RNAi has prompted discussion about whether RNAi can be used as an antiviral therapy. In mammalian cells, siRNAs, rather than long dsRNAs, are required to induce RNAi as an antiviral therapy because these cells possess interferon and other antiviral innate immune pathways that are triggered by dsRNA >30 bp in length. Numerous studies in cell culture have shown that HIV replication can be halted when cells are treated with siRNAs that target the viral genome. West Nile virus (WNV) replication also can be reduced in cultured cells by treatment with siRNAs targeting the virus RNA. These studies showed a significant reduction in levels of WNV RNA if cells were pretreated with siRNAs; however, the cells that were treated subsequent to the establishment of viral replication did not show the same reduction in viral RNA, suggesting that the RNA may be sequestered from the RNAi machinery after replication is established in the cell. Studies investigating RNAi as a therapy for hepatitis C virus (HCV) infection have used siRNAs to effectively target HCV replicon RNAs in cultured human cells as well as in a mouse model.

Model Systems for Studying Role of RNA in Virus Infections

While the genetics and biochemistry of RNAi in *C. elegans* and *D. melanogaster* have been investigated in detail, there have been no virus infection models of these animals until very recently. RNAi-based innate immunity has been detected in *C. elegans*-derived cultured cells infected with vesicular stomatitis virus (VSV; *Rhabdoviridae*). FHV replication in *C. elegans* triggered potent antiviral silencing that required RDE-1, an argonaute protein essential for RNAi mediated by siRNAs. This antiviral innate immunity was capable of rapid virus clearance in *C. elegans* in the absence of FHV RNAi suppressor protein B2. Two recent papers have shown that successful infection and killing of *Drosophila* by FHV was strictly dependent on expression of the viral suppressor protein B2. *Drosophila* with a knockout mutation in the gene encoding Dicer-2 showed enhanced susceptibility to infection by FHV, cricket paralysis, and *Drosophila* C viruses (*Dicistroviridae*) and Sindbis virus (*Alphavirus*, *Togaviridae*). These data demonstrate the importance of RNAi for controlling virus replication *in vivo* and establish *dcr2* as a drosophila susceptibility locus for virus infections. *C. elegans* and drosophila are important models for studying virus–RNAi interactions. The drosophila model system allows advanced genetic approaches such as generation of null mutants of RNAi components in a system that has few other dsRNA-triggered defensive responses to complicate mechanistic studies. The availability of the annotated drosophila genome sequence and established genetic approaches is critical for understanding RNAi mechanisms and continues to make the drosophila model important to our understanding of innate immune responses to viruses.

Mosquitoes, RNAi, and Arboviruses

Mosquitoes and arboviruses provide an important naturally occurring insect–virus system to study the potential role of RNAi in host defense. Arboviruses are RNA viruses that must replicate in their arthropod vector for amplification before they can be transmitted to a vertebrate host, such as humans. Obviously, arboviruses must somehow evade the RNAi pathway to successfully replicate in the mosquito prior to transmission. There are several advantages to studying RNAi in mosquitoes. First, the complete genome sequence is now available for at least two medically important vectors, *A. gambiae* and *A. aegypti*. Second, the RNAi pathway of mosquitoes is similar in structure and function to the pathway in drosophila and many of the component genes of RNAi have been identified. Third, new genetic approaches are allowing researchers to manipulate genes that affect innate immune responses in the mosquito. Fourth, infectious cDNA clones of arbovirus genomes from at least three virus families are available to allow manipulation of the viral genes and identify determinants of RNAi modulation. Fifth, RNAi–virus interactions studies can occur in systems directly relating to medically important pathogens.

Mosquitoes, like drosophila, do not have responses comparable to interferon induction, so they can be injected with long dsRNAs (300–500 bp) to trigger the RNAi pathway and efficiently silence specific vector genes that may participate in innate immune pathways, including RNAi. As an example, to determine whether RNAi conditions the vector competence of *A. gambiae* for O'nyong-nyong virus (ONNV; *Alphavirus*), a genetically modified ONNV expressing GFP (eGFP) was developed to readily track virus infection. After intrathoracic injection, ONNV-eGFP slowly spread to other *A. gambiae* tissues over a 9 day period. Mosquitoes were co-injected with virus and dsRNA derived from the ONNV nsP3 gene. Treatment with nsP3 dsRNA inhibited virus spread significantly, as determined by GFP expression patterns. ONNV-GFP titers from mosquitoes co-injected with nsP3 dsRNA also were significantly lower at 3 and 6 days after injection than in mosquitoes co-injected with non-virus-related β-galactosidase (β-gal) dsRNA. However, mosquitoes co-injected with ONNV-GFP and dsRNA derived from the *A. gambiae ago2* gene displayed widespread GFP expression and virus titers 16-fold higher than β-gal dsRNA controls at 3 or 6 days after injection. These observations provided direct evidence that RNAi is an antagonist of ONNV replication in *A. gambiae* and suggest that this innate immune response plays a role in conditioning vector competence. These types of experiments could be vital to understanding why some mosquito species are excellent vectors of disease and others are not.

The Biological Arms Race between Viruses and Hosts

RNA virus–host interactions are often characterized as an escalating arms race between two mortal enemies. The host evolves innate and acquired immune pathways to counter the destructive effects of virus invasion and the virus adapts to these defense measures by rapidly evolving new ways of evading the host's attempts at pathogen control. RNA–virus interactions with the host's RNAi pathway exemplify this struggle for dominance. Many RNA viruses evolve rapidly because their RNA-directed RNA polymerases are error prone, providing significant variation in virus genome populations needed to probe weaknesses in the host's defense and allow selection of new virus variants that have an advantage in their replication. As we have described earlier, a number of families of plant RNA viruses have evolved RNAi suppressors. Each family has developed a different strategy to downregulate RNAi, as shown by the fact that their suppressors attack different steps in the pathway. Still other viruses may have adapted to host RNAi by sequestering their replicative intermediate dsRNA triggers in double-membrane structures derived from host endoplasmic reticulum. Rapid evolution of RNA viruses also produces significant challenges to developing therapeutic strategies for humans using siRNAs to target and destroy viruses. Strategies that use siRNAs to trigger RNAi can be thwarted by point mutations in the target sequence. This has prompted development of siRNAs that target multiple regions of the viral RNA, highly conserved regions of the viral genome, or host genes essential for virus infection. However, RNA viruses are constrained in the amount of genetic variation they can tolerate and remain genetically fit for cell entry, replication, packaging, and egress.

Viruses that have complex life cycles, such as arthropod-borne viruses that must replicate in both vertebrate and invertebrate cells, are further constrained in their evolutionary potential. Finally, there is evidence that the host can evolve to counter the threat posed by RNA viruses. A recent finding emphasized the critical role of RNAi as an innate immune mechanism in *Drosophila* when researchers showed that RNAi pathway genes (*dcr2, ago2, r2d2*) are among the 3% fastest evolving genes among drosophila species, confirming the biological arms race between viruses and insects. The antiviral role of RNAi has been studied for less than a decade and is only now being exploited as a mechanism to fight viral diseases. Knowledge gained since the discovery of RNAi in 1998 should allow researchers to fully exploit RNAi as a means of controlling a number of infectious agents that cause disease in plants, animals, and humans.

Acknowledgments

Our research is funded by NIH grants AI34014 and AI48740 and the Grand Challenges in Global Health through the foundation for NIH.

See also: Satellite Nucleic Acids and Viruses.

Further Reading

Akira S, Uematsu S, and Takeuchi O (2006) Pathogen recognition and innate immunity. *Cell* 124: 783–801.

Baulcombe D (2004) RNA silencing in plants. *Nature* 431: 356–363.

Franz AWE, Sanchez-Vargas I, Adelman ZN, et al. (2006) Engineering RNA interference-based resistance to dengue virus type 2 in genetically modified *Aedes aegypti*. *Proceedings of the National Academy of Sciences, USA* 103: 4198–4203.

Galiana-Arnoux D, Dostert C, Schneemann A, Hoffmann JA, and Imler JL (2006) Essential function *in vivo* for dicer-2 in host defense against RNA viruses in *Drosophila*. *Nature Immunology* 7: 590–597.

Hammond SM, Boettcher S, Caudy AA, Kobayashi R, and Hannon GJ (2001) Argonaute2, a link between genetic and biochemical analyses of RNAi. *Science* 293: 1146–1150.

Kavi HH, Fernandez HR, Xie W, and Birchler JA (2005) RNA silencing in *Drosophila*. *FEBS Letters* 579: 5940–5949.

Keene KM, Foy BD, Sanchez-Vargas I, Beaty BJ, Blair CD, and Olson KE (2004) RNA interference as a natural antiviral response to O'nyong-nyong virus (*Alphavirus; Togaviridae*) infection of *Anopheles gambiae*. *Proceedings of the National Academy of Sciences, USA* 101: 17240–17245.

Leonard JN and Schaffer DV (2006) Antiviral RNAi therapy: Emerging approaches for hitting a moving target. *Gene Therapy* 13: 532–540.

Li WX, Li H, Lu R, et al. (2004) Interferon antagonist proteins of influenza and vaccinia viruses are suppressors of RNA silencing. *Proceedings of the National Academy of Sciences, USA* 101(5): 1350–1355.

Voinnet O (2005) Induction and suppression of RNA silencing: Insights from viral infections. *Nature Reviews Genetics* 6: 206–220.

Zamore PD (2002) Ancient pathways programmed by small RNAs. *Science* 296: 1265–1269.

Iridoviruses of Vertebrates

A D Hyatt, Australian Animal Health Laboratory, Geelong, VIC, Australia
V G Chinchar, University of Mississippi Medical Center, Jackson, MS, USA

© 2008 Elsevier Ltd. All rights reserved.

Glossary

Anemia Reduced number (below normal) of erythrocytes (red blood cells).

Ectothemic Animals whose temperature varies with the surrounding environment. Also known as poikliotherms.

Epitheliotropic Having a special affinity for epithelial cells.

Hyperplasia Abnormal increase in the volume of a tissue or organ caused by an increase in the number of normal cells.

Karyolysis Dissolution of a cell nucleus.

Karyorrhexis Rupture of the cell nucleus in which the chromatin disintegrates and is extruded from the cell.

Pyknosis Degeneration of a cell in which the nucleus shrinks in size and the chromatin appears as a solid, structureless feature.

Urodeles Amphibians belonging to the order Caudata, including the salamanders and newts, in which the larval tail persists in adult life.

Introduction

The family *Iridoviridae* encompasses five recognized genera, two of which infect invertebrates (*Iridovirus, Chloriridovirus*), and three of which infect ectothermic vertebrates (*Ranavirus, Lymphocystivirus,* and *Megalocytivirus*). In addition, two other viruses that infect cold-blooded vertebrates (*Erthrocytic necrosis virus* and *White sturgeon iridovirus* (WSIV) remain unassigned members of the family. Iridoviruses that infect 'cold-blooded' vertebrates (fish, amphibians, and reptiles) have become the focus of recent interest. These viruses are being identified and isolated with increasing frequency and their importance is being measured in terms of their impact on farmed production and trade in fish and amphibians. There are also significant impacts of iridoviruses on biodiversity, most notably the decline of local amphibian populations.

The history of research into these viruses extends back to the nineteenth century for lymphocystis virus, the 1940s for *Frog virus 3* (FV3, genus *Ranavirus*) and 1980s for the first description of a highly infectious fresh water piscine ranavirus, *Epizootic haematopoietic necrosis virus* (EHNV, genus *Ranavirus*) and saltwater megalocytiviruses such as red sea bream iridovirus (RSIV, species *Infectious spleen and kidney necrosis virus*, genus *Megalocystivirus*). Recent interest in the ranaviruses and megalocytiviruses is associated with disease epizootics of freshwater and salt water finfish, die-offs of frogs and salamanders, and illegal trade in wildlife. Fish, amphibians, and reptiles are being bought and sold illegally and transported across national and/or international borders without appropriate certification and quarantine. Increasing reports of disease involving vertebrate iridoviruses suggests expansion of the geographic distribution and host range, and, as such, they are considered to represent a significant group of emerging viruses.

Structure

Each of the three recognized genera of chordate iridoviruses shares common structural, replicative, genomic, and protein characteristics, and each genus contains several distinct species and isolates/strains. The virions of vertebrate iridoviruses range in size from 100–300 nm. They are comprised of four concentric layers: an outer envelope composed of a lipid bilayer and virus-encoded transmembrane proteins; an icosahedral protein shell comprised of the major capsid protein (MCP), an inner lipid membrane; and a central dsDNA core (approximately 170 kbp) and associated proteins. The outer membrane is acquired as the virus buds through the plasma membrane (**Figure 1(a)**). Lymphocystiviruses and megalocytiviruses differ from the above in that they are seldom observed budding from the host cell plasma membranes. Lymphocystiviruses differ from ranaviruses, megalocytiviruses, and erythrocytic necrosis viruses in that the capsids contain an outer fringe of external, fibril-like protusions (**Figure 1(g)**). Megalocytiviruses also differ from ranaviruses in that assembly sites within cells infected *in vivo* are membrane bound (**Figure 1(e)**).

Geographical Distribution and Host Range

Table 1 illustrates the diversity of hosts and geographic locations from which vertebrate iridoviruses have been identified and/or isolated. It is important to note that while many iridoviruses have been observed and isolated, they cannot be assigned to a specific taxonomic group until specific demarcation criteria relating to their molecular, structural, and host-disease characteristics are determined and interpreted as prescribed by the International Committee on Taxonomy of Viruses (ICTV).

Differences in virion size and topography, host range, and geographic locations distribution are generally restricted to specific isolates. Lymphocystis viruses have been reported worldwide and infect both freshwater and marine finfish. Similarly, ranaviruses have been isolated from most continents and have been reported to infect freshwater finfish, anurans (frogs and toads), urodeles (salamanders), and reptiles (turtles and snakes). Megalocytiviruses infect more than 30 species of cultured marine and freshwater finfish belonging mainly to the orders Perciformes and Pleuroneciformes (**Table 1**).

Hosts for vertebrate iridoviruses include all classes of ectothermic vertebrates. However, to date, only one vertebrate iridovirus, namely Bohle iridovirus (BIV), has been shown, under experimental conditions, to infect multiple vertebrate species, genera, and classes. While cross-species/class transmission has been demonstrated following experimental exposure, BIV is yet to be isolated from any epizootic. Moreover, the maximum permissive temperature for vertebrate iridovirus replication (approximately 15–25 °C) precludes replication in mammals or in mammalian cells incubated at 37 °C.

Phylogeny

At the genomic level, megalocytiviruses, ranaviruses, and lymphocystiviruses differ markedly in genomic organization, GC content, and sequence identity. The GC content of lymphocystiviruses (27–29%) is markedly lower than that of ranaviruses or megalocytiviruses (49–55%). Construction of a phylogenetic tree based on the inferred amino acid sequence of the major capsid protein indicates that three genera of vertebrate iridoviruses form separate clusters that are distinct from each other and from the invertebrate viruses (**Figure 2**). Sequence similarity/identity between different vertebrate iridovirus genera is typically <50%, whereas it is greater than 70% within a genus. Moreover, there is more sequence diversity among ranaviruses, which infect all classes of ectotherms, than among megalocytiviruses, which infect only teleost fish.

Clinical Features and Pathology

Iridoviruses of fish, amphibians, and reptiles are often highly virulent and can cause fatal infections in their hosts. Epizootics leading to mass mortalities with death rates approaching 100% have been reported in fish and amphibians.

Lymphocystosis in fish is, in the main, chronic and benign and not life-threatening; its impact is mainly

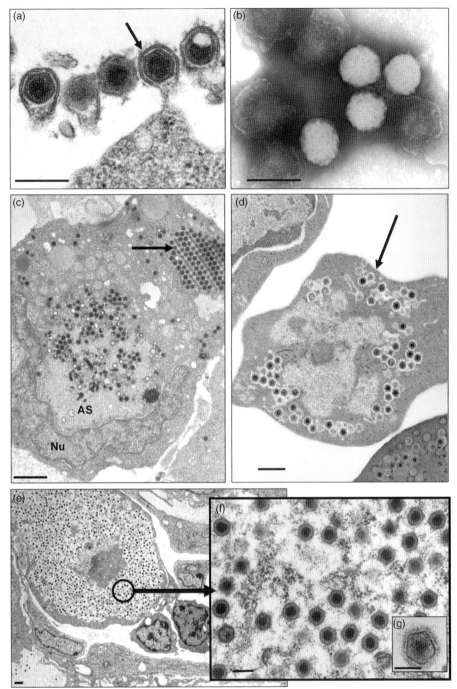

Figure 1 Electron micrographs of iridoviruses identified within or isolated from fish, amphibians, and reptiles. (a) Transmission electron micrograph of an ultrathin section from an EHNV-infected cell. The host-derived membrane is indicated (arrow). (b) Micrograph of a negatively stained EHNV illustrating the difference in appearance when the whole virus is imaged. (c) Low magnification image of a single cultured cell (Chinook salmon embryo cell) infected with EHNV. The image is from an ultrathin section and illustrates the presence of paracrystalline arrays (arrow), virus assembly sites (AS), and a distorted nucleus (Nu). (d) Image of an ultrathin section of a reptilian erythrocyte (arrow) infected with erythrocytic necrosis iridovirus. Bar represents 1 μm. (e) An ultrathin section of an unknown hypertrophied cell from a dwarf gourami infected with an unknown megalocytivrus. The enlarged cell is apparent. (f) Enlargement of the indicated region from panel (e). Large icosahedral viruses are apparent. (g) Image of a single lymphocystivirus from an ultrathin section of an infected cell. The virus differs from the others due to the presence of surface-associated fibrils (arrow). Scale = 200 nm (a, b, g); 1 μm (c–e); 300 μm (f).

Table 1 Vertebrate iridoviruses[a]

Host	Virus	Country or region where isolated
Fish		
Examples of ranaviruses		
Red-fin perch (*Perca fluviatilis*) and rainbow trout (*Onchorhynchus mykiss*)	Epizootic hemotopoietic necrosis virus (EHNV)	Australia
Catfish (*Ictalurus melas*)	European catfish virus (ECV)	Europe (France)
Largemouth bass (*Micropterus salmonides*)	Largemouth bass virus (LMBV)	North America (USA)
Guppy fish (*Poecilia reticlata*)	Guppy fish iridovirus (GV6)	Southeast Asia
Examples of meglacytiviruses		
Red sea bream (*Pagrus major*)	Red seabream iridovirus (RSIV)	Japan
Sea bass (*Lateolabrax sp.*)		Japan
Brown spotted grouper (*Epinephelus tauvina*)	Grouper sleepy disease virus (GSIV)	Southeast Asia
Cultured mandarin fish (*Siniperca chuatsi*)	Infectious spleen and kidney necrosis virus (ISKNV)	Southeast Asia
Examples of lymphocystiviruses		
Infect a large range of fish including flounder (*Platichthys flesus*), plaice (*Pleuronectes platessa*), and dab (*Limanda limanda*)	Lymphocystis disease virus 1 (LCDV-1) (also referred to as Flounder lymphocystis disease virus, FLDV, Dab lymphocystis disease virus (LCDV-2, tentative species of genus *Lymphocystivius*)	Ubiquitous
Amphibians		
Examples of ranaviruses		
Leopard frog (*Rana pipiens*)	Frog virus 3	North America (USA)
Leopard frog (*Rana pipiens*)	Leopard frog iridoviruses (LT1-LT4)	North America (USA)
Red eft (*Diemictylus viridescens*)	T6–20	North America (USA)
North American bullfrog (*Rana catesbeiana*)	Tadpole edema virus (TEV)	North America (USA)
Edible frog (*Rana esculenta*)	*Rana esculenta* iridovirus (REIR)	Europe (Croatia)
Ornate burrowing frog (*Limnodynastis ornatus*)	Bohle iridovirus (BIV)	Australia
Cane toad (*Bufo marinus*)	*Bufo marinus* Venezuelan iridovirus 1 (GV)	South America (Venezuela)
Common frog (*Rana temporaria*)	*Rana temporaria* United Kingdom iridovirus 1 (RUK 11)*	Europe (UK)
Tiger salamander (*Ambystoma tigrinum stebbensi*)	*Ambystoma tigrinum* iridovirus (ATV)	North America (USA)
Reptiles		
Examples of ranaviruses		
Box turtle (*Terrapene c. Carolina*)	Tortoise virus 3 (TV3)	North America (USA)
Central Asian tortoise (*Testudo horsefieldi*)	Tortoise virus 5 (TV5)	North America (USA)
Gopher tortoise (*Gopherus polyphemus*)		North America (USA)
Green tree python (*Morelia viridis*)	Wamena virus (WV)*	Australia (origin Irian Jaya)

[a]Listed viruses are examples of vertebrate iridoviruses; the list is not exhaustive. Not all listed viruses are included within the *Eighth Report of the International Committee on Taxonomy of Viruses* (Fauquet *et al*.); those which are not assigned to genera are indicated (*).

cosmetic. Infection causes cellular hypertrophy with individual cells reaching 100 μm to 1 mm in diameter. Infected cells develop a thick hyline capsule, a central, enlarged nucleus and basophilic inclusions which correspond to the comparatively large size (up to 300 nm) of the viruses. Megalocytiviruses, on the other hand, cause a darkening of body color and lethargy which has led to the disease being referred to as 'sleepy disease'. Infected animals also exhibit severe anemia, petechia of the gill, and enlargement of the spleen. Enlarged, inclusion body-bearing cells which function as viral assembly sites are characteristic structures of the spleen, kidney, liver, and other internal organs. WSIV, which is currently an unassigned member of the family, causes significant mortalities among farm-raised juvenile white sturgeon (*Acipenser transmontanus*) in North America. The virus is epitheliotropic, infecting the skin, gills, and upper alimentary tract. Affected tissues display hyperplasia with characteristic amphophilic to basophilic enlarged Malpighian cells filled with virus particles. Erthrocytic necrosis iridoviruses of fish, amphibians, and reptiles are associated with intracytoplasmic inclusion bodies within erythrocytes. Infection is characterized by nuclear degeneration, margination of chromatin, pyknosis, karyorrhexis, and karyolysis. A major clinical feature of such animals is anemia.

The clinical outcome of ranavirus infections varies from benign to fatal. Infections can lead to ulceration and/or systemic hematopoietic necrosis in amphibians and fish, and skin polyps, skin sloughing, and systemic hematopoietic necrosis in urodeles. The pathology of

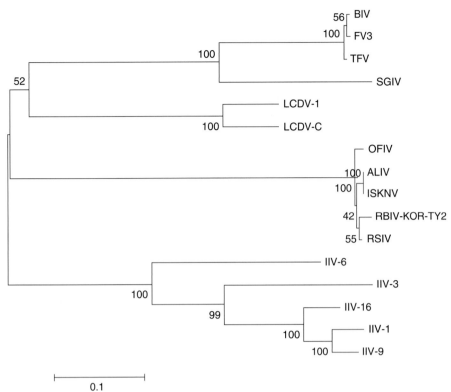

Figure 2 Phylogenetic relationships among iridoviruses. The inferred amino acid sequences of the MCP of 16 iridoviruses, representing all five currently recognized genera, were aligned using the CLUSTAL W program and used to construct a phylogenetic tree using the Neighbor-Joining algorithm and Poisson correction within MEGA version 3.1. The tree was validated by 1000 bootstrap repetitions. Branch lengths are drawn to scale and a scale bar is shown. The number at each node indicates bootstrapped percentage values. The sequences used to construct the tree were obtained from the following viruses: genus *Megalocytivirus* – ISKNV, infectious skin and kidney necrosis virus (AF370008); ALIV, African lampeye iridovirus (AB109368); OFIV, olive flounder iridovirus (AY661546); RSIV, red sea bream iridovirus (AY310918); RBIV, rock bream iridovirus (AY533035); genus *Ranavirus* – SGIV, Singapore grouper iridovirus (AF364593); TFV, tiger frog virus (AY033630); BIV, Bohle iridovirus (AY187046); FV3, frog virus 3 (U36913); genus *Lymphocystivirus* – LCDV-1, lymphocystis disease virus (L63545); LCDV-C, lymphocystis disease virus – China (AAS47819.1); genus *Iridovirus* – IIV-6, invertebrate iridescent virus 6 (AAK82135.1); IIV-16 (AF025775), IIV-1 (M33542), and IIV-9 (AF025774); genus *Chloriridovirus* – IIV-3 (DQ643392).

vertebrate iridovirus infections is best described for the ranavirus EHNV. The associated disease is referred to as epizootic hemaopoietic necrosis and this designation is applicable to most ranavirus infections. Infection results in the degeneration of hematopoietic cells and damage to the vascular endothelium within most organs. For example, in infected kidneys (**Figure 3**) destruction of the blood-forming cells, termed acute hematopoietic necrosis, occurs. Within the liver and spleen, multifocal necrosis is common. Other organs are also affected, including the pancreas and the vascular endothelium within the liver, spleen, kidney, gill, and heart.

Emerging Infectious Pathogens

Although lymphocystis disease has been known since the nineteenth century, other vertebrate iridoviruses are more recently recognized and the incidence of reports of ranavirus and meglocytivirus infections has increased in recent years with respect to the number of new hosts and new geographic locations. Ranavirus and megalocytivirus epizootics have been reported in finfish in Asia, North America, South America, the United Kingdom, and Australia. However, long-term declines in finfish populations have been attributed to over-fishing rather than infectious disease. Dramatic fluctuations of amphibian populations due to ranavirus infections have been reported but these have not been recognized as causative agents for reported global amphibian population declines; these are due, in the main, to the fungus Batrachochytrium dendrobatidis.

Transmission and Control

Lymphocystiviruses are transmitted horizontally via abraded lesions, *Aedes aegypti*, *Aedes albopictus* (C6/36),

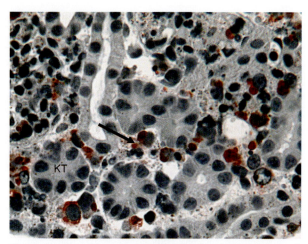

Figure 3 Photomicrograph (light microscopy) of the kidney of an infected *Bufo marinus* tadpole infected with a South American ranavirus. Extensive necrosis of the hematopoietic cells is indicated by the presence of chromogen (brick-red color) following an immunoperoxidase procedure using a primary antibody against EHNV. KT, kidney tubules. The image was viewed at ×100 magnification.

and *Drosophila melanogaster*. The principal mode of transmission of megalocytiviruses (e.g., RSIV) and ranaviruses is horizontally via virus-containing water. Ranaviruses remain viable in water, as dried culture medium, in frozen carcasses, and at various temperatures (4 °C, −20 °C, and −70 °C) for prolonged periods, indicating that they can exist between epizootics both within and outside their biological host(s). The resistant nature of these viruses also indicates their potential for translocation via fomites such as boots, boat hulls, and fishing tackle, in live fish used for stocking aquaculture ponds, via bait fish, and on the skin surfaces of predatory animals such as birds. The international trade in wildlife, a considerable portion of which is illegally performed, is also a recognized mechanism for the effective global transport of vertebrate iridoviruses.

To date, vaccines have only been developed against the megalocytiviruses of Asia. For RSIV, a formalin-inactivated vaccine and a DNA vaccine have been used for the protection of finfish including red sea bream, yellowtail, and amberjack. Control of other diseases such as epizootic hematopoietic necrosis is by containment using diagnosis, surveillance, and management strategies documented by the Office Internationale des Epizooties (OIE). No control measures exist for other ranaviruses or erythrocytic necrosis viruses.

Concluding Remarks

Iridoviruses of fish, amphibians, and reptiles represent a potential health risk to both free-ranging and captured ectothermic vertebrates and are recognized as agents of economical importance by commercial fishing and aquaculture industries; they are not a direct health risk to humans. Over the past two decades, there has been an increase in the number of reported disease incidents involving these large dsDNA viruses. In some reports, surveillance has indicated that these viruses have increased their geographical range and may, therefore, be regarded as emerging viruses. While iridovirus-associated diseases can cause mass mortalities of some species, they are not recognized as a primary cause of the reported global population declines of amphibians or finfish.

The patterns of disease and details on viral structure for the recognized genera are known. However, detailed knowledge of the cellular pathways associated with the pathogenesis of the various diseases is limited.

See also: Iridoviruses of Invertebrates.

Further Reading

Chinchar VG (2002) Ranaviruses (family *Iridoviridae*): Cold-blooded killers. *Archives of Virology* 147: 447–470.

Chinchar VG, Essbauer S, He JG, et al. (2005) Family *Iridoviridae*. In: Fauquat C, Mayo MA, Maniloff M, Desselberger U, and Ball LA (eds.) *Virus Taxonomy: Eighth Report of the International Committee on Taxonomy of Viruses*, pp. 145–162. San Diego, CA: Elsevier Academic Press.

Chinchar VG, Hyatt AD, Miyazaki T, and Williams T (in press) Family *Iridoviridae*: Poor viral relations no longer. *Current Topics Microbiology and Immunology*.

Daszak P, Berger L, Cunningham AA, Hyatt AD, Green DE, and Speare R (1999) Emerging infectious diseases and amphibian population declines. *Emerging Infectious Diseases* 5: 735–748.

Daszak P, Cunningham AA, and Hyatt AD (2000) Emerging infectious diseases of wildlife – threats to biodiversity and human health. *Science* 287: 443–449.

Eaton HE, Metcalf J, Penny E, Tcherepanov V, Upton C, and Brunetti CR (2007) Comparative genomic analsis of the family *Iridoviridae*: Re-annotating and defining the core set of iridovirus genes. *Virology Journal* 4: 11.

Fauquet CM, Mayo MA, Maniloff J, Desselberger U, and Ball LA (eds.) (2005) *Virus Taxonomy. Classification and Nomenclature of Viruses. Eighth Report of the International Committee on the Taxonomy of Viruses*, San Diego, CA: Elsevier Academic Press.

Hyatt AD, Gould AR, Zupanovic Z, et al. (2000) Comparative studies of piscine and amphibian iridovirus. *Archives of Virology* 145: 301–331.

Office Internationale des Epizooties (2006) Epizootic haematopoietic necrosis. In: *Aquatic Animal Health Code*, p. 2.1.1. Paris: OIE.

Robert J, Morales H, Buck W, Cohen N, Marr S, and Gantress J (2005) Adaptive immunity and histopathology in frog virus 3-infected Xenopus. *Virology* 332: 667–675.

Schock DM, Bollinger TK, Chinchar VG, Jancovich JK, and Collins JP (2008) Experimental evidence that amphibian ranaviruses are multi-host pathogens. *Copeia* 2008: 133–143.

Tan WGH, Barkman TJ, Chinchar VG, and Essani K (2004) Comparative genomic analyses of frog virus 3, type species of the genus *Ranavirus* (family *Iridoviridae*). *Virology* 323: 70–84.

Williams T, Barbosa-Solomieu V, and Chinchar VG (2005) A decade of advances in iridovirus research. *Advances in Virus Research* 65: 173–248.

Iridoviruses of Invertebrates

T Williams, Instituto de Ecología A.C., Xalapa, Mexico

© 2008 Elsevier Ltd. All rights reserved.

Glossary

Circular permution Genome is a linear molecule of DNA with different terminal sequences at the population level. Physical maps of such genomes are circular.

Covert infection A sublethal infection that causes no obvious changes in the appearance or behavior of the host.

Cytopathic effect Alteration in the microscopic appearance of cultured cells following virus infection.

MCP Major capsid protein. A highly conserved structural polypeptide of *c.* 50 kDa that represents about 40% of the total protein content of the virion.

Patent infection Invertebrates that display iridescent hues due to an abundance of virus particles in crystalline arrays in infected cells.

Introduction

Invertebrate iridescent viruses (IIVs) (aka invertebrate iridoviruses) are icosahedral particles of approximately 120–200 nm in diameter that infect invertebrates, mostly insects, in damp or aquatic habitats. These viruses cause two types of disease: one patent and the other covert (inapparent). An abundance of virus particles in the cells of patently infected insects causes them to develop an obvious iridescent color that typically ranges from violet, blue, green, or orange. Patent disease is usually fatal in the larval or pupal stages. In contrast, covert infections are not lethal; covertly infected insects appear healthy and may develop to the adult stage and reproduce. Interest in these viruses has been limited by the perception that they have little potential as biological control agents against insect pests. However, there is now growing awareness of the potential impact of sublethal IIV disease on the dynamics of insect populations, including insect vectors of medical importance worldwide.

History

Originally discovered in 1954 infecting soil-dwelling populations of cranefly larvae (*Tipula paludosa*) in England, IIVs were subsequently reported in insects and other invertebrates worldwide. Detailed electron microscope observations in the late 1960s and early 1970s confirmed the icosahedral nature of the particle and revealed a complex ultrastructure including an internal lipid membrane and an external fringe of fibrils extending from the capsid of certain isolates. Serological relationships among IIVs and with iridoviruses from vertebrates were revealed during the 1970s and the genome of IIV-6 was shown to be circularly permuted and terminally redundant in 1984. Abundant covert IIV infections in insect populations were detected using molecular techniques and comparative genetic studies broadly supported previous serological findings on the relationships among these viruses in the 1990s. High-resolution ultrastructural studies in 2000 built on the previous model and the first complete genome of an IIV was sequenced in 2001.

Classification

IIVs are currently assigned to one of two genera in the family based on particle size and genetic characteristics. Small IIVs with dehydrated particle sizes in ultrathin section typically around 120 nm diameter have been isolated from several different orders of insects and terrestrial isopods and are assigned to the genus *Iridovirus* (**Table 1**). Due to the limited genome sequence data available, only two IIVs have been assigned species status, *Invertebrate iridescent virus 1* (IIV-1) and *Invertebrate iridescent virus 6* (IIV-6), that is the type species of the genus. Tipula iridescent virus and chilo iridescent virus are recognized synonyms of each of these names, respectively. An additional 11 viruses are presently considered as tentative species in the genus. The viruses of this genus can be further divided into three distinct complexes based on genetic and serological characteristics: one large complex containing IIV-1 and at least nine other IIVs, and two smaller complexes, one containing IIV-6, and the other containing IIV-31 and an IIV from a beetle (*Popillia japonica*).

The genus *Chloriridovirus* is comprised of a single species, *Invertebrate iridescent virus 3* (IIV-3), with a particle size of ~180 nm diameter in ultrathin section, isolated from a mosquito. There are a great many additional records of iridoviruses from invertebrate hosts but these have not been characterized.

Geographical Distribution

IIVs have been observed infecting invertebrates in tropical and temperate regions on every continent except Antarctica. A number of marine invertebrates have also been reported as

Table 1 Classification of iridoviruses isolated from invertebrates

Genus, virus (alternative name)	Abbreviation	Host species (order)[a]	Location	Accession numbers
Iridovirus				
Invertebrate iridescent virus 1 (Tipula iridescent virus)	IIV-1 (TIV)	*Tipula paludosa* (D)	UK	M33542, M62953
Invertebrate iridescent virus 6 (Chilo iridescent virus)	IIV-6 (CIV)	*Chilo suppressalis* (L)	Japan	AF303741, NC_003038
Anticarsia gemmatalis iridescent virus[b]	AGIV	*Anticarsia gemmatalis* (L)	USA	AF042343
Invertebrate iridescent virus 2[b]	IIV-2	*Sericesthis pruinosa* (C)	Australia	AF042335
Invertebrate iridescent virus 9[b]	IIV-9	*Wiseana cervinata* (L)[c]	New Zealand	AF025774, AY873793
Invertebrate iridescent virus 16[b]	IIV-16	*Costelytra zealandica* (C)	New Zealand	AF025775, AY873794
Invertebrate iridescent virus 21[b]	IIV-21	*Helicoverpa armigera* (L)[d]	Malawi	
Invertebrate iridescent virus 22[b]	IIV-22	*Simulium variegatum* (D)	UK	AF042341, M32799
Invertebrate iridescent virus 23[b]	IIV-23	*Heteronychus arator* (C)	South Africa	AF042342
Invertebrate iridescent virus 24[b]	IIV-24	*Apis cerana* (Hy)	Kashmir	AF042340
Invertebrate iridescent virus 29[b]	IIV-29	*Tenebrio molitor* (C)	USA	AF042339
Invertebrate iridescent virus 30[b]	IIV-30	*Helicoverpa zea* (L)	USA	AF042336
Invertebrate iridescent virus 31[b]	IIV-31	*Armadillidium vulgare* (Is)[e]	USA	AF042337, AJ279821, AF297060
Chloriridovirus				
Invertebrate iridescent virus 3	IIV-3	*Ochlerotatus (Aedes) taeniorhynchus* (D)	USA	AJ312708

[a]Insect orders Coleoptera (C), Diptera (D), Homoptera (H), Hymenoptera (Hy), Lepidoptera (L), and terrestrial isopods (Is) (Crustacea).
[b]Tentative member.
[c]Also isolated from sympatric insect species *Witlesia sabulosella* (L) and *Opogonia* sp. (C).
[d]Also isolated from *Lethocerus colombiae* (H) in Lake Victoria, Uganda, but may have been contaminated.
[e]Also isolated from *Porcellio dilatatus* and probably several other species of terrestrial isopods.

hosts to iridoviruses. Humidity appears to be the principal factor limiting the distribution of these viruses. Records of IIV infections are most common in aquatic or soil-dwelling arthropods during periods of rainfall and absent in species that inhabit arid or desiccated habitats.

Host Range and Virus Propagation

IIVs replicate in many types of insect cells and may even replicate in reptilian cells *in vitro*. Host range *in vivo* depends very much on the route of infection. Most IIVs show a remarkably broad host range when the inoculum is injected compared to a reduced host range following oral administration of inoculum. For example, injection of particles of IIV-6 results in patent infections in species from all major insect orders and a number of other arthropods isopods and a centipede. Other IIVs, such as IIV-3, IIV-16, or IIV-24, appear restricted to one or two closely related host species. Certain IIVs are capable of infecting multiple host species in their natural habitat, including IIV-9 that infects soil-dwelling species of Lepidoptera and Coleoptera in New Zealand, and IIV-31 that infects several species of terrestrial isopods in the USA. IIVs are also capable of replication in host species that do not develop signs of disease, but the range of species susceptible to such asymptomatic infections is largely unknown.

Virus propagation *in vitro* is most readily achieved in cell lines from dipteran (*Aedes aegypti*, *Ae. albopictus*, *Drosophila* DR1, DL2, etc.) and lepidopteran species (Sf-9, Sf-21, Cf-124, etc.), although recently, cell lines from species of Homoptera and Coleoptera have also been successfully used. Most IIVs can be grown in massive quantities in the standard laboratory host, *Galleria mellonella* (Lepidoptera: Pyralidae). The mosquito virus IIV-3 can only be produced *in vivo* in larvae of *Ochlerotatus taeniorhynchus*.

Properties of the Virion

IIV particles comprise an electron dense core of DNA and associated proteins, surrounded by a lipid membrane encased by an exterior protein capsid (**Figure 1**). Virions released by budding may have an additional outer envelope but this is not essential for infectivity. Detailed studies using cryo-electron microscopy and three-dimensional image reconstruction have examined particles of IIV-6 in closely packed quasi-crystalline hexagonal arrays with an interparticle distance of 40–60 nm. Particle diameter was calculated to be 162 nm along the two- and threefold axes of symmetry and 185 nm along the fivefold axis, considerably larger than the diameter in ultrathin section. The outer capsid is composed of trimeric capsomers, each approximately 8 nm diameter and 7.5 nm high, arranged in a pseudo-hexagonal array. A thin fiber projects radially from the surface of each capsomer that probably regulates interparticle distance, a key characteristic for the iridescence of infected hosts. At the base, the

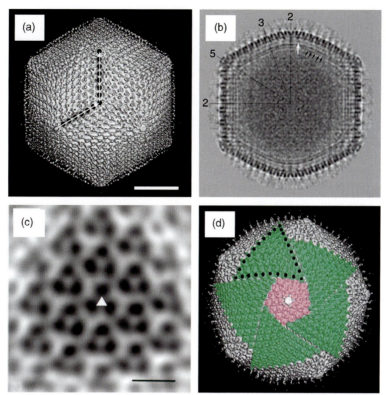

Figure 1 (a) 3-D reconstruction of an IIV-6 particle viewed down threefold axes of symmetry by cryoelectron microscopy. Circles indicate the position of capsomers along the h and k lattice used to calculate the triangulation number (*T*). The proximal end of surface fibers are visible, whereas the flexible distal ends have become lost during the reconstruction process (white bar = 50 nm applies to images (a)–(c)). (b) Central section of the reconstruction density map viewed along twofold axes indicating two-, three-, and fivefold axes of symmetry. A lipid bilayer is observed beneath the capsid shell (black arrows) with numerous connections to an additional shell (white arrow) beneath the outer shell. (c) Close-up view of the trimers comprising each capsomer shown as a planar section through a reconstruction density map viewed along threefold axes (triangle) (black bar = 10 nm). (d) Facets of capsid with five trisymmetrons (highlighted in green) arranged around one pentasymmetron (in pink). The central capsomer of the pentasymmetron (uncolored) is pentavalent. The edge of each trisymmetron comprises 10 capsomers (black dots). Reprinted by permission from Macmillan Publishers Ltd: Nature Structural Biology, Yan X, Olson NH, Van Etten JL, Bergoin M, Rossmann MG, Baker TS (2000) Structure and assembly of large lipid-containing dsDNA viruses. *Nature Structural Biology* 7: 101–103, copyright (2000).

interconnected capsomers form a contiguous icosahedral shell ∼2.5 nm thick. The major capsid protein (MCP) exists externally as a noncovalent trimer and internally as a trimer linked by disulfide bonds. The capsomers are arranged into trisymmetron and pentasymmetron facets. Each particle consists of 20 trisymmetrons, composed of 55 capsomers, and 12 pentasymmetrons composed of 30 capsomers and one hexavalent capsomer of uncertain composition. Pentavalent capsomers are located at the vertices of the particle in the center of each pentasymmetron. This gives a total of 1460 capsomers + 12 pentavalent capsomers per particle. The triangulation number (T) is 147. Larger IIVs that infect mosquitoes and midges have larger trisymmetrons, probably comprising 78 subunits giving a likely 1560 subunits per particle. A lipid bilayer, 4 nm thick, surrounds the DNA core, and is intimately associated with an additional inner shell beneath the fused layer of the capsid. Core and capsid polypeptides are likely interconnected by intermembrane proteins passing through the lipid layer. The core is a highly hydrated entity in which the DNA–protein complex appears to be arranged in a long coiled filament of some 10 nm diameter.

IIVs are structurally complex: one-dimensional PAGE resolves 20–32 polypeptides with weights typically from 11 to 200 kDa. Much of the polypeptide diversity of IVs appears to be associated with the core and lipid membrane. At least six polypeptide species are associated with the DNA within the core, the major component being a 12.5 kDa species in IV-6. The MCP comprises about 470 amino acids (∼50 kDa) and represents 40–45% of the total particle polypeptide.

Properties of the Genome

Each IIV genome is comprised of a linear molecule of DNA that ranges in size from 140 to 210 kbp, and is circularly permuted and terminally redundant. Circular permution means that the terminal sequences differ for each genome in a population, whereas terminal redundancy means that part of the sequence at one end of the DNA molecule is repeated at the other end. For example, if a complete genome is represented by the numbers 0–9, the DNA molecules from individual virus particles may be represented by the following combinations: 012345678901, 234567890123, 4567890123456, etc., where terminal redundancy is indicated by the underlined numbers. In IIV-6, the degree of terminal redundancy has been estimated as 12% and the genome contains six origins of replication.

The IIV genome is either not methylated, or methylated at very low or undetectable levels. In contrast, high levels of methylation of cytosine residues are seen in virtually all vertebrate iridoviruses. Currently, only two IIV genomes have been sequenced in their entirety: IIV-3 and IIV-6. The genome of IIV-6 is 212 kbp (unique portion) with 28.6% G+C content and comprises 468 open reading frames (ORFs), of which 234 are nonoverlapping. The genome of IIV-3 is 191 kbp (unique portion) with a 48% G+C content and comprises 453 ORFs, of which 126 are nonoverlapping. No collinearity is observed between the genomes of IIV-3 and IIV-6.

Core IIV genes include those involved in (1) replication, including DNA polymerase (037L), RNA polymerase II (α-subunit 176R, β-subunit 428L), Rnase III (142R), a helicase (161L), and a DNA topoisomerase II (045L); (2) nucleotide metabolism, such as ribonucleotide reductase (α-subunit 085L, β-subunit 376L), dUTPase (438L), thymidylate synthase (225R), thymidylate kinase (251L), and thymidine kinase (143R); and (3) other proteins of known function including IAP inhibitor of apoptosis (157L, 193R), PCNA (436R), and the MCP (274L).

Other notable putative genes identified in IIV-6 include an NAD+-dependent DNA ligase (205R) and a putative homolog of sillucin (160L), a cysteine-rich peptide antibiotic. The promoter regions of the MCP and DNA polymerase genes have been located to essential sequences at 29–53 and 6–27 positions upstream of the transcriptional start site, respectively (**Figure 2**). An ORF (100L) has been detected in IIV-6 with truncated homology to the nuclear polymerizing (ADP-ribosyl) transferase from eukaryotic organisms. Interestingly, the large subunit of the IIV-6 ribonucleotide reductase appears to contain an intein, a form of selfish genetic element that removes itself from the protein by post-translational autocatalytic splicing. Genes unique to IIV-3 include two putative transmembrane proteins, a protein similar to fungal DNA polymerase, and a protein similar to a fungi RNA Pol II subunit.

IIVs have been shown to have extensive regions of repetitive DNA that account for 20% (IIV-3) to over 25% (IIV-9) of the genome. The coding function of these regions is unknown although transcription of these regions has been detected late in the infection cycle.

The pattern of repetitive DNA in the genome of IIV-6 is complex and involves boxes of tandem repeat sequences and others with a number of different interdigitated repeat sequences of variable size and homology.

Replication

The model for iridovirus replication is that of frog virus 3 (FV-3). Virions display cytotoxic properties and are capable of nongenetic reactivation. Like all other members of the family, IIVs do not replicate at temperatures above 30 °C.

Evolution

Iridoviruses are members of a monophyletic clade of large, nucleocytoplasmic DNA viruses that includes the families *Poxviridae*, *Phycodnaviridae*, *Asfarviridae* and the recently discovered giant icosahedral mimivirus from an ameba. The clade shares a total of 41 ancestral genes including structural components, and those involved in DNA packaging, replication, transcription and RNA modification including subunits of RNA polymerase and transcription factors, many of which appear to have been acquired from eukaryotic host cells.

Sequence comparisons of the virus-encoded δ DNA polymerase and MCP indicated putative evolutionary relationships between IIVs and ascoviruses of lepidopteran insects. Homologs to about 40% of the proteins encoded by *Spodoptera frugiperda ascovirus 1a* are found in IIV-6 with lower percentages seen among vertebrate iridoviruses, phycodnaviruses, and African swine fever virus (ASFV). Additional analyses based on *bro* genes (Baculovirus repeated ORFs), a multigene family of unknown function, support the conclusion that ascoviruses are more closely related to invertebrate iridoviruses than to vertebrate iridoviruses. Like IIVs, lepidopteran ascoviruses have very low oral infectivity but are highly infectious by injection and depend on parasitoid wasps for transmission. Structural similarities between iridoviruses and the allantoid particles of ascoviruses are not immediately apparent although molecular evidence indicates clear relationships between these viruses.

Signs and Characteristics of Disease

The principal sign of patent IV infection is the iridescent hues which arise from the paracrystalline arrangement of virus particles in host cells. Light is reflected from the surface of close packed particles and causes interference with incident light known as 'Bragg reflections'.

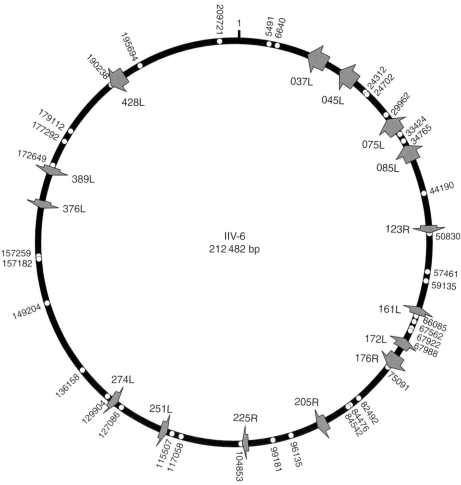

Figure 2 Genetic map of IIV-6. Position of selected ORFs (arrows) identified in the complete genome sequence of IIV-6 that encode the following putative proteins: DNA polymerase (037L), topoisomerase II (045L), ATPase (075L), ribonucleoside diphosphate reductase large subunit (085L), protein-tyrosine phosphatase (123R), helicase (161L), global transactivator homolog (172L), DNA-dependent RNA polymerase 1 (176R), DNA ligase (205R), thymidylate synthase (225R), thymidylate kinase (251L), major capsid protein (274L), ribonucleoside diphosphate reductase small subunit (376L), serine-threonine protein kinase (389L), DNA-dependent RNA polymerase 2 (428L). White dots accompanied by figures indicate the nucleotide positions of *Eco*RI cleavage sites. Reproduced from Jakob N, Darai G, and Williams T (2002) Genus *Iridovirus*. In: Christian T and Darai G (eds.) *Springer Index of Viruses*, Berlin: Springer, with kind permission of Springer Science and Business Media.

The small IIVs of the genus *Iridovirus* usually display violet, blue, or turquoise colors, whereas large IIVs from mosquitoes (genus *Chloriridovirus*) commonly display colors such as green, yellow, or orange. Purified pellets of IIVs also iridesce. Some isolates have unusually long external fibrils attached to the capsid and these isolates do not iridesce.

Covert infections have been detected in natural populations of blackflies (*Simulium variegatum*) and a mayfly (*Ecdyonurus torrentis*), and in laboratory populations of a mosquito (*Ae. aegypti*) and a lepidopteran (*G. mellonella*). Covert infections have been detected by polymerase chain reaction (PCR) amplification of the MCP gene, electron microscope observations, and insect bioassay techniques (**Figure 3**). Studies on IIV-6 in *Ae. aegypti* have revealed clear costs of covert infection including an increase in larval development time and reductions in adult body size, longevity, and fecundity. Overall the reproductive capacity of covertly infected mosquitoes is reduced by 22–50% compared to healthy mosquitoes, depending on the number of cycles of blood meals followed by oviposition.

Pathology

IIVs replicate extensively in most host tissues, especially the epidermis, muscles, fat body, nerves, hemocytes, and areas of the gut. IIV-1 caused the formation of epidermal

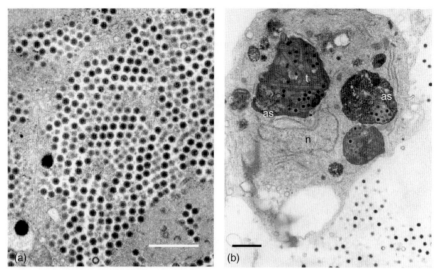

Figure 3 IIV particles in cytoplasm of cells from insects with patent or covert infection. (a) Arrays of closely packed particles of IIV-3 in an epidermal cell from a patently infected mosquito larva. Scale = 1 μm. (b) Low density of particles in a hemocyte from a covertly infected mayfly larva. Cells from covertly infected insects show the characteristic cytoplasmic virus assembly sites (as) close to the nucleus (n), and also the presence of tubular structures (t) likely to be aberrant forms of virus capsids. Scale = 1 μm. (a) Photo courtesy of J. J. Becnel. (b) Reproduced from Tonka T and Weiser J (2000) Iridovirus infection in mayfly larvae. *Journal of Invertebrate Pathology* 76: 229–231, with permission from Elsevier.

tumors in silkworm larvae but such pathology is not observed in other hosts. In mosquito larvae infected by IIV-3, the fat body, epidermis, imaginal disks, hemocytes, trachea, muscle, visceral nerves, gonads, and esophagus were infected but not the remaining gut or malpighian tubules. Individuals with patent infections that survive to pupate frequently show marked deformations of the pupa, particularly of the wing buds.

Pathological changes at the cellular level include cell rounding and the appearance of extensive areas of finely granulated material devoid of cell organelles, the cytoplasmic virus assembly sites. Marked contraction of the cells followed by cell detachment are also common cytopathic effects. Rapid cell–cell fusion is influenced by the multiplicity of infection in cells infected by IIV-6. A virion-associated protein appears to be responsible. The formation of numerous vesicles arising from blebbing of cell membranes followed by loss of cell adhesion and cell–cell fusion has also been observed in lepidopteran cells infected by IIV-1. Changes in the position and morphology of mitochondria have been reported.

Ecology

Ecological studies of IIVs are sparse, probably because the incidence of patent disease is typically very low. The majority of studies have used iridescence as the sole criterion for diagnosing infection, although PCR and insect bioassay have also been employed successfully to detect and quantify covert infections. Studies on backflies and Lepidoptera have indicated that there exists considerable genotypic variation in IIV populations such that individual insects collected at the same place and time may harbor genetically distinct variants.

Studies on transmission have been hindered because the route of infection is unknown or uncertain for most IIV–host systems. Cannibalism or predation of infected individuals involves the consumption of massive doses of particles and appears to be the principal mechanism of transmission in populations of mosquitoes, isopods, tipulids, mole crickets, and cannibalistic species of Lepidoptera. Hymenopteran parasitoids and entomopathogenic nematodes have been shown capable of transmitting IIV infections during the act of oviposition or host penetration, respectively. Survival of IIV-3 in mosquito populations appears to depend on alternating cycles of horizontal transmission between cannibalistic larvae and vertical transmission from adult female mosquitoes that acquire infection shortly before pupating. Horizontal transmission is also favored in high-density populations of some hosts wherein the frequency of aggressive encounters between conspecifics and the probability of wounding is greater than at low densities.

There is clear evidence of seasonality in many IIV–host associations due to seasonal fluctuations in precipitation and/or host densities. The persistence of IIV-6 in soil depends on moisture, whereas persistence in water is markedly affected by solar ultraviolet radiation.

Occasional epizootics of patent disease have been reported in lepidopteran species, *Helicoverpa zea* and *Anticarsia gemmatalis*, the cricket *Scapteriscus borellii*, as well as tipulid, mosquito, and blackfly larvae.

Economic Importance

IIVs have been observed to infect natural populations of major insect pests and numerous species of insect vectors of medical or veterinary importance (mosquitoes, midges, and blackflies). However, the low prevalence of patent infections and relatively broad host range of most IIVs means that these viruses are not considered as likely agents for programs of biological control. An IIV is believed to be responsible for periodically devastating populations of mopane worms (*Gonimbrasia belina*, Lepidoptera) that represent a multimillion dollar food industry in several southern African countries. Iridovirus infections are also associated with severe diseases and mass mortalities in oyster populations, but the relationship between these marine iridovirus and those infecting terrestrial and freshwater arthropods is unknown.

See also: Ascoviruses; Baculoviruses: General Features; Iridoviruses: General Features; Iridoviruses of Vertebrates.

Further Reading

Darai G (ed.) (1990) *Molecular Biology of Iridoviruses.* Dordrecht, The Netherlands: Kluwer.

Delhon G, Tulman ER, Afonso CL, et al. (2006) Genome of Invertebrate iridescent virus type 3 (mosquito iridescent virus). *Journal of Virology* 80: 8439–8449.

Jakob N, Darai G, and Williams T (2002) Genus *Iridovirus*. In: Christian T and Darai G (eds.) *Springer Index of Viruses*, Berlin: Springer.

Jakob NJ, Muller K, Bahr U, and Darai G (2001) Analysis of the first complete DNA sequence of an invertebrate iridovirus: Coding strategy of the genome of Chilo iridescent virus. *Virology* 286: 182–196.

Tonka T and Weiser J (2000) Iridovirus infection in mayfly larvae. *Journal of Invertebrate Pathology* 76: 229–231.

Webby R and Kalmakoff J (1998) Sequence comparison of the major capsid protein gene from 18 diverse iridoviruses. *Archives of Virology* 143: 1949–1966.

Williams T (1998) Invertebrate iridescent viruses. In: Miller LK and Ball LA (eds.) *The Insect Viruses*, pp. 31–68. pp. 31–68. New York: Plenum.

Williams T, Barbosa-Solomieu V, and Chinchar VG (2005) A decade of advances in iridovirus research. *Advances in Virus Research* 65: 173–248.

Yan X, Olson NH, Van Etten JL, Bergoin M, Rossmann MG, and Baker TS (2000) Structure and assembly of large lipid-containing dsDNA viruses. *Nature Structural Biology* 7: 101–103.

Iridoviruses: General Features

V G Chinchar, University of Mississippi Medical Center, Jackson, MS, USA
A D Hyatt, Australian Animal Health Laboratory, Geelong, VIC, Australia

© 2008 Elsevier Ltd. All rights reserved.

Glossary

Anuran An amphibian of the order Salientia (formerly Anura or Batrachia) which includes frogs and toads. Also called, salientians.

Introduction

Members of the family *Iridoviridae*, hereafter referred to as iridovirids to distinguish them from members of the genus *Iridovirus*, are large (120–200 nm), double-stranded DNA viruses that utilize both the nucleus and the cytoplasm in the synthesis of viral macromolecules, but confine virion formation to morphologically distinct, cytoplasmic assembly sites. Virus particles display icosahedral symmetry but, unlike other virus families, infectious virions can be either nonenveloped (i.e., naked) or enveloped, although the latter show a higher specific infectivity. The viral capsid is composed primarily of the major capsid protein (MCP), a ~50 kDa protein that is highly conserved among all members of the family. An internal lipid membrane, that is essential for infectivity, underlies the capsid and encloses the viral DNA core. Approximately 30 virion-associated proteins have been identified by gel electrophoresis, but the functions of most of these proteins are unknown. The viral genome is linear, double-stranded DNA and ranges in size from 103 to 212 kbp, depending upon the viral species. As a likely consequence of its mode of packaging, viral DNA is terminally redundant and circularly permuted. The size of the repeat regions range from 5% to 50% of the genome and, like the genome size, appears to vary with the specific viral species.

Iridovirus Taxonomy

Members of the family *Iridoviridae* are classified into five genera, two of which infect invertebrates (*Iridovirus*, *Chloriridovirus*), and three that infect ectothermic vertebrates (*Ranavirus*, *Lymphocystivirus*, and *Megalocytivirus*). In addition to differences in host range, viruses within the three vertebrate iridovirus genera, with one known exception, contain highly methylated genomes in which every cytosine present within a CpG motif is methylated.

Viral isolates within the genus *Megalocytivirus* show high levels of sequence identity and it is not clear if they represent strains of the same viral species, or a small number of closely related species. To a lesser extent, the same is true within the genus *Ranavirus*, although here differences in host range and clinical presentation, along with a higher level of sequence variation, allows identification of individual viral species (**Table 1**).

Recently, genomic analysis validated taxonomic divisions made earlier on the basis of physical characteristics, clinical presentation, and host range. A phylogenetic tree constructed using the inferred amino acid sequences of the major capsid proteins (MCPs) of 16 iridovirids, representing all five genera, supports the division of the family into three genera of vertebrate iridoviruses (*Megalocytivirus*, *Ranavirus*, and *Lymphocystivirus*), and a phylogenetically diverse cluster of invertebrate iridescent viruses (IIVs) composed of the existing *Chloriridovirus* and *Iridovirus* genera (**Figure 1**). Recently, a maximum likelihood (M-L) tree, constructed using a concatenated set of 11 protein sequences conserved among all iridoviruses, confirmed this finding. Moreover, a M-L tree constructed using MCP sequences from 14 invertebrate iridescent viruses (many of which had been tentatively classified as members of the genus *Iridovirus*) has shown that IIV-6 clusters with two tentative members of the genus *Iridovirus* on one branch, whereas IIV-3 is linked with the remaining 12 isolates on a separate branch. Given this admixture of putative chloriridovirus and iridovirus species, it has been suggested that the status of the existing invertebrate genera may require revision. However, the lack of collinearity between IIV-3 and IIV-6, the low levels of amino acid identity between IIV-3 and IIV-6, along with differences in host range, virion size, and GC content, indicate that IIV-3 and IIV-6 are likely members of different viral genera. On a larger taxonomic scale, recent work has linked the family *Iridoviridae* with other large DNA virus families such as the *Poxviridae*, *Asfarviridae*, *Phycodnaviridae*, the newly discovered mimiviruses, and the *Ascoviridae*. However, although it is clear that these large DNA viruses share a set of common genes, it is not known if iridoviruses are more closely related to ascoviruses, the recently discovered mimivirus, or another virus family.

Viral Genome

The 12 completely sequenced iridovirid genomes range in size from 103 to 212 kbp (**Table 2**). The genomes of vertebrate iridovirids are found at the lower end of this size range, whereas the genomes of IIV-3 and IIV-6 (invertebrate iridovirids) occupy the high end. Consistent with the differences in size, vertebrate iridovirids encode ~100 putative ORFs, whereas invertebrate iridovirids code for

Table 1 Taxonomy of the family *Iridoviridae*

Genus	Viral species [strains][a]	Tentative species
Iridovirus	*Invertebrate iridescent virus 6* (IIV-6) [Gryllus iridovirus, Chilo iridescent virus]	Anticarsia gemmatalis iridescent virus (AGIV), IIV-2, -9, -16, -21, -22, -23, -24, 29, -30, -31
	Invertebrate iridescent virus 1 (IIV-1) [Tipula iridescent virus]	
Chloriridovirus	*Invertebrate iridescent virus-3* (IIV-3) [Aedes taeniorhynchus iridescent virus, mosquito iridescent virus]	
Ranavirus	*Frog virus 3* (FV3), [tadpole edema virus, TEV; tiger frog virus, TFV]	Singapore grouper iridovirus (SGIV), grouper iridovirus (GIV), Rana esculenta iridovirus, Testudo iridovirus, Rana catesbeiana virus Z (RCV-Z)
	Ambystoma tigrinum virus (ATV), [Regina ranavirus, RRV]	
	Bohle iridovirus (BIV)	
	Epizootic haematopoietic necrosis virus (EHNV)	
	European catfish virus (ECV), [European sheatfish virus, ESV]	
	Santee-Cooper ranavirus, [largemouth bass virus, LMBV; doctor fish virus, DFV; guppy virus 6, GV-6]	
Megalocytivirus	*Infectious spleen and kidney necrosis virus* (ISKNV) [Red Sea bream iridovirus, RSIV; African lampeye iridovirus, ALIV; orange spotted grouper iridovirus, OSGIV; rock bream iridovirus, RBIV]	
Lymphocystivirus	*Lymphocystis disease virus 1* (LCDV-1) [flounder lymphocystis disease virus, flounder virus]	LCDV-2, LCDV-C, LCDV-RF; [Dab lymphocystis disease virus]
Unclassified		White sturgeon iridovirus (WSIV)
		Erythrocytic necrosis virus (ENV)

[a]Recognized viral species are italicized and likely virus strains or isolates are enclosed within brackets. Tentative species are listed, as are commonly used abbreviations for species, strains, isolates, and tentative species.

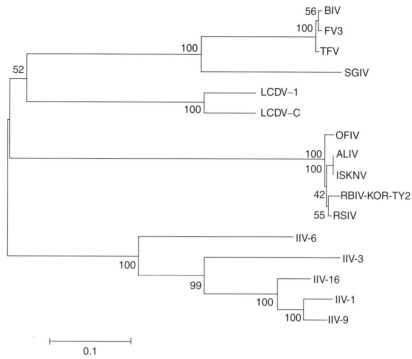

Figure 1 Phylogenetic relationships among iridovirids. The inferred amino acid sequences of the MCP of 16 iridovirids, representing all five currently recognized genera, were aligned using the CLUSTAL W program and used to construct a phylogenetic tree using the neighbor-joining algorithm and Poisson correction within MEGA version 3.1. The tree was validated by 1000 bootstrap repetitions. Branch lengths are drawn to scale. The number at each node indicates bootstrapped percentage values. The sequences used to construct the tree were obtained from the following viruses: genus *Megalocytivirus* – ISKNV, infectious skin and kidney necrosis virus (AF370008); ALIV, African lampeye iridovirus (AB109368); OFIV, olive flounder iridovirus (AY661546); RSIV, red sea bream iridovirus (AY310918); RBIV, rock bream iridovirus (AY533035); genus *Ranavirus* – SGIV, Singapore grouper iridovirus (AF364593); TFV, tiger frog virus (AY033630); BIV, Bohle iridovirus (AY187046); FV3, frog virus 3 (U36913); genus *Lymphocystivirus* – LCDV-1, lymphocystis disease virus (L63545); LCDV-C, lymphocystis disease virus (China) (AAS47819.1); genus *Iridovirus* – IIV-6, invertebrate iridescent virus 6 (AAK82135.1); IIV-16 (AF025775), IIV-1 (M33542), and IIV-9 (AF025774); genus *Chloriridovirus* – IIV-3 (DQ643392).

Table 2 Coding potential of iridovirid genomes

Genus	Virus	bp[a]	No. of genes[b]	%GC	GenBank Acc. No.
Iridovirus	IIV-6	212 482	234	29	AF303741
Chloriridovirus	IIV-3	191 132	126	48	DQ643392
Lymphocystivirus	LCDV-1	102 653	110	29	L63545
	LCDV-C	186 250	176	27	AY380826
Megalocytivirus	ISKNV	111 362	105	55	AF371960
	OSGIV	112 636	121	54	AY894343
	RBIV	112 080	118	53	AY532606
Ranavirus	FV3	105 903	98	55	AY548484
	TFV	105 057	105	55	AF389451
	ATV	106 332	96	54	AY150217
	SGIV	140 131	162	49	AY521625
	GIV	139 793	120	49	AY666015

[a]The value shown represents the unique genome size in base pairs (bp) minus the length of the terminal repeats.
[b]The value shown is an estimate of the total number of nonoverlapping genes encoded by a given virus. It is generally lower than the total number of putative ORFs which includes putative genes encoded on opposing DNA strands and overlapping genes.

126–234 putative proteins. BLAST analysis of the various viral genomes indicates that about a quarter to a third of the viral ORFs share sequence identity/similarity to eukaryotic and viral proteins of known function including the two largest subunits of RNA polymerase II, a viral DNA polymerase, the two subunits of ribonucleotide reductase, dUTPase, thymidylate synthase, thymidylate kinase, a major capsid protein, and other proteins likely to be directly involved in viral replication. In addition to these replicative proteins, iridovirids also encode one or more putative 'immune evasion' proteins such as a viral homolog of eukaryotic translational initiation factor 2α (vIF-2α), a steroid oxidoreductase, a caspase activation and recruitment domain (CARD)-containing protein, and a homolog of the TNF receptor. Dot plot analyses have shown that virus species within the genus *Ranavirus* display some degree of collinearity in their gene order, but that inversions occur even between closely related viruses such as Ambystoma tigrinum virus (ATV) and tiger frog virus. Moreover, between more distantly related ranaviruses (e.g., Singapore grouper iridovirus and FV3) and between viruses from different genera, collinearity appears to break down completely, indicating that, although iridovirids share many genes in common, their precise arrangement within the viral genome does not appear to be critical for successful virus replication. Furthermore, although viral genes are temporally expressed, there is no clustering of immediate early (IE), delayed early (DE), or late (L) viral genes. These observations suggest that the marked differences in gene order between different viral species may be due to the propensity of iridovirids to undergo high levels of recombination.

DNA repeats are present within the intergenic regions of all iridovirids, but differ between genera in their extent, arrangement, and sequence motifs. For example, 15 distinct repeats, comprising two distinct groups, 0.8–4.6 kbp in length, are found within IIV-3. The repeats are composed of a ~100 bp sequence that is present in 2–10 copies/genome. They show 80–100% identity within a group and ~60% identity between groups. The function(s) of the repeat regions is unknown but, by analogy to similar regions in other DNA viruses, they may play roles in genome replication and gene expression. Among members of individual vertebrate iridovirid genera, sequence identity within particular coding regions, for example, the MCP, is relatively high (>70%). However, there is considerable diversity among the MCPs of invertebrate viruses, with some members of the genus *Iridovirus* showing only 50% identity to others. As expected, between genera amino acid identity/similarity is only ~50%.

Virus Replication Cycle

Most of what is known about the replication of iridovirids is based on studies conducted with *Frog virus 3* (FV3), the type species of the genus *Ranavirus*. Thus, any discussion of iridovirid life cycles will have a strongly 'ranacentric' orientation. By necessity, that bias will continue here, but with the caveat that other members of the family, especially viruses from different genera, may not follow the FV3 pathway. **Figure 2** summarizes the salient features of FV3 replication.

Virus Entry

The cellular receptor for FV3 is apparently a quite common and highly conserved molecule as a range of mammalian, avian, piscine, and amphibian cell lines can be infected *in vitro*. However, since the host range *in vivo* is restricted mainly to anurans, the receptor molecule may not be expressed as widely in whole animals, at least in cells in which infection may be initiated. In addition, since FV3 does not replicate at temperatures above 32 °C, mammals and birds are nonpermissive hosts. The host range varies with the particular virus. Some iridovirids, such as RSIV and LCDV, possess a broad host range and infect a wide variety of different fish species, whereas other members of the family, such as ATV, infect a limited number of host species. Although both naked and enveloped virions are infectious, early work suggests that these two forms of the virus enter target cells via different mechanisms. Non-enveloped virions are thought to bind to the plasma membrane and inject their viral cores into the cytoplasm whence it is transported to the nucleus. In contrast, enveloped virions enter the cell by receptor-mediated endocytosis.

Early Viral RNA Synthesis

Following entry of the viral DNA into the nucleus, IE viral RNA synthesis begins in a process catalyzed by host cell RNA polymerase II (Pol II). Naked viral DNA is not infectious and IE transcription is dependent upon one or more virion-associated proteins, referred to as virion-associated transcriptional transactivators (VATTs). However, it is not known if VATTs modify the viral DNA template or interact with cellular RNA polymerase II (Pol II) itself. Subsequently, one or more newly synthesized IE proteins, designated virus-induced transcriptional transactivators (VITTs), are required for the transcription of delayed early (DE) viral genes. Although its mechanism of action is unknown, VITTs are likely required to overcome the inhibitory effect of high levels of DNA methylation on Pol II-catalyzed transcription. Like other DNA viruses, transcription of both invertebrate and vertebrate iridovirid genes is a highly regulated process in which the three classes of viral genes (IE, DE, and late) are synthesized in a coordinated manner. In general, IE transcripts likely encode regulatory and catalytic proteins required for the synthesis of DE and late viral gene products, whereas late mRNAs are translated to yield viral structural proteins.

Viral DNA Synthesis and Methylation

Following its synthesis in the cytoplasm and translocation to the nucleus, viral DNA polymerase synthesizes genome-sized DNA molecules. Subsequently, viral DNA is transported to the cytoplasm where the second stage of viral DNA replication takes place. In this stage, large,

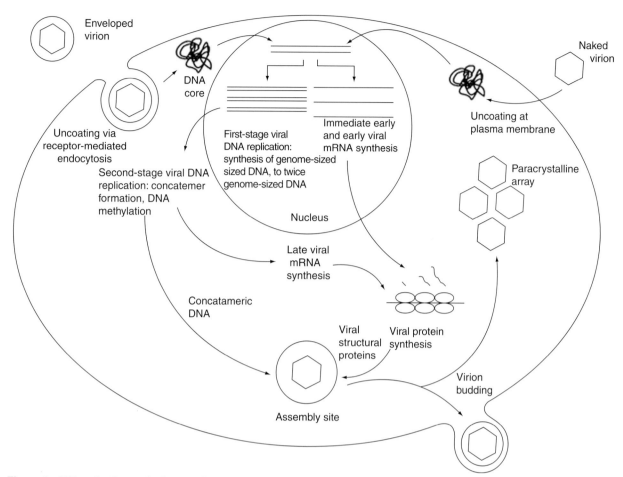

Figure 2 FV3 replication cycle. See text for details. Reproduced from Chinchar VG (2002) Ranaviruses (family *Iridoviridae*): Cold-blooded killers. *Archives of Virology* 147: 447–470, with permission.

concatameric structures are generated that contain more than 10 times the unit length of viral DNA. The mechanics of concatamer formation are not known, but likely involve multiple recombination events and may occur within viral assembly sites.

Following transport to the cytoplasm, viral DNA is methylated by a virus-encoded cytosine DNA methyltransferase. It is not known whether unit length or concatameric DNA is the template for methylation, or whether both serve equally well. The precise role of DNA methylation in the virus life cycle has not yet been elucidated. It has been postulated that methylation protects viral DNA from degradation mediated by a virus-encoded endonuclease similar to those found in bacterial restriction-modification systems. While this scenario is plausible, the rationale for encoding the endonuclease is not clear, unless one postulates that breakdown products of host cell DNA are used to synthesize viral DNA. Regardless of the reason, methylation is important for successful virus replication since treatment of cells with 5-azacytidine, an inhibitor of DNA methylation, reduces virus yields 100-fold. Analysis of viral DNA following sucrose gradient centrifugation suggests that the reduction in viral yields is likely due to the increased sensitivity of unmethylated DNA to nucleolytic attack and the inability of nicked DNA to be properly packaged into virions. An alternative hypothesis that is currently being tested suggests that viral DNA is methylated to block activation of a Toll-like receptor 9 (TLR-9) response. According to this hypothesis, unmethylated FV3 DNA, like that of herpes simplex virus DNA, binds TLR9 and triggers a pro-inflammatory response. In contrast, methylated DNA may not trigger as rapid an immune response and thus permit higher levels of virus replication *in vivo*.

Late Viral RNA Synthesis

Although early viral transcription is catalyzed by host Pol II, late viral transcription is likely catalyzed by a virus-encoded or virus-modified enzyme, designated vPol II. Supporting this view is the observation that all iridovirids sequenced to date encode homologs of the two largest subunits of Pol II which either form a unique, wholly virus-encoded polymerase, or a chimeric polymerase composed of host and viral components. It is likely that vPol II catalyzes the synthesis of late viral messages within viral assembly sites. However, it is not known if some vPol II returns to the nucleus where it contributes to the synthesis of early transcripts, or if viral transcription remains spatially and temporally separated into nuclear (i.e., early transcription catalyzed by host Pol II) and cytoplasmic (i.e., late transcripts synthesized by vPol II) compartments. Recent studies indicate that cells treated with an antisense morpholino oligonucleotide (asMO) targeted to the largest subunit of vPol II show ~80% reduction in late, but not early, viral message expression, and >95% drop in virus yields. These results support the hypothesis that late viral gene transcription is likely catalyzed by a virus-encoded or virus-modified Pol II-like molecule.

Virion Assembly

Following synthesis, late viral proteins, which are likely structural components of the virion, are transported into assembly sites where they participate in virion formation. Unfortunately, the mechanics of virion assembly are poorly understood. Packaging of viral DNA appears to occur via a headful mechanism in which a unit length of viral DNA, plus an additional variable amount (depending upon the virus species), is inserted into the developing virion. As a consequence of this mechanism, viral DNA is terminally redundant and circularly permuted. Completed (but not yet enveloped) virions are often seen scattered within assembly sites or present within paracrystalline arrays (**Figure 3**). It is unclear whether virions are transported from assembly sites to the arrays, or whether, as virion morphogenesis continues, the original assembly site is transformed into a paracrystalline array. In contrast to the accumulation of naked virions within the cytoplasm, enveloped virions form by budding at the plasma membrane. However, most virions remain cell associated and are only released from the cell upon lysis.

Virus assembly sites are one of the most striking features seen in virus-infected cells following visualization by either fluorescent antibody staining or transmission electron microscopy (**Figure 3**). In ranavirus-infected cells, assembly sites are electrolucent areas within the

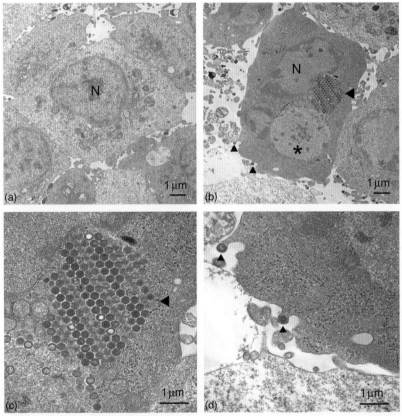

Figure 3 Electronmicrographic analysis of RCV-Z-infected FHM cells. FHM cells were infected with RCV-Z, a tentative member of the genus *Ranavirus*, at an MOI ~20 PFU/cell and fixed for electron microscopy at 9 h post infection. (a) Mock-infected FHM cells; (b) RCV-Z-infected cells showing chromatin condensation within the nucleus (N), a single large paracrystalline array within the cytoplasm (arrowhead), and a viral assembly site (asterisk); (c) enlargement of the paracrystalline array seen within (b); (d) enlargement of a virion in the process of budding from the plasma membrane. Arrowheads in (c) and (d) indicate virions within a paracrystalline array (c) or in the process of budding (d). Reproduced from Majji S, LaPatra S, Long SM, *et al.* (2006) Rana catesbeiana virus Z (RCV-Z): A novel pathogenic ranavirus. *Diseases of Aquatic Organisms* 73: 1–11, with permission from Inter-Research Science Center.

cytoplasm that are devoid of cellular organelles such as mitochondria, endoplasmic reticulum, ribosomes, or cytoskeletal elements. Assembly sites contain viruses in various stages of formation ranging from partially formed capsids to complete, nonenveloped capsids with electron-dense cores. The matrix is granular and frequently contains smooth-surfaced vesicles and diffuse electron-dense material which may be nucleic acid. FV3 assembly sites are surrounded by intermediate filaments. It has been suggested that the filaments may play a direct role in virus replication, perhaps by anchoring assembly sites and facilitating entry of viral components, or by excluding cellular organelles that might interfere with virion morphogenesis. Furthermore, the association of ranaviruses with the cytoskeleton suggests that intermediate filaments may be involved in virion envelopment and release. Disruption of intermediate filaments by treatment of FV3-infected cells with taxol or microinjection of anti-vimentin antibodies has been shown to interfere with the ability of intermediate filaments to encompass assembly sites. It has also been shown to lead to intrusion of cellular components into the assembly site, reduced accumulation of viral proteins, and 70–80% reduction in virus yields. Moreover, assembly sites are observed in the absence of late gene expression, suggesting that one or more early viral gene products and the viral DNA are sufficient for their formation.

The morphological appearance of cells from fish infected *in vivo* with megalocytiviruses such as RSIV is markedly different from the assembly site formation observed in ranavirus-infected cells. RSIV-infected splenocytes contain large inclusion bodies that are surrounded by a membrane and enclose not only the viral assembly site, but also mitochondria, rough and smooth endoplasmic reticulum, and electron-dense amorphous structures that likely contain viral DNA. Inclusion body-bearing cells are also found within the kidney, liver, heart, and gills, and are a hallmark of megalocytivirus infection. Interestingly, *in vitro* infection of cultured cells with RSIV does not result in the formation of membrane-bound inclusions, and the resulting histology is similar to that seen following ranavirus infection.

Effects of Virus Infection on Host Cell Function and Viability

FV3-infection results in a marked inhibition of cellular DNA, RNA, and protein synthesis and culminates in apoptotic cell death. The mechanisms responsible for these outcomes are slowly being resolved. Whereas the inhibition of cellular DNA synthesis appears to be a secondary consequence of the prior inhibition of host RNA and protein synthesis, the latter are likely to be the direct effect of virus infection. Inhibition of translation appears to be due to a series of events that include the synthesis of abundant amounts of highly efficient viral mRNAs, degradation of preexisting host transcripts, inhibition of host transcription, and the induction of a dsRNA-activated protein kinase (PKR) that phosphorylates and thereby inactivates eIF-2α. However, as viral translation persists in the face of host translational shutdown, FV3, like many other viruses, may encode factors that antagonize PKR-mediated effects. Most ranaviruses encode an eIF-2α homolog (vIF-2α) that, like its vaccinia virus counterpart (K3L), is thought to act as a pseudosubstrate that binds PKR, preventing phosphorylation of eIF-2α. Surprisingly, in at least one strain of FV3, vIF-2α is truncated by a deletion that removes the region homologous to eIF-2α and K3L, rendering it unable to bind to PKR. However, these strains maintain high levels of viral protein synthesis, suggesting that, as in the case of vaccinia virus, other viral proteins also play key roles in maintaining viral translation.

Apoptosis, as evidenced by DNA fragmentation, chromatin condensation, and membrane reversal, takes place in FV3-infected cells from ~6 to 9 h post infection. The immediate trigger for apoptosis is not known, but may involve PKR activation, translational shutdown, or another viral insult. Interestingly, both host cell shutdown and apoptosis can be triggered by productive infections as well as nonproductive infections with heat- or ultraviolet (UV)-inactivated virus, suggesting that a virion-associated protein is responsible. Receptor binding by virions may be sufficient to trigger these effects.

Concluding Remarks

For most of the 40 years since its discovery, FV3 has been studied, not for its impact on the infected host, but because it is the type species of a novel virus family that possesses certain features that can best (or only) be studied in this system. However, as other iridovirids are now emerging as important agents of disease in commercially and ecologically important animal species, it is likely that we will see concerted efforts to study their life cycles and, in the process, uncover novel aspects of iridovirus biogenesis. It is anticipated that the pioneering studies of the 1970s and 1980s will be extended using contemporary techniques such as asMOs and small interfering RNAs to elucidate the function of key viral proteins. In addition, study of the role of the host immune system in combating iridovirus infections in fish and amphibians may provide a way, not only to uncover potential virus-encoded immune evasion proteins, but also to elucidate key elements in antiviral immunity in lower vertebrates.

See also: Iridoviruses of Invertebrates; Iridoviruses of Vertebrates.

Further Reading

Allen MJ, Schroeder DC, Holden MTG, and Wilson WH (2005) Evolutionary history of the *Coccolithoviridae*. *Molecular Biology and Evolution* 23: 86–92.

Chinchar VG (2002) Ranaviruses (family *Iridoviridae*): Cold-blooded killers. *Archives of Virology* 147: 447–470.

Chinchar VG, Hyatt AD, Miyazaki T, and Williams T (in press) Family *Iridoviridae*: Poor viral relations no longer. *Current Topics in Microbiology and Immunology*.

Delhon G, Tulman ER, Afonso CL, et al. (2006) Genome of invertebrate iridescent virus type 3 (mosquito iridescent virus). *Journal of Virology* 80: 8439–8449.

Hyatt AD, Gould AR, Zupanovic Z, et al. (2000) Comparative studies of piscine and amphibian iridoviruses. *Archives of Virology* 145: 301–331.

Iyer LM, Aravind L, and Koonin EV (2001) Common origin of four diverse families of large eukaryotic DNA viruses. *Journal of Virology* 75: 11720–11734.

Jancovich JK, Mao J, Chinchar VG, et al. (2003) Genomic sequence of a ranavirus (family *Iridoviridae*) associated with salamander mortalities in North America. *Virology* 316: 90–113.

Lua DT, Yasuike M, Hirono I, and Aoki T (2005) Transcription program of Red Sea bream iridovirus as revealed by DNA microarrays. *Journal of Virology* 79: 15151–15164.

Majji S, LaPatra S, Long SM, et al. (2006) Rana catesbeiana virus Z (RCV-Z): A novel pathogenic ranavirus. *Diseases of Aquatic Organisms* 73: 1–11.

Murti KG, Goorha R, and Chen M (1985) Interaction of frog virus 3 with the cytoskeleton. *Current Topics in Microbiology and Immunology* 116: 107–131.

Sample RC, Bryan L, Long S, et al. (2007) Inhibition of iridovirus protein synthesis and virus replication by antisense morpholino oligonucleotides targeted to the major capsid protein, the 18 kDa immediate early protein, and a viral homologue of RNA polymerase II. *Virology* 358: 311–320.

Stasiak K, Renault S, Demattei M-V, Bigot Y, and Federici BA (2003) Evidence for the evolution of ascoviruses from iridoviruses. *Journal of General Virology* 84: 2999–3009.

Tan WGH, Barkman TJ, Chinchar VG, and Essani K (2004) Comparative genomic analyses of frog virus 3, type species of the genus *Ranavirus* (family *Iridoviridae*). *Virology* 323: 70–84.

Williams T, Barbosa-Solomieu V, and Chinchar VG (2005) A decade of advances in iridovirus research. *Advances in Virus Research* 65: 173–248.

J

Jaagsiekte Sheep Retrovirus

J M Sharp, Veterinary Laboratories Agency, Penicuik, UK
M de las Heras, University of Glasgow Veterinary School, Glasgow, UK
T E Spencer, Texas A&M University, College Station, TX, USA
M Palmarini, University of Glasgow Veterinary School, Glasgow, UK

© 2008 Elsevier Ltd. All rights reserved.

History

Jaagsiekte sheep retrovirus (JSRV) is the causative agent of a naturally occurring lung adenocarcinoma of sheep known as ovine pulmonary adenocarcinoma (OPA, also known as jaagsiekte or sheep pulmonary adenomatosis). OPA was recognized for the first time in South Africa in the nineteenth century as a cause of dyspnea in herded sheep, hence the origin of the Afrikaans name 'jaagsiekte', meaning driving (=jaagt) sickness (=ziekte). OPA is one of the original 'slow diseases' of sheep (along with scrapie, maedi-visna and paratuberculosis) originally described in the 1930s by the Icelandic physician Björn Sigurdsson. The slow diseases of sheep were of great biological importance as they allowed, for the first time, the recognition that an infectious agent could cause clinical disease many months or years after the initial infection of the host.

Studies on JSRV were hampered for years by the lack of a suitable tissue culture system for the cultivation of JSRV. The isolation of full-length JSRV molecular clones (JSRV$_{21}$ and JS$_7$) allowed the *in vitro* generation of infectious viral particles by a transient transfection system, which has sparked a variety of studies that have elucidated many aspects of the molecular biology of JSRV. JSRV is a remarkable virus in many respects. It is the only known virus that induces a naturally occurring lung adenocarcinoma. In addition, the JSRV envelope glycoprotein is an oncoprotein. This is a unique example among retroviruses and oncogenic viruses in general in which a structural protein functions also as a dominant oncoprotein.

Classification

JSRV belongs to the genus *Betaretrovirus* within the family of the *Retroviridae*. Retroviruses are divided on the basis of their modality of transmission as 'exogenous' and 'endogenous' viruses. Exogenous retroviruses are horizontally transmitted between infected and uninfected hosts. Endogenous retroviruses are stably integrated in the genome of the host species from which they derive, are usually defective, and are transmitted vertically like any other Mendelian gene. JSRV is an exogenous retrovirus and it is highly related to enzootic nasal tumor virus (ENTV) of sheep (ENTV-1) and goats (ENTV-2). JSRV and ENTV have common pathogenic characteristics, and both cause low-grade adenocarcinomas of secretory cells in different portions of the respiratory tract of small ruminants. Interestingly, sheep, goats, and other members of the *Caprinae* have several copies of nonpathogenic JSRV-related 'endogenous' retroviruses (commonly referred as enJSRVs) stably integrated in their genome. The phylogenetic relationship between JSRV, ENTV, and enJSRVs is indicated in **Figure 1**.

Genetic Organization and Virion Proteins

The JSRV virions are enveloped particles of approximately 100 nm in diameter and a density by isopycnic centrifugation in sucrose gradients of 1.15 g ml^{-1} for particles obtained from cell cultures. Virions purified from lung secretions of OPA-affected sheep have a slightly higher density (1.16–1.18 g ml^{-1}).

JSRV has the typical genomic organization of a simple retrovirus; the genomic RNA of 7455 nt (in the JSRV$_{21}$ infectious molecular clone) contains the canonical retroviral genes *gag, pro, pol,* and *env* (**Figure 2**). Apart from Env, few studies have been undertaken to assign functions to the JSRV proteins and these are assumed to be the same as other retroviruses. The *gag* gene encodes the structural proteins of the viral core. Gag is expressed as an immature polyprotein that is cleaved upon exit by the viral protease. The JSRV Gag is cleaved into at least five proteins: MA

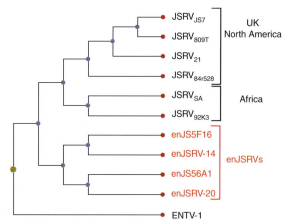

Figure 1 Phylogenetic analysis of sheep betaretroviruses. A phylogenetic tree of representative small ruminant betaretroviruses was derived by the neighbor-joining method using JSRV, ENTV-1, and enJSRVs *env* aligned using Clustal W. JSRV, enJSRVs, and ENTV-1 cluster in three distinct phylogenetic groups. Note that JSRV isolates from Africa cluster in separate branches from European and North American isolates.

(p23), p15, CA (p26), NC, and p4. MA is myristoylated and presumably interacts with the cell membrane during viral egress. CA is the major capsid protein, while NC interacts tightly with the genomic RNA although no specific studies have been conducted on the JSRV NC. The *pro* gene overlaps *gag* and encodes most probably a dUTPase (DU, deoxyuridine tryphosphatase) and the viral protease (PR). The main function of DU is to avoid misincorporation of uracil into DNA during reverse transcription (see below). As mentioned above, PR cleaves the Gag polyprotein upon exit and it is absolutely required in order to obtain mature infectious viral particles. The *pol* gene overlaps *pro* and is predicted to encode the viral reverse transcriptase (RT) and integrase (IN). Both RT and IN are virion-associated enzymes; RT copies the viral single-stranded RNA genome into double-stranded DNA, while IN serves to join the proviral DNA into the host genome. RNaseH is a subdomain of RT and serves to degrade the viral genomic RNA once has been copied into DNA.

An additional open reading frame (*orf-x*) overlaps *pol* and has some unusual features, including a codon usage different from other genes within JSRV, and a very hydrophobic predicted amino-acid sequence that shows no strong similarities to any known protein and only weak homology to a member of the G protein coupled receptor family. The role of this open reading frame is unknown. Orf-x is conserved among all of the exogenous JSRVs examined to date, but it does not seem to be required for JSRV replication *in vitro* nor for cell transformation either *in vitro* or *in vivo*.

The *env* gene encodes the glycoproteins of the viral envelope (Env). The JSRV Env is formed by two subunits, the surface domain (SU) which interacts with the cellular receptor and mediates viral entry, and the transmembrane domain (TM), which fixes SU to the lipid bilayer. A unique feature of the JSRV *env* is that it functions essentially as a dominant oncogene, and its sole expression is sufficient to induce cell transformation *in vitro* and *in vivo* (see below).

Noncoding regions are present at the 5′ and 3′ end of the genome. The R region is repeated at the 5′ and 3′ end of the genome. U5 is present uniquely at the 5′ end, and the U3 is present at the 3′ end of the genome. The viral promoter and enhancer regions are present in the U3 (see below).

Genomic sequence variability among JSRV strains is very low. For instance, the infectious molecular clones $JSRV_{21}$ and $JSRV_{JS7}$ are 99.3% identical along the entire genomic sequence, and they were derived from naturally occurring OPA cases from Scotland collected many years apart. $JSRV_{21}$ and $JSRV_{JS7}$ Env proteins are 100% identical. JSRV isolates from Africa can be distinguished phylogenetically from the UK and North American isolates, although there is still a high degree of homology (∼93% along the entire nucleotide genomic sequence) among the two groups.

Replication Cycle

In general, the replication cycle of JSRV is not thought to be markedly different from other betaretroviruses. The lack of a suitable tissue culture system for the propagation of JSRV has not allowed detailed studies on the replication of this virus.

JSRV interacts with a specific cellular receptor to enter the cell. The use of retroviral pseudotypes identified hyaluronidase 2 (HYAL2) as the cellular receptor for JSRV entry. HYAL2 is a glycosylphosphatidylinositol (GPI) linked membrane protein with a low hyaluronidase activity and is widely expressed on many different cell types. Besides ovine HYAL2, the human ortholog also allows JSRV entry, while mouse and rat Hyal2 do not. Detailed steps of JSRV entry and post-entry, such as membrane fusion and the dependence or independence from acidic pH have not been investigated, but are likely to be similar to those of other retroviruses including uncoating, reverse transcription of the viral genome into double-stranded DNA, entry into the nucleus of the pre-integration complex, and stable integration of the proviral DNA into the host DNA. During the process of reverse transcription, the noncoding regions at the 5′ and 3′ ends of the genome (R, U5, and U3) are duplicated and give origin to the viral long terminal repeats (LTRs).

The retroviral LTRs are major determinants of retrovirus tropism, as the 5′ LTR of the provirus initiates transcription and the U3 region contains the majority of *cis*-acting sequences interacting with the cellular RNA polymerase II and with cellular transcription factors. The exogenous JSRV LTRs are particularly active in reporter

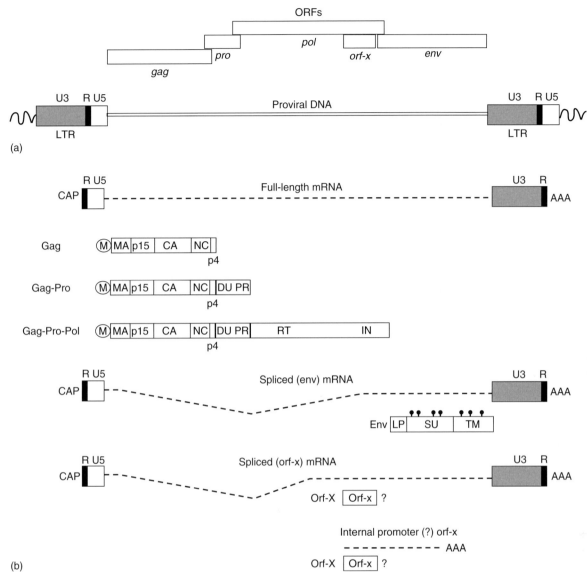

Figure 2 Genomic organization, mRNAs and viral proteins of JSRV. (a) Schematic organization of the JSRV open reading frames (ORFs) and their relative position in the JSRV provirus. The JSRV LTR, repeated at both ends of the provirus genome, is divided into U3, R and U5. (b) Major JSRV RNAs and proteins. The JSRV provirus transcribes various mRNAs. A full-length mRNA encodes for the Gag, Pro and Pol proteins and serves as genome for the newly synthesized viral particles. A spliced mRNA encodes the viral Env. The Env glycoprotein is glycosylated, and putative glycosylated sites are indicated by full circles. Two additional mRNAs have been detected and encompass the *orf-x* reading frame, but their biological significance is unknown at present. MA, matrix; CA, capsid; NC, nucleocapsid; DU, dUTPase; PR, protease; RT, reverse transcriptase; IN, integrase; LP, leader peptide; SU, surface; TM, transmembrane.

assays using type II pneumocytes/Clara cell lines and interact with lung-specific transcription factors such as forkhead box A2 (FOXA2; alias HNF-3β). These features may explain the preferential expression of exogenous JSRV in the transformed cells of the lungs, which have phenotypic characteristics of type II pneumocytes and Clara cells.

The expression of JSRV, as in all retroviruses, is believed to follow the basic transcriptional events of cellular mRNAs including capping of the 5′ end and polyadenylation at the 3′ end. JSRV encodes a full-length mRNA that serves as genome of the viral progeny and is also translated into the Gag polyprotein and the proteins encoded by *pro* and *pol*. The latter are expressed as fusion proteins with Gag. As in all betaretroviruses, *pro* and *pol* are in different open reading frames to *gag* and are expressed by ribosomal frameshifting. The viral Env is produced from a mRNA which is spliced using the splice donor in the untranslated *gag* region and a splice acceptor immediately before *env*.

Another spliced mRNA has been found to use the same splice donor of the *env* mRNA and a splice acceptor before the *orf-x* reading frame. This mRNA presumably expresses a protein encoded by *orf-x*, but there are no published data supporting this assumption. A possible mRNA expressing *orf-x* has been found to lack the R-U5 regions (typical of all other mRNAs starting from the U3) and is thought to derive from the expression of an internal promoter

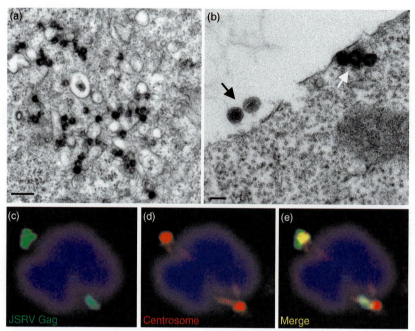

Figure 3 Electron and confocal microscopy of cells expressing JSRV. Electron microscopy in 293T cells transiently transfected with the JSRV$_{21}$ infectious molecular clone showing intracytoplasmic particles (a; scale = 200 μm). In panel b, intracellular particles in the vicinity of the cell membrane (white arrow) and extracellular particles complete with viral envelope (black arrow) are also visible (scale = 100 μm). (c–e) Confocal microscopy in HeLa cells transiently transfected with JSRV and probed with an antiserum towards the JSRV matrix (c) and γ-tubulin (d). The JSRV Gag concentrates in the pericentrosomal area (e).

that has not been characterized. Other mRNAs deriving from the use of secondary splice acceptors and nonconventional polyadenylated sites have been detected but their biological significance, if any, is unknown.

JSRV assembles in the cytoplasm (**Figures 3(a)** and **3(b)**) like all betaretroviruses, most likely in the pericentriolar region (**Figures 3(c)–3(e)**). Other retroviruses such as lentiviruses and gammaretroviruses assemble instead mostly at the cell membrane. The JSRV intracellular particles interact with Env at the cell membrane or in a cellular compartment not yet fully identified and egress from the cell. Upon exit, the Gag polyprotein is cleaved by the viral protease into the mature proteins described above. A more detailed description of the replication cycle of retroviruses is discussed elsewhere in this encyclopedia.

enJSRVs

The sheep genome contains at least 27 copies of endogenous retroviruses highly related to JSRV (hence the name enJSRVs). Endogenous retroviruses are believed to derive from integration events of ancestral exogenous retroviruses into the germline of the host. enJSRVs have a high degree of homology with JSRV. For example, JSRV$_{21}$ and enJS56A1 are 92% identical at the nucleotide level along the entire genome. Major differences are located in the U3 region, in two regions in Gag (termed variable regions 1 and 2), and in the cytoplasmic tail of the transmembrane domain of the Env (termed variable region 3). At least some of these highly divergent regions are the basis of important biological differences between JSRV and enJSRVs.

enJSRV transcripts have been detected in most tissues by sensitive reverse transcriptase polymerase chain reaction (RT-PCR) assays. However, high levels of enJSRVs (both mRNA and proteins) are found specifically in the epithelia of the genital tract of the ewe and, particularly, in the epithelia of the uterus (**Figure 4**). In the placenta, enJSRVs are expressed in the mononuclear trophectoderm cells of the conceptus (embryo/fetus and associated extra-embryonic membranes), but are most abundant in the trophoblast giant binucleate cells (BNCs) and multinucleated syncytial plaques of the placentomes. The temporal expression of enJSRVs envelope (*env*) gene in the trophectoderm is coincident with key events in the development of the sheep conceptus. Indeed, using a morpholino antisense oligonucleotide to induce loss-of-function *in utero*, enJSRVs Env knockdown caused a reduction in trophoblast outgrowth and inhibited/prevented trophoblast giant binucleate cell differentiation during blastocyst elongation and formation of the conceptus. Thus, enJSRVs have been proposed to be essential in sheep for peri-implantion growth and differentiation.

enJSRVs expression appears to be regulated by progesterone and expression of the progesterone receptor. However, the specific enJSRV loci that are transcriptionally active are not known at present. enJSRVs are also expressed in the sheep fetus, supporting the idea that sheep are tolerized toward JSRV. Indeed, JSRV-infected naïve sheep (with or without lung adenocarcinoma) have

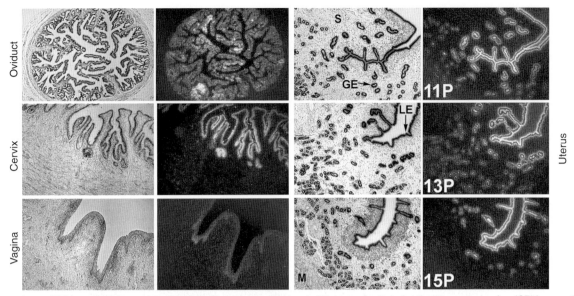

Figure 4 enJSRVs are highly expressed in epithelia of the genital tract of the ewe. In situ hybridization analysis of enJSRVs env mRNA in the oviduct, uterus, cervix, and vagina. Note the specific expression of enJSRVs env mRNA in the epithelia lining the oviduct, cervix, and vagina of the cycling ewe. During the estrous cycle and pregnancy (P), enJSRVs mRNA is particularly abundant in the luminal and glandular epithelia of the uterus. LE, luminal epithelium; GE, glandular epithelium; M, myometrium; S, stroma. Modified from Palmarini M, Gray CA, Carpenter K, Fan H, Bazer FW, and Spencer TE (2001) Expression of endogenous betaretroviruses in the ovine uterus: Effects of neonatal age, estrous cycle, pregnancy, and progesterone. *Journal of Virology* 75(23): 11319–11327.

no detectable specific humoral or cellular immune responses, although recombinant JSRV CA, in the presence of adjuvants, can induce antibody production and specific T-cell responses in vaccinated sheep.

Three full-length enJSRV loci have been cloned and characterized (enJS56A1, enJS5F16, and enJS59A1). All three loci have deletions or stop codons that make them replication incompetent. However, enJS56A1 and enJS5F16 maintain intact ORFs for *gag* and *env*. By using retroviral vectors pseudotyped by the enJSRVs Env, it was found that they too use HYAL2 as a cellular receptor and interfere by receptor competition with JSRV entry.

One of the enJSRVs loci, enJS56A1, is defective for viral exit when overexpressed in transfected cells, although abundant intracytoplasmic Gag is detected and intracytoplasmic viral particles are visible by electron microscopy. The replication defect of enJS56A1 is determined by its Gag protein, which is *trans*-dominant over the exogenous JSRV Gag if co-expressed in the same cell. A tryptophan residue in position 21 of the enJS56A1 Gag (replacing an arginine in JSRV) is the main determinant for the block induced by enJS56A1. Thus, enJS56A1 exerts a unique mechanism of retroviral interference, which occurs at a late step of the replication cycle. The mechanism and timing of the block induced by enJS56A1 are not yet understood. However, the observation of viral particles by electron microscopy in cells expressing enJS56A1 (or co-expressing enJS56A1 and JSRV) suggests that enJS56A1-induced interference depends on a defect in Gag trafficking. Recent studies suggest that enJS56A1 appears to block JSRV, most likely in *trans*, by hampering the ability of the latter to reach the centrosome, the proposed site of assembly for betaretroviruses.

enJSRVs are not expressed in the differentiated epithelial cells of the lungs but are expressed in the epithelium of the genital tract. As mentioned above, enhancer regions are located in the U3 and indeed enJSRVs transcription is regulated by progesterone via the progesterone receptor. Interestingly, the enJSRVs LTRs do not respond to lung transcription factors (such as FOXA2) unlike the exogenous JSRV LTRs. Thus, the LTR appears to be a major contributor to the different tropisms (genital tract vs. respiratory tract) shown by enJSRVs and JSRV.

The enJSRVs Env (or at least the Env of those enJSRVs loci cloned so far) is not able to transform cells *in vitro*, unlike the highly related JSRV Env. Indeed, the cytoplasmic tail of the JSRV Env is where major determinants of transformation are located. In particular, an SH2 binding domain is present in the exogenous JSRV Env, but is absent in the enJSRVs Env sequenced to date. It is hypothesized that enJSRVs have been selected in the sheep for their ability to confer resistance to infection of the host by the related exogenous betaretroviruses. This innate resistance could have provided a selection pressure for betaretroviruses with tropism towards the respiratory tract (i.e., the current JSRV) rather than the genital tract.

Ovine Pulmonary Adenocarcinoma

OPA is a naturally occurring lung cancer of sheep that has been reported in most sheep-rearing countries. It is absent from Australia and New Zealand and has been eradicated from Iceland. The disease is characterized by a progressive respiratory condition caused by the growth of the lung adenocarcinoma. The tumor appears to originate from two types of differentiated epithelial secretory cells of the distal respiratory tract, the type II pneumocyte and the Clara cell, which retain their phenotype (**Figure 5**). OPA is invariably fatal once the disease is diagnosed, and affected sheep die as a result of compromised respiratory function caused by tumor enlargement or from secondary bacterial infections. The incubation period of the naturally occurring disease can be very long lasting several months to years.

Susceptibility/resistance to JSRV of some breeds has been suggested but not proved. OPA can be transmitted experimentally only with material that contains JSRV, such as lung secretions collected from affected animals or virions obtained by transiently transfecting cells with JSRV infectious molecular clones. The experimental OPA model is highly reproducible. JSRV infection can be induced in most, if not all, inoculated lambs aged 1–6 months at the time of inoculation and a high proportion of them develop clinical signs and OPA lesions. Neonatal lambs are most susceptible and, in contrast to naturally occurring OPA, the incubation period for experimentally induced OPA can be as short as a few weeks.

Although most experimentally inoculated lambs develop clinical signs and OPA lesions, under natural conditions the majority of JSRV-infected sheep do not develop OPA during their commercial lifespan. Within an endemically infected flock, lambs appear to become infected at a very early age although the routes of transmission have not been determined yet. Interestingly, most infected animals harbor the virus as a disseminated infection of their lymphoid tissues and do not show detectable pulmonary lesions. Although lymphoreticular cells appear to serve as the principal reservoir of virus infection, viral antigens are detected only rarely in this compartment where sensitive PCR assays are necessary to detect viral RNA or proviral DNA.

The short incubation period in young lambs experimentally infected with JSRV may be explained by the combination of high virus infectious doses present in the inoculum and the higher abundance of the permissive target cells (type 2 pneumocytes and Clara cells) in lambs compared to adult animals. In natural infections, the mechanisms involved in converting the stable persistent lymphoid infection in peripheral tissues to a progressive pulmonary epithelial tumor are not clear. In addition, it is not known whether and how efficiently sheep infected by JSRV, but with no neoplastic lesions, are able to transmit the virus to uninfected sheep. In an affected flock, control of OPA is very difficult as no vaccines or effective diagnostic tests are available.

Mechanisms of Virus-Induced Cell Transformation

The mechanisms used by JSRV to induce cell transformation are different from those followed by the majority of oncogenic retroviruses. Most of the tumors caused by

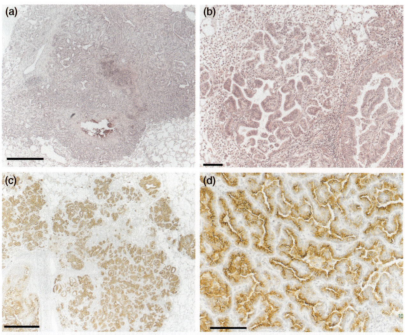

Figure 5 JSRV-induced tumours *in vivo*. (a, b) Histology from a lung tumor section of a naturally occurring case of OPA. Sections were stained with hematoxylin and eosin. Note the presence of papillary to acinar neoplastic lesions that replace the normal alveolar structure of the lungs. (c, d) Immunohistochemistry in lung sections from a JSRV experimentally inoculated lamb show expression of the viral Env in tumor cells (characterized by the intracytoplasmic brown color). Scale = 500 μm (a, c); 100 μm (b, d).

retroviruses are due to insertional activation of cellular oncogenes nearby the integrated proviruses. As a result, tumors originated by retroviral insertional activation are monoclonal or oligoclonal and common proviral integration sites are detected in tumors from different animals. Another mechanism of retrovirus transformation is transduction of cellular proto-oncogenes by the retroviral genome that are transmitted as a mixture of replication competent 'helper' viruses and replication defective viruses that have captured the oncogene. Tumors caused by retroviral transduction are in general multiclonal.

In contrast, JSRV follows neither of the mechanisms mentioned above. OPA tumors are multiclonal, and common proviral integration sites have been rarely observed. However, no cellular oncogenes have been found to be transduced by JSRV. The oncogenic properties of JSRV are due directly to one of its structural proteins.

Transfection of a variety of fibroblast and epithelial cell lines with expression plasmids for the JSRV Env leads to efficient cell transformation (**Figure 6**). Moreover, experimental infection of mice and lambs, with replication incompetent vectors expressing JSRV Env, induced tumors in the inoculated animals. Thus, the JSRV Env is a dominant oncoprotein both *in vitro* and *in vivo* and viral spread is not necessary for tumorigenesis.

The mechanisms involved in JSRV Env-induced transformation have not been fully elucidated; however, signal transduction involving the PI3K-Akt and H/N Ras-MEK-MAPK pathways are important. Conflicting reports are available on the involvement of other cellular oncogenes such as the Stk/Ron tyrosine kinase in the onset of JSRV Env-induced cell transformation.

One of the major determinants of JSRV Env transformation is the transmembrane domain (TM), although other regions may be important. In particular, a putative docking site (Y-X-X-M) for phosphatidylinositol 3-kinase (PI-3K) is critical for JSRV-Env induced cell transformation. Within this motif, Y590 is crucial for JSRV Env-induced cell transformation, although Y590 mutants maintain a reduced ability to induce cell transformation in some cell lines. In summary, the JSRV Env acts as a dominant oncogene *in vitro* and *in vivo* and its expression is sufficient to induce lung adenocarcinoma in the target species.

Enzootic Nasal Tumor Virus

ENTVs of sheep and goats are distinct betaretroviruses highly related to each other and JSRV. ENTV causes a contagious tumor of the mucosal nasal glands, as well as respiratory and olfactory mucosa, in their respective target species known as enzootic nasal adenocarcinoma (ENT). Clinically, ENT is characterized by respiratory distress caused by the enlargement of the tumor, nasal discharge, and skull deformation. Like OPA, ENT has a long incubation period.

The molecular biology of ENTV has not been studied in great detail, but the virus displays many features in common

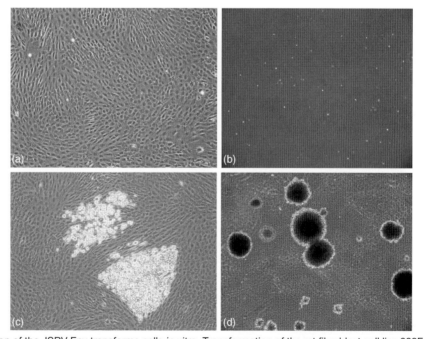

Figure 6 Expression of the JSRV Env transforms cells *in vitro*. Transformation of the rat fibroblast cell line 208F by the JSRV envelope glycoprotein. 208F cells were mock-transfected (a) or transfected with an expression plasmid for the JSRV Env (c). Mock-transfected 208F cells are morphologically flat, possess a strong contact-inhibition and do not grow in soft agar (b). On the other hand, 208F cells transfected with an expression plasmid for the JSRV Env show the onset of foci of transformed cells (c), which are able to form colonies in soft agar (d).

with JSRV. The cellular receptor for ENTV is also HYAL2 and the ENTV Env, like its JSRV counterpart, is a dominant oncogene that appears to follow the same mechanisms of cell transformation used by JSRV. The main differences between the ENTVs and JSRV appear to be concentrated in the U3 region of the viral LTRs. The ENTV LTR, unlike the JSRV LTR, does not bind lung-specific transcription factors such as FOXA2, which is likely the basis of tropism differences between the two viruses.

Future Perspectives

Small ruminant betaretroviruses are a fascinating group of viruses with unique characteristics that are of broad interest through their veterinary, comparative medical and biological importance. The veterinary importance arises from the economic impact in many sheep rearing countries of the diseases induced by JSRV and ENTV combined with the absence of any effective control tools or mechanisms. Their comparative medical interest stems from the striking similarity of some forms of human lung adenocarcinoma to OPA, which is considered an excellent outbred large animal model for these tumors with opportunities to investigate issues that are not available from other systems. Lung cancer is the main cause of death among cancer patients and effective therapeutic strategies are greatly needed to improve patient survival and well-being. OPA is a large animal model that can identify and test the efficacy of new therapeutic interventions in a highly reproducible system.

enJSRVs are an especially active group of endogenous retroviruses and offer insights into several areas of general biological interest, such as viral replication, interference, and reproductive biology. Understanding the mechanisms of the enJS56A1-induced block could inspire the design of novel anti-retroviral strategies and shed light on early events in retroviral assembly and/or trafficking. In particular, this unique viral block provides additional clues on the variety of mechanisms shaping co-evolution of endogenous/exogenous retroviruses and their hosts.

enJSRVs are highly expressed in the genital tract of the ewe and are intimately involved in early placental development in this animal species. The sheep/enJSRVs model can be useful to experimentally address the hypothesis that endogenous retroviruses have shaped and are essential for mammalian biology.

See also: Retroviruses: General Features; Simian Retrovirus D.

Further Reading

Fan H (ed.) (2003) *Jaagsiekte Sheep Retrovirus and Lung Cancer.* Berlin: Springer.

Maeda N, Palmarini M, Murgia C, and Fan H (2001) Direct transformation of rodent fibroblasts by jaagsiekte sheep retrovirus DNA. *Proceedings of the National Academy of Sciences, USA* 98: 4449–4454.

Palmarini M, Gray CA, Carpenter K, Fan H, Bazer FW, and Spencer TE (2001) Expression of endogenous betaretroviruses in the ovine uterus: Effects of neonatal age, estrous cycle, pregnancy, and progesterone. *Journal of Virology* 75(23): 11319–11327.

Palmarini M, Sharp JM, De las Heras M, and Fan H (1999) Jaagsiekte sheep retrovirus is necessary and sufficient to induce a contagious lung cancer in sheep. *Journal of Virology* 73: 6964–6972.

Wootton SK, Halbert CL, and Miller AD (2005) Sheep retrovirus structural protein induces lung tumours. *Nature* 434: 904–907.

Japanese Encephalitis Virus

A D T Barrett, University of Texas Medical Branch, Galveston, TX, USA

© 2008 Elsevier Ltd. All rights reserved.

Glossary

Arbovirus Virus which replicates in hematophagous insects and which then may be transmitted to vertebrates.
Flavivirus Any virus in the genus *Flavivirus*, family *Flaviviridae*.
Japanese encephalitis Disease caused by Japanese encephalitis virus.
Viremia Multiplication of virus in the blood of animals.

Introduction

Japanese encephalitis (JE) is a rural, zoonotic viral disease and is a major public health problem in many Asian countries. It is the most important of the arthropod-borne virus encephalitides and has replaced polioviruses as the major cause of human epidemic encephalitis in some parts of the world. Although a disease resembling JE was described in the late nineteenth century, the virus causing it, Japanese encephalitis virus (JEV), was first isolated in Japan in 1935. This prototype strain is known as Nakayama. Originally, JEV was termed 'Japanese B encephalitis virus' and was classified on the basis of antigenic characteristics as a member of the group B arboviruses. Subsequently it was reclassified as JEV in the family *Togaviridae*, before being reclassified as a member of the family *Flaviviridae* (species *Japanese encephalitis virus*). The virus is transmitted between vertebrate hosts by mosquitoes. Humans, as are most vertebrates, are 'dead-end' hosts, due to production of a low viremia, such that mosquitoes cannot be infected while feeding. JE is characterized by infection of the central nervous system. There are at least 50 000 clinical cases of JE reported each year but it is thought that the actual number is much higher. The majority of cases occur in children below the age of 10 years. The case–fatality rate is 15–25% and up to 70% of those who survive infection develop neurological sequelae. In addition, JE is the most important flavivirus disease of livestock. Horses and pigs are considered to be of veterinary importance. Horses become encephalitic and are dead-end hosts, whereas JEV can induce abortion in pigs, which are considered a major amplifying host.

Virus

Japanese encephalitis virus is a species in the genus *Flavivirus*, family *Flaviviridae*. The genus contains approximately 70 viruses. JEV is a member of the JEV complex, viruses that are closely related on the basis of cross-reactivity in neutralization tests and at the nucleotide level. The JEV complex includes Cacipacore, JEV, Koutango, Kunjin, Murray Valley encephalitis, St. Louis encephalitis, Usutu, West Nile, and Yaounde viruses. JEV replicates in a wide range of cell cultures derived from various animals and mosquitoes. Monkey kidney-derived Vero and LLC-MK2 cells usually are used for infectivity titrations, and these cell lines as well as C6-36 cells from *Aedes albopictus* mosquitoes usually are used to grow the virus.

Physical Properties

JEV virions are approximately 50 nm in diameter, icosahedral in shape, and have a lipid envelope. The envelope is derived from the host cell. Infectivity is lost after treatment with heat, detergents, lipid solvents, or acidic pH. Lipid solvents (e.g., ether and chloroform) and ionic detergents (e.g., sodium dodecyl sulfate) inactivate both infectivity and hemagglutination activity, while the milder nonionic detergents (e.g., nonidet P40, triton X-100, or X-113) only destroy infectivity. The sedimentation coefficient of the virion is approximately 200S and it sediments at a density of $1.20–1.23 \, \text{g cm}^{-3}$ in potassium tartrate-glycerol or sucrose gradients.

Virions contain three structural proteins. The small capsid (C) protein surrounds the genome of the virus and the envelope contains two proteins known as envelope (E) and membrane (M). The E protein is the viral hemagglutinin (i.e., the protein that binds to red blood cells) and contains most of the epitopes recognized by neutralizing antibodies. The E protein is the major virion protein and has one glycosylation site, at residue 154. Two types of virions are recognized; mature extracellular virions containing M protein, and immature intracellular virions containing precursor M (prM) protein, which is proteolytically cleaved during maturation to yield M protein. The genome comprises one positive-sense, single-stranded RNA molecule of *c.* 11 000 nt and is infectious. The 5′ terminus of the genome possesses a type I cap (m-^{7}GpppAmp) followed by the conserved dinucleotide AG. There is no terminal poly(A) tract at the 3′ terminus. The gene order is C-prM-E-NS1-NS2A-NS2B-NS3-NS4A-NS4B-NS5. There are 95 nucleotides in the 5′ noncoding region, and the single open reading frame has 10 296 nucleotides. The 3′ noncoding region is variable in length, depending on the strain.

In addition to the mature virion, two additional physical entities have been described, namely the slowly sedimenting hemagglutinin and soluble complement-fixing antigen. The former is associated with immature particles and the latter with secreted NS1 protein.

Replication Cycle

The replication cycle involves virion binding to cell receptor(s), mediated by the viral E protein. Uptake of the virion into cells is via receptor-mediated endocytosis followed by pH-dependent membrane fusion to release the virus nucleocapsid into the cytoplasm. The input virus does not contain viral RNA-dependent RNA polymerase. Thus, the positive-sense genomic RNA is translated to generate the nonstructural proteins required for replication of the virus, including the RNA-dependent RNA polymerase. RNA replication is associated with membranes and begins with transcription of the input genomic RNA to synthesize complementary negative-sense strands, which are then used as templates to transcribe positive-sense genomic RNA. The genomic RNA is synthesized by a semiconservative mechanism involving replicative intermediates (i.e., containing double-stranded regions as well as nascent single-stranded molecules) and replicative forms (i.e., duplex RNA molecules). Synthesis of negative-sense RNA continues throughout the replication cycle. All viral proteins are produced as a single polyprotein that is co- and post-translationally processed by cellular proteases and by the viral NS2B-NS3 serine protease to generate individual structural and nonstructural proteins. In addition to the three structural proteins C, prM, and E, seven nonstructural (NS) proteins are found in virus-infected cells: NS1, NS2A, NS2B, NS3, NS4A, NS4B, and NS5. Few of the nonstructural proteins have been studied in detail. NS3 is a multifunctional protein whose N-terminal one-third forms the viral serine proteinase complex together with NS2B. The C-terminal portion of NS3 contains an RNA helicase domain involved in RNA replication, as well as an RNA triphosphatase activity involved in the formation of the 5′ terminal cap structure of the viral RNA. Two enzymatic activities have been assigned to NS5: the RNA-dependent RNA polymerase and the methyltransferase activity necessary for methylation of the 5′ cap structure. NS1 is an unusual nonstructural protein as it is glycosylated at two sites: residues 130 and 207. The functions of NS2A, NS4A, and NS4B are poorly understood, but current evidence suggests that NS2A, NS2B, NS4A, and NS5 are all part of the replication complex, and that NS1 is involved in RNA synthesis and virus assembly. The function in NS4B is not clear but may be an interferon antagonist. Other studies suggest that NS5 is involved in this function.

Virions are first observed in the rough endoplasmic reticulum, which is believed to be the site of virus assembly (i.e., interaction of genomic positive-sense RNA molecules with structural proteins C, prM, and E). Progeny virions assemble by budding through intracellular membranes into cytoplasmic vesicles. These immature virions (i.e., containing prM (which is thought to act as a chaperone) rather than M protein) are then transported through the membrane systems of the host secretory pathway to the cell surface where exocytosis occurs. Shortly before virion release, the prM protein is cleaved by furin or a furin-like cellular protease to generate mature virions that contain M protein. Immature virions have low infectivity compared to mature virions. Host-cell macromolecular synthesis is not shut-off during virus replication and is not decreased until cytopathic effect is evident late in the infection process.

Geographic Distribution

JEV is found throughout much of Asia. It is epidemic in temperate regions of Asia (e.g., Japan, Taiwan, People's Republic of China, Korea, eastern Russia, northern Vietnam, northern Thailand, Burma, Nepal, Sri Lanka, and India) and endemic in tropical regions (e.g., Malaysia, Indonesia, southern Vietnam, southern Thailand, and the Philippines). JE was not described in Australia until 1995, when cases of JE were reported on the island of Badu in the Torres Strait. The first human case of JE was reported on mainland Australia in 1998. This recent introduction into the Torres Straits and northern Australia is thought to be due to wind-blown mosquitoes.

Epidemiology

JE is reported in nearly all Asian countries. It is endemic in tropical areas of Asia and epidemic in temperate regions of Asia. Endemicity in tropical areas is thought to be due to the availability of mosquito vectors throughout the year, while there is seasonal incidence in temperate climates. Specifically, the disease is reported annually with seasonal (peaking in June and July during the rainy season) and age (mainly children between the ages of 1 and 15 but peaking in those 3–5 years of age) distributions. Some endemic countries that utilize childhood immunization have seen the age distribution of cases move toward older children and adults. Human infections are related to increased vector densities associated with rainfall or with irrigation practices. It is estimated that 3 billion people live in JE endemic areas. Of this population, approximately 700 million are children under the age of 15 years with an annual birth cohort of 70 million per year. The disease incidence varies greatly between countries, ranging from <10 to >100 per 100 000 population. On the basis of these figures, it is estimated that there are approximately 175 000 cases of JE per year, of which approximately 50 000 have clinical encephalitis. Only 1 in 1000 JEV infections are symptomatic. The estimated case–fatality rate is 25% (5–40% have been reported), and approximately 45% (plus reports of up to 70%) of surviving patients have neurological and/or psychiatric sequelae. These include neurological and psychomotor retardation, motor deficits, convulsions, memory impairment, optic nerve atrophy, limb paralysis, parkinsonism, and also psychological and behavioral disorders. Much of the information on neurological sequelae comes from case reports rather than detailed studies incorporating controls. Nonetheless, it appears that a large proportion of patients with neurological sequelae do recover completely over time.

The effect of concurrent infection with human immunodeficiency virus on JEV infection is unknown and there are few reports of JEV infection of pregnant women, given that the virus predominantly causes disease in children. However, there are reports from India that JEV infection can cause abortion during the first two trimesters of pregnancy.

JE appears to be age related, with the majority of patients being children and the elderly tending to have encephalitis. It is thought that this is due to the high prevalence of antibody in adults. All evidence indicates that there is one serotype of JEV and that infection provides lifelong protection from reinfection against all known antigenic and genetic variants of the virus.

Swine are important amplifying hosts of JEV. Infected pigs develop viremias sufficient to infect mosquitoes that subsequently feed on them. Importantly, adult pigs do not show clinical signs of JE. In terms of veterinary disease, infection of pregnant sows results in abortion and stillbirth due to transplacental infection and causes aspermia in boars. Infected equids can succumb to JE and cattle seroconvert; however, both are considered dead-end hosts.

Molecular Epidemiology

Strain variation has been recognized for many years. Initially antigenic variation was detected with polyclonal antisera in neutralization, complement fixation, agar gel diffusion, and hemagglutination inhibition tests and subsequently with monoclonal antibodies. Overall, two major immunotypes of JEV have been differentiated: Nakayama and Beijing-1/JaGAr-01. Other antigenically distinct groups of strains have been identified, including those from northern Thailand and Malaysia, while studies with monoclonal antibodies suggested that at least five antigenic groups could be identified. Oligonucleotide mapping provided the first evidence for genetic variation among strains of JEV. Subsequent studies used direct nucleotide sequencing of portions of the genome to identify four genotypes (genetic clusters) of JEV. Genotype I includes isolates from Cambodia and northern Thailand; genotype II includes isolates from Indonesia, southern Thailand, and Malaysia; and genotype III includes isolates from temperate regions of Asia (Japan, China, Taiwan, India, Nepal, Sri Lanka, and the Philippines). Subsequent studies showed that isolates from Japan, Korea, and India clustered within genotype III. Genotype IV includes certain isolates from Indonesia. Interestingly, recent studies have shown that during the past 25 years, genotype I is replacing genotype III in China, Japan, Korea, and Vietnam.

Overall, molecular epidemiologic studies of JEV have identified at least five antigenic groups and four genotypes. The relationship of the antigenic groups to the genotypes is not clear at present. The practical significance of these differences has been the subject of much debate. In particular, should strains from one or more antigenic groups/genotypes be included in vaccines to give effective protection against all JEV strains found in nature (see below)? Also, there is very little information on the molecular basis of neurovirulence of JEV or on biological differences between genotypes of JEV (see below). However, it is noteworthy that JEV isolates from Thailand can be distinguished genetically: isolates from northern Thailand are of genotype I while isolates from southern Thailand are of genotype II. The incidence of encephalitis in humans in northern Thailand is 12.2 per 100 000 while it is only 0.3 per 100 000 in southern Thailand. There are many potential explanations for this situation, but genetic differences between the viruses may contribute to these epidemiologic differences. Finally, the entire genome of a number of strains of JEV has been sequenced, including representatives of all four genotypes.

Pathogenicity and Virulence

Vertebrates of many species are susceptible to JEV. The virus is lethal for newborn mice and rats by all routes of inoculation. As the animals get older, age-related resistance to disease is observed following inoculation by peripheral routes, while adult mice are still susceptible following direct inoculation of virus into the brain, that is, neuroinvasiveness decreases with age while the virus is still neurovirulent for all ages of mice. The virus causes a lethal disease in primates inoculated intracerebrally, while some strains are lethal also by intranasal inoculation; peripheral inoculation of JEV causes only asymptomatic infection.

In nature, the virus readily infects individuals of a large number of vertebrate and invertebrate species. Many vertebrates are considered 'dead-end' hosts. Horses and humans develop encephalitis, and swine have inapparent infections, while the virus causes aspermia in boars and stillbirth or abortion in pregnant swine. Persistent infections have been reported in experimental infection of pregnant mice and virus has been passed vertically to offspring.

Although it is known that the virus appears in the blood before invading the central nervous system, the host and viral determinants of disease are poorly understood. It is not known whether the virus directly crosses the blood–brain barrier or uses the olfactory nerve route to invade the brain, nor is it known why the virus targets particular regions of the brain. The cell receptor for the virus has not been identified.

Strains of the virus differ in neuroinvasiveness and this appears to be related to the level of viremia, but the exact molecular determinants of JEV that control neuroinvasion have not been identified. However, studies on the attenuated vaccine strain SA14-14-2 have shown that the E protein is a major determinant of neurovirulence in the mouse model, with residue 138 dominant, and residues 107, 176, and 279 also contributing to neurovirulence.

Clinical Symptoms

Following the bite of a JEV-infected mosquito, there is an incubation period of 4–16 days before clinical symptoms are observed. These take the form of a febrile illness with headache, aseptic meningitis, or encephalitis. The most important form of the disease is acute meningomyeloencephalitis. Onset is rapid, with 1–4 days of fever, headache, chills, drowsiness, mental confusion, stupor, anorexia, nausea and vomiting. Subsequently, patients show symptoms of nuchal rigidity, photophobia, tremors, involuntary movements, focal motor nerve impairment involving the central and peripheral nervous systems, or coma. Generalized motor seizures are seen in children.

Examination of the cerebrospinal fluids of patients may provide indicators of infection, including one or more of increased pressure, increased concentration of protein, an increased number of lymphocytes, anti-JEV IgM antibody, and virus. Fatal cases usually have respiratory complications, seizures, virus in cerebrospinal fluid, and low levels of anti-JEV IgM antibody. Pathological examination of brain tissue from fatal cases and imaging studies of patients show that infection of neurons is widespread in the central nervous system, although the thalamus, basal ganglia, and anterior horns of the spinal cord appear to be particularly involved. Microscopic lesions include perivascular inflammation, neuronal degeneration, and necrosis, rather than apoptosis. Nonfatal cases recover in 1 to 2 weeks. However, neurological and/or psychiatric sequelae are seen in a large proportion of patients who survive the acute disease (see the section titled 'Epidemiology').

Diagnosis

The virus is rarely isolated from peripheral blood during the acute stage of the disease in humans. This is thought to be due to a combination of a low viremia and clinical symptoms of the disease, which are not normally seen until after the virus has invaded the central nervous system by which time the viremia has finished. Virus can be isolated from cerebrospinal fluid early in the course of acute encephalitis, but this is consistent with a poor prognosis. Most virus isolates have been obtained from brains of patients at autopsy, or from mosquito pools. Viral antigen can also be detected by immunohistochemical techniques applied to neurons of patients at autopsy.

Many procedures have been used to detect serum antibodies against JEV: hemagglutination-inhibition, complement fixation, immunofluoresence, and enzyme-linked immunosorbent assay (ELISA). It is necessary to show at least a fourfold increase or decrease in titers of antibody to JEV between paired serum samples for any of these tests to be used to make a presumptive diagnosis of JE. Such a 'presumptive' diagnosis is required because of the extensive serological cross-reactions with antibodies against other flaviviruses, and all such serodiagnoses must be confirmed by neutralization tests. Many other flaviviruses overlap geographically with JEV, including the dengue and West Nile viruses, and can result in misinterpretation of test results. Detection of IgM antibodies is considered a relatively (JE complex-specific) specific test for JEV infection and is the serologic method of choice. An IgM-capture ELISA is usually used to detect IgM antibody to JEV in blood or cerebrospinal fluid within 7 days of onset of clinical symptoms. Recently, a dot-blot IgM assay has become available. Detection of IgM antibody in cerebrospinal fluid is associated with clinical JE and is predictive of a poor outcome.

Transmission

The virus is transmitted between vertebrate hosts by mosquitoes. The natural cycle involves rice-field-breeding mosquitoes and domestic pigs or wading ardeids (e.g., egrets and herons). The most important vector is *Culex tritaeniorhynchus*, found in most parts of Asia and which breeds in water pools and flooded rice fields. The virus also has been demonstrated to infect mosquitoes of other species, including *Cx. fuscocephala*, *Cx. gelidus*, *Cx. pipiens*, *Cx. bitaeniorhynchus*, *Cx. epidesmus*, *Cx. vishnui*, *Mansonia uniformis*, *M. bonneae/dives*, *Aedes curtipes*, *Ae. albopictus*, *Armigeres obturans*, *Anopheles hyrcanus*, and *An. barbirostris*. In temperate regions, the 'JE season' is considered to start in April–June with detection of virus in mosquitoes; this peaks in July. During July and August, virus is detected with increasing frequency in pig and bird amplifying hosts. The 'season' usually ends in September–October. Human infections are concurrent with the increased frequency in amplifying hosts. The exact timing of the 'JE season' will vary depending on geographic location in Asia, rainfall, and migration of birds. It is not clear how JEV survives between 'JE seasons' in temperate areas. There is no evidence that JE epidemics follow heavy rains, major floods, etc.; intervals between outbreaks vary from 2 to 15 years. There is evidence to support vertical transmission by *Culex* and *Aedes* spp. mosquitoes, sexual transmission between male and female mosquitoes, and a potential role for migratory birds in long-distance movement of JEV. However, none of these possibilities has been conclusively demonstrated.

Culex tritaeniorhynchus preferentially feeds on animals other than humans. Consequently, high seroprevalence rates are seen in a wide range of animals including dogs, ducks, chickens, cattle, bats, water buffaloes, donkeys, monkeys, snakes, and frogs. The role, if any, that these animals play in the ecology of JEV is not known. However, birds and pigs are considered to be the major viremic amplifying hosts of the virus. Among birds, ducks, chickens, water hens, egrets, and herons seroconvert.

Swine are important amplifying hosts in the epidemiology of the virus. Infected pigs develop sufficiently high viremias to infect mosquitoes. Importantly, adult pigs do not show clinical signs of JEV infection.

Infected equids can succumb to JE, while cattle seroconvert; however, both are considered dead-end hosts as viremias are too low to infect mosquitoes.

Treatment

There are no antiviral treatments to control flavivirus infections; rather, supportive therapy is the norm.

Control

As with most mosquito-borne diseases, transmission of JEV can be blocked by mosquito control measures. In addition, immunization of pigs, a major vertebrate amplifying host, blocks the transmission cycle. However, given that JE is usually found in rural areas, human immunization is the method of choice. Although vaccination is used to control JE, socioeconomic development of Asian countries is also contributing to control. Smaller areas of rice fields, centralized pig production, use of agricultural pesticides, and reduction of the rural population at the expense of increased urbanization have all contributed to a decrease in cases of JE.

There are licensed vaccines available to control JE. Inactivated vaccines are based on strains Nakayama, Beijing-1, or P3. The former two are based on formalin-inactivated mouse brain preparations that are semipurified to remove brain materials while strain P3 is also formalin-inactivated, but grown in primary hamster kidney cell cultures in the People's Republic of China. Inactivated mouse brain-derived vaccines are approved by the World Health Organization for international use. The above vaccines require two doses given on days 0 and 7–28 to induce protective immunity with a booster dose at 1 year, and subsequently every 3–4 years to maintain immunity. Initial formalin-inactivated mouse brain vaccines were based on strain Nakayama; however, in 1989, this strain was replaced with strain Beijing-1 in most markets because the latter is antigenically closer to recent Japanese isolates of JEV, a high potency being retained following purification, and immunogenicity is superior to that of strain Nakayama. This resulted in a vaccine that required a dose equivalent to half that needed for strain Nakayama vaccines. Inactivated vaccines are manufactured in India, Japan, Republic of Korea, People's Republic of China, Thailand, Taiwan, and Vietnam. The mouse-brain inactivated vaccine has been reported to cause occasional adverse events (estimated to be one in a million doses), including allergic mucocutaneous reactions. The majority of the reactions take place after the second or subsequent dose of vaccine. There were a number of adverse events reported during the period 1989–92 of which 15% were hospitalized and two-thirds required medical treatment. In addition, there have been occasional reported cases of acute disseminated encephalomyelitis since 1983. Studies to date suggest there is no reduced seroconversion rate or an increase in adverse events when mouse brain-derived JE vaccine is given simultaneously with vaccines against measles, DPT, and polio. Due to the adverse events, consideration is being given in Japan to employ an inactivated vaccine produced in Vero cells, and this second-generation vaccine is undergoing clinical trials to demonstrate noninferiority of immunogenicity compared to the mouse brain product.

There are no contraindications to the use of this vaccine, other than a history of hypersensitivity reactions to previous doses. However, vaccination is not recommended during pregnancy and pregnant women are only vaccinated when at high risk of exposure to the infection. Mouse brain-derived vaccine has been given safely in various states of immunodeficiency, including HIV infection.

There is also a live Japanese encephalitis vaccine based on strain SA14-14-2 grown in primary hamster kidney cell culture. Until recently, this vaccine was only licensed for use in the People's Republic of China. However, the World Health Organization has now developed criteria for the production of the vaccine, and it is now licensed in Nepal, India, and Republic of Korea, and clinical trials are being undertaken in Sri Lanka, the Philippines, Thailand, and Indonesia. This vaccine appears to be very efficacious and over 300 million doses of this vaccine have been administered in the People's Republic of China since 1988 with no known reports of serious adverse events.

The success of human vaccination has been demonstrated by the decrease in the number of cases in Japan. Prior to 1966, there were 1000–5000 cases per year with mortality up to 50%. Following the introduction of vaccination, the number of cases has steadily decreased to the point where fewer than 10 cases have been reported each year during the 1990s. A similar situation has taken place in the Republic of Korea and Taiwan, and all three countries have reported a shift from cases in children to cases in adults, particularly in the elderly. Since the vast majority of JE cases are children, vaccination has focused on children.

Vaccination is used to reduce the incidence of abortions in JEV-infected pigs. A live vaccine has been used in the People's Republic of China to protect horses.

The mechanism of protective immunity is poorly understood for most flaviviruses, although production of neutralizing antibodies ($\geq 1:10$) appears to correlate with immunity. Most neutralizing antibodies recognize epitopes of the E protein. Significantly, in passively immunized mice, detectable neutralizing antibodies will protect against a wild-type challenge administered by the intracerebral route. Cell-mediated immunity, in particular T-cell epitopes, have been mapped to nonstructural proteins, in particular NS3.

See also: Flaviviruses: General Features; Flaviviruses of Veterinary Importance; Vaccine Safety; Vaccine Strategies.

Further Reading

Burke DS and Leake CJ (1988) Japanese encephalitis. In: Monath TP (ed.) *The Arboviruses: Epidemiology and Ecology,* vol. 3, pp. 63–92. Boca Raton, FL: CRC Press.

Burke DS, Tingpalapong M, Ward GS, Andre R, and Leake CJ (1986) Intense transmission of Japanese encephalitis to pigs in a region free of epidemic encephalitis. *Japanese Encephalitis and Haemorrhagic Renal Syndrome Bulletin* 1: 17–26.

Chen W-R, Tesh RB, and Rico-Hesse R (1990) Genetic variation of Japanese encephalitis virus in nature. *Journal of General Virology* 71: 2915–2922.

Halstead HB and Jacobson J (2003) Japanese encephalitis. *Advances in Virus Research* 61: 103–138.

Halstead SB and Tsai TF (2004) Japanese encephalitis vaccines. In: Plotkin SA and Orenstein WA (eds.) *Vaccines,* 4th edn., pp. 919–958. Philadelphia: W B Saunders.

Huang C (1982) Studies of Japanese encephalitis in China. *Advances in Virus Research* 27: 72–100.

Mackenzie JS, Deubel V and Barrett ADT (eds.) (2002) *Japanese Encephalitis and West Nile Viruses, Vol. 267*: Current Topics in Microbiology and Immunology, pp. 1–416. Vienna: Springer.

Mackenzie JS, Gubler DJ, and Petersen LR (2004) Emerging flaviviruses: The spread and resurgence of Japanese encephalitis, West Nile and dengue viruses. *Nature Medicine* 10(supplement): S98–S109.

Solomon T (2003) Recent advances in Japanese encephalitis. *Journal of Neurovirology* 9: 274–283.

Solomon T and Winter PM (2004) Neurovirulence and host factors in flavivirus encephalitis – Evidence from clinical epidemiology. *Archives of Virology Supplement* 18: 161–170.

Kaposi's Sarcoma-Associated Herpesvirus: General Features

Y Chang and **P S Moore**, University of Pittsburgh Cancer Institute, Pittsburgh, PA, USA

© 2008 Elsevier Ltd. All rights reserved.

History

Kaposi's sarcoma (KS) was first described by Moritz Kaposi in 1872 as a rapidly fatal, 'idiopathic, pigmented sarcoma' of the skin. Although initially associated with men of Eastern European, Ashkenazi Jewish and Mediterranean ancestry, by the 1950s KS was recognized to be the third most common malignancy in sub-Saharan Africa. The unusual geographic distribution of this cancer led to speculation about a viral cause, speculation that intensified with the onset of the acquired immune deficiency syndrome (AIDS) epidemic in the early 1980s. AIDS patients were 100 000-fold more likely to develop KS than the general population and the tumor took on a more aggressive and frequently lethal form in AIDS patients.

Two epidemiologists, Harold Jaffe and Valerie Beral, examined patterns among AIDS surveillance data and concluded that KS is caused by a viral agent other than human immunodeficiency virus (HIV) and that the agent is uncommon in the general population but sexually transmitted efficiently among gay men. These predictions were later found to be strikingly accurate. Over 20 possible culprits were described as 'the KS agent'. As early as 1972, Giraldo and colleagues suggested that human cytomegalovirus might cause KS. These findings were discounted after further study, but in 1984 Walter and colleagues succeeded in directly observing herpesvirus particles in KS tumor specimens, thus reviving the possibility that KS is a herpesvirus disease.

In 1993, Chang and Moore used representational difference analysis to isolate two small DNA fragments from a KS lesion similar to, but distinct from, known herpesvirus sequences. Collaborating with Ethel Cesarman, they were then able to show that this putative new virus was present in virtually all KS tumors, was more likely to be found at the tumor site than distal tissues, and was not present in control tissues from patients without AIDS. The virus was given the common name Kaposi's sarcoma-associated herpesvirus (KSHV) and recognized to be the eighth human herpesvirus (HHV-8).

A B-cell lymphoma, called primary effusion lymphoma (PEL), was quickly identified as harboring KSHV. These lymphomas were known to be common secondary malignancies among KS patients, and isolation of PEL cells allowed the first *in vitro* culture of KSHV. These early cell lines were coinfected with both Epstein–Barr virus (EBV) and KSHV, but nonetheless accurate first generation serologic assays were developed using them. A second neoplastic disorder called multicentric Castleman's disease (MCD) was also recognized to be frequently infected with KSHV by Soulier and colleagues.

Once the virus was identified, its characterization proceeded rapidly. Studies demonstrated that all forms of KS, both HIV-positive and HIV-negative, are uniformly infected with KSHV. Extensive HIV/AIDS cohort databanks were used for serologic test development and showed that KSHV infection preceded onset of disease. These early studies also revealed that KSHV infection is uncommon in the general population of the USA and Europe – despite technical confusion resulting from highly cross-reactive assays – and patterns of infection mirror KS tumor rates among high risk populations, following closely the Beral-Jaffe predictions for the KS agent. Parravincini and colleagues showed that KS arising among transplant patients could either be due to reactivation of preexisting infection or to *de novo* infection from the organ allograft, helping to explain the association of this tumor with transplantation.

Virologic characterization was equally rapid. In 1996, Ganem's group isolated a KSHV-infected PEL cell line free of EBV infection, and performed the first studies on viral replication. While KSHV had been previously cultured *in vitro* in EBV-coinfected cell lines, this was a major breakthrough allowing unambiguous characterization of the virus. That same year, Chang and Moore finished sequencing the 165 kbp viral genome, almost exactly 2 years after reports of its initial discovery. These studies revealed that the virus makes extensive use of molecular piracy, encoding homologs to multiple known cellular oncogenes. Shortly thereafter, research groups identified KSHV inhibitory proteins targeting p53 and retinoblastoma protein, interferon regulatory proteins, and KSHV paracrine-signaling molecules regulating cell growth control, providing key molecular clues on how this virus might initiate tumorigenesis.

Shortly after its initial description, KSHV was widely recognized to be the infectious cofactor causing KS and related diseases. With the onset of the AIDS epidemic, KS is now the most common tumor reported in sub-Saharan Africa, resulting in an enormous, unappreciated morbidity and mortality. In contrast, the introduction of highly effective retroviral therapy into developed countries in the mid-1990s has reduced KS incidence among AIDS patients by nearly 90% – a remarkable public health success story.

Taxonomy, Classification and Evolution

KSHV is a gammaherpesvirus (subfamily *Gammaherpesvirinae*) belonging to the genus *Rhadinovirus*, which is distinct from the genus *Lymphocryptovirus* containing EBV (**Figure 1**). While rhadinoviruses are distributed among both primates and nonprimates (e.g., mice, cattle, and horses), EBV-like viruses are found exclusively among primates.

The genus *Rhadinovirus* contains three evolutionarily distinct lineages of primate viruses. One lineage is found among New World monkeys and two parallel but distinct lineages (including KSHV) infect Old World primates including humans. KSHV was first recognized to be closely related to the New World squirrel monkey virus, herpesvirus saimiri. In the search for KSHV-like viruses among nonhuman primates, similar viruses were found among rhesus macaques as well as baboons, gorillas, and chimps, suggesting an evolutionary codivergence for these viruses with their hosts.

A second rhadinovirus lineage found in rhesus macaques (and hence commonly called rhesus rhadinovirus, RRV) was quickly extended to other Old World primate hosts. The importance of this finding is that no human RRV-like virus has been described. If it exists, and evolutionary evidence strongly suggests that it might, then it would become HHV-9.

Molecular epidemiology studies based on variation of the K1 gene and the right-hand end of the genome reveal at least four unique viral clades called A–D that largely match patterns of human migration from Africa. There are several conserved alleles of the right-hand end of the genome (called M, N, O, and P) found across clades suggesting that these viruses are chimeric and have arisen by recombination.

Geographic Distribution

Unlike most human herpesviruses, KSHV is not a ubiquitous human infection. KSHV infection is hyperendemic throughout sub-Saharan Africa with estimated adult infection rates of 30–60%. Other isolated populations, including indigenous Amazonian tribes, may also have infection rates at this level or higher. The prevalence of infection declines among Mediterranean basin countries such that general adult infection rates in Sicily average about 10% and continue to decline at more northern latitudes in Italy. In the USA and Northern Europe, approximately 2% of blood donors are KSHV infected. While general infection rates are low in the USA and Europe, gay men are recognized to have a rate of infection 10–20 times higher than heterosexuals.

Transmission

In the absence of underlying immunosuppression, KSHV infection appears to be asymptomatic, or perhaps associated with transient fever and rash. KSHV is recognized to have several patterns of transmission, and modes of transmission differ dramatically between developed and developing countries. It is secreted from the oral mucosa through salivary secretions. In developed countries, there is a strong sexual component to transmission among gay and bisexual men but the precise sexual behavior responsible for transmission is unknown. Two possibilities include deep kissing and use of saliva as a sexual lubricant. In contrast, heterosexual transmission is apparently rare.

In developing countries, and possibly endemic regions of Europe and the Middle East, casual transmission is common. Again precise mechanisms of transmission are

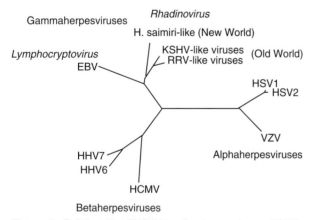

Figure 1 Relationship of KSHV to other herpesviruses. KSHV belongs to a lineage of rhadinoviruses infecting Old World primates. A second rhadinovirus lineage, exemplified by rhesus rhadinovirus (RRV), has been discovered infecting Old World primates. No human infections with an RRV-like virus have been described.

unknown, but KSHV transmission is much more efficient in sub-Saharan Africa where, unlike Northern Europe and the USA, significant rates of infection occur among children and adolescents. While intrauterine transmission and transmission through breast milk is uncommon, maternal-child transmission does occur during infancy and early childhood.

Of increasing concern, KSHV is also transmitted both through organ allografts and through blood transfusion. KS arising during transplantation is an important clinical problem with rates among all transplant patients between 0.2% and 6% depending on the geographic locale and background endemicity for viral infection. KS in this setting has a high mortality and morbidity since primary treatment involves removing immunosuppression. The bulk of KS among transplant recipients derives from reactivation of preexisting, quiescent infection which subsequently flares during immunosuppression, but transmission from donor allografts has been documented. Interestingly, one study has demonstrated that KS tumor cells rather than virus itself can be transmitted during transplantation.

Blood transfusion has in the past been considered a minor component of KSHV transmission. Several studies have shown a relationship between non-AIDS KS and past transfusion history, but this relationship has been obscured by the overwhelming risk of sexual transmission among HIV/AIDS patients. Transfusion transmission has recently been documented to occur among a fraction of infected blood samples. Whether a public health response to this is needed remains unknown.

Host Range and Virus Propagation

KSHV is grown in the laboratory using PEL-derived cell lines and can be activated into lytic virus production by treatment with chemicals such as protein kinase C activators or histone deacetylase inhibitors. The virus cannot be maintained in cultured cells from KS tumor explants and is rapidly lost within a few passages.

KSHV can, however, be readily transmitted to a range of human and nonhuman cell lines in culture, although it is lost unless it is kept under some form of selection. The virus is commonly transmitted in culture to endothelial cell lines, which undergo transformation into a spindle growth pattern resembling KS. Receptors for virus binding and entry include extracellular heparan, integrin $\alpha 3\beta 1$, the cystine transporter xCT, and DC-SIGN. It is not yet clear which of these receptors act in concert or independently of each other to allow different steps of binding, internalization, and entry. Transmission to nonobese diabetic/severe combined immunodeficiency (NOD/SCID) mice has been documented, although the consequences of infection have not been fully examined.

Although whole virus transmission to animal models is limited, single gene expression studies have proved valuable. The K1 protein, for example, has been placed into a herpesvirus saimiri backbone and caused lymphatic tumors in rhesus macaques. Many KSHV genes, including those encoding viral interleukin 6 (vIL6), viral interferon regulatory factor 1 (vIRF1), and kaposin, transform rodent cell lines and cause tumors in mice. In addition, expression of the viral G protein-coupled receptor (vGPCR) in transgenic mice results in paracrine development of endothelial cell tumors that closely resemble human KS tumors.

Virus Genetics and Molecular Biology

KSHV is a ~165 kbp double-stranded DNA virus with a long unique region (LUR) containing all protein-coding sequences that is flanked by high $G + C$ content direct terminal repeats (TRs) on both ends. The TRs are 801 bp in length each, with 20 or more units in a typical virus, and are the site for circularization (latency) or linearization (lytic replication). One KSHV protein, latency-associated nuclear antigen 1 (LANA1), acts similarly to a corresponding EBV protein, EBNA1, to bridge between the viral episome and host cell chromosomes during latency, allowing equal segregation of virus to daughter cells. During latent replication, in which the virus is in a circular state, the origin of DNA replication is located within the TR region.

The LUR encodes over 90 independent protein-coding open reading frames (ORFs), and alternative splicing may markedly increase the expressed number of viral proteins. During latency, the KSHV genome is largely quiescent with only a few genes constitutively expressed, primarily from the right-hand latency locus and including LANA1, viral cyclin (vCYC), viral FLICE-inhibitory protein (vFLIP), and approximately 12 micro-RNAs.

KSHV, like other herpesviruses, has a highly choreographed cascade of gene expression during lytic replication. Lytic replication is initiated by a single transactivator protein, RTA, expressed from ORF50. This leads to sequential early, delayed-early, and late gene expression. Ultimately, the viral polymerase is expressed and linear KSHV chromosomes are replicated through a rolling circle mechanism.

KSHV genes are commonly divided into 'latent' and 'lytic' classes based on whether or not they are activated during the lytic replication cycle. This is clearly too simplistic a categorization to be meaningful. A large group of nonstructural protein genes have long been known to be expressed at low levels during viral latency but are upregulated during lytic replication (so called type II genes). Notch signaling, which activates the transcription factor RBP-Jk that acts in concert with RTA to initiate lytic replication, has recently been found to activate expression of these genes without late gene expression or virion production. Increased notch signaling is present in PEL cells and appears to be responsible for low level expression of type II genes during viral latency.

Molecular Piracy by KSHV

Rhadinoviruses show the most extensive piracy of host genes among the herpesviruses. KSHV genes that have been directly acquired from the host genome include those encoding proteins involved in DNA synthesis (thymidine kinase, dihydrofolate reductase, thymidylate synthase, ribonucleotide reductase subunits, and formyglycinamide ribotide amidotransferase (FGARAT)), deoxyuridine metabolism (dUTPase), paracrine signaling (viral complement control protein, vIL6, viral chemokine ligands 1, 2 and 3 (vCCL1, vCCL2, and vCCL3)) and vGPCR, interferon signaling (vIRF1, vIRF2, and LANA2), major histocompatibility class (MHC) regulation (modulator of immune recognition 1 and 2 (MIR1 and MIR2)), and cell cycle (vCYC) and apoptosis control (vBCL-2, vFLIP, and viral inhibitor of apoptosis (vIAP)). These proteins frequently fall into the type II expression pattern described above.

All of these proteins have recognizable sequence similarity to their human counterparts and appear to act in the same regulatory pathways. The primary difference between the viral and human homologs is at the level of protein regulation, in which the viral protein generally escapes from normal regulatory controls imposed on the human protein. vCYC, for example, targets the retinoblastoma protein for phosphorylation by recruiting cellular cyclin-dependent kinases. This allows the viral protein to initiate unscheduled S-phase entry of the cell cycle. vCYC resembles cellular D-type cyclins but differs in a key characteristic in that it is resistant to cyclin-dependent kinase inhibitors that control cellular cyclin function. Expression of these proteins, together with viral proteins that have no cellular counterparts, such as LANA1, is thought to be responsible for the tumorigenic capacity of this virus.

While the range of these homologous nonstructural proteins is unique to KSHV among the human herpesviruses, careful inspection reveals that the functions of these proteins are similar to those found in other viruses. There is a close functional correspondence between latent EBV proteins and latent KSHV proteins despite the fact that there is little or no sequence similarity. In general, EBV nonstructural proteins act as signaling molecules to activate specific cellular pathways, whereas KSHV has pirated members of these pathways to act in a similar fashion.

Immune Response

Infection with KSHV generates a strong antibody response that can be measured by indirect immunofluorescence assays (IFA), ELISA, and western blotting. In the absence of lytic stimulation, the nuclear antigen LANA1 commonly elicits antibody titers exceeding 1:50 000. This protein migrates aberrantly as a 223–234 kDa doublet band on western blotting. While LANA1 immunoreactivity is highly specific for KSHV infection, approximately 20% of AIDS-KS patients do not have immunoreactivity. For this reason, attempts to increase test sensitivity using lytic genes have been made. Whole cell lytic immunoassays can be highly sensitive but have demonstrated cross-reactivity that can be measured using adsorption with formalin-fixed KSHV-negative cell line antigen. Use of lytic cell assays led to considerable confusion in early studies about the prevalence of KSHV infection in various low-risk populations.

Two recombinant structural antigens, the ORF65 and K8.1 proteins, have been valuable in maintaining high sensitivity with low levels of cross-reactivity. Assays based on these antigens have been further refined using peptide epitopes in ELISA assays. Together with a recombinant LANA1, ELISA sensitivity and specificity of greater than 90% for KSHV detection can be routinely achieved. Complicating the serologic detection of KSHV is the fact that much of this work is based on patients with end-stage AIDS-KS. These patients routinely lose antibody reactivity due to loss of CD4+ helper activity.

The surge in KS among AIDS patients revealed the importance of cellular immunity, particularly CD4+ based immunity, to KS tumor control. Cell-mediated immunity assays have measured cellular immune responses against a number of KSHV protein epitopes, including those found on both structural and latent antigens. Among AIDS patients, these responses closely follow regression of KS tumors during highly active antiretroviral therapy (HAART).

Surprisingly, cellular immunity to the constitutively expressed LANA1 protein is generally low. The corresponding EBV protein, EBNA1, escapes immune surveillance by inhibiting its own proteosomal degradation and retarding its own translational synthesis. The latter effect appears to inhibit the formation of misfolded EBNA1 defective ribosomal products that are rapidly processed into MHC class I (MHC I) presented peptides. A similar mechanism seems to be at work for KSHV LANA1. Although the two proteins are not homologous, the nucleotide sequence of a central repeat region of both viral proteins is similar but is frameshifted between the two viruses. It remains to be seen whether the nucleotide sequence or the individual protein sequences are responsible for this immunoevasion effect.

KSHV also evades MHC I immune processing by expressing two proteins, MIR1 and MIR2, that ubiquitinate MHC I and other immune accessory proteins, causing them to be internalized and degraded. Although MIR1 and MIR2 are generally expressed during lytic replication, MIR1 is expressed at low levels during latency and can be upregulated by notch signaling.

In addition to cell-mediated immune response evasion, KSHV encodes a number of proteins that act to overcome innate immune signaling pathways. vIRF1 and the ORF45 protein act to sequester interferon-regulated transcription factors, effectively shutting off interferon signaling during infection. vIL6 signals in a similar manner to human IL-6 but it can bind directly to the signal transducer molecule gp130 to initiate signaling. One consequence of this effect is to abrogate signaling from interferon receptors to downstream signaling machinery.

KSHV-Associated Diseases

Kaposi's Sarcoma

KS is an endothelial tumor having molecular markers suggesting that it originates from lymphatic endothelium (**Figure 2**). Microscopically, the tumor is characterized by tumor spindle cells forming disorganized vascular clefts that fill with blood cells, giving the tumor a bruised like appearance. KSHV is detected in virtually all spindle cells, but tumors also show active neoangiogenesis in which more organized vessels lace the tumor mass. KSHV is generally absent from these cells and the intrusion of neovascular and immune cell components give the tumors a mixed cellular origin. Thus, the pathology of KS strongly suggests that the tumor has both neoplastic and hyperplastic components. Conflicting data exist as to whether or not KS tumors are cellularly monoclonal, and studies of viral TRs suggest that KS tumors may originate from both polyclonal and monoclonal infection. It is likely that tumor cells evolve from polyclonal populations and gradually evolve into monoclonal or oligoclonal entities over time, helping to explain conflicting studies on the origin of the tumor.

Similarly, there is mixed evidence for the virus contributing to KS tumor growth through virus-transforming and paracrine activities. In tumors, virtually all tumor spindle cells are infected with KSHV, suggesting that the virus drives cell multiplication through intracellular expression of specific oncoproteins while the overwhelming majority of tumor cells are in a tight latency expression pattern. In line with this, antiherpesvirus drugs, such as ganciclovir, are effective in preventing KS but, once tumors are established, have little impact on disease. In contrast, KS tumors show marked hyperplasia and cellular cytokine overexpression. Since the virus is rapidly lost from tumor cells once the tumor is disaggregated and put into culture, there is strong evidence that the tumor environment maintains KSHV infection through paracrine signaling mechanisms. The best animal model for KS involves transgenic expression of vGPCR, which is known to induce factors such as VEGF in T cells, resulting in disorganized proliferation of endothelial cells. Taken together, these data suggest that the virus contributes to tumorigenesis through both neoplastic and hyperplastic mechanisms.

KS tumors from all sites and geographic settings are nearly universally infected with KSHV. While KS in the past has been split into categories depending on the epidemiologic setting, such as endemic KS, classical KS, iatrogenic KS, and epidemic KS, all forms of the tumor are pathologically indistinguishable. Cellular immune status appears to be key to the clinical expression of the tumor among infected persons. In the absence of HIV, KS primarily occurs in those with immune compromise such as the extreme elderly and patients on immunosuppressive therapy. The incubation period from infection to KS has been reported to average 3 years among HIV/AIDS patients, but this appears to be determined by degree of underlying immunosuppression at time of infection. In one case, transient KS was detected in an AIDS patient weeks after initial infection. While KSHV is largely asymptomatic in immunologically intact individuals, it has a very high degree of tumor expression among those who are immunosuppressed. Up to half of early AIDS patients develop KS and among transplant patients infected with KSHV, and 20–60% have been reported to develop clinical disease.

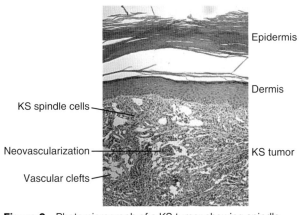

Figure 2 Photomicrograph of a KS tumor showing spindle tumor cells forming irregular vascular clefts that fill with blood. While near-universal KSHV infection of spindle cells is generally seen, the virus is generally absent from the neovascularization

Primary Effusion Lymphoma

PEL is a rare, postgerminal center B-cell tumor accounting for less than 3% of B-cell lymphomas from various clinical series in developed countries. The tumor can form malignant effusions in body cavities, and among AIDS patients it generally has an aggressive course. Unlike KS, PELs are monoclonal. Tumor cells lack most B-cell markers, although V(D)J rearrangement studies confirm their B-cell origin and they express syndecan-1. PEL tumor cells have been reported to be autocrine-dependent on human IL10 and viral IL6. The latter cytokine is peculiar since it appears to form a 'xenocrine' loop in which vIL6 signaling abrogates cellular interferon responses resulting from virus infection that would otherwise stop growth of the tumor cell. PEL cells are uniformly infected with KSHV and can be grown stably in culture with each cell maintaining 40–150 copies of the viral episome. Cell lines derived from PEL tumors, particularly from AIDS patients, frequently have EBV coinfection, but EBV-negative, KSHV-positive tumors are also found.

Multicentric Castleman's Disease

Unlike PEL, MCD is considered a hyperproliferative lymphoid disorder, with affected individuals displaying a high rate for subsequent development of other lymphoid malignancies. Among HIV-negative persons, only about 50% of MCD tumors show evidence of KSHV infection whereas >90% of HIV-positive MCD patients have KSHV present.

The primary characteristic of MCD is the formation of adventitious germinal centers with marked hyperplasia in lymphoid organs. Examination of these KSHV-positive tumors reveals scattered KSHV infected cells in the marginal zone of the germinal centers, with the bulk of the cells being uninfected. Aberrant IL6 signaling has been demonstrated to be central to the pathogenesis of MCD, and KSHV-positive tumors have marked expression of vIL6, with little human IL6 expression, while KSHV-negative tumors show extensive overexpression of the human cytokine.

Unlike KS and PEL, KSHV-infected tumor cells in MCD show a broad range of viral gene expression and perhaps full lytic viral replication. While standard treatment of KSHV-positive tumors has a high mortality, recent studies have reported marked success in treatment using anti-CD20 antibodies together with ganciclovir.

Other KSHV-Related Disorders

Transplant patients have been reported to develop bone marrow failure after graft infection with KSHV. The extent and consequences of this nonmalignant complication are unknown. In addition, a plasmacytic post-transplant lymphoproliferative disorder from KSHV has been reported to occur among transplant patients.

KSHV has been reported to be associated with a number of other malignant and autoimmune diseases, including multiple myeloma, sarcoidosis, post-transplant skin tumors, and idiopathic pulmonary hypertension. Despite initial enthusiasm, subsequent studies have largely failed to demonstrate any association between the virus and these diseases. It remains formally possible that KSHV causes a subset of one or more of these conditions with the majority of disease cases caused by other agents or conditions.

Diagnosis, Treatment, and Prevention

At present, there are no clinically certified tests to detect KSHV infection, although individual research laboratories have the capacity to test samples under special conditions. This presents a major public health problem, particularly in the setting of transplantation and in blood transfusions in highly endemic areas of the world. Treatment of KSHV-related diseases also generally does not directly target underlying viral infection, instead relying on traditional surgical and chemotherapy. An exception to this are ongoing clinical studies of MCD in which the lytic form of the virus is controlled with ganciclovir. Similarly, blinded clinical trials among high-risk AIDS patients show that antiherpesviral treatment can be highly effective in preventing KS. The introduction of HAART therapy has dramatically reduced the risk of KS among these patients, and thus attempts to extend these results have been limited. Since HAART does not target KSHV, it is expected that a resurgence of KS may occur with aging of the AIDS population.

There is currently no development of vaccines that might be useful to developing country populations. The exquisite sensitivity of KSHV-associated tumors to immune reconstitution and the loss of KSHV from developed populations throughout the world provides evidence that an appropriate vaccine might be extremely effective in controlling this virus infection.

In summary, the public health importance of this virus is clear: KS is the most common malignancy in sub-Saharan Africa and remains an important cause of cancer death among transplant and immunocompromised patients. It remains distressing that the wealth of basic science data available for this virus has yet to be used in a significant fashion to prevent or treat diseases caused by KSHV.

See also: Epstein–Barr Virus: General Features; Herpesviruses: Discovery; Herpesviruses: General Features; Kaposi's Sarcoma-Associated Herpesvirus: Molecular Biology; Simian Gammaherpesviruses.

Further Reading

Beral V, Peterman TA, Berkelman RL, and Jaffe HW (1990) Kaposi's sarcoma among persons with AIDS: A sexually transmitted infection? *Lancet* 335: 123–128.

Chang Y, Cesarman E, Pessin MS, *et al.* (1994) Identification of herpesvirus-like DNA sequences in AIDS-associated Kaposi's sarcoma. *Science* 265: 1865–1869.

Ganem D (2006) KSHV-induced oncogenesis. In: Arvin A, Campadelli-Fiume G, Moore PS, Mocarski E, Roizman B, Whitley R, and Yamanishi K (eds.) *The Human Herpesviruses: Biology, Therapy and Immunoprophylaxis*, ch. 56, pp. 1007–1030. Cambridge, UK: Cambridge University Press.

Jarviluoma A and Ojala PM (2006) Cell signaling pathways engaged by KSHV. *Biochimica Biophysica Acta* 1766: 140–158.

Kaposi M (1872) Idiopathic multiple pigmented sarcoma of the skin. *Archives of Dermatology and Syphilology* 4: 265–273.

Martin J (2006) Epidemiology of KSHV. In: Arvin A, Campadelli-Fiume G, Moore PS, Mocarski E, Roizman B, Whitley R, and Yamanishi K (eds.) *The Human Herpesviruses: Biology, Therapy and Immunoprophylaxis*, ch. 5, pp. 960–985. Cambridge, UK: Cambridge University Press.

Mendez JC and Paya CV (2000) Kaposi's sarcoma and transplantation. *Herpes* 7: 18–23.

Russo JJ, Bohenzky RA, Chien MC, et al. (1996) Nucleotide sequence of the Kaposi sarcoma-associated herpesvirus (HHV8). *Proceedings of the National Academy of Sciences, USA* 93: 14862–14867.

Kaposi's Sarcoma-Associated Herpesvirus: Molecular Biology

E Gellermann and T F Schulz, Hannover Medical School, Hannover, Germany

© 2008 Elsevier Ltd. All rights reserved.

History

Kaposi's sarcoma (KS) is an endothelial cell-derived tumor. It is common in acquired immunodeficiency syndrome (AIDS) patients, and is known to have occurred before the arrival of human immunodeficiency virus (HIV). Today, it is one of the most frequently occurring tumors in Africa.

In 1994, following epidemiological studies that suggested the involvement of a sexually transmissible agent other than HIV in the pathogenesis of epidemic (AIDS-associated) KS, Y. Chang, and P. S. Moore used representational difference analysis to identify two sequence fragments of a then unknown rhadinovirus (or γ_2-herpesvirus). Epidemiological studies have now established KS-associated herpesvirus (KSHV, also known as human herpesvirus 8, HHV-8) as a necessary and indispensable causative factor in the pathogenesis of all forms of KS. In addition to 'epidemic' KS, these include the 'classic' form that is more common mainly in HIV-negative elderly men in some Mediterranean countries, the 'endemic' form known to have existed in East and Central Africa long before the spread of HIV, and 'iatrogenic' KS as seen, for example, in transplant recipients. KSHV is also essential in the pathogenesis of primary effusion lymphoma (PEL) and frequently found in the plasma cell variant of multicentric Castleman's disease (MCD). PEL is a rare lymphoma characterized by effusions in the pleural or abdominal cavity. MCD is a lymphoproliferative disorder characterized by multiple lesions that involve lymph nodes and spleen and featuring expanded germinal centers with B-cell proliferation and vascular proliferation.

KSHV (species *Human herpesvirus 8*) is a member of the genus *Rhadinovirus* in subfamily *Gammaherpesvirinae* of the family *Herpesviridae*. It was the first rhadinovirus discovered in humans and thereby also the first in an Old World primate. In the 12 years since this discovery, it has become apparent that most, if not all, Old World primate species, including the great apes (chimpanzee, gorilla and orangutan) harbor related rhadinoviruses. Of these, the genomes of two distinct rhesus macaque viruses have been sequenced completely (or nearly completely), and found to represent two major rhadinovirus lineages. Short-sequence fragments of rhadinovirus genomes representing these lineages have been found in most Old World primates, with the exception of humans, who appear to harbor only KSHV.

Virion Structure

KSHV has the characteristic morphological appearance of a herpesvirus, with a core containing the double-stranded, linear viral DNA, an icosahedral capsid (100–110 nm in diameter) containing 162 capsomers, the tegument that surrounds the capsid, and the envelope, in which glycoproteins are embedded. The virion contains at least 24 proteins, among which are five capsid proteins, eight envelope proteins, six tegument proteins, and five other proteins such as the regulator of transcriptional activation (RTA) and replication-associated protein (RAP or K-bZIP) that are not usually considered as structural virion proteins. The presence of RTA may account for the transient lytic gene expression program that occurs soon after virus entry into the cell. Among the envelope glycoproteins are gB and gpK8.1, which interact with cellular receptors like heparan sulphate or integrin molecules, thereby mediating entry of the virus into the host cell.

Genome Organization

The KSHV genome consists of a long unique region of approximately 140 kbp flanked by multiple, variable numbers of a tandemly repeated 801 bp terminal repeat (TR) (**Figure 1**). All the protein-coding open reading frames (ORFs) are located in the long unique region, and around 90 have been described so far. These are named ORF4 through ORF75 and K1 through K15. In addition, a noncoding RNA and an alternatively spliced RNA that is processed into 12 microRNAs (miRNAs) has been reported. miRNAs are a class of evolutionarily conserved, noncoding RNAs of about 22 nucleotides that are presumed to have regulatory functions.

The majority of KSHV genes are homologous to genes in the alpha- and betaherpesviruses. Their functions as structural virion proteins or nonstructural players during the lytic (productive) replication cycle are therefore assumed to be similar to those of their counterparts. These genes are not discussed in further detail in this article.

In addition, several genes are only found in KSHV or closely related rhadinoviruses of Old World primates. These are shaded gray in **Figure 1** and include genes encoding latency-associated nuclear antigen 1 (LANA-1; encoded by ORF73) and several homologs of cellular proteins, including D-type cyclin (vcyc; encoded by ORF72), a homolog of a cellular FLICE-inhibitory protein (vFLIP; encoded by ORF71), interleukin 6 (vIL6; encoded by K2), macrophage inflammatory proteins (vMIP1, vMIP2, and vMIP3; encoded by K6, K4.1, and K4.2), interferon regulatory factors (vIRF1, vIRF2, vIRF3, and vIRF4; encoded by K9, K11, K10.5, and K10), complement regulatory protein (KCP; encoded by ORF4), and two members of a class of membrane-associated ubiquitin E3 ligases involved in the downregulation of major histocompatability class I (MHC I) and adhesion molecules from the cell surface (MIR1 and MIR2; encoded by K3 and K5). Ten of the miRNAs are located in the intergenic region between K12 and ORF71 and two are within K12.

The KSHV genome contains two lytic origins of DNA replication (Ori-Lyt), which show extensive sequence similarity to each other but are oriented in opposite directions. Each copy of TR contains a latent origin of DNA replication (Ori-Lat) that consists of two binding sites for LANA-1 and an adjacent region that is essential for replication.

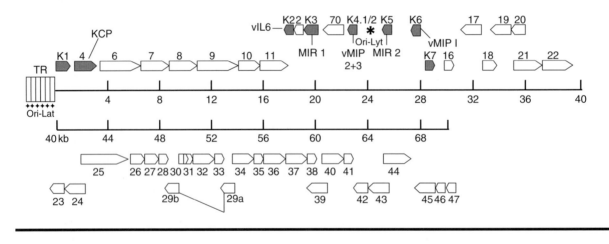

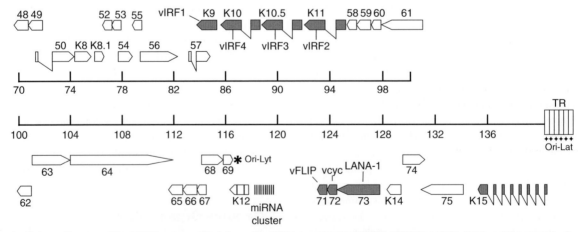

Figure 1 Schematic map of the KSHV genome. Protein-coding ORFs are denoted ORF4–ORF75 and K1–K15, and the names of encoded proteins are given for some (see RefSeq NC_003409 and NC_009333). ORFs typical of Old World or New World Primate rhadinoviruses are shaded gray and other marked features are described in the text. The scale is in kbp.

Evolution

Rhadinoviruses closely related to KSHV appear to exist in many, if not all, Old World primate species. The phylogenetic relationships between DNA fragments from these viruses suggest the co-evolution of primate rhadinoviruses with their host species. Modern KSHV isolates can be grouped into five subtypes (A–E), three of which are still associated with particular human populations (B, Africa; D, Australasia; E, Native Americans), while for two (A and C) such an association is no longer easily discernable, presumably as a result of extensive population mixing in Europe and in Near and Middle Eastern countries. However, one variant of subtype A (A5) is often seen in populations with links to Africa. With the exception of the K1 gene and the right end of the viral genome, variability between subtypes does not exceed a few percent.

Extensive recombination has evidently occurred between KSHV subtypes, and many modern isolates have chimeric genomes. At the right end of the genome, which includes the K15 gene, two types of very divergent and slowly evolving sequences have been found, suggesting that recombination events with rhadinoviruses other than KSHV could have taken place. By contrast, at the left end of the genome, the K1 gene shows evidence of extensive variation and evolution under selective pressure. It is possible that selective pressure exerted by T cells recognizing epitopes may have shaped the marked variability of the encoded membrane glycoprotein, but other explanations are also conceivable.

Tropism and Cell Entry

In vivo, KSHV has been found in human peripheral blood B and T cells, monocytes/macrophages, and endothelial cells. *In vitro*, a wide range of cell lines and primary cell cultures have been infected with KSHV, including human primary and immortalized endothelial cell cultures, epithelial cell lines and primary epithelial cell cultures, fibroblast cultures, monocytes/macrophages, and dendritic cells. Cell cultures from some other host species can also be infected *in vitro*, as can a strain of non-obese diabetic/severe combined immunodeficiency (NOD/SCID) mouse. These observations indicate that the restriction of KSHV for humans in the context of 'natural' infections is not determined at the level of cell entry.

The viral membrane glycoproteins gB (encoded by ORF8) and gpK8.1 (K8.1) are involved in entry of KSHV into target cells. Both interact with heparan sulfate, and gB also binds to $\alpha_3\beta_1$ integrin via an arginine-glycine-aspartate (RGD) motif. In addition, DC-SIGN (CD209) serves as a receptor for KSHV on macrophages and dendritic cells. xCT, the 12-transmembrane light chain of the human cystine/glutamate exchange transporter system x-c, serves as a receptor for KSHV fusion and entry.

The interaction of KSHV with heparan sulfate is thought to mediate the first, 'low-affinity' contact with the target cell, and subsequent binding of individual viral proteins to particular receptors then initiates the next stages of fusion and endocytosis of viral particles. Different entry routes may predominate, depending on the lineage of the infected cell. The importance of the interaction of gB with $\alpha_3\beta_1$ integrin may lie in the integrin-mediated activation of focal adhesion kinase (FAK), whose phosphorylation has been shown to play an important role during the early stages of KSHV entry, such as cellular uptake of viral DNA. The related kinase Pyk2 may also contribute to this process. The subsequent activation of Src, phosphatidylinositol 3-kinase (PI-3K), protein kinase C (PKC), RhoGTPase, mitogen-activated protein kinase (MEK), and extracellular signal-regulated kinase 1/2 (ERK1/2) facilitates virus uptake and immediate-early gene expression. KSHV-activated RhoGTPase in turn contributes to microtubular acetylation and leads to the modulation of microtubule dynamics required for movement of KSHV in the cytoplasm, and for delivery of viral DNA into the cell nucleus.

Latent Persistence and Replication

One of the characteristic features of KSHV is its propensity of quickly establishing a nonproductive, persistent state of infection following cell entry. Akin to its gammaherpesvirus (lymphocryptovirus) cousin, Epstein–Barr virus (EBV), but in contrast to alpha- and betaherpesviruses, this latent state of infection appears to be the default option for KSHV in most cell types, and presumably reflects its mode of long-term persistence *in vivo*. Latency is characterized by a restricted pattern of viral gene expression. Latent viral genes are depicted in **Figure 2** and include those for LANA-1, vcyc, vFLIP, and kaposin A and B, as well as the miRNAs. Several promoters, indicated in **Figure 2**, direct expression of cognate transcripts, of which two (P1 and P3) are strictly latent whereas two others (P2 and P4) become active during the lytic replication cycle. In addition, vIRF3 (also known as LANA-2) is part of the latent gene expression pattern in latently infected B cells. *In vivo*, expression of many of these latent genes has been demonstrated in the majority of KSHV-infected endothelial and spindle cells of KS, and the neoplastic B cells of PEL and MCD, using immunohistochemical or *in situ* hybridization techniques.

During latency, the KSHV genome persists as a circular episome, of which multiple copies (up to 100 or more) have been detected in latently infected PEL cell lines, although it is likely that the copy number varies widely

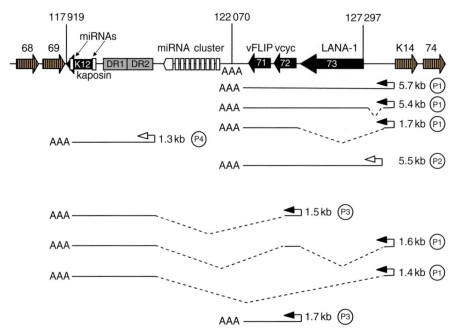

Figure 2 The latency gene cluster in the KSHV genome. The KSHV latency locus encodes four proteins indicated by black arrows (K12/kaposin, ORF71/vFLIP, ORF72/vcyc, and ORF73/LANA-1), as well as 12 miRNAs (white boxes), and is flanked by lytic genes (striped arrows). Lytic promoters are indicated by white arrows and latent promoters by black arrows. Readthrough of a viral polyadenylation signal located at nucleotide 122 070 leads to miRNA expression. P1 (127 880/86) gives rise to a precursor RNA that is spliced to produce 5.7 and 5.4 kb tricistronic (ORF71, ORF72, and ORF73) and 1.7 kb bicistronic (ORF71 and ORF72) mRNAs. Additionally, P1 drives two transcripts that could function as precursors for the miRNAs. As soon as RTA is expressed, P2 (127 610) is induced and gives rise to a 5.5 kb mRNA encompassing all three ORFs. Lytic transcription from P4 (118 758) leads to a 1.3 kb transcript coding for K12/kaposin. P3 (123 751/60) controls a 1.7 kb transcript encoding vFLIP and vcyc and a 1.5 kb transcript that has the potential to encode the miRNAs. Sequence coordinates are derived from GenBank NC_003409.

in other cell types. Replication of the genome during S-phase is ensured by LANA-1, which binds to viral episomes via its C-terminal region and recruits a range of cellular factors, including the 'origin-recognition complex', which is involved in replication of cellular DNA. Two binding sites for LANA-1, spaced 39 bp apart, are located in each of the multiple 801 bp TRs that flank the viral genome. In addition, LANA-1 attaches viral episomes to mitotic chromosomes via its N-terminal domain, thus ensuring their segregation to daughter cells during mitosis as well as their replication during S-phase.

Possibly linked to its role in the replication of viral episomes is the ability of LANA-1 to promote the transition of the G1/S cell cycle checkpoint, thereby enabling viral DNA synthesis. The physical interaction of LANA-1 with the retinoblastoma protein pRB and with GSK-3β is thought to contribute to this property. In addition, LANA-1 binds to p53 and may thereby inactivate cell cycle checkpoint functions that could interfere with successful viral DNA replication, although this can also lead to genomic instability in the infected cell.

A further facet of the role of LANA-1 during latent persistence is its inhibition of the central activator of the lytic replication cycle, RTA, thereby contributing to maintenance of the latent infection. The ability of LANA-1 to interact with components of silenced and transcriptionally active chromatin, such as MeCP2 (which binds to methylated CpG motifs and thereby promotes chromatin silencing) or brd2/RING3, brd4/MCAP, brd4/HUNK, and brd3/orfX (which bind to acetylated histones in transcriptionally active chromatin regions), may be related to its role as a transcriptional repressor and activator, but such a link is currently still tenuous.

Among the other latent viral proteins is vcyc, which, although a homolog of a D-type cyclin, has a more pleiotropic range of functions. The ability of vcyc to promote the transition of the G1/S cell cycle checkpoint is linked to its phosphorylation of pRB (in concert with cdk4/6) and perhaps its inactivation of p27. This property may be important in the context of establishing S-phase conditions required for viral episome replication. However, recent findings clearly indicate that vcyc also contributes to the enhanced proliferation of virus-infected cells.

Translated from the same bicistronic mRNA as vcyc, vFLIP has anti-apoptotic functions and is a potent activator of the NFκB pathway. It is currently thought that the contribution of this protein to maintaining latency may involve protection against apoptosis, induced either by intracellular surveillance mechanisms detecting the replication of an extraneous agent or by cytotoxic T cells directed against latent viral proteins. Another role could be the repression of the lytic cycle via inhibition of the RTA promoter, which has been shown in several rhadinoviruses to be downregulated by the NFκB pathway. In PEL-derived cell lines, continuous activation of the NFκB pathway is required for the survival of these latently infected cells, and in endothelial cells NFκB activation may contribute to the formation of spindle cells, one of the histological hallmarks of KS lesions.

The fourth strictly latent protein is vIRF-3 (or LANA-2), which is expressed in a strictly B-cell-specific manner by K10.5 (**Figure 1**). Its contribution to latency is incompletely understood, but may involve its ability to interact with p53.

Unlike the latency proteins described above, expression of the transcript encoding kaposin A and B increases after initiation of the lytic replication cycle. It is currently not clear whether these two proteins play a role during latent persistence or whether they contribute to early stages of the lytic cycle. Both proteins influence intracellular signaling pathways, kaposin A by associating with cytohesin-1, a guanine nucleotide exchange factor for ARF GTPases, and regulator of integrin signaling that triggers the Erk/MAPK signaling cascade, and kaposin B by binding to MK2 and increasing the stability of cellular cytokine mRNAs that are regulated by the p38 MAPK pathway. Likewise, the roles of the KSHV miRNAs during latency, if any, are not known precisely, but may involve regulation of cellular and viral mRNAs.

Lytic Replication

Lytic replication results in the production of new KSHV virions. The central regulator of the lytic replication cycle is RTA, which is encoded by ORF50. Overexpression of RTA in latently infected cells is sufficient to trigger reactivation from latency. However, in many experimental systems additional inhibition of histone deacetylases will augment this process. As in other herpesviruses, lytic replication involves a 'rolling circle' mechanism to synthesize new viral DNA concatamers, from which individual linear viral genomes are cleaved during packaging into newly assembled capsids. This cleavage occurs in the TR region, thus ensuring the packaging of the complete long unique region of the genome flanked by a variable number (within limits) of copies of TR. Synthesis of new viral DNA occurs in nuclear replication compartments in the vicinity of ND10/POD domains, which undergo morphological changes during this process. As in other herpesviruses, the exact role of ND10/POD domains in productive herpesviral replication is not fully understood. KSHV proteins recruited to lytic replication compartments include the single-stranded DNA-binding protein (encoded by ORF6), the polymerase processivity factor (ORF59), the DNA polymerase (ORF9), the primase-associated factor (ORF40/41), the primase (ORF56), and the DNA helicase (ORF44), which together mediate viral DNA replication. Other viral proteins are involved in aspects of this process. Thus RAP (K-bZIP; encoded by K8), a structural but not functional homolog of EBV ZTA, is also associated with viral replication compartments and blocks the cell cycle at the G1/S transition by increasing expression levels of p21 and C/EBPα as well as by associating with and downmodulating the activity of cyclin-dependent kinase 2 (cdk-2). The ability to block the cell cycle at G1/S appears to be an important early step in the lytic cycle for most herpesviruses, presumably because it facilitates diversion of the nucleotide pools from cellular to viral DNA replication.

Immune Evasion

To establish an infection in higher vertebrates, viruses must interfere with the host cell immune system. Mechanisms of immune evasion are especially important for viruses that establish long-term infections (e.g., herpesviruses). Like other herpesviruses, KSHV employs several mechanisms to protect infected cells from attacks by the immune system.

Latency represents one strategy to escape from the immune system. During latency only a minimal number of viral proteins are expressed, thus reducing the number of antigens that are presented to the immune system. Another strategy favored by herpesviruses is to downregulate MHC I proteins on the surface of infected cells, thereby decreasing antigen presentation and the recognition of infected cells by cytotoxic T lymphocytes.

KSHV encodes two proteins, modulator of immune recognition 1 and 2 (MIR1 and MIR2; encoded by K3 and K5), which are involved in protecting virus-infected cells against natural killer (NK) cells or cytotoxic T lymphocytes. They are both expressed immediately after viral reactivation. While K3 efficiently modulates multiple MHC alleles, K5 affects the expression not only of HLA-A, but also of ICAM-1 and B7.2. These membrane proteins downregulate MHC I molecules by increasing their endocytosis and degradation rate. This mechanism differs from that of other known viral inhibitors of MHC I expression, which either interfere with the synthesis of MHC I chains or retain them in the endoplasmic reticulum. MIR1 and MIR2 are members of the family of enzyme type 3 (E3) ubiquitin ligases, which regulate the last step of ubiquitination, and thereby increase the degradation rate of MHC I proteins.

The complement system represents a potentially important antiviral mechanism. KSHV encodes a lytic protein called KSHV complement-control protein (KCP; encoded by ORF4), which is incorporated into virions and inhibits activation of the complement cascade, thereby protecting virions and virus-infected cells from complement-mediated opsonisation or lysis.

Kaposin B, noted above for its ability to modulate the p38 MAPK pathway through its interaction with MK2, has an immunomodulatory function. It is able to increase the expression of certain cytokines by stabilizing their mRNAs. The turnover of some cytokine transcripts is regulated via AU-rich elements (AREs) present in the 3′ nontranslated region, which target the RNAs for degradation. The p38 pathway, via MK2, regulates this turnover. By binding to MK2, kaposin B inhibits the decay of cytokine mRNAs containing AREs. This leads to the enhanced production of certain cytokines, such as IL6 and granulocyte macrophage colony-stimulating factor (GM-CSF).

The viral homolog of IL6, vIL6, binds directly to gp130, the signal-transducing chain of the cellular IL6 receptor and a variety of other cellular cytokine receptors. Unlike cellular IL6, which requires binding to the IL6 receptor α-chain specifically to trigger only the α-chain/gp130 receptor complex, vIL6 has a much broader range of cellular targets and more pleiotropic effects. It is a potent stimulant for B-cell proliferation and also acts on other bone-marrow-derived cells.

Finally, three other KSHV genes show significant homologies to cellular chemokines. Their encoded proteins (vMIP1, vMIP2, and vMIP3) modulate the immune system. They are able to bind to chemokine receptors, act as chemo-attractants, and presumably affect the composition of leukocyte infiltrates around KSHV-infected cells. vMIP1 and vMIP2 have also been shown to have angiogenic properties and may therefore contribute to angiogenesis in KS lesions.

Inhibition of Apoptosis

Apoptosis is necessary for the elimination of cells that are no longer required or have become damaged. Many viruses modulate apoptotic pathways to prevent the premature death of an infected cell, which must be avoided in order to complete the replication cycle.

In PEL cells, pharmacological inhibition of the NFκB pathway leads to apoptosis, suggesting that NFκB, which is activated by several viral proteins, plays an important role in preventing apoptosis in KSHV-infected cells. Several viral proteins have been shown to have an anti-apoptotic function, including vFLIP, which blocks Fas- and Fas-ligand-induced apoptosis at the level of procaspase 8 activation and which is also a potent inducer of NFκB (see above).

Among the KSHV vIRFs, vIRF-1 and vIRF-2 both inhibit interferon signaling and subsequently prevent induction of apoptosis. Both vIRF-1 and vIRF-3 inhibit the activation of p53-dependent promoters.

The product of K7, which is the viral inhibitor of apoptosis (vIAP), is thought to operate as a molecular adaptor, bringing together Bcl-2 and effector caspases. The viral homolog of human Bcl-2, vBcl-2 (encoded by ORF16) is expressed only in late-stage KS lesions. It inhibits virus-induced apoptosis and the pro-death protein BAX.

KSHV-Associated Diseases

KSHV is responsible for KS, a tumor of endothelial origin, as well as for PEL, a rare B-cell lymphoma. It is also found in the plasma cell variant of MCD and thought to play an essential role in the pathogenesis of this B-cell-derived lymphoproliferative disorder. Additional sightings of KSHV in a variety of other proliferative or neoplastic disorders have not been confirmed. However, KSHV may play a role in the occasional case of hemophagocytic syndrome and bone marrow failure in immunosuppressed individuals.

Kaposi's Sarcoma

With very few exceptions, KSHV sequences have been detected by PCR in all KS cases. The few exceptions are held to reflect technical problems or the variable load of KSHV DNA in this pleomorphic tumor, which, in addition to KSHV-infected spindle and endothelial cells, contains many other (uninfected) cellular lineages. KSHV adopts a latent gene-expression pattern in the majority of infected endothelial or spindle cells. As shown by immunohistochemistry and *in situ* hybridization studies, the latent genes discussed above (i.e., LANA-1, vcyc, vFLIP, and kaposin) are expressed in the majority of cells, whereas immediate-early, early, or late genes or proteins are expressed only in a small percentage. Most KS tumors also show a uniform pattern of TR lengths in the circular viral episome, consistent with viral latency. This could suggest that viral latent proteins play a key role in the atypical differentiation of endothelial cells into spindle cells, shown by gene expression array studies to represent an intermediate endothelial cell differentiation stage between vascular and lymphatic endothelial cells. However, it is conceivable that certain viral proteins of the lytic cycle, although apparently only expressed in a few cells, could also contribute to pathogenesis by either exerting a paracrine effect (if secreted, as in the case of vIL6 or vMIPs) or inducing the expression of cellular growth factors that could then act on neighboring latently infected cells (e.g., the viral chemokine receptor homolog encoded by ORF74 induces the secretion of VEGF). Given the inefficient persistence of KSHV in latently infected endothelial cell cultures *in vitro*, it is also conceivable that periodic lytic reactivation of KSHV is required to infect new endothelial cells and thereby maintain KSHV in this population. On the other hand, *in vitro* experiments suggest that, in spite of being lost rapidly from most infected dividing cells in culture, a small percentage of such cells can remain stably infected, possibly as a result of epigenetic modifications of the viral genome. It is therefore equally conceivable that the latently infected spindle cells in KS tumors are derived from cells that harbor such epigenetically modified latent viral genomes.

Primary Effusion Lymphoma

The presence of KSHV DNA is a defining feature of PEL, a rare lymphoma entity characterized by a lymphomatous effusion in the pleural or abdominal cavity containing malignant B cells that have a rearranged immunoglobulin gene and often express syndecan-1/CD138 and MUM1/IRF4, two markers characteristic for postgerminal center B cells. The majority of PEL samples are dually infected with KSHV and EBV, but 'KSHV-only' cases also occur. KSHV adopts a latent gene-expression pattern in the majority of PEL cells and in permanent cell lines derived from PEL samples. In addition to the latent proteins found in KS (LANA-1, vcyc, vFLIP; see above and **Figure 2**), vIRF3 is expressed in latently infected PEL cells, indicating a B-cell-specific pattern of latent gene expression. A minority of PEL cells, or PEL-derived cell lines, can spontaneously undergo productive viral replication, or be induced to do so by treatment with phorbol esters, histone deacetylase inhibitors such as sodium butyrate, or overexpression of RTA, the central regulator of the productive viral replication cycle. Such PEL-derived cell lines have for a long time been a source of virus for *in vitro* experiments.

Multicentric Castleman's Disease

KSHV is frequently found in the B-cell population of the plasma cell variant of MCD. As in KS and PEL, the majority of virus-infected cells are in a latent stage of infection, expressing characteristic latency proteins such as LANA-1, but also the B-cell-specific latent protein vIRF3. However, unlike KS and PEL, in which only a very small proportion of virus-infected cells appear to undergo productive viral replication at any point in time (see above), a sizable number of KSHV-infected B cells in MCD spontaneously express viral proteins that are thought to reflect productive viral replication. In addition, case reports suggest that MCD activity is reflected by the level of peripheral blood viral load, indicating that MCD may be the manifestation of, or be accompanied by, productive viral reactivation in B cells. Among the 'lytic' viral proteins expressed in MCD B cells is vIL6, a potent stimulator of B-cell growth (see above).

See also: Herpesviruses: General Features; Kaposi's Sarcoma-Associated Herpesvirus: General Features.

Further Reading

Ablashi DV, Chatlynne LG, Whitman JE, Jr., and Cesarman E (2002) Spectrum of Kaposi's sarcoma-associated herpesvirus, or human herpesvirus 8, diseases. *Clinical Microbiology Reviews* 15: 439–464.

Brinkmann MM and Schulz TF (2006) Regulation of intracellular signalling by the terminal membrane proteins of members of the *Gammaherpesvirinae*. *Journal of General Virology* 87: 1047–1074.

Damania B (2004) Oncogenic gamma-herpesviruses: Comparison of viral proteins involved in tumorigenesis. *Nature Reviews Microbiology* 2: 656–668.

Dourmishev LA, Dourmishev AL, Palmeri D, Schwartz RA, and Lukac DM (2003) Molecular genetics of Kaposi's sarcoma-associated herpesvirus (human herpesvirus-8) epidemiology and pathogenesis. *Microbiology and Molecular Biology Reviews* 67: 175–212.

Edelman DC (2005) Human herpesvirus 8 – A novel human pathogen. *Virology Journal* 2: 78.

Kempf W and Adams V (1996) Viruses in the pathogenesis of Kaposi's sarcoma – A review. *Biochemical and Molecular Medicine* 58: 1–12.

Moore PS and Chang Y (2001) Kaposi's sarcoma-associated herpesvirus. In: Knipe DM and Howley PM (eds.) *Fields Virology*, 4th edn., pp. 2803–2833. Philadelphia, PA: Lippincott Williams and Wilkins.

Rezaee SA, Cunningham C, Davison AJ, and Blackbourn DJ (2006) Kaposi's sarcoma-associated herpesvirus immune modulation: An overview. *Journal of General Virology* 87: 1781–1804.

Schulz TF (2006) The pleiotropic effects of Kaposi's sarcoma herpesvirus. *Journal of Pathology* 208: 187–198.

L

Lassa, Junin, Machupo and Guanarito Viruses

J B McCormick, University of Texas, School of Public Health, Brownsville, TX, USA

© 2008 Elsevier Ltd. All rights reserved.

This article is reproduced from the previous edition, volume 2, pp 887–897, © 1999, Elsevier Ltd., with an update by the Editor.

History

Lymphocytic choriomeningitis (LCM) virus, the first identified arenavirus, was isolated by Lillie and Armstrong in 1933 from the cerebrospinal fluid of a patient suspected of having St Louis encephalitis. The virus was again isolated in 1935 from patients with aseptic meningitis, and finally by Traub in 1935 from laboratory mice. More than 20 years passed before Junin virus, the next member of this taxon to be identified, was isolated from patients with Argentine hemorrhagic fever in 1957. Machupo virus from patients with Bolivian hemorrhagic fever was similarly identified in 1964. A third arenavirus from South America which is pathogenic for humans is Guanarito virus, isolated from patients in Venezuela in 1991. Yet another pathogenic arenavirus, Sabia virus, this time from Brazil, was isolated from a fatally ill individual. Several other arenaviruses have also recently been isolated from rodent species in South America, but none of these have yet been associated with human illness. Lassa virus was the first arenavirus isolated from Africa, in 1969, and remains the only arenavirus pathogenic for humans from Africa, although several other arenaviruses have also been identified from Africa. In total, 23 arenaviruses have been identified worldwide, but only six are known to be pathogenic for humans.

Classification

The 23 members in the *Arenavirus* genus of the *Arenaviridae* have traditionally been placed in categories of Old and New World arenaviruses (**Table 1**), based on geographic locations. More recently, Tacaribe complex viruses (in the Western hemisphere) have been more completely placed in phylogenetic relationship with each other based on a sequence of about 600+nucleotides in the nucleoprotein. The present information suggests three related groups: lineage A contains Pichinde, Parana, Flexal and Tamiami, Allpahuayo, Pirital, Whitewater Arroyo, and Bear Canyou viruses; lineage B contains Junin, Machupo, Amapari, Guanarito, Sabia, Cupixi, Chapare, and Tacaribe viruses; lineage C contains Latino and Oliveros viruses. More recently still, further genetic analysis has led to the suggestion that Pichinde and Oliveros viruses are most closely associated with Old World arenaviruses. The likelihood that arenaviruses and their rodent hosts have coevolved was suggested sometime ago, and now seems a possible hypothesis to begin to test through the parallel use of rodent genetics, alongside the understanding of the genetic relationships between arenaviruses.

The six arenaviruses presently known to be pathogenic for humans (LCM, Lassa, Junin, Machupo, Guanarito, and Sabia viruses) will be the primary subjects of this article.

Properties of the Virion

The arenaviruses are enveloped, pleomorphic, membrane viruses ranging in diameter from 50 to 300 nm, with a mean diameter of 110–130 nm. The virion density in sucrose is $1.17\,g\,ml^{-1}$. They contain two segments of single-stranded RNA tightly associated with a nucleocapsid protein. This is enclosed in a membrane consisting of two glycosylated proteins (or in some cases a single glycosylated protein). The genome consists of two segments of single-stranded RNA both containing two genes encoded in an ambisense structure. The small RNA segments encode for the glycoprotein precursor (GPC) and for the nucleoprotein (NP). The NP and GPC genes are encoded in nonoverlapping reading frames with origins at the $3'$ and $5'$ ends of the molecule, respectively. The N gene is encoded by the $5'$ half of the viral complementary RNA sequence corresponding to the $3'$ half of the viral RNA molecule. The GPC gene is encoded by the $5'$ half of the viral RNA

Table 1 Arenaviruses: basic biological information

Virus	Human disease	Geographic distribution
LCM	Choriomeningitis	Europe, Asia, Western Hemisphere
Lassa	Lassa fever	West Africa
Mopeia	Human infection, no disease known	Southern Africa/Mozambique, Zimbabwe, Rep. of South Africa
Mobala	Human infection, no known disease	Central African Republic
Ippy	Unknown	Central African Republic
Lineage B[a]		
Cupixi	None	Brazil
Chapare	One human hemorrhagic fever	Bolivia
Junin	Argentine hemorrhagic fever	Argentina
Machupo	Bolivian hemorrhagic fever	Bolivia
Guanarito	Venezuelan hemorrhagic fever	Venezuela
Sabia	Hemorrhagic fever	Brazil
Amapari	None	Brazil
Tacaribe	None	Trinidad
Lineage A[a]		
Parana	None	Paraguay
Tamiami	None	Florida
Pichinde	None	Colombia
Flexal	None	Brazil
Pirital	None known	Venezuela
Whitewater Arroyo	None known	North America
Allpahuayo	None	Peru
Bear Canyou	None	California
Lineage C[a]		
Latino	None	Bolivia
Oliveros	None known	Argentina

[a]In current phylogenetic scheme suggested for Tacaribe complex viruses.

molecule. Similarly, the large strand of RNA codes for the RNA-dependent viral RNA polymerase and a smaller ring-finger protein involved in replication. The arenaviruses are virtually indistinguishable from each other morphologically, and all share the characteristic granules noted in electron micrographs. These granules appear to be the results of the binding of the zinc-binding ring-finger protein of arenaviruses with the nuclear fraction of ribosomal proportion. The *nuclear ribosomal* protein (PO) appears in the virion, while other ribosomal proteins do not. This suggests that the granules, believed to be nonspecific inclusion of ribosomes into the virion, may rather be a specific process related to virion replication and assembly. These granules give rise to the family name *Arenaviridae*, derived from *arenos*, the Latin word for sand, based on their electron micrographic appearance as grains of sand.

Geographic and Seasonal Distribution

Junin, Machupo, and Guanarito viruses occur in Argentina, Bolivia, and Venezuela, respectively. Junin and Guanarito are endemic in their respective areas. Junin virus primarily infects workers during the corn harvesting season, by disturbance of the rodent host, *Calomys callosus*, which lives in the corn fields. Guanarito virus infection is endemic, and its epidemiology may resemble that of Lassa virus in Africa, which occurs throughout the year; however, insufficient data presently exist to confirm this impression. Machupo virus occurred in epidemic fashion in the 1960s in a circumscribed area of Bolivia. It was associated with the transient marked increase in the population of *Calomys* rodents, which are normally field rodents but because of overpopulation moved into human dwellings in search of food. Elimination of the rodents in the towns stopped the epidemic, and few further cases have since been reported (though a few cases were reported in 1996). Chapare virus has been isolated from a human fatal hemorrhagic fever patient in Bolivia in 2004. No reservoir has been identified and nothing is known about its epidemiology or geographic distribution. Lassa virus infection is endemic in West Africa from Senegal to Cameroon and perhaps other areas not yet explored. There are increases in Lassa infection during the dry seasons, perhaps because of increased virus stability in lower humidity, but other, as yet unknown, factors may also be involved.

Host Range and Virus Propagation

All of the New World arenaviruses have rodent reservoir hosts with the exception of Tacaribe virus, which was isolated from bats in Trinidad (**Table 1**). A hallmark of the arenaviruses is their intimate biological relationship with rodents, resulting in lifetime infection and chronic virus excretion. Many arenaviruses have more than one rodent host, although usually a single species will predominate as the reservoir in nature.

The hosts of LCM virus (LCMV) have included *Mus* species and hamsters. Guinea pigs are also capable of transmitting the virus in laboratory settings. Machupo virus often renders its major natural host, *Calomys callosus*, essentially sterile by causing the young to die *in utero*. Machupo virus also induces a hemolytic anemia in its rodent host, with significant splenomegaly, often an important identifier of infected rodents in the field. The major rodent hosts for Junin virus are *Calomys* species. Transmission of Junin virus from rodent to rodent is generally horizontal, and not vertical, and is believed to occur through contaminated saliva and urine. The *Calomys* rodents are affected by the virus, with up to 50% fatality among infected suckling animals, and stunted growth in many others. Both Junin and Machupo viruses induce a humoral immune response when transmitted to their suckling natural rodent hosts, which may have neutralizing antibody in the face of persistent infection. Guanarito virus has been isolated from *Zygodontomys brevicauda,* though its detailed biology remains to be learned.

The only known reservoir of Lassa virus in West Africa is *Mastomys natalensis,* one of the most commonly occurring rodents in Africa. At least two species of *Mastomys* (diploid types with 32 and 38 chromosomes) inhabit West Africa, and both have been found to harbor the virus. All species are equally susceptible to silent persistent infection, as seen when LCMV infects mice. This presumably occurs as in LCM infection, from a selective deletion of the thymic T cell response to the virus. All of the arenaviruses pathogenic for humans will also infect and produce illness in a wide range of primates. However, it is not known whether such infections occur in nature, as is known for Ebola virus for example. In addition, human infection plays no biological role in the life cycle and ecology of the arenaviruses.

Virus Propagation

The original isolation of LCMV was made in suckling mice, which have been important in isolation and characterization of several of the arenaviruses. The arenaviruses are, however, easily cultivated in a wide variety of mammalian cell monolayers. The Vero E6 remains the cell of choice for primary isolation and cultivation, but arenaviruses also replicate in baby hamster kidney cells, as well as in a number of specialized cells such as continuous macrophage lines, endothelial cells, fibroblasts and a variety of mouse cell lines, with specific MHC markers used as targets for immunological studies. The infected cells may produce a cytopathic effect (CPE) beginning on days 4–7 of incubation. However, not all arenaviruses produce CPE, especially on primary isolation. For diagnosis, cells may be harvested after 48–72 h and assayed for antigen by immunofluorescent antibody (IFA) or ELISA. Virus plaquing techniques may also be used for the arenaviruses.

Genetics

While advances have been made in determining the genetic relationship between different arenaviruses (see Classification), the level of genetic variability within species of arenaviruses is not well characterized, though it must be added that little genetic data on field isolates exist on which to make this judgment. It would appear that the frequency of variability at the amino acid level, as judged by B cell epitope variability, is not high. Thus the variability in B cell epitopes among Lassa viruses isolated from humans or rodents over a 10-year period in a circumscribed area was not substantial, suggesting that B cell epitopes are under limited immune pressure in their rodent hosts. Knowledge of the epitopes recognized by T cells may be crucial for the development of recombinant Lassa virus vaccines. A human T-helper cell epitope, highly conserved between Old and New World arenaviruses, as well as HLA-A2-restricted CD8 T cell protective Lassa virus epitope have been described. The South American arenaviruses may be under both B and T cell immune pressure in their rodent hosts, though no data are available on this issue. Reassortment, demonstrated only in the laboratory, may also be a means of genetic variability, but its occurrence in nature and therefore its importance is unknown.

Evolution

The *Arenaviridae* are distributed over five continents and can be divided into three 'coevolutionary' groups: Lassa complex in Africa, LCM in North America, Europe, and South America, and the Tacaribe complex in South America. Today's arenaviruses probably descended from an ancestral virus which subsequently differentiated in parallel with the evolution of the Cricetid rodents persistently infected by arena viruses. It seems likely, therefore, that the present distribution and evolution of these viruses are directly related to the distribution and evolution of the earliest Cricetid rodents and their descendants, which now make up the natural hosts of most of this family of viruses (Tacaribe virus has a bat host). The coevolution of these viruses will undoubtedly continue and depend primarily on mutations, selected by the persistently infected host's immune pressure, and perhaps on reassortment in the rodent host.

Epidemiology

The fundamental determinant of the ecology of hemorrhagic fevers is the occurrence of persistent virus infection in rodents. Who becomes infected and what are the functions of the behavior of the persistently infected rodent, and the cultural and occupational patterns of human populations. The arenavirus hemorrhagic fevers are primarily rural and semirural diseases; however, some evidence exists that under certain conditions (poverty, overcrowding), their rodent reservoirs may also establish urban habits. The rodents infect the environment via urine, feces and saliva. Interactions between rodents and humans are peridomestic or in agricultural areas. However, details of the rodent population dynamics, behavior, and the natural history of the persistent virus infection in the feral rodent hosts are only poorly understood. The Old World arenaviruses, LCM and Lassa, produce persistent infection in their rodent hosts without significant detrimental effects. The South American viruses, in contrast, may cause illness and death in newborn rodents, or may induce persistence. The modes of transmission from rodents to humans are not precisely known. Direct contact by humans, with cuts and scratches on hands and feet, with articles and surfaces contaminated by virus may be a more important and consistent mode of transmission. Transmission through mucosal surfaces may also occur under some circumstances.

Lassa Fever

Lassa fever occurs in West Africa, but with a wide geographic area from Northern Nigeria to Guinea, encompassing perhaps 100 million population. At least two species of *Mastomys* occupy West Africa, and both have been found to harbor the virus. These rodents, especially the species with 32 chromosomes, are highly commensal with humans. The movement of *Mastomys* within a village is very limited, and their average lifespan is about 6 months, with little seasonal fluctuation in their breeding pattern. From 5% to as many as 70% have been found to be infected with virus in some village houses. Therefore, most virus transmission takes place in and around the homes. All age groups and both sexes are affected and antibody prevalence increases with age. Risk factors for human infection include contact with rodents, direct contact with ill persons infected with Lassa virus, presence of a large household rodent population, and human practices such as catching, cooking and eating rodents, and indiscriminant storage of food. In many endemic areas, Lassa fever is a common cause of hospitalization. The death rate in systematically studied, untreated hospitalized patients with Lassa fever is 16%, very similar to that described for Junin and Machupo infections. Nosocomial transmission occurs in Africa from contact with infected patients, or through improper use or sterilization of needles, sometimes leading to an outbreak. Person-to-person spread of Lassa virus in households is common, a unique characteristic of Lassa virus in relation to other arenaviruses (no data are yet available on Guanarito virus from Venezuela, or Sabia virus from Brazil), and is usually associated with direct contact or care of someone with a febrile illness, or possibly with sexual contact with the spouse during the incubation or convalescent phases of illness. Illness to infection ratios are 10–25% in some endemic areas of West Africa. Antibody prevalence ranges from less than 1% to over 40% in some villages.

Argentine Hemorrhagic Fever (AHF)

AHF, caused by Junin virus, was recognized in the 1950s in the northwestern part of the Buenos Aires Province in Argentina, an area of very fertile farmland and therefore of great economic importance. The total number of cases reported over a 30 year period is about 21 000. AHF is a seasonal disease with peak yearly incidence in May. The average number of cases from 1981 to 1986 was 360 per year. Although all ages and sexes are susceptible, nevertheless the major group affected is the male working population, explained by the habits of the rodent hosts for Junin virus, *Calomys musculinus* and *Calomys laucha*. These animals are not peridomestic, but rather occupy grain fields, and this is the major reason for the affected population of field workers. Infection also occurs infrequently in other rodents: *Mus musculus*, *Akodon azarae*, and *Oryzomys flavecens*. Transmission from rodent to rodent is horizontal, not vertical, and is thought to occur via contaminated saliva and urine. It is believed that the major routes of virus transmission from humans is through contact with virus-infected dust and grain products and subsequent infection through cuts and abrasions on the skin, or through airborne dust generated primarily by killing and scattering of rodents during mechanized farming. The disease has spread over the 30 years or so since its recognition from an area of 16 000 km^2 and a quarter of a million persons to an area greater than 120 000 km^2 containing a population of over 1 million persons. Furthermore, the incidence in the older affected areas appears to wane after 5–10 years. Overall antibody prevalence is 12%, with a typical predominance in agricultural workers. One-third of the seropositive individuals have no history of typical illness, suggesting that the case: infection ratio is about 2:3. The incidence of AHF is now very low due to the extensive vaccination campaign in the endemic area.

Bolivian Hemorrhagic Fever (BHF)

The only known reservoir for the virus is *Callomys callosus*, a Cricetid rodent that is found in the highest density at the borders of tropical grassland and forest. The distribution of this rodent appears to include the eastern Bolivian plains, as well as northern Paraguay and adjacent areas of western Brazil. The disease was recognized in 1959, and by 1962 more than 1000 cases had been identified in a confined area of two provinces, with a 22% case fatality ratio. The largest known epidemic of BHF, involving several hundred cases, occurred in the town of San Joaquin in 1963 and 1964. This outbreak occurred because of a marked increase in the *Callomys* population, and the subsequent invasion of homes in the town by these rodents. Although the *Callomys* appears to be capable of living both in the areas surrounding the towns and in the towns themselves (where the most efficient transmission of virus would appear to occur), they favor a nonperidomestic habitat where contact with humans is much reduced. It appears that the situation in San Joaquin was unusual, and such an event has not been observed again after nearly 25 years. There has not appeared to be any increase in the geographic areas affected by BHF recently, and few cases have been reported.

Venezuelan Hemorrhagic Fever

The epidemiologic pattern of this disease has not yet been well characterized. In one survey the antibody prevalence in the population in the endemic area in central Venezuela was 2.6%. Guanarito virus was isolated from the cane rat *Zygodontomys brevicauda*. The occurrence of person-to-person transmission has not yet been demonstrated. The low frequency of infection in family contacts and lack of disease in hospital workers caring for patients suggest that person-to-person spread is uncommon. The pattern of infection includes all ages and sexes, suggesting that transmission occurs in and around houses, similar to Lassa fever and BHF and unlike AHF.

Transmission and Tissue Tropism

The primary mode of arenavirus transmission is from human contact with rodent urine or blood. This probably occurs primarily when individuals come into contact with surfaces or materials recently contaminated by rodent urine, or when they trap a rodent and handle the carcass. Some evidence suggests that rodent blood or urine might be aerosolized by machinery during mechanized harvesting of corn in Argentina, with consequent transmission of Junin virus to people working near the machinery; however, no detailed studies of specific risk factors exist. In Africa, the rodents are highly commensal and probably deposit virus-laden urine in many areas of the houses they inhabit. The people tend to walk barefooted, and, based on the epidemiologic pattern of somewhat sporadic infection in households, it seems likely that contact with infectious urine is the primary source of contamination. In addition, in parts of Africa people catch and eat rodents, putting them in direct contact with rodent tissue, blood and secretions during the preparation. Finally, for some of the viruses, particularly Lassa virus, there is person-to-person transmission primarily through contact with the blood or secretions of an infected, ill person. Whatever the mode of transmission, it would seem that the reticuloendothelial system is probably a primary target of replication of the arenaviruses, though they replicate well in many organs, including liver, adrenal gland, placenta, lung and many other organs.

Pathogenesis

The most common sites of initiation of human infection by the arenaviruses are not yet known, although they seem likely to be cuts and abrasions in the skin. Following the initiation of infection, all arenaviruses progress to generalized multiorgan infections, especially of the reticuloendothelial system. Thereafter, however, the pathogenesis of the different infections is variable. Cellular receptors for both New and Old world arenaviruses have recently been described.

Lassa fever

The 1–3 week incubation that follows infection suggests an unknown primary replication site, probably within the reticuloendothelial system. Route and titer of infecting dose may be important determinants of outcome, as may the virus strain. For example, death rates in nosocomial outbreaks where parenteral exposure is substantial are usually higher than community-acquired infections.

The degree of organ damage in fatal human infections is mild, which is sharply at variance with the clinical course and collapse of the patient. Indeed, there are few clues to the pathogenesis of Lassa fever in standard pathological studies. Liver damage is variable, with concomitant cellular injury, necrosis and regeneration. Nevertheless, serum aspartate aminotransferase (AST) levels over 150 iu ml^{-1} are correlated with poor outcome, and an ever-increasing level is also associated with increased risk of death. Alanine aminotransferase (ALT) is only marginally raised, and the ratio of AST:ALT in natural infections and in experimentally infected primates is as high as 11:1. Furthermore, prothrombin times, glucose and bilirubin levels are near normal, excluding biochemical hepatic failure. An increasing Lassa viremia is also associated with increasing case fatality. In addition to the liver, high virus titers occur in many other organs without significant pathologic or functional lesions, perhaps reflecting blood rather than parenchymal levels of virus.

Some patients develop severe pulmonary edema and adult respiratory distress syndrome, gross head and neck edema, pharyngeal stridor and hypovolemic shock. This pattern is consistent with edema due to capillary leakage, rather than cardiac failure and impaired venous return. Endothelial cell dysfunction has been demonstrated in primates experimentally infected by Lassa fever, in that there is apparently a marked decrease in prostacyclin production by endothelial cells. Loss of integrity of the capillary bed presumably causes the leakage of fluids and macromolecules into the extravascular spaces and the subsequent hemoconcentration, hypoalbuminemia, and hypovolemic shock. Proteinuria is common, occurring in two-thirds of patients.

Edema and bleeding may occur together or independently. Since there is a minimal disturbance of the intrinsic, and almost none of the extrinsic, coagulation system, and there is no increase in fibrinogen breakdown products, disseminated intravascular coagulation (DIC) is excluded. Furthermore, platelet and fibrinogen turnover in experimental primate infections are normal. Though platelet numbers are only moderately depressed, in severe disease platelet aggregation is almost completely abolished by a circulating inhibitor. The origin of this inhibitor is not known; however, it cannot be reproduced with viral material nor can it be blocked by antibodies to Lassa virus. In the platelet, it blocks dense granule and ATP release and thus abolishes the secondary wave of *in vitro* aggregation, while sparing the arachidonic acid metabolite-dependent primary wave. The inhibitor of platelet function also interferes with the generation of the FMLP-induced superoxide generation in polymorphonuclear leukocytes probably through a similar mechanism.

AHF and BHF

Despite the different degrees of bleeding, there are sufficient similarities between the course of disease in AHF, BHF, and Lassa fever to speculate that there exists a similar pathophysiologic pathway underlying all of the diseases. Organ function, other than the endothelial system, appears to remain intact, and the critical period of shock is brief, lasting only 24–48 h. Hepatitis is mild, and renal function is also well maintained. Bleeding is more pronounced with AHF and BHF than Lassa fever, but it is not the cause of shock and death. Capillary leakage is significant, with loss of protein and intravascular volume being much more pronounced than loss of red cells. Proteinuria is significant, and dehydration with hemoconcentration appears to be an important process. The shock is not associated with evidence of disseminated intravascular coagulation, and even though there are petechiae suggesting some direct endothelial damage, no clear evidence of virus replication in, and damage of, endothelium has been demonstrated. Thus, clinical observations suggest that vascular endothelial dysfunction may be the basis of subsequent circulatory failure in AHF and BHF. Persistent hypovolemic shock in the face of intravascular volume expanders suggests that it is due to the loss of endothelial function and leakage of fluid into extravascular spaces. This is supported by the tissue edema frequently observed, and more directly by the pulmonary edema which may result from vigorous fluid therapy of the shock. These events lead to irreversible shock and death in the most severely ill patients. Two other observations have been made in AHF: high levels of interferon in severely/fatally ill patients, and a decrease in complement. These are general phenomena observed in other severe infectious processes and are consistent with the events described above. Neither would dictate a substantially different pathophysiologic explanation of these diseases.

Clinical Features

Lassa Fever

Lassa fever begins after 7–18 days of incubation, with fever, headache, and malaise. Aching in the large joints, pain in the lower back, a nonproductive cough, severe headache, and sore throat are common. Many patients also develop severe retrosternal or epigastric pain. Vomiting and diarrhea occurs in 50–70% of patients. In more severely ill patients, complete prostration may occur by the 6th to 8th day of illness. Patients with Lassa fever appear toxic and anxious, and in the absence of shock, the skin is usually moist from diaphoresis. There is an elevated respiratory rate and pulse. The systolic blood pressure may be low. There is no characteristic skin rash; petechiae and ecchymoses are rare, nor is jaundice a feature of Lassa fever. Conjunctivitis is common, but rare conjunctival hemorrhages portend a poor prognosis. Seventy percent of patients have pharyngitis, often exudative, but few if any petechiae, and ulcers are rare. Mucosal bleeding occurs in 15–20% of all patients, and although associated with severe disease it is almost never of a magnitude to produce shock by itself. Edema of the face and neck are commonly seen in severe disease. About 20% of patients have pleural or pericardial rubs heard late in disease at the beginning of convalescence. The abdomen may be diffusely tender. The ECG may be abnormal, primarily with elevated T waves, but without a characteristic pattern. Moderate or severe diffuse encephalopathy with or without general seizures is characteristic in severe disease. Severe Lassa fever can progress rapidly between the 6th and 10th day to *respiratory distress* with stridor due to laryngeal edema, central cyanosis, hypovolemic shock and clinical signs of encephalopathy, sometimes with coma and seizures. Tremors are often seen in the hours just before death. Acute, focal neurological signs rarely occur, with the exception of VIII nerve deafness seen in convalescence. Residual ataxia is common. Lassa fever is also a pediatric disease affecting all ages of children. The disease appears to be even more difficult to diagnose by clinical criteria in children than in adults because its manifestations are so general. In very young babies marked edema may be seen, associated with very severe disease. In older children the disease may manifest as a primary diarrheal disease or as pneumonia or simply as an unexplained prolonged fever. The case fatality in children is 12–14%.

The mean white count in early Lassa fever may be low or normal, with a relative lymphopenia, but late neutrophilia may supervene in severe disease. Proteinuria is very characteristic. A serum AST (SGOT) level of >150 iu ml^{-1} is associated with a case fatality of 50%, and there is a correlation between an ever-increasing level and an increased risk of death. A viremia of $>1 \times 10^3$ TCID$_{50}$ ml^{-1} is associated with an increasing case fatality. Both factors together carry a risk of death of 80%.

Pharyngitis, proteinuria, and retrosternal chest pain have a predictive value for Lassa fever in febrile hospitalized patients of 81% and a specificity of 89%. Likewise, a triad of pharyngitis, retrosternal chest pain, and proteinuria (in a febrile patient) correctly predicted Lassa fever 80% of the time (in an endemic area). Both triads have sensitivities of 50% for detecting cases of Lassa fever. Significant complications of Lassa fever include a two- to threefold increased risk of maternal death from infection in the third trimester, and a fetal/perinatal loss of 84% that does not seem to vary by trimester. Another significant complication is that of acute VIII nerve deafness, with nearly 30% of patients suffering an acute loss of hearing in one or both ears. Other complications which appear to occur much less frequently are uveitis, pericarditis, orchitis, pleural effusion and ascites.

AHF and BHF

These diseases have an insidious onset of malaise, fever, general myalgia, and anorexia. Lumbar pain, epigastric pain, retro-orbital pain, often with photophobia, and constipation occur commonly. Nausea and vomiting frequently occur. Temperature is high, reaching 40 °C or above. Unlike LCM and Lassa fever, AHF and BHF do not usually lead to respiratory symptoms and sore throat. On physical examination, patients appear toxic. Conjunctivitis, erythema of the face, neck, and thorax are prominent. Petechiae may be observed in the axillae by the 4th or 5th day of the illness. There may be a pharyngeal enanthem, but pharyngitis is uncommon. Relative bradycardia is often observed. The disease may begin to subside after 6 days of illness; if not, the second stage of illness supervenes, and most commonly manifests as epistaxis and/or hematemesis or acute neurological disease. The bleeding may be from mucosal surfaces or into the skin, with petechiae and hemorrhagic rash, with preceding increase in packed red cell volume. Pulmonary edema is common in severely ill patients. The appearance of intractable shock is a serious sign which becomes irreversible in some patients, and accounts for the majority of deaths from AHF and BHF. Fifty percent of AHF and BHF patients also have neurologic symptoms during the second stage of illness, such as tremors of the hands and tongue, progressing in some patients to delirium, oculogyrus, and strabismus.

A low white blood cell count (under 1000 mm^{-1}) and a platelet count under 100 000 are invariable. Proteinuria is common and microscopic hematuria also occurs. Alterations in clotting functions are minor, and DIC is not a significant part of the diseases. Liver and renal function tests are only mildly abnormal.

Venezuelan Hemorrhagic Fever

Little information is available on the spectrum of disease caused by Guanarito virus infection. Hospitalized patients with severe diseases are febrile on admission, with prostration, headache, arthralgia, cough, sore throat, nausea/vomiting, diarrhea, and hemorrhage. The bleeding includes epistaxis, bleeding gums, menorrhagia, and melena. On physical examination, patients with severe disease appear toxic, and usually dehydrated. They are described as having one or more of a series of signs, including pharyngitis, conjunctivitis, cervical lymphadenopathy, facial edema, and petechiae. Thrombocytopenia and neutropenia are common at admission. Case fatality of a single group of 15 hospitalized patients was over 60%; however, the single serum survey available suggests that overall mortality:infection ratio is much lower.

Pathology and Histopathology

Lassa Fever

The most frequently and consistently observed microscopic lesions in fatal human Lassa fever are focal necrosis of the liver, adrenal glands, and spleen. The liver damage is variable in the degree of hepatocytic necrosis. The liver demonstrates cellular injury, necrosis, and regeneration, with any or all present at death. A substantial macrophage response occurs, with little if any lymphocytic inflammatory response. Nevertheless, fatal cases do not exhibit sufficient hepatic damage to implicate hepatic failure as a cause of death. Similarly, moderate splenic necrosis is a consistent finding, primarily involving the marginal zone of the periarteriolar lymphocytic sheath. Diffuse focal adrenocortical cellular necrosis has been less frequently observed. Although high virus titers occur in other organs, such as brain, ovary, pancreas, uterus, and placenta, no significant lesions have been found. Thus, few clues to the pathogenesis of Lassa fever are found in standard pathological studies. It is clear that the outcome in Lassa fever is associated with the degree of virus replication; nevertheless the effect of replication is not major tissue destruction, but a more subtle biological effect on vascular endothelium, and perhaps other key cells or organs.

AHF and BHF

AHF and BHF are typical hemorrhagic diseases and have very similar pathologic features, some of which differ from Lassa fever. Patients with AHF manifest a skin rash and petechiae, and gross examination of organs at necropsy also show petechiae on the organ surfaces. Ulcerations of the digestive tract have been described, but the bleeding is not massive. Microscopic examination shows a general alteration in endothelial cells, mild edema of the vascular walls, with capillary swelling, and perivascular hemorrhage. Large areas of intra-alveolar or bronchial hemorrhage are often seen with no evidence of inflammatory process. Pneumonia with necrotizing bronchitis or pulmonary emboli is observed in half of the fatal necropsied cases. Hemorrhage and a lymphocytic infiltrate have been observed in the pericardium, occasionally with interstitial myocarditis. The lymphnodes are enlarged and congested with reticular cell hyperplasia. Splenic hemorrhage is common. Medullary congestion with pericapsular and pelvic hemorrhages are frequently seen. Acute tubular necrosis occurs in about half of the fatal cases, but adrenal necrosis has not been reported.

Venezuelan Hemorrhagic Fever

Only a limited number of post-mortem dissections have been performed. Observations included pulmonary edema with diffuse hemorrhages in the parenchyma, and subpleural, focal hepatic hemorrhages with congestion and yellow discoloration, cardiomegaly with epicardial hemorrhages, splenic and renal swelling, and bleeding into numerous cavities, including stomach, intestines, bladder, and uterus.

Immune Response

There is a brisk B cell response to Lassa virus, with a classic primary IgG and IgM antibody response to virus early in the illness. This event does not, however, coincide with virus clearance, and high viremia and high IgG and IgM titers often coexist in both humans and primates. Indeed, virus may persist in the serum and urine of humans for several months after infection, and possibly longer in occult sites, such as renal tissue. Neutralizing antibodies to Lassa virus are absent in the serum of patients at the beginning of convalescence, and in most people they are never detectable. In a minority of patients some low titer serum neutralizing activity may be observed several months after resolution of the disease. Passive protection with antibody to Lassa virus has been demonstrated in animals given selected antiserum at the time or soon after inoculation with virus, but clinical trials of human plasma given after onset of illness have shown little or no protective effect. Thus, the clearance of Lassa virus in acute infection appears to be less dependent of antibody formation, and presumably depends more on the cell immune response. This is supported by recent experience with experimental Lassa vaccines in primates. In recent studies of LCM infection there is clearly apoptosis of T cells and probably B cells during acute infection in the mouse model. Such a scenario must be entertained for acute Lassa infection, and could help explain the fulminant course in some patients. Such work will need to be performed in primate models or during studies of human disease in endemic areas. Neutralizing and complement fixing antibody to Junin and Machupo are usually detectable 3–4 weeks after the onset of illness. Indirect fluorescent antibodies may be detected at the end of the second week of illness. The efficacy of convalescent plasma has been demonstrated in the therapy of Junin infection (see Therapy). The effectiveness of the plasma has been demonstrated to be associated with the level of Junin virus-neutralizing antibodies. The IFA test is the most commonly performed test for diagnosis by antibody detection for AHF and BHF. Little data are available on Guanarito virus, which appear to induce an antibody response. However, more like AHF and BHF, antibodies to Guanarito virus seem to appear later in illness.

Prevention and Control

Rodent Control

The ideal method of prevention for these rodent-borne diseases is to prevent contact between rodents and humans. The effectiveness of this was shown in the outbreaks of BHF in the 1960s. However, the prospects of rodent control in preventing AHF are not so bright. The human–rodent encounter resulting in AHF occurs during the crop harvests, and with the present technology it is difficult to imagine how control of noncommensal feral rodents could be accomplished. The best choice may be better protection of the agricultural worker from contact with rodent secretions and blood. Similarly, the control of rodents as a broad approach to preventing Lassa fever is not realistic. The improvement of housing and food storage could reduce the domestic rodent population, but such changes are not easily made. Rodent trapping in an individual village where transmission is high has demonstrated as much as a fivefold reduction in the rate of virus transmission. However, such a program is only applicable in villages with exceptional transmission rates, and would certainly not be applicable to large areas.

Vaccines

The live attenuated Junin vaccine has now been shown to be not only effective but also has had a dramatic effect in reducing the number of cases of AHF seen each year in the endemic area. A vaccine against Lassa fever has been made by cloning and expressing the Lassa virus glycoprotein gene into vaccinia virus. This vaccine has proved highly successful in preventing severe disease and death in challenged monkeys. Only the glycoprotein gene is protective in the two species of monkeys thus far tested, and the basis of protection is not neutralizing antibody, which does not develop, but more likely the cell-mediated immune response. Vaccines using the glycoprotein vectored by VSV prevent disease and death in the non-human primates model.

Drug Prophylaxis

In the event of identifiable exposure to Lassa virus (and possibly other pathogenic arenaviruses) in a hospital or laboratory setting, the prophylactic use of orally administered ribavirin is recommended, although efficacy data are missing.

Therapy

Significant advances have been made in the therapy of Junin and Lassa virus infections since the early 1990s. Convalescent plasma is effective in the treatment of Junin virus during the first 8 days of illness, but not Lassa virus. Ribavirin is effective in reducing the viremia and mortality of Lassa fever particularly when given in the first week of illness.

Lassa Fever

Ribavirin can prevent death in Lassa fever when given at any point in the illness, but it is more effective when given early and intravenously. Thus, patients with risk factors for severe disease who were treated within the first 6 days of illness experienced a 5–9% case fatality. Those with the same risk factors, in whom treatment was initiated more than 6 days after the onset of illness, had a case fatality of 20–47%. Regardless, the case fatality was significantly less for ribavirin-treated patients in all categories than for either nontreated patients or for those treated with plasma alone. As such it seems reasonable to assume that it would also be an effective measure in the event of laboratory or hospital exposure to the disease. The pathogenesis of the infection is less reversible later in illness. Furthermore, patients treated with ribavirin had significant declines in viremia regardless of outcome, whereas patients who were untreated or treated with plasma and who died showed no decrease in viremia, consistent with the observation that outcome, and presumably the result of therapy, is closely related to the inhibition of virus replication. Therefore, patients coming late in disease will require more effective clinical management of physiologic dysfunction, and perhaps other drugs which may be used to stabilize the shock state sufficiently long to allow recovery and improve survival (see Pathogenesis).

AHF

A randomized trial of patients with AHF demonstrated that convalescent-phase plasma reduced the mortality from 16% to 1% in the patients who were treated in the first 8 days of illness. The efficacy of the plasma was directly related to the concentration of neutralizing antibodies of the plasma.

The success of this therapy was not without a price, however, which was the development of a late neurological syndrome in about 10% of cases. Thus far, there is no correlation between either the day of therapy or the dose of neutralizing antibodies given and the occurrence of late neurological syndrome. Passive antibody therapy depends on collection of plasma from persons known to have had the disease, testing the plasma (or screening the donor) for antibodies to blood-borne agents such as hepatitis, and proper storage until its use. In addition, the advent of the acquired immunodeficiency syndrome (AIDS) and the other diseases transmissible by blood products means that further screening is required before use.

BHF

Recent experience of successful treatment of two patients with intravenous ribavirin suggests that further efforts to evaluate the effectiveness of ribavirin in this disease will be worthwhile.

Future Perspectives

An important area of future research includes a more through understanding of the viral and host components of the virus-clearing and protective immune response in humans. A second area of substantial interest is the nature of the persistent infection in the rodent host. A more complete understanding of the pathogenesis of Lassa fever may provide insight not only to that disease, but to basic elements of how the host response may be detrimental as well as beneficial to the host. Finally, a more through comprehension of how to control the more widespread arenavirus diseases, either through vaccination, or preferably rodent control, is essential.

See also: Lymphocytic Choriomeningitis Virus: General Features.

Further Reading

Buchmeier MJ, Bowen MD, and Peters CJ (2001) Arenaviridae: The viruses and their replication. In: Knipe DM and Howley PM (eds.) *Fields Virology,* 4th edn., pp. 1635–1668. Philadelphia: Lippincott Williams and Wilkins.

Cajimat MNB and Fulhorst CF (2004) Phylogeny of the Venezuelan arenaviruses. *Virus Research* 102: 199–206.
Clegg JCS (2002) Molecular Phylogeny of the Arenaviruses. *Current Topics in Microbiology and Immunology* 262: 1–24.
Damonte EB and Coto CE (2002) Treatment of arenavirus infections: from basic studies to the challenge of antiviral therapy. *Advances in Virus Research* 58: 125–155.
Enria D, Mills JA, Flick R, *et al.* (2006) Arenavirus infections. In: Guerrant RL, Walker DH, and Weller PF (eds.) *Tropical Infectious Diseases. Principles, Pathogens and Practice*, 2nd edn., pp. 734–755. Philadelphia, PA: Elsevier Churchill Livingstone.
Fisher-Hoch SP and McCormick JB (2004) Lassa fever vaccine. *Expert Review of Vaccines* 3: 189–197.
Gunther S and Lenz O (2004) Lassa virus. *Critical Reviews in Clinical Laboratory Sciences* 41: 339–390.
Kunz S, Borrow P, and Oldstone MBA (2002) Receptor structure, binding, and cell entry of arenaviruses. *Current Topics in Microbiology and Immunology* 262: 111–137.
McCormick JB, Webb PA, Scribner CL, *et al.* (1986) Lassa fever: Effective therapy with ribavirin. *New England Journal of Medicine* 314: 20.
McCormick JB, Webb PA, Krebs JW, Johnson KM, and Smith K (1986) A prospective study of the epidemiology and ecology of Lassa fever. *Journal of Infectious Diseases* 155: 437.
Rojek JM, Spiropoulou CF, and Kunz S (2006) Characterization of the cellular receptors for the South American hemorrhagic fever viruses Junin, Guanarito, and Machupo. *Virology* 349: 476–491.
Salazar-Bravo J, Ruedas LA, and Yates TL (2002) Mammalian Reservoirs of Arenaviruses. *Current Topics in Microbiology and Immunology* 262: 25–63.

Legume Viruses

L Bos, Wageningen University and Research Centre (WUR), Wageningen, The Netherlands

© 2008 Elsevier Ltd. All rights reserved.

Glossary

Ecology Highly complex interaction of organisms with each other and with their physical environment and its study.

Epidemiology (quantitative ecology) Rapid quantitative spread of parasites and pathogens in populations of their host(s) and its study.

Vector Organism or other agent that can spread a virus or any parasite and introduce it to or into other organisms including plants so that attack or infection results.

Virus identification The acts of (1) describing the particular identity or independence of a new virus (i.e., virus characterization) and (2) recognizing a virus as an entity earlier described, classified, and named, with the aim of etiological disease diagnosis.

Introduction

Legume viruses and those of any other type of crop must be dealt with in relation to the crop concerned. Initially, plant viruses drew attention in crops because they were injurious to plants and reduced crop productivity. Legumes were among the first crops studied for virus infections. Also, plants were for long the sole medium to study the viruses and distinguish between them from the view point of the species they infected the symptoms they produce. Most plant viruses still derive their names from crops on which they were first detected and from the symptoms they caused, for example, bean yellow mosaic virus (BYMV). Plants were also used as differential hosts for separating viruses from mixtures by selective passage. Natural hosts affect the evolution of the viruses that infect them. Selected test or indicator plants soon helped for detecting viruses in plants and in vector organisms, and later in fractions during purification. Laboratory studies of viruses for their intrinsic properties made virological interest gradually shift from their effects on plants to molecular biology. But plants remain indispensable for propagating most plant viruses, and basic information on plant viruses must merge in relation to crops such as legumes for elucidating the biological and societal impact of viruses, and showing how to combat them in agricultural practice.

Legumes

Legumes are botanical species in the pod-bearing family Leguminosae or Fabaceae (order Fabales) of which the subfamily Papilionoideae is the largest and economically most important. Root-nodule bacteria (*Rhizobium* spp.) that bind nitrogen from the air make legumes do well under poor growing conditions and improve soil fertility. Legumes are protein-rich sources of food and fodder of high nutritive value to man and animals. The seeds of grain legumes serve as a meat substitute in developing countries. Those of soybean and groundnut contain valuable oil for industrial processing, and they are increasingly produced on large holdings as cash crops for export, as of soybean in Brazil. Green seeds or whole pods and sometimes green leaves or sprouted seeds are also eaten as a vegetable. Some tropical legume species are grown for their edible swollen roots. Clovers are forage or fodder crops, or serve as green manure and cover crops used in crop rotation for maintaining soil fertility. Leguminous tree species are often grown in the tropics for shading and sometimes for timber.

Plants, including leguminous species, keep playing important roles in studying plant viruses for their effects on plants and crops, and also for their separation from natural mixtures, their propagation, and even their detection. Common bean (French bean or bush bean; *Phaseolus vulgaris*) and cowpea (*Vigna unguiculata*) are still used as test plants for detecting several viruses or measuring their infectivity by the local lesions they produce on uniform opposite primary leaves either on plants or detached.

History of Legume Virology

The history of research on legume viruses illustrates how, through trial and error, plant virology has evolved from its middle ages concentrating on their effects on plants and crops into the present high-technology era dominated by molecular biology of the viruses themselves. Legumes were among the first crops studied for virus diseases. Some early examples were bean common mosaic (1917, 1921) (**Figure 1**) and soybean mosaic (1921). Viruses were originally distinguished by their hosts or host ranges and by symptoms, and later also by their ways of spread, as in seed (bean common mosaic potyvirus, BCMV, 1919; **Figure 2**) and by specific insects such as aphids or beetles. Later, further distinction was by the so-called 'physicochemical properties' studied in expressed sap, especially since 1930 when sap transmission was facilitated with abrasives. Viruses of annual legumes, such as peas and beans, were also found in perennial legumes such as clovers, and viruses described from legumes and those from nonlegumes increasingly appeared to cross-infect their respective hosts. For example, beet curly-top curtovirus (BCTV) was found to cause severe disease in beans, and BYMV often infected gladiolus. With information accruing in the 1930s, efforts to devise keys for the recognition of viruses of legumes were first made in 1939 and 1945, and order slowly began to emerge.

After Wold War II, viruses were increasingly recognized as worldwide production constraints, and legume virus workers were among the first to standardize the techniques for identifying the viruses dealt with (1960). An International Working Group on Legume Viruses (IWGLV) was established in 1961 for the exchange of seeds of test plants and antisera and of information. New techniques such as serology and electron microscopy were advocated for further laboratory characterization of the viruses themselves, and collaboration between researchers across national barriers was stimulated. A resulting tentative list of viruses reported from naturally infected leguminous plants (1964) animated further worldwide assemblage of information in computerized form as by the Australian Virus Identification Data Exchange project (VIDE). Its microfiche publication on *VIDE Viruses of Legumes* was soon followed by a printed version (1983). Similar books on viruses of plants in Australia (1988) and of tropical plants (1990) were succeeded by *Viruses of Plants* (1996), which was also distributed on the Internet. It later contributed to the database of the International Committee on Taxonomy of Viruses (ICTV), but it all began with IWGLV.

The IWGLV has in 2001 merged with the International Working Group on Vegetable Viruses, set up in 1970 with similar means and objectives and partially dealing with the same crops. The major aim remains surveying of the crops for viruses and their economic importance, understanding their ecology, and learning

Figure 1 Dramatic effect of common mosaic, caused by bean common mosaic potyvirus (BCMV) in some cultivars of common bean (*Phaseolus vulgaris*). The disease was described during the decade starting 1910 as one of the first recognized, now globally distributed legume virus diseases. (Left, resistant cultivar.)

Figure 2 BCMV in common bean with primary infection from the seed in plant at left and secondary infection by aphids in adjacent plant at right.

how they can be controlled for safeguarding crop productivity. But first of all comes virus identification.

Legume Viruses

In early plant virology viruses were thought to be rather specific in host range, for example, limited to Leguminosae or even to single species. Examples of narrow host specialization were BCMV and soybean mosaic virus (SMV) on beans (**Figure 1**) and soybean, respectively. They are highly seed transmitted and are introduced into crops every year by the seed that is sown, whereupon they are further spread from infected seedlings by aphids (**Figure 2**). Alternative hosts then are not needed for survival from one growing season to another. The aphid-transmitted wisteria vein mosaic potyvirus (WVMV) has so far been found worldwide in a few ornamental *Wistera* spp. only, possibly from a single vegetative origin.

Several viruses, of legumes, soon appeared to be much more polyphagous than supposed before. BYMV, for example, infects several annual legumes, but is prevalent in ornamental gladiolus as well, and occurs in other Iridaceae and some 14 other plant families. The related cowpea aphid-borne potyvirus (CABMV) also causes woodiness disease in passionfruit, and BCMV has meanwhile been found naturally in at least 26 leguminous species.

Some 'nonlegume viruses', other than the beet virus BCTV, also infecting legumes are the potyviruses beet mosaic virus (BtMV), lettuce mosaic virus (LMV), and turnip mosaic virus (TuMV). Extremely polyphagous viruses like the aphid-borne cucumber mosaic cucumovirus (CMV) and the soil-borne and fungus-transmitted tobacco necrosis necrovirus (TNV) include legumes among their many natural hosts (**Figure 3**). Presumed host barriers are increasingly transgressed, and novel types of disease emerge in cases of often haphazard new encounters between plants and viruses – as through the global movement of plant species and cultivars and of viruses and their strains distributed worldwide in plant propagation materials. Legume viruses are therefore increasingly difficult to define as a special category, but over 14% of the plant viruses still derive their name from legumes.

Most taxonomic groups of plant viruses are represented in legumes, and a 1991 monograph by Edwardson and Christie lists 279 different legume species reported naturally infected by 171 viruses then thought to be distinct. Perennial legumes, such as clovers, host many viruses, often in mixtures, and in plants the viruses may interact, either directly or via their host. BYMV, for example, helps multiply CMV and increase disease severity as in cowpea stunt. Groundnut rosette in Africa is caused by a complex of groundnut rosette umbravirus (GRV) together with an associated satellite RNA and groundnut rosette assistor luteovirus (GRAV), that helps aphid transmission. In nature, leguminous hosts and the viruses infecting them vary greatly and the result of infection may range from symptomless infection to severe disease up to premature plant death. Symptoms may often be mistaken for mere physiological disturbances as by mineral deficiency (**Figure 4**). Concise literature on virus diseases of specific legume crops is given by the compendia

Figure 3 Severe pod necrosis in common bean caused by the soil-borne and sap- and fungus-transmitted tobacco necrosis necrovirus (TNV).

Figure 4 Beginning chlorosis in top of faba-bean plants with bean leafroll luteovirus (BLRV), often mistaken for a mere effect of mineral deficiency.

of crop diseases published by the American Phytopathological Society, St. Paul, MN, USA. These and other reviews up till 1996 have been listed by Bos in 1996.

Variation of Legume Viruses

BCMV was one of the first viruses of which 'strains' were detected on new cultivars of common bean bred for resistance. The same was later found for SMV on soybean where resistance breeding was also prominent. Genetic change of crops is known to exert selection pressure on the viruses allowing these to co-evolve. Of BCMV, at least seven pathogenicity genes have been postulated and four of them interact with the bean's susceptibility or resistance genes in a gene-for-gene relationship.

Genetic adaptation to plant species and host cultivars where the viruses happen to land may also account for the wide variation that exists within the virus genera, explaining the evolution of virus species. Legume potyviruses (genus *Potyvirus*, now part of the family *Potyviridae*) have many variants and intermediates as between and around BCMV, SMV, and BYMV. Their expanding number made biological properties, such as host ranges, symptoms, cross-protection, and serology, increasingly inadequate for distinction and specific detection, and a semblance of continuity started to blur the originally postulated boundaries. It suggested progressive evolution through mutation, recombination, and selective adaptation in niches often created by man. Molecular genetics, recording percentages of nucleotide and amino acid sequence identities, has helped artificially to draw boundaries between different 'viruses' and devise phylogenetic dendrograms (**Figure 5**). This allows more final identification but entails increasing reliance on molecular techniques. Similar evolutionary developments have been observed among, for example, legume luteoviruses and begomoviruses.

Ecology

What happens to the legume viruses in legume crops is a matter of an intriguingly complex ecology. Factors to consider are: (1) the viruses already mentioned, (2) the sources of infection, (3) the vectors that spread the viruses, (4) the crops that are subject to infection and differ in genetic vulnerability, and finally (5) the growing conditions, including soil, climate, and the cropping system that influence host plants and vectors. These factors may capriciously interact, and the outcome is often hard to predict. Several of the factors listed largely depend on the grower's decisions, as for his/her choice of place and time of cultivation and the choice of crop and cultivar and of cultural practices. Legume viruses illustrate the widest known array of ecological relationships between viruses and plants. The quantitative involvement

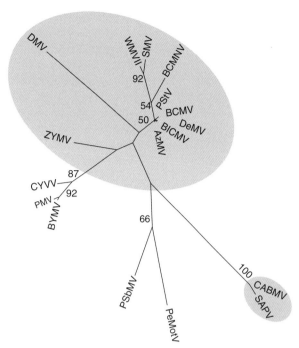

Figure 5 Phylogenetic tree of the legume-infecting viruses in the family *Potyviridae* based on analysis of the nucleotide sequences of the 3′-end noncoding region (NCR) of their ss (+) RNA genome. Branch lengths are proportional to genetic distance. Reproduced from Berger PH, Wyatt SD, Shiel PJ, Silbernagel MJ, Druffel K, and Mink GI (1997) Phylogenetic analysis of the *Potyviridae* with emphasis on legume-infecting potyviruses. *Archives of Virology* 142: 1979–1999, with permission from Springer-Verlag.

of the factors mentioned is the subject of epidemiology. Comprehension of the ecology of the viruses and of their epidemiology explaining to what extent and with what speed the viruses spread is essential for devising ways and means to control the viruses or limit their effects.

Sources of Infection

Many legume viruses are seed-borne, and commercial seed lots, even when usually only partially infested, provide infected seedlings throughout a newly sown crop. For BCMV (**Figure 2**) and SMV in common bean and soybean, respectively, these are the almost only, but efficient and early, within-crop sources of infection. BYMV that is seed-transmitted in yellow lupin and at low rates in pea and faba bean, has other sources of infection as well. Pea seed-borne mosaic potyvirus (PSbMV), which caused considerable concern in the pea-growing and -processing industry in the USA during the late 1960s and 1970s, is now known to often occur in seed of several other legumes worldwide. Notorious and vast other sources of infection for annual legumes are nearby perennial clovers and medics in pastures or grown as pure crops. BYMV readily moves from clovers (**Figure 6**) and also from nonleguminous ornamental gladiolus to bean when grown nearby.

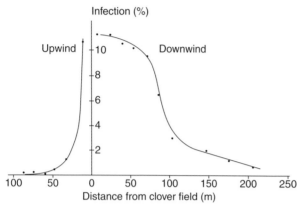

Figure 6 Spread of bean yellow mosaic potyvirus (BYMV) that is nonpersistently transmitted by aphids from red clover into field of common bean, influenced by prevailing wind. Reproduced from Hampton RO (1967) Natural spread of viruses infectious to beans. *Phytopathology* 57: 476–481, with permission from American Physopathological Society.

Infection of wild plants, as by CMV, pea early-browning tobravirus (PEBV) and TNV, is often symptomless and the wild sources of infection may then be hard to find.

Means and Ways of Spread

By contact

For legumes, spread from one plant to another when plants touch or are touched, seems to be limited to white clover mosaic potexvirus (WClMV), TNV, and the beetle-transmitted red clover mottle comovirus (RCMV). They occur in plants in high concentration and have very stable virions. RCMV may even be spread with a lawn mower. Some stable plant viruses, such as the fungus-transmitted red clover necrotic mosaic dianthovirus (RCNMV) may also be released into soil and water from decaying roots and may then get mechanically into plants when these are wounded. Tomato bushy stunt tombusvirus (TBSV), reported from *Robinia pseudoacacia* and *Wisteria floribunda*, may be transmitted directly from living roots.

Via seed

Many legume viruses (69 of 171 listed) have been reported to be distributed via seed, although often in such low percentages (below 1% or less) that infected seedlings easily escape notice. Such seed may still spread viruses over long distances to places where they did not occur before. Many legume viruses have already been spread worldwide in seed, and they continue to threaten legume crop improvement in developing countries. However, rates of seed-lot infestation with infected seeds may be high as of PSbMV where up to 90% of the seeds of infected mother plants may contain the virus, so that crops grown from such seed will directly suffer economically. Most seed-transmitted viruses of legumes are carried in the embryo, which remains infective for the seed's life. The legume viruses that in their plant hosts are limited to the phloem, as a rule, do not pass via the seed. Most mechanically transmissible viruses are detectable for a while in the seed coats from where they usually cannot reach the seedling.

By living organisms

Most spread of legume viruses is by living organisms, called vectors. It usually is rather, if not highly, specific, and most viruses have complicated relationships with their vectors so that viruses transmitted by one type of vector usually are not transmitted by any other type, and the vectors have complex ecologies of their own as well. Seed-borne viruses, with the exception of cryptoviruses which have no known vector, have either an aerial or subsoil vector for further dissemination.

Aboveground, most legume viruses are readily insect transmitted.

1. 'Nonpersistent and noncirculative transmission', as of many carla-, caulimo-, cucumo-, and fabaviruses, and the large group of potyviruses that are all readily transmissible in expressed sap, is by mere mouthpart contamination and usually is not specific. It is rapid but over short distances only (**Figure 6**). These viruses usually move in zones from a nearby other crop or do so spotwise from infected seedlings or weeds.

2. 'Persistent transmission' particularly holds for phloem-limited viruses that usually are not sap-transmissible. Examples are the luteoviruses of bean leafroll (BLRV) (**Figure 4**) and soybean dwarf (SbDV) by aphids, the mastrevirus of chickpea chlorotic dwarf (CpCDV) and the curtovirus BCTV in common beans by leafhoppers, the bean golden mosaic begomovirus (BGMV) by whiteflies, and the tospoviruses of tomato spotted wilt (TSWV) and groundnut bud necrosis (PBNV) in groundnut by thrips. Their relationship with the vectors is highly specific, that is, circulative. Some viruses even multiply in their vector (propagative transmission) and transovarially pass to offspring. The vectors then remain viruliferous for extended periods or for life. Propagative viruses are viruses of animals (insects) as well as of plants. Persistently transmitted plant viruses may be transferred over very long distances. Leafhoppers are strong flyers and they may cover hundreds of kilometres in high-level wind currents.

3. 'Semipersistent transmission' as of clover yellows closterovirus (CYV) and related criniviruses often is by whiteflies. Virus uptake is from the sieve tubes, and the virus is only adsorbed in the insect's foregut.

4. A variety of viruses that are artificially transmissible in plant sap is naturally transmitted by beetles (*Coleoptera*) including blister beetles (bean pod mottle comovirus; BPMV), lady beetles (southern bean mosaic sobemovirus; SBMV), leaf beetles (cowpea mosaic comovirus; CPMV; and BPMV) and some weevils (broad bean mottle

bromovirus; BBMV). These have biting mouthparts, and virus acquisition and introduction by these vectors is immediate and after a single bite. Virus retention in the beetles is in the hemolymph for days or weeks, and specificity of transmission is due to specific inactivation of most of the viruses in the regurgitant fluid that is produced by the beetles during feeding.

5. Exceptional among legume viruses is the possibly persistent transmission of pigeon pea sterility mosaic virus (PPSMV) in India by the eryophyid mite *Aceria cajani* that is invisible to the naked eye and mostly moves passively as by wind.

Within the soil, a number of legume viruses are transmitted by either nematodes or fungi.

1. The trichodorid nematodes that spread PEBV are of low mobility. Patches of infected plants in fields of pea and common bean hardly enlarge. The transmission resembles semipersistence, and retention in the vector may be for months especially at low temperatures. Disease recurrence the next season, as well as long-distance dissemination of the viruses, is especially explained by high rates of seed transmission and by occurrence in weeds. Trichodorid nematodes and the viruses they transmit do best on sandy soils or in sandy patches.

2. (a) Chytrid fungi (*Olpidium brassicae* and *O. bornovanus*) transmit the necrovirus TNV (**Figure 3**) and the dianthovirus RCNMV, respectively. These viruses are carried externally on the fungal zoospores, and transmission is nonpersistent. The virus-carrying spores move for short distances in soil water or over longer distances in irrigation water. (b) The plasmodiophorid fungus *Polymyxa graminis* spreads peanut clump pecluvirus (PCV) and broad bean necrosis pomovirus (BBNV). The viruses are carried within the hard-walled resting spores of the fungus and may persist there for many years. These spores may be transferred over long distances in drainage water and much farther in dry soil on seeds, vegetative propagation material, tools and transport vehicles, and in wind-blown dry soil and seeds. Fungus-transmitted viruses are favored by rainfall and irrigation.

The Role of Man

Viruses are nearly always and often invisibly present in or around crops or in wild species. Whether disease results greatly depends on susceptibility and sensitivity (vulnerability) of the crop and thus on the crop species or cultivar that the growers choose. A change of farming system may also create new niches for viruses and their vectors. Large-scale commercial cropping of groundnut in Northern Nigeria, for example, led to a dramatic epidemic of groundnut rosette in 1975. Dense populations of the vectoring aphids resulted from their survival and multiplication on volunteer groundnut plants and on numerous other plant species available throughout the year due to increased irrigation. In Brazil, the acreage of soybean for the world market tremendously increased and its growing period extended, creating vast areas of an excellent food and breeding plant for the whitefly *Bemisia tabaci*. This led to the enormous upsurge of bean golden mosaic virus (BGMV) in common bean. Where soybean cultivation had also expanded in traditional bean-producing areas, the seed-borne SMV, not usually infecting bean, also showed up in that crop because of high infection pressure from soybean and the cultivation of susceptible bean varieties. In the Latin Americas and other parts of the world, whitefly-transmitted viruses have also increased enormously in soybean and a range of crops including many nonlegumes such as cotton, tobacco and tomato. New niches may also result from the introduction of alien crops or new cultivars, and alien viruses inadvertently introduced in germplasm for breeding programs and in commercial seed may well be pathogenic to local crops or crop cultivars.

Most of the many seed-borne viruses of legumes have already shown up everywhere. Legume improvement programs in developing countries often entailed new diseases by viruses introduced in germplasm. In developed countries the introduction of virus-resistant cultivars, as of bean and soybean, has evoked the emergence of resistance-breaking virus strains. Continuing agricultural modernization and further internationalization of trade and traffic are therefore bound to lead to newly emerging diseases. In fact, most if not all virus diseases of crops are 'man made'.

Epidemiology

Accurate quantitative information about epidemic development of virus infection in crops usually is scarce. Epidemic disease development in crops when beginning with a limited number of infected plants is known to be a polycyclic process and to proceed according to an S-curve. This also holds for legume viruses (**Figure 7**; curve *V*). The speed of development greatly depends on distance between source and subject of infection, on the number of infection sources, on vector population density (**Figure 7**; curve *A*) and efficiency, and on crop and cultivar susceptibility (or resistance). When seeds are major if not sole primary sources of infection as of BCMV and SMV in bean and soybean, respectively, initial disease incidence determined by seed infestation below 0.1% mostly does not lead to incidences at the time of harvest that cause appreciable loss. In resistant cultivars, virus multiplication in individual plants usually is lower than in susceptible ones so that it takes longer for plants to become efficient secondary sources of infection. Epidemic development in crops of such cultivars is then delayed and crops are resistant as a whole.

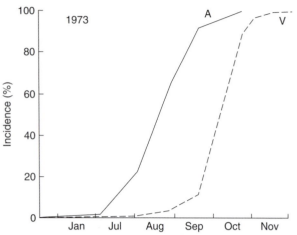

Figure 7 The increase in incidence with time of subterranean clover red leaf virus (a strain of soybean dwarf virus, SbDV, an unassigned virus of the family *Luteoviridae*) (**V**), and of its aphid vector *Aulacorthum solani* (**A**) in crop of faba bean in Tasmania. Reproduced from Thresh JM (1983) Progress curves of plant virus disease. *Advances in Applied Biology* 8: 1–85, with permission from Elsevier.

Figure 8 Total crop loss in faba bean (*Vicia faba*) over vast area in Egypt in 1992 caused by faba bean necrotic yellows nanovirus (FBNYV).

Persistently transmitted viruses such as the luteoviruses usually move in from far away and many plants may become infected around the site where the viruliferous aphids land and move about. Then, infection may soon evenly invade an entire crop. Nematode-transmitted viruses like PEBV in pea and bean occur spotwise depending on irregularity of soil type and nematode distribution. Plants usually are invaded with such viruses early when the nematodes become active, and the spots may slowly enlarge over the years if the soil permits the nematodes to extend their territory.

Economic Importance

Crop loss depends on incidence of diseased plants, on time of infection during crop development, and on symptom severity. Many viruses of legumes are highly contagious and often rapidly reach all plants of a crop. Losses then are high if the symptoms are severe as of faba bean with faba bean necrotic yellows nanovirus (FBNYV) where plant death often leads to total crop failure (e.g., **Figure 8**). Early infections usually are most severe, particularly when symptoms accumulate as of the phloem-limited luteoviruses viruses in chickpea, faba bean (**Figure 4**), pea, and several other legumes. These viruses primarily cause degeneration of the sieve tubes and impede transport in vascular bundles. This then is followed by a cascade of secondary and tertiary symptoms such as poor root growth and leaf yellowing, plant stunting, and even premature death. Late infections usually are less harmful than early ones because plant susceptibility and sensitivity often decrease toward plant maturity. Necrosis directly affects yield and quality particularly when the pods are involved. Examples are certain genetic host/virus combinations of soybean with SMV and of bean with BCMV and BCMNV, of bean with TNV (**Figure 3**), and of the same and pea with PEBV.

The effects of viruses on yield are often overlooked or neglected. Plant yellowing (**Figure 4**) may be mistaken for noninfectious mineral deficiency, but final losses may still be dramatic. Chlorophyl defects, as in various mosaics and yellows diseases (**Figures 1** and **4**), always reduce assimilation capacity and adversely affect yield. Even when visible symptoms are absent, root nodulation, plant vitality and productivity may considerably be diminished, and plant susceptibility and sensitivity to other pathogens increased. In perennial clovers, longevity of stands may be reduced by increased infection by root rot fungi and by raised sensitivity to drought and winter injury. One virus may also support another virus. In cowpea stunt, caused by a mixture of CMV and blackeye cowpea mosaic virus (BCMV or a variant), experiments have shown that together they caused a reduction in seed yield of 86%, whereas single infections only led to seed losses of 14 and 2.5%, respectively.

In practice, severe crop losses have been reported in legumes. The dramatic 1975 epidemic of groundnut rosette in Northern Nigeria occurred on over 1 million ha of groundnut and destroyed an estimated 0.7 million ha worth over $250 million. The same year, the whitefly-transmitted BGMV had become the main production constraint in the traditional cultivation of bean in Brazil. In Latin America, where beans is one of the main staple foods, more than 2.5 million ha was under attack by the virus during the 1990s. The related bean dwarf mosaic begomovirus (BDMV) showed up in Argentina in 1981 and at least one million ha previously planted with bean in South America was abandoned because of the risk of total yield loss. In 1992 in the Nile Valley, Egypt, the then newly recognized FBNYV wiped out entire crops of faba bean on *c.* 16 000 ha in the Beni Suef Governorate (**Figure 8**).

Financial losses that mostly are not taken into account but may be substantial are the costs of hygienic measures to be taken by growers, and of the more expensive price of seed of resistant cultivars and of seed certified for low rates of infection. Nationwide, costs of research, education, extension, and quarantine must also be considered.

Control

Virus-diseased plants cannot be cured once they are infected, and control nearly always is preventive. Either one or, mostly, a combination of mainly hygienic measures are to be taken based on profound knowledge of the ecology and epidemiology of the viruses in their relation to crop ecology. But this, first of all, requires proper disease diagnosis, that is, reliable identification of the causal virus for allowing a goal-directed strategy of control.

Often highly effective, even when applied singly, is the avoidance or removal of the sources of infection. For example, soil infested with soil-borne and nematode- or fungus-transmitted viruses, such as PEBV, TNV, or PCV, should not be used for growing peas, bean, and groundnut. Commercial seed used for growing crops must be relatively free of seed infected with seed-borne viruses, and possibly be certified for low rates of infection. Certification schemes have been developed for the production of commercial seed of legume crops that often harbor aphid-transmitted seed-borne viruses such as BCMV and SMV. The commercial seed must be produced under relatively vector-free conditions away from sources of infection, while infected plants must be regularly removed. The harvested product must be tested and certified for low rates of virus infection so that infection of crops to be grown from the seed is likely to remain below damage thresholds. For most legumes, commercial seed should not contain more than 0.1–0.5% of infected seeds. Germplasm used for breeding purposes and as basic material for the production of commercial seed, must be multiplied under vector-proof conditions from seed of plants that have been tested individually for virus freedom, and suspected plants must be removed. Annual legume crops must not overlap with nearby older crops or with other perennial crops such as clovers that may harbor several legume viruses (**Figure 6**).

Epidemic buildup must be prevented by avoiding high densities of vector insects by choosing growing conditions or a season with less vector insects. Systemic insecticides are helpful to reduce the spread of persistently insect-transmitted viruses such as the legume luteoviruses. Virus dissemination over long distances must be avoided or prevented also by the use of clean seed, which is especially important for legumes because of the many viruses that are seed-borne in them.

The use of resistant cultivars, and breeding for resistance, has been the most widely used means of managing virus disease incidence and severity, and has a long tradition for legumes. It should be an ongoing effort because of the selection pressure that new cultivars exert on viruses for developing new resistance-breaking strains. Breeding for resistance to seed transmission is also possible as found in beans and lupin for BCMV and CMV, respectively. For developing countries, International Agricultural Research Institutes (IARCs) as CIAT in Cali, Colombia, ICARDA in Aleppo, Syria, ICRISAT in Hyderabad, India, and IITA in Ibadan, Nigeria, that are mandated for legume crops, are instrumental in improving resistance of important leguminous food and fodder crops by breeding, often in concert with national institutes in their respective regions. They also provide information and survey the respective crops in their regions of outreach, study the identity and ecology of the viruses that are important there, and develop techniques and means for their proper identification and for testing of plant genetic materials for resistances to the viruses concerned. The institutes have immense collections of germplasm in their gene banks that are also sources of genes for resistance to viruses. Genetic engineering is increasingly replacing original methods of gene transfer by pollination.

The ever-modernizing agriculture, including the continuing change of crop genetic makeup as well as of farming systems, however, keeps entailing new risks. Even resistance of cultivars practically never means immunity, and resistance hardly ever is durable. For the many seed-borne viruses of legumes, in particular, the continuing globalization of trade and traffic create a special hazard. Certification of commercial seed, which must be grown in the open with a view to the immense quantities needed, and quarantine of germplasm, which cannot handle individual seeds of the large quantities that are distributed worldwide, is unlikely to ever guarantee absolute freedom from the many known and still unknown and often latent seed-borne viruses.

Most virus control in crops, therefore, involves interference through crop management with the ecology of the viruses. Most measures are imperfect and are likely to entail new problems. Hence, man will never be ready to combat viruses in legumes and whatever other crops. Since the mostly preventive measures of control must be taken outside the field-grown crops, for example, at the source of infection, and what individual growers are doing or neglecting has a bearing on the health of crops of other growers, public interests are at stake. That is why public institutions, governments, and international organizations must remain involved as for regulation, quarantine, certification, advice, teaching, and continuing research. Such research, as on legume viruses, is of great relevance to human society. The study of legume viruses provides an outstanding example of the dynamics and complexity of viruses in their relation to crops.

See also: Bean Common Mosaic Virus and Bean Common Mosaic Necrosis Virus; Luteoviruses; Plant Resistance to Viruses: Engineered Resistance; Plant Resistance to Viruses: Geminiviruses; Plant Resistance to Viruses: Natural Resistance Associated with Dominant Genes; Plant Resistance to Viruses: Natural Resistance Associated with Recessive Genes; Potyviruses; Virus Species.

Further Reading

Berger PH, Wyatt SD, Shiel PJ, *et al.* (1999) Phylogenetic analysis of the *Potyviridae* with emphasis on legume-infecting potyviruses. *Archives of Virology* 142: 1979.

Berger PH, Wyatt SD, Shiel PJ, Silbernagel MJ, Druffel K, and Mink GI (1997) Phylogenetic analysis of the *Potyviridae* with emphasis on legume-infecting potyviruses. *Archives of Virology* 142: 1979–1999.

Bos L (1970) The identification of three new viruses isolated from *Wisteria* and *Pisum* in the Netherlands, and the problem of variation within the potato virus Y group. *Netherlands Journal of Plant Pathology* 76: 8.

Bos L (1983) Plant virus ecology: The role of man, and the involvement of governments and international organizations. In: Plumb RT and Thresh JM (eds.) *Plant Virus Epidemiology; the Spread and Control of Aphid-Borne Viruses*, 7pp. Oxford: Blackwell.

Bos L (1992) New plant virus problems in developing countries: A corollary of agricultural modernization. *Advances in Virus Research* 41: 349.

Bos L (1996) *Research on Viruses of Legume Crops and the International Working Group on Legume Viruses; Historical Facts and Personal Reminiscences*, 151pp. Aleppo, Syria: International Working Group Legume Viruses.

Bos L, Hagedorn DJ, and Quantz L (1960) Suggested procedures for international identification of legume viruses. *Tijdschr. Plantenziekten* 66: 328.

Boswell KF and Gibbs AJ (1983) *Viruses of Legumes 1983. Descriptions and Keys from VIDE*, 139pp. Canberra: Research School of Biological Sciences, The Australian National University.

Edwardson RE and Christie RG (1986) *Viruses Infecting Forage Legumes*, monograph No.14, 742pp + appendices. Gainesville: University of Florida Agricultural Experiment Station.

Edwardson RE and Christie RG (1991) *CRC Handbook of Viruses Infecting Legumes*, 504pp. Boca Raton, FL: CRC Press.

Hampton RO (1967) Natural spread of viruses infectious to beans. *Phytopathology* 57: 476.

Jones RAC (2000) Determining 'threshold' levels for seed-borne virus infection in seed stocks. *Virus Research* 71: 171.

Summerfield RJ and Roberts EH (eds.) (1985) *Grain Legume Crops*. 859pp. London: Collins.

Thresh JM (1983) Progress curves of plant virus disease. *Advances in Applied Biology* 8: 1–85.

Van der Maesen LJG and Sadikin S (eds.) (1989) *Pulses. Plant Resources of South-East Asia No. 1*. 105pp. Wageningen: Pudoc.

Weiss F (1945) Viruses described primarily on leguminous vegetable and forage crops. *Plant Disease Reporter, Supplement* 155: 32.

Leishmaniaviruses

R Carrion Jr., Southwest Foundation for Biomedical Research, San Antonio, TX, USA
Y-T Ro, Konkuk University, Seoul, South Korea
J L Patterson, Southwest Foundation for Biomedical Research, San Antonio, TX, USA

© 2008 Elsevier Ltd. All rights reserved.

Glossary

Polyprotein An expressed protein comprising two functionally distinct domains.
Ribosomal frameshift Translation of a protein from an alternate reading frame.
Shine–Dalgarno sequence A consensus sequence (AGGAGG) that helps recruit the ribosome to the mRNA to initiate protein synthesis by aligning it with the start codon.

Introduction

Leishmania RNA virus (LRV) is the type species of the genus *Leishmaniavirus* in the family *Totiviridae*. The virus persistently infects the protozoan *Leishmania*. It possesses an icosahedral capsid of approximately 32–33 nm in diameter. The double-stranded (ds) RNA genome of leishmaniaviruses ranges in size from 5.2 to 5.3 kbp in length and encodes two major proteins – the capsid protein and the RNA-dependent RNA polymerase (RdRp). Its capsid protein has a unique endoribonuclease activity that targets the viral genome in a site-specific manner. In addition to primary structure, a hairpin loop upstream of the cleavage site is essential for processing of the viral RNA. The polymerase is expressed as a polyprotein that is processed by a host-encoded cysteine protease. Over time, this ancient virus has elaborated a variety of mechanisms to control copy number that has allowed it to co-evolve with its protozoan host.

History

The first documentation that some strains of *Leishmania* were infected with a virus came in 1974 when virus-like particles (VLPs) were observed in the cytoplasm of *Leishmania hertigi*. However, over a decade would pass before molecular data were reported. Initially, a 6.0 kbp band was visualized on an ethidium bromide-stained gel of parasite extracts from *Leishmania braziliensis guyanensis*. A probe generated from cDNA of nucleic acid moiety, termed LR1, failed to hybridize to the parasites genomic DNA confirming its existence as an extrachromosomal entity. Electron microscope examination of sucrose gradient fractions of parasite lysates confirmed the existence of 32 nm VLPs in fractions that contained LR1 RNA (**Figure 1**).

Additional virus-infected isolates were identified on the basis of RdRp activity. Those isolates exhibiting RdRp activity also harbored a 6 kbp nucleic acid that was determined to be dsRNA. The products generated from polymerase assays showed no hybridization to *Leishmania* genomic DNA, but did hybridize to the 6 kbp dsRNA, thus providing conclusive evidence that the dsRNA served as template. Soon after, electron microscopy correlated the polymerase activity with $CsCl_2$ purified virus particles. To date, 14 leishmaniavirus species comprising both New World and Old World viruses have been identified (**Table 1**).

Initial studies to characterize the LRV polymerase revealed that the RdRp was associated with viral particles. *In vitro* polymerase assays resulted in the production of genome length RNA as well as a faster migrating RNA species. RNase protection assays confirmed that the genome length was dsRNA and the faster migrating species ssRNA. This evidence supports a dual function for the LRV polymerase as both replicase and transcriptase. The LRV replication cycle appears to undergo a strategy similar to the dsRNA totivirus that infects yeast. LRV replication involves generation of a full-length positive-sense RNA species that is encapsidated and serves as a template for negative-sense RNA synthesis. Positive-sense RNA transcripts are then produced from the parental dsRNA in a conservative manner and extruded into the cytoplasm where they are translated into viral proteins, or encapsidated to serve as template for viral replication.

Genome Organization

The genome of leishmaniaviruses ranges from 5.2 kbp (**Figure 2**) for the Old World viruses to 5.3 kbp for New World viruses. Open reading frames (ORFs) are encoded on the positive-sense strand of New World viruses (**Figure 3**). ORF1 and ORF X encode small ORFs that share no homology to any known protein sequence. These small ORFs correspond to the 5′-untranslated region (UTR), which is 449 nt in length in LRV1-4 showing greater than 90% identity to all known leishmaniaviruses at the nucleotide level. ORF2 encodes the 82 kDa viral capsid protein that has been shown to spontaneously assemble into VLPs when expressed

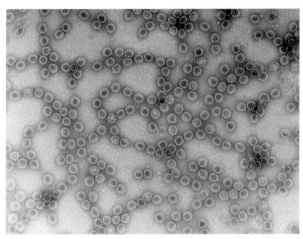

Figure 1 An electron micrograph of LRV1-4 particles isolated from a sucrose gradient of *Leishmania* lysate. Reproduced from Cadd TL and Patterson JL (1994) Synthesis of viruslike particles by expression of the putative capsid protein of Leishmania RNA virus in a recombinant baculovirus expression system. *Journal of Virology* 68: 358–365, with permission from American Society for Microbiology.

Table 1 Known leishmaniaviruses and associated sequences

Viruses	Host range	Sequences available	Accession numbers
LRV1-1	*Leishmania* CUMC1	Complete genome (1–5284)	M92355
LRV1-2	*Leishmania* CUMC3	Partial 5′-UTR (1–253)	AF230881
LRV1-3	*Leishmania* M2904		
LRV1-4	*Leishmania* M4147	Complete genome (1–5283)	U01899
LRV1-5	*Leishmania* M1142		
LRV1-6	*Leishmania* M1176		
LRV1-7	*Leishmania* BOS12	Partial 5′-UTR (1–251)	AF230882
LRV1-8	*Leishmania* BOS16	Partial 5′-UTR (1–253)	AF230883
LRV1-9	*Leishmania* M6200	Partial 5′-UTR (1–253)	AF230884
LRV1-10	*Leishmania* LC76	Partial 5′-UTR (1–251)	AF230885
LRV1-11	*Leishmania* LH77	Partial 5′-UTR (1–251)	AF230886
LRV1-12	*Leishmania* LC56		
LRV1-13	*Leishmania* NC	Partial 5′-UTR/RdRp	U23810/U39069
LRV2-1	*Leishmania* 5ASKH	Complete genome (1–5241)	U32108

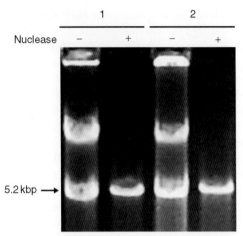

Figure 2 New World parasite strains containing a dsRNA moeity. Equal concentrations of total RNA from late-log phase parasites were incubated in the presence (+) or absence (−) of mung bean nuclease/DNase then resolved on a 1% agarose gel. The gel was stained with ethidium bromide then visualized on a UV transilluminator. Note the presence of dsRNA at approximately 5.2 kbp.

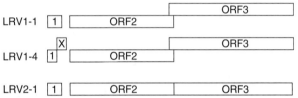

Figure 3 Schematic representation of genome organization of New World and Old World leishmaniaviruses. Reproduced from Scheffter S, Widmer G, and Patterson JL (1994) Complete sequence of Leishmania RNA virus 1-4 and identification of conserved sequences. 199: 479–483, with permission from Elsevier.

in vitro. The viral polymerase is encoded by ORF3, which contains the conserved six consensus RdRp motifs and overlaps the capsid domain by 71 nt. *In vitro* data and sequence analysis support the notion that the viral polymerase is expressed by a ribosomal frameshift to yield a capsid–polymerase polyprotein in a manner similar to several other members of the family *Totiviridae*.

Endoribonuclease Activity

A unique feature of LRV1-4 is that it encodes endoribonuclease activity within its assembled capsid particle. The cleavage event is site specific requiring a minimal essential sequence, and is mediated by both purified wild-type virus particles, as well as *in vitro* expressed VLPs. Later it was shown that the Old World LRV2-1 virus particles also possessed site-specific endoribonuclease activity.

Mutational analysis of the viral genome suggests that the minimal essential sequence required for site-specific targeting of the capsid endoribonuclease resides within viral nucleotides 249–342 in the 5′-UTR. This region contains conserved secondary structure, which was hypothesized to contain determinants for capsid-mediated cleavage. Nuclease mapping and site-specific mutagenesis identified a hairpin structure 40 nt upstream of the cleavage site that, when eliminated, abrogated cleavage. Reconstituting the hairpin through compensatory mutations allowed for precise cleavage of the substrate. Characterization of endoribonuclease activity associated with baculovirus expressed LRV1-4 VLPs reveals an absolute requirement for divalent cations such as Mg^{2+} and Ca^{2+}. This requirement is probably for stabilizing substrate secondary structure since electrophoretic mobility shift assay (EMSA) showed that the divalent cation itself changed the substrate RNA mobility. Furthermore, endoribonuclease activity was greatly inhibited by salt concentrations greater than 25 mM, indicating that high concentration of salts may interfere with enzyme–substrate interactions.

Processing of a Viral Polyprotein by a Cellular Protease

The inability to observe a capsid–polymerase fusion protein *in vivo* is explained by a proteoytic cleavage event that separates the two domains. Host cell extract from either virus infected or uninfected *Leishmania* strains produced a cleavage event in an *in vitro* expressed capsid–polymerase polyprotein. In both cases, a specific cleavage event was observed to yield both 95 and 82 kDa fragments, which matched well the predicted sizes of the individual polymerase and capsid products, respectively. The presence of specific proteolytic cleavage activities in both virus-infected and -uninfected parasite extracts indicates that the protease is host-encoded. It has been recently demonstrated that the protease responsible for cleavage is a unique trypsin-like cysteine protease. The purified protease is related to the cysteine proteases of *Leishamnia mexicana* found to be important for survival within macrophages. They are part of a large array of cathepsin L-like proteases that are differentially expressed. That processing of the Gag-Pol polyprotein is linked to an essential gene for parasite survival suggests that it plays a role in ensuring maintenance the LRV infection. The role of the proteolytic processing in generating functional viral proteins and/or its role in the virus replication has yet to be deciphered.

It is hypothesized that excessive production of viral proteins may adversely affect host macromolecule synthesis potentially leading to cell death, and ultimately a selective pressure to eliminate the virus. Targeting a fusion protein is an effective approach to maintain a low copy number. It is believed that the LRV replication cycle is similar to the yeast LA virus. From *gag-pol* overlapping reading frames, capsid protein is produced primarily until a ribosomal frameshifting event occurs. The Gag-Pol fusion proteins dimerize, enabling the binding of positive-sense viral RNA, which primes Gag polymerization and genome packaging. It can be envisioned that by cleaving the fusion protein at the junction of the Gag-Pol domains, packaging would ultimately be aborted. This scenario would require tight regulation of protease activity, since indiscriminate proteolysis would lead to loss of virus.

Mechanisms of Viral Persistence

Expression of proteins in *Leishmania*, as well as other trypanosomes, requires a highly conserved 39 nt splice leader. Early studies have shown that LRV1-4 transcripts do not possess the leader sequence and also lack a cap structure. Therefore, LRV1-4 must translate its proteins by a cap-independent mechanism. Computer sequence analysis of LRV1-4 clones has revealed that the 5′-UTR contains extensive secondary structure similar to those that promote internal initiation in poliovirus. In fact, it has been demonstrated that the 5′-UTR can function as a *cis*-element in promoting ribosomal entry. These observations, in addition to its similarity to the internal initiation sites of picornaviruses, suggest that the 5′-UTR functions as an internal ribosome entry site (IRES) in LRV1-4.

If the 5′-UTR of LRV1-4 functions as an IRES, how does one reconcile the observed removal of the UTR with its presumed role as a *cis*-acting internal initiation element? The conundrum may be explained by the hypothesis that the endoribonuclease activity plays a regulatory role in maintaining a persistent infection. Phylogenetic evidence has confirmed that LRV1-4 is an ancient virus that co-evolved with its protozoan host. Since the parasite's reproductive cycle is predominately asexual, and there is no evidence horizontal transfer between parasites, viral persistence is dependent on maintaining a copy number compatible with host survival. Excessive production of viral proteins may adversely affect host macromolecule synthesis potentially leading to cell death, and ultimately, a selective pressure to eliminate the virus. Cleavage of the LRV1-4 transcripts could alter the functionality of the message by affecting translation, RNA stability, RNA packaging, and/or viral replication. These events are not exclusive as LRV1-4 is likely to employ a variety of regulatory mechanisms to maintain a persistent infection. It has been demonstrated that cleavage of LRV1-4 RNA correlates with the growth of the parasite. In addition, it has been shown that overexpression of capsid protein results in a suppression of LRV replication. These observations support a tightly regulated model of viral translation in which viral transcripts are being translated efficiently by cap-independent mechanisms involving the 5′-UTR. As the molar concentration of VLPs increases, endoribonuclease cleavage increases, thus maintaining a low-level viral persistence by decreasing efficiency of viral translation.

Alternatively, cleavage of the 5′-UTR may represent removal of translation-attenuating *cis*-elements. Other viral systems, such as coronaviruses, have translation-attenuating elements which are selected for during persistent infection. The extensive secondary structure present in the LRV1-4 UTR may function in the same respect. Cleavage of the LRV1-4 5′-UTR exposes a cryptic protein-binding site on transcripts that permits complexing with *Leishmania* cytoplasmic proteins. Although the translation machinery of *Leishmania* has not been well characterized, the speculation is that these factors are accessory proteins which may facilitate translation. Analysis of LRV1-4 UTR sequences has identified a pyrimidine-rich region complementary to a pyrimidine-rich sequence of *Leishmania* 18S rRNA. It is postulated that binding of ribosomal RNA would facilitate translation analogous to the Shine–Dalgarno sequence in prokaryotes. In this case, the cleavage of the 5′-UTR would be removing translation attenuation factors and noninitiating AUGs to allow for efficient viral gene expression through the binding of translation accessory proteins as well as rRNA to cleaved transcripts.

Another manner in which the virus may interact with its host is through the cleavage of host RNA. Sequence analysis of *Leishmania* rRNA and the endoribonuclease cleavage site of both New World and Old World viruses have identified several regions of the rRNA sequence that possesses similarity to the cleavage site. If the hypothesis were validated, cleavage of rRNA would likely have an effect on host translation machinery. Since the translation of host genes is predicated upon the presence of a splice-leader sequence, host genes may have an advantage over viral transcripts in recruiting translation machinery. LRV has adapted by allowing for efficient internal initiation via an IRES; the endoribonuclease may have evolved to provide an additional role in promoting viral gene expression by decreasing translation of cellular mRNAs.

It is entirely possible that the cleavage event benefits the virus in multiple ways. When virus copy number is low, translation is allowed to proceed via a cap-independent mechanism using the LRV IRES. As copy number increases, transcripts are cleaved in addition to *Leishmania* rRNA and a slower more regulated form of translation occurs. Cryptic sites are exposed on the cleaved transcript and bind host translation accessory proteins. The short 5′-cleaved fragment also binds the cleaved transcript blocking assembly of translation accessory proteins, thus decreasing the

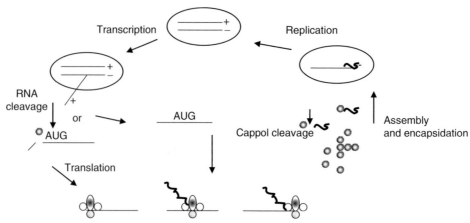

Figure 4 A model of LRV replication. Reproduced from MacBeth KJ and Patterson JL (1998) Overview of the leishmaniavirus endoribonuclease and functions of other endoribonucleases affecting viral gene expression. *Journal of Experimental Zoology* 282(1–2): 254–260 (review), with permission from Research Trends.

rate of viral protein synthesis. The result is a tightly regulated persistent infection compatible with parasite survival.

Leishmaniavirus: A Model

Recent advances in studies on leishmaniavirus have permitted the creation of a model for the LRV-replication cycle (**Figure 4**). As described above, there are many opportunities for regulation that have potentially aided in maintaining a persistent infection as the virus co-evolved with its host. Beginning with a mature particle containing dsRNA and viral transcriptase, dsRNA is transcribed in a conservative fashion and the resultant transcript is extruded from capsid. At a low molar abundance of assembled particles, viral proteins are translated by a cap-independent mechanism utilizing the LRV IRES located within the 5′-UTR. As the number of viral particles increases, the endoribonuclease activity of the LRV capsid cleaves the viral transcripts slowing the rate of translation. The cleaved 5′-UTR and the 3′-cleaved product compete for host proteins, which may be essential for translation of viral transcripts lacking a cap structure and now an IRES. Translation results in the production of a large number of capsid proteins, then occasionally through a ribosomal frameshift, a less abundant capsid–polymerase polyprotein is expressed. A host encoded-cysteine protease quickly cleaves the polyprotein in a process that is probably required for regulating viral copy number or generating functional viral proteins. Full-length viral RNA then presumably binds to a yet-to-be-determined binding site on the RdRp ensuring encapsidation of the genome. Replication follows through an unknown mechanism, thus reconstituting the dsRNA genome.

See also: Giardiaviruses; Totiviruses; Yeast L-A Virus.

Further Reading

Cadd TL and Patterson JL (1994) Synthesis of virus like particles by expression of the putative capsid protein of Leishmania RNA virus in a recombinant baculovirus expression system. *Journal of Virology* 68: 358–365.

Carrion R, Jr., Ro YT, and Patterson JL (2003) Purification, identification, and biochemical characterization of a host-encoded cysteine protease that cleaves a leishmaniavirus gag-pol polyprotein. *Journal of Virology* 77(19): 10448–10455.

Lee SE, Suh JM, Scheffter S, Patterson JL, and Chung IK (1996) Identification of a ribosomal frameshift in Leishmania RNA virus 1-4. *Journal of Biochemistry (Tokyo)* 120(1): 22–25.

MacBeth KJ and Patterson JL (1995) Single-site cleavage in the 5′-untranslated region of Leishmaniavirus RNA is mediated by the viral capsid protein. *Proceedings of the National Academy of Sciences, USA* 92(19): 8994–8998.

MacBeth KJ and Patterson JL (1995) The short transcript of Leishmania RNA virus is generated by RNA cleavage. *Journal of Virology* 69(6): 3458–3464.

MacBeth KJ and Patterson JL (1998) Overview of the leishmaniavirus endoribonuclease and functions of other endoribonucleases affecting viral gene expression. *Journal of Experimental Zoology* 282(1–2): 254–260 (review).

Ro YT and Patterson JL (2000) Identification of the minimal essential RNA sequences responsible for site-specific targeting of the Leishmania RNA virus 1-4 capsid endoribonuclease. *Journal of Virology* 74(1): 130–138.

Ro YT, Scheffter SM, and Patterson JL (1997) Specific *in vitro* cleavage of a Leishmania virus capsid-RNA-dependent RNA polymerase polyprotein by a host cysteine-like protease. *Journal of Virology* 71(12): 8983–8990.

Scheffter S, Widmer G, and Patterson JL (1994) Complete sequence of Leishmania RNA virus 1-4 and identification of conserved sequences. 199: 479–483.

Weeks RS, Patterson JL, Stuart K, and Widmer G (1992) Transcribing and replicating particles in a double-stranded RNA virus from Leishmania. *Molecular and Biochemical Parasitology* 52(2): 207–213.

Widmer G, Comeau AM, Furlong DB, Wirth DF, and Patterson JL (1989) Characterization of a RNA virus from the parasite Leishmania. *Proceedings of the National Academy of Sciences, USA* 86(15): 5979–5982.

Widmer G and Dooley S (1995) Phylogenetic analysis of Leishmania RNA virus and Leishmania suggests ancient virus–parasite association. *Nucleic Acids Research* 23(12): 2300–2304.

Leporipoviruses and Suipoxviruses

G McFadden, University of Florida, Gainesville, FL, USA

© 2008 Elsevier Ltd. All rights reserved.

History

Poxviruses of leporids and swine cause a broad range of symptoms varying from mild lesions of the skin right up to the lethal systemic diseases (**Table 1**). The agent of myxomatosis, a virulent disease of domestic rabbits described originally by G. Sanarelli in 1896, was in fact the first viral pathogen discovered for a laboratory animal. The close similarity of myxoma virus (MYX) with other members of the poxvirus family, such as variola and fowlpox viruses, was first recognized by Aragão in 1927. MYX is notable because, although it causes rather benign lesions in the native *Sylvilagus* rabbit (the brush rabbit in North America and the tropical forest rabbit in South America), when introduced to the European (*Oryctolagus*) rabbit it causes an invasive disease syndrome called myxomatosis with up to 100% mortality. MYX was the first viral agent ever introduced into the wild for the purpose of eradicating a vertebrate pest, namely the feral European rabbit population in Australia in 1950 and, 2 years later, in Europe. The resulting genetic selection of virus isolates with lesser pathogenicity and upsurgence of rabbits with greater resistance to the viral disease was studied intensively by Frank Fenner and his colleagues as a model system to investigate the ecological consequences of virus/host evolution in an outbred population.

Also, of interest to the history of animal virology is the fact that the first DNA virus associated with transmissible tumors was Shope fibroma virus (SFV), described in 1932 by Richard Shope as an infectious agent of fibroma-like hyperplasia in cottontail rabbits (*Sylvilagus floridanus*) in the eastern USA. It is likely that the agent of 'hare sarcoma', described first in Germany in 1909, was also a poxvirus, now called hare fibroma virus (HFV). HFV remains the only leporipoxvirus to have arisen outside the Americas but its biology closely resembles that of SFV.

Very little is known about the remaining leporipoxviruses. Subcutaneous fibromatosis in gray squirrels of the eastern USA and western gray squirrels in California, caused by poxviruses now collectively called squirrel fibroma virus (SqFV), has been observed since 1936, but their rigorous classification with the MYX–SFV group was not made until 1951 by L. Kilham. Similarly, HFV, described first in 1959 in the European hare (*Lepus europaeus*), was also shown to be a closely related poxvirus in 1961. In 1983, an outbreak of a disease resembling

Table 1 Members of the genera *Leporipoxvirus* and *Suipoxvirus*

Member	Abbreviation	Natural host	Major arthropod vector	Natural host disease	Disease in domesticated European rabbit (Oryctolagus cuniculus)
Leporipoxvirus					
Myxoma	MYX	California brush rabbit[a], S. American tapeti[b] (*Sylvilagus* sp.)	Mosquito, flea	Localized benign fibroma	Systemic lethal myxomatosis
Rabbit fibroma (Shope fibroma)	SFV	N. American cottontail rabbit (*Sylvilagus floridans*)	Mosquito, flea	Localized benign fibroma	Localized benign fibroma
Malignant rabbit fibroma[c]	MRV	Lab. rabbit[d] (*Oryctolagus cuniculus*)		Not observed in wild	Systemic lethal syndrome similar to myxomatosis
Squirrel fibroma	SqFV	Gray squirrel (*Sciurus* sp.)	Probably mosquito	Localized or multiple fibromas	Occasional nodular dermal lesions
Hare fibroma	HFV	Wild hares (*Lepus* sp.)	Probably mosquito	Localized benign fibroma	Localized benign fibroma
Suipoxvirus					
Swinepox	SPV	Domestic pigs (Suidae sp.)	Hog lice	Localized cutaneous lesions	Intradermal lesions but no serial propagation

[a]Also called Marshall–Regnery myxoma.
[b]Also called Aragão's (or Brazilian) myxoma.
[c]Laboratory recombinant between MYX and SFV.
[d]MRV has been propagated only by serial inoculation of laboratory rabbits and in cultured cells.

myxomatosis in laboratory rabbits in San Diego was caused by a novel leporipoxvirus later shown to be a genetic recombinant between SFV and a still-undefined strain of MYX. This virus, called malignant rabbit fibroma virus (MRV), has never been observed in wild rabbit populations but is of interest as an experimental model for poxvirus-induced immunosuppression and tumorigenesis. In most respects, MRV can be considered to be a substrain of MYX.

Based on landmark experiments with pneumococcus in the 1920s, the very first example of what was believed to be genetic interaction between viruses was reported in 1936 with the discovery that heat-inactivated myxoma could be reactivated with live SFV (Berry–Dedrick transformation), but later work showed this to be a genome rescue phenomenon rather than true recombination.

The only known member of the *Suipoxvirus* genus, swinepox virus (SPV), has been observed sporadically in pig populations throughout the world, but is not considered a serious pathogen because infected animals usually have only moderate symptoms and recover completely.

Taxonomy and Classification

The genera *Leporipoxvirus* and *Suipoxvirus* are in the subfamily *Chordopoxvirinae* of the family *Poxviridae*. The prefix 'lepori' comes from Latin *lepus* or *leporis* ('hare') and 'sui' from Latin *sus* ('swine'), to denote the relatively restricted host range of these viruses. All the viruses in the genus *Leporipoxvirus* can be shown to be closely related to each other by serology, immunodiffusion, and fluorescent antibody tests, although antigenic differences can be detected in strains of MYX. SPV (genus *Suipoxvirus*) is antigenically unique and is not known to have any closely related members. In terms of broad features, all are typical poxviruses, with characteristic brick-shaped virions containing a double-stranded DNA (dsDNA) genome with covalently closed hairpin termini and terminal inverted repeat (TIR) sequences. Like other poxviruses, viral macromolecular synthesis takes place exclusively in the cytoplasm of infected cells.

Properties of the Virion

As for all other members of the poxvirus family, the virions have a characteristic brick-shaped morphology with dimensions of approximately 250–300 nm × 250 nm × 200 nm. The leporipoxviruses are uniquely sensitive to ether and chloroform but otherwise the virions are very stable at ambient temperatures and in skin lesions. In all other respects, such as chemical composition and physical properties, the virus particles are very similar to those of vaccinia virus (VACV).

Properties of the Viral DNA and Protein

Complete genomic sequences are available for SFV, MYX, and SPV. The leporipoxviruses have dsDNA genomes of 160–163 kbp, with hairpin termini and TIR sequences of 10–13 kbp. SPV DNA is somewhat smaller (146.5 kbp) but otherwise the genome has similar characteristics. Each virus encodes from 150 to 160 proteins. A web-based resource is available that records the predicted proteins expressed by these three poxviruses (see 'Relevant website' section). Viral DNA of leporipoxviruses cross-hybridize at moderate stringencies only with other members of the genus and SPV DNA is unique and is not known to cross-hybridize with any other poxvirus DNA. The MRV DNA genome is 95% identical to that of MYX, except that it possesses five genes derived from SFV plus three SFV/MYX fusion genes.

The nucleotide composition of the leporipoxviruses (44% G + C for MYX) is higher than that of the orthopoxviruses (35% for VACV) but there is evidence that many of the viral genes important for virus replication, gene expression, and viral assembly are conserved between the genera. These conserved genes are clustered near the central regions of the viral genome. In contrast, viral genes mapping near the genomic termini show considerable variability, and are believed to encode many of the specific determinants of pathogenesis, host range, and disease characteristics.

The protein complexities of these viruses are comparable to those of most poxviruses, although the profiles are unique for each member. In general, about 80–90 poxvirus genes are relatively well conserved among the various member poxviruses, whereas the remainder (usually in excess of 50–60 genes) are more diverged and specify the unique features of virus–host interactions, such as host tropism, immunomodulation, and virulence.

DNA Replication, Transcription, and Translation

All of the major features of macromolecular synthesis by these viruses are very similar to those deduced for the prototype poxvirus, VACV. Viral DNA synthesis is restricted to cytoplasmic sites, although replication for leporipoxviruses tends to be initiated somewhat more slowly than for VACV. The virus-encoded transcriptional apparatus is well conserved between the poxvirus genera, and many of the important regulatory signals that are utilized by VACV, such as promoters and transcription termination sequences, are also utilized with comparable efficiency in the leporipoxviruses. Thus, viral genes from one genus can be introduced to another by recombination or by DNA transfection technologies to generate chimeric virus constructs that maintain the correct regulation of the new genetic information. As in the case of VACV, transcriptional units can be of different kinetic classes (early/intermediate/late) and there is no splicing of viral mRNA.

The leporipoxviruses replicate in cytoplasmic factories that appear by microscopic analysis as eosinophilic B-type inclusion bodies. These factories, also called virosomes, can be visualized by Feulgen, Giemsa, or fluorescent antibody staining. SPV produces nuclear inclusions and vacuolations in addition to cytoplasmic bodies but these nuclear alterations are not believed to be sites of viral replication.

Molecular Mechanisms of Pathogenesis

Since these viruses are of only minor veterinary importance, recent research has focused on the elucidation of the determinants for viral virulence, particularly with respect to the strategies that these viruses employ to subvert the immune system of the infected host. Particular attention has been paid to the mechanism(s) underlying the immune dysfunction caused by MYX infection in *Oryctolagus* rabbits. To date, at least two classes of viral gene products have been directly implicated in the immunomodulation induced by these viruses:

1. 'Virokines' are secreted virus-encoded proteins that are targeted to host-specific pathways outside the infected cell. For example, SFV and MYX encode growth factors related to epidermal growth factor and transforming growth factor α that participate in stimulating fibroblastic proliferation at primary and secondary tumors.
2. 'Viroceptors' are viral proteins that mimic cellular receptors and function by sequestering important host cytokines that normally participate in the antiviral immune response. Leporipoxviral-encoded receptor-like molecules have been discovered for tumor necrosis factor (TNF) and interferon γ (IFN-γ), and may exist for other antiviral lymphokines as well. SPV encodes a novel homolog of cellular chemokine receptors, and the leporipoxviruses express secreted chemokine-binding proteins that are important for virus pathogenesis.

Interference with antigen presentation by MYX is also believed to play a role in circumventing T-cell recognition during early stages of virus infection. One MYX gene product responsible for evading immune clearance, designated Serp1, is an extracellular inhibitor of cellular serine proteinases, and the purified protein exhibits potent anti-inflammatory properties in a variety of animal model systems of inflammatory disease.

Geographic and Seasonal Distribution

All three major species of *Sylvilagus* rabbits in the Americas have endemic fibroma-like poxviruses, and myxomatosis is now established in wild *Oryctolagus* rabbit populations of South America, Europe, and Australia. SqFV and HFV have been reported to date only in North America and Europe, respectively. The leporipoxviruses in the wild undergo seasonal fluctuations that correlate well with increased populations of arthropod vectors in summer and autumn, most prominently mosquitoes. An exception to this is found in Britain, where the major vector of MYX is the flea, which is not as seasonally variable.

In the case of SPV, outbreaks are not tied to seasonal cycles but are generally associated with the degree of hog lice infestation.

Host Range and Virus Propagation

These viruses demonstrate a very restricted host range in terms of ability to cause disease, although viral replication can also occur in cultured cells from some nonsusceptible hosts as well. In some cases, viral replication in tissue culture monolayers or chicken chorioallantoic membranes produces 'foci' in which infected cells manifest minimal cytopathic effects, thus permitting macroscopic cell aggregations to develop. The extent of cytopathology is markedly influenced by both the cell type and the virus strain, and in some instances the infected cells may detach from the monolayer to produce visible plaques. When viral replication is relatively slow and the toxicity to the target cell sufficiently moderate, a chronically infected carrier culture can be established in which progeny virus production persists for extended passages. Although poxviruses cannot permanently transform primary cells into an immortalized state, cells persistently infected with the fibroma-inducing leporipoxviruses can assume many of the phenotypic characteristics associated with the transformed phenotype, such as novel morphology, growth in reduced serum, and ability to form colonies in soft agar. It is likely that some of these phenotypic characteristics are facilitated by secreted poxviral proteins (virokines) that mimic cellular mitogens, such as epidermal growth factor, and trigger neighboring cells into excessive proliferation.

In the cases of the benign leporipoxviruses and SPV, replication is restricted to dermal and subcutaneous sites, with occasional involvement of draining lymph nodes. However, MYX and MRV are unique in that they also replicate efficiently in lymphoid cells, such as macrophages, B cells, and T cells. MYX, like human immunodeficiency virus type 1, replicates in either resting or stimulated T cells, and can be readily isolated from splenocyte cultures. The molecular basis for the uniquely permissive nature of MYX replication in lymphocytes is unknown, but is unquestionably an important factor in the extreme virulence of myxomatosis. Several MYX genes have been identified (e.g., M-T2, M-T4, M-T5, and M11L) that express host range determinants that block the cellular apoptosis response to infection of lymphocytes.

Evolution and Genetic Variability

The deliberate release of MYX into rabbit populations of Australia, France, and Britain in the early 1950s provided a unique opportunity to study the natural selection pressures exerted on a particularly virulent virus/host interaction. There is an extensive literature on the ecological consequences of the feral rabbit eradication program, and the rapid evolution of myxomatosis in the wild is well documented. Although the original South American MYX virus strain that was introduced left very few survivors in selected populations, within a few years attenuated viral strains with reduced virulence took over and more resistant rabbits became predominant.

In terms of the categories of viral virulence, some strains of MYX are classified as highly virulent (e.g., Moses and Lausanne), and attenuated variants exist down to relatively nonpathogenic (e.g., neuromyxoma and the Nottingham strains). Little is known about the extent of genetic variation in other leporipoxviruses, although different isolates of SFV show marked variations in tumorigenicity. Generally, leporids that recover from infection with one member either become resistant or undergo partial protection from infection by another member.

SPV shares some antigenic crossreactivity with VACV, but neutralizing antibody does not confer cross-protection for secondary infection by members of different genera.

Transmission and Tissue Tropism

The principal mode of transmission is by biting arthropod vectors, and the major inoculation route is dermal. Since these viruses do not replicate in the vector, the transmission is purely mechanical and hence virus spread can be readily accomplished by alternative routes. Thus, mosquitoes, fleas, blackflies, ticks, lice, mites, and even thistles and the claws of predatory birds have all been implicated in leporipoxvirus transmission. The efficiency of transmission by arthropods is quite variable, and is related to viral titers in skin lesions as well as the size of the vector populations. There are no known respiratory or oral routes of infection with members of either genus, but in some infections, such as MYX in domestic rabbits, the disease can be transmitted by direct contact with ocular discharges or open cutaneous lesions.

The sui- and leporipoxviruses in their native hosts are specific for the epidermis or subdermis and usually do not progress to secondary sites, although draining lymph nodes can be affected. However, in the specific case of MYX infection of the domestic rabbit, the virus can propagate efficiently in lymphocytes and migrate via infected leukocytes through lymphatic channels to establish secondary sites of infection. Recently, it was shown that MYX can productively infect and kill a variety of human cancer cells, likely due to cell signaling changes associated with cellular transformation in the tumor cells.

Pathogenicity

The leporipoxviruses are restricted to rabbits, squirrels, and hares, and swinepox is found only in domestic pigs. For SFV infection of *Sylvilagus* rabbits, tumors can last for many months before regressing, whereas in *Oryctolagus* rabbits recovery is usually complete within a few weeks. Only MYX manifests dramatic alterations in pathogenicity when the European rabbit is infected. For all of these viruses, the immune status of the host rabbit plays a critical role; for example, in adult rabbits, SFV rarely causes disease symptoms except for the primary fibroblastic lesion, but in newborn or immunocompromised animals the infection can lead to invasive tumors and much higher titers of infectious virus in infected tissues. Agents such as cortisone, X-rays, or immunosuppressants can dramatically increase SFV tumor development, and chemical promoters like 3,4-benzopyrene or methylcholanthrene can predispose progression to invasive fibromatosis or even metastatic fibrosarcoma.

The ability to cause collapse of the host immune response, replicate in lymphocytes, and spread efficiently to secondary sites is a unique property of MYX in *Oryctolagus* rabbits. The myxomatosis syndrome can be associated with multiple external signs (e.g., South American MYX) or may have relatively fewer gross symptoms (e.g., California MYX) and mortalities can range up to 100%. Supervening Gram-negative bacterial infections in the respiratory tract and conjunctiva are often observed concomitantly during myxomatosis, particularly by the adventitious pathogens *Pasteurella multocida* or *Bordetella bronchoseptica*, and contribute to the lethality of the disease.

SPV is only mildly pathogenic in pigs although it can cause a minor level of mortality, usually associated with milk-feeding reduction in younger animals.

Clinical Features of Infection

The cutaneous tumors induced by the different leporipoxviruses in their natural hosts are clinically very similar to each other. The fibromas are rarely associated with any other symptoms, such as fever or appetite loss, and invariably regress as long as the animal is not otherwise immunocompromised. In the case of MYX in *Oryctolagus* rabbits, however, the symptoms rapidly become severe as the tumors fail to regress and the concomitant immunosuppression contributes to the lethal myxomatosis syndrome. The clinical features of myxomatosis are influenced by the genetic background of both the virus strain and the rabbit host. In the preacute form of the disease caused by California MYX, the rabbits succumb in less than 1 week, and often have only minor external symptoms, such as inflammation and edema of the eyelids. Skin hemorrhages can be observed in some cases and convulsions often precede death. In the acute form caused by South American strains of MYX, the rabbits survive 1–2 weeks and develop more distinctive symptoms. The primary tumor can be either flat and diffuse or protuberant, and secondary site tumors around the nose, eyes, and ears become prominent by 6–7 days, at which time purulent exudates from the nose and eyes frequently develop. The cutaneous tumors often become necrotic and a generalized immune dysfunction exacerbates the progressive secondary bacterial infestation of the respiratory tract. In the case of the more attenuated MYX isolates, such as neuromyxoma, the disease course is less severe and may be associated with little or no mortality.

The disease course of SPV in pigs is rather different, and resembles that of VACV in humans. Inoculation results in localized dermal papules, which progress to vesicles and pustules, after which the lesions crust and scab over. The only clinical symptom is occasional minor fever and the animals recover within 3 weeks.

Pathology and Histopathology

The primary tumors caused by leporipoxviruses in *Sylvilagus* rabbits, squirrels, and hares all closely resemble proliferant fibromas. Following inoculation, an acute inflammatory reaction occurs with infiltration of polymorphonuclear and mononuclear cells and proliferation of fibroblast-like cells of uncertain origin. The 'tumor' consists of pleomorphic cells imbedded in a matrix of intercellular fibrils of collagen. Unlike the transformed cells induced by other DNA tumor viruses, cells from poxviral tumors are not immortalized and cannot be propagated independently. Instead they appear to require secreted virus-encoded proteins in order to sustain the hyperproliferative state. Inclusion bodies characteristic of poxviral replication can be observed in the cytoplasms of epithelial and some fibroma cells. As the tumor develops, mononuclear leukocyte cuffing of adjacent vessels is observed and at the base of the tumor there is accumulation of lymphocytes, plasma cells, macrophages, and neutrophils. The ratio between influx of inflammatory cells and fibroblast proliferation is variable but generally there is little or no necrosis. The speed with which immune cells clear the viral infection and reverse the hyperproliferation can range from 1–2 weeks to 6 months, depending on both the virus and the host.

The principal difference between the benign fibroma syndrome described above and the devastating disease caused by MYX in *Oryctolagus* rabbits is that the latter virus efficiently propagates in host lymphocytes and is able to circumvent the cell-mediated immune response to the viral infection. The subcutaneous tumors consist of proliferating undifferentiated mesenchymal cells, which become large and stellate with prominent nuclei ('myxoma' cells). In surrounding tissue there can be extensive proliferation of endothelial cells of the local capillaries and venules, often to the point where complete occusion leads to extensive necrosis of the infected site. The overlying epithelial cells can show hyperplasia or degeneration, depending on the virus strain, and poxviral inclusion bodies are frequently observed in the prickle-cell layer. In some MYX strains, primary and secondary skin tumors can undergo extensive hemorrhage and internal lesions may be found in the stomach, intestines, and heart. The virus readily migrates to secondary sites within infected immune cells and concomitant cellular proliferation can be detected in the reticulum cells of lymph nodes and spleen, as well as the conjunctival and pulmonary alveolar epithelium. The nasal mucosa and conjunctiva overlying secondary tumors undergo squamous metaplasia such that the epithelia become nonciliated and nonkeratizing. Disruption of the ciliary architecture may be one of the factors that facilitate the extensive Gram-negative bacterial infections of the eyes, nose, and respiratory tract. Varying degrees of inflammatory cell infiltration by polymorphonuclear heterophils occur soon after infection but there is only a limited and ineffective cellular immune response. The lymph nodes and spleen show evidence of aberrant T-cell activation and hyperplasia, and infectious virus can be isolated from all lymphoid organs except the thymus. Death is believed to be caused by a combination of tissue damage from the increasing tumor burden, generalized immunosuppression, and debilitating bacterial colonization of the respiratory tract.

Little is known about SPV pathogenesis but gross features closely resemble those of the noninvasive orthopoxviruses in their native hosts.

Immune Response

The benign fibromas caused by SFV/SqFV/HFV regress, albeit slowly, due to a combined cellular and humoral immune response. These viruses are excellent antigens, and neutralizing antibody produced during recovery will also cross-react with other members of the genus. All of the leporipoxviruses are strongly cell associated, and cell-mediated immunity is probably the single most important mechanism of viral clearance. Other immune mechanisms are also activated, including interferon production, antibody-mediated cell lysis, sensitized macrophages, and natural killer cells. Neutralizing antibody can last for many months after viral clearance and immunity is usually cross-protective to the other leporipoxviruses.

In the unique case of MYX in *Oryctolagus* rabbits the picture is very different. Although circulating antibody can be detected against virions, as determined by neutralization or agglutination, and against soluble antigens, as determined by complement fixation and precipitin tests, the antibody provides little protection against the disease progression. Instead, cellular immunity is severely compromised, and by day 6–7 lymphocytes (especially splenocytes) are demonstrably dysfunctional in their response to mitogens and lose the ability to secrete critical cytokines such as interleukin-2. Unlike the case of SFV, there is a notable absence of virus-specific T cells in either the spleen or draining lymph nodes. Immune dysfunction is common for viruses that replicate in lymphocytes, but the precise levels at which MYX intervenes in cellular immunity remain to be clarified. There is some evidence that these viruses interfere with the function of cell surface major histocompatibility complex (MHC) class I molecules, which could prevent proper viral antigen presentation and hence interfere with immune recognition of infected cells. Also, several virus-specific gene products have been shown to be secreted homologs of the cellular receptors for TNF and IFN-γ that are believed to bind and sequester these extracellular ligands in the vicinity of virus-infected cells and thus short-circuit immune pathways dependent on TNF and IFN-γ.

SPV-infected pigs generally recover from the infection and become immune to secondary challenge. There are few data on the nature of this immunity, but it bears close resemblance to that of VACV immunization in humans.

Prevention and Control

Since these viruses are spread principally by biting arthropods, vector control is the single most effective method of disease prevention. The viruses are susceptible to standard anti-poxvirus chemical agents, such as phosphonoacetic acid, arabinosyl cytosine, and rifampicin, but these are of limited utility in infected animals. Immunization against myxomatosis can be accomplished with live SFV or attenuated strains of MYX.

Future Perspectives

Now that DNA sequencing studies have revealed the genomic repertoire of so many poxviruses, it is likely that more viral proteins which determine the clinical characteristics of their diseases will be characterized. Studies on viral gene products that stimulate fibroblastic and endothelial cells to proliferate will likely provide information on how mitogenesis is regulated by surface receptors on these target cells. Some of the secreted virokines and viroceptors have the potential to be used as drugs to treat inflammatory diseases. The ability of MYX to replicate in lymphocytes offers an important system in which to elucidate the mechanisms of cellular tropism by which these viruses suppress the innate apoptosis response to virus infection. Furthermore, the analysis of virus-induced immunosuppression should shed light on the various immune strategies used by the host to combat viral infections in general. The restricted host ranges of the lepori- and suipoxviruses suggest the potential for the genetic manipulation of these viruses such that heterologous foreign antigen genes can be expressed for the purpose of developing novel vaccines against important pathogens of domestic leporids and swine. Finally, the ability of MYX to infect and kill many human tumor cells offers the potential as a therapy against human cancer.

See also: History of Virology: Vertebrate Viruses; Poxviruses; Vaccinia Virus; Viral Pathogenesis.

Further Reading

Barrett JW, Cao J-X, Hota-Mitchell S, and McFadden G (2001) Immunomodulatory proteins of myxoma virus. *Seminars in Immunology* 13: 73–84.

DiGiacomo RF and Maré CJ (1994) Viral diseases. In: Manning P, Ringler DH, and Newcomer CE (eds.) *The Biology of the Laboratory Rabbit,* 2nd edn., p. 171. San Diego, CA: Academic Press.

Fenner F and Radcliffe FN (1965) *Myxomatosis.* Cambridge: Cambridge University Press.

Johnston JB and McFadden G (2004) Technical knockout: Understanding poxvirus pathogenesis by selectively deleting viral immunomodulatory genes. *Cellular Microbiology* 6: 695–705.

Kerr PJ and Best SM (1998) Myxoma virus in rabbits. *Revue Scientifique et Technique de l'office International des Épizooties* 17: 256–268.

Kerr P and McFadden G (2002) Immune responses to myxoma virus. *Viral Immunology* 15: 229–246.

Lucas A and McFadden G (2004) Secreted immunomodulatory proteins as novel biotherapeutics. *Journal of Immunology* 173: 4765–4774.

Nazarian SH and McFadden G (2006) Immune evasion by poxviruses. *Future Medicine* 1: 129–132.

Seet BT, Johnston JB, Brunetti CR, *et al.* (2003) Poxviruses and immune evasion. *Annual Review of Immunology* 21: 377–423.

Sypula J, Wang F, Ma Y, Bell J, and McFadden G (2004) Myxoma virus tropism in human tumor cells. *Gene Therapy and Molecular Biology* 8: 103–114.

Relevant Website

http://www.poxvirus.org – Poxvirus Bioinformatics Resource Center.

Luteoviruses

L L Domier, USDA–ARS, Urbana, IL, USA
C J D'Arcy, University of Illinois at Urbana-Champaign, Urbana, IL, USA

© 2008 Elsevier Ltd. All rights reserved.

Glossary

Hemocoel The primary body cavity of most arthropods that contains most of the major organs and through which the hemolymph circulates.
Hemolymph A circulatory fluid in the body cavities (hemocoels) and tissues of arthropods that is analogous to blood and/or lymph of vertebrates.

Introduction

Viruses of the family *Luteoviridae* (luteovirids) cause economically important diseases in many monocotyledonous and dicotyledonous crop plants, including barley, wheat, potatoes, lettuce, legumes, and sugar beets. Yield reductions as high as 30% have been reported in epidemic years, although in some cases crops can be totally destroyed. Diseases caused by the viruses were recorded decades and even centuries before they were associated with the causal viruses. In many cases, the stunted, deformed, and discolored plants that result from luteovirid infection were thought to be the result of abiotic factors, such as mineral imbalances or stressful environmental conditions, or of other biotic agents. This, along with their inabilities to be transmitted mechanically, delayed the initial association of the symptoms with plant viruses. For example, curling of potato leaves was first described in Lancashire, UK, in the 1760s, but was not recognized as a specific disease of potato until 1905 and to be caused by an aphid-transmitted virus until the 1920s. The causal agent, potato leaf roll virus (PLRV), was not purified until the 1960s. Similarly, widespread disease outbreaks in cereals, probably caused by barley yellow dwarf virus (BYDV), were noted in the United States in 1907 and 1949. In 1951, a virus was proposed as the cause. Other diseases caused by luteovirids, like sugarcane yellow leaf, which is caused by sugarcane yellow leaf virus (ScYLV), were not described until the 1990s.

Taxonomy and Classification

Members of the family *Luteoviridae* were first grouped because of their common biological properties. These properties included persistent transmission by aphid vectors and the induction of yellowing symptoms in many infected host plants. 'Luteo' comes from the Latin *luteus*, which translates as yellowish. All luteovirids have small (*c.* 25 nm diameter) icosahedral particles, composed of one major and one minor protein component and a single molecule of positive-sense single-stranded RNA of approximately 5600 nt in length.

The family *Luteoviridae* is divided into three genera – *Luteovirus*, *Polerovirus* (derived from potato leaf roll), and *Enamovirus* (derived from pea enation mosaic) – based on the arrangements, sizes, and phylogenetic relationships of the predicted amino acid sequences of the open reading frames (ORFs). In some plant virus families, a single gene can be used to infer taxonomic and phylogenetic relationships. Within the family *Luteoviridae*, however, different taxonomic relationships can be predicted depending on whether sequences of the replicase (ORF2) or coat protein (CP; ORF3) genes are analyzed (**Figure 1**). ORFs 1 and 2 of the luteoviruses are most closely related to the polymerase genes of viruses of the family *Tombusviridae*, while ORFs 1 and 2 of the poleroviruses and enamoviruses are related to those of the genus *Sobemovirus*. These polymerase types are distantly related in evolutionary terms. Consequently, it has been suggested that luteovirid genomic RNAs arose by recombination between ancestral genomes containing the CP genes characteristic of the family *Luteoviridae* and genomes containing either of the two polymerase types. For taxonomic purposes, the polymerase type has been the primary determinant in assigning a virus to a genus. For this reason, viruses for which only CP sequences have been determined have not been assigned to a genus. The current members of the family are listed in **Table 1**. The genus *Luteovirus* contains five species, and the *Polerovirus* genus has nine species. The genus *Enamovirus* contains a single virus, pea enation mosaic virus 1 (PEMV-1). The family also contains 11 virus species that have not been assigned to a genus. Of these, recently determined sequences of genomic RNAs of BYDV-GAV and carrot red leaf virus (CtRLV) suggest that BYDV-GAV is a strain of BYDV-PAV and that CtRLV is a unique species in the genus *Polerovirus*.

Virion Properties

The sedimentation coefficients $S_{20,w}$ (in Svedberg units) for luteoviruses and poleroviruses range from 106S to 127S. Buoyant densities in CsCl are approximately $1.40 \, g \, cm^{-3}$. The particles formed as result of the mixed

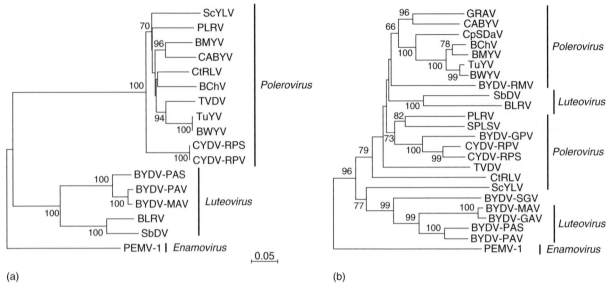

Figure 1 Phylogenetic relationships of the predicted amino acid sequences of the (a) RNA-dependent RNA polymerase (ORF2) and (b) major capsid protein (ORF3). When predicted amino acid sequences from ORF2 are used to group virus species the genera form three distinct groups. Using predicted amino acid sequences from ORF3, species of the genera *Luteovirus* and *Polerovirus* are intermingled in the tree. The resulting consensus trees from 1000 bootstrap replications are shown. The numbers above each node indicate the percentage of bootstrap replicates in which that node was recovered. For virus abbreviations, see **Table 1**.

Table 1 Virus members in the family *Luteoviridae*

Genus	Virus	Abbreviation	Accession number
Luteovirus	Barley yellow dwarf virus – MAV	BYDV-MAV	NC_003680[a]
	Barley yellow dwarf virus – PAS	BYDV-PAS	NC_002160
	Barley yellow dwarf virus – PAV	BYDV-PAV	NC_004750
	Bean leafroll virus	BLRV	NC_003369
	Soybean dwarf virus	SbDV	NC_003056
Polerovirus	Beet chlorosis virus	BChV	NC_002766
	Beet mild yellowing virus	BMYV	NC_003491
	Beet western yellows virus	BWYV	NC_004756
	Cereal yellow dwarf virus – RPS	CYDV-RPS	NC_002198
	Cereal yellow dwarf virus – RPV	CYDV-RPV	NC_004751
	Cucurbit aphid-borne yellows virus	CABYV	NC_003688
	Potato leafroll virus	PLRV	NC_001747
	Turnip yellows virus	TuYV	NC_003743
	Sugarcane yellow leaf virus	ScYLV	NC_000874
Enamovirus	Pea enation mosaic virus 1	PEMV-1	NC_003629
Unassigned	Barley yellow dwarf virus – GAV	BYDV-GAV	NC_004666
	Barley yellow dwarf virus – GPV	BYDV-GPV	L10356
	Barley yellow dwarf virus – RMV	BYDV-RMV	Z14123
	Barley yellow dwarf virus – SGV	BYDV-SGV	U06865
	Carrot red leaf virus	CtRLV	NC_006265
	Chickpea stunt disease associated virus	CpSDaV	Y11530
	Groundnut rosette assistor virus	GRAV	Z68894
	Indonesian soybean dwarf virus	ISDV	
	Sweet potato leaf speckling virus	SPLSV	
	Tobacco necrotic dwarf virus	TNDV	
	Tobacco vein distorting virus	TVDV	AJ575129

[a]Accession numbers beginning with NC_ represent complete genomic sequences.

infections by PEMV-1 and PEMV-2 sediment as two components. The $S_{20,w}$ are 107–122S for B components (PEMV-1) and 91–106S for T components (PEMV-2, an umbravirus). Virions are moderately stable and are insensitive to treatment with chloroform or nonionic detergents, but are disrupted by prolonged treatment with high concentrations of salts. Luteovirus and polerovirus particles are insensitive to freezing.

Virion Structure and Composition

All members of the *Luteoviridae* have nonenveloped icosahedral particles with diameters of 25–28 nm (**Figure 2**). Capsids are composed of major (21–23 kDa) and minor (54–76 kDa) CPs, which contain a C-terminal extension to the major CP called the readthrough domain (RTD). According to X-ray diffraction and molecular mass analysis, virions consist of 180 protein subunits, arranged in $T = 3$ icosahedra. Virus particles do not contain lipids or carbohydrates.

Virions contain a single molecule of single-stranded positive-sense RNA of 5300–5900 nt. The RNAs do not have a 3′ terminal poly(A) tract. A small protein (VPg) is covalently linked to the 5′ end of polerovirus and enamovirus genomic RNAs. Cereal yellow dwarf virus RPV (CYDV-RPV) also encapsidates a 322 nt satellite RNA that accumulates to high levels in the presence of the helper virus. Complete genome sequences have been determined for 17 members of the *Luteoviridae* (**Table 1**). For several viruses, genome sequences have been determined from multiple isolates.

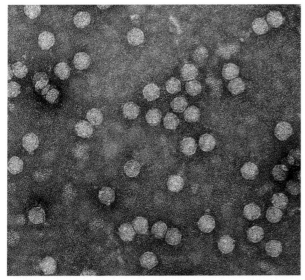

Figure 2 Transmission electron micrograph of soybean dwarf virus particles magnified 240 000×. Virions (stained with uranyl acetate) are *c*. 25 nm in diameter, hexagonal in appearance, and have no envelope.

Genome Organization and Expression

Genomic RNAs of luteovirids contain five to eight ORFs (**Figure 3**). ORFs 1, 2, 3, and 5 are shared among all members of the *Luteoviridae*. Luteoviruses lack ORF0. Enamoviruses lack ORF4. The luteo- and polerovirus genomes contain one or two small ORFs, ORFs 6 and 7, within or downstream of ORF5. An additional ORF, ORF8, has been discovered in ORF1 of PLRV. In the enamo- and poleroviruses ORF0 overlaps ORF1 by more than 600 nt, which also overlaps ORF2 by more than 600 nt. In the luteoviruses, ORF1 overlaps ORF2 by less than 50 nt. In all luteo- and polerovirus genome sequences (except for cucurbit aphid-borne yellows virus (CABYV) and GRAV), ORF4 is contained within ORF3. A single, in-frame amber (UAG) termination codon separates ORF5 from ORF3.

Luteovirids have relatively short 5′ and intergenic noncoding sequences. The first ORF is preceded by 21 nt in CABYV RNA and 142 nt in soybean dwarf virus (SbDV) RNA. ORFs 2 and 3 are separated by 112–200 nt of noncoding RNA. There is considerable variation in the length of sequence downstream of ORF5, which ranges from 125 nt for CYDV-RPV to 650 nt for SbDV.

Luteovirids employ an almost bewildering array of strategies to express their compact genomes. ORFs 0, 1, 2, and 8 are expressed directly from genomic RNA. Downstream ORFs are expressed from subgenomic RNAs (sgRNAs) that are transcribed from internal initiation sites by virus-encoded RNA-dependent RNA polymerases (RdRps) from negative-strand RNAs and are 3′-co-terminal with the genomic RNA. Since the initiation codon for ORF0 of polero- and enamoviruses is upstream of that of ORF1, translation of ORF1 is initiated by 'leaky scanning' in which ribosomes bypass the AUG of ORF0 and continue to scan the genomic RNA until they reach the ORF1 AUG. The protein products of ORF2 are expressed as a translational fusion with the product of ORF1. At a low but significant frequency during the expression of ORF1, translation continues into ORF2 through a −1 frameshift that produces a large protein containing sequences encoded by both ORFs 1 and 2 in a single polypeptide. ORF8, which has only been identified in PLRV, resides entirely within ORF1 in a different reading frame and encodes a 5 kDa replication-associated protein. To express ORF8, sequences within the ORF fold into a structure called an internal ribosome entry site (IRES), which recruits ribosomes to initiate translation about 1600 nt downstream of the 5′ terminus of PLRV RNA.

ORFs 3, 4, and 5 are expressed from sgRNA1, the 5′ terminus of which is located about 200 nt upstream of ORF3 at the end of ORF2, and extends to the 3′ terminus of the genome. Luteo- and poleroviruses produce a second sgRNA that expresses ORFs 6 and 7. Luteoviruses produce a third sgRNA, which does not appear to encode

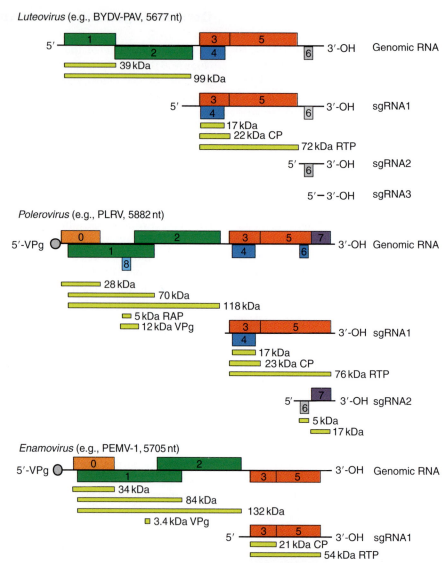

Figure 3 Maps of the virus genomes of genera in the family *Luteoviridae*. Individual ORFs are shown with open boxes. The ORFs are staggered vertically to show the different reading frames occupied by each ORF. The yellow boxes indicate protein products with the predicted sizes listed to the right of each. The polyproteins encoded by ORF1 of enamo- and poleroviruses contain the protease and the genome-linked protein (VPg). The predicted amino acid sequences of proteins encoded by ORF2 are similar to RNA-dependent RNA polymerases. ORF3, which encodes the major coat protein, is separated from ORF5 by an amber termination codon. ORF4, when present, is contained within ORF3 and encodes a protein required for virus cell-to-cell movement.

a protein. ORF3 is translated from the 5′ terminus of sgRNA1. ORF4 of luteo- and poleroviruses, which encodes a 17 kDa protein, is contained within ORF3, and is expressed from the same sgRNA as ORF3 through a leaky scanning mechanism much like that used to express ORF1 in polero- and enamoviruses. In all luteovirids, ORF5 is expressed only as a translational fusion with the products of ORF3 by readthrough of the UAG stop codon at the end of ORF3. This produces a protein with the product of ORF3 at its N-terminus and the product of ORF5 at its C-terminus.

While enamo- and polerovirus RNAs contain 5′ VPgs that interact with translation initiation factors, luteovirus RNAs contain only a 5′ phosphate. Unmodified 5′ termini are recognized poorly for translation initiation. To circumvent this problem, a short sequence located in the noncoding region just downstream of ORF5 in the BYDV-PAV genome acts as a potent enhancer of cap-independent translation by interacting with sequences near the 5′ termini of the genomic and sgRNAs to promote efficient translation initiation.

Research into the functions of the proteins encoded by luteovirids has shown that the 28–34 kDa proteins encoded by ORF0 are effective inhibitors of post-transcriptional gene silencing (PTGS). PTGS is an innate and highly adaptive antiviral defense found in all eukaryotes that is activated by double-stranded RNAs (dsRNAs), which are produced during virus replication. Consequently, viruses that contain mutations in ORF0 show greatly reduced accumulations in infected plants.

The ORF1-encoded proteins of enamo- and poleroviruses contain the VPg and a chymotrypsin-like serine protease that is responsible for the proteolytic processing of ORF1-encoded polyproteins. The protease cleaves the ORF1 proteins in *trans* to liberate the VPg, which is covalently attached to genomic RNA. The protein expressed by ORF8 of PLRV is required for virus replication. Luteovirid ORF2s have a coding capacity of 59–67 kDa for proteins that are very similar to known RdRps and hence likely represent the catalytic portion of the viral replicase.

ORF3 encodes the major CP of the luteovirids, which ranges in size from 21 to 23 kDa. ORF5 has a coding capacity of 29–56 kDa. However, ORF5 is expressed only as a translational fusion with the product of ORF3 when, about 10% of the time, translation does not stop at the end of ORF3 and continues through to the end of ORF5. The ORF5 portion of this readthrough protein has been implicated in aphid transmission and virus stability. Experiments with PLRV and BYDV-PAV have shown that the N-terminal region of the ORF5 readthrough protein determines the ability of virus particles to bind to proteins produced by endosymbiotic bacteria of aphid vectors. Interactions of virus particles with these proteins seem to be essential for persistence of the viruses in aphids. Nucleotide sequence changes within ORF5 of PEMV-1 abolish aphid transmissibility. The N-terminal portions of ORF5 proteins are highly conserved among luteovirids while the C-termini are much more variable.

The luteo- and polerovirus genomes possess an ORF4 that is contained within ORF3 and encodes proteins of 17–21 kDa. Viruses that contain mutations in ORF4 are able to replicate in isolated plant protoplasts, but are deficient or delayed in systemic movement in whole plants. Hence, the product of ORF4 seems to be required for movement of the virus within infected plants. This hypothesis is supported by the observation that enamoviruses lack ORF4. While luteo- and poleroviruses are limited to phloem and associated tissues, the enamovirus PEMV-1 is able to move systemically through other plant tissues in the presence of PEMV-2, which under natural conditions invariably coexists with PEMV-1.

Some luteo- and polerovirus genomes contain small ORFs within and/or downstream of ORF5. In luteoviruses, no protein products have been detected from these ORFs in infected cells. BYDV-PAV genomes that do not express ORF6 are still able to replicate in protoplasts. The predicted sizes of the proteins expressed by ORFs 6 and 7 of PLRV are 4 and 14 kDa, respectively. Based on mutational studies, it has been proposed that these genome regions may regulate transcription late in infection.

Evolutionary Relationships among Members of the *Luteoviridae*

Viruses in the family *Luteoviridae* have replication-related proteins that are similar to those in other plant virus families and genera. The luteovirus replication proteins encoded by ORFs 1 and 2 resemble those of members of the family *Tombusviridae*. In contrast, polymerases of poleroviruses and enamoviruses resemble those of viruses in the genus *Sobemovirus*. The structural proteins of some sobemoviruses also are similar to the major CP of luteovirids. Using an X-ray crystallography-derived structure of virions of the sobemovirus rice yellow mottle virus, which shares a CP amino acid sequence similarity of 33% with PLRV, it was possible to predict the virion structure of PLRV and other luteovirids.

Host Range and Transmission

Several luteovirids have natural host ranges largely restricted to one plant family. For example, BYDV and CYDV infect many grasses, BLRV infects mainly legumes, and CtRLV infects mainly plants in the family Apiaceae. Other luteovirids infect plants in several or many different families. For example, beet western yellows virus (BWYV) infects more than 150 species of plants in more than 20 families. As techniques for infecting plants with recombinant viruses have improved, the experimental host ranges of viruses have been expanded to include plants on which aphid vectors would not normally feed. For example, BYDV, CYDV, PLRV, and SbDV have been shown to infect *Nicotiana* species that had not been described previously as experimental hosts for the viruses when inoculated biolistically with viral RNA or using *Agrobacterium tumefaciens* harboring binary plasmids containing infectious copies of the viruses. These results suggest that feeding preferences of vector aphids play important roles in defining luteovirid host ranges.

Luteovirids are transmitted in a circulative manner with varying efficiencies by at least 25 aphid species. With the exception of the enamovirus PEMV-1, members of the family *Luteoviridae* are transmitted from infected plants to healthy plants in nature only by the feeding activities of specific species of aphids. There is no evidence for replication of the viruses within aphid vectors. *Myzus persicae* is the most common aphid vector of luteovirids that infect dicots. Several different species of aphids transmit luteovirids that infect monocotyledenous plants (BYDV and CYDV) in a species-specific manner.

Circulative transmission of the viruses is initiated when aphids acquire viruses from sieve tubes of infected plants during feeding. The viruses travel up the stylet, through the food canal, and into the foregut (**Figure 4**). The viruses then are actively transported across the cells

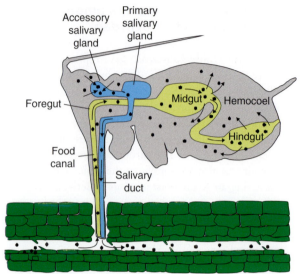

Figure 4 Circulative transmission of viruses of the family *Luteoviridae* by vector aphids. While feeding from sieve tubes of an infected plant, an aphid (shown in cross section) acquires virus particles, which travel up the stylet, through the food canal, and into the foregut. The virions are actively transported across cells of the posterior midgut and/or hindgut into the hemocoel in a process that involves receptor-mediated endocytosis. Virions then passively migrate through the hemolymph to the accessory salivary gland where they are again transported by a receptor-mediated process to reach the lumen of the gland. Once in the salivary gland lumen, the virions are expelled with the saliva into the vascular tissue of host plants. Aphids can retain the ability to transmit virus for several weeks. Hindgut membranes usually are much less selective than those of the accessory salivary glands, which is why viruses that are not transmitted by a particular species of aphid often accumulate in the hemocoel, but do not traverse the membranes of the accessory salivary gland.

of the alimentary tract into the hemocoel in a process that involves receptor-mediated endocytosis of the viruses and the formation of tubular vesicles that transport the viruses through the epithelial cells and into the hemocoel. Luteovirids are acquired at different sites within the gut of vector aphids. PLRV and BWYV are acquired in the posterior midgut. BYDV, CYDV, and SbDV are acquired in the hindgut. CABYV is taken up at both sites. Viruses then passively migrate through the hemolymph to the accessory salivary gland where the viruses must pass through the membranes of the accessory salivary gland cells in a similar type of receptor-mediated transport process to reach the lumen of the gland. Once in the salivary gland lumen, viruses are expelled with saliva into vascular tissues of host plants. Since large amounts of virus can accumulate in the hemocoel of aphids, they may retain the ability to transmit virus for several weeks. Typically hindgut membranes are much less selective than those of the accessory salivary glands. Consequently, viruses that are not transmitted by a particular species of aphid often are transported across gut membranes and accumulate in the hemocoel, but do not traverse the membranes of the accessory salivary gland.

The RTD of the minor capsid protein plays a major role in aphid transmission of luteovirids. The RTD interacts with symbionin produced by endosymbiotic aphid-borne bacteria, which may protect virions from degradation by the aphid immune system. The specificity of aphid transmission and gut tropism has been linked to the RTD in multiple luteovirids.

Unlike other luteovirids, PEMV-1 can be transmitted by rubbing sap taken from an infected plant on a healthy plant, in addition to being transmitted by aphids. This difference in transmissibility is dependent on its multiplication in cells co-infected with PEMV-2, but aphid transmissibility can be lost after several mechanical passages.

Replication

Luteovirids infect and replicate in sieve elements and companion cells of the phloem and occasionally are found in phloem parenchyma cells. PEMV-1 is able to move systemically into other tissues in the presence of PEMV-2. Virus infections commonly result in cytopathological changes in cells that include formation of vesicles containing filaments and inclusions that contain viral RNA and virions. The subcellular location of viral RNA replication has not been determined unequivocally. However, early in infection, negative-strand RNAs of BYDV-PAV are first detected in the nucleus and later in the cytoplasm, which suggests that at least a portion of luteovirus replication occurs in the nucleus. Synthesis of negative-strand RNA, which requires tetraloop structures at the $3'$ end of BYDV-PAV genomic RNAs, is detected in infected cells before the formation of virus particles. Late in infection, BYDV-PAV sgRNA2 inhibits translation from genomic RNA, which may promote a switch from translation to replication and packaging of genomic RNAs.

Virus–Host Relationships

While some infected plants display no obvious symptoms, most luteovirids induce characteristic symptoms that include stunting, leaves that become thickened, curled or brittle, and yellow, orange, or red leaf discoloration, particularly of older leaves of infected plants. These symptoms result from phloem necrosis that spreads from inoculated sieve elements and causes symptoms by inhibiting translocation, slowing plant growth, and inducing the loss of chlorophyll. Symptoms may persist, may vary seasonally, or may disappear soon after infection. Temperature and light intensity often affect symptom severity and development. In addition, symptoms can vary greatly with different virus isolates or strains and with different host cultivars.

Yield losses caused by luteovirids are difficult to estimate because the symptoms often are overlooked or attributed to other agents. US Department of Agriculture specialists estimated that yield losses from BWYV, BYDV, and PLRV infections were over $65 million during the period 1951–60. Plants infected at early stages of development by luteovirids suffer the most significant yield losses, which often are linearly correlated with the incidence of virus infection.

Epidemiology

Luteovirid infections have been reported from temperate, subtropical, and tropical regions of the world. Some of the viruses are found worldwide, such as BWYV, BYDV, and PLRV. Others have more restricted distribution, such as tobacco necrotic dwarf virus, which has been reported only from Japan, and groundnut rosette assistor virus, which has been reported only in African countries south of the Sahara.

Most luteovirids infect annual crops and must be reintroduced each year by their aphid vectors. Some viruses are disseminated in infected planting material, like PLRV where infected potato tubers are the principal source of inoculum for new epidemics. Consequently, programs to produce clean stock are operated around the world to control these viruses. Alate, that is, winged, aphid vectors may transmit viruses from local cultivated, volunteer, or weed hosts. Alternatively, alate aphids may be transported into crops from distant locations by wind currents. These vectors may bring the virus with them, or they may first have to acquire virus from locally infected hosts. The agronomic impact of luteovirid diseases depends both on meteorological events that favor movement and reproduction of vector aphids and susceptibility of the crop at the time of aphid arrival. Only aphid species that feed on a particular crop plant can transmit virus. Aphids that merely probe briefly to determine a plant's suitability will not transmit the viruses. Secondary spread of the viruses is often primarily by apterous, that is, wingless, aphids. The relative importance of primary introduction of virus by alate aphids and of secondary spread of virus by apterous aphids in disease severity varies with the virus, aphid species, crop, and environmental conditions.

Some members of the family *Luteoviridae* occur in complexes with other members of the family or with other plant viruses. For example, BYDV and CYDV often are found co-infecting cereals; BWYV and SbDV are often found together in legumes; and PLRV is often found co-infecting potatoes with potato virus Y and/or potato virus X. Some other plant viruses depend on luteovirids for their aphid transmission, such as the groundnut rosette virus, carrot mottle virus, and bean yellow vein banding virus (all umbraviruses), which depend on groundnut rosette assistor virus, carrot red leaf virus, and PEMV-1, respectively.

Diagnosis

An integral part of controlling luteovirid diseases is accurate diagnosis of infection. Because symptoms caused by luteovirids often resemble those caused by other biotic and abiotic factors, visual diagnosis is unreliable and other methods have been developed. Initially, infectivity, or biological, assays were used to diagnose infections. These techniques also have been used to identify species of vector aphids and vector preferences. In bioassays, aphids are allowed to feed on infected plants and then are transferred to indicator plants. These techniques are very sensitive, but can require several weeks for symptoms to develop on indicator plants. The strong immunogenicity of luteovirids has facilitated development of very specific and highly sensitive serological tests that can discriminate different luteovirids and sometimes even strains of a single virus species. Poly- and monoclonal antibodies for virus detection are produced by immunizing rabbits and/or mice with virus particles purified from infected plants. Techniques also have been developed to detect viral RNAs from infected plant tissues by reverse transcription-polymerase chain reaction (RT-PCR), which can be more sensitive and discriminatory than serological diagnostic techniques. Even so, serological tests are the most commonly used techniques for the detection of infections because of their simplicity and speed.

Control

Because methods are not available to cure luteovirid infections after diagnosis, emphasis has been placed on reducing losses through the use of tolerant or resistant plant cultivars and/or on reducing the spread of viruses by controlling aphid populations. Many luteovirids are transmitted by migrating populations of aphids that occur at similar times each year. For those virus–aphid combinations, it is sometimes possible to plant crops so that young, highly susceptible plants are not in the field when the seasonal aphid migrations occur. Insecticides have been used in a prophylactic manner to reduce crop losses. While insecticide treatments do not prevent initial infections, they can greatly limit secondary spread of aphids and therefore of viruses. In some instances biological control agents such as predatory insects and parasites have reduced aphid populations significantly. Genes for resistance or tolerance to infection by luteovirids have been identified in most agronomically important plant species infected by the viruses. For BYDV, PLRV, and SbDV, transgenic plants that express portions of the virus genomes have been produced through DNA-mediated transformation. In some cases, the expression of these virus genes in transgenic plants confers higher levels of virus resistance than resistance genes from plants.

See also: Barley Yellow Dwarf Viruses; Cereal Viruses: Wheat and Barley; Sobemovirus; Tombusviruses.

Further Reading

Brault V, Perigon S, Reinbold C, et al. (2005) The polerovirus minor capsid protein determines vector specificity and intestinal tropism in the aphid. *Journal of Virology* 79: 9685–9693.

Falk BW, Tian T, and Yeh HH (1999) Luteovirus-associated viruses and subviral RNAs. *Current Topics in Microbiology and Immunology* 239: 159–175.

Gray S and Gildow FE (2003) Luteovirus–aphid interactions. *Annual Review of Phytopathology* 41: 539–566.

Hogenhout SA, van der Wilk F, Verbeek M, Goldbach RW, and van den Heuvel JF (2000) Identifying the determinants in the equatorial domain of Buchnera GroEL implicated in binding potato leafroll virus. *Journal of Virology* 74: 4541–4548.

Lee L, Palukaitis P, and Gray SM (2002) Host-dependent requirement for the potato leafroll virus 17-kDa protein in virus movement. *Molecular Plant–Microbe Interactions* 10: 1086–1094.

Mayo M, Ryabov E, Fraser G, and Taliansky M (2000) Mechanical transmission of potato leafroll virus. *Journal of General Virology* 81: 2791–2795.

Miller WA and White KA (2006) Long-distance RNA–RNA interactions in plant virus gene expression and replication. *Annual Review of Phytopathology* 44: 447–467.

Moonan F, Molina J, and Mirkov TE (2000) Sugarcane yellow leaf virus: An emerging virus that has evolved by recombination between luteoviral and poleroviral ancestors. *Virology* 269: 156–171.

Nass PH, Domier LL, Jakstys BP, and D'Arcy CJ (1998) In situ localization of barley yellow dwarf virus PAV 17-kDa protein and nucleic acids in oats. *Phytopathology* 88: 1031–1039.

Nixon PL, Cornish PV, Suram SV, and Giedroc DP (2002) Thermodynamic analysis of conserved loopstem interactions in P1–P2 frameshifting RNA pseudoknots from plant *Luteoviridae*. *Biochemistry* 41: 10665–10674.

Pfeffer S, Dunoyer P, Heim F, et al. (2002) P0 of beet western yellows virus is a suppressor of posttranscriptional gene silencing. *Journal of Virology* 76: 6815–6824.

Robert Y, Woodford JA, and Ducray-Bourdin DG (2000) Some epidemiological approaches to the control of aphid-borne virus diseases in seed potato crops in Northern Europe. *Virus Research* 71: 33–47.

Taliansky M, Mayo MA, and Barker H (2003) *Potato leafroll virus*: A classic pathogen shows some new tricks. *Molecular Plant Pathology* 4: 81–89.

Terradot L, Souchet M, Tran V, and Giblot Ducray-Bourdin D (2001) Analysis of a three-dimensional structure of potato leafroll virus coat protein obtained by homology modeling. *Virology* 286: 72–82.

Thomas PE, Lawson EC, Zalewski JC, Reed GL, and Kaniewski WK (2002) Extreme resistance to potato leafroll virus in potato cv. Russet Burbank mediated by the viral replicase gene. *Virus Research* 71: 49–62.

Lymphocytic Choriomeningitis Virus: General Features

R M Welsh, University of Massachusetts Medical School, Worcester, MA, USA

© 2008 Elsevier Ltd. All rights reserved.

History

Lymphocytic choriomeningitis virus (LCMV), of the family *Arenaviridae*, is an etiological agent for human acute aseptic meningitis and grippe-like infections and is maintained in nature by lifelong persistent infections of mice (*Mus musculus*). Three strains still being studied today were initially isolated in the USA in the 1930s. During an attempt to recover and passage virus from a suspected human case of St. Louis encephalitis, Armstrong and Lillie isolated LCMV (Armstrong strain) from a monkey that developed a lymphocytic choriomeningitis – hence the name. Traub then reported the isolation of a serologically indistinguishable virus contaminating his mouse colony (Traub strain), and Rivers and Scott isolated similar viruses from human meningitis patients (one being the WE strain). Subsequently, many other isolations from man and animals were made, and a clear cause–effect relationship between LCMV and about 8% of the US cases of human nonbacterial meningitis was established.

The LCMV infection of the mouse soon became an important model system for studying viral immunology. A distinguishing feature of the LCMV infection was that it established a long-term persistent infection in mice infected *in utero* or shortly after birth, whereas adult mice inoculated with LCMV either cleared the virus with lasting immunity or, in the case of an intracranial infection, died of a lethal meningoencephalitis (**Figure 1**). Antiviral antibody was difficult to detect in the persistently infected mice, and this led Burnet and Fenner to postulate that exposure to viral antigens before the maturation of the immune system resulted in mice becoming immunologically tolerant to LCMV and thus unable to clear the infection. This was one of the systems that provided the basis for Burnet's Nobel Prize-winning theories of immunological tolerance. Subsequent work by Oldstone and Dixon, however, demonstrated that persistently infected mice did make antiviral antibody, but it was difficult to detect because of excess viral antigen. The antiviral antibody traveled in the circulation complexed

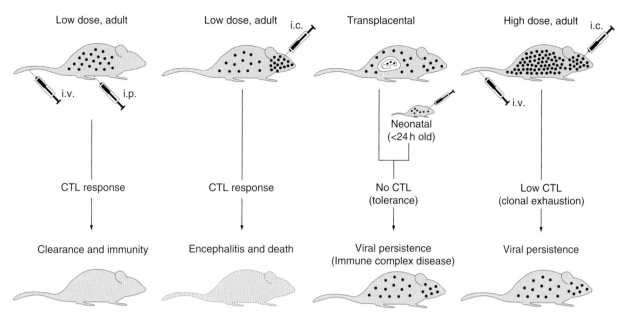

Figure 1 Pathogenesis of the LCMV infection is dependent on the age of the host, the route of infection, and the dose of the virus.

to virus and complement, and these circulating as well as tissue-bound immune complexes contributed to a progressive degenerative disease involving glomerulonephritis, arteritis, and chronic inflammatory lesions. In the acute infection both the clearance of the virus and the lethal meningoencephalitis were shown to be due mostly to the cytotoxic T lymphocyte (CTL) response. Antiviral CTLs were first demonstrated in the LCMV model, and Zinkernagel and Doherty used this model to demonstrate the Nobel Prize-winning concept of major histocompatibility complex (MHC) restriction in CTL recognition. Persistently infected mice were eventually shown to have split-tolerance to LCMV, in that, although they could mount an antibody response to LCMV, they could not generate the LCMV-specific CTLs which were needed to clear the infection.

Taxonomy and Classification

LCMV is the prototype virus of the *Arenaviridae* family of RNA viruses, and, although it has some homology with all arenaviruses, it is most closely related to the African virus *Lassa* and is classified as an Old World arenavirus. Its ambisense genome consists of two single-stranded RNAs, each encoding two genes of opposite polarity and separated by intergenic regions with strong RNA secondary structure. The small S RNA (1.1×10^3 kDa) in the virion encodes in the negative-sense a 63 kDa nucleoprotein (NP) and in the positive- or message-sense a 75 kDa cell-associated glycoprotein (GPC), which is cleaved into two virion glycoproteins, GP1 (44 kDa) and GP2 (35 kDa). The large L RNA (2.3×10^3 kDa) in the virion encodes in the negative-sense a 200 kDa RNA-dependent RNA polymerase (L) and in the positive-sense a smaller 11 kDa zinc-binding protein (Z) which binds to the ribonucleoprotein complex and is involved in transcriptional regulation. LCMV replicates in the cytoplasm and buds from the plasma membrane, incorporating host lipids into the viral membrane. It is pleomorphic, with sizes reported from 50 to 300 nm. Some of the virions, which may or may not be infectious, contain ribosome-like structures, giving the virion the characteristic arenavirus appearance.

Geographic and Seasonal Distribution

LCMV infections of mice and man have been well established in Europe and in North and South America, but they are not well documented elsewhere. LCMV is considered an Old World arenavirus brought to the Americas by its host, *Mus musculus*. There is some evidence that human infections occur more commonly in the winter and spring.

Host Range and Viral Propagation

The natural host for LCMV is the mouse, but it can be transmitted to man, hamsters, guinea pigs, rats, dogs, swine, monkeys, chimpanzees, and chick embryos. It commonly causes long-term persistent infections in mice and hamsters, and this has provided a source for human infections. The receptor for LCMV is alpha-dystroglycan, and LCMV strains binding to this receptor at high affinity tend to disseminate *in vivo* more than those that bind at lower affinity. LCMV grows in a wide variety of tissues *in vivo* and in most cells tested in culture. For instance, LCMV has been propagated and plaque assayed in 3T3, baby hamster kidney

(BHK), Detroit-98, HeLa, JLSV-9, L-929, MDBK, MDCK, Vero, and vole cell lines. Vero, L-929, and BHK cells are most commonly used for plaque assays, and BHK cells have been the choice for most biochemical analyses because of the relatively high yields of virus ($c.\ 10^8$ plaque-forming units (PFU) ml^{-1}) and the lack of secreted endogenous retrovirus contaminants. LCMV also grows in lymphocytes and macrophages, and the latter can be used as stimulator or target cells in T-cell proliferation or cytotoxicity assays, respectively.

Genetics

Like other single-stranded RNA viruses, LCMV mutates frequently, and these mutants vary in their tropism and disease-producing potential. A single passage of a cloned LCMV variant into mice will soon segregate into clear neurotropic and turbid viscerotropic plaque variants, which can be recovered from the brain and spleen, respectively. A single amino acid change in the LCMV glycoprotein (residue 260) can convert the immunostimulatory Armstrong strain of LCMV into an immunosuppressive (clone 13) variant, and these genotypes rapidly intraconvert during *in vivo* passage. Several strains of LCMV have been sequenced, and the highest level of sequence homology is at the 5′ and 3′ termini of the S and L virion RNA. These are presumed polymerase-binding sites well conserved throughout the arenavirus family. Different arenaviruses cross-interfere via a defective-interfering virus mechanism. The preservation of these polymerase-binding sites may allow for this heterotypic interference. The NP and GP of the Armstrong and WE strains share 90% amino acid homology.

The presence of two virion RNAs allows for high-frequency recombination due to reassortment of viral genomes. The technique of generating reassortants has led to the assignment of viral-encoded proteins to the appropriate RNA and has facilitated the mapping of genes required for disease-producing potential. The ease of producing reassortants in the laboratory suggests that they also occur in nature and probably play roles in enhancing the genetic diversity of arenaviruses.

Evolution

Arenaviruses do not show substantial sequence homology with any other virus group, but the homology within members of this family suggests a common origin for all arenaviruses. LCMV is most closely related to Lassa virus, which, like LCMV, has its origins in the Eastern Hemisphere. Each arenavirus favors a specific rodent host, and selective evolutional pressure on these viruses must have been conferred by their adaptation to their respective hosts in forms that established persistent infections. Of interest is that LCMV and other arenaviruses undergo rapid evolution as they form persistent infections. These persistent infections *in vitro* and *in vivo* result in the extensive production of defective-interfering virus and of attenuated relatively noncytopathic plaque variants which may help in the maintenance of persistent infections by preventing virus-induced cell death.

Serological Relationships and Variability

Antisera to Lassa virus and to all members of the Tacaribe complex cross-react to some extent with LCMV by complement fixation and immunofluorescence assays but not at the level of viral neutralization. Few monoclonal antibodies show cross-reactivity between LCMV and other Tacaribe complex viruses, but several cross-react with the more closely related Lassa virus. Infection of guinea pigs with LCMV immunizes them against a lethal dose of Lassa virus. Different strains of LCMV are not easily distinguishable by antisera but can be distinguished by some monoclonal antibodies and molecular methods.

Epidemiology

Human LCMV infections occur without sexual bias and at all ages, but most frequently in the 20–30-year age group. A longitudinal study in the USA from 1941 to 1958 implicated LCMV infections in about 8% of patients diagnosed with suspected viral meningitis, and serological studies have suggested up to a 10–15% incidence of LCMV infection in the general population. Most of these infections are probably mild or subclinical. Laboratory infections with LCMV are relatively common, and several cases have occurred in laboratories working with the WE strain, which was re-isolated from one laboratory worker and identified serologically with monoclonal antibodies.

Transmission and Tissue Tropism

LCMV has been experimentally transferred to man by intramuscular injection, but the normal route of infection is probably via the respiratory tract after exposure to mouse secretions. LCMV is shed at high titer in mouse feces and urine and is probably not transmitted by arthropod vectors. Another source of infection is the Syrian hamster (*Mesocricetus auratus*), which, like the mouse, can harbor a long-term persistent infection. Several cases of LCMV in different geographic areas have been linked to a colony of persistently infected hamsters distributed throughout the

USA. Recently, several patients developed severe LCMV infections after receiving transplanted organs from a deceased individual who had contracted LCMV from a pet hamster. Horizontal man-to-man transmission is rare, but LCMV can cross the placenta and infect the fetus.

Pathogenicity

LCMV strains appear to differ in their pathogenicity in man, but conclusive analyses of strain virulence differences in humans have not been carried out. Several human infections have occurred in laboratories working with the WE strain, and the WE strain has been reisolated from a laboratory worker with meningitis and clearly identified. There appear to be fewer (if any) anecdotal reports of human infection with the parent Armstrong strain. However, the more rapidly disseminating clone 13 variant of the Armstrong strain of LCMV has also been linked to human infection. The WE strain is much more virulent than the parent Armstrong strain in hamsters and guinea pigs, and reassortant analyses have mapped the ability to cause lethal infections in guinea pigs to the L RNA of the WE strain. The United States Centers for Disease Control recommends Biosafety Level 2 practices for most studies with LCMV in mice but Biosafety Level 3 practices for work with hamsters. This is based on the presumption that LCMV becomes more virulent for humans as it passes through hamsters. Although this has not been formally proven, this precaution would appear necessary due to the high number of clinical infections in individuals exposed to persistently infected hamsters.

LCMV has been used in a number of pathogenesis studies in mice. Viral strain variant differences exist in the encephalitis model, in which 'docile' or viscerotropic variants fail to kill mice, whereas 'aggressive' or more exclusively neurotropic variants do. However, a docile variant for one strain of mouse may be aggressive to another strain of mouse, and the susceptibility of mice to these variants seems to be linked to both MHC- and non-MHC genes. One reason for 'docility' is the immunosuppressive nature of some LCMV strains, particularly when inoculated into mice at high dose. Lethal encephalitis does not occur when the T-cell response is severely compromised. Immunosuppressive variants of LCMV tend to replicate to very high levels in the visceral organs. This high-dose antigen load can clonally exhaust the T cells by either driving them into apoptosis or into a functional anergy. This clonal exhaustion may limit the encephalitis but can result in a long-term persistent infection (**Figure 1**).

One interesting pathogenic feature of LCMV is its ability to cause a loss in cellular specialized or 'luxury' functions required not for cell survival but for homeostasis of the whole organism. Persistent LCMV infection results in reduced neurotransmitter enzyme activity in cultured neuroblastoma cells *in vitro* and reduced levels of growth and thyroid hormones in mice. Reduced growth hormone synthesis is associated with a runting syndrome in young infected mice. LCMV infects *in vivo* the cells that produce growth hormone and thyroid hormone and causes significant reductions in levels of mRNA encoding these hormones but not other hormones and proteins such as thyroid-stimulating hormone, actin, and collagen.

Clinical Features of Infection

The LCMV infection of man can be an inapparent or subclinical infection, or it can present as a nonmeningeal grippe-like ailment, an aseptic meningitis, or a more severe meningoencephalitis. The incubation period is 1–2 weeks or longer, and the disease may come in two or three waves. The grippal type is characterized by fever, malaise, lethargy, weakness, myalgia, arthralgia, fever, headache, photophobia, anorexia, and nausea. Some patients develop a rash, arthritis, parotitis, or orchitis. The grippal type is likely the most common form of the disease. The meningeal type is often preceded by the grippal type and presents with stiff neck, vomiting, irritability, and Brudzinski's and Kernig's signs. The meningoencephalomyelitic form, which is relatively rare, is associated with confusion, hallucinations, papilledema, and weakness progressing to paralysis. Patients usually recover without lasting sequelae. Death is very rare. There have been only nine fatal cases documented between 1942 and 1992, but two deaths were more recently seen in infected transplant recipients. There are also some reports of human transplacental infection, resulting in fetal abortion or malformation.

Pathology and Histopathology

LCMV can be recovered from the blood, cerebrospinal fluid (CSF), urine, and nasopharyngeal secretions during the human LCMV disease. Leukopenia is a common feature of the infection, but there is a pleocytosis in the CSF in the meningeal stages, and histological analyses of diseased brain tissue in lethal cases of LCMV have revealed meningeal perivascular inflammation and many lymphocytes and monocytes in the arachnoid. This is consistent with studies in the mouse which have demonstrated a lymphocytic infiltration of the meninges. Extensive studies in the mouse model of LCMV-induced acute meningoencephalitis have indicated that virus-specific MHC class I-restricted CD8+ CTLs are the major mediators of the lethal disease, and this may also be the mechanism for the human disease. The mouse model has also shown that antibody–virus–complement immune complexes can be pathological entities, and these could be involved in some of the human arthritic symptoms and the rash, when present.

Immune Response

Human cases of LCMV are characterized by a rise in antibody titer after infection, followed by lasting immunity. Little work has been done on human cellular immunity to infection. However, the LCMV infection of the mouse has provided the most extensive analyses of the cellular immune response to any virus disease. The infection is characterized by an early stage (1–5 days post infection) and later stage (6–10 days post infection). The early stage involves the log-phase replication of the virus and a variety of antigen-nonspecific responses including the liberation of type 1 interferon (IFN)-α and -β and tumor necrosis factor (TNF)-α, the activation and proliferation of natural killer (NK) cells, and a depression in bone marrow hematopoiesis. The type 1 IFN and probably some of the TNFα is directly and rapidly induced in cells by the LCMV infection. Type 1 IFN and IL-15 stimulate the activation and proliferation of NK cells, which substantially increase in number and accumulate in virus-infected tissue. There is initially a type 1 IFN-induced attrition of memory T cells and other cell types, resulting in a leukopenia. Depressed bone marrow function, which is likely due to the effects of inhibitory cytokines like IFN and TNFα and to the ability of activated NK cells to lyse or suppress hematopoietic precursor cells, may contribute to the leukopenia. Antibodies to IFN-α and -β (as well as γ) enhance the synthesis of LCMV, but antibodies to NK cells do not alter the course of the infection. During this early stage of infection there is a pronounced type 1 IFN-induced increase in class I MHC antigen expression throughout the body, an increase in the susceptibility of the cells to CTLs, and a decrease in their susceptibility to NK cells. IFN-mediated protection of uninfected and LCMV-infected target cells may make them resistant to surveillance by NK cells.

The second phase of the response is associated with the expansion of clones of virus-specific T and B cells, a decrease in the production of virus, an increase in the production of IFN-γ and interleukin (IL)-2, and an increase in the activation of macrophages. The major factor in the control of the LCMV infection and in the development of encephalitis and immunopathological lesions throughout the body is the generation of CD8+ CTLs, which can develop in CD4+ T-cell-depleted mice. CD4 T cells and CD40/CD40L interactions, however, seem to be needed for the maintenance of long-term memory CD8 T-cell responses. The IFN-induced upregulation of class I MHC antigens conditions cells in the host to be good targets for CD8+ T cells, which recognize viral peptides in the context of class I antigens. Many LCMV-encoded immunodominant and subdominant T-cell epitopes have been identified, and, in the context of synthetic peptides or vaccinia virus recombinants, they can immunize mice against LCMV.

LCMV infection is so profound at inducing CD8 T-cell responses in the mouse that over 20% of the CD8 T cells can remain LCMV-specific in the memory pool. This has enabled the LCMV infection to be very useful in the discovery and analysis of T-cell-dependent heterologous immunity, where T cells specific for one pathogen are recruited into the response to an unrelated pathogen and alter protective immunity and immunopathology. For example, a complex network of T-cell cross-reactivities between LCMV and vaccinia virus causes LCMV-immune mice to partially resist vaccinia virus infection, while at the same time developing severe T-cell-associated immunopathological lesions.

Antiviral antibody plays only a minor role in the acute LCMV infection and is not needed to clear the infection. However, in the absence of antibody, virus may eventually recrudesce in immune mice. In the absence of a CTL response, such as in congenitally infected mice, antibody–virus–complement immune complexes mediate the development of glomerulonephritis, arteritis, and other inflammatory lesions. Usually the LCMV infection *per se* is relatively mild, and it is the immune response to the virus that causes the damage in the infected mouse.

Prevention and Control of LCMV

LCMV has not been a sufficiently significant human pathogen to warrant special public health measures for its control. However, the disease can be reduced in frequency by ridding houses of mice, by ensuring that pet hamsters are not infected, and by adhering at least to Biosafety Level 2 procedures in the laboratory. Supportive therapy without antiviral chemotherapy is recommended for patients experiencing LCMV infection.

Future Perspectives

Studies with LCMV have in the past provided much of the conceptual basis for viral immunology and pathogenesis, including (1) the theory of immunological tolerance, (2) the concept of virus-induced immunopathology, (3) the analysis of immune complex disease, (4) the discovery of virus-specific CTLs and demonstration of their roles in viral clearance and immunopathology, (5) the discovery of MHC restriction in T-cell recognition, (6) the discovery and analysis of NK cell activation and proliferation, (7) the concept that sublethal viral infections can abrogate specialized 'luxury' functions of cells and lead to metabolic disturbances, (8) the observation that high antigen loads can cause T-cell clonal exhaustion and lead to persistent infection, and (9) the phenomenon of T-cell-dependent heterologous immunity. In the future, the LCMV infection will likely continue to be an important model used for the development of vaccines directed against T-cell epitopes, for examining the clearance of virus by T-cell immunotherapy, for elucidating mechanisms of virus-induced immunosuppression and T-cell tolerance that contribute to persistent infections, for analyzing the regulatory roles of cytokines in the development of the NK-cell and T-cell responses, and for clarifying regulatory mechanisms of heterologous immunity. LCMV variants can produce in mice syndromes that very closely resemble acquired immunodeficiency syndrome (AIDS), and it is possible to manipulate this system to get a persistent infection without a CTL response at all, a persistent infection associated with immunosuppression, and a low-level CTL response, or an acute infection with a strong CTL response which clears the infection. Exploiting the biology of these systems should continue to provide basic concepts fundamental to viral immunology and pathogenesis.

See also: Defective-Interfering Viruses; History of Virology: Vertebrate Viruses; Immune Response to viruses: Antibody-Mediated Immunity; Lassa, Junin, Machupo and Guanarito Viruses; Persistent and Latent Viral Infection.

Further Reading

Buchmeier MJ, Welsh RM, Dutko FJ, and Oldstone MB (1980) The virology and immunobiology of lymphocytic choriomeningitis virus infection. *Advances in Immunology* 30: 275–331.

Lehmann-Grube F (1986) Lymphocytic choriomeningitis virus. In: Braude AL, Davis CE, and Fierer J (eds.) *Infectious Diseases and Medical Microbiology*, 1076pp. Philadelphia, PA: W.B. Saunders.

Oldstone MBA (2002) Biology and pathogenesis of lymphocytic choriomeningitis virus infection. *Current Topics in Microbiology and Immunology* 263: 83–117.

Sevilla N and de la Torre JC (2006) Arenavirus diversity and evolution: Quasispecies *in vivo*. *Current Topics in Microbiology and Immunology* 299: 315–335.

Welsh RM (1999) Lymphocytic choriomeningitis virus as a model for the study of cellular immunology. In: Cunningham MW and Fujinami RS (eds.) *Effects of Microbes on the Immune System*, pp. 280–312. Philadelphia, PA: Lippincott Williams and Wilkins.

Lymphocytic Choriomeningitis Virus: Molecular Biology

J C de la Torre, The Scripps Research Institute, La Jolla, CA, USA

© 2008 Elsevier Ltd. All rights reserved.

Genome Organization

Arenaviruses have a bi-segmented negative single-stranded RNA genome and a life cycle restricted to the cell cytoplasm. Individual arenaviruses exhibit some variability in the lengths of the two genomic RNA segments, L ($c.$ 7.2 kbp) and S ($c.$ 3.5 kbp), but their overall organization is well conserved across the virus family. As with other negative-strand (NS) RNA viruses, arenaviruses are characterized by a lack of infectivity of their purified genome RNA species and the presence of a virion-associated RNA-dependent-RNA polymerase (RdRp). However, the arenavirus coding strategy has unique features compared to prototypical NS RNA viruses. Each arenavirus genome segment uses an ambisense coding strategy to direct the synthesis of two polypeptides in opposite orientation, separated by a non-coding intergenic region (IGR) (**Figure 1**). The S RNA encodes the viral glycoprotein precursor, GPC ($c.$ 75 kDa), and the nucleoprotein, NP ($c.$ 63 kDa), whereas the L RNA encodes the viral RdRp, or L polymerase ($c.$ 200 kDa), and a small RING finger protein Z ($c.$ 11 kDa). The NP and L coding regions are transcribed into a genomic complementary mRNA, whereas the GPC and Z coding regions are not translated directly from genomic RNA, but rather from genomic sense mRNAs that are transcribed using as templates the corresponding antigenome RNA species, which also function as replicative intermediates.

Terminal Nucleotide Sequences

Arenaviruses exhibit high degree of sequence conservation at the 3′-end of the L and S RNA segments (17 out of 19 nucleotides (nt) are identical). Likewise, similar to other NS RNA viruses, arenaviruses also exhibit complementarity between the 5′- and 3′-ends of their genomes and antigenomes. The almost exact inverted complement of the 3′-end 19 nt is found at the 5′-termini of genomes and antigenomes of arenaviruses. Thus, the 5′- and 3′-ends of both L and S genome segments are predicted to form panhandle structures. This prediction is supported by electron microscopy data showing the existence of circular ribonucleoprotein (RNP) complexes within arenavirus virion particles. This terminal complementarity may reflect the presence at the 5′-ends of *cis*-acting signal sequences that provide a nucleation site for RNA encapsidation, required to generate the nucleocapsid (NC) templates recognized by the virus polymerase. Terminal complementarity may also be a consequence of strong similarities between the genome

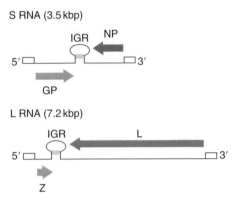

Figure 1 Scheme of the LCMV genome organization.

and antigenome promoters used by the virus polymerases. Studies using reverse genetics approaches have shown that both conserved terminal sequence elements and the panhandle structure are essential for the activity of the virus genome. This terminal complementarity has been proposed to favor the formation of both intra- and intermolecular L and S duplexes that might be part of the replication initiation complex, but its mechanistic aspects remain undefined. For several arenaviruses, an additional nontemplated G residue has been detected on the 5′-end of their genome and antigenome RNAs.

Intergenic Regions

Arenavirus IGRs are predicted to fold into a stable hairpin structure. Transcription termination of the S-derived NP and GP occurs at multiple sites within the predicted stem of the IGR suggesting that a structural motif rather than a sequence-specific signal promotes the release of the arenavirus polymerase from the template RNA. Some arenaviruses, including lymphocytic choriomeningitis virus (LCMV), Lassa fever virus (LFV), and Pichinde virus (PCV), contain one single predicted stem loop within the S IGR, whereas the S IGR of others like Tacaribe virus (TACV) and Mopeia are predicted to contain two distinct stem loops located downstream to the translation termination codons from NP and GPC.

Virion Structure

Virions are pleomorphic but typically spherical with an average diameter of 90–110 nm, and surrounded by a lipid envelope covered with surface glycoprotein spikes. Virions contain the L and S genomic RNAs as helical NC structures that are organized into circular configurations, with lengths ranging from 400 to 1300 nm. The L and S genomic RNA species are not present in equimolar amounts within virions (L:S ratios ∼ 1:2), and low levels of both L and S antigenomic RNA species are also present within virions. Host ribosomes can be incorporated into virions, but the biological implications of this remain to be determined. Virus particles have a 'sandy' appearance from which the family name is derived.

Arenavirus Gene Products

Viral Nucleoprotein

The NP (M_r 60–68 kDa) is the most abundant viral polypeptide both in infected cells and virions (about 1530 NP molecules/virion particle). The NP is the main structural element of the viral NC and associates with the genome RNA to form beadlike structures. Phosphorylated forms of the NP are usually detected at late stages of acute infection and their abundance increases in persistently infected cells, but the functional implications of these changes in the stage of NP phosphorylation have not been established.

Viral Glycoproteins

The viral GPC (70–80 kDa) is post-translationally proteolytically processed to generate the two mature virion glycoproteins GP-1 (40–46 kDa) and GP-2 (35 kDa), and a stable 58-amino-acid signal peptide (SP). Evidence indicates that in addition to its predicted role in targeting the nascent polypetide to the endoplasmic reticulum, the SP of the arenavirus GPC likely serves additional roles in the biosynthesis, trafficking, and function of the viral envelope glycoproteins. Cleavage of GPC into GP-1 and GP-2 is mediated by the SKI-1/S1P cellular protease. GP-1 is located at the top of the spike, away from the membrane, and is held in place by ionic interactions with the N-terminus of the transmembrane GP-2, that forms the stalk of the spike. GP-1 is the virion attachment protein that mediates virus interaction with host cell-surface receptors and subsequent virus cell entry via receptor mediated endocytosis. Alpha-dystroglycan (α-DG) has been identified as the receptor for LCMV, LFV, and several other arenaviruses. Evidence indicates that unprocessed arenavirus GPC can traffic to the cell surface, but the correct processing of the virus GPC into GP1 and GP2 is strictly required for the production of infectious particles that bud from the plasma membrane. Accordingly, growth of LCMV was severely impaired in cells deficient in SK1-1/S1P. Notably, proteolytic processing of GP requires the structural integrity of its GP-2 cytoplasmic tail.

L Polymerase

The arenavirus L protein has the characteristic sequence motifs conserved among the RdRp (L proteins), of NS RNA viruses. The proposed polymerase module of L is located within domain III, which contains highly conserved amino acids within motifs designated A and C. Mutation-function analysis showed that sequence serine-aspartic acid-aspartic acid (SDD) characteristic of motif C of segmented NS RNA viruses, as well as the presence of the highly conserved D residue within motif A of L proteins, are strictly required for the function of the LCMV L polymerase. Several mutant L proteins exhibit a strong dominant negative (DN) under assay conditions where wt and mutant L proteins do not compete for template RNA or other *trans*-acting proteins, viral or cellular, required for polymerase activity. These results suggested that L–L interaction is required for LCMV polymerase activity. Intragenic complementation has been documented for the L genes of several NS RNA viruses. Consistent with this, direct L–L physical interaction has been demonstrated for the paramyxoviruses Sendai and parainfluenza virus 3 (PIV3), and this L oligomerization was required for polymerase activity. Likewise, biochemical evidence supports that the LCMV L protein has also the property of forming oligomeric structures.

Detailed sequence analysis and secondary structure predictions have been documented for the LFV L polymerase. These studies unequivocally identified only one functional domain corresponding to the putative RNA polymerase domain that exhibited a similar folding as those found in the corresponding domains of determined crystal structures of viral RdRp. In addition, several regions, outside the polymerase region, of strong alpha-helical content were identified, as well as a putative coiled-coil domain at the N-terminus. Secondary structure-assisted alignment of the RNA polymerase region indicated that arenaviruses are most closely related to nairoviruses.

Z Protein

The arenavirus small RING finger protein Z (*c.* 11 kDa) has no homolog among other known NS RNA viruses. Z is a structural component of the virion, and in LCMV-infected cells Z has been shown to interact with several cellular proteins including the promyelocytic leukemia (PML) protein, and the eukaryote translation initiation factor 4E (eIF4E). The former has been proposed to contribute to the noncytolytic nature of LCMV infection, whereas the latter has been proposed to repress cap-dependent translation. Early studies suggested a role of Z in arenavirus transcriptional regulation. However, more recent studies using reverse genetic approaches have shown that Z is not required for virus RNA replication and transcription, but rather it exhibits a dose-dependent inhibitory effect on RNA synthesis mediated by the arenavirus polymerase. The linear amino acid structure of Z could be seen as composed of three modules: the N-terminal part, the RING finger domain RD, and the C-terminal part including conserved proline motifs that constitute canonical late (L) domains. Despite some minor differences on protein size, the same overall structure is also seen among different arenavirus Z proteins. The Z RD domain exhibits the highest levels of conservation. The N-terminus contains a conserved canonical mirystoylation motif, whereas the C-terminus of Z contains conserved proline motifs that are similar to those present within the L domains of several Gag and matrix viral proteins and that have been identified as key functional elements during budding of enveloped viruses from the plasma membrane.

Arenavirus Life Cycle

Cell Attachment and Entry

Consistent with a broad host range and cell type tropism, a highly conserved and widely expressed cell-surface protein, alpha-dystroglycan (aDG) has been identified as a main receptor for LCMV, Lassa fever, and several other arenaviruses. Upon receptor binding, arenavirus virions are internalized by uncoated vesicles and released into the cytoplasm by a pH-dependent membrane fusion step. The pH-dependent fusion between viral and cell membranes in the acidic environment of the endosome is mediated by the GP2 portion of arenavirus GP, which is structurally similar to the fusion active membrane proximal portions of the GP of other enveloped viruses. Recent molecular modeling studies revealed the presence of two antiparallel alpha-helices separated by a glycosylated peptide loop within the ectodomain of arenavirus GP2, which is a characteristic feature of the proposed viral transmembrane protein superfamily.

RNA Replication and Transcription

The fusion between viral and cellular membranes results in the release of the viral RNP into the cytoplasm, which is followed by viral RNA synthesis. The encapsidated vRNA present in the vRNP does not serve as a direct template for protein synthesis, but rather as a template for both transcription and replication. The viral NP and L polypeptides were identified as the minimal viral *trans*-acting factors required for efficient RNA synthesis, both transcription and replication, mediated by the polymerase of LCMV, as well as LFV and the New World arenavirus TACV.

Z is not required for intracellular transcription and replication of an LCMV MG, but rather Z exhibits a dose-dependent inhibitory effect on both transcription and replication of LCMV MG. This inhibitory effect of Z has also been reported for TAV. Moreover, cells transduced with a recombinant, replication-deficient, adenovirus expressing Z (rAd-Z) become resistant to LCMV and

LFV. Z-mediated resistance to virus infection was not due to a blockade of virus entry, but rather to a strong inhibitory effect of Z on both viral transcription and RNA replication that in the case of TAV appears to involve Z–L interaction. Mutation-function studies showed that the N- and C-terminal residues of Z were not required for its inhibitory activity on RNA synthesis. In contrast, the structural integrity of the Z RING domain (RD) was strictly required, but not sufficient for the Z inhibitory activity. Many RING finger proteins play a key role in the control of intracellular protein degradation by acting as ubiquitin (Ub)–protein ligases (or E3s). A subset of RING proteins implicated in protein degradation has been shown to have a conserved tryptophan (W) where other hydrophobic amino acids are otherwise found. Notably, arenavirus Z proteins contain a highly conserved W residue located next to the second conserved cystein (C) residue of its RD. Substitution of this W residue by A resulted in a Z protein with reduced inhibitory activity, an effect that might be related to its potential E3 activity. Nevertheless, experimental evidence has failed to support, yet, an E3 activity associated with Z.

It is plausible that increased expression of Z protein during the virus life cycle might contribute to block replication of an additional infection by a genetically closely related arenavirus, which is known as homotypic viral interference. Superinfection exclusion could also influence arenavirus evolution and contribute to explain the observed population partitioning in the field, resulting in the maintenance of independent evolutionary lineages of the same strain within a small geographic range.

A widely accepted model for the control of arenavirus RNA replication and gene transcription (**Figure 2**) proposes that at early times of infection, low levels of NP prevent the virus polymerase to go across the IGR, and hence transcription is favored over replication. Viral mRNAs have extra nontemplated nucleotides (nt) and a cap structure at their 5′-ends, but the origin of both the cap and 5′-non-templated nt extensions remain to be determined. The 3′-termini of the subgenomic nonpolyadenylated viral mRNAs has been mapped to multiple sites within the distal side of the IGR, which is reminiscent of 'intrinsic' or 'rho-independent' termination in prokaryotes and represents strong support for the IGR as termination signal for transcription complexes. Direct experimental data has confirmed that the IGR is a *bona fide* transcription termination signal. Intriguingly, the IGR appears to also play a key role in packaging or cell egress, or both, of infectious viral particles. Increasing intracellular NP levels during the course of the infection would unfold secondary RNA structures within the IGR. This would attenuate structure-dependent transcription termination at the IGR, thus promoting replication of genome and antigenome RNA species. However, studies using a helper virus-free minigenome (MG) rescue

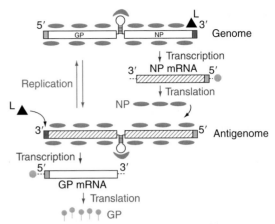

Figure 2 Basic aspects of arenavirus RNA replication and gene transcription illustrated for the S segment. Once the virus RNP is delivered in to the cytoplasm of the infected cell, the polymerase associated with the virus RNP initiates transcription from the genome promoter located at the genome 3′-end. Primary transcription results in synthesis of NP and L mRNA from the S and L, respectively, segments. Subsequently, the virus polymerase can adopt a replicase mode and moves across of the IGR to generate a copy of the full length antigenome RNA (agRNA). This agRNA will serve as template for the synthesis of the GP (agS) and Z (agL) mRNAs. The agRNA species serve also as templates for the amplification of the corresponding genome RNA species.

system showed that although transcription and replication were strictly dependent on NP, both were however equally enhanced by incrementally increasing amounts of NP up to levels in the range of LCMV-infected cells. These findings, similar to those described for the paramyxovirus respiratory syncytial virus, are consistent with a central role for NP in transcription and replication of the LCMV genome, but they do not support a participation of NP levels in balancing the two processes. These observations would be consistent with a model where two distinct polymerase complexes may exist, which are committed to either transcription or replication. This view is supported by the finding that LCMV mRNAs are capped while genomes and replicative intermediates are not, suggesting that the replicase is committed to either process already when initiating at the 3′-end of the template using either a cap-oligonucleotide primer or a prime and realign mechanism.

The activity of the genomic promoter recognized by virus polymerase requires both sequence specificity within the highly conserved 3′-terminal 19 nt of arenavirus genomes, and the integrity of the predicted panhandle structure formed via sequence complementarity between the 5′- and 3′-termini of viral genome RNAs. A prime and realign mechanism has been proposed for the initiation of arenavirus RNA replication. This model accounts for the presence of a nontemplated G at the 5′-ends of the arenavirus genomic and antigenomic RNAs. This model assumes that arenavirus polymerases, like many other viral RdRps, can initiate RNA synthesis *de novo* only with ATP or GTP. Therefore, arenavirus RNA initiation is thought to take place from an internal (+2) templated cytidilate. Once the first phosphodiester bond has been formed, the pppGpC will slip backwards on the template and realign, creating a nascent chain whose 5′-end is at position −1 with respect to the template, before the polymerase can continue the downstream synthesis. The length of the RNA is kept constant despite a nontemplated nucleotide being added at the 5′-end, because the polymerase is supposed to terminate RNA synthesis by removing the last base. A similar mechanism has been proposed for RNA initiation of Hantaviruses (*Bunyaviridae*).

Genomic and antigenomic viral RNA species with terminal deletions, or nontemplated nucleotide extensions, have been documented during persistent LCMV infections. These truncated RNAs have been proposed to be a new type of defective interfering (DI) genome that contributes to the establishment and maintenance of LCMV persistence. It has been hypothesized that truncated RNAs are replication competent but are not used as templates for transcription. Increased ratio of truncated to full-length RNAs would be expected to cause a decrease in viral gene expression. On the other hand, addition of nontemplated nucleotides could repair some truncated RNAs to full-length sequence, thus leading to a transient increase in virus gene expression. This hypothesis is compatible with the observations that LCMV persistence is characterized by cyclical changes in numbers of antigen-positive cells and production of infectious virus. A corollary of this hypothesis would be that the arenavirus' highly conserved 3′-terminal 19 nt sequence is not strictly required for initiation of replication, which appears to be in conflict with mutation-function analysis of the arenavirus genomic promoter using reverse genetic approaches. One possible explanation for this apparent discrepancy would be that truncated RNAs detected in LCMV persistence are continuously generated *de novo* from full-length templates by an unknown mechanism. On the other hand, cellular factors induced or modified by LCMV persistence might influence the activity of the virus polymerase.

Assembly and Budding

Z has been postulated to participate in virion morphogenesis. For most enveloped viruses, a matrix (M) protein is involved in organizing the virion components prior to assembly. Interestingly, arenaviruses do not have an obvious counterpart of M. However, cross-linking studies showed complex formation between NP and Z, suggesting a possible role of Z in virion morphogenesis. Recent data have shown that Z is the main driving force of arenavirus budding and that this process is mediated by the Z proline-rich late (L) domain motifs PT/SAP and PpxY, similar to those known to control budding of several other viruses including HIV and Ebola, via interaction with specific host cell proteins. Consistent with this observation Z exhibits features characteristic of bona fide budding proteins: (1) Z has ability to bud from cells by itself, and (2) Z can efficiently substitute for the RSV L domain. Targeting of Z to the plasma membrane, the location of arenavirus budding, strictly required its myristoylation. Accordingly, the myristoylation inhibitor 2-hydroxymyristic acid (2-OHM) inhibited Z-mediated arenavirus budding without affecting RNA synthesis mediated by the LCMV polymerase. Intriguingly, the IGR appears to also play a key role in packaging or cell egress, or both, of infectious viral particles.

Arenavirus Reverse Genetics

The development of reverse genetic systems for several arenaviruses including the prototypic LCMV has provided investigators with a novel and powerful approach for the investigation of the *trans*-acting factors and *cis*-acting sequences that control arenavirus replication and gene expression, as well as assembly, and budding. Likewise, the implementation of reverse genetic approaches for arenaviruses has opened new avenues for the development of novel vaccine and antiviral strategies to combat arenavirus infections. Moreover, the ability to generate recombinant LCM viruses with predetermined specific mutations and analyze their phenotypic expression in its natural host, the mouse, would significantly contribute to the elucidation of the molecular mechanisms underlying LCMV–host interactions including virus persistence and associated disease.

Molecular Phylogeny

The family *Arenaviridae* comprises two distinct complexes: the LCMV–Lassa complex, which includes the Old World arenaviruses; and the TACV complex, which includes all known New World arenaviruses. Early sequence analysis of laboratory adapted arenaviruses revealed a significant degree of genetic stability with amino acid sequence homologies of 90–95% among different strains of the prototypic arenavirus LCMV, whereas significant higher levels of genetic diversity (37–56%) were observed for homologous proteins of different arenaviruses species. More recent, genetic studies on arenavirus field isolates (including Lassa, Junin, Guanarito (GTO), Pirital (PIR), and Whitewater Arroyo) have revealed a high degree of genetic variation among geographical and temporal isolates of the same virus species. Notably, a remarkably high level of genetic divergence (26% and 16% at the nt and amino acid level, respectively) has been documented among PIR isolates within very small geographic regions. The substantial degree of inter- and intraspecies genetic variation among arenaviruses appear to have important biological correlates, as suggested by the significant variation in biological properties observed among LCMV strains. Dramatic phenotypic differences among genetically closely related LCMV isolates indicate that a few amino acid replacements in LCMV proteins could suffice to produce important alterations in the virus biological properties.

See also: Lymphocytic Choriomeningitis Virus: General Features.

Further Reading

Bowen MD, Rollin PE, Ksiazek TG, et al. (2000) Genetic diversity among Lassa virus strains. *Journal of Virology* 74: 6992–7004.

Buchmeier MJ, de la Torre JC, and Peters CJ (2006) *Arenaviridae*: The virus and their replication. In: Knipe DM, Howley PM, Griffin DE, et al. (eds.) *Fields Virology,* 5th edn., pp. 1791–1827. Philadelphia, PA: Lippincott Williams and Wilkins.

Cornu TI, Feldmann H, and de la Torre JC (2004) Cells expressing the RING finger Z protein are resistant to arenavirus infection. *Journal of Virology* 78: 2979–2983.

Eichler R, Lenz O, Strecker T, Eickmann M, Klenk HD, and Garten W (2003) Identification of Lassa virus glycoprotein signal peptide as a *trans*-acting maturation factor. *EMBO Reports* 4: 1084–1088.

Hass M, Golnitz U, Muller S, Becker-Ziaja B, and Gunther S (2004) Replicon system for Lassa virus. *Journal of Virology* 78: 13793–13803.

Kunz S, Borrow P, and Oldstone MB (2002) Receptor structure, binding, and cell entry of arenaviruses. *Current Topics in Microbiology and Immunology* 262: 111–137.

Lee KJ, Novella IS, Teng MN, Oldstone MB, and de la Torre JC (2000) NP and L proteins of lymphocytic choriomeningitis virus (LCMV) are sufficient for efficient transcription and replication of LCMV genomic RNA analogs. *Journal of Virology* 74: 3470–3477.

Meyer BJ, de la Torre JC, and Southern PJ (2002) Arenaviruses: Genomic RNAs, transcription, and replication. In: Oldstone MB (ed.) *Arenaviruses I*, vol. 262, pp. 139–149. Berlin: Springer.

Perez M, Craven RC, and de la Torre JC (2003) The small RING finger protein Z drives arenavirus budding: Implications for antiviral strategies. *Proceedings of the National Academy of Sciences, USA* 100: 12978–12983.

Perez M and de la Torre JC (2003) Characterization of the genomic promoter of the prototypic arenavirus lymphocytic choriomeningitis virus (LCMV). *Journal of Virology* 77: 1184–1194.

Pinschewer DD, Perez M, Sanchez AB, and de la Torre JC (2003) Recombinant lymphocytic choriomeningitis virus expressing vesicular stomatitis virus glycoprotein. *Proceedings of the National Academy of Sciences, USA* 100: 7895–7900.

Sanchez AB and de la Torre JC (2005) Genetic and biochemical evidence for an oligomeric structure of the functional L polymerase of the prototypic arenavirus lymphocytic choriomeningitis virus. *Journal of Virology* 79: 7262–7268.

Sanchez AB and de la Torre JC (2006) Rescue of the prototypic arenavirus LCMV entirely from plasmid. *Virology* 350: 370–380.

Strecker T, Eichler R, Meulen J, et al. (2003) Lassa virus Z protein is a matrix protein and sufficient for the release of virus-like particles [corrected]. *Journal of Virology* 77: 10700–10705.

York J, Romanowski V, Lu M, and Nunberg JH (2004) The signal peptide of the Junin arenavirus envelope glycoprotein is myristoylated and forms an essential subunit of the mature G1–G2 complex. *Journal of Virology* 78: 10783–10792.

Lysis of the Host by Bacteriophage

R F YoungIII and R L White, Texas A&M University, College Station, TX, USA

© 2008 Elsevier Ltd. All rights reserved.

Glossary

Antiholin Phage-encoded protein that inhibits the holin by directly interacting with it; may be produced from the same coding sequence as the holin, as in the case of the lambda dual-start motif, or may be encoded by a completely separate gene, as in the case of the T4 holin *t* and antiholin *rl* genes.

Dual-start motif Motif at the beginning of a holin gene consisting of two start codons separated by one or two codons, at least one of which specifies a positively charged residue; from each of these two start codons, a separate protein product is translated, and the products differ only by two or three N-terminal residues but possess opposing functions (the shorter polypeptide is the active holin and the longer polypeptide is the antiholin).

Endolysin Enzyme with muralytic activity; that is, it degrades the host cell wall by cleaving one or more of three chemical bonds in the peptidoglycan.

Holin Small phage-encoded membrane protein responsible for the release or activation of the endolysin to accomplish lysis and for the timing of the lysis event.

Holin triggering The onset of lysis as caused by lesion formation by the holin.

Holin–endolysin lysis This two-component lysis strategy relies on a small, phage-encoded membrane protein (the holin) to time the lytic event and activate or release second component, the endolysin, a phage-encoded muralytic enzyme that destroys the host cell wall. It is used by essentially all phages with dsNA genomes.

Lysis Destruction of host cells to release progeny virions at the end of the phage vegetative cycle; two

fundamental lysis strategies are the 'holin–endolysin' and the 'single-gene' lysis systems.
Pinholin Holin, such as S^{21}, that makes small holes that depolarize the membrane but are too small to allow a fully folded, soluble endolysin to escape into the periplasm.
Protein antibiotic Single-gene lysis system, such as the φX174 E protein or the Qβ A_2 protein, which causes lysis in a manner that is indistinguishable from lysis induced by antibiotics inhibiting any of the steps in the pathway of cell wall biogenesis.
Rz–Rz1 Type II cytoplasmic membrane protein and outer-membrane lipoprotein, respectively, encoded by most phages with Gram-negative hosts. Through specific C-terminal interactions, Rz and Rz1 form a complex that spans the periplasm. In some cases, the Rz–Rz1 complex is required for lysis under conditions that stabilize the outer membrane.
SAR endolysin Signal anchor release endolysin; the endolysin is initially secreted as a type II integral membrane protein (single transmembrane domain or TMD; N-in, C-out topology), with the TMD acting as a signal anchor domain. The SAR domain is not cleaved like a normal signal sequence but is able to escape from the membrane. The exit of the SAR domain from the bilayer is accelerated when the membrane is depolarized. SAR domains are rich in nonhydrophobic residues, especially glycine and alanine. These endolysins are inactive in the membrane tethered form and become active upon release from the membrane, which occurs slowly and spontaneously in the absence of holin function or rapidly upon holin-mediated membrane depolarization.
Single-gene lysis This strategy effects lysis with a single protein by inhibiting a step in cell wall synthesis and without inducing any detectable muralytic enzyme activity; it is used by non-filamentous lytic ssDNA and ssRNA phages.
Spanin A phage-encoded outer membrane lipoprotein that contains a C-terminal TMD and thus spans the entire periplasm. It replaces the Rz–Rz1 proteins in some phages and can complement an *Rz–Rz1* defect.

Preface: Host Lysis and Release of Progeny Virions

In eubacterial systems, there are two modes by which bacteriophages effect release of the progeny from the infected host: 'extrusion' and 'lysis'. Extrusion is limited to the male-specific filamentous phages and can be viewed as a chronic process that is not fatal to the host cell. Essentially, each filamentous phage particle is extruded through the envelope as its DNA is coated with coat proteins that have accumulated in the cytoplasmic membrane. There is no concomitant disruption of the membrane and thus the process is compatible with continued cell growth. This process is intimately related to the morphogenesis of the virion and is thus treated in detail elsewhere in this encyclopedia. All nonfilamentous phages cause complete 'lysis' of the host, requiring that the murein component of the cell envelope be at least partially destroyed. Thus, this article is devoted to the molecular basis of host lysis.

Lysis versus 'Rotting'

It is important at the outset to make a distinction between phage lysis and the cell lysis that can occur as a result of antibiotic intoxication, developmental processes, or fatal cellular dysfunction. Phage lysis is generally a much more rapid and saltatory event. Under ideal physiological conditions in the laboratory, where a culture is uniformly and synchronously infected, lysis of the entire culture can be observed within a time interval that is usually small compared to the length of the total infection cycle. In most cases, cell lysis not deriving from phage infection is essentially a passive event; the cell wall deteriorates as a result of a failure or lack of coordination of murein synthesis and turnover processes. An exception to this may be in the autolysis of bacteria that occurs as a result of a developmental pathway, such as lysis of the mother cell during sporulation, where murein degradation is actively promoted by expression of muralytic enzymes. There is some evidence that lysis systems were evolutionarily late additions to the genetic architecture of phages (see below). It seems likely that primitive phages did not directly promote lysis of the host; instead, liberation of the progeny virions was eventually effected as a result of the general derangement of cellular growth and maintenance systems caused by massive viral-specific reprogramming of gene expression. However, the rapid, actively programmed lysis now universally characteristic of all nonfilamentous phages presumably conferred a great competitive advantage, allowing the progeny virions an early opportunity to diffuse into the environment and initiate new infection cycles, instead of being trapped in a dead host and relying on 'rotting' of the corpse for eventual release.

Implicit in this argument, however, is an intriguing question; if lysis is an active programmed process, what determines when it will be initiated? The simplest expectation might be that lysis would be triggered when the cell was exhausted in terms of its macromolecular synthesis capacity, so that the assembly of virions was significantly diminished or halted. This turns out not to be the case.

Instead, at least for almost all dsDNA and, perhaps, dsRNA phages (hereafter: dsNA) the timing of lysis is determined by a small phage-encoded membrane protein called a 'holin' (see below). The holin triggers lysis at a time built into its primary structure, irrespective of the ability of the host cell to maintain or even increase the rate of intracellular assembly of virions. Holin-mediated lysis can be changed by mutations in the holin sequence or expression signals or by the direct inhibitory effects of another phage-encoded protein called the 'antiholin'. Irrespective of the molecular details, it must be kept in mind that there is no direct link between the progress of the morphogenetic program of the infecting phage and lysis timing. Lysis timing is an independent fitness parameter. In fact, the only decision made during the vegetative cycle of any phage is when to terminate the infection and effect lytic release of the progeny virions. It should be noted that this is not a purely basic science matter; phage-mediated lysis has a direct role in many disease processes. Some bacterial toxins are actually proteins or polypeptides encoded by prophages in the pathogenic bacterium. The expression of genes encoding products like the Shiga-like toxins of hemorraghic *Escherichia coli* and release of the toxin itself occurs as a result of the lysogenic induction of a fraction of the bacterial population and the subsequent phage-mediated lysis of these host cells. The study of phage lysis may thus reveal opportunities for therapeutic intervention in these diseases.

Modes of Phage Lysis

There are at least two fundamental strategies by which host lysis is accomplished by bacteriophages: 'holin–endolysin' lysis and 'single gene' lysis. The former strategy relies on a viral-encoded muralytic enzyme to destroy the host cell wall and is used by phages with genome sizes above about ~10 kb; that is, essentially all phages with dsNA genomes. The latter mode causes lysis without inducing any detectable muralytic enzyme activity and is used by nonfilamentous phages with genomes under this limit, including the lytic ssDNA and ssRNA phages. The molecular bases of these two fundamental strategies are the substance of the remainder of this article.

Holin–Endolysin Lysis

The lytic strategy adapted by most phages other than the smallest ones involves, at a minimum, two proteins: a membrane protein called a 'holin', and a muralytic enzyme called an 'endolysin'. The endolysin is an enzyme that degrades the cell wall (also known as murein or peptidoglycan). The 'holin', mentioned above, is a small membrane protein that controls the enzymatic activity of the endolysin or its access to its murein substrate; the timing of the onset, or 'triggering', of lysis is entirely determined by the holin's primary structure. In addition, most dsDNA phages of Gram-negative bacteria also either a pair of interacting secreted proteins called Rz and Rz1, or a 'spanin' protein, which has both cytoplasmic membrane and OM localization signals. The Rz–Rz1 complex and the spanin protein both span the entire periplasm. Finally, in many phages, a regulatory protein called an 'antiholin' is also produced. As its name implies, the antiholin binds to and inhibits the function of the holin and can delay, or, in some circumstances, completely block lysis.

The holin–endolysin systems thus involve a minimum of two genes, a holin and an endolysin. However, some phage lysis systems use as many as five proteins; one example is the well-studied coliphage P2, which encodes a holin, endolysin, antiholin, and Rz–Rz1 proteins. It must be noted here, however, that there has been little study of the lysis systems of phages with Gram-positive hosts, although the existence of the simplest two-gene holin–endolysin systems has been demonstrated, as in the case of phage φ29, which is the prototype of a very diverse and widespread family of phages of Gram-positive bacteria. In addition, the emerging field of phage genomics suggests that some phages of Gram-positive bacteria may have lysis components that have no counterpart in the better-studied Gram-negative systems. This is especially likely in phages of mycobacteria and similar microbes, where there is a unique waxy outer surface that in some ways serves the purpose of an outer membrane for these important Gram-positive organisms.

Holins

Holins may be the most diverse functional family in biology. The last comprehensive survey, more than 5 years ago, found more than 250 holin or putative holin genes among the sequenced phages and in the prophages found in bacterial genomes. Homology analysis found that these sequences could be grouped into 50 unrelated families. Primary structure analysis by membrane topology algorithms indicates that most holins can be grouped into three topology classes, based on the number of transmembrane helical domains (TMDs): class I: 3 TMDs, with the N-terminus out and C-terminus in (the cytoplasm); class II: 2 TMDs, N-in, C-in; and class III, 1 TMD, with N-in and C-out. In many cases, the membrane topology is not obvious from the primary structure, and there is no reason to doubt that there may be more topological classes in this large and diverse functional group. The prototype class I, II, and III members are the holins from phages lambda, phage 21, and phage T4, respectively (**Figure 1**); all have been subjected to genetic analysis, and the holin from lambda has also been studied by biochemical means. Only one other holin has had extensive genetic study, the class I holin of the tectivirus PRD1.

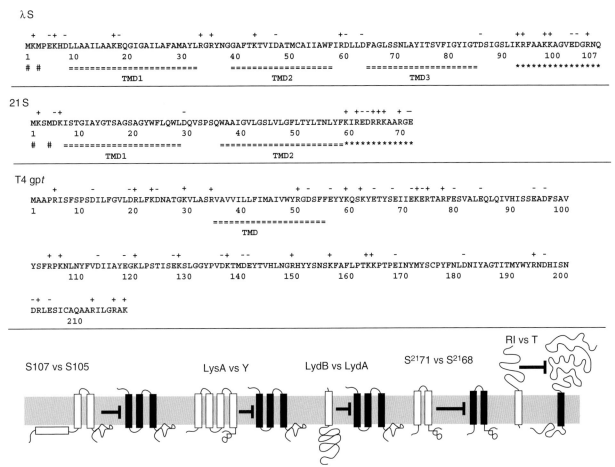

Figure 1 Holins and antiholins: sequence and topology. Upper 3 panels: Primary structure and membrane topology of the products of lambda S, phage 21 S^{21}, and T4 t; ===, transmembrane domains (numbered for each protein) as predicted by TMHMM program (http://www.cbs.dtu.dk/services/TMHMM-2.0/); ***, the highly charged, dispensable C-terminal region of lambda S and S^{21} gene products; #, the two start codons of lambda S and S^{21}. Lower panel: Membrane topologies of coliphage holins and their cognate antiholins; interaction between holin/antiholin pair is indicated. From left to right: antiholin S107 and holin S105 of lambda; antiholin LysA and holin Y of P2; antiholin LydB and holin LydA of P1; antiholin S^{21}71 and holin S^{21}68 of lambdoid phage 21; antiholin RI and holin gpt of T4.

Endolysins

There are three chemical bonds that constitute the covalent polymer of the peptidoglycan: glycosidic bonds between the amino-sugars, amide bonds between between the linking oligopeptide and the lactyl group of MurNAc, and peptide bonds within the linking oligopeptide. Lambda R is a 17 kDa muralytic enzyme that cleaves the MurNac–GlcNac glycosidic linkage of the peptidoglycan, releasing a disaccharide digestion product lacking a reducing end. Three other different enzyme activities are associated with other endolysins, each associated with the breakage of one of the three different types of covalent bonds that link the subunits of the murein polymer. (Note the historical term lysozyme is also associated with three different enzymatic entities.)

Glycosidases. Also known as true lysozymes or, less precisely, as muramidases, these enzymes cleave the same peptidoglycan bonds as the transglycosylases but do so by hydrolysis, freeing disaccharide products with reducing ends. Strictly, glycosidases can also be called either muraminidases or glucosaminidases, depending on whether they cleave after MurNac or GlcNac residues. The phage T4 lysozyme, E (product of the T4 *e* gene), is perhaps the most intensively studied protein structure, in terms of the energetics and mechanics of its tertiary structure, and is a 164 residue monomeric, soluble globular enzyme.

Amidases. These enzymes hydrolyze the amide bond. The best-characterized phage amidase is the T7 lysozyme, or gp3.5. Gene *3.5* is expressed as an early gene because it has a second important function beyond the endolysin role: it complexes with and inhibits T7 RNA polymerase.

Endopeptidases. These enzymes hydrolyze a peptide bond or isopeptide bond in the cross-linking polypeptide moieties. For example, the *S. aureus* phage A500 has an L-alanyl-D-glutamate endopeptidase as its endolysin. Currently, no endolysin with endopeptidase activity has been characterized structurally.

There does not appear to be any linkage between the enzymatic type of endolysin and the topology class of the holin. For example, the holins of lambda and the *Salmonella* phage P22 are nearly identical class I holins, but the P22 endolysin is a T4 E glycosidase homolog, whereas lambda R is a transglycosylase. Moreover, the lambda holin has been shown to function with the T4 glycosidase E and the T7 amidase gp*3.5*, at least to the extent that co-expression of the holin and either endolysin results in host lysis.

Holin–Endolysin Lysis: The Lambda Paradigm

The paradigm system for studying lysis has been the classic phage lambda, and the functions of the holin, antiholin, endolysin, and Rz–Rz1/spanin are best described with reference to this system. For studying lysis, the main advantage of lambda, besides its unparalleled genetics, is that one can use thermo-inducible lysogens of *E. coli* carrying a single lambda prophage to obtain the synchronous initiation of the vegetative cycle in all the cells of a culture in balanced growth. The extent and timing of lysis can then easily be monitored by following the turbidity of the culture. Under carefully controlled physiological conditions, such lysogenic inductions result in a synchronous and saltatory lysis of all the cells in the culture at about 50 min after induction, releasing ∼100 progeny virions per induced cell. Noninvasive methods to assess the energy state of the cell have shown that, up to the instant of lysis triggering, the induced cells have normal energy metabolism and an intact, energized membrane. This rigidly scheduled lysis program of lambda can be subverted by any disturbance of the energy metabolism of the cells; for example, addition of an energy poison like cyanide or the uncoupler dinitrophenol (DNP) triggers 'premature' lysis, indicating that holin–endolysin systems rely on the energized state of the membrane to achieve temporal scheduling.

The lambda lysis genes, *SRRzRz1*, are clustered in the 'lysis cassette' immediately downstream of the lambda late gene promoter (**Figure 2**). This promoter is activated about 8 min after induction and serves not only the lysis

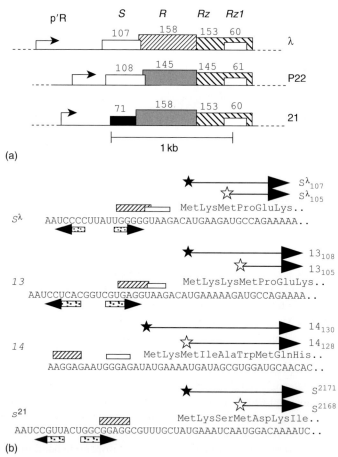

Figure 2 The lambda lysis cassette and the dual start motif. (a) The late promoter and *S, R, Rz* and *Rz1* genes of lambda, P22 and phage 21; amino acid lengths of each gene product are indicated above each gene. (b) The dual start motifs of the *S* gene of coliphage lambda, gene *13* of *Salmonella* phage P22, gene *14* of *B. subtilis* phage ϕ29, and the *S* gene of lambdoid coliphage 21; the boxed sequences (diagonal striped box for the antiholin, white box for the holin) indicate the Shine–Dalgarno sequences for the dual translational starts; the start codon for each protein is marked with a star and arrow (filled stars indicate antiholin start codons, white stars indicate holin start codons) in the direction of translation above the sequence and the length of both protein products is given in amino acid (aa) residues. Stem–loop structures (if present) are marked with arrows below the sequence in the direction of base of the stem–loop.

genes but also the approximately 25 essential genes downstream of the lysis cassette, which are required for DNA packaging and the synthesis and assembly of heads, tails, and tail fibers. Thus, the lysis genes are expressed simultaneously with all the other genes needed for viral morphogenesis. Genes *S* and *R* are required for lysis under all conditions; *R* encodes the endolysin, a transglycosylase, which accumulates fully folded and active in the cytoplasm. *S* is a 107 codon reading frame which encodes two products, named S105 and S107 for their length in amino acid residues, as a consequence of dual translational starts at Met codons 1 and 3 (**Figures 1** and **2**). S105 is the essential holin; the properties of S107, which is nonessential and functions as an antiholin, will be described below. In the absence of functional S105 protein, the lambda vegetative cycle, begun either from an induction of a lysogen or from infection, continues unabated for several hours, far beyond the 50 min lysis time observed for phages carrying the parental *S* allele. During this extended latent period, progeny virions and lambda endolysin accumulate in the cytoplasm. Lysis can be elicited almost instantaneously if the host membranes are disrupted by chloroform or a detergent, reflecting the endolysin's requirement for the *S* holin to achieve access to the cell wall. In contrast, null mutations in *R* do not result in an extended latent period, if the *S* gene is functional; instead, cellular respiration, ATP synthesis, all macromolecular events, and phage assembly stop at the normal time of lysis, but the progeny virions are left within the dead cell.

The lambda holin S105

These physiological effects and other observations indicate that the *S* holin functions by forming a hole in the membrane. This conclusion is reinforced by the ability of purified S105 protein to form holes in artificial membranes; moreover, missense mutations that block hole formation *in vivo* also block *in vitro* hole formation. The holes that are formed appear to be nonspecific, since null alleles in the lambda *S* gene can be complemented by completely unrelated cloned holin genes from other phages of Gram-negative hosts, such as T4, P2, and PRD1 and also from phages of Gram-positive hosts. Nothing has been published about the structure of the S holes except that they must be very large; fusions of *R* to β-galactosidase (resulting in a tetrameric *R*-βgal chimera with a molecular mass exceeding 500 kDa) are fully functional as endolysins and support plaque formation.

Genetic and biochemical studies focused on the unique ability of S105, an integral membrane protein, to form large nonspecific holes in the membrane, and on the equally unique capacity to accomplish this at a programmed time. The S105 protein has three transmembrane domains, with an N-out, C-in topology (**Figure 1**). The C-terminal cytoplasmic domain is nonessential, although the number of basic residues in this sequence is important for lysis timing. *S* represents an extreme case of genetic malleability. Missense mutations in all three TMDs, including in series of positions all the way around each helical domain, affect lysis timing unpredictably and dramatically. For example, substitutions of Gly or Leu at Ala52 result in catastrophically early lysis at about 20 min, before the first progeny virion is assembled, whereas Ala52 to His or Trp result in delayed lysis (and increased burst size) and Ala52 to Val, Ile, or Thr are functionally nonlytic, null alleles. The fact that all surfaces of all three TMDs are sensitive to substitution suggests that helical packing, both intramolecular and intermolecular, is involved in the *S* holin function and timing. This notion is reinforced by cross-linking studies that show S oligomerizes in the membrane. Some of the mutants that block lysis also block oligomerization, such as A52V, which yields only dimers, but others do not.

Based on these properties, a model for S105 function has been proposed in which it assembles into large multimeric 'death rafts' within which lipid is completely or largely excluded. At some point dependent on the overall helical packing properties of the particular S105 allele, the raft develops a lesion, leading to a local collapse of the membrane potential. In this normal triggering event, and also in the case of premature lysis triggering with an energy poison, this local collapse of membrane potential triggers a conformational shift at the tertiary or quaternary level, leading to immediate hole formation. It is thought that the remarkable malleability of S105 for its lysis timing phenotype reflects the effects of changing side chain mass and shape on the helical packing involved in formation and stability of the rafts. In support of this notion is the interesting observation that some defective *S* alleles have an unusual form of partial dominant character. In lysogens carrying two prophages, one wild type for *S* and the other an absolute-defective missense *S* allele, lysis timing is actually earlier than with a lysogen carrying two S^+ prophages.

The antiholin S107

Compared to S105, without which lysis does not occur, S107 normally has a much more subtle null phenotype for lysis. Elimination of the start codon for S107 advances the timing of lysis by 5–7 min. By altering RNA secondary structures flanking the translation initiation region, alleles with different S105:S107 ratios can be constructed. Strikingly, any allele generating a S105:S107 ratio <1 does not spontaneously trigger until very late after induction and has dominant-negative character.

The only distinction between S105 and S107 is the latter's N-terminal additional dipeptide Met_1–Lys_2, and the critical feature is the Lys residue. It has been shown that the N-terminal TMD of S107 is prevented from integrating into the membrane as long as the bilayer is energized; thus, S107 and S105 have different topologies. Depolarization by an energy poison, or by holin-dependent hole formation, relaxes the barrier, allows the

N-terminal TMD of the S107 molecule to integrate into the bilayer, and converts the antiholin into a functional holin. S107 binds S105, and it is thought that S107:S105 complexes are inactive for hole formation; the net effect is that, when triggering does occur, thereby collapsing the membrane potential, all the S105:S107 complexes are converted into active holin complexes, which may increase the size or number of the holes.

It is important to note that the dependence of the inhibitory activity of S107 on the energized state of the membrane is distinct from the susceptibility of S105 to being triggered by the collapse of the membrane potential. In the former case, a topological change is caused by the de-energization of the bilayer, whereas S105 has all three TMDs integrated into the membrane from the outset.

Rz–Rz1 encodes the spanin complex

The most distal genes of the lambda lysis cassette, *Rz* and *Rz1*, have an unusual relationship in that the smaller gene, *Rz1*, is entirely embedded within *Rz* in the +1 reading frame. Although there are a few other examples of such out-of-frame nested genes (see below for two examples in other phage lysis systems), *Rz* and *Rz1* are unique in biology because the two genes are required for the same biological function. The products of the two reading frames are components of a complex that spans the entire periplasm, from the cytoplasmic membrane to the outer membrane (OM). This function of this complex is unknown, but it appears to be important for efficient disruption of the envelope. The evidence for this is that, in the absence of either *Rz* or *Rz1*, lysis is blocked if the OM is stabilized by ~10 mM divalent cations in the medium. The terminal phenotype is a mechanically fragile spherical cell, presumably bounded by the stabilized OM.

Rz–Rz1 equivalents are found in almost all phages of Gram-negative bacteria. The genes are highly diverse, both in terms of sequence and gene arrangement. In many cases, the *Rz1* gene begins within the *Rz* gene but extends beyond its end. In other cases, the *Rz* and *Rz1* genes are completely separated, although always in the order *Rz–Rz1*. Rz proteins are type II 'signal anchor' proteins, with an N-terminal TMD tethering a periplasmic domain to the bilayer. Rz1 proteins are lipoproteins that are attached to the inner surface of the OM by the lipid and fatty acid groups of the modified N-terminal Cys residue. Rz and Rz1 form complexes by a specific C-terminal interaction. In the classic coliphage T1, the *Rz–Rz1* genes are replaced by a single gene encoding a larger OM lipoprotein with the unique feature of a C-terminal TMD. Because this lipoprotein spans the entire envelope, it has been designated as a 'spanin'. The more common Rz–Rz1 pairs comprise a functional equivalent to the spanin, albeit with different topology.

A New Paradigm: Signal Sequences, Pinholins, and SAR Endolysins

Until recently, the lambda lysis paradigm, where the holin controls lysis by gating the escape of the muralytic endolysin from the cytoplasm, was thought to apply to all dsNA phages. However, recent findings have established that there are completely different modes of holin–endolysin lysis. The first hint of this came from experiments with the lysis system of a phage, fOg44, which grows on the Gram-positive host *Oenococcus oeni*. The lysozyme of this phage was shown to have a bona fide signal sequence, such that it is exported by the *sec* translocon of the host, concomitant with proteolytic removal of the signal. Nevertheless, fOg44 also encodes what appears to be a canonical holin, suggesting that the holin plays a different role in the control of lysis in this phage system. Studies from the more genetically accessible phages of *E. coli* have provided unequivocal evidence of a new paradigm. Genetic and molecular analysis revealed that the holin of lambdoid phage 21 does not form holes large enough to allow exit of a soluble endolysin. Instead, the class II holin, called S^{21} because it is the product of the *S* gene of phage 21, forms small holes that cause depolarization of the membrane but are not large enough to allow an endolysin to cross the membrane. For lambda, this result would be the worst possible outcome of a lysis program; the host cell would be dead, stopping the further accumulation of progeny virions and leaving the existing progeny, along with the cytoplasmic endolysin, trapped in a corpse. Phage 21, and many other phages, however, encode an endolysin with a *sec*-specific secretory signal in the form of an N-terminal TMD appended to a glycosidase catalytic domain. The TMD functions as 'signal anchor' domain, so that the endolysin is initially secreted as type II integral membrane proteins (single TMD, N-in, C-out). This would seem to be inconsistent with the notion of specifically timed lysis, since the exported endolysin activity would begin appearing in the periplasm immediately after the onset of late protein synthesis. Remarkably, however, these enzymes are inactive in the membrane-tethered form, so there is no catastrophically early lysis. In addition, the signal anchor TMDs have a special character, in that they can completely escape from the membrane, although they are not cleaved off like normal signal sequences. When this happens, the endolysins become active and murein degradation ensues. Endolysins with this feature are called SAR (signal anchor release) endolysins. It is not known yet what allows these membrane-embedded helical domains to have the unprecedented ability to be extracted efficiently from the bilayer; at least from primary structure analysis, it is clear that most SAR domains are rich in Ala, Gly, Ser, and other minimally hydrophobic neutral residues. This feature suggests that the SAR domains are localized to the membrane by 'tricking' the signal recognition machinery, but are metastable in the environment of the bilayer, requiring the polarization of the membrane to remain integrated.

In support of this notion, SAR endolysins are found to spontaneously release from the membrane at a low rate, even when the membrane remains energized. Thus, the induction of a cloned SAR endolysin gene can eventually cause lysis as a result of spontaneous release and activation of the induced protein. Moreover, in the unrelated phages P1 and 21, both with SAR endolysins, neither holin gene is required for plaque formation, since, in the absence of holin triggering, the infected cell will eventually lyse and release progeny virions. Even so, the release of the SAR endolysin from the membrane is quantitative and rapid if the membrane is de-energized, as happens when the S^{21} holin triggers to form its small holes. This elegant variation on the classic lambda paradigm serves to emphasize the real role of the holin in lysis – to provide timing. In the lambda/T4 system, the holin effects timing by controlling the escape of the already active phage endolysin from the cytoplasm. In contrast, in the phage 21 system and many others, the holin determines timing by depolarizing the membrane, thus causing the concerted activation of all the membrane-tethered, inactive endolysin molecules.

To distinguish holins like lambda S105 and T4 T, which form holes sufficiently large to let fully folded enzymes transit the membrane nonspecifically, from holins like S^{21}, which make small holes that serve to depolarize the bilayer, the latter type of holin has been termed a 'pinholin'. Phage genomes are highly mosaic, and the lysis cassettes are no exception; in general, then, there is no significant barrier to recombination events that pair different holin and endolysin genes. However, pinholin genes are obviously more limited in this respect, compared to the canonical holins, in that they can only function with SAR endolysins, which do not require large holes to allow escape from the cytoplasm.

Diversity in Antiholin Strategies

The dual start motif which allows the S gene to produce both the S105 holin and S107 antiholin has been found in a number of holin genes, including the unrelated class I holin gene from the *B. subtilis* phage ϕ29 and in the class II holin genes of the S^{21} family (**Figures 1** and **2**). In the latter, the extra positive charge in the longer antiholin form of the protein still influences the membrane topology of the first TMD, as in lambda S107, but in this case, it inhibits TMD1 from exiting the membrane, rather than entering it. Remarkably, S^{21} TMD1 has the properties of an SAR domain, spontaneously leaving the bilayer at a certain rate, which is accelerated if the membrane is depolarized. As long as it stays within the bilayer, TMD1 binds TMD2 and prevents it from oligomerizing to form the pinhole. In effect, TMD1 is an internal antiholin of the S^{21} class II holin, and in the longer form produced from the dual start, the internal antiholin sequence is retarded from escaping the membrane.

In other lysis systems, the holin and antiholin are encoded by separate genes, with a striking number of different topologies (**Figure 1**). The well-studied coliphage P2 has a putative antiholin gene, *lysA*, sandwiched in its lysis gene cluster between the holin and endolysin genes, Y and K, and the Rz–Rz1 equivalents, *lysB–lysC*. LysA is predicted to be an integral membrane protein with four predicted TMDs. In the classic phage T4, the antiholin gene is *rI*, one of the first genetic markers identified in storied history of phage biology; the RI protein is a type II signal anchor protein with a 72 residue periplasmic domain. The T4 holin, encoded by gene *t*, is the prototype class III holin, with a single TMD and a large periplasmic domain. RI inhibits gp*t* through interactions of the periplasmic domains. In infected cells, the RI protein is activated to be an inhibitor of gp*t* by secondary infections that occur about 2 min after the primary infection. T4 gene product, gp*imm*, blocks passage of the secondary phage's DNA into the cytoplasm, resulting in ectopic localization of the capsid contents in the periplasm. It is not known whether it is the mislocalized DNA or internal capsid proteins that activate RI, or what the activating event is. Interestingly, RI is also an SAR protein, releasing from the membrane at a certain rate, and its N-terminal SAR TMD confers instability on the protein. This instability ensures that RI inhibition of the holin is not permanent. The concept is that holin triggering should be postponed as long as the infected cell continues to be superinfected with more T4 phage particles, which is an indication that the current environment is host poor. When superinfection ceases, the instability of RI ensures that the gp*t* holin will be allowed to trigger and cause the release of the progeny virions. Overexpression of *rI* bypasses the need for activation, just as altering the ratio of S105 to S107 in favor of S107 blocks lysis in the lambda system.

Antiholins have also been identified in P1, where the *lydB* gene encodes a small type I integral membrane protein, that inhibits the LydA holin, a class I holin unrelated to lambda S105. Interestingly, in P1, the holin is not required for lysis or plaque-formation, since the P1 Lyz is a SAR endolysin. Instead, the antiholin LydB is essential, because in its absence, the LydA holin triggers lysis too early, before the first virion has been completed (on average). LydB binds to LydA, but other details of its inhibitory function have not been elucidated. In general, except for the T4 system, nothing is known about how antiholin function is regulated during infection. Indeed, it is not clear why antiholins exist at all in cases like the lambda and phage 21 systems. The loss of the antiholin product in these cases has only a slight effect – the acceleration of lysis by a few minutes. Considering the genetic malleability of the lambda S gene, it is easy to isolate mutations which would cause S105 to trigger at almost any time, so the presence of the elaborate dual start system to generate a slight delay in lysis would seem to be unnecessary. As noted above, altering the 5′-end of the S-gene to cause synthesis of S107 in excess of S105 results in a block in lysis; it is possible that some physiological conditions may exist which cause the S107:S105 ratio to change, which would thus achieve real-time control of lysis timing.

'Single-Gene' Lysis

Although the holin–endolysin system is nearly universal among dsNA phages, it is not found in all lytic phages. A second well-studied mode for host lysis is found in two other major classes of icosahedral phages in Gram-negative bacteria, the ssDNA *Microviridae* and the male-specific ssRNA *Alloleviviridae*. In both of these cases, a single gene is necessary and sufficient for lysis – gene E in the prototype *Microvirus* φX174 and gene A_2 in the prototype *Allolevivirus*, Qβ. Moreover, both lytic proteins cause lysis in the same way, by inhibiting an enzyme in the synthesis pathway of peptidoglycan precursors. In eubacteria, the committed step for synthesis of murein is the reaction catalyzed by MurA, which adds the 3C backbone of PEP to the amino-sugar ring of the sugar nucleotide UDP-GlcNac (**Figure 3**). A series of further modifications to this sugar nucleotide yields UDP-MurNac-pentapeptide. The aminosugar-pentapeptide is transferred to a C55 lipid carrier by a membrane-embedded enzyme, MraY, converted to a disaccharide-pentapeptide by the addition of a GlcNac moiety, and then flipped to the outside surface of the cytoplasmic membrane by a yet unidentified 'flippase' component. Linkage of the disaccharide-pentapeptide to the murein is catalyzed by the transglycosylase activity of a multicomponent machinery involving the penicillin-binding proteins (PBPs). The formation of the new glycosidic bond results in the liberation of the undecaprenol carrier and is followed, in Gram-negative bacteria, by the formation of isopeptide cross-links between the peptidoglycan chains, a process mediated by transpeptidase activities of the PBP complex. The two lysis proteins E and A_2 were found to be specific inhibitors of MraY and MurA, respectively. Thus, infection with φX174 or Qβ results in the cessation of the flow of precursors to the periplasmic murein synthesis complex. It is not completely clear why this results in lysis. Most cells are seen to undergo lysis during septation, when new cell wall material is required to separate the daughter cells. Apparently, bacterial cells are unable to stop the septation process if the flow of murein precursors is interrupted. In any case, E- and A_2-mediated lysis is indistinguishable from lysis induced by antibiotics that inhibit any of the steps in the pathway

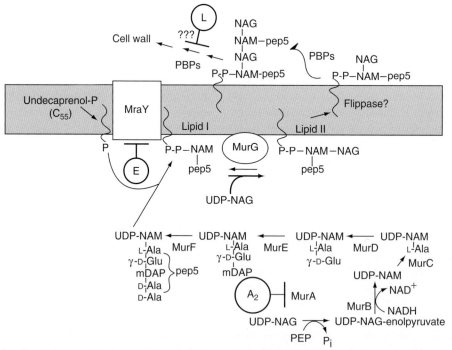

Figure 3 Phage-encoded 'protein antibiotics' inhibit the synthesis of cell wall precursors. The murein synthesis pathway beginning with the first committed step to peptidoglycan synthesis, which is catalyzed by MurA. Inhibition by E and A_2 is shown at the step that they inhibit, and inhibition by the L protein is indicated after the incorporation of the disaccharide oligopeptide subunit into the peptidoglycan (the specific target of L-mediated lysis is not currently known).

for precursor synthesis or by β-lactams, which inhibit the PBPs. E and A_2 have been called 'protein antibiotics' because of this similarity of action.

E: The First Embedded Gene

The molecular mechanisms of E and A_2 function are of interest at least in part because both MraY and MurA are universally conserved in bacteria and are thus attractive antibiotic targets. The *E* gene of φX174 was originally defined as an essential cistron in φX174 that gave rise to nonsense mutants with defects in lysis but not in virion production. Strangely, mutations in an essential morphogenesis gene, *D*, mapped to both sides of the *E* mutations. φX174 was the first DNA genome to be sequenced, and part of the interest in this ground-breaking work was the revelation that *E* was entirely embedded within *D*, in the +1 reading frame. (Note, however, that the *D–E* embedded relationship is very different from that of *Rz–Rz1*, because in the latter case, the two protein products are required for the same biological function.) *E* encodes a 91 residue type I integral membrane protein. The C-terminal cytoplasmic domain is highly basic (predicted +11 net charge) and completely dispensable. Fusions of the N-terminal domain of *E* to heterologous protein domains such as β-galactosidase and GFP result in chimeras with unimpaired lytic function. Mutations conferring resistance to *E* have been mapped to two of the ten TMDs of MraY. In *in vitro* assays, MraY activity has been shown to be inhibited in membranes containing E. However, to date, the inhibition has not been demonstrated in a purified system. A host gene, *slyD*, encoding a cytoplasmic *cis–trans* peptidylprolyl *cis–trans* isomerase (PPIase), is required for E-mediated lysis. In the absence of SlyD, the E protein is extremely unstable, suggesting that the PPIase is needed for proper folding of E. Mutations that restore the ability of the phage to plate on *slyD* hosts have been mapped to N-terminal residues of E. These turn out to be translation-up mutations that bypass the *slyD* block by increasing the rate of synthesis without affecting the proteolytic instability. While it is unclear if there is a regulatory relationship between E-mediated lysis and the host *slyD* gene, the strict dependence of φX174 plating on the *slyD* function remains the strongest ligand-independent phenotype ever found for PPIases, which, despite their ubiquity in most compartments of both eukaryotic and prokaryotic cells, have not been associated with essential functions.

A_2: A Virion-Borne Lysis Protein

Unlike *E*, the A_2 gene occupies a unique reading frame, in this case the first cistron of the Qβ genome (**Figure 4**), encoding a 420 residue polypeptide. However, the lysis function is just one of several essential functions of A_2, which is present in one copy in the Qβ virion. In addition to lysis, A_2 is required for adsorption to the male sex pilus and protection of the 5′-end of the ssRNA genome from external RNase attack. Unlike the membrane-embedded MraY, MurA, the target of A_2, is a soluble enzyme that has been extensively studied at both the structural and mechanistic levels. Thus far, however, the only soluble form of A_2 that has been obtained has been in the form of purified Qβ virions, and these have been shown to inhibit MurA *in vitro*. A misssense mutation in MurA, *rat1* (resistance to A-two), that blocks the inhibitory action of A_2 and abolishes Qβ plaque formation, has been found to be a Leu to Gln change in residue 138; this residue is on the surface of MurA near the active site cleft. Qβ *por* (plates on rat) mutations that overcome the *rat1* block and restore

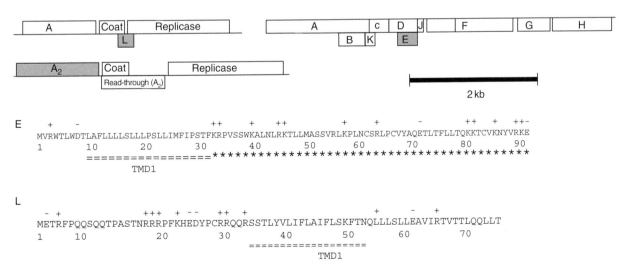

Figure 4 Single gene lysis systems. Upper panel: Genome maps of phages MS2, φX174, and Qβ; the lysis gene of each phage is colored gray. Lower panel: Primary sequence and membrane topology of the lysis proteins L, E; ===, transmembrane domains as predicted by TMHMM (see above.); ***, the highly charged, dispensable C-terminal region of E.

plaque-forming ability map to several positions in the N-terminal domain of A_2. Among the issues not yet resolved, besides the molecular details of the MurA-A_2 interaction, is whether or not the actual *in vivo* inhibitor of MurA is A_2 on the Qβ virion or free A_2 protein.

L: Another Embedded Gene and an Enduring Mystery

Operationally, the fundamentals of how the holin-endolysin and the two 'protein antibiotic' systems described above effect lysis are now established, even though the molecular mechanisms are still a matter of investigation. The remaining mystery in phage lysis, in terms of the general mode of action, is that of the male-specific ssRNA *Leviviridae*. These viruses have nearly the same genome organization as the *Alloleviviridae*, except that the lysis gene is a separate, small (75 codon) reading frame designated as *L* (**Figure 4**). Expression of *L* is necessary and sufficient for lysis, but, unlike the case with E and A_2, synthesis of murein is undisturbed, at least up to the initial incorporation of the disaccharide oligopeptide subunit into the peptidoglycan. The simplest notion would be that L interferes with cell wall synthesis or maturation in some step downstream of the initial glycosidic bond formation. Alternatively, L may be triggering an autolytic response analogous to those triggered in many Gram-positive systems as a result of starvation or developmental processes.

The *L* gene is yet another example of an embedded gene; in this case, the *L* cistron overlaps the end of the *coat* gene and the beginning of the *rep* gene encoding the RNA-dependent RNA polymerase. The predicted L protein spans only 75 residues, with a highly charged basic N-terminal domain and a hydrophobic C-terminal domain. Although no topological studies have been done, it has been reported that the L protein accumulates in zones of adhesion between the inner and outer membranes. The expression of *L* is regulated by RNA structures that control access to its translation initiation site and are influenced by ribosomes transiting the end of the *coat* gene.

Perspective

Lysis was once considered a rather trivial and inevitable consequence of bacteriophage infection, an indirect consequence of the extensive re-direction of the host biosynthetic and maintenance functions to the purposes of the virus. However, it is now clear that lysis is a carefully orchestrated and regulated process programmed by the phage. Moreover, the molecular basis of the lytic event and its regulation are compellingly interesting, involving fundamental issues of the behavior of membrane proteins and also of the function of the bacterial cell wall biosynthesis pathway. The availability of the traditionally powerful genetic facility of phages, coupled with modern genomic, proteomic, and structural approaches, gives promise of rapid progress in our understanding of this fundamental biological process.

See also: Ecology of Viruses Infecting bacteria; Virus Entry to Bacterial Cells.

Further Reading

Bernhardt TG, Wang IN, Struck DK, and Young R (2002) Breaking free: 'Protein antibiotics' and phage lysis. *Research in Micriobiology* 153: 493–501.

Bläsi U and Young R (1996) Two beginnings for a single purpose: The dual-start holins in the regulation of phage lysis. *Molecular Microbiology* 21: 675–682.

Rydman PS and Bamford DH (2003) Identifcation and mutational analysis of bacteriophage PRD1 holin protein P35. *Journal of Bacteriology* 185: 3795–3803.

Sao-Jose C, Parreira R, and Santos MA (2003) Triggering of host-cell lysis by double-stranded DNA bacteriophages: Fundamental concepts, recent developments and emerging applications. In: Pandalais G (ed.) *Recent Research Developments in Bacteriology*, vol. 1, pp. 103–130. Kerala, India: Transworld Research Network.

Sao-Jose C, Parreira R, Vieira G, and Santos MA (2000) The N-terminal region of the *Oenococcus oeni* bacteriophage fOg44 lysin behaves as a bona fide signal peptide in *Escherichia coli* and as a cis-inhibitory element, preventing lytic activity on oenococcal cells. *Journal of Bacteriology* 182: 5823–5831.

Steiner M, Lubitz W, and Bläsi U (1993) The missing link in phage lysis of Gram-positive bacteria: Gene 14 of *Bacillus subtilis* phage φ29 encodes the functional homolog of lambda S protein. *Journal of Bacteriology* 175: 1038–1042.

Tran TAT, Struck DK, and Young R (2005) Periplasmic comains define holin–antiholin interactions in T4 lysis inhibition. *Journal of Bacteriology* 187: 6631–6640.

Wang IN, Deaton J, and Young R (2003) Sizing the holin lesion with endolysin-beta-galactosidase fusion. *Journal of Bacteriology* 185: 779–787.

Wang IN, Smith DL, and Young R (2000) Holins: The protein clocks of bacteriophage infections. *Annual Reviews of Microbiology* 54: 799–825.

Xu M, Struck DK, Deaton J, Wang IN, and Young R (2004) A signal-arrest-release sequence mediates export and control of the phage P1 endolysin. *Proceedings of the National Academy of Sciences* 101: 6415–6420.

Young R (2005) Phage lysis. In: Waldor MK, Friedman DI, and Adhya SA (eds.) *Phages: Their Role in Pathogenesis and Biotechnology*, pp. 92–127. Washington, DC: ASM Press.

Young R and Wang IN (2006) Phage lysis. In: Calendar R (ed.) *The Bacteriophages*, pp. 104–126. Oxford: Oxford University Press.

Young R, Wang IN, and Roof WD (2000) Phages will out: Strategies in host cell lysis. *Trends in Microbiology* 8: 120–128.

Machlomovirus

K Scheets, Oklahoma State University, Stillwater, OK, USA

© 2008 Elsevier Ltd. All rights reserved.

History

Maize chlorotic mottle virus (MCMV) was initially found in maize (*Zea mays*) and sorghum (*Sorghum bicolor*) fields in Peru in 1973. In 1976, the virus appeared in Kansas (USA) maize fields, alone and as part of a synergistic disease. In 1978, the second component of the synergistic disease, corn lethal necrosis (CLN), was identified as any maize-infecting virus of the family *Potyviridae*. MCMV and CLN spread to Nebraska, and both MCMV and CLN appeared in Mexico in 1982. On the island of Kauai, Hawaii (USA), a severe outbreak of MCMV appeared in the winter seed nurseries in 1989–90. MCMV is endemic in Peru and along the Kansas–Nebraska border. In 2004, MCMV and CLN were first detected in Thailand where MCMV continues to spread.

Taxonomy and Classification

The only known member of the genus *Machlomovirus* is the species *Maize chlorotic mottle virus*. Its inclusion in the family *Tombusviridae* is based on the high degree of homology of the encoded viral replicase. The genome organization of the single viral RNA of MCMV is most similar to panicum mosaic virus (PMV), species *Panicum mosaic virus*, genus *Panicovirus*, family *Tombusviridae*. Key taxonomic features of the genus *Machlomovirus* are a unique open reading frame (ORF) at the 5′ end of the genome that largely overlaps the pre-readthrough portion of the viral replicase gene, and a readthrough ORF preceding and overlapping the capsid protein (CP) gene near the 3′ end of the genome.

Virion Structure and Properties

Transmission electron microscopy of negatively stained MCMV virions shows a smooth sphere or hexagonal structure approximately 28 nm in diameter (**Figure 1**). The virion consists of a 4437 nt single-stranded RNA surrounded by 25.1 kDa CP subunits that lack the protruding domain found on CPs of viruses from many genera in the family *Tombusviridae*. Sequence similarity to the CPs of PMV, tobacco necrosis virus genus *Necrovirus*, family *Tombusviridae*, and southern bean mosaic virus genus *Sobemovirus* suggest that MCMV is a $T = 3$ icosahedral virion with 180 copies of its CP in the viral shell. The estimated weight of the virion is 5.95 mDa. Purified MCMV has a sedimentation coefficient ($s_{20,w}$) of 109S and a buoyant density in CsCl of 1.365 g ml^{-1}. The thermal inactivation point in maize sap is 85–90 °C. The virion is stable at pH 5.

Genome Structure and Organization

The genome organization of the genus *Machlomovirus* is based on the complete sequence of MCMV. The plus-sense RNA is 4437 nt long and contains seven overlapping ORFs that encode proteins of 7 kDa or larger (**Figure 2**). The RNA was reported to have no poly(A) tail and an ^{m7}G cap at the 5′ end. However, in common with other viruses in the family, the encoded MCMV replicase does not have any motifs characteristic of the methyltransferase domain found in viral replicases of capped RNA viruses. None of the other MCMV-encoded proteins contain a methyltransferase domain, so it is likely that the RNA in MCMV, like in other members of the family, is in fact uncapped. The 5′ untranslated region (UTR) is 117 nt long, and ORF1 encodes a 32 kDa highly acidic protein (estimated pI = 3.83). ORF2 begins 19 nt downstream and encodes a 50 kDa highly basic protein (estimated pI = 10.59) so the migration of these two proteins in sodium dodecyl sulfate (SDS) polyacrylamide gels is likely to be anomalous. Suppression of the UAG stop codon of ORF2 would produce a 111 kDa protein. A cluster of four ORFs is encoded in the 3′ third of the viral RNA downstream of the transcription start site for the 1467-nt-long subgenomic RNA1 (sgRNA1). ORF4 encodes a 7.5 kDa protein (p7a), and suppression of its UGA stop codon would produce a 31 kDa protein. The second AUG of sgRNA1 begins ORF7 which encodes the viral CP. ORF6 was identified by similarity of its gene product p7b to small peptides encoded in similar locations on carmoviruses, necroviruses, and PMV, and it begins with a noncanonical start codon. *In vitro* translation of MCMV virion RNA in rabbit reticulocyte lysate produces p32, p50, p111, and p25. The two 7 kDa peptides and p31 were not detected in rabbit reticulocyte lysate translations. The 3′ UTR is 343 nt long and encodes a 337-nt-long sgRNA (sgRNA2).

Replication

The replication strategy of MCMV has not been completely determined, but inoculation of maize protoplasts with transcripts from wild type and mutant versions of an infectious cDNA has provided some information. Transcripts with mutations in the 3′ third of the genome that stop expression of one or more of the proteins encoded on sgRNA1 are capable of replication. Additionally, mutations just upstream of the sgRNA1 transcription start site that stop expression of sgRNA1 but do not alter the sequence of p111 are capable of replication, indicating that none of the proteins encoded on sgRNA1 are necessary for replication. Based on the replication mechanisms of other tombusvirus family members it is likely that after virion disassembly, MCMV viral RNA is translated to produce the viral replicase which then synthesizes the negative strand of genomic RNA after recognizing sequences and structures located at the viral 3′ terminus that have sequence and structural similarities to the promoters of carmoviruses. The complementary strand is then used as template for synthesis of progeny viral RNA strands. sgRNA synthesis mechanisms differ between genera in the family *Tombusviridae*, and it is not known which mechanism is used by MCMV. sgRNA1 synthesis may initiate by replicase binding internally to the sgRNA promoter on the genomic complementary strand. Alternatively, occasional premature termination

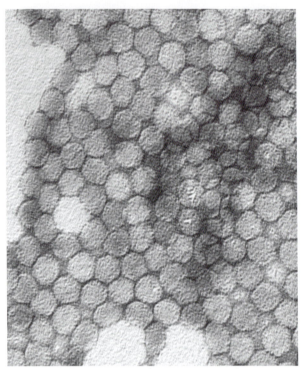

Figure 1 Purified maize chlorotic mottle virus negatively stained with uranyl acetate.

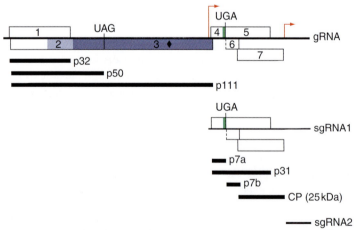

Figure 2 Genome organization and protein products of maize chlorotic mottle virus. The seven ORFs are marked as boxes on the genomic RNA and subgenomic RNA1 (sgRNA1) in each reading frame. The two suppressible stop codons are indicated (UAG and UGA), and the noncanonical start codon for ORF6 is marked with a dashed line. The proteins are indicated as heavy black bars beneath their mRNAs. The dark blue box marks the region encoding high protein sequence similarity to replicases of other members of the family *Tombusviridae* with a monopartite genome, and the light blue box indicates the area encoding protein similarity only to PMV replicase. The diamond marks the location encoding the 'GDD' motif. The green box in ORF4 marks the coding region for a conserved peptide sequence also found in PMV, carmoviruses, and necroviruses. Bent red arrows mark the sgRNA transcription start sites. sgRNA2 does not contain any significant ORFs.

of viral complementary strand synthesis at a specific location may produce separate complementary strand copies of sgRNA1 that are used as templates to synthesize many copies of sgRNA1. Although sgRNA2 accumulates in infected maize plants and inoculated protoplasts, its function and method of transcription are not known.

Geographic and Seasonal Distribution

Originally MCMV was found only in the Western Hemisphere. Peru, Argentina, Mexico, and the USA have reported infections, and it is possible that MCMV is present in other countries in the Western Hemisphere anywhere the beetle vectors are found in maize-growing areas. In Peru, MCMV is found in northern coastal areas where maize is grown as a seasonal crop as well as in the southern coastal areas where it is grown year-round. MCMV is endemic in several valleys in the Department of Lima. In the continental USA, where maize is a seasonal crop, MCMV was initially found in Kansas near the Nebraska border, then spread along the Republican and Big Blue River Valleys. The virus continued spreading along river valleys and is endemic in Nebraska and Kansas. MCMV initially appeared in the state of Guanajuato, Mexico, and has spread from there to additional nearby states. After its appearance in Kauai, Hawaii, MCMV has not remained as a problem. In Thailand, maize production is year-round, and MCMV was first detected in the province of Saraburi. It is now found in Saraburi, Nakohn Ratchasrima, Tak, and Petchaboon provinces.

Host Range and Virus Propagation

Susceptibility has been tested for 73 plant species (19 dicots) representing 35 genera, and only members of the family Poaceae are hosts. Although some cultivars of hexaploid wheat (*Triticum aestivum*), durum wheat (*Triticum durum*), barley (*Hordeum vulgare*), and sorghum (*S. bicolor*) are susceptible to infection, maize is the only crop in which MCMV infection is of economical importance, with many susceptible cultivars in all maize types (sweet, floury, field, and popcorn). MCMV can be readily propagated by mechanical inoculation in the inbred lines N28Ht, N28, and Oh28, but no reliable local lesion host has been identified.

Genetics

Viral RNA or uncapped transcripts of a full-length cDNA are sufficient for replication in protoplasts and infection of plants. The first ORF encodes a highly acidic 32 kDa protein with no similarity to any other protein in databases. The location of ORF1 suggests that it is a protein needed early in the viral life cycle, but its function is not known. ORF2 and ORF3 comprise a 50 kDa protein and a readthrough protein of 111 kDa containing the 'GDD' motif found in almost all RNA-dependent RNA polymerases (**Figure 2**). Sequence similarities of the carboxyl end of p50 and the p111 readthrough region to viral replicase proteins of other members of the family *Tombusviridae* suggest that these are similarly involved in replication of MCMV. p50 is much larger than the pre-readthrough proteins encoded by most of the monopartite viruses in the same family (27–33 kDa) and the sequence similarity only extends to the carboxyl third of p50. The similarity between p50 and the 48 kDa protein from PMV encompasses about 60% of the carboxyl end. Along with other members of the family, no regions identifiable as a helicase domain or a methyltransferase domain are found in p111. Like PMV and oat chlorotic stunt virus (genus *Avenavirus*), MCMV produces a single sgRNA to express a cluster of ORFs in the 3′ third of the genome, and none of the proteins encoded in this region are required for replication. The genes for p7a and p7b are upstream of the CP ORF, similar to the location of small ORFs in PMV, necroviruses, and carmoviruses. p7a and p7b have similar hydrophobic/hydrophilic characteristics and some sequence similarity to the peptides encoded in the corresponding locations in PMV, necroviruses, and carmoviruses, with the greatest sequence similarity occurring in the C-terminal region of p7a. These small peptides are required for cell-to-cell movement for PMV, necroviruses, and carmoviruses, so it is likely that p7a and p7b have similar functions. Database searches with the protein encoded in ORF5 do not identify any proteins with related sequence. The most 3′ ORF encodes the viral CP.

Serologic Relationships and Variability

The complete relationship of MCMV isolates from countries where MCMV has been found on a noticeable scale has not been determined. Two serotypes from Kansas (K1 and K2) and one from Peru have been compared, and they can be differentiated by agar double-diffusion assays. Most isolates from Nebraska and Kansas are similar to MCMV-K1, which is the source used for the MCMV sequence in GenBank (X14736) and the infectious transcript cDNA. Sequence comparisons have been done using reverse transcription-polymerase chain reaction (RT-PCR) amplification of the CP shell domain ORF from 47 isolates collected from Nebraska/Kansas, Hawaii, and Peru. Alignment of the 200 bp fragments indicates that each geographic region can be identified by a predominant unique genotype. The Hawaiian isolates show the least intrapopulational divergence, consistent with the recent appearance of the virus in Hawaii before the sample collection period. The Hawaiian isolates are more closely related to Peru isolates than to those from Nebraska/Kansas. The complete sequence of the CP gene of MCMV isolated in Thailand (AY587605) shows 96%

identity to MCMV-K1 at the nucleotide level and 97% identity at the protein level. When the corresponding 200 nt region is compared to the data in the larger study, the Thai sequence groups with Nebraska/Kansas sequences. These data suggest that Peru was the source for MCMV in Hawaii while the source of virus in Thailand was Nebraska/Kansas.

Transmission

MCMV is readily transmitted by mechanical inoculation of leaves or roots and by vascular puncture inoculation of seeds. Six species of beetles of the family *Chrysomelidae* found in the continental USA can transmit MCMV: the southern corn rootworm beetle (*Diabrotica undecimpunctata howardi*), the northern corn rootworm beetle (*Diabrotica barberi*), the western corn rootworm beetle (*Diabrotica virgifera virgifera*), the cereal leaf beetle and larvae (*Oulema melanopus*), the corn flea beetle (*Chaetocnema pulicaria*), and the flea beetle (*Systena frontalis*). The southern corn rootworm beetle is the vector in Mexico. In Peru, *Diabrotica viridula* (adult and larvae) and *Diabrotica decempunctata sparsella* transmitted MCMV-P. MCMV was spread by a large infestation of thrips (*Frankliniella williamsi*) in Hawaii. MCMV-K1 epidemiology indicates a soil and water connection, so it is possible that MCMV is soil borne, water borne, or transmitted by a fungus as is seen with various other viruses in the tombusvirus family. It is possible that the different major isolates of MCMV are predominantly transmitted by either a soil route or insects. Seed transmission occurs at a very low rate (0.008–0.04%).

Epidemiology

MCMV reappears in previously infected fields in Nebraska and Kansas, and there is little indication that the virus spreads very rapidly during the maize-growing season, suggesting that beetles are not the major source of viral spread there. MCMV does not overwinter in grasses surrounding the fields, suggesting that soil is the reservoir for future infections. Although it was hypothesized that infected crop residue might be a food source for beetle larvae which would infect new seedlings, it was later shown that MCMV loses its infectivity over the winter, and its loss in infectivity correlates with drying. Additionally, larvae do not feed on dead infected plant material. Infected fields are in river valleys and likely to be irrigated suggesting a water-borne mechanism. In the regions of Peru where maize is planted year-round, infection patterns correlate with emergence and spread of the beetle vectors. The mechanism of spread in Thailand has not yet been determined. The large geographic jumps in MCMV outbreaks suggest that initial infections occur via seed transmission. Although all the known beetle vectors are native to the New World, the western corn rootworm beetle was introduced to Serbia in 1992 and has since spread to additional European countries. This increases the potential for an outbreak of MCMV in Europe if it is introduced there via seeds.

Pathogenicity

MCMV infections can reduce yields of maize in experimental plots by up to 59%, but natural infections are seldom as severe. Average losses of 10–15% of sweet and floury maize crops occur in Peru. The more devastating disease CLN is caused by a synergistic interaction with potyviruses. CLN has been reported in Peru, Kansas, Nebraska, Mexico, and Thailand. In Kansas and Nebraska, maize dwarf mosaic virus-A (MDMV-A; genus *Potyvirus*), sugarcane mosaic virus MD-B (SCMV MD-B) (genus *Potyvirus*), and wheat streak mosaic virus (WSMV) (genus *Tritimovirus*) are found in natural CLN infections. In Mexico and Thailand, the potyvirus is usually SCMV MD-B. CLN has caused crop losses as high as 91%. If both viruses are present when plants are young, they initially develop a chlorotic mottle, then become extremely chlorotic and stunted. Leaves become necrotic starting at the leaf margins, then necrosis spreads inward causing rapid plant death. If the viral combination infects the plants at later stages, plants may appear green and healthy except for early drying of husks, producing ears that are fully developed except for wrinkled and shriveled seeds. As well as the synergistic increase in symptoms in CLN, the concentration of MCMV increases dramatically similarly to most synergisms involving potyviruses. With SCMV MD-B present, MCMV concentrations averaged more than fivefold higher than in singly infected plants, and in WSMV/MCMV synergisms the MCMV levels increased up to 11-fold higher than in singly infected plants. Although most potyviral synergisms do not cause an increase in potyvirus concentrations, the WSMV/MCMV synergism increased the WSMV infection rate and caused a two- to threefold increase in WSMV concentrations.

Pathology and Histopathology

Disease symptoms vary depending on maize genotype and age at infection. The mildest symptoms are a light chlorotic mottle on leaves with little effect on corn yield. In more susceptible varieties, these symptoms progress to chlorotic stripes parallel to the midvein that may coalesce forming elongated chlorotic blotches that may further turn necrotic. Severe symptoms include necrosis, stunting, decreased male inflorescence with few spikes, decreased number of ears, and malformed and partially filled ears. Plants infected at earlier stages usually have stronger symptoms and lower yields than later infections. MCMV has been found in all parts of the plant including leaves, stem, roots, cob, silk, pollen, ear sheaves, bracts, and in all seed parts (embryo, cotyledon, endosperm, and pericarp). Electron microscopy of cross sections of MCMV-infected maize showed virus-like particles in xylem vessels, often filling the entire vessel lumen. Some xylem tubes were filled with an electron dense matrix that appeared to be virus-like particles embedded in viroplasm. Some parenchyma cells contained vacuolated viroplasms. Chloroplasts were highly disorganized.

Prevention and Control

MCMV-resistant maize lines are being developed, and perennial diploid teosinte (*Zea diploperennis*) is a source for additional resistance genes. If maize lines resistant to either MCMV or the potyviral components of CLN are used, the potential for CLN outbreaks is diminished. Interestingly, a transgenic maize line expressing the CP gene from SCMV MD-B showed fewer disease symptoms when inoculated singly with SCMV MD-B, MDMV-A, and MCMV, as well as showing fewer symptoms when inoculated with MCMV + SCMV-MD-B or MCMV + MDMV-A. In Kansas and Nebraska, crop rotation with soybeans or MCMV-resistant sorghum in fields previously infected with MCMV decreases the incidence of MCMV in the following year. In Kauai, Hawaii, MCMV was essentially eliminated as a detectable disease by completely destroying all maize on the island and not replanting for 6 months. This was followed by including a 90 day fallow period each year for 2 years, and has continued with a 60 day fallow period every year.

Evolution

Based on genome organization and protein sequence similarities, MCMV is most closely related to PMV and vice versa. MCMV CP and the putative viral replicase proteins are markedly most similar to the PMV equivalents, and p7b is most similar to the 6.6 kDa peptide encoded by PMV. Both MCMV and PMV encode four overlapping ORFs on large sgRNAs that have similar start codon characteristics; the second ORFs begin with noncanonical start codons, the CP ORFs begin with the second AUGs, and a fourth ORF overlaps the CP ORFs. Additionally, the host range for both viruses is restricted to monocots. Since there is no similarity in the proteins encoded by the ORFs overlapping the CP ORFs, it suggests that MCMV and PMV evolved from a virus lacking those ORFs and the 5′ ORF encoding p32. The origins of the p32 ORF and p31 readthrough region are unknown. MCMV and the other members of the family *Tombusviridae* belong to virus supergroup II based on their viral replicases along with umbraviruses, a subgroup of luteoviruses, hepatitis C virus, pestiviruses, flaviviruses, and the positive-strand RNA coliphages.

See also: *Carmovirus*; Cereal Viruses: Maize/Corn; *Necrovirus*; Tombusviruses.

Further Reading

Batten JS and Scholthof K-BG (2004) Genus *Panicovirus*. In: Lapierre H and Signoret PA (eds.) *Viruses and Virus Diseases of Poaceae (Gramineae)* pp. 411–412. Paris: INRA.

Castillo J and Hebert TT (1974) A new virus disease of maize in Peru. *Fitopatologia* 9: 79–84.

Chiemsombat P, Larprom A, See-Tou W, and Patarapuwadol S (2006) Mixed infection of maize chlorotic mottle virus and sugarcane mosaic virus in sweet corn. In: Pohsoong T (ed.) *Proceedings of the Second Kasetsart University Corn and Sorghum Research Program Workshop*, pp. 214–219. Nakhon Nayok: Thailand Kasetsart University.

Gordon DT, Bradfute OE, Gingery RE, Nault LR, and Uyemoto JK (1984) Maize chlorotic mottle virus. In: *CMI/AAB Description of Plant Viruses*, No. 284. Kew, UK: Commonwealth Mycological Institute.

Lommel SA, Martelli GP, Rubino L, and Russo M (2005) Genus *Machlomovirus*. In: Fauquet CM, Mayo MA, Maniloff J, Desselberger U, and Ball LA (eds.) *Virus Taxonomy: Eighth Report of the International Committee on Taxonomy of Viruses*, pp. 932–934. San Diego, CA: Academic Press.

Scheets K (2000) Maize chlorotic mottle machlomovirus expresses its coat protein from a 1.47-kb subgenomic RNA and makes a 0.34-kb subgenomic RNA. *Virology* 267: 90–101.

Scheets K (2004) Maize chlorotic mottle. In: Lapierre H and Signoret PA (eds.) *Viruses & Virus Diseases of Poaceae (Gramineae)*, pp. 642–644. Paris: INRA.

Shafer KS (1992) *Molecular Evolution of Maize Chlorotic Mottle Virus: Isolates from Nebraska, Hawaii and Peru*. MD Thesis, University of Nebraska, Lincoln.

Maize Streak Virus

D P Martin, D N Shepherd, and E P Rybicki, University of Cape Town, Cape Town, South Africa

© 2008 Published by Elsevier Ltd.

Glossary

Agroinfection Technique used for the infection of host plants with cloned virus genomes involving transfer of virus DNA into the nuclei of host cells by the bacterium *Agrobacterium tumefaciens*.

Agroinoculation See agroinfection.

Bicistronic Contains two protein-coding regions within a single mRNA transcript.

Capsomer A subunit of the mature virus particle containing an ordered series of polymerized coat protein molecules.
Cicadulina spp. A group of leafhopper species involved in transmission of MSV.
C-ori Complementary strand (i.e., the half of the DNA duplex that is not packaged into virus particles) origin of replication.
CP Coat protein. The only protein component of the virus particle also believed to be involved in nuclear trafficking and cell-to-cell movement of viral DNA.
LIR Long or large intergenic region containing the origin of virion strand replication and gene promoters.
MP Movement protein. A small (c. 10 kDa) protein believed to be involved in intercellular virus movement via plasmodesmata.
Oviposition Egg-laying. To oviposit means to lay eggs.
Plastochron The time interval between successive leaf primordia, or the attainment of a certain stage of leaf development.
Rep Replication-associated protein involved in the initiation of virion strand replication.
RepA A truncated version of Rep with a unique C-terminal domain believed to be involved in regulation of host and/or virus gene expression.
SIR Short or small intergenic region containing gene polyadenylation signals and the origin of complementary strand replication.
Viruliferous A state in which an MSV vector species is carrying and is capable of transmitting the virus.
V-ori Virion strand (i.e., the half of the DNA duplex that is packaged into virus particles) origin of replication.

Introduction

Maize streak virus (MSV) is the causal agent of maize streak disease (MSD), the most serious viral disease of maize in Africa. It is a major contributor to the continent's food security problems and is endemic throughout Africa south of the Sahara. It is also found on the Indian Ocean islands of Madagascar, Mauritius, and La Réunion. There is no obvious barrier to spread of the virus outside of this region; hence it should be considered as a serious potential problem for other as yet unaffected maize-growing areas.

History and Taxonomy

"The disorder of the mealie plant, locally described as 'Mealie Blight', 'Mealie Yellows', or 'Striped Leaf Disease', belongs to a group of plant troubles arising from obscure causes..." was how MSD was first described by Claude Fuller in 1901 in Natal, South Africa. Fuller mistakenly attributed the disease to a soil disorder, but in retrospect it is quite clear that the 'mealie (a local word for maize or corn) variegation' he described and drew in minute detail can be attributed to MSV.

The first milestone in MSD research was reached in 1924, when H. H. Storey determined that a virus obligately transmitted by leafhopper species of the genus *Cicadulina* (**Figure 1**) was the causal agent of MSD. Storey named the virus 'maize streak virus', and was also the first to determine both the genetic basis of MSV transmission by *Cicadulina mbila*, and that resistance to MSD in maize could be inherited.

When MSV particles were first purified in 1974 they were found to have a novel twinned quasi-icosahedral (geminate) shape (**Figure 2**), from which the name 'geminivirus' was derived. This was followed by the unexpected discovery in 1977 that geminivirus particles contain circular single-stranded DNA (ssDNA), a genome type never before observed in plant viruses. These novel characteristics

Figure 1 The leafhopper vector of MSV, *Cicadulina mbila* Naudé. Photograph courtesy of Dr. Benjamin Odhiambo, Kenyan Agricultural Research Institute (KARI).

Figure 2 MSD symptoms on a maize leaf: note characteristic veinal streaks. Photograph courtesy of Dr. Frederik Kloppers, PANNAR (Pty) Ltd., Greytown, KwaZulu-Natal, South Africa.

led to the proposal of a new virus group – the geminiviruses – consisting of MSV and other viruses with geminate particle morphology and ssDNA genomes. *Maize streak virus* is now recognized as the type species of the genus *Mastrevirus*, in the family *Geminiviridae*.

Host Range and Symptoms

While most notorious for the yield losses it causes when infecting maize, MSV also infects over 80 other grass species including the economically important crops wheat, barley, and rye. In susceptible maize and grass genotypes, the virus first causes symptoms between 3 and 7 days after inoculation. These first appear as almost circular pale spots of 0.5–2 mm diameter in the lowest exposed portions of the youngest leaves. Later, fully emerged symptomatic leaves show veinal streaks from a few millimeters long to the entire length of the leaf and between 0.5 and 3 mm wide. These streaks often fuse laterally and symptomatic leaves may become >95% chlorotic (**Figure 2**).

Plants are worst affected when infected within a few days of coleoptile emergence; symptoms only develop above the site of inoculation on newly emerging leaves. Susceptible varieties may display severe stunting as well as very severe streaking, and cob development may be abolished. Yield losses can reach 100% when susceptible varieties are infected early.

Of the nine major MSV strains so far identified (designated MSV-A through MSV-I; (**Figure 3**)), only MSV-A produces economically significant infections in maize. The 'grass-adapted' MSV strains (MSV-B, -C, -D, -E, -F, -G, -H, and -I) differ from MSV-A types by 5–25% in nucleotide sequence, and produce substantially milder symptoms in maize than do MSV-A viruses and are often incapable of producing symptomatic infections in MSV-resistant maize genotypes. They may also have distinct but overlapping host ranges.

Diversity and Evolution

MSV is closely related to the other distinct 'African streak' mastreviruses, panicum streak virus, sugarcane streak virus, sugarcane streak Egypt virus, and sugarcane streak Reunion virus, with which it shares ~65% genome sequence identity. It is, however, most similar to, although not necessarily more closely related to, an isolate of digitaria streak virus from the Pacific island of Vanuatu, with which it shares ~67% genome sequence identity (**Figure 3**).

The full genome nucleotide sequences of MSV-A isolates display relatively low degrees of diversity, with any two MSV-A isolates obtained from anywhere in Africa invariably having genome sequences that are more than 97% identical. MSV isolates from La Réunion share

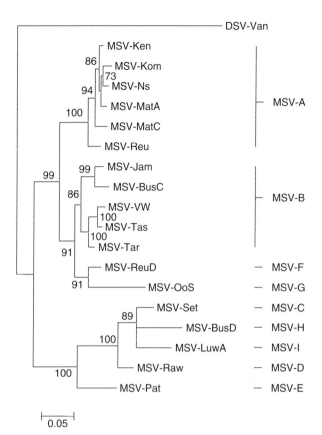

Figure 3 Phylogenetic relationships between the full genome sequences of different MSV strains. The tree is constructed using the maximum-likelihood method (HKY model transition and transversion weight determined from the data and 100 bootstrap replicates) and numbers associated with branches indicate degrees of bootstrap support for those branches. Branches with less than 70% support have been collapsed and the genome sequence of a digitaria streak virus (DSV) from Vanuatu is included as an outgroup. Only viruses in the MSV-A group have been isolated from maize. All viruses in the other groups have been isolated from wheat, barley, or wild grass species.

~95% identity with mainland isolates. Given that maize is a crop introduced at multiple points into Africa and its neighboring islands less than 400 years ago, and that the virus is not seed-borne, this may indicate either that the rate at which MSV-A is evolving is fairly slow, or that continent-wide spread and dominance of new MSV genotypes with enhanced fitness is very rapid. Experimental assessments of the rates at which the MSV-A-type isolate MSV-Reu evolves when maintained in a susceptible maize genotype, a resistant maize genotype, and a non-maize host (*Coix lacryma-jobi*) are, respectively, 9.5×10^{-5}, 17.3×10^{-5}, and 26.5×10^{-5} nucleotide substitutions per site per year. These evolutionary rates are relatively low when compared with RNA viruses and imply that genome-wide only one nucleotide becomes fixed every ~1.5–4 years, depending on the selection pressures exerted by the host species. Despite these low nucleotide fixation rates, it has been demonstrated that certain artificially induced mutations will revert to wild-type states at an unexpectedly high rate of approximately 36 substitutions per site per year. This combination of a low evolution rate and high mutation rate implies that MSV is currently occupying a fitness peak in maize. Despite the virus' apparent evolutionary sluggishness it may therefore be able to adapt very quickly and break any inbred or transgenic MSV resistance traits it is exposed to. In any case, the fact that there is only a very narrow range of sequence variants that can cause severe disease in maize over the whole geographical range of the virus, compared to the wide range of sequence variation in viruses adapted to other grasses, indicates that there is a high degree of sequence selection by the host genotype.

Transmission

In nature, MSV and other African streak viruses are neither seed nor contact transmissible and rely instead on transmission by cicadellid leafhoppers in the genus *Cicadulina* (including among others *C. mbila*, *C. storeyi*, *C. bipunctella zeae*, *C. latens*, and *C. parazeae*). Of these *C. mbila* is considered the most important MSV vector as it is the most widely distributed. Also, a greater proportion of *C. mbila* individuals are capable of transmitting the virus than is found with other *Cicadulina* species. The virus may be acquired by leafhoppers at any developmental stage in less than 1 h of feeding with a minimum acquisition time of 15 s. A latent period within the vector during which the virus cannot be transmitted lasts between 12 and 30 h at 30 °C. Once this latent period is over (signaled by the appearance of virus within the leafhopper's body fluids) the virus can be transmitted within 5 min of feeding. The so-called viruliferous leafhoppers can then transmit the virus for the rest of their lives.

Particle Structure

MSV particles (**Figure 4**) consist of two incomplete icosohedra with a $T=1$ surface lattice, comprising 22 pentameric capsomers each containing five coat protein (CP) molecules. Particle dimensions are 38 nm × 22 nm with the 110 CP molecules in each virion packaging a ~2690 nt covalently closed mostly single-stranded circular DNA genome. The packaged DNA has annealed to it a complementary ~80 nt sequence believed to act as a primer for complementary strand synthesis following infection and uncoating.

Genome Organization

As with all other mastreviruses discovered to date, the MSV genome encodes four proteins: a movement protein

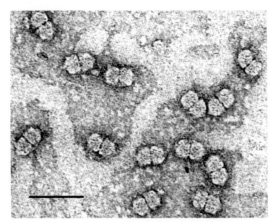

Figure 4 Electron micrograph of MSV purified from infected maize, showing particles of 18 nm × 30 nm stained with uranyl acetate. Scale = 50 nm. Photograph courtesy of Kassie Kasdorf; copyright EP Rybicki.

(MP) and a CP in the virion sense, and a replication-associated protein (Rep) and a regulatory protein (RepA) that is expressed from the same transcript as Rep in the complementary sense. MSV genomes also contain two intergenic regions: these are a short or small one (SIR), and a long or large one (LIR), which are the complementary sense strand and virion sense strand origins of replication, respectively (**Figure 5**).

The Long Intergenic Region

Besides containing divergent RNA polymerase II-type promoters and other transcriptional regulatory features necessary for the expression of the complementary and virion sense genes, the LIR also contains sequence elements that are essential for replication. The most striking of these is an inverted repeat sequence that is capable of forming a stable hairpin loop structure. All geminiviruses sequenced to date have the highly conserved nonanucleotide sequence TAATATT↓AC within the loop sequence of similar hairpin structures: this sequence contains the virion sense strand origin of replication (V-*ori*;↓).

A sequence 6–12 nt long occurring in all known mastreviruses between the TATA box that directs *rep/repA* transcription and the *repA* initiation codon is directly repeated in the stem near the V-*ori* hairpin, and is probably involved in Rep and/or RepA binding during replication.

The hairpin and two GC boxes on the 5′ side of the stem also forms part of an upstream activator sequence (UAS) required for efficient CP expression. The GC boxes bind nuclear factors to the UAS and are known as the rightward promoter element (*rpe1*).

The Short Intergenic Region

The MSV SIR occurs between the termination codons of the *CP* and *rep* genes (**Figure 5**) and contains the

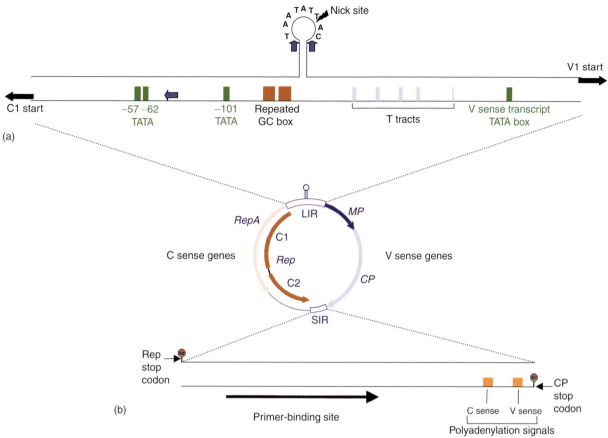

Figure 5 A schematic representation of the MSV LIR (a) and SIR (b), shown in context with the MSV genome. In (a) the main features of the MSV LIR are shown. These include a stem–loop structure with the loop's nonanucleotide sequence conserved among all geminiviruses and other rolling-circle systems. The site at which Rep introduces an endonucleolytic nick to initiate virion strand replication is shown. Iterated sequences (iterons) are shown in the V sense, with blue arrows indicating their location in the LIR. Iterons are potentially specific Rep-recognition sequences via which Rep may bind to the LIR. 5′ of the stem–loop is a repeated GC-box, which binds host transcription factors. A series of T tracts 3′ of the stem–loop may be involved in DNA bending of this region of the LIR. TATA boxes 5′ and 3′ of the stem–loop are potential C sense and V sense transcription initiation sites, respectively. In (b), the main features of the SIR include polyadenylation signals for V and C sense transcripts, and a primer-binding site on the plus strand. An ~80 bp DNA primer-like molecule, encapsidated with the viral genome and annealed to this site, is thought to be involved in initiating complementary strand replication. Both the MSV LIR and SIR are essential for viral replication.

polyadenylation and termination signals of the virion and complementary sense transcripts. The SIR also contains the origin of complementary strand synthesis (C-*ori*). A small 80-nt-long primer-like molecule is bound to the SIR of encapsidated MSV DNA and, at the onset of an infection, probably enables synthesis of double-stranded DNA (dsDNA) replicative forms (RFs) of the genome from newly uncoated virion strand DNA.

The Complementary Sense Genes (*rep* and *repA*)

The replication initiator/associated protein (Rep) is the only MSV gene product that is absolutely required for virus replication. In mastreviruses, *repA* (the *C1* open reading frame (ORF)) and the *C2* ORF (also called *repB*), respectively, encode the N- and C-terminal portions of Rep (**Figure 5**). Beginning at the same transcription initiation site, two C sense transcripts (1.5 and 1.2 kbp in size) are produced during MSV infections. Splicing of the larger transcript removes an intron, which permits expression of full-length Rep from the two ORFs.

It is very probable, although as yet unproven *in vivo*, that RepA is translated from both the unspliced 1.5 and 1.2 kbp C sense transcripts. If expressed in infected cells, MSV RepA would have the same N-terminal 214-amino-acid sequence as Rep, but would have a different C-terminus. RepA is possibly a multifunctional protein that modifies the nuclear environment to favor viral replication.

The N-terminal portions of Rep and RepA contain three conserved amino acid sequence motifs commonly found in replication-associated proteins of many extremely diverse rolling-circle replicons. Other significant landmarks include a plant retinoblastoma-related protein (pRBR) interaction motif, via which RepA but not Rep binds to the host pRBR which is involved in host cell cycle regulation, and oligomerization domains via which Rep and RepA bind to other Rep/RepA molecules (Rep activities are influenced by the aggregation state of Rep and/or RepA homo- and heterooligomers). It is also likely that the approximately 100 N-terminal amino acids of both Rep and RepA are involved in binding of the proteins to the viral LIR.

The C-terminal portion of Rep contains a dNTP-binding domain with motifs similar to those found in proteins with kinase and helicase activities. The dNTP-binding domain also sits within a region with similarity to the DNA-binding domains of the *myb*-related class of plant transcription factors: this domain may be functional in the induction of virus and/or host gene transcription.

The C-terminal portion of RepA, which is different from that of Rep, contains another potential transactivation domain also possibly involved in regulation of virus and/or host gene expression. A second domain within the C-terminal portion of RepA possibly interacts with host proteins involved in developmental regulation.

The Virion Sense Genes (MP and CP)

Transcription of the MSV virion (V) sense genes is directed by two TATA boxes within the LIR 26 and 214 nucleotides 5′ of the *MP* start codon. Each TATA box directs the production of different-sized transcripts, both of which terminate at the same place. Splicing of an intron within the *MP* portion of V sense transcripts appears to be an important determinant of relative MP and CP expression levels. Whereas CP is expressed from both long and short, spliced and unspliced V sense transcripts, MP is most likely only expressed from unspliced long transcripts.

The MP is post-translationally modified and contains a hydrophobic domain that may either facilitate its interaction with host cell membranes or be involved in homo- or heterooligomerization with the CP.

The N-terminal ~100 amino acids of the CP contain both nuclear localization signals and a sequence-nonspecific dsDNA- and ssDNA-binding domain. The CP and MP interact with one another and it is possible that this interaction is involved in trafficking of naked and/or packaged virus DNA from nuclei through nuclear pores, to the cell periphery, through plasmodesmata, and into the nuclei of neighboring cells.

Molecular Biology of MSV

Replication

As with other geminiviruses, MSV replicates by a rolling-circle mechanism (rolling-circle replication, or RCR; (**Figure 6**)) within the nuclei of infected cells. Replication may also occur by recombination-dependent mechanisms but this has not been conclusively demonstrated for MSV. As with other rolling-circle replicons, MSV replication is discontinuous with virion strand replication being initiated from the hairpin structure in the LIR and complementary strand synthesis being initiated from a short 80 nt primer-like molecule synthesized on the SIR of newly replicated virion strands.

Particle Assembly and Movement

Besides being the primary location of replication, the nucleus is also the site of virus particle assembly. CP molecules in the nucleus nonspecifically bind virion strands released during RCR (there is no known encapsidation signal in mastrevirus genomes), arresting the synthesis of new RF DNAs. Viral ssDNA molecules are packaged into particles that aggregate to form large paracrystalline nuclear inclusions. Crystalline arrays of MSV particles have also been detected outside nuclei within physiologically active phloem companion cells, and inside the vacuoles of dead and dying cells within chlorotic lesions. These lesions are caused by an as yet unexplained degeneration of chloroplasts in infected cells.

The mechanistic details of MSV cell-to-cell movement are still obscure, but it seems to involve an interaction between the CP, MP, and viral DNA. Besides requiring the coordinated interactions of viral gene products and DNA, the successful movement of MSV genomes from infected to uninfected cells is strongly dependent on the extent of plasmodesmatal connections between neighboring cells. Also, in maize it appears as though certain cell types are more sensitive to MSV infection than others. For example, in maize leaves the virus infects all photosynthetic cell types (e.g., mesophyll and bundle sheath cells) but despite abundant plasmodesmatal connections between photosynthetic, epidermal, and parenchyma cells, MSV is only rarely detectable in the latter two cell types.

It is unknown whether systemic movement of geminiviruses within plants simply relies on normal cell-to-cell movement to deliver genomic DNA into the phloem, or whether viral DNA is specifically packaged for long-distance transport. It is possible that cell-to-cell movement might involve unencapsidated ss- or dsDNA but that long-distance movement in the phloem might require encapsidation.

Long-distance movement of MSV within infected plants occurs via phloem elements and it is believed that MSV is incapable of invading the root apical, shoot apical, and reproductive meristems due to the absence of developed vasculature in these tissues. Thus, the virus is not found in tissue which develops into gametes and is not seed-borne. It also does not appear to travel down plants from the site of inoculation to older tissue.

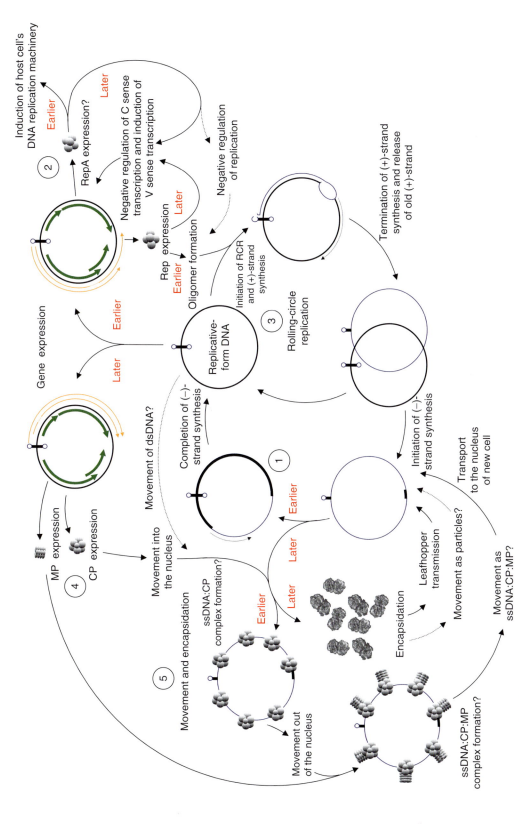

Figure 6 Summary of the MSV infection process. Early during an infection following the synthesis of a dsDNA replicative form (RF; 1) RepA is most likely expressed and induces a cellular state in which viral DNA replication can occur (2). Rep is also expressed early and RCR begins (3). At a later point in the infection process, following genome amplification and possibly Rep and/or RepA induction of the V sense promoter, MP and CP are expressed (4), and movement and encapsidation occur (5). Represented here is movement of unencapsidated ssDNA, but it should be noted that it is possible that dsDNA and/or encapsidated ssDNA may also be moved either cell to cell or systemically within the phloem of plants. Whereas the involvement of MSV CP and MP in movement has been demonstrated, the mechanics of the process are obscure. While the probable timing of events is indicated, it is unlikely, for example, that absolutely no MP and CP expression occurs during the earlier stages of the infection process. ssDNA is represented by blue lines, dsDNA by bold black lines, and RNA by orange lines.

Within the shoot apex where most productive MSV replication occurs, MSV first enters developing leaves at approximately plastochron five. While the virus is restricted to the developing leaf vasculature before plastochron 12, it is likely that the development of metaphloem elements at approximately plastochron 12 provides an opportunity for the virus to escape the vasculature into the photosynthetic cells of the leaf. Metaphloem develops with the abundant plasmodesmatal connections required for efficient loading of photoassimilates once the leaf emerges from the whorl. Before emergence, however, the developing photosynthetic tissues are still net importers of photoassimilates and the virus most likely moves into these cells through their plasmodesmatal connections with the metaphloem.

On the leaves, the pattern of chlorotic streak-like lesions that characterizes MSV infections is directly correlated with the pattern of virus accumulation within the leaves and the virus can only be acquired by leafhoppers from these lesions. The degree of chlorosis that occurs within lesions can differ between MSV isolates and is related to the severity of chloroplast malformation that occurs in infected photosynthetic cells.

Control of MSD

Although effective control of MSD in cultivated crops is possible with the use of carbamate insecticides, and it is possible to avoid the worst of leafhopper infestations by varying planting dates, the fact that small farmers cannot generally use these options means that the development and use of MSV-resistant crop genotypes is probably the best way to minimize the impact of MSD on African agriculture. MSV resistance is associated with up to five separate alleles conferring a mixture of both dominant and recessive traits, none of which is sufficient by itself. Despite great successes achieved in the development of MSV-resistant maize genotypes that tolerate infection without significant yield loss, there has been only limited success in the field. For example, severe infections of the so-called MSV-tolerant genotypes can occur when they are grown under environmental conditions different from those in which the plants were selected, meaning each distinct geographical growing area requires specific varieties to be bred for maximum resistance. Another problem facing breeders is that natural genetic resistance is not usually associated with desirable agronomic traits such as good yield and it can therefore be difficult to transfer resistance traits without also transferring undesirable characteristics. Moreover, the number of alleles involved means that successful breeding takes years for each release. Even in the absence of any predictive modeling of sporadic MSD outbreaks, most farmers would still prefer to gamble on the use of higher-yielding MSV-sensitive genotypes. Efforts are currently underway to introduce MSV resistance traits into commercial maize genotypes by genetic engineering. This technique has the advantage of enabling the direct transfer of single-gene resistance, without linkage to undesirable characteristics, to many different breeding lines suited to different environmental conditions. However, up till now this strategy has been limited by negative public perception of genetically modified organisms, and the expensive and time-consuming risk assessment necessary to ensure a safe feed and food product.

MSD Epidemiology

There are loose correlations between MSD incidence and both environmental conditions and agricultural practices. Environmental influences on MSD epidemiology are mostly driven by a strong correlation between rainfall and leafhopper population densities. For example, drought conditions followed by irregular rains at the beginning of growing seasons tend to be associated with severe MSD outbreaks. Also, maize planted later in the growing season tends to get more severely infected than that planted at the beginning of the season, probably due to steady increases in leafhopper numbers and inoculum sources over the course of the season. As is the case with most insect-borne virus diseases, however, the incidence of MSD is erratic. Whereas MSD can devastate maize production in some years, in others it has only a negligible effect. The reason for this is that apart from MSD epidemiology being strongly dependent on environmental variables it is also the product of extremely complex interactions between the various MSV leafhopper vector and host species, and an as yet unknown number of virus strains.

Leafhopper Vectors

Serious MSD outbreaks are absolutely governed by leafhopper acquisition and movement of severe MSV isolates from infected plants (wild grasses or crop plants) to sensitive, uninfected crop plants. The distance that MSV spreads from a source of inoculum is determined by the movement behavior of leafhoppers. Distinct long- and short-distance flight morphs have been detected among certain *Cicadulina* populations. It is believed that the long-flight morphs are a migratory form and as such they may play an important part in the rapid long-distance spread of virulent MSV variants. Migratory movement is more common in certain *Cicadulina* species than in others and it is probably influenced by environmental conditions.

The dynamics of primary infection following leafhopper invasion of a susceptible maize crop are influenced by leafhopper population densities, the proportions of viruliferous individuals in populations, and the virus titer within these individuals. Disease spread within individual maize fields is apparently linear when only a few viruliferous leafhoppers are involved in transmission, but becomes exponential once the number of insects exceeds one individual per three plants.

Plant Hosts

Although attempts to understand the dynamics of MSD epidemics have focused primarily on vector population dynamics and behavior, an important component of MSD epidemiology is the population density, turnover, and demographics of the over 80 grass species that are both MSV and vector hosts. Because *Cicadulina* species favor certain annual grass hosts for mating and oviposition, the species composition of grass populations that vary seasonally in any particular area will directly influence leafhopper population densities and feeding behaviors in that area.

The species composition and age distribution of grasses (including cultivated crops) in an area may also affect the amount of MSV inoculum available for transmission in that area. While MSV infects at least 80 of the 138 grass species that leafhoppers feed on, both the susceptibility of these grasses to MSV infection and the severity of symptoms that occur following their infection may be strongly influenced by a number of factors. While sensitivity to infection can vary substantially from species to species, it can also vary within a species with genotype and plant age at the time of inoculation: for example, plants from many species, including maize, generally become more resistant to MSV infection with age, thereby reducing the inoculum available for transmission to other plants.

The Virus

While efforts are underway to promote the widespread cultivation of MSV-resistant maize in Africa, surprisingly little is known about the MSV populations that will confront these new genotypes. Although to date nine major MSV strain groupings have been discovered, it is unknown whether any other than the maize-adapted MSV-A strain play an important role in the epidemiology of MSD. MSV-B, -C, -D, and -E isolates only produce very mild symptoms in MSV-sensitive maize genotypes and are therefore unlikely to pose any significant direct threat to maize production. Mixed MSV-A and -B infections have, however, been detected in nature and there is also strong evidence of recombination occurring between these strains. It is therefore possible that MSV-B, and possibly other MSV strains, may indirectly influence MSD epidemiology through recombination with MSV-A-type viruses. Recombination has been linked with the emergence of a number of geminivirus diseases and it is quite conceivable that it may have already contributed to the emergence of MSV and may also eventually contribute to the evolution of MSV genotypes with elevated virulence in resistant maize varieties.

It seems highly probable that MSV isolates, strains, and even distinct MSV-related mastreviruses travel with the leafhoppers as a 'swarm' of virus types through a variety of grasses, both perennial and annual, each of which has a virus genotype or group of genotypes most suited to replication in it. Thus, the dominant virus in any particular host at any one time will probably be different, but the swarm diversity is preserved – in part because the dominant type in any case may facilitate the replication of other, less fit virus types in any one host. Consequently, it is possible to see very different dominant viruses in maize and wheat grown in consecutive summer and winter growing seasons in the same field, as a result of independent selection by the host genotype of the most fit virus genotype.

Future Threat

MSV is rightly regarded as a significant potential threat to maize production outside of Africa: while the vectors do not occur outside of the current range, there is no obvious reason that they would not survive in other, climatically similar areas such as the southern USA and South America and Eurasia, and it is a distinct possibility that they could inadvertently spread or be deliberately taken there. If one or more vector species did become established, and were viruliferous, spread of MSV and its relatives into native grasses and cultivated maize and other cereals would be inevitable. It is worth noting here that as none of the maize varieties grown outside of Africa has any but the weakest resistance to MSV, the probability of severe economic consequences would be very high.

See also: Plant Resistance to Viruses: Geminiviruses.

Further Reading

Bosque-Pérez NA (2000) Eight decades of maize streak virus research. *Virus Research* 71: 107–121.

Boulton MI (2002) Functions and interactions of mastrevirus gene products. *Physiological and Molecular Plant Pathology* 60: 243–255.

Damsteegt V (1983) Maize streak virus. Part I: Host range and vulnerability of maize germ plasm. *Plant Disease* 67: 734–737.

Efron Y, Kim SK, Fajemisin JM, et al. (1989) Breeding for resistance to maize streak virus – A multidisciplinary team-approach. *Plant Breeding* 103: 1–36.

Fuller C (1901) *First Report of the Government Entomologist, Natal, 1899–1900.* http://www.mcb.uct.ac.za//msv/fuller.htm (accessed February 2008).

Harrison BD, Barker I, Bock K, et al. (1977) Plant viruses with circular single-stranded DNA. *Nature* 270: 760.

Martin DP, Willment J, Billharz R, *et al.* (2001) Sequence diversity and virulence in *Zea mays* of maize streak virus isolates. *Virology* 288: 247.

McLean AP (1947) Some forms of streak virus occurring in maize, sugarcane and wild grasses. *Science Bulletin of Department of Agriculture for Union of South Africa* 265: 1–39.

Palmer KE and Rybicki EP (1998) The molecular biology of mastreviruses. *Advances in Virus Research* 50: 183–234.

Storey HH (1925) The transmission of streak disease of maize by the leafhopper. *Balclutha mbila naudé. Annals of Applied Biology* 12: 422–443.

Zhang W, Olson NH, Baker TS, *et al.* (2001) Structure of the maize streak virus germinate particle. *Virology* 279: 471–477.

Relevant Website

http://www.mcb.uct.ac.za – The Maize Streak Virus Home Page at Online Resources of the University of Cape Town Department of Molecular and Cell Biology.

Marburg Virus

D Falzarano, University of Manitoba, Winnipeg, MB, Canada
H Feldmann, Public Health Agency of Canada, Winnipeg, MB, Canada

© 2008 Elsevier Ltd. All rights reserved.

History

In 1967 the first noted cases of a viral hemorrhagic fever (VHF) caused by a new family of viruses, the *Filoviridae*, occurred when laboratory workers in Germany contracted severe VHF after handling tissues from African green monkeys (*Cercopithecus aethiops*) imported from Uganda. Shortly after, two cases were identified in the former Yugoslavia, where a veterinarian was infected during the necropsy of a dead monkey. In total there were 32 cases, including six secondary infections and a single retrospective case. Overall, there were seven fatalities in the primary infections resulting in a case–fatality rate of 23%. The new virus was named Marburg virus (MARV) after the German city which had reported the first cases (see **Figure 1** for a map of all known occurrences of MARV).

Following these initial cases of Marburg hemorrhagic fever (MHF) there were only a small number of isolated cases noted until 1998. In 1975, three cases of MHF were reported in Johannesburg, South Africa. The index case, who did not survive, had traveled to Zimbabwe immediately before becoming ill. Shortly afterward, his travel companion and a nurse who cared for them also became ill but both later recovered. Cases of MHF also occurred in Kenya in 1980 and 1987. In 1980 the index case became ill in western Kenya and died in Nairobi where a physician also got infected but survived. In 1987, a fatal case occurred in the same region of western Kenya. The index cases in both 1980 and 1987 had traveled to the Mt. Elgon region, which is located close to Lake Victoria and was the source of the monkeys that initiated the original 1967 outbreak (trapped near Lake Kyogo, Uganda).

The first large community outbreak of MHF occurred in 1998 in central Africa. The community of Durba/Watsa, located in the northeastern region of the Democratic Republic of the Congo (DRC), had 149 cases with an 83% fatality rate. The response to this outbreak was limited due in part to the remote location and an ongoing conflict in the region. Sporadic cases continued after the end of the outbreak with most cases being linked either directly or indirectly to illegal mining in an underground gold mine. Typically, the index cases were gold miners who initiated multiple, short chains of human-to-human transmission within their families. The Durba outbreak was somewhat unusual compared to previous MARV and Ebola virus (EBOV) outbreaks in that it continued for almost 2 years, during which multiple distinct genetic lineages were found to be circulating, indicating several independent introductions of virus into the human population from the unknown natural reservoir.

To date, the largest outbreak of MHF occurred in the Uíge Province of Angola in 2004–05. At the conclusion of the outbreak, the Ministry of Health in Angola reported a total of 252 cases, including 227 deaths. This case–fatality rate of 90% is the highest observed during an MHF outbreak thus far and is more typical of past severe Ebola hemorrhagic fever (EHF) outbreaks. A large proportion of the cases were found among young children, which is unusual for filovirus outbreaks. The high case–fatality rate and fast progression of the disease were suggestive of a more virulent strain. In contrast to the Durba/Watsa outbreak, the viruses isolated from patients were highly conserved, indicating a single introduction with little evolution of the virus during the outbreak similar to what has been reported for previous EHF outbreaks. The most recent occurrence of MHF was in 2007 in Uganda, where there were three confirmed cases.

In addition to the initial outbreak and those that have occurred in the natural setting there have also been at least three laboratory-acquired infections of MARV since the mid-1980s, with one fatality occurring in Russia.

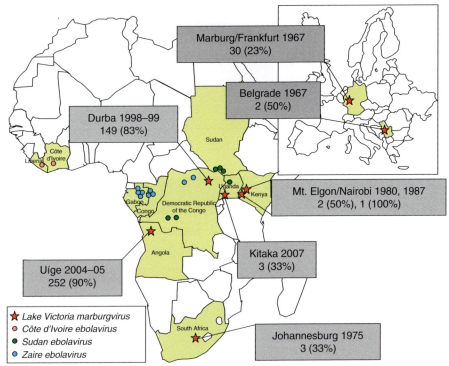

Figure 1 Location of known filovirus outbreaks. Countries that have experienced occurrences of filoviruses are indicated in yellow. For MARVs, the location of the outbreak is indicated by a red star. The main center (city/village) of the outbreak is indicated, along with the year of the outbreak, the number of cases, and the case–fatality rate (indicated in brackets). Incidents of the various EBOV species are also indicated.

Taxonomy and Classification

All MARVs are classified as a single species, *Lake Victoria marburgvirus*, which comprise the genus *Marburgvirus*, family *Filoviridae*, within the order *Mononegavirales*. They share unique morphologic, physicochemical, genetic, and biological features with members of the genus *Ebolavirus*. Due to their high fatality rates, frequency of person-to-person transmission, the potential for aerosol infectivity, and the absence of vaccines and treatments, these viruses are considered to be 'biosafety level 4' (BCL-4) agents and have been placed on the 'category A' and 'select agent' list.

Virus Structure and Composition

Electron microscopy of MARV reveals distinctive pleomorphic filamentous particles (**Figure 2(a)**) that can appear in U-shaped, 6-shaped, or circular (torus) configurations, or as elongated filamentous forms of varying length (up to 14 000 nm). Filamentous particles may also form branched structures. The length of peak infectivity for MARV particles is between 790 and 860 nm with a uniform diameter of 80 nm for all particles. Virus particles contain a helical ribonucleoprotein (RNP) complex, which contains the negative-sense viral RNA genome, the polymerase (L) protein, the nucleoprotein (NP), and viron protein (VP) 35 and 30. The RNP is surrounded by the matrix protein (VP40) and a closely apposed outer envelope derived from the host cell plasma membrane (**Figure 2(b)**). The surface of the particle has membrane-anchored protein spikes, made up by the virus glycoprotein (GP), which gives the virus particle a rough appearance. These spikes are approximately 7 nm in length and are spaced at approximately 10 nm intervals. Virus particles have a molecular weight of approximately $3-6 \times 10^8$ Da and a density of $1.14 \mathrm{~g~ml}^{-1}$ as determined by centrifugation in a potassium tartrate gradient; uniform, bacilliform particles have a sedimentation coefficient of 1300–1400 S. Virus infectivity is rather stable at room temperature. Inactivation can be performed by ultraviolet (UV) light or gamma irradiation, 1% formalin, β-propiolactone or brief exposure to phenolic disinfectants and lipid solvents, like deoxycholate and ether, as well as ionic detergents such as sodium dodecyl sulfate.

Genome Organization and Expression

The MARV genome consists of a single, linear molecule of negative-stranded RNA that is just over 19 000 bp in length. The genome is slightly larger than that of EBOV but is organized in a similar manner (**Figure 2(b)**). The

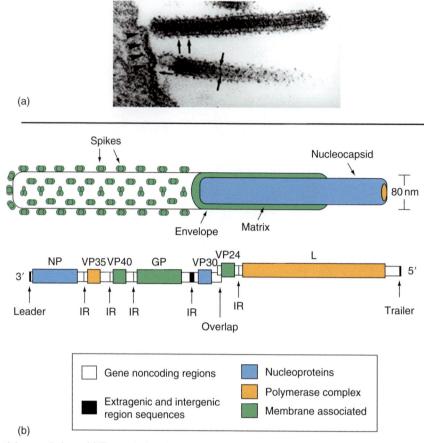

Figure 2 Virus particle morphology. (a) Transmission electron micrograph (negative stain) of MARV particles budding from an infected cell. The arrows indicate the glycoprotein spikes. (b) Schematic representation of the MARV particle structure (upper portion) and genome (lower portion) indicating open reading frames, noncoding regions, a single gene overlap (VP30/VP24), and intergenic regions (IRs).

MARV genome is noninfectious, rich in adenosine and uridine residues, not polyadenylated and complementary to viral-specific messenger RNA. Complete genome sequences are available from isolates covering six episodes of MHF (1967, 1975, 1980, 1987, 1999, and 2005). The genome amounts to 1.1% of the total virion weight with a sedimentation coefficient of 46 S (0.15 M NaCl, pH 7.4).

The genes are arranged in a linear fashion in the following order: 3′ leader; NP; viral structural protein (VP)-35; VP40; GP; VP30; VP24; polymerase (L); 5′ trailer (**Figure 2(b)**). All genes are flanked at their 3′ and 5′ ends by highly conserved transcriptional start and stop signal sequences that almost always include the pentamer 3′-UAAUU-5′. Most genes are separated by intergenic sequences that are variable in length and nucleotide composition. An unusual feature of the MARV genome is the presence of a gene overlap, between VP30 and VP24, a feature that is shared with EBOV, which have two or three gene overlaps (VP35–VP40, GP–VP30, and or VP24–L). Extragenic sequences that are complementary at their very extremities are present at the 3′ and 5′ end of the genome. This complementarity favors formation of a panhandle structure between the genomic termini, but it is unclear if such a structure can actually form due to co-transcriptional encapsidation. Both genomic ends are self-complementary and modeling suggests that almost identical hairpin-like structures are present at the 3′ ends of the genome and antigenome.

All of the MARV proteins are incorporated into the virus and are either part of the RNP complex or associated with the envelope. For a summary of protein functions, see **Table 1**. RNP-associated proteins are involved in transcription and replication while envelope-associated proteins are primarily involved in assembly/budding or entry. NP and VP30 are phosphoproteins and are considered the major and minor nucleoproteins, respectively. They interact strongly with the genomic RNA molecule, forming the viral RNP complex, in combination with VP35 and L which form the polymerase complex. EBOV VP30 is thought to be a transcriptional activator that is regulated by phosphorylation but MARV VP30 does not appear to function in this manner.

The polymerase complex transcribes and replicates the MARV genome. L is the RNA-dependent RNA

Table 1 MARV proteins: Function(s) and localization

Gene order [a]	Marburg virus proteins	Protein function (localization)
1	Nucleoprotein (NP)	Major nucleoprotein, RNA genome encapsidation (component of RNP[b])
2	Virion protein 35 (VP35)	Polymerase complex cofactor, type I interferon antagonist (component of RNP)
3	Virion protein 40 (VP40)	Matrix protein, virion assembly and budding (membrane associated)
4	Glycoprotein (GP)	Receptor binding and membrane fusion (membrane associated)
5	Virion protein 30 (VP30)	Minor nucleoprotein (component of RNP)
6	Virion protein 24 (VP24)	Minor matrix protein, virion assembly (membrane associated)
7	Polymerase (L)	RNA-dependent RNA polymerase, enzymatic portion of polymerase complex (component of RNP)

[a]Gene order refers to 3′–5′ gene arrangement as shown in **Figure 2(b)**.
[b]Ribonucleoprotein complex.

polymerase and contains motifs that are linked to RNA (template), phosphodiester (catalytic site), and ribonucleotide triphosphate binding. MARV L is larger than EBOV L and while there are areas of conservation between the two, nearly 25% of the C-terminal quarter is not conserved, in addition to stretches of sequences that are totally unique to MARV L. VP35 is thought to act as an essential cofactor of L that affects the mode of RNA synthesis (transcription or replication) similar to that of the P proteins of other negative-stranded viruses. In addition, VP35 also acts as a linker between L and NP as well as having an antagonistic effect on the type I interferon pathway.

The surface of MARV particles is covered with spike structures (**Figures 2(a)** and **2(b)**) that are composed of the structural glycoprotein, GP, which is anchored in the membrane as a timer in a type I orientation. These spike proteins play a role in entry and are thought to influence pathogenesis. MARV GP is encoded in a single open reading frame in contrast to EBOV GP, which is encoded in two open reading frames and is only expressed after RNA editing. Sequence analysis of MARV GP genes indicates that a nucleotide sequence that corresponds to the editing region of EBOV GP genes is absent. MARV GP is processed similar to EBOV GP but in contrast to EBOV GP the glycans on MARV GP lack terminal sialic acids if propagated in specific cell lines. This may be caused by differences in processing as the protein is directed through the *trans*-Golgi apparatus.

VP40 functions as the matrix protein in combination with VP24, which acts as a minor matrix protein. VP40 is the most abundant protein in the virion; however, only small amounts of VP24 are incorporated into virus particles. Both VP40 and VP24 are hydrophobic, have an affinity for membranes, and are associated with the virion envelope. VP40 is essential for virus budding as it initiates and drives the envelopment of the RNP by the plasma membrane. The role of VP24 in the replication of MARV is still unclear and direct interactions with other virus proteins have not been described.

Replication

MARVs are thought to replicate in a similar manner to EBOV; however, most of the research on replication has been performed on EBOV. Only the NP, VP35, and L proteins are required for transcription and replication of MARV, in contrast to EBOV, which also requires VP30. This indicates that there may be differences in the factors required for transcription/replication. It also appears that there might be differences in receptor binding for MARV, which appear to have a higher affinity for asialoglycoprotein, whereas this interaction does not seem to be important for EBOV infection. For a detailed description of MARV replication, the reader is referred to the replication section of Zaire ebolovirus (*Filoviridae*) which is covered elsewhere in this encyclopedia.

Evolutionary Relationship between Viruses

MARV and EBOV are clearly related in sequence, morphology, and the general disease that they cause; however, it is also clear that they represent different genera of filoviruses. The nucleotide and amino acid differences between MARV and EBOV are both *c.* 55%. Phylogenetic analysis indicates that within MARV there are two distinct lineages of viruses (**Figure 3**) that show a diversity of up to 21%, with the 1987 isolate from Kenya (Ravn) being the most divergent from the rest. This distinct lineage was also seen again in the Durba/Watsa outbreak, along with other isolates that cover most, if not all, of the MARV repertoire.

The outbreak in 2005 in the Uíge Province of northern Angola was a surprise as the origins of all earlier outbreaks were linked directly to East Africa. What is somewhat surprising is that the Angola strain of MARV was less different (~7%) from the main group of East African MARV than one would expect given the large geographic separation (**Figure 3**). This suggests that the

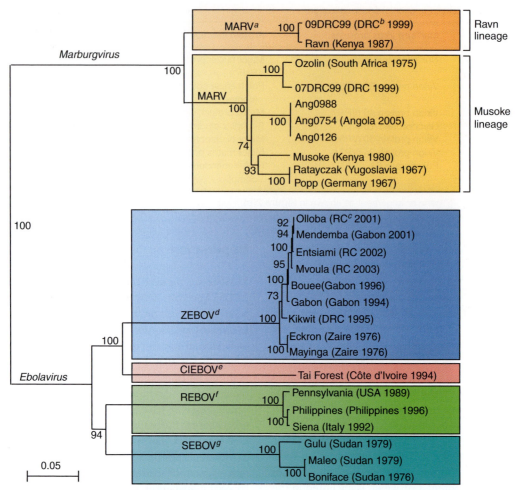

Figure 3 Phylogeny of the family *Filoviridae*. Neighbor-joining analysis of the nucleotide sequence of the GP gene of Marburg and Ebola virus isolates, indicating the two lineages of MARV and the different species of EBOV. Sequences were obtained from GenBank. Confidence values at branch point were obtained from 1000 bootstraps. The bar length equals 5% nucleotide difference. [a]MARV – *Lake Victoria marburgvirus*, [b]Democratic Republic of the Congo, [c]Republic of the Cango, [d]ZEBOV – *Zaire ebolavirus*, [e]CIEBOV – *Côte d'Ivore ebolavirus*, [f]REBOV – *Reston ebolavirus*, [g]SEBOV – *Sudan ebolavirus*.

virus reservoir in these regions may not be substantially different. However, an index case was never identified, and thus the importation of the virus from East Africa cannot be ruled out. Given the length of the outbreak (from fall of 2004 to July 2005) it seemed possible that multiple introductions of MARV could be possible. However, remarkably few nucleotide differences were found among the Angolan clinical specimens (0–0.07%), which is consistent with a single introduction of virus into the human population, followed by person-to-person transmission with little accumulation of mutations. Typically, RNA viruses evolve rapidly due to positive selection in combination with the large number of errors that are made and cannot be corrected during replication (missing proof-reading activity of the polymerase). However, disease progression is so rapid that most individuals seem to die before an effective immune response can be mounted. Therefore, the positive selection of viruses may not occur.

Transmission and Host Range

The reservoir for MARV remains unknown, but its emergence in Angola extends the scope of the reservoir search beyond East Africa. Humans and nonhuman primates serve as natural hosts and it is unclear if other animals are infected. In Durba/Watsa, epidemiological data linked over 70% of the cases with mines or caves, suggesting that the natural reservoir could well be associated with such environments and that bats have been a favorable species for a reservoir. With the Angola outbreak, difficulties in surveillance and contact tracing, combined with the delay in the identification of the outbreak, led to poor epidemiological linkage of MARV cases and ultimately to a lack of success in identifying a point source or mounting any ecological study. Filovirus outbreaks in general are relatively rare events. If the natural reservoir of MARV is similar to that of EBOV, the emergence of MARV in western Africa should not be surprising, as the sites of multiple large EHF outbreaks are less than 500 miles away (**Figure 1**), including areas which have experienced almost yearly activity over the last decade. EBOV has recently been linked to fruit bats in Gabon and the Republic of the Congo. It is thought that MHF outbreaks start with the rare introduction of the virus into the human population followed by waves of human-to-human transmission (usually through close contact with infected individuals or their body fluids), with little if any evolution of the virus during the course of the outbreak.

Clinical Features

MARVs cause severe hemorrhagic fever in both humans and nonhuman primates. The incubation period lasts from 2 to 21 days (average 4–10), after which there is a sudden onset of nonspecific flu-like symptoms that can include fever, chills, malaise, headache, and myalgia (**Figure 4**). This is followed 2–10 days later by the development of more severe symptoms that include systemic (i.e., prostration, anorexia), gastrointestinal (nausea, vomiting, abdominal pain, diarrhea), respiratory (chest pain, shortness of breath, cough), vascular (conjunctival injection, postural hypotension, edema), and neurologic (headache, confusion, coma) manifestations. The presence of a macropapular rash associated with varying degrees of erythema is also frequently observed and is a useful differential diagnostic.

In cases where coagulation abnormalities develop, hemorrhagic manifestations can include petechiae, ecchymoses, bleeding from venipuncture sites, mucosal bleeding (typically involving the gastrointestinal tract), and visceral hemorrhagic effusions. Patients that develop coagulation abnormalities usually have a bad prognosis. The late stages of the disease are characterized by the development of shock with convulsions, severe metabolic disturbances, and coagulopathy. The onset of shock, with or without obvious bleeding, leads to multiple organ failure with death typically occurring between days 7 and 16. Nonfatal cases have fever for approximately 5–9 days with improvement typically occurring around days 7–11 – the time that the humoral antibody response is noted (**Figure 4**). In patients who survive, convalescence is prolonged and is sometimes associated with myelitis, recurrent hepatitis, psychosis, or uveitis in addition to psycho-social difficulties integrating back into their community.

The mortality for MARV seems to average around 70–85% with the exception of the outbreak in Europe (only 23%). The reason for the large difference in mortality in the European versus the African cases is unknown and may be the result of important host-genetic differences, genetic difference between virus strains, or the standard of care available to infected individuals.

Pathogenesis and Pathology

Clinical investigations of outbreaks of human MARV infections have provided most of the descriptive information on the pathology and pathogenesis of these viruses. However, studies in laboratory animals are much more

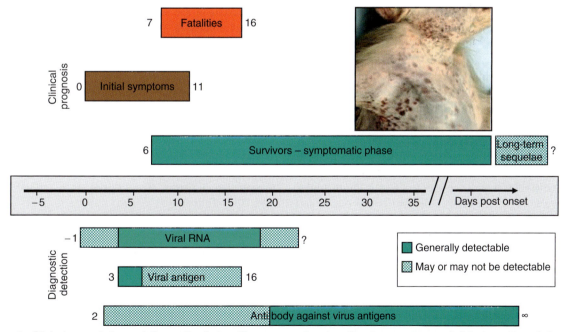

Figure 4 Clinical presentation and diagnostic window. The upper portion describes the appearance of clinical symptoms in humans. Also shown are the characteristic petechiae, seen here on the neck and upper body, of a nonhuman primate with MHF (inset). The lower portion illustrates the timeframe for the appearance of diagnostic targets (viral RNA, viral antigen, and host antibody response).

comprehensive and consistent. Guinea pigs have been employed to study MHF. While guinea pigs have served as effective early screens for evaluating antiviral drugs and candidate vaccines, the disease pathogenesis seen in these animals is not nearly as representative of the human clinical picture as nonhuman primates are.

The different strains of MARV appear to have different levels of virulence. Initially, it appeared that MARV strains were more comparable to *Sudan ebolavirus*; however, virulence of the recently isolated Angola strain appears to be similar to *Zaire ebolavirus*. Most strains of MARV produce uniformly lethal infections in cynomolgus and rhesus macaques. Infections of macaques with the Angola strain appear to progress more rapidly than infections with other strains. For example, challenge of rhesus macaques by intramuscular injection with 1000 pfu of the Musoke or Angola strains produces a uniformly lethal infection, but death occurs within 10–12 days versus 6–8 days, respectively.

While the pathology of MARV is less extensively studied than that of EBOV, it appears to be similar. The pathological changes seen in patients fatally infected with any of the filoviruses include extensive necrosis of parenchymal cells of many organs, including the liver, spleen, kidney, and gonads with little infiltration in infected tissues. However, no single organ is sufficiently damaged to explain the fatal outcome. Infection of cells leads to intracytoplasmic vesiculation and mitochondrial swelling which is followed by a breakdown of organelles and terminal cytoplasmic rarification or condensation. MARV appears to cause more severe liver damage than EBOV. Hepatocellular necrosis is widespread with extremely high infectivity titers present in infected liver samples. Elevations in liver enzymes are prominent findings in most filovirus infections and the hepatocellular degeneration and necrosis observed during MARV infections is extensive. Liver function impairment could contribute to the overall pathogenesis because hemorrhagic tendencies in some cases may be related to decreased synthesis of coagulation factors and other plasma proteins as a result of hepatocellular necrosis. In the late stages of the disease, hemorrhage occurs in the gastrointestinal tract, pleural, pericardial, and peritoneal spaces, as well as the renal tubules with deposition of fibrin. Abnormalities in coagulation parameters include fibrin split products and prolonged prothrombin and partial thromboplastin times. Clinical and biochemical findings support anatomical observations of extensive liver involvement, renal damage, changes in vascular permeability, and activation of the clotting cascade.

Fluid distribution and platelet abnormalities indicate dysfunction of endothelial cells and platelets. In addition to direct vascular involvement in infected hosts, active host mediator molecules probably play a significant role in these disorders. Infected monocytes/macrophages are probably responsible for producing different proteases, peroxide, and other mediators, such as tumor necrosis factor alpha (TNF-α) that may have a negative effect on the infected host. The increased production of TNF-α can result in secondary activation of mediators with important protective as well as deleterious effects. For example, supernatants of MARV-infected monocytes/macrophages cultures are capable of increasing paraendothelial permeability, thus exacerbating the development of the shock syndrome seen in severe and lethal cases. The endothelium is also directly targeted by the virus and endothelial cells support cytolytic MARV replication in culture. The bleeding disorders observed during infection could be due to direct endothelial damage caused directly by virus replication or indirectly by cytokine-mediated processes. The combination of viral replication in endothelial cells and virus-induced mediator release from infected leukocytes may also promote a distinct pro-inflammatory phenotype of the endothelium that triggers, most likely via tissue factor, the coagulation cascade.

Diagnosis

Diagnostics for MARV and EBOV infections use the same principles. As these infections typically occur in isolated regions of Africa that do not have the diagnostic capabilities to identify filovirus infections, the initial diagnosis of MHF and EHF will most likely be based on clinical symptoms. Diagnosis of single cases is very difficult due to similarity of symptoms to other diseases also present in the endemic areas. Due to the rarity of filovirus infections, diagnostic testing is usually performed at national and/or international reference laboratories that are capable of performing the required tests under suitable containment conditions. During outbreaks, healthcare workers who have direct contact with patients are at high risk for infection and adequate barrier nursing precautions need to be implemented during the collection of samples and treatment of patients. Laboratory diagnosis is based on either detection of virus-specific antibodies, virus particles, or particle components. Inactivation of samples is necessary when testing is not done under BSL-4 conditions.

Currently, reverse transcriptase-polymerase chain reaction (RT-PCR) and antigen detection enzyme-linked immunosorbent assay (ELISA) are the primary test systems to diagnose acute infections; RT-PCR, however, is more sensitive than antigen detection ELISA. Viral antigen and/or nucleic acid can be detected in blood from 3 to 18 days post onset of symptoms (**Figure 4**). Most laboratories currently favor RT-PCR because of its sensitivity, specificity, and rapidity. RT-PCR is also readily used in a mobile lab setting and has proved to be accurate and effective in both MHF and EHF outbreaks. Due to the seriousness of a positive test for filoviruses, the diagnosis of index cases or of single imported cases should not be solely based on a single technology. Confirmation by an independent assay and/or laboratory should always be attempted.

Serological assays are second choice for acute diagnosis as patients often succumb to the disease before an antibody response is generated. Alternatively, immunohistochemistry, direct immunofluorescence on tissues, or electron microscopy can be used for diagnostics, but these methods lack sensitivity, are time consuming, or require expensive equipment. Virus isolation from clinical specimens in tissue culture and/or animals is easily achieved but takes time and requires BSL-4. Filoviruses grow well in a number of cell lines, including Vero and Vero E6, which are the most frequently used. The most commonly used laboratory animals for virus isolation are inbred guinea pig strains; however, it should be kept in mind that often several passages are required to obtain a lethal infection.

Treatment

The current treatment of MARV infections mainly involves supportive therapy, which is directed toward the maintenance of effective blood volume and electrolyte balance. Shock, cerebral edema, renal failure, coagulation disorders, and secondary bacterial infections may be life threatening and have to be managed. Most experimental treatment strategies have been studied for EBOV, but it is expected that several of these approaches would have similar effects on MARV.

Therapeutic antibodies are still considered a valuable short-term solution. This strategy might perhaps be more realistic for the treatment of MHF than EHF based on the observation that humoral responses seemed to be more effective against MARV. The use of recombinant nematode anticoagulant protein c2 (rNAPc2), which seems to be a promising approach for EBOV, did not result in a similar positive effect for MARV infections (Angola strain). Studies on novel antiviral strategies such as viral gene silencing through specific siRNA or functional domain interference with small peptides have been very limited for MARV. In contrast, more focus has been given to strategies targeting host responses. Treatment with TNF-α neutralizing antibodies has been partially successful in guinea pig models of MHF, but has yet to be evaluated in nonhuman primates. MARV-infected guinea pigs are partially protected by Desferal, an interleukin 1 (IL-1) and TNF-α antagonist. In a separate study, treatment of animals with IL-1 receptor antagonist (IL-1RA) or anti-TNF-α decreased the concentration of circulating TNF-α and allowed 50% survival.

As with EBOV, ribavirin is not indicated for MHF treatment. In general, it seems plausible that combination therapy for MHF and EHF will be superior over any single treatment form.

Prevention

Protective MARV vaccines would be extremely valuable for at-risk medical personnel, first responders, military personnel, researchers, and high-risk contact groups during outbreaks (such as family members). Past vaccine approaches were based on either inactivated virus preparations, which were of limited protective efficacy and considered unsafe, or subunit vaccines, which showed efficacy in the rodent but not nonhuman primate model. Protective efficacy in nonhuman primates could first be demonstrated with a system based on Venezuelan equine encephalitis virus replicons expressing the MARV GP and/or NP. Currently, the most promising vaccine approach seems to be a live-attenuated vector based on vesicular stomatis virus expressing the MARV GP (strain Musoke). The protective efficacy and safety of this vaccine vector has been demonstrated in two animal models, the guinea pig and nonhuman primate. Protective efficacy could also be achieved against challenge with heterologous MARV strains as well as homologous aerosol challenge. In addition, the vector showed astonishing efficacy in postexposure treatment of rhesus macaques; single, high-dose treatment 30 min after high-dose challenge protected all animals from lethal disease. Despite the success it remains questionable if a replication-competent vector will be granted approval for human use.

Acknowledgments

The Public Health Agency of Canada (PHAC), Canadian Institutes of Health Research (CIHR), and CBRNE (chemical, biological, radiological, and nuclear) Research and Technology Initiative (CRTI), Canada, supported work on filoviruses at the National Microbiology Laboratory of the Public Health Agency of Canada.

See also: Ebolavirus.

Further Reading

Bausch DG and Geisbert TW (2007) Development of vaccines for Marburg hemorrhagic fever. *Expert Reviews of Vaccines* 6: 57–74.

Bausch DG, Nichol ST, Muyembe-Tamfum JJ, et al. (2006) Marburg hemorrhagic fever associated with multiple genetic lineages of virus. *New England Journal of Medicine* 355: 909–919.

Becker S and Muhlberger E (1999) Co- and posttranslational modifications and functions of Marburg virus proteins. *Current Topics in Microbiology and Immunology* 235: 23–34.

Bray M and Paragas J (2002) Experimental therapy of filovirus infections. *Antiviral Research* 54: 1–17.

Feldmann H, Geisbert TW, Jahrling PB, *et al*. (2005) *Filoviridae*. In: Fauquet CM, Mayo MA, Maniloff J, Desselberger U, and Ball LA (eds.) *Virus Taxonomy: Eighth Report of the International Committee on Taxonomy of Viruses*, pp. 645–653. San Diego, CA: Elsevier Academic Press.

Hensley LE, Jones SM, Feldmann H, Jahrling PB, and Geisbert TW (2005) Ebola and Marburg viruses: Pathogenesis and development of countermeasures. *Current Molecular Medicine* 5: 761–772.

Martini GA and Siegert R (eds.) (1971) *Marburg Virus Disease*. New York: Springer.

Mohamadzadeh M, Chen L, Olinger GG, Pratt WD, and Schmaljohn AL (2006) Filoviruses and the balance of innate, adaptive, and inflammatory responses. *Viral Immunology* 19: 602–612.

Paragas J and Geisbert TW (2006) Development of treatment strategies to combat Ebola and Marburg viruses. *Expert Reviews of Anti-Infective Therapy* 4: 67–76.

Sanchez A, Geisbert TW, and Feldmann H (2007) Marburg and Ebola viruses. In: Knipe DM, Howley PM, Griffin DE, *et al*. (eds.) *Fields Virology*, 5th edn., pp. 1409–1448. Philadelphia, PA: Lippincott Williams and Wilkins.

Siegert R, Shu HL, Slenczka W, Peters D, and Muller G (1967) On the etiology of an unknown human infection originating from monkeys. *Deutsche Medizinische Wochenschrift* 92: 2341–2343.

Slenczka WG (1999) The Marburg virus outbreak of 1967 and subsequent episodes. *Current Topics in Microbiology and Immunology* 235: 49–75.

Zaki SR and Goldsmith CS (1999) Pathologic features of filovirus infections in humans. *Current Topics in Microbiology and Immunology* 235: 97–116.

Marnaviruses

A S Lang, Memorial University of Newfoundland, St. John's, NL, Canada
C A Suttle, University of British Columbia, Vancouver, BC, Canada

© 2008 Elsevier Ltd. All rights reserved.

The first, and currently the only, virus species classified within the family *Marnaviridae* is *Heterosigma akashiwo* RNA virus (HaRNAV). This virus was isolated from the Strait of Georgia in coastal British Columbia and infects and causes the lysis of the marine, toxic bloom-forming, unicellular, photosynthetic alga *Heterosigma akashiwo*. The host organism, *H. akashiwo*, forms blooms in the Strait of Georgia waters every year, and the blooms of this organism are toxic to fish.

When the virus is added to a growing culture of susceptible *H. akashiwo*, complete lysis of the culture generally occurs within 1 week or less. Only some strains of *H. akashiwo* are susceptible to infection by HaRNAV, and the pattern of susceptibility does not appear to relate to geographic origin of the strain in question. Therefore, HaRNAV causes lysis of some, but not all, *H. akashiwo* strains isolated from coastal British Columbia waters as well as some, but not all, strains from other locations in the Pacific such as coastal Japan.

HaRNAV particles are small (approximately 25 nm in diameter) and spherical, and appear to have an icosahedral symmetry (**Figure 1**). Signs of viral-induced intracellular cytopathology are visible within 48 h of infection of *H. akashiwo* by HaRNAV. One notable effect is the appearance of membranous vesicles within the cytoplasm; such structures are the site of RNA replication for other positive-stranded ssRNA viruses and are well characterized for picornaviruses such as poliovirus. This is followed by the appearance of assembled viral particles throughout the cytoplasm, and these can be at sufficient density for the particles to form crystalline arrays.

The genome of HaRNAV (strain SOG263) has been completely sequenced. The single-stranded RNA (ssRNA) genome is 8587 nucleotides long with positive polarity and a poly (A) tail. The 5' and 3' untranslated regions (UTRs) total 9.8% of the genome sequence and the 5' UTR contains sequence features believed to be important for protein translation; for example, there is a

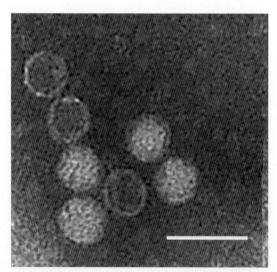

Figure 1 Transmission electron micrograph of HaRNAV particles negatively stained with phosphotungstic acid. Scale = 50 nm. Reprinted from Tai V, Lawrence JE, Lang AS, *et al*. (2003) Characterization of HaRNAV, a single-stranded RNA virus causing lysis of *Heterosigma akashiwo* (Raphidophyceae). *Journal of Phycology* 39: 343–352, with permission from Blackwell Publishing.

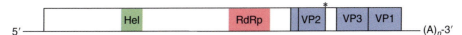

Figure 2 Map of the HaRNAV genome. The approximate predicted location of the single open reading frame is shown as a box on top of the positive-polarity ssRNA genome. Regions within the polyprotein sequence that have homology to known ssRNA viral protein domains are colored and annotated: Hel, 'RNA helicase'; RdRp, RNA-dependent RNA polymerase; VP1, VP2, and VP3, capsid protein subunits. For the structural proteins, the locations of N-termini found by sequencing are shown as vertical lines. The N-terminus marked with an asterisk (*) was only found in immature virus particles. Modified from Lang AS, Culley AI, and Suttle CA (2004) Genome sequence and characterization of a virus (HaRNAV) related to picorna-like viruses that infects the marine toxic bloom-forming alga *Heterosigma akashiwo*. *Virology* 320: 206–217, with permission from Elsevier.

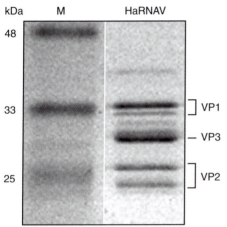

Figure 3 SDS-PAGE analysis of HaRNAV particles. The identities of the major bands are labeled, where known, corresponding to the designations in **Figure 2**. The lane marked M contains molecular weight markers with sizes indicated in kilodaltons on the left. Modified from Lang AS, Culley AI, and Suttle CA (2004) Genome sequence and characterization of a virus (HaRNAV) related to picorna-like viruses that infects the marine toxic bloom-forming alga *Heterosigma akashiwo*. *Virology* 320: 206–217, with permission from Elsevier.

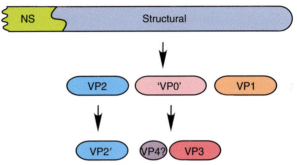

Figure 4 Organization of the structural protein subunits within the polyprotein. NS, nonstructural protein domains.

pyrimidine-rich tract of sequence immediately preceding the predicted start codon and it is also predicted to contain extensive secondary structure. The genome sequence predicts a single open reading frame (ORF) encoding a protein that contains conserved +ssRNA virus protein domains, as indicated by comparisons with sequence databases. Conserved 'RNA helicase', RNA-dependent RNA polymerase (RdRp), and structural protein domains are recognizable by these comparisons (**Figure 2**).

When subjected to SDS-PAGE, HaRNAV particles appear to contain at least six major protein bands varying in size between 24 and 39 kDa (**Figure 3**). Sequence analyses of these protein bands indicate that they represent multiple 'versions' of only three different polypeptides. Other than the two bands corresponding to the VP2 protein which have different N-termini (**Figures 2–4**), the differences between protein bands that migrate to different positions in the gel but that appear to contain the same polypeptide sequences are not known. All three of the known structural protein sequences (VP1, 2, and 3) have significant similarity with the structural proteins of viruses classified in the family *Dicistroviridae*, including cricket paralysis virus (CrPV) for which a crystal structure has been determined. Also, the HaRNAV organization of the structural protein subunits within the larger (structural portion of the) polyprotein is the same as found in the family *Dicistroviridae* (**Figure 4**). Note however that a VP4-like protein has not been identified associated with HaRNAV particles.

The N-terminal sequences of several structural proteins were determined and this showed that two of the proteins have a conserved sequence at the cleavage site (ST–SEI). Beyond this, there is surprisingly little recognizable conservation at the protein processing sites. Based on computer searches of the sequence databases, there is no conserved putative protease domain recognizable in the HaRNAV polyprotein sequence, but it is reasonable to expect that there is one, as found in well-studied members of the proposed viral order *Picornavirales*.

Phylogenetic analyses that have been performed with the recognizable conserved domains from the HaRNAV polyprotein sequence and the corresponding regions of other related virus sequences clearly indicate that HaRNAV does not fall within the previously characterized groups within the proposed order *Picornavirales*. A phylogenetic analysis of viral (putative) RdRp domains is shown in **Figure 4**. All defined groups are found to cluster in these analyses.

In the time since the characterization of the HaRNAV genome, other positive ssRNA viruses that infect marine protists have been isolated and these viruses (HcRNAV, RsRNAV, and SssRNAV) all differ from HaRNAV in that their genomes are dicistronic. Phylogenetic analyses indicate that RsRNAV and SssRNAV likely belong in the proposed order *Picornavirales* while HcRNAV does not belong in this group. Our analyses with the RdRp domains (the complete sequence spanning regions I–VIII) do not suggest that these viruses are closely related to the family *Marnaviridae* (**Figure 5**, **Table 1**), although HaRNAV does give the highest scoring match of any cultured virus when the putative RdRp sequences from RsRNAV were used in a BLAST search against the GenBank database. (The overall top-scoring matches to the RsRNAV sequence were from unidentified picorna-like RdRp sequences amplified from marine environments by degenerate RT-PCR.) The same search with the putative RdRp sequence from SssRNAV gave viruses in the family *Dicistroviridae* as the top-scoring matches.

Analyses performed with the VP3-like capsid protein sequences from these viruses tell a different story. BLAST searches with the VP3-like sequence from any of the three cultured marine viruses (HaRNAV, RsRNAV, and SssRNAV) give the other two viruses as the top-scoring matches. Pairwise alignments show that they are all >30% identical: 34% for HaRNAV versus RsRNAV; 33% for HaRNAV versus SssRNAV; and 32% for RsRNAV versus SssRNAV. Phylogenetic analyses performed with the capsid protein sequences (**Figure 6**, **Table 1**) clearly indicate a closer relationship between these marine RNA viruses than the analyses with the RdRp domain. If we extend the amount of sequence used for the phylogenetic analyses and use the (putative) RNA helicase, RdRp, and VP3-like sequences concatenated into one sequence, the same groupings are strongly supported (**Figure 6**). However, there are other significant differences between HaRNAV and these viruses, such as monocistronic (HaRNAV) versus dicistronic (RsRNAV and SssRNAV) genome organization. Furthermore, although all analyses we have conducted support the idea that RdRp phylogenies are robust at discriminating established genus and family-level relationships between picorna-like viruses, a relationship between these marine viruses is not supported (**Figure 5**). Clearly more work is required and more related viruses need to be cultured and characterized to allow

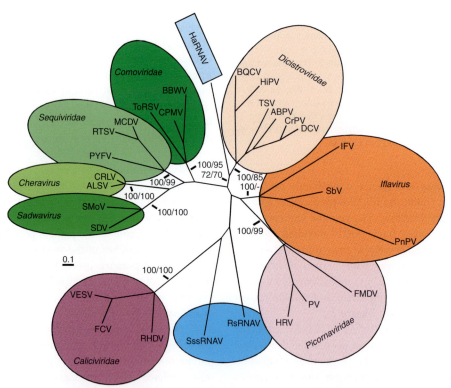

Figure 5 Phylogenetic analyses of RdRp sequences. Bayesian and neighbor-joining analyses were performed with aligned complete (putative) RdRp domain regions I through VIII. The Bayesian maximum likelihood tree is shown with the Bayesian scale bar. Important branches are labeled with the Bayesian clade credibility values (percentages based on 250 000 generations), followed by the neighbor-joining bootstrap values (percentages based on 10 000 replicates). A dash indicates that the branch was not supported by the neighbor-joining analysis. Defined viral groups are labeled by their family or genus designation where appropriate and highlighted together by color. The colors reflect the host organism species as follows: orange, insects/arthropods; pink, mammals; green, plants; blue, marine protists. HaRNAV represents the family *Marnaviridae* and the other two marine viruses (RsRNAV and SssRNAV) are not yet classified. Full names of the viruses and sequence database accession numbers are in **Table 1**.

Table 1 Viruses used in genome comparisons with HaRNAV

Virus	Abbreviation	Accession number
Acute bee paralysis virus	ABPV	NC_002548
Apple latent spherical virus	ALSV	NC_003787
Black queen cell virus	BQCV	NC_003784
Broad bean wilt virus 2	BBWV	AF225953, AF225954
Cherry rasp leaf virus	CRLV	NC_006271
Cowpea mosaic virus	CPMV	NC_003549
Cricket paralysis virus	CrPV	NC_003924
Drosophila C virus	DCV	NC_001834
Feline calicivirus	FCV	NC_001481
Foot-and-mouth disease virus	FMDV	NC_004004
Heterosigma akashiwo RNA virus	HaRNAV	AY337486
Himetobi P virus	HiPV	NC_003782
Human rhinovirus 14	HRV	NC_001490
Human poliovirus	PV	NC_002058
Infectious flacherie virus	IFV	NC_003781
Maize chlorotic dwarf virus	MCDV	NC_003626
Parsnip yellow fleck virus	PYFV	NC_003628
Perina nuda picorna-like virus	PnPV	NC_003113
Rabbit hemorrhagic disease virus	RHDV	NC_001543
Rhizosolenia setigera RNA virus	RsRNAV	AB243297
Rice tungro spherical virus	RTSV	NC_001632
Sacbrood virus	SbV	NC_002066
Satsuma dwarf virus	SDV	NC_003785
Schizochytrium single-stranded RNA virus	SssRNAV	AB193726
Strawberry mottle virus	SMoV	NP_733954
Taura syndrome virus	TSV	NC_003005
Tomato ringspot virus	ToRSV	NC_003840
Vesicular exanthema of swine virus	VESV	NC_002551

the phylogenetic the relationships among these viruses to be understood better. Given the extreme evolutionary distance between the hosts of these viruses (and the viruses themselves based on RdRp analyses), it seems unlikely that they would fall within the same family.

Although HaRNAV (strain SOG263) is the only marnavirus to be cultured and fully described so far, there is clear evidence that closely related viruses occur in marine systems. A degenerate RT-PCR approach that targeted picorna-like virus RdRp genes in marine virus communities produced 11 distinct RdRp sequences that grouped very closely with HaRNAV in phylogenetic analyses. These 11 sequences show variation at a total of 17 sites in the amplified 450-nucleotide region and each different sequence has between 1 and 5 nucleotide polymorphisms relative to the HaRNAV sequence (**Table 2**). The differences across all of these sequences (relative to HaRNAV) at the nucleotide level result in four polymorphic sites in the amino acid sequences, and two of these sites have two different polymorphisms at the same site in different clones (**Table 2**). Although these variations are in one of the most conserved regions of all +ssRNA viruses, the sequences have a very high level of identity with HaRNAV suggesting that these different sequences represent different strains of HaRNAV.

The RT-PCR probing of natural marine virus community RdRp genes further showed that there are many more unknown picorna-like viruses present in these communities. These virus sequences are more closely related to RsRNAV than to any of the other currently known and cultured marine viruses. We speculate that these sequences represent viruses from completely distinct genera, if not families, relative to marnaviruses, and so are beyond the scope of this article. One notable point about the sequences recovered with this approach is the vast diversity of these types of viruses that are likely to be present in the marine environment, even within a single water sample.

Work in the ocean shows that marine systems harbor a whole new world of ssRNA viruses. Culturing of some of these marine ssRNA viruses has shown that several have features that recognizably link them with known groups, such as picornaviruses. HaRNAV, representing the family *Marnaviridae*, shares many features with known groups of ssRNA viruses, particularly with viruses in the proposed order *Picornavirales*. Despite these shared characteristics, however, a large part of the HaRNAV genome has no recognizable similarity to any other known sequences, viral or otherwise. More work is clearly needed with the individual marine RNA viruses, and even more viruses need to be brought into culture, sequenced, and characterized.

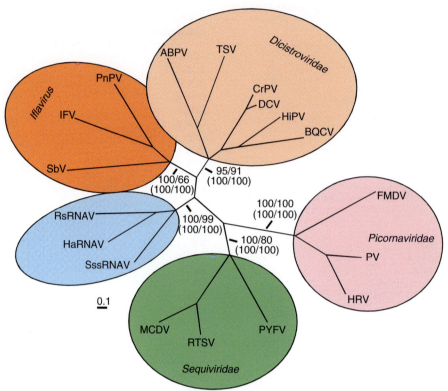

Figure 6 Phylogenetic analyses of VP3-like capsid sequences. Bayesian and neighbor-joining analyses were performed with aligned complete (putative) VP3-like sequences. The Bayesian maximum likelihood tree is shown with the Bayesian scale bar. Important branches are labeled with the Bayesian clade credibility values (percentages based on 175 000 generations), followed by the neighbor-joining bootstrap values (percentages based on 10 000 replicates). The corresponding support values for the branches from analyses of the (putative) 'RNA Helicase', RdRp, and VP3-like sequences concatenated into one long sequence are shown in brackets with Bayesian clade credibility values (percentages based on 175 000 generations). Defined viral groups are labeled by their family or genus designation where appropriate and highlighted together by color. The colors reflect the host organism species as follows: orange, insects/arthropods; pink, mammals; green, plants; blue, marine protists. HaRNAV represents the family *Marnaviridae* and the other two marine viruses (RsRNAV and SssRNAV) are not yet classified. Full names of the viruses and sequence database accession numbers are in **Table 1**.

Table 2 Polymorphisms, relative to HaRNAV, in RdRp sequence fragments amplified from natural marine virus communities

Sequence	Position[a]																
	63	72	78	134	212	223	258	267	285	316	317	363	379	380	396	432	438
HaRNAV[b]	A	C	G	T	C	A	A	C	T	A	A	C	T	T	A	G	A
JP500-1	.[c]	.	.	.	**T**[d]	T	.	.	.	.	.	.	.	.	.	.	.
JP800-1	.	.	.	**C**	**T**	.	.	.	.	.	.	.	.	**C**	.	A	.
JP800-4	.	.	.	.	**T**	.	.	.	.	.	.	.	.	.	.	.	.
JP800-8	.	.	.	.	**T**	.	.	.	.	.	.	.	T	.	**G**	.	.
JP800-11	.	.	.	.	**T**	.	**G**	.	.	**C**	.	.	**C**	.	.	.	.
FRP896-2	.	.	.	.	**T**	.	.	.	.	.	**C**	.	.	.	**G**	.	T
FRP896-3	G	.	.	.	**T**	.	.	.	.	.	.	.	.	.	**G**	.	T
FRP896-4	.	.	**A**	.	**T**	.	.	.	.	.	.	.	.	.	**G**	.	T
FRP896-5	.	.	.	.	**T**	.	.	.	**C**	.	.	.	.	.	**G**	.	T
FRP896-6	.	T	.	.	**T**	.	.	T	.	.	.	.	.	.	**G**	.	T
FRP896-7	.	.	.	.	**T**	.	.	.	.	.	.	**G**	.	.	**G**	.	T

[a]Sequences are 450 bases long.
[b]The cultured and sequenced strain HaRNAV SOG263.
[c]'.' indicates the sequence is the same as the sequenced strain HaRNAV SOG263 at this position.
[d]Changes that result in different amino acid sequences are in bold.

The ssRNA viruses isolated from marine systems infect basal eukaryotes, protists, and so these viruses may be ancestral to the related viruses infecting higher organisms such as mammals. Study of these viruses will further our understanding of RNA virus evolution.

See also: Algal Viruses.

Further Reading

Christian P, Fauquet CM, Gorbalenya AE, et al. (2005) A proposed *Picornavirales* order. In: Fauquet CM (ed.) *Microbes in a Changing World*. San Francisco: International Unions of Microbiological Societies.

Culley AI, Lang AS, and Suttle CA (2003) High diversity of unknown picorna-like viruses in the sea. *Nature* 424: 1054–1057.

Culley AI, Lang AS, and Suttle CA (2005) Taxonomic structure of the family *Marnaviridae*. In: Fauquet CM, Mayo MA, Maniloff J, Desselberger U, and Ball LA (eds.) *Virus Taxonomy: Eighth Report of the International Committee on Taxonomy of Viruses*, pp. 789–792. San Diego, CA: Elsevier Academic Press.

Culley AI, Lang AS, and Suttle CA (2006) Metagenomic analysis of coastal RNA virus communities. *Science* 312: 1795–1798.

Hellen CUT and Sarnow P (2001) Internal ribosome entry sites in eukaryotic mRNA molecules. *Genes and Development* 15(13): 1593–1612.

Jackson WT, Giddings TH, Taylor MP, et al. (2005) Subversion of cellular autophagosomal machinery by RNA viruses. *PLoS Biology* 3(5): e156.

Koonin EV and Dolja VV (1993) Evolution and taxonomy of positive-strand RNA viruses: Implication of comparative analysis of amino acid sequences. *Critical Reviews in Biochemistry and Molecular Biology* 28: 375–430.

Lang AS, Culley AI, and Suttle CA (2004) Genome sequence and characterization of a virus (HaRNAV) related to picorna-like viruses that infects the marine toxic bloom-forming alga *Heterosigma akashiwo*. *Virology* 320: 206–217.

Nagasaki K, Shirai Y, Takao Y, et al. (2005) Comparison of genome sequences of single-stranded RNA viruses infecting the bivalve-killing Dinoflagellate *Heterocapsa circularisquama*. *Applied and Environmental Microbiology* 71(12): 8888–8894.

Nagasaki K, Tomaru Y, Katanozaka N, et al. (2004) Isolation and characterization of a novel single-stranded RNA virus infecting the bloom-forming diatom *Rhizosolenia setigera*. *Applied and Environmental Microbiology* 70(2): 704–711.

Shirai Y, Takao Y, Mizumoto H, Tomaru Y, Honda D, and Nagasaki K (2006) Genomic and phylogenetic analysis of a single-stranded RNA virus infecting *Rhizosolenia setigera* (Stramenopiles: Bacillariophyceae). *Journal of the Marine Biological Association of the UK* 86: 475–483.

Tai V, Lawrence JE, Lang AS, et al. (2003) Characterization of HaRNAV, a single-stranded RNA virus causing lysis of *Heterosigma akashiwo* (Raphidophyceae). *Journal of Phycology* 39: 343–352.

Takao Y, Mise K, Nagasaki K, Okuno T, and Honda D (2006) Complete nucleotide sequence and genome organization of a single-stranded RNA virus infecting the marine fungoid protist *Schizochytrium* sp. *Journal of General Viology* 87: 723–733.

Takao Y, Nagasaki K, Mise K, Okuno T, and Honda D (2005) Isolation and Characterization of a novel single-stranded RNA virus infectious to a marine fungoid protist *Schizochytrium* sp. (Thraustochytriaceae, Labyrinthulea). *Applied and Environmental Microbiology* 71(8): 4516–4522.

Tate J, Liljas L, Scotti P, et al. (1999) The crystal structure of cricket paralysis virus: The first view of a new virus family. *Nature Structural Biology* 6: 765–774.

Measles Virus

R Cattaneo, Mayo Clinic College of Medicine, Rochester, MN, USA
M McChesney, University of California, Davis, Davis, CA, USA

© 2008 Elsevier Ltd. All rights reserved.

Glossary

Oncolytic virotherapy The experimental treatment of cancer patients based on the administration of replication-competent viruses that selectively destroy tumor cells but leave healthy tissue unaffected.

RNA editing The introduction into a RNA molecule of nucleotides that are not specified by the gene; the measles virus polymerase introduces a single G nucleotide in the middle of the phosphoprotein messenger RNA by reading twice over a C template (polymerase stuttering).

Subacute sclerosing panencephalitis (SSPE) A rare but always lethal brain disease caused by measles virus.

Syncytia Fused cells with multiple nuclei characteristic of measles virus infection.

Introduction and Classification

Measles virus (MV) is an enveloped nonsegmented negative-strand RNA virus of the family *Paramyxoviridae*, genus *Morbillivirus*. The *Paramyxoviridae* are important agents of disease, causing age-old diseases of human and animals (measles, mumps, respiratory syncytial virus (RSV), the parainfluenza viruses), and newly recognized emerging diseases (Nipah, Hendra, morbilliviruses of aquatic mammals).

Among negative-strand RNA viruses, the *Paramyxoviridae* are defined by having a protein (F) that causes fusion of viral and cell membranes at neutral pH. The organization and expression strategy of the nonsegmented genome of *Paramyxoviridae* including MV is similar to that of the *Rhabdoviridae*.

The defining characteristics of the genus *Morbilliviruses* are the lack of neuraminidase activity, and cell entry through the primary receptor signaling lymphocyte activation molecule (SLAM, CD150): MV, canine distemper virus, and rinderpest virus all enter cells through SLAM (human, canine, or bovine, respectively). The cellular distribution of SLAM overlaps with the sensitivity of different cell types to wild type MV infection, and explains immunosuppression.

Viral Particle Structure and Components

MV particles are enveloped by a lipid bilayer derived from the plasma membrane of the cell in which the virus was grown. They have been visualized as pleomorphic or spherical, depending on the methods used for their purification. Their diameter ranges from 120 to 300–1000 nm, implying that their cargo volume may differ by a factor 30 and that the large particles are polyploid. Inserted into the envelope are glycoprotein spikes that extend about 10 nm from the surface of the membrane and can be visualized by electron microscopy (**Figure 1**). A schematic diagram of an MV particle is shown in **Figure 2(a)**.

Inside the viral membrane is the nucleocapsid core, typically including several genomes. Each genome has 15 894 nucleotides tightly encapsidated by a helically arranged nucleocapsid (N) protein (**Figure 2(a)**). Two other proteins, a polymerase (L for large) and a polymerase cofactor (P for phosphoprotein), are associated with the RNA and the N protein to form a replicationally active ribonucleoprotein (RNP) complex. The MV RNP is condensed by the matrix (M) protein and then selectively covered by the two envelope spikes, consisting of oligomers of the F and H proteins.

The F protein spike is trimeric, whereas the H protein forms covalently linked dimers that may form noncovalently linked tetramers. The H protein contacts the cellular receptors, whereas F executes fusion. This protein is cleaved and activated by the ubiquitous intracellular protease furin. In the assembly process viral proteins are preferentially incorporated into nascent viral particles, whereas the majority of host proteins are excluded.

Genome, Replication Complex, and Replication Strategy

The MV negative-strand genome begins with a 56 nt 3′ extracistronic region known as leader, and ends with a 40 nt extracistronic region known as trailer. These control regions are essential for transcription and replication and flank the six genes. The term gene refers here to contiguous, nonoverlapping transcription units separated by three untranscribed nucleotides. There are six genes coding for eight viral proteins, in the order (positive strand): 5′-N-P/V/C-M-F-H-L-3′ (**Figure 2(b)**). The P gene uses overlapping reading frames to code for three proteins, P, V, and C. Complete sequences of several MV wild type and vaccine strains have been obtained.

The first gene codes for the N protein. Each N protein covers 6 nt, and about 13 N proteins may constitute a turn in the RNP helix. RNPs are formed when N is expressed in the absence of other viral components, suggesting that N–N interactions drive RNP assembly. Two domains have been identified in the N protein: a conserved N-terminal N core (about 400 amino acids) and a variable C-terminal N tail (about 100 amino acids). N core is essential for self-assembly, RNA binding, and replication activity, whereas N tail interacts with a C-terminal domain of the P protein. N protein exists in at least two forms in infected cells: one associated with RNA in a RNP structure and a second unassembled soluble form named N^0 that may encapsidate the nascent RNA strand during genome and antigenome replication.

The second gene codes for three proteins implicated in transcription or innate immunity control: P, V, and C. Two of these proteins have a modular structure: the 231 N-terminal residues of V are identical with those of P, but its 68 C-terminal residues are translated from a reading frame accessed by insertion of a nontemplated G residue through co-transcriptional editing (polymerase stuttering). This V domain is highly conserved in paramyxoviruses, with cysteine and histidine residues binding two zinc molecules per protein. The main function of the V protein is to counteract the innate immune response; V interferes with intracellular signaling pathways supporting the interferon response and sustains virus spread in the host immune system.

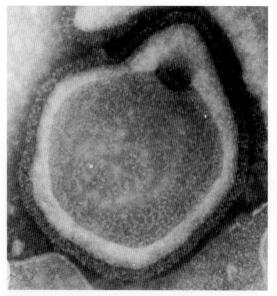

Figure 1 MV particle. Electron micrograph from Claire Moore and Shmuel Rozenblatt, Tel-Aviv University, Israel.

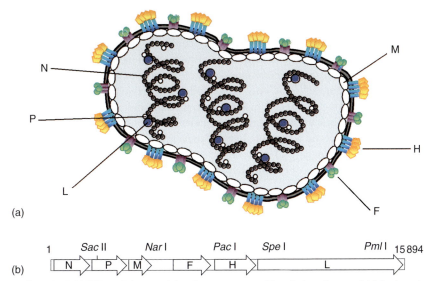

Figure 2 (a) Diagram of a polyploid MV particle containing three genomes. The viral nucleocapsid (N), phosphoprotein (P), polymerase (large, L), matrix (M), fusion (F), and hemagglutinin (H) proteins are indicated with different symbols. (b) Schematic representation of the MV antigenome (plus strand). The open reading frames of the six largest proteins are indicated with open arrows. The P gene codes for three proteins, P, V, and C. Unique restriction enzyme sites used for reverse genetics are indicated above the genome.

The main function of the P protein is to support viral replication and transcription; it is an essential component of the polymerase, and of the protein complex mediating RNA encapsidation by N^0 protein. The P protein N-terminal segment, identical with the V protein N-terminus, is phosphorylated on serines and threonines and contains regions of high intrinsic disorder, possibly facilitating interactions with multiple viral and cellular proteins. P self-assembles as a tetramer through a central region in its unique domain, and interacts with N tail through its C-terminus.

The third protein expressed from the P gene is named C. Its reading frame is accessed by ribosomes initiating translation 22 bases downstream of the P AUG codon. The MV C protein not only inhibits the interferon response but also has infectivity factor function, and a role in virus particle release has been demonstrated for the C protein of the related Sendai virus.

The third, fourth, and fifth genes code for envelope-associated proteins that are discussed below in the context of receptor recognition and membrane fusion. The large (L) sixth and last gene codes for the RNA-dependent RNA polymerase, believed to possess all enzymatic activities necessary to synthesize mRNA: nucleotide polymerization, capping and methylation, and polyadenylation. L adds poly-A tails to nascent viral mRNAs cotranscriptionally, by stuttering on a stretch of U residues occurring at the end of each viral gene. Sequence comparison identified six highly conserved domains in the L protein that were tentatively assigned different catalytic functions, whereas interaction of this 200 kDa protein with the P and other viral proteins was mapped to nonconserved regions.

The N, P, and L proteins are associated with the RNA to form the replicationally active RNP complex that starts primary transcription after cell entry. The negative-strand genome is then used to synthesize positive-strand 'antigenomes' that produce more genomes, completing one amplification cycle. Amplification produces the genomic templates for secondary transcription.

The polymerase transcribes the viral genome with a sequential 'stop–start' mechanism. It accesses it through an entry site located near its $3'$ end, transcribes the first gene (N) with high processivity, polyadenylates the N mRNA, and reinitiates P mRNA synthesis. The frequency of reinitiation is less than 100%, resulting in a gradient of transcript levels; N is transcribed at the highest levels, the most promoter-distal L gene (coding for a catalytic enzyme) at the lowest. The gene order and transcription strategy are fundamental characteristics that MV shares with all other paramyxoviruses.

Envelope Proteins and Cellular Receptors

The M protein, coded by the third gene, is a somewhat hydrophobic protein visualized as an electron-dense layer underlying the lipid bilayer. M is not an intrinsic membrane protein but can associate with membranes. It also binds RNPs, associates with the F and H protein cytoplasmic tails and modulates cell fusion. Thus, it is considered the assembly organizer and may also drive virus release/budding.

The two spike glycoproteins F and H are primarily responsible for membrane fusion and receptor attachment, respectively. An interesting peculiarity of the F gene is the long (almost 500 nt) 5′ untranslated region, whose function has not yet been characterized. The trimeric F protein, which is cleaved and activated by the ubiquitous intracellular protease furin, executes membrane fusion. This process is necessary for MV not only to enter cells, but also to spread through cell–cell fusion forming giant cells (syncytia). Membrane fusion activity depends not only on F protein proteolytic activation, but also on receptor recognition by the cognate H protein, which transmits a signal to F eliciting a conformational change and membrane fusion.

Since the MV H protein can hemagglutinate certain nonhuman primate red cells but lacks neuraminidase activity, it is named H and not HN. Other *Paramyxoviridae* use sialic acid as receptor and need neuraminidase to destroy receptor activity while budding from a host cell. Morbilliviruses do not use sialic acid as a receptor and do not need neuraminidase.

Two MV receptors have been characterized: Wild-type and vaccine MV target SLAM, whose expression is limited to immune cells. This protein was originally identified on activated B and T lymphocytes, but it is also expressed constitutively on immature thymocytes, memory T cells, and certain B cells. Other cell types including monocytes and dendritic cells (DCs) express SLAM only following activation. This cellular distribution overlaps with the sensitivity of different cell types to wild-type MV infection. Another strong argument for the central role of SLAM in MV tropism is the fact that three morbilliviruses (MV, canine distemper virus, and rinderpest virus) enter cells through SLAM (human, canine, or bovine, respectively).

The live attenuated MV vaccine strain Edmonston can also use the regulator of complement activation membrane cofactor protein (MCP; CD46) as a receptor. CD46's primary function is to bind and promote inactivation of the C3b and C4b complement products, a process protecting human cells from lysis by autologous complement, a function that requires ubiquitous expression. The question of the relevance of an ubiquitous receptor for the pathogenesis of a lymphotropic virus was raised when it was shown that the H protein of the attenuated Edmonston strain interacts much more efficiently with CD46 than the proteins of certain wild-type MV strains, and then again when the MV receptor function of human SLAM was discovered. The relevance of CD46 has been discussed in the context of both MV virulence and attenuation.

Clinical Features

Measles, one of the most contagious infectious diseases of man, was recognized clinically by the rash and other signs from early historical times and it was differentiated from small pox by Rhazes in the tenth century, who noted that it could be more fatal than small pox. Transmitted by aerosol droplets, the infection has an incubation period of 7–10 days, with onset of fever, cough, and coryza, followed in about 4 days by the skin rash which begins on the face and spreads to the whole body. The skin rash fades after about 5 days and clinical resolution is usually uneventful. Measles is typically an infection of childhood and protective immunity is lifelong, such that a second case of measles in a child or adult would be highly unusual. Prior to widespread vaccination against measles in the 1960s, the infection had a case–fatality rate of less than 5% in children, higher in infants and in children in developing countries, where fatality rates of up to 20% can still occur. Complications include pneumonia, encephalitis, otitis media, blindness, and secondary infections by common bacteria and viruses. In developing countries, common complications are diarrheal illness and wasting. 'Atypical measles', a severe MV infection that occurred following the use of whole, inactivated virus vaccine from 1963 to 1967, is characterized by high and persistent fever, a different body rash that resembles Rocky Mountain spotted fever, and lobar pneumonia with effusions. This unusual clinical manifestation is discussed in the section titled 'Pathology and histopathology'.

Compared to infections by other viruses of the *Paramyxoviridae*, measles has more serious clinical potential than infection by RSV or the parainfluenza viruses, possibly because the latter infections are usually confined to the respiratory tract and not systemic; but measles in man causes less disease and death than infections due to the most closely related morbilliviruses, rinderpest in cattle, or canine distemper in dogs.

Latent tuberculosis may be reactivated following a case of measles. A regular complication of measles virus-induced immunosuppression was initially studied by von Pirquet, who observed, in 1908, that the tuberculin skin test response is suppressed during measles. MV-induced immunosuppression has an onset near to the peak of viremia, such that the primary immune responses to MV are not impaired, but the immunosuppression is global and lasts for several months. Several distinct mechanisms of virus–cell interaction have been elucidated to account for impaired functions of lymphocytes and antigen presenting cells by MV *in vitro*, but the global and sustained nature of immunosuppression *in vivo* is not understood. Infections by other morbilliviruses are also associated with immunosuppression and this complication was a hallmark of epidemic disease in harbor seals that resulted in the discovery of a new morbillivirus, the phocine distemper virus. In the extreme case, immunosuppression with massive lymphocyte depletion is fatal in canine distemper virus-infected ferrets.

Immune Response and Complications

The primary immune responses to MV, initially IgM antibody, type 1 CD4 and CD8+ T-cell responses, followed by neutralizing IgG antibody, are completely effective in controlling viral replication and resolution of the infectious process. Both primary and secondary immunodeficiencies that impair T-cell responses, for example, the DiGeorge syndrome or advanced HIV infection, are a significant risk for failure of the host to control MV infection, resulting in persistent infection and death or serious disease in the lower respiratory tract and central nervous system. In contrast, deficiency of the antibody response does not impair the immune control of MV replication.

These fundamental insights about antiviral immunity, based upon the tragic consequences of immunodeficiency diseases or childhood leukemia, were confirmed by the experimental infection of rhesus monkeys with MV. CD8+ T-cell depletion of monkeys at the time of viral inoculation resulted in prolonged viremia until the T cells repopulated, but depletion of B cells had no significant effect on MV infection. Persistent and/or fatal MV infection has occurred in HIV-infected children who contracted measles and this happened in a young, HIV-infected man who was vaccinated with live, attenuated MV. However, the risk of serious disease with measles in the immunocompromised host (50% or greater case fatality) outweighs the risk of vaccination, and measles vaccination is recommended for HIV-infected infants and children unless they have severe immunodeficiency with low, age-adjusted CD4+ T-cell counts in blood.

Persistent MV infection can result in giant cell pneumonia or two neurological diseases: measles inclusion body encephalitis and subacute sclerosing panencephalitis (SSPE). Another rare neurologic complication, acute demyelinating encephalomyelitis, not due to continuing viral replication in the brain, is associated with autoimmunity to myelin. Several other diseases, including multiple sclerosis, inflammatory bowel disease, and autism, have been linked to MV infection, either anecdotally or by inference or deduction, but causal connections to MV infection or vaccination have never been established.

Pathology and Histopathology

The histopathological hallmark of MV infection is the formation of syncytia, or multinucleated giant cells. Mediated by the viral fusion protein, syncytial cells are not unique to MV infection but they are characteristic. MV infects cells of ectodermal, endodermal, and mesenchymal origin and synctial cells have been observed in all of these cell types. Multinucleated giant cells are readily observed in the organized lymphoid organs and in mononuclear cell aggregates in many tissues, including the inflamed lung, where they are referred to as Warthin–Finkeldy giant cells. These syncytia are of lymphocyte, macrophage, dendritic, or reticular cell origin. Syncytial cells are also readily observed in many epithelia, including the columnar epithelieum of the trachea and bronchi, the stratified squamous epithelium of the skin and buccal mucosa, and the transitional epithelium of the urinary bladder and urethra. Endothelial syncytial cells were observed in small pulmonary arteries of monkeys infected with MV. Eosinophilic cytoplasmic and nuclear inclusion bodies can be seen in measles giant cells. The syncytial cells of measles are not long lived and they disappear with resolution of the infection.

The major pathologic changes of measles are due to inflammation and necrosis followed by tissue repair without fibrosis. Secondary infections by bacterial or other viral agents are common and they alter the pathologic process accordingly, especially in the respiratory and gastrointestinal tracts. Pathology in the lower respiratory tract is mainly peribronchiolar inflammation and necrosis with a mild exudate, but interstitial pneumonitis with mononuclear cell infiltrates may occur. In the brain and central nervous system, perivascular mononuclear infiltrates can occur with a few necrotic endothelial cells, microglia, and neurons. In lymph nodes, spleen, and thymus, the major changes are mild to moderate lymphocyte depletion and multinucleated giant cells. Lymph node and splenic follicular hyperplasia are not seen in primary measles but are present if the host was previously infected or vaccinated against measles.

The unexpected occurrence of atypical measles following vaccination with whole-inactivated virus and subsequent infection with live, attenuated vaccine virus or wild-type virus was thought to be due to an aberrant, anamnestic host immune response resulting in an Arthus reaction or delayed hypersensitivity. Similar immunopathology was seen in children exposed to RSV following vaccination with whole-inactivated RSV. Atypical measles was experimentally induced in monkeys vaccinated with whole-inactivated MV and then challenged with wild type virus. Compared to monkeys vaccinated with live, attenuated MV, the aberrant immune responses associated with atypical measles resulted in immune complex deposition and eosinophilia. A marked skewing of T-cell cytokine responses toward type 2, with abnormal interleukin 4 production was also observed.

Diagnosis, Prevention and Control

As measles becomes a rare illness in regions with high vaccine coverage where the virus has been eliminated or controlled, the diagnosis of MV infection will be a challenge. The typical punctate, maculopapular skin rash, and the prodromal signs and symptoms are not specific for this viral infection and measles diagnosis requires differentiation from other pathogens causing exanthems of children, including infectious mononucleosis (Epstein–Barr virus) and cytomegalovirus infections, rubella, scarlet fever, typhus, toxoplasmosis, meningococcus, and staphylococcus. An enanthem with small white lesions on the buccal, labial, or gingival mucosal (Koplik's spots) is considered specific for MV infection and it typically occurs a few days before the skin rash. A buccal swab smeared on a glass slide, fixed and stained (e.g., Wright–Giemsa), may show multinucleated syncytia of epithelial cells. Secondary infections due to MV-induced immunosuppression are caused by the common pathogens in a geographic region. Detection of the early IgM response in blood by antibody-capture ELISA is diagnostic. MV can be cultured from peripheral blood mononuclear cells and from oral or nasal aspirates. Rescue of wild-type MV is readily done using B95a cells or Vero cells transfected with human SLAM. The virus can be detected by reverse transcription polymerase chain reaction (RT-PCR) from these samples and from urine. Filter paper spot assays of blood, buccal, or nasopharyngeal swabs have recently been developed for the detection of viral antibodies or viral RNA. The World Health Organization (WHO) hosts a website with descriptions of measles diagnostic samples and assays.

Measles remains the leading cause of vaccine-preventable illness and death in the world. The WHO and other international bodies have proposed the global eradication of MV in the twenty-first century, following the current campaign to eradicate poliovirus. This is an achievable goal as humans are the only known host species for this virus, although nonhuman primates are susceptible hosts. Regional elimination of MV has been achieved. Endogenous, circulating MV has been eliminated from the Americas in the last 20 years by the vaccination of children with the live, attenuated vaccines in current usage.

Many countries are planning to increase vaccination coverage as they progress toward targeted reductions in measles mortality or elimination of transmission. Recent experience in both developed and developing countries has shown that the maintenance of effective herd immunity against MV requires a two-dose vaccination strategy. Because of the highly contagious nature of this infection, about 95% of susceptibles will require vaccination to prevent the circulation of MV in a population of more than several hundred thousand. Greater than 90% measles vaccine coverage has been achieved in most countries but more than 25% of countries have not achieved this level of vaccination (**Figure 3**). The HIV pandemic

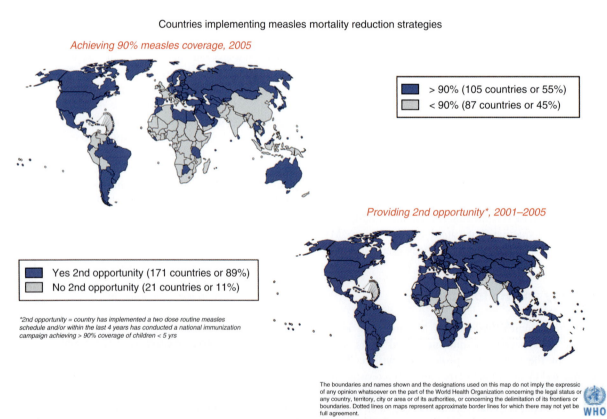

Figure 3 Global MV vaccine coverage, 2005. Reproduced from World Health Organization (2005) *Progress Towards Global Immunization Goals – 2005: Summary presentation of key indicators*, with permission.

presents a potential obstacle to MV eradication globally, especially in Africa and Asia, but HIV-infected children that are not immunocompromised can be safely vaccinated with live, attenuated MV (see above). The importation of MV into countries where elimination has been achieved is a very real challenge for diagnosis and control which has been successfully met by molecular genetic methods of rapid viral genomic sequencing and taxonomy.

The New Frontiers: MV-Based Multivalent Vaccines and Oncolysis

The live attenuated measles vaccine, which has an outstanding efficacy and safety record, is being developed as pediatric vaccine eliciting immunity against additional pathogens. An infectious cDNA with the identical coding capacity of a vaccine strain but with regularly spaced restriction sites has been engineered, and vectored MV expressing the hepatitis B surface antigen (HBsAg) generated by reverse genetics. One of these vectored MV vaccines induced protective levels of HBsAg antibodies while protecting Rhesus macaques against measles challenge. Another vectored MV expressing the West Nile virus envelope protein protected mice from West Nile virus encephalitis. A distinctive advantage of immunization with a di- or multivalent MV-based vaccine is delivery of an additional immunization safely without additional cost.

The knowledge gained from basic research has also been applied to the development of vectors for targeting and eliminating cancer cells. Oncolytic virotherapy is the experimental treatment of cancer patients based on the administration of replication-competent viruses that selectively destroy tumor cells but leave healthy tissue unaffected. MV is one of several human vaccine strains, or apathogenic animal viruses, currently being genetically modified to improve oncolytic specificity and efficacy. These modifications include targeting cell entry through designated receptors expressed on cancer cells like CD20 (lymphoma), CD38 (myeloma), or carcinoembryonic antigen (colon cancer).

Moreover, cancer cell specificity can be enhanced by silencing the expression of proteins that counteract the interferon system, often inactivated in cancer cells. Finally, proteolytic activation of the F protein has been engineered for exclusive cleavage through matrix metalloproteinases, enzymes degrading the extracellular matrix and promoting cancer invasiveness. Clinical trials of MV-based oncolysis for ovarian cancer, myeloma, and glioma are ongoing. These trials will soon profit from second generation targeted oncolytic MV with enhanced cancer specificity.

See also: Paramyxoviruses of Animals; Parainfluenza Viruses of Humans; Mumps Virus; Paramyxoviruses.

Further Reading

Cathomen T, Mrkic B, Spehner D, et al. (1998) A matrix-less measles virus is infectious and elicits extensive cell fusion: consequences for propagation in the brain. *EMBO Journal* 17: 3899–3908.

Cattaneo R (2004) Four viruses, two bacteria, and one receptor: Membrane cofactor protein (CD46) as pathogens' magnet. *Journal of Virology* 78: 4385–4388.

Griffin DE (2007) Measles virus. In: Knipe DM, Howley PM, Griffin DE, et al. (eds.) *Fields Virology,* 5th edn., vol. 2, pp. 1551–1586. Philadelphia, PA: Lippincott Williams and Wilkins.

Katz SL (2004) Measles (rubeola). In: Gershon AA, Hotez PJ, and Katz SL (eds.) *Krugman's Infectious Diseases of Children*, 11th edn., pp. 353–372. Philadelphia: Mosby.

Lamb RA and Parks GD (2007) *Paramyxoviridae*: The viruses and their replication. In: Knipe DM, Howley PM, Griffin DE, et al. (eds.) *Fields Virology,* 5th edn., vol. 1, pp. 1449–1496. Philadelphia, PA: Lippincott Williams and Wilkins.

McChesney MB, Miller CJ, Rota PA, et al. (1997) Experimental measles. Part I. Pathogenesis in the normal and the immunized host. *Virology* 233: 74–84.

Moss WJ and Griffin DE (2006) Global measles elimination. *Nature Reviews Microbiology* 4: 900–908.

Panum P (1939) Observations made during the epidemic of measles on the Faroe Islands in the year 1846. *Medical Classics* 3: 829–886.

Rager M, Vongpunsawad S, Duprex WP, and Cattaneo R (2002) Polyploid measles virus with hexameric genome length. *EMBO Journal* 21: 2364–2372.

Rima BK and Duprex WP (2006) Morbilliviruses and human disease. *Journal of Pathology* 208: 199–214.

Springfeld C and Cattaneo R (in press) Oncolytic virotherapy. In: Schwab M (ed.) *Encyclopedia of Cancer*, 2nd edn. Heidelberg: Springer.

von Messling V and Cattaneo R (2004) Toward novel vaccines and therapies based on negative-strand RNA viruses. *Current Topics in Microbiology and Immunology* 283: 281–312.

von Messling V, Svitek N, and Cattaneo R (2006) Receptor (SLAM [CD150]) recognition and the V protein sustain swift lymphocyte-based invasion of mucosal tissue and lymphatic organs by a morbillivirus. *Journal of Virology* 80: 6084–6092.

von Pirquet C (1908) Das Verhalten der kutanen Tuberkulinreaktion während der Masern. *Deutsche medizinische Wochenschrift* 34: 1297–1300.

World Health Organization (2005) *Progress Towards Global Immunization Goals – 2005: Summary Presentation of Key Indicators.* http://www.who.int/immunization_monitoring/data/SlidesGlobalImmunization.pdf.

World Health Organization (2007) *Manual for the Laboratory diagnosis of Measles and Rubella virus Infections*, 2nd edn., http://www.who.int/immunization_monitoring/LabManualFinal.pdf (accessed December 2007).

Yanagi Y, Takeda M, and Ohno S (2006) Measles virus: Cellular receptors, tropism and pathogenesis. *Journal of General Virology* 87: 2767–2779.

Relevant Website

http://www.ncbi.nlm.nih.gov – National Center for Biotechnological Information.

Membrane Fusion

A Hinz and W Weissenhorn, UMR 5233 UJF-EMBL-CNRS, Grenoble, France

© 2008 Elsevier Ltd. All rights reserved.

Glossary

Fusion pore Small opening at the site of two merged lipid bilayers, which allows the exchange of fluids. Fusion pores expand gradually to complete membrane fusion.

Hemifusion Membrane fusion intermediate state with the two proximal leaflets of two opposed bilayers merged to one.

Lipid raft Small membrane microdomain enriched in cholesterol and glycosphingolipids. These domains are resistant to solubilization by Triton X-100.

Type 1 TM protein A glycoprotein composed of an N-terminal external domain and a single transmembrane region followed by a cytoplasmic domain.

Introduction

Enveloped viruses contain a lipid bilayer that serves as an anchor for viral glycoproteins and protects the nucleocapsid containing the genetic information from the environment. The lipid bilayer is derived from host cell membranes during the process of virus assembly and budding. Consequently, infection of host cells requires that enveloped viruses fuse their membrane with cellular membranes to release the nucleocapsid and accessory proteins into the host cell in order to establish a new infectious cycle. Glycoproteins from enveloped viruses evolved to combine two main features. Firstly, they contain a receptor-binding function, which attaches the virus to the host cell. Secondly, they include a fusion protein function that can be activated to mediate fusion of viral and cellular membranes. Both tasks can be encoded by a single glycoprotein or by separate glycoproteins, which act in concert.

Three different classes of viral fusion proteins have been identified to date based on common structural motifs. These include class I fusion proteins, characterized by trimers of hairpins containing a central alpha-helical coiled-coil structure, class II fusion proteins, characterized by trimers of hairpins composed of beta structures, and class III proteins, forming trimers of hairpins by combining structural elements of both class I and class II fusion proteins (**Table 1**).

Viral glycoproteins interact with distinct cellular receptors by initiating conformational changes in the fusion protein leading to membrane fusion. Fusion occurs either at the plasma membrane, where receptor binding triggers conformational changes in the glycoprotein, or in endosomes upon virus uptake by endocytosis. In the latter case the low pH environment of the endosome leads to protonation (key histidine residues have been specifically implicated in the process), which induces conformational changes that lead to fusion of viral and cellular membranes.

The biophysics of membrane fusion is dominated by the stalk hypothesis. According to this view, fusion of two lipid bilayers in an aqueous environment requires that they come into close contact associated with a significant energy barrier. This process involves local membrane bending creating a first site of contact. Complete dehydration of the initial contact site induces monolayer rupture that allows mixing of lipids from the two outer leaflets, resulting in a hemifusion stalk. In a next step, the model predicts that radial expansion of the stalk leads to either direct fusion pore opening or to the formation of another intermediate, the hemifusion diaphragm, an extended bilayer connecting both membranes. The hemifusion diaphragm may also expand into a fusion pore. Fusion pore formation, which is characterized by an initial opening and closing ('flickering') of the pore may be mediated by several factors such as lateral tension in the hemifusion stalk or bilayer and the curvature at the edges of the hemifusion state. Finally the fusion pore extends laterally until both membranes form a new extended lipid bilayer (**Figure 1**).

The applicability of the stalk model to viral membrane fusion processes is supported by a number of observations. Labeling techniques allow to distinguish between merging of lipid bilayers and content mixing thus visualizing intermediate steps in membrane fusion. This has been applied to several liposome fusion systems demonstrating that membrane fusion steps can be arrested at different stages. Furthermore, certain lipids such as inverted cone-shaped lysophospholipids induce spontaneous positive bilayer curvature and inhibit hemifusion, while cone-shaped phosphatidylethanolamines induce negative curvature and promote hemifusion. In contrast, the lipid effect on the opening of the fusion pore is the opposite. Finally, electron microscopy images of influenza virus particles fused with liposomes reveal structures resembling stalk intermediates.

These observations are consistent with the hypothesis that viral fusion proteins generate initial contacts between two opposing membranes and their extensive refolding

Table 1 Crystal structures of viral fusion proteins

Virus family	Virus species	PDB code
Class I		
Orthomyxoviridae	Influenza A virus HA	1HA0, 3HMG, 1HTM, 1QU1
	Influenza C virus HEF	1FLC
Paramyxoviridae	Simian parainfluenza virus 5 F	2B9B, 1SVF
	Human Parainfluenza virus F	1ZTM
	Newcastle disease virus F	1G5G
	Respiratory syncytial F	1G2C
Filoviridae	Ebola virus gp2	1EBO, 2EBO
Retroviridae	Moloney Murine leukemia virus TM	1AOL
	Human immunodeficiency virus 1 gp41	1ENV, 1AIK
	Simian immunodeficiency virus gp41	2SIV, 2EZO
	Human T cell leukemia virus 1 gp21	1MG1
	Human syncytin-2 TM	1Y4M
	Visna virus TM	1JEK
Coronaviridae	Mouse hepatitis virus S2	1WDG
	Sars corona virus E2	2BEQ, 1WYY
Class II		
Flaviviridae	Tick-borne encephalitis virus E	1URZ, 1SVB
	Dengue 2, and 3 virus E	1OK8 IUZG, 1OAN, 1TG8
Togaviridae	Semliki forest virus E1	1E9W 1RER
Class III		
Rhabdoviridae	Vesicular stomatitis virus G	2GUM
Herpesviridae	Herpes simplex virus gB	2CMZ

regulates and facilitates fusion via lipidic intermediate states by lowering the energy to form stalk-like intermediate structures.

Class I Fusion Glycoproteins

Biosynthesis of Fusion Proteins

Class I fusion proteins are expressed as trimeric precursor glycoproteins that are activated by proteolytic cleavage with subtilisin-like enzymes such as furin. This produces a receptor-binding subunit that is either covalently or noncovalently attached to the membrane fusion protein subunit, which anchors the heterotrimer to the viral membrane. The endoproteolytic cleavage positions a hydrophobic fusion peptide at or close to the N-terminus of the fusion domain. Subtilisin-like proteases recognize a conserved multibasic recognition sequence R-X-K/R-R or a monobasic cleavage site present in various glycoproteins. The nature of the cleavage site and its efficient cleavage (e.g., influenza virus hemagglutinin) has been associated with pathogenicity. The multibasic recognition sequences present in influenza virus HA, *SV5* F protein, *HIV*-1 gp160, and Ebola virus GP lead to mostly intracellular processing, whereas monobasic cleavage sites in Sendai virus F protein or influenza virus HA are efficiently cleaved extracellularly, resulting in a more tissue-restricted distribution of these viruses.

Cleavage activates the fusion potential of the viral -glycoproteins and is required for most class I glycoprotein-mediated fusion events. Although some evidence suggests Ebola virus processing by furin is not required for entry, it still requires the activity of endosomal cysteine proteases for efficient entry. Proteolytic cleavage thus generates in most cases a metastable glycoprotein structure that can switch into a more stable structure upon cellular receptor interaction including proton binding in the acidic environment of endosomes. This metastability was first recognized to play an important role in influenza virus hemagglutinin-mediated entry and has since been associated with all class I glycoproteins.

Structure of Native Influenza Virus Hemagglutinin

Since the structure solution of influenza virus hemagglutinin (HA), HA has served as the prototype of a class I fusion protein. The HA_1 domain, which contains the receptor-binding domain, folds into a beta structure that binds sialic acid-containing cellular receptors at the top of the molecule. In addition, both N- and C-termini of HA_1 interact with the stem of fusion domain HA_2 in an extended conformation. HA_2 anchors hemagglutinin to the viral membrane and folds into a central triple-stranded coiled-coil structure that is followed by a loop region and an antiparallel helix, which extends towards the N-terminal fusion peptide that is buried within the trimer interface (**Figure 2**).

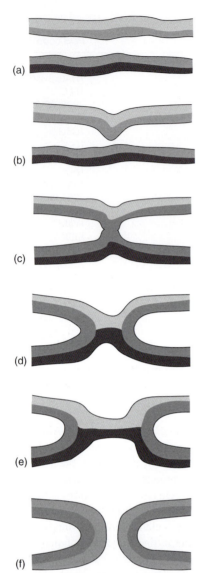

Figure 1 Fusion of two lipid bilayers. (a) Two parallel lipid bilayers do not approach closely. (b) Close contact mediated by local membrane bending. Hemifusion stalks with contact of outer leaflets (c) and inner leaflets (d). Fusion pore opening (f) may proceed directly from the stalk structure (d) or via a hemifusion diaphragm (e).

Structure of the Precursor Influenza Virus Hemagglutinin HA$_0$

The structure of uncleaved influenza virus HA$_0$ shows that only 19 residues around the cleavage site are in a conformation, which is different from the one seen in the native cleaved HA structure. This difference entails an outward projection of the last residues of HA$_1$ (323–328) and the N-terminal residues of HA$_2$ (1-12), resulting in the exposure of the proteolytic cleavage site (**Figure 2**). Upon cleavage, HA$_2$ residues 1–10 fill a mostly negatively charged cavity adjacent to the cleavage site, which leads to the sequestering of the fusion peptide within the trimeric structure (**Figure 2**).

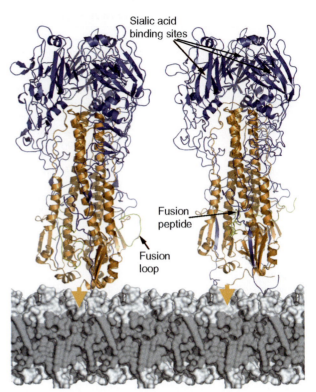

Figure 2 Ribbon diagram of trimeric conformations of influenza A virus hemagglutinin before (left, pdb code 1HA0) and after (right, pdb code 3HMG) proteolytic processing. The HA$_1$ receptor-binding domain is shown in blue and the positions of the sialic acid-binding sites are indicated. The fusion protein subunit HA$_2$ is shown in orange. The orientation towards the lipid bilayer is indicated by the orange triangle.

Structure of the Low pH-Activated Conformation of Hemagglutinin HA$_2$

Low pH destabilizes the HA$_1$ trimer contacts, which causes the globular head domains to dissociate. This movement facilitates two major conformational changes. (1) A loop region (residues 55–76) refolds into a helix (segment B in HA) and extends the central triple-stranded coiled-coil in a process that projects the fusion peptide approximately 100 Å away from its buried position in native HA. (2) Another dramatic change occurs toward the end of the central triple-stranded coiled-coil, where a short fragment unfolds to form a reverse turn which positions a short helix antiparallel against the central core. This chain reversal also repositions a β-hairpin and the extended conformation that leads to the transmembrane region (**Figure 3**). Although its orientation changes, the core structure of the receptor-binding domain HA$_1$ does not change upon acidification.

Since both the neutral pH structure and the core of the low pH structure from hemagglutinin have been solved, a number of class I fusion protein structures have been determined and a common picture has emerged for their mode of action (**Table 1**). The characteristic of all class I

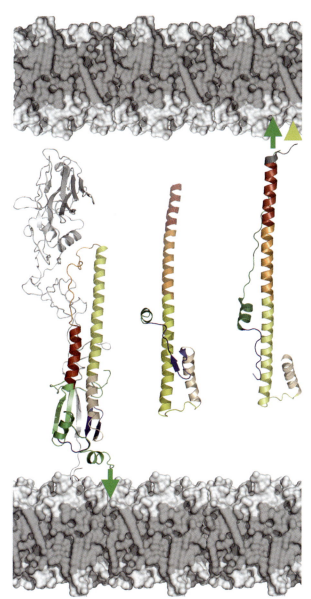

Figure 3 Ribbon representations of the conformational changes in HA₂ upon low pH exposure. Only one monomer is shown for clarity. Left panel: Native cleaved HA (pdb code 3HMG), HA$_1$ in gray, the secondary structure elements for HA$_2$ that change are shown in different colors. Middle panel: Low pH HA$_2$ (pdb code 1HTM) projecting the N-terminus leading to the fusion peptide towards the target cell membrane. Right panel: The C-terminal region has completely zipped up against the N-terminal coiled-coil domain (pdb code 1QU1). The membrane orientations of the TM region and the fusion peptide are indicated by green arrows and yellow triangles, respectively.

assumed that they all represent the postfusion conformation. Although there is only structural evidence for extensive conformational rearrangement of the fusion protein subunit in case of hemagglutinin HA and the paramyxovirus F protein (**Figure 4**), it is assumed that all known class I fusion protein core structures are the product of conformational rearrangements induced by receptor binding.

Class I-Mediated Membrane Fusion

The positioning of the N- and C-terminal ends containing the fusion peptide and the transmembrane region at the same end of a core structure, which was first established for the *HIV*-1 gp41 core structure, led to the proposal of the following general fusion model. (1) Proteolytic cleavage activation transforms the glycoprotein into a metastable conformation. (2) Receptor binding induces conformational changes in the fusion protein that exposes the fusion peptide and allows fusion peptide interactions with the target membrane (**Figures 5(a)** and **5(b)**). This generates a prehairpin intermediate structure that can be targeted by fusion inhibitors such as the HIV-1-specific T-20 peptide (**Figure 5(b)**). (3) Extensive refolding of the fusion domain most likely requires the dissociation of the C-terminal regions (**Figures 5(b)** and **5(c)**, indicated by blue lines) and leads to the apposition of the two membranes, concomitantly with the zipping up of the C-terminal region against the N-terminal coiled-coil domain ultimately forming the hairpin structure (**Figures 5(c)** and **5(d)**). The complete refolding process is thought to pull the two membranes into close-enough proximity to concomitantly allow membrane fusion. The extensive rearrangement of the fusion protein is thought to control the formation of different intermediate bilayer structures such as the hemifusion stalk (**Figure 5(d)**), and or the hemifusion diaphragm, followed by fusion pore opening and expansion (**Figure 5(e)**). It is generally assumed that membrane fusion occurs while the helical hairpin structure is formed and the core fusion protein structures represent postfusion conformations. Refolding of the fusion protein might produce defined stable intermediate structures, as suggested by the two low pH structures of influenza virus HA$_2$. One indicates that most of the outer layer has not yet zipped up to form the hairpin structure (**Figure 3**, middle panel: the C-terminal ends could extend back to the transmembrane region), while the other one reveals the extended conformation of the outer layer which forms together with the N-terminal coiled-coil a stable N-capped structure (**Figure 3**, right panel). Stepwise refolding may thus lock the fusion process irreversibly at distinct steps in agreement with a general irreversibility of class I-mediated fusion processes. The two membrane anchors, which are not present in the fusion conformation structures, also play an active

fusion protein cores is their high thermostability suggesting that they represent the lowest energy state of the fusion protein. Secondly, they all contain a central triple-stranded coiled-coil region with outer C-terminal anti-parallel layers that are either mostly helical or adopt extended conformations, thus forming trimers of helical hairpins. Since they resemble the low pH form of HA, it is

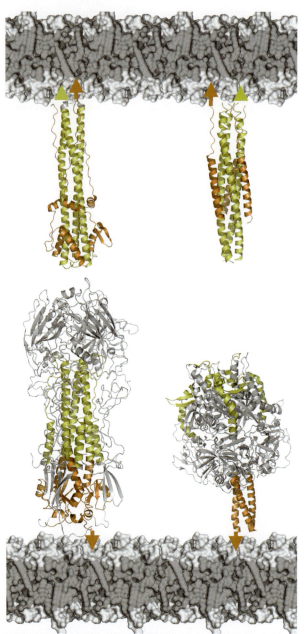

Figure 4 Comparison of the conformational changes induced upon receptor binding of two class I fusion proteins, influenza A virus hemagglutinin (left panel, pdb codes 3HMG and 1QU1) and paramyxovirus F protein (right panel, Simian parainfluenza virus 5 F, pdb codes 2B9B and 1SVF). The lower panel shows ribbon diagrams of native HA and F with secondary structure elements that change conformation upon activation highlighted in two colors (the inner triple-stranded coiled-coil region of the postfusion conformation in yellow and the C-terminal layers in orange). Although both native structures differ quite substantially, the conformational changes result in similar hairpin structures (upper panel) orienting both membrane anchors toward the target cell membrane. The membrane orientations of the TM region and the fusion peptide are indicated by orange arrows and yellow triangles, respectively.

role in the fusion reaction. Replacement of the transmembrane region by a glycophosphatidylinositol anchor leads to a hemifusion phenotype in case of hemagglutinin-driven membrane fusion, highlighting the role of the transmembrane region. Furthermore the C-terminal membrane proximal region plays an important role in fusion as shown in the case of *HIV*-1 gp41-mediated fusion.

Fusion Peptide

Class I fusion peptide sequences vary between virus families and are usually characterized by their hydrophobicity and a general preference for the presence of glycine residues. Although most fusion peptides of class I fusion proteins locate to the very N-terminus of the fusion protein, a few are found within internal disulfide-linked loops (e.g., filovirus Gp2 and the Avian sarcoma virus fusion proteins). NMR studies on the isolated influenza virus hemagglutinin fusion peptide revealed a kinked helical arrangement, which was suggested to insert into one lipid bilayer leaflet. This mode of bilayer interaction was proposed not only to mediate membrane attachment but also to destabilize the lipid bilayer. A further important function of fusion peptide sequences might be their specific oligomerization at the membrane contact site, which might constitute sites of initial membrane curvature.

Cooperativity of Fusion Proteins

A number of studies suggest that more than one class I fusion protein trimer is required to promote class I-driven membrane fusion. It has been suggested that activated hemagglutinin glycoproteins interact with each other in a synchronized manner and cooperativity of refolding allows the synchronized release of free energy required for the fusion process. This implies that activated fusion proteins assemble into a protein coat-like structure that helps to induce membrane curvature, possibly also by inserting the fusion peptides into the viral membrane outside of the direct virus–cell contact site. However, it should be noted that no clearly ordered arrays of activated class I glycoproteins have yet been observed experimentally.

Role of Lipids in Fusion

The lipid content of a viral envelope such as that of *HIV*-1 was shown to contain mostly lipids normally present in lipid raft microdomains at the plasma membrane. Lipid rafts are small ordered lipid domains that are enriched in cholesterol, sphingomyelin, and glycosphingolipids. A number of enveloped viruses (e.g., influenza virus, HIV-1, Ebola virus, measles virus) use these platforms for assembly and budding, and some evidence suggest

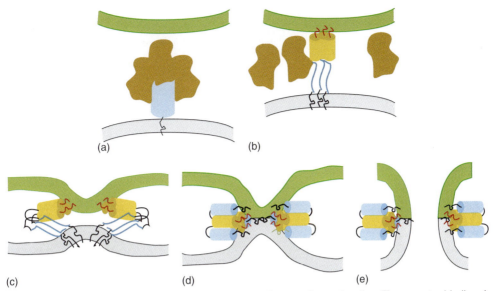

Figure 5 Model for class I glycoprotein-mediated membrane fusion. See text for explanation. The receptor-binding domains are indicated in brown and the fusion protein as cylinders. Note that some fusion proteins such as F from paramyxoviruses associate with an attachment protein (HN, H, or G). The latter interacts with F and cellular receptors triggering F-mediated fusion at the plasma membrane.

that raft platforms are also required for virus entry. Since the fusion activity of viral glycoproteins such as *HIV-1* Env is affected by cellular receptor density as well as Env glycoprotein density, it has been suggested that both ligands have to be clustered efficiently to cooperatively trigger productive Env-mediated membrane fusion. This observation is consistent with the sensitivity of *HIV-1* entry to cholesterol depletion.

Class II Fusion Proteins

Biosynthesis of Fusion Proteins

Class II fusion proteins comprise the fusion proteins from positive-strand RNA viruses such as the *Togaviridae* family, genus *Alphaviruses* (e.g., *Semliki Forest virus* (SFV)), and the *Flaviviridae* (e.g., *Dengue, Yellow fever,* and *Tick-borne encephalitis* virus (TBE)) (**Table 1**). Flaviviruses express the glycoprotein E that associates with a second precursor glycoprotein prM, while alphaviruses express two glycoproteins, the fusion protein E1 and the receptor-binding protein E2. E1 associates with the regulatory precursor protein p62. Both E-prM and E1-p62 heterodimerization are important for folding and transport of the fusion proteins. Cleavage of the fusion protein chaperones p62 and prM by the cellular protease furin in the secretory pathway is a crucial step in the activation of E and E1 fusion proteins.

Structure of the Native Fusion Protein

The native conformations of the flavivirus E glycoproteins and that of the alphavirus E1 glycoproteins are similar and fold into three domains primarily composed of β-sheets, with a central domain I, flanked by domain III connecting to the transmembrane region on one side and domain II on the other side (**Figure 6**, lower panel). Domain II harbors the fusion loop that is stabilized by a disulfide bridge and mostly sequestered within the antiparallel flavivirus E glycoprotein homodimer. In analogy, the fusion loop might be sequestered within the *SFV* E1–E2 heterodimer. Dimeric E–E and E1–E2 interactions keep the glycoproteins in an inactive, membrane-parallel conformation that covers the viral membrane. *SFV* E1–E2 heterodimers form an icosahedral scaffold with $T=4$ symmetry. Similarly, flavivirus E homodimers completely cover the viral membrane surface. The arrangement of the class II glycoproteins is thus completely different from the appearance of class I glycoprotein spikes, which do not form a specific symmetrical protein coat. In addition to forming the outer protein shell, flavivirus E and alphavirus E2 interact with cellular receptors, which direct the virion to the endocytotic pathway.

Structure of the Activated Fusion Protein

There are only minimal changes in secondary structure during the low pH-induced rearrangement of *TBE* E and *SFV* E1. However, the conformational changes result in an approximate 35–40 Å movement of domain III and a rotation of domain II around the hinge axis connecting domains I and II. This rearrangement produces a hairpin-like structure with a similar functional architecture as class I fusion proteins (**Figure 6**, upper panel). The outside of the trimer reveals a groove that was suggested to

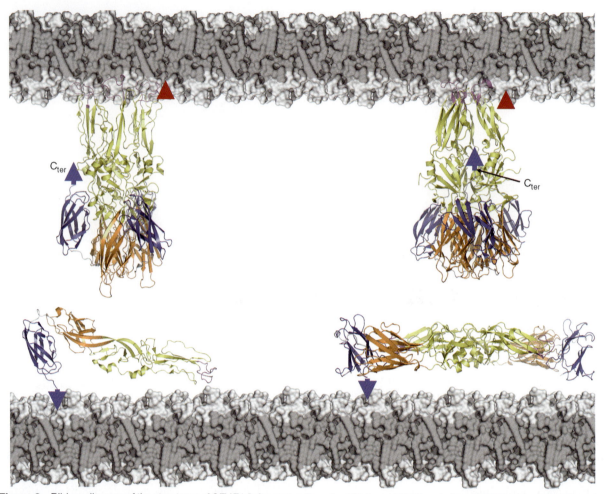

Figure 6 Ribbon diagram of the structures of *SFV* E1 (left panel, pdb codes 2ALA and 1RER) and of the *TBE* E (right panel, pdb codes 1URZ and 1SVB) in their native dimeric state (lower panel) and low-pH-activated trimeric conformation (upper panel). The three main domains of E1 and E are colored differently: domain I in blue, domain II in orange, and domain III in yellow. In both cases, activation of the conformational changes leads to trimeric hairpin structures. The membrane orientations of the TM region and the fusion peptide are indicated by blue arrows and red triangles, respectively.

accommodate the segment, which connects to the transmembrane anchor and thus positions the fusion loops next to the membrane anchors. One significant difference between the *TBE* E and *SFV* E1 low pH conformations are the orientations of the fusion loops. *TBE* E fusion loops undergo homotrimer interactions, while *SFV* E1 fusion loops do not interact within trimers (**Figure 6**).

Class II-Mediated Membrane Fusion

At the low pH of endosomes E and E1 undergo conformational rearrangements that involve three major steps. Firstly, the homo- or heterodimers dissociate from the membrane-parallel conformation in a reversible manner assuming monomeric fusion proteins that expose their fusion loop to the target membrane (**Figures 7(a)** and **7(a′)**). This seems to be a main difference between class I and class II fusion, since trimer dissociation into monomers has not been implicated in any class I fusion pathway.

Secondly, fusion loop membrane interaction leads to the formation of homotrimers with an extended conformation. Trimerization is irreversible and tethers the fusion protein to the target membrane (**Figure 7(b)**). It is comparable to the postulated prehairpin structure of class I fusion proteins such as *HIV*-1 gp41 (**Figure 7(b)**). Notably, both fusion intermediates can be targeted by either fusion protein peptides (T-20 in case of *HIV*-1 gp41) or recombinant fusion protein domains (such as the E3 domain in case of *TBE*), to block membrane fusion. Further refolding, namely the reorientation of domain I, then pulls the two membranes into closer apposition that ultimately leads to the formation of a hemifusion stalk-like structure (**Figures 7(c)** and **7(d)**). Finally complete zipping up of the C-terminal ends against the N-terminal core domains allows fusion pore opening and its expansion (**Figure 7(e)**). Similar to the case of class I fusion protein-driven fusion, refolding is thought to provide the energy for fusion (**Figure 7**).

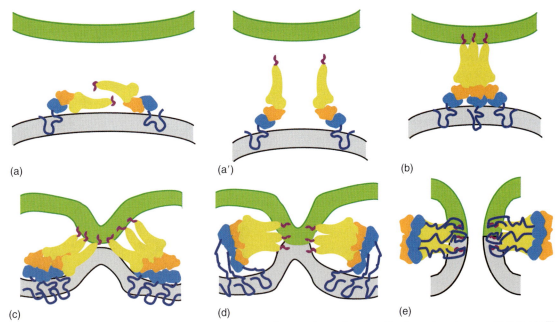

Figure 7 Model for membrane fusion of class II fusion proteins. See text for explanation. The three domains are colored as in **Figure 6**.

Fusion Peptide

The native and low-pH-induced crystal structures of the *TBE* virus E, dengue fever virus E, and *SFV* E1 proteins reveal that the conformation of the fusion loop changes upon acidification. The low-pH structures indicate that only hydrophobic side chains of the loop insert into the hydrocarbon chains of the outer leaflet of a target membrane. This is sufficient to anchor the fusion protein to the host cell membrane. Further oligomerization of fusion loops, as shown in the case of the low pH form of *SFV* E1, where crystal packing analysis revealed fusion loop interactions between trimers, was suggested to induce local membrane deformation, such as induction of a nipple-like membrane deformation (**Figure 7(c)**). This has been predicted in the stalk model to play an important role in the generation of lipidic intermediates during membrane fusion. Therefore the *SFV* E1 conformation might reflect an intermediate fusion state preceding the suggested post-fusion conformation of flavivirus E trimers with homotrimeric fusion loop interactions. *In vivo*, the latter conformation might be induced by the final refolding of the C-terminal membrane proximal region and thus determining 'open' and 'closed' conformations of *SFV* E1 trimers and *TBE virus* E trimers, respectively (**Figure 6**).

Fusion Protein Cooperativity in Membrane Fusion

Homo- or heterodimeric class II fusion proteins already form a protein shell covering the complete viral membrane in the native state. Upon activation *in vitro*, both, soluble *SFV* E1 protein and flavivirus E protein insert their fusion loops into liposomes and form arrays of trimers organized in a lattice composed of rings of five or six. The E protein lattice on liposomes contains preferentially rings of five, which seems to affect the curvature of coated liposomes. In contrast rings of six form mostly flat hexagonal arrays *in vitro*. E1 pentameric rings can also be reconstructed from the crystal packing of E1 trimers. This strongly suggests that formation of a distinct fusion protein lattice might exert a cooperative effect on the fusion process.

Role of Lipids in Fusion

Although heterodimer dissociation exposes the *SFV* E1 fusion loop, its insertion into a lipid bilayer requires low pH triggering and cholesterol, which is consistent with the observation that *SFV* fusion depends on cholesterol and sphingolipids. The lipids required for E1 activation and fusion imply indirectly that lipid raft microdomains might be targeted for fusion. Flavivirus fusion, however, seems to be less dependent on cholesterol than alphavirus fusion.

Class III Fusion Proteins

Biosynthesis of Fusion Proteins

The glycoprotein G from vesicular stomatitis virus (VSV), a member of the *Rhabdoviridae* (e.g., *VSV* and *Rabies virus*), negative-strand RNA viruses, and gB from Herpesvirus, a member of the *Herpesviridae*, double-stranded DNA viruses, constitute a third class of viral fusion proteins based on the structural similarity of the postfusion conformation of their respective glycoproteins. Unlike class I and II envelope proteins both, *VSV* G and herpesvirus gB, are neither expressed as precursor proteins nor are they proteolytically activated.

Rhabdoviruses express a single trimeric glycoprotein G, which acts as receptor-binding domain to induce endocytosis and as the fusion protein that controls fusion with endosomal membranes upon acidification. However, different from class I and class II fusion machines the conformational changes induced by low pH are reversible. Changes in pH can easily revert the three proposed conformations of G, the native conformation as detected on virions, an activated state that is required for membrane interaction, and an inactive postfusion conformation.

Herpesvirus entry and fusion is more complex since it requires four glycoproteins, namely gD, gH, gL, and gB. Glycoprotein gD contains the receptor-binding activity and associates with gB as well as gH and gL. While gB seems to constitute the main fusion protein, the others are thought to be required for activation of the fusion potential of gB, which is pH independent.

Structure of the Low-pH-Activated VSV Glycoprotein and Herpesvirus gB

VSV G is composed of four domains that, interestingly, show similarities to both class I and class II fusion protein structures. It contains a β-sheet-rich lateral domain at the top, a central α-helical domain that mediates trimerization, and resembles the α-helical hairpin structure of class I fusion molecules, a neck domain containing a pleckstrin homologous (PH) domain, and the fusion loop domain that builds the trimeric stem of G. This stem-like domain exposes two loops at its very tip containing aromatic residues constituting the membrane-interacting motif of G. The stem domain resembles that of class II fusion proteins, albeit its different strand topology, which could be the result of convergent evolution. Although the complete C-terminus is not present in the structure, it points towards the tip of the fusion domain, indicating that both membrane anchors, the fusion loops and the transmembrane region, could be positioned at the same end of an elongated hairpin structure (**Figure 8**). The overall similarity of the structural organization of *VSV* G with that of herpesvirus gB indicates a strong evolutionary relationship between the rhabdovirus G and herpesvirus gB fusion proteins (**Figure 8**).

The Fusion Loops

The fusion loops extending from the stem-like domain of *VSV*G is similar to those observed in class II fusion proteins. The architecture is such that only few hydrophobic

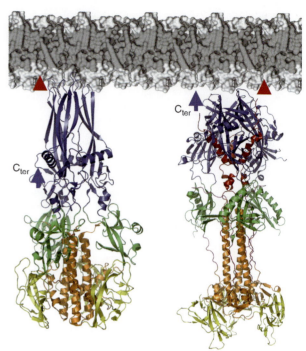

Figure 8 Ribbon diagram of class III fusion glycoproteins from *VSV* G (left panel, pdb code 2CMZ) and Herpesvirus gB (right panel, pdb code 2GUM). The individual domains are colored as follows: domain I in yellow, domain II in orange, domain III in green, and domain IV in blue. Their orientation toward the target membrane is indicated and shows the attachment of the putative fusion loops to one leaflet of the bilayer (red triangles) and the putative orientation of the TM region toward the same end of the fusion loop region (blue arrow).

side chains can intercalate into one lipid bilayer leaflet, potentially up to 8.5 Å. Intercalation of side chains into one leaflet may induce curvature of the outer leaflet with respect to the inner leaflet, which would satisfy the stalk model. The role of lipidic intermediate states, including the hemifusion state, has been confirmed experimentally in case of rhabdovirus G-mediated fusion.

Since both *VSV* G and herpesvirus gB resemble class I and class II fusion proteins which adopt a hairpin conformation with both membrane anchors at the same end of the molecule, it is most likely that they follow very similar paths in membrane fusion as suggested for class I and class II fusion proteins.

Summary

Although accumulating structural evidence suggests that the structural motifs used by viral fusion proteins and the mode of their extensive refolding varies substantially, the final product, namely the generation of a hairpin-like structure with two membrane anchors at the same end of an elongated structure, is maintained in all known postfusion conformations of viral glycoproteins. Thus the overall membrane fusion process is predicted to be the same for class I, II, and III fusion proteins, although the kinetics of refolding and fusion might vary to a large extent due to the involvement of different structural motifs to solve the problem of close apposition of two membranes.

See also: Metaviruses.

Further Reading

Chernomordik LV and Kozlov MM (2005) Membrane hemifusion: Crossing a chasm in two leaps. *Cell* 123(3): 375–382.

Earp LJ, Delos SE, Park HE, and White JM (2005) The many mechanisms of viral membrane fusion. *Current Topics in Microbiology and Immunology* 285: 25–66.

Gallo SA, Finnegan CM, Viard M, et al. (2003) The HIV Env-mediated fusion reaction. *Biochimica Biophysica Acta* 1614: 36–50.

Harrison SC (2005) Mechanism of membrane fusion by viral envelope proteins. *Advances in Virus Research* 64: 231–261.

Kelian M and Rey FA (2006) Virus membrane-fusion proteins: More than one way to make a hairpin. *Nature Reviews Microbiology* 4(1): 67–76.

Lamb RA, Paterson RG, and Jardetzky TS (2006) Paramyxovirus membrane fusion: Lessons from the F and HN atomic structures. *Virology* 344(1): 30–37.

Roche S and Gaudin Y (2003) Pathway of virus-induced membrane fusion studied with liposomes. *Methods in Enzymology* 372: 392–407.

Skehel JJ and Wiley DC (2000) Receptor binding and membrane fusion in virus entry: The influenza hemagglutinin. *Annual Review of Biochemistry* 69: 531–569.

Metaviruses

H L Levin, National Institutes of Health, Bethesda, MD, USA

Published by Elsevier Ltd.

Glossary

Erranti Comes from the Latin *errans*, meaning to wander.
Integration The insertion of cDNA into the genome of a host cell.
Long terminal repeats (LTRs) Sequence repeats on both ends of retroviruses and many retrotransposons that play a critical role in reverse transcription.
LTR-retrotransposon A form of transposable element that propagates by the reverse transcription of an RNA intermediate and the subsequent integration of the cDNA.
LTR-retroelements Include any genetic element with LTRs. They are retroviruses and LTR-retrotransposons regardless of whether they possess env or env-like proteins.
Meta Comes from the Greek metathesis for 'transposition'. Also to indicate uncertainty regarding whether these are true viruses.
Semoti Comes from the Latin *semotus*, meaning 'distant' or 'removed'. This refers to the large evolutionary distance between semotiviruses and members of the other two genera of the family *Metaviridae*.

Introduction

The family *Metaviridae* includes a vast number of genetic elements that populate the genomes of eukaryotes. They possess two long terminal repeats (LTRs) that flank coding sequences for the capsid protein Gag, and the enzymes protease (PR), reverse transcriptase (RT), and integrase (IN) (**Figure 1**). The assignment of elements to the *Metaviridae* versus other families with the same structure is based on phylogenetic relationships. Seven regions of RT sequence with strong homology are aligned and elements with similar sequence patterns are grouped into families such as the *Metaviridae*. In addition, elements belonging to the *Metaviridae* are distinguished from those in the family *Pseudoviridae* by the order of the coding sequences for RT and IN. While any element belonging to the *Metaviridae* contains RT sequence upstream of IN (**Figure 1**), members of the family *Pseudoviridae* encode IN before RT. Before the nomenclature for viruses and retrotransposons was standardized by the ICTV, the *Metaviridae* was named after two of its founding elements, gypsy/Ty3.

It is of great interest that many members of the *Metaviridae* possess envelope (env) proteins with similar structure to retrovirus env's, known for their role in particle release and infection (**Figure 1**). This presents the possibility that many elements first thought to be retrotransposons may actually be infectious viruses. These potential

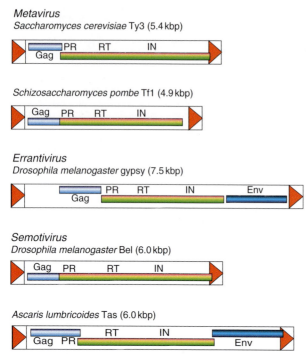

Figure 1 Genome structure of representative members of the family *Metaviridae*. The genomes in their integrated forms are shown with LTRs (red triangles) and the coding sequences for their proteins (colored rectangles). Overlap between two coding sequences indicates the position of a ribosomal frameshift.

the synthesis of a full-length cDNA that is then inserted by IN into the host genome. This life cycle as described for members of the metavirus family is general and applies to the propagation of all LTR-containing transposons and viruses.

The early discovery of LTR-retrotransposons and their env-containing derivatives resulted from genetic studies of model organisms such as *Saccharomyces cerevisiae* and *Drosophila melanogaster*. Gypsy, Tom, 17.6, 297, Idefix, Tirant, and Zam were discovered in *D. melanogaster* and form the basis of the errantivirus lineage (**Table 1**). The retrotransposons 412, Blastopia, Mdg1, Mdg3, Micropia, and 412 were also identified in *D. melanogaster*, and these, based on the phylogeny of their RT sequences, are classified as metaviruses. The study of yeast sequences led to the discovery of other metaviruses such as Ty3 of *Sa. cerevisiae* and Tf1 of *Schizosaccharomyces pombe*. The original members of the genus *Semotivirus* were Pao of *Bombyx mori* and TAS of *Ascaris lumbricoides*. Although these contain mutations and are poorly described, other semotiviruses have been identified including BEL, an abundant and active element of *D. melanogaster*.

The completion of genome sequence projects has contributed considerably to the discovery of transposons that belong to the family *Metaviridae*. For example, the completed sequence of the *D. melanogaster* genome revealed the presence of at least 20 clades of retroelements within the family *Metaviridae*; 15 are clearly errantiviruses. Similarly, analyses of the genome sequences of *Arabidopsis thaliana* and *Caenorhabditis elegans* revealed nine and six metavirus lineages, respectively. For the most part, the increased sequence data show that the members of any given clade propagate in a limited range of host species. For example, the nine clades of metaviruses in *A. thaliana* are only observed in other plant species. However, exceptions to this pattern have recently been observed. Although the clade of metaviruses that contains Tf1 was only found in fungi, Sushi and Sushi II were discovered in the puffer fish *Fugu rubripes*. Another example of a metavirus clade that exists in distant host species is that of Osvaldo, an element of *Drosophila buzzatii*. The genome of the Atlantic cod *Gadus morhua* contains a metavirus that clearly falls in the same lineage as Osvaldo. These examples of clades that bridge distantly related host species indicate that our current collection of sequenced genomes is still insufficient to document the breadth of hosts that possess transposon lineages.

viruses might have been classified as members of the family *Retroviridae* but their env proteins are unrelated to those of *Retroviridae* and their RT sequences firmly place them within the *Metaviridae*. Many of the env-containing elements belong to a single clade that constitutes the genus *Errantivirus*, within the family *Metaviridae*. The two other genera in the *Metaviridae*, the genera *Metavirus* and *Semotivirus* (**Figure 2**), were defined based on the separate phylogenies of their RT sequences. The semotiviruses are notable for their extensive divergence from the other genera of *Metaviridae*. LTR-retrotransposons and their env-containing relatives are collectively referred to as LTR-retroelements.

The life cycle of members of the family *Metaviridae* relies on the reverse transcription of their mRNA followed by the integration of this cDNA into the genome of the host cell (**Figure 3**). Integrated copies of the transposons are transcribed by RNA pol II and the mRNA is translated into the Gag, PR, RT, and IN proteins. Gag assembles into particles that encapsidate the RT, IN, and the transposon mRNA. The particles produced by errantiviruses, and some of the metaviruses and semotiviruses, are unique in their ability to interact with an env-like protein and bud from the plasma membrane. Once these particles have escaped they are similar to retroviruses in their ability to infect other cells. All the members of *Metaviridae* form a complex between RT and the mRNA that results in

Expression

The survival and propagation of retrotransposons depends on a carefully balanced relationship with their host. High levels of transposition would damage the host genome and jeopardize the existence of the element. In general, the

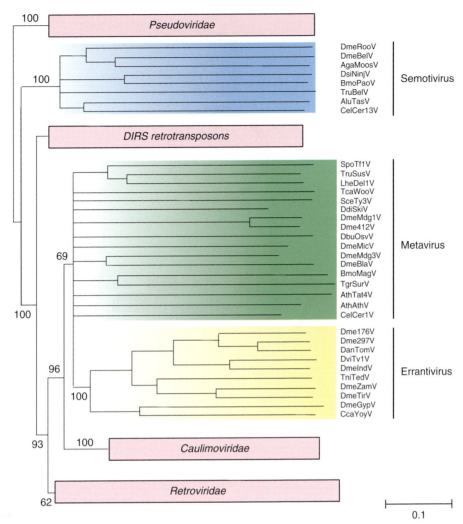

Figure 2 Phylogeny of the family *Metaviridae* and related groups based on their reverse transcriptase domain. The portion of the reverse transcriptase domain used in this analysis spans approximately 250 aa and includes the most conserved residues found in all retroelements. The phylogram is a 50% consensus tree of the transposons based on neighbor-joining distance algorithms and was rooted using sequences of members of the family *Pseudoviridae*. Bootstrap values (percentage of the time all elements are located on that branch) are shown for the major branches only. Transposons that are included in the various divisions of *Metaviridae* are indicated by the vertical lines to the right of each systematic name. Each group of elements that are not part of the family *Metaviridae* is represented by a box with the length of the box related to the sequence diversity within that group. DIRS retrotransposons are mobile elements that utilize a reverse transcriptase closely related to that of the family *Metaviridae* but lack many structural features of this group and integrate by a different mechanism. Scale bar at bottom represents divergence per site. Reproduced from Eickbush T, Boeke JD, Sandmeyer SB, and Voytas DF (2005) *Metaviridae*. In: Fauquet CM, Mayo MA, Maniloff J, Desselberger U, and Ball LA (eds.) *Virus Taxonomy: Eighth Report of the International Committee on Taxonomy of Viruses*, pp. 409–420. San Diego, CA: Elsevier Academic Press, with permission from Elsevier.

LTR-retrotransposons of yeast and *D. melanogaster* are dormant and exhibit very low levels of transposition. These low frequencies of spontaneous insertions suggest that the transcription of LTR-retrotransposons is heavily controlled. For example, the expression of most LTR-retrotransposons is induced by stress or specific stages of development. The gypsy element in wild-type stocks has very low transposition activity. However, stocks with mutant versions of the transcription factor *flamenco* possess substantially higher levels of transposition. The high rates of integration are only observed in female progeny of homozygous *flamenco* females. The mutations in *flamenco* cause 10- to 20-fold increases in full-length gypsy transcripts and this RNA occurs only in the follicle cells surrounding oocytes at stages 8–10 of oogenesis. A hypothesis that accounts for how increased transcription in follicle cells results in germ-line transposition in the progeny is that the gypsy env allows particles assembled in the follicle cells to infect the oocyte. Thus, transposition rates of gypsy in the progeny appear to be regulated at the level of transcription in the follicle cells of the mother. This program of regulated expression benefits gypsy by promoting transmission into the germ line

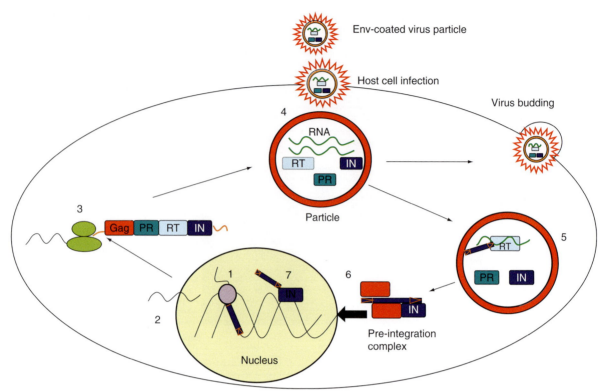

Figure 3 Life cycle of retrotransposons. The retrotransposon life cycle is contained entirely within the host cell. A seven-step life cycle is shown. First, cellular RNA polymerase transcribes an endogenous DNA copy of the transposon (blue) (1), producing an RNA copy (green). Next, that RNA transcript is exported from the nucleus (2) and then translated by host-cell ribosomes (green ovals) (3). The proteins of the retrotransposon then assemble into a virus-like particle (4). Reverse transcription takes place within this particle (5), producing a new DNA copy of the transposon. The pre-integration complex then enters the nucleus (6). Finally, the integrase protein mediates the integration of the transposon into a new location in the genome (7). The members of the family *Metaviridae* that encode an env-like protein have a life cycle that is quite similar to that of retroviruses. They have the additional ability to enter and leave the host cell. The ruffled circles indicate the particles with env-like proteins. Reproduced from **figure 1** of Kelly FD and Levin HL (2005) The evolution of transposons in *Schizosaccharomyces pombe*. *Cytogenetic and Genome Research*. 110: 566–574, with permission from S. Karger AG. Basel.

while minimizing the damage to the fly that would result from somatic events.

Perhaps the most extensively studied regulation of expression is that of Ty3. Its transcription is low in diploid cells but is dramatically induced in haploids exposed to mating factor. In cells with the mating type *a*, Ty3 is induced by α-factor treatment while cells with the mating type α are induced by a-factor. This developmental control is mediated by regulatory sequences within the LTR. Deletion analyses of the Ty3 LTR in a plasmid system identified a 101 nt segment that restricts transcription in diploids and is required for the mating-type regulation. Additional studies with a *lacZ* reporter fused to the *CYC1* promoter found that nucleotides 56–97 of the Ty3 LTR conferred a tenfold increase in expression in *a* cells treated with α-factor and a 100-fold decrease in expression in diploid cells. Inspection of the sequence between nucleotides 56 and 97 identified two matches to the binding site for Ste12p, a transcription factor that binds to pheromone response elements. This highly regulated expression of Ty3 is likely a mechanism that restricts transposition activity to cells in the process of mating. The benefit of this strategy for the transposon is not immediately clear.

Microarray studies of *Sc. pombe* revealed that Tf2, a species of the genus *Metavirus* closely related to Tf1, is a typical example of a transposon that is induced by conditions of stress. Cells challenged with hydrogen peroxide to induce oxidative stress produced as much as 20-fold more Tf2 RNA. This increase requires Atf1p, the transcription factor responsible for inducing a large number of environmental stress response genes. Other experiments showed that cells exposed to elevated temperatures (39 °C) induced Tf2 RNA by tenfold. The increase in Tf2 RNA that results from environmental stress is typical of many transposons. Although the function of this response to stress is not clear, an increase in transposition may help the elements to survive conditions that damage DNA.

Once the RNA of LTR-containing elements is produced, its translation presents particular challenges that must be overcome to allow the assembly of virus-like particles (VLPs). For all classes of LTR-containing elements, the coding sequences for Gag, PR, RT, and IN are

Table 1 Virus names, their abbreviations, the accession numbers (in square brackets), and discerning features (sor some cases)

Metaviruses			
Arabidopsis thaliana Athila virus	AthAthV	[AC007209]	Has env-like protein
Arabidopsis thaliana Tat4	AthTat4V	[AB005247]	Has env-like protein
Bombyx mori Mag virus	BmoMagV	[X17219]	
Caenorhabditis elegans Cer1 virus	CelCer1V	[U15406]	Gag, PR, RT, and IN are in single ORF
Cladosporium fulvum T-1 virus	CfuT1V	[Z11866]	Has self-priming structure
Dictyostelium discoideum Skipper virus	DdiSkiV	[AF049230]	PR is encoded in a separate ORF
Drosophila buzzatii Osvaldo virus	DbuOsvV	[AJ133521]	Has env-like protein
Drosophila melanogaster Blastopia virus	DmeBlaV	[Z27119]	
Drosophila melanogaster Mdg1 virus	DmeMdg1V	[X59545]	
Drosophila melanogaster Mdg3 virus	DmeMdg3V	[X95908]	
Drosophila melanogaster Micropia virus	DmeMicV	[X14037]	
Drosophila melanogaster412 virus	Dme412V	[X04132]	
Drosophila virilis Ulysses virus	DviUlyV	[X56645]	
Fusarium oxysporum Skippy virus	FoxSkiV	[L34658]	Has self-priming structure
Lilium henryi Del1 virus	LheDel1V	[X13886]	
Saccharomyces cerevisiae Ty3 virus	SceTy3V	[M34549]	Incerts 1 to 4 nt upstream of pol III transcribed genes
Schizosaccharomyces pombe Tf1 virus	SpoTf1V	[M38526]	Has self-primed reverse transcription
Schizosaccharomyces pombe Tf2 virus	SpoTf2V	[L10324]	
Takifugu rubripes Sushi virus	TruSusV	[AF030881]	Has self-priming structure
Tribolium castaneum Woot virus	TcaWooV	[U09586]	
Tripneustis gratilla SURL virus	TgrSurV	[M75723]	
Errantiviruses			
Ceratitis capitata Yoyo virus	CcaYoyV	[U60529]	Has env-like protein
Drosophila ananassae Tom virus	DanTomV	[Z24451]	Has env-like protein
Drosophila melanogaster virus	DmeZamV	[AJ000387]	Has env-like protein
Drosophila melanogaster Gypsy virus	DmeGypV	[M12927]	Is infectious and has env-like protein
Drosophila melanogaster Idefix virus	DmeIdeV	[AJ009736]	Has env-like protein
Drosophila melanogaster Tirant virus	DmeTirV	[X93507]	Has env-like protein
Drosophila melanogaster 17.6 virus	Dme176V	[X01472]	Has env-like protein
Drosophila melanogaster 297 virus	Dme297V	[X03431]	Has env-like protein
Drosophila virilis Tv1 virus	DviTv1V	[AF056940]	Has env-like protein
Trichoplusia ni TED virus	TniTedV	[M32662]	Has env-like protein
Semotiviruses			
Anopheles gambiae Moose virus	AgaMooV	[AF060859]	
Ascaris lumbricoides Tas virus	AluTasV	[Z29712]	Env-like protein
Bombyx mori Pao virus	BmoPaoV	[Z79443]	
Caenorhabditis elegans Cer13 virus	CelCer13V	[Z81510]	Has env-like protein
Drosophila melanogaster Bel virus	DmeBelV	[U23420]	
Drosophila melanogaster Roo virus	DmeRooV	[AY180917]	
Drosophila simulans Ninja virus	DsiNinV	[D83207]	
Fugu rubripes Suzu virus	FruSuzV	[AF537216]	

contained within a single transcript. The proteins are first expressed as polyproteins that are then cleaved by PR to produce the mature polypeptides. However, the Gag protein functions as the particle capsid and must be present in substantial excess to allow functional particles to assemble. Typically, the high levels of Gag relative to RT and IN result from programmed frameshifts positioned after Gag but before the coding sequences for RT and IN. The frameshifts are relatively inefficient and allow translation of RT and IN at levels 10- to 20-fold lower than Gag. The expression of the Ty3 proteins is a good example of this regulation. The coding sequence for Gag is separated from the ORF for PR, RT, and IN by a +1 frameshift. Fusions of these coding sequences to *lacZ* show that 20 times more Gag is translated than RT and IN. The +1 frameshift is also observed in the translation of retroviruses such as human immunodeficiency virus 1 (HIV-1) and Rous sarcoma virus (RSV). However, detailed studies of the translation mechanisms reveal fundamental differences in the frameshift of Ty3. The programmed frameshift of HIV-1 and RSV depends on a so-called slippery site where the ribonucleotides in a codon are repeated allowing the charged tRNA to rebind in the +1 frame. The sequence that causes frameshifting in Ty3 is GCG AGU U. The first codon is GCG (Ala), a sequence that is not repeated so the tRNA is not predicted to rebind in the +1 frame. Instead evidence indicates that the AGU is decoded by a low-abundance tRNA and this allows a near-cognate tRNA$^{Ala}_{IGC}$ to decode the GCG. The resulting I:G pairing in the wobble position of the peptidyl-tRNA coupled with the ribosomal pausing caused by low availability of the tRNA decoding AGU, mispositions the incoming tRNA^{Val} in the +1 frame.

Although most elements in the family *Metaviridae* have +1 or −1 frameshifts that regulate the ratios of Gag versus RT and IN, the semotiviruses BEL1 and Cer13 and the metavirus Tf1 are examples that encode Gag, PR, RT, and IN all within a single ORF. Extensive studies of the Tf1 proteins in *Sc. pombe* found that they are all expressed within a single primary translation product. Cells in log phase cultures contain equal amounts of Gag and IN. But when cells become stationary in saturated cultures, most of the IN is degraded. This regulated degradation of IN provides the excess levels of Gag that are typically required for assembly of particles. Once the excess IN is degraded, VLPs form that contain 25 times more Gag than IN.

All three genera of the family *Metaviridae* contain elements that have the additional challenge of expressing an env protein. The source of the env proteins appears to be spliced copies of the retroelement RNA that lack the sequences for Gag, PR, RT, and IN. In the case of gypsy, Zam, and Tom, experimental data show correlations between the expression of spliced RNAs and transposition.

Reverse Transcription

After VLPs mature, the encapsidated RNA is reverse transcribed into double-stranded DNA copies of the retroelement. This process of reverse transcription is a complex series of reactions that rely on two catalytic activities of RT, the synthesis of DNA and the degradation of RNA annealed to DNA known as RNase H activity. Despite the great differences between LTR-retroelements of the *Retroviridae*, *Pseudoviridae*, and *Metaviridae*, their mechanisms of reverse transcription are extremely similar (**Figure 4**). A full-length RNA begins and ends with sequence from the R portion of the LTRs. It is called R because this sequence is repeated on each end of the transcript. Upstream of R is U3 (unique on the 3′ end), a segment of the LTR that is only found on the 3′ end of the transcript. U5 (unique on the 5′ end) is the segment of the LTR present only on the 5′ end of the transcript. Reverse transcription is primed by specific tRNA species that are also packaged in the particles. The tRNA anneals to the retroelement RNA just after the upstream LTR at a position called the primer-binding site (PBS). The tRNA primes the reverse transcription of a short minus-strand species of DNA referred to as the minus-strand-strong-stop (−SSS). As this cDNA is being synthesized, the RNase H activity of RT degrades the RNA after it is reverse transcribed. Degradation of the 5′ end of the RNA releases the −SSS from the 5′ end of the RNA and allows it to anneal to the R sequence at the 3′ end of the RNA. This positions the −SSS so it can be extended by RT to generate the minus-strand of the LTR-retroelement. During the synthesis of the minus-strand cDNA, the remainder of the RNA is degraded by RNase H. The exception to this is RT does not degrade a polypurine tract (PPT) that functions as the plus-strand primer for reverse transcription. A short segment of plus-strand cDNA is produced and RNase H removes the RNA sequence from the 5′ end of the minus-strand DNA. Circularization is then mediated by the annealing of PBS sequences. The formation of a circle allows RT to extend the plus-strand intermediate into a full-length DNA. The final section of the minus strand is also generated.

Priming of Reverse Transcription

The features of reverse transcription as described above have been observed experimentally during the replication of elements in the *Metaviridae*. Based on sequence complementarity, the primer for gypsy is thought to be tRNA^{Lys} while the majority of the other errantiviruses are predicted to use tRNA^{Ser}. The metavirus Ty3 is thought to be primed by the initiator tRNA^{Met} because there are 8 nt after the upstream LTR which are complementary to 8 nt at the 3′ end of the tRNA. This prediction was demonstrated experimentally in strains of *Sa. cerevisiae* lacking the four genomic genes for the initiator tRNA^{Met}. The strain contains a plasmid copy of the initiator tRNA^{Met} that supports the initiation of translation. Mutations in the plasmid-encoded tRNA^{Met} that specifically alter the complementarity with Ty3 restricted reverse transcription but not protein translation.

Another mechanism of priming that has been studied extensively is that of the metavirus Tf1. In this case, there is a surprising difference in the nature of the primer. No tRNA candidate with complementarity to the Tf1 mRNA is known. Instead, the first 11 nt of the Tf1 mRNA itself has perfect complementarity to its PBS. This observation led to the discovery of an unusual self-priming mechanism in which the first 11 nt of the Tf1 mRNA function as the primer of the minus-strand strong stop cDNA (**Figure 5(a)**). Evidence for this mechanism includes the result that single nucleotide mutations in the PBS and in the 11 nt at the 5′ end of the mRNA cause substantial reductions in Tf1 transposition. Importantly, the combination of mutations in the PBS and the self-primer that reestablish complementarity returns transposition frequencies to near wild-type levels. Further evidence that the first 11 nt prime reverse transcription is the Tf1 mRNA in VLPs is cleaved between nucleotides 11 and 12. The cleavage depends on the complementarity with the PBS and the cleavage is required for production of −SSS. Another surprising feature of the self-priming mechanism is that the cleavage requires the RNase H activity of RT. One initial question that the self-priming mechanism poses is how can the 11 nt sequence removed from the Tf1 mRNA be restored during reverse transcription. The answer is that the sequence of

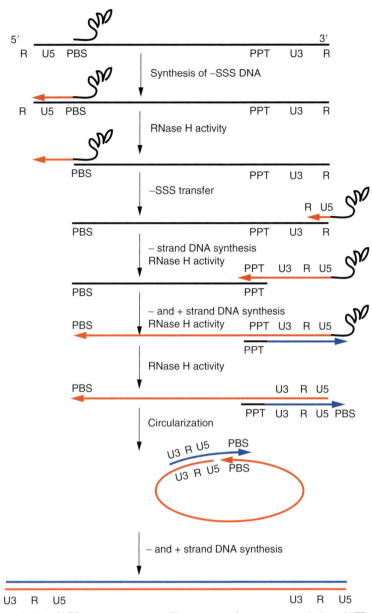

Figure 4 The reverse transcription of LTR-retrotransposons. The process of reverse transcription of LTR-retrotransposons is the same as for retroviruses. RNA is shown in black. The plus-strand and minus-strand cDNAs are shown in blue and red, respectively. The first intermediate of cDNA produced is the minus-strand strong-stop (−SSS). RNase H degrades the R and U5 regions from the 5′ end of the transcript and the −SSS transfers to the 3′ end of the RNA. The bulk of the minus-strand cDNA is then produced while RNase H degrades the rest of the transcript except for the plus-strand primer (PPT). The plus-strand strong-stop is produced and RNase H removes the tRNA primer from the 5′ end of the minus-strand cDNA. Once circularized, the plus-strand and minus-strand cDNAs are extended to generate their full-length forms.

the first 11 nt is present on both ends of the Tf1 mRNA and during reverse transcription of the downstream LTR, the 11 nt are restored (**Figure 4**). Although the self-priming mechanism appears quite different from the priming of other LTR-retroelements, experiments showed the cleavage of the Tf1 mRNA depends on its formation of a complex structure surrounding the PBS that is quite similar to the priming structure found in the RNA of RSV (**Figures 5(b)** and **5(c)**). The characterization of the complex structure in the Tf1 mRNA made it possible to identify many other LTR-retrotransposons that likely use self-priming to initiate reverse transcription.

Integration

The INs of retroviruses catalyze two reactions critical for integration, 3′ processing and strand transfer. The processing step removes two or three nucleotides from the

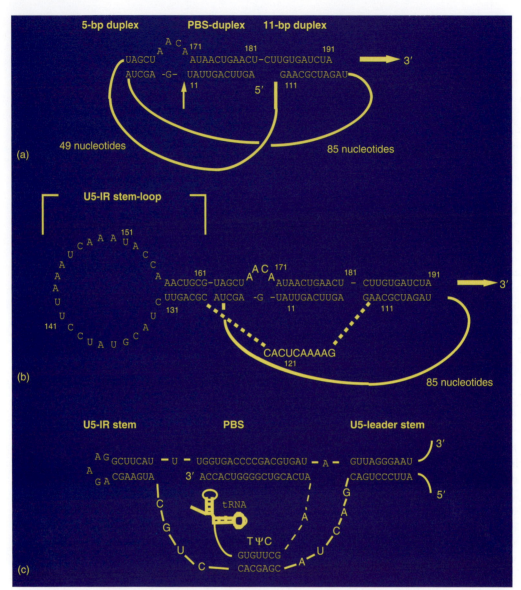

Figure 5 The self-priming structure of the Tf1 mRNA. (a) The self-priming mechanism of Tf1 requires an RNA structure consisting of three duplexes. The regions of duplex structure are labeled 5-bp duplex, PBS-duplex, and 11-bp duplex. The 5′ end of the RNA is labeled and the yellow lines represent extended stretches of sequence. The arrow labeled 3′ indicates that the majority of the transcript is 3′ of the sequence shown. The numbers shown in small type indicate the nucleotide position relative to the first base of the transcript. (b) Additional base pairing exists between nucleotides (nt 124–130 and nt 162–156) within a 49 nt loop. The resulting structure is a U5-IR stem–loop with a 7 bp stem and a 25 nt loop. The dashed line indicates that the nucleotides shown connect directly to the indicated regions of the diagram. (c) The U5-IR stem–loop of RSV. The three regions of duplex structure in the transcript of RSV are U5-IR, PBS, and U5-leader stem. These structures show considerable similarity to the self-priming RNA of Tf1. The tRNA primer is shown in diagrammatic form and all the lines between nucleotides indicate a direct connection to adjacent nucleotides. The 3′ and 5′ labels indicate the polarity of the RNA and the position of the transcript sequences that are not shown. Reproduced from Lin JH and Levin HL (1998) Reverse transcription of a self-primed retrotransposon requires an RNA structure similar to the U5-IR stem-loop of retroviruses. *Molecular and Cell Biology* 18: 6859-6869, with permission from American Society for Microbiology.

3′ ends of the cDNA in order to position the highly conserved CA dinucleotide at the 3′ end of the cDNA. The processing step allows the 3′ hydroxyls of the terminal 'A' to act as nucleophiles in a pair of strand-transfer reactions that cleave the phosphodiester bonds in the target DNA and make a covalent bond between the 3′ ends of the viral DNA and the 5′ ends of the target DNA. In retroviruses, and some LTR-retrotransposons, the tRNA primers initiate reverse transcription two or three nucleotides from the LTR and this results in the additional deoxynucleotides at the cDNA termini that IN removes during the processing reaction. This is the case for some metaviruses such as Ty3. Direct support for 3′ processing is that cDNA isolated from Ty3 particles terminates with the conserved

CA dinucleotide. However, mutations that inactivate IN cause cDNA to accumulate that retains two additional nucleotides 3' of the CA.

In the case of other LTR-retrotransposons including Tf1, the minus-strand primers initiate reverse transcription with no extra nucleotides added before the LTR. These LTR-retrotransposons were thought to proceed to the strand-transfer reaction without a processing step. This assumption was tested for Tf1 and was found to be incorrect. Despite the position of the self-primer adjacent to the upstream LTR, cDNA in Tf1 particles has nucleotides 3' of the CA dinucleotide. The sequences of these 1–4 nt is untemplated and thought to be added by the terminal transferase activity of RT. Because these extra nucleotides must be removed before strand transfer, Tf1 IN was tested for 3' processing activity. Experiments with oligonucleotide substrates show that Tf1 IN as a recombinant protein does possess 3' processing activity that specifically removes the nucleotides 3' of the CA. The finding that Tf1 IN has processing activity brings into question the predicted lack of processing activity in the INs of the other LTR-retrotransposons.

Target Site Specificity

The integration of cDNA can disrupt gene function and has the potential to reduce the viability of host cells. Since the survival of transposons depends entirely on the well being of the host, mechanisms have evolved that specifically target integration to noncoding portions of the host genome. The integration of Ty3 is one of the best-studied examples of this. Ty3 expressed from the Gal promoter generates insertions 1–4 nt upstream of tRNA, 5S, and U6 genes, all of which are expressed by pol III RNA polymerase. The promoters of tRNA, U6, and 5S genes bind transcription factors TFIIIC downstream and TFIIIB upstream of the position that transcription initiates. Both of these factors do play a role in targeting Ty3 integration. In *in vitro* assays that measure integration of Ty3 into a plasmid copy of a tRNA gene, fractions containing TFIIIB and TFIIIC are required. However, the role of TFIIIC in transcription is to load TFIIIB onto the DNA of the tRNA gene. This suggests the function of TFIIIC in integration could be indirect. In *in vitro* reactions with purified factors, the U6 gene binds TFIIIB without the requirement of TFIIIC. Thus, the role of TFIIIC in Ty3 integration can be tested independently from its function in loading TFIIIB. The results of integration assays using a plasmid copy of the U6 gene as a target show that TFIIIB, and not TFIIIC, is necessary for insertion. TFIIIB is composed of three subunits, TBP, Brf, and B". Further analysis reveals that TBP and Brf are the minimum number of subunits that support Ty3 integration. When the *in vitro* integration experiments are performed using truncated segments of Brf, just the N-terminal 282 amino acids are necessary.

Taken together, these results implicate Brf as a key transcription factor responsible for positioning Ty3 insertions.

Tf1 is another metavirus that has a strong preference for a specific type of target site. The genome of *Sc. pombe* contains 186 inserts of Tf1 or transposons closely related to Tf1. Ninety-six percent of the inserts occur in intergenic sequences that contain a pol II promoter. It is significant that the inserts cluster within 400 nt of the 5' end of ORFs. This accumulation of insertions near the 5' end of genes could be the result of a mechanism that targets integration to these locations or, conversely, they could represent an evolutionary equilibrium where insertions into the ORFs or near the 3' end occur but are lost due to selective pressure. However, when a *neo* marked version of Tf1 is expressed from a plasmid, the insertions also cluster within 400 nt of the start of ORFs. Insertions do not occur within ORFs. These data confirm that the positioning of Tf1 near the 5' end of ORFs is due to an insertion mechanism. The preference for the 5' end of genes suggests that the integration mechanism may recognize components of the pol II transcription machinery.

The location of Ty3 and Tf1 in chromatin that is actively expressed is representative of transposons in compact genomes such as yeast. The transposons of complex eukaryotes commonly cluster in heterochromatin where the bulk of transcription is off. For example, the great majority of gypsy elements are located in the pericentromeric regions of the *D. melanogaster* genome. The location of gypsy elements is conserved in strains from different geographic origins. This suggests that a high number of gypsy elements are nonmobile and were already present in the pericentromeric sequences of the common ancestor of the *D. melanogaster* strains.

The accumulation of metaviruses in centromeric sequences is also observed in plants. Transposons belonging to the *Metaviridae* in *A. thaliana* form a diverse set of retroelements that fall into three distinct lineages, *Athila*, *Tat*, and *Metavirus*. Together the *Metaviridae* of *A. thaliana* constitute as much as 2.34% of the genome and the retroelements of each of the three lineages cluster in the pericentromeric heterochromatin. Interestingly, *Tat* and *Athila* exhibited significantly stronger associations with heterochromatin than metaviruses. In addition, members of the family *Pseudoviridae* make up 1.25% of the *A. thaliana* genome and these elements are only loosely associated with pericentromeric regions. The pronounced differences in distribution patterns of the retroelements in *A. thaliana* indicate that targeted integration plays an important role in positioning the elements.

Env-Mediated Infection

The vertebrate retroviruses make up a relatively young clade that has a close relationship to members of the family

Metaviridae. They are thought to have originated from a member of the *Metaviridae* that acquired a third ORF, the *env* gene. Env is a transmembrane protein that confers infectivity by mediating receptor recognition and membrane fusion. It is glycosylated and has an N-terminal signal peptidase cleavage site. It is surprising that among the members of the families *Pseudoviridae* and *Metaviridae*, six lineages have independently acquired *env*-like genes. In each of these cases the third ORFs have structural features typical of the vertebrate retroviruses. The *env*-like proteins are predicted to possess N-terminal leader peptides, N-glycosylation sites, and transmembrane domains. Each of the three genera of the *Metaviridae* contains *env*-like lineages. Two lineages in the metavirus family with *env*-like genes are the clades represented by Athila and Osvaldo. One lineage with *env*-like genes in the errantiviruses is represented by gypsy and two lineages in the semotiviruses are represented by Cer13 and TAS.

Efforts to identify the source of the retroviral *env* genes have been difficult because their sequences have no similarity to the *env*-like genes of other retroelements or to any other genes known. Fortunately, sequences similar to the *env*-like proteins of members of the family *Metaviridae* have been identified. More importantly, the sequence similarities suggest that the env-like proteins were acquired from unrelated viruses and these proteins in their original viruses already had env activities. The *env*-like genes of the errantiviruses have highly significant homology to genes from three insect baculoviruses, indicating that the *env* genes from errantiviruses and the baculoviral ORFs share common ancestry. Although the function of the baculoviral proteins is not known, their structural features indicate they are analogous to the env proteins of vertebrate retroviruses.

The env-like protein of the semotiviruses also has homology to proteins of unrelated viruses. The third ORF of Cer13 has significant homology to a group of proteins (G2) from phleboviruses, a class of single-stranded RNA viruses. The G2 proteins are thought to have env-like function because they are glycoproteins with a predicted transmembrane domain. It is interesting that despite the similarity between the semotivirus Cer13 and Tas, their env-like proteins are unrelated. It appears that the env-like protein of TAS resulted from a different acquisition. The Tas protein has a modest similarity to gB glycoproteins from viruses in the family *Herpesviridae*, a class of double-stranded DNA viruses. The glycoprotein gB has been implicated in the viral attachment and fusion of herpesviruses. It is significant that the segment of gB glycoproteins thought to be responsible for viral attachment to the cell surface is exactly the segment that has similarity to Tas.

Despite the many examples of env-like proteins encoded by LTR-retroelements, only the VLPs of gypsy are known to be infectious. Stocks of *D. melanogaster* that produce high levels of gypsy transposition can be used to make preparations of VLPs. Extracts of pupae from these stocks can be fed to strains that lack active gypsy elements. The result is the horizontal transmission of active gypsy. In addition, extracts from the stocks with active gypsy can be used to fractionate VLPs on sucrose gradients. The fractions contain a clear peak with high levels of RT activity and immunoelectron micrographs show these fractions contain gypsy VLPs. When these fractions are fed to larvae that lacked active gypsy, 1.5% of the flies that emerged contained new insertions of gypsy. These results show that gypsy has the properties expected for an infectious virus. One surprising result is that mutations that inactivate env in an expressed copy of gypsy do not reduce transposition activity. One explanation is that the infectious behavior of the particles is due to env-like protein expressed from a related transposon in the genome of *D. melanogaster*.

Conclusions

LTR-retroelements have been highly successful throughout the evolution of eukaryotes. Analyses of the available genome sequences indicate that members of the family *Metaviridae* are the most abundant class of LTR-retroelements. One significant distinction of the family *Metaviridae* is its close relationship to the disease-causing retroviruses such as HIV-1. As a result, various members of the *Metaviridae* are actively being studied as models for the vertebrate retroviruses. Perhaps the most significant result from the study of elements belonging to the *Metaviridae* is the discovery that structures responsible for specific mechanisms are modular and can be readily substituted within relatively short periods of evolution. For example, LTR-retroelements that use the self-priming mechanism exist in the same genus as elements that use tRNA priming. Just a limited modification of transposon sequence rendered them able to self-prime. The *env*-like genes are an example of how modules that promote infection can be acquired multiple times during independent events. Whether self-priming PBSs and *env*-like genes are the only forms of modular evolution for LTR-retroelements is unlikely. As increasing numbers of genomes are sequenced from a broad swath of organisms, we are likely to learn much more about modular evolution and the diversity of LTR-retroelements.

See also: Pseudoviruses; Retroviruses: General Features.

Further Reading

Bowen NJ, Jordan I, Epstein J, Wood V, and Levin HL (2003) Retrotransposons and their recognition of pol II promoters: A comprehensive survey of the transposable elements derived from the complete genome sequence of *Schizosaccharomyces pombe*. *Genome Research* 13: 1984–1997.

Eickbush T, Boeke JD, Sandmeyer SB, and Voytas DF (2005) Metaviridae. In: Fauquet CM, Mayo MA, Maniloff J, Desselberger U, and Ball LA (eds.) *Virus Taxonomy: Eighth Report of International Committee on Taxonomy of Viruses*, pp. 409–420. San Diego, CA: Elsevier Academic Press.

Eickbush TH and Malik HS (2002) Newly identified retrotransposons of the gypsy/Ty3 class in fungi, plants, and vertebrates. In: Craig NL, Craigie R, Gellert M, and Lambowitz AL (eds.) *Mobile DNA II*, pp. 1111–1144. Washington, DC: American Society of Microbiology.

Kelly FD and Levin HL (2005) The evolution of transporons in *Schizosaccharomyces pombe*. *Cytogenetic and Genome Research*. 110: 566–574.

Kirchner J, Connolly CM, and Sandmeyer SB (1995) Requirement of RNA polymerase III transcription factors for *in vitro* position-specific integration of a retroviruslike element. *Science* 267: 1488–1491.

Le Grice SFJ (2003) 'In the beginning': Initiation of minus strand DNA synthesis in retroviruses and LTR-containing retrotransposons. *Biochemistry* 42: 14349–14355.

Levin H (2002) Newly identified retrotransposons of the gypsy/Ty3 class in fungi, plants, and vertebrates. In: Craig NL, Craigie R, Gellert M, and Lambowitz AL (eds.) *Mobile DNA II*, pp. 684–704. Washington, DC: American Society of Microbiology.

Levin HL (1995) A novel mechanism of self-primed reverse transcription defines a new family of retroelements. *Molecular and Cell Biology* 15: 3310–3317.

Lin JH and Levin HL (1998) Reverse transcription of a self-primed retrotransposon requires an RNA structure similar to the U5-IR stem-loop of retroviruses. *Molecular and Cell Biology* 18: 6859–6869.

Malik HS, Henikoff S, and Eickbush TH (2000) Poised for contagion: Evolutionary origins of the infectious abilities of invertebrate retroviruses. *Genome Research* 10: 1307–1318.

Pelisson A, Mejlumian L, Robert V, Terzian C, and Bucheton A (2002) Drosophila germline invasion by the endogenous retrovirus gypsy: Involvement of the viral *env* gene. *Insect Biochemistry and Molecular Biology* 32: 1249–1256.

Sandmeyer S (2003) Integration by design. *Proceedings of the National Academy of Sciences, USA* 100: 5586–5588.

Sandmeyer SB, Aye M, and Menees T (2002) Newly identified retrotransposons of the gypsy/Ty3 class in fungi, plants, and vertebrates. In: Craig NL, Craigie R, Gellert M, and Lambowitz AL (eds.) *Mobile DNA II*, pp. 663–683. Washington, DC: American Society of Microbiology.

Song SU, Gerasimova T, Kurkulos M, Boeke JD, and Corces VG (1994) An env-like protein encoded by a *Drosophila* retroelement – Evidence that gypsy is an infectious retrovirus. *Genes and Development* 8: 2046–2057.

Mimivirus

J-M Claverie, Université de la Méditerranée, Marseille, France

© 2008 Elsevier Ltd. All rights reserved.

Glossary

16S rDNA The gene encoding the small ribosomal RNA molecule, a universal component of all cellular prokaryotic organisms, the sequence of which is used for identification and classification purposes.

Aminoacyl-tRNA synthetases The highly specific enzymes responsible for the loading of a given amino acid onto its cognate tRNA(s). These enzymes are at the center of the use of the genetic code, together with the specific tRNA/mRNA recognition mediated by the anticodon/codon pairing on the ribosome.

COG Cluster of orthologous groups. Families of evolutionarily conserved genes, usually associated with well-defined protein functions and 3-D structures.

NCLDV Nucleocytoplasmic large DNA viruses, a group of large dsDNA viruses, the replication of which involves a stage the host cell cytoplasm and the host cell nucleus. Although presently very diverse in size, host range, and morphology, these are suspected to have originated from a common ancestor, contemporary to the emergence of the eukaryote lineage.

Paralogs Copies of similar genes (originated by duplication) found in the same genome.

Proteomics New methodological approaches allowing the whole protein complement of an organism (or a virus) to be identified.

Introduction

The recent discovery in 2003 of acanthamoeba polyphaga mimivirus and the analysis of its complete genome sequence sent a shock wave through the community of virologists and evolutionists. Its particle size (750 nm), genome length (1.2 million bp), and large gene repertoire (911 protein coding genes) blur the established boundaries between viruses and parasitic cellular organisms. In addition, the analysis of its genome sequence identified many types of genes never before encountered in a virus, including aminoacyl-tRNA synthetases and other central components of the translation machinery previously thought to be the signature of cellular organisms. The information available on this giant double-stranded DNA (dsDNA) virus mostly consists of electron microscopy (EM) images, its genome sequence, and proteomic data. Very little is yet known about its pathogenicity.

History and Classification

In 1992, following a pneumonia outbreak in the West Yorkshire mill town of Bradford (England), a routine investigation for legionella (a pneumonia causing intracellular parasitic bacteria) within the amoeba colonizing the water of a cooling tower led Timothy Rowbotham from the Britain's Public Health Laboratory Service to discover a microorganism resembling a small Gram-positive coccus (initially called Bradford coccus). Cultivation attempts failed, and no amplification product was obtained with universal 16S rDNA bacterial primers at that time (**Figure 1**). The mysterious sample was stored in a freezer for about 10 years, until it reached the laboratory of Prof. Didier Raoult, at the school of medicine in Marseilles, France. There, EM of infected *Acanthamoeba polyphaga* cells provided the first hints that Bradford coccus was in fact a giant virus, with mature icosahedral particles *c*. 0.7 μm in diameter, a size comparable to that of mycoplasma cells (**Figure 2**). The viral nature of the agent was definitively established by the demonstration of an eclipse phase during its replication, and the analysis of several gene sequences exhibiting a clear phylogenetic affinity with nucleocytoplasmic large DNA viruses (NCLDVs), a group of viruses including members of the *Poxviridae*, the *Iridoviridae*, the *Phycodnaviridae*, and the *Asfarviridae*. This new virus was named mimivirus (for mimicking microbe) and is now classified by ICTV as the first member of the *Mimiviridae*, a new family within the NCLDV. The size of its particles makes mimivirus the largest virus ever described.

Host Range and Pathogenicity

Among a large number of primary or established cell lines from vertebrates or invertebrates that were tested for their ability to support mimivirus infection and replication, only cells from the species *Acanthamoeba polyphaga*, *A. castellanii*, and *A. mauritaniensis* could be productively infected by a cell-free viral suspension. This narrow range of target cell specificity, restricted to protozoans, apparently conflicts with other reports suggesting that mimivirus might be an amoebal pathogen-like legionella, causing pneumonia in human. Numerous seroconversions to mimivirus have been documented in patients with both community- and hospital-acquired pneumonia. In addition, mimivirus DNA was found in respiratory samples of a patient with hospital-acquired pneumonia and a laboratory infection of a technician by mimivirus has also been reported. The patient's serum sample was found to strongly react with many mimivirus proteins, most of them unique to the virus, making cross-reactions with other pathogens unlikely. Isolation of the mimivirus from an infected patient is yet to be accomplished in order to formally establish that it is indeed pathogenic to human. In the meantime, as a precautionary measure, mimivirus should be considered a potential pneumonia agent and manipulated as a class 2 pathogen. In this context, it is worth noticing that mimivirus particles have been shown to remain infectious for at least 1 year at temperature of 4–32 °C in a neutral buffer.

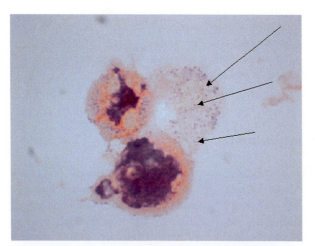

Figure 1 Coccus-like appearance of mimivirus particles (indicated by arrows) under the light microscope after Gram coloration in infected amoeba. Reproduced from La Scola B, Audic S, Robert C, *et al*. (2003) A giant virus in amoebae. *Science* 299: 2033, with permission from American Association for the Advancement of Science.

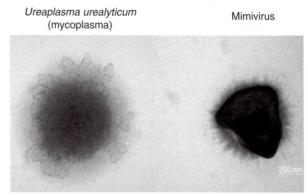

Figure 2 Comparable size of a mimivirus particle and a mycoplasma cell. EM picture of a co-culture. Reproduced from La Scola B, Audic S, Robert C, *et al*. (2003) A giant virus in amoebae. *Science* 299: 2033, with permission from American Association for the Advancement of Science.

Replication Cycle

Upon infection of *A. polyphaga* cells, mimivirus exhibits a typical viral replication cycle with an eclipse phase of 4 h post infection (p.i.), followed by the appearance of newly synthesized virions at 8 h p.i. in the cytoplasm, leading to the clustered accumulation of viral particles filling up most of the intracellular space, until infected amoebae are lysed at 24 h p.i. Little more is known about the details of the various stages of the replication cycle. Combining various hints from genomic and proteomic analysis as well as from EM suggests the following infection scenario (**Figure 3**):

1. free virus particles mimicking bacteria (by their micron size and perhaps the lipopolysaccharide (LPS)-like layer surrounding the capsid) are taken up as putative food by the amoeba;
2. the LPS-like layer is digested within the amoeba endocytic vacuole, making protein at the surface of the capsid accessible for a specific interaction at the vacuole membrane;
3. the content of the capsid is then injected into the amoeba cytoplasm, eventually using a specialized apparatus visible as a 600 Å 'vertex' (see below) leaving the empty capsid in the endocytic vacuole (documented by EM); and
4. early transcription events occur in the cytoplasm, as suggested by the proteomic content of the capsid.

It is not clear if, how, and when the virus genome gets into the amoeba nucleus (for its replication) or if the synthesis of particle components and capsid assembly only proceeds from 'virus factories' located in the cytoplasm.

Genome Organization

The mimivirus genome (**Figure 4**) consists of a single dsDNA, 1 181 404 bp long, within which 911 protein-coding genes are predicted. Two inverted repeats of about 900 nt are found near both extremities of the sequence, suggesting that the genome might adopt a circular topology as result of their annealing. The genome nucleotide composition is 72% A + T and exhibits some level of local strand asymmetry revealed by plotting the A + C strand excess as well as the gene excess (number of genes expressed from one strand minus the number of genes expressed from the

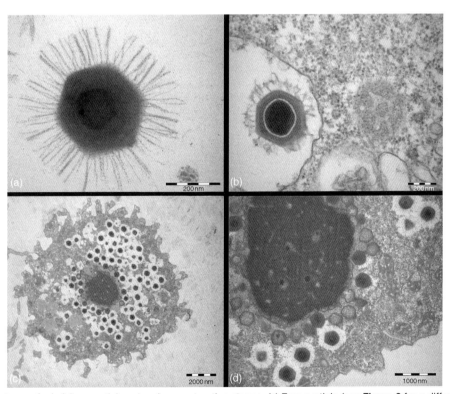

Figure 3 EM pictures of mimivirus particles at various maturation stages. (a) Free particle (see **Figure 6** for a different aspect of the fibers recovering the capsid). (b) Mimivirus particle in an amoeba vacuole, exhibiting a damaged (digested) fiber layer. (c) Clustered particles in the cytoplasm of an amoeba, 8 h p.i. (d) Close-up of a virus factory surrounded by particles at various stages of maturation: empty particles, filled particle (with a dark central core) without their translucent capsular material, mature particles (central dark core, white halo of cross-linked fibers). Reproduced from Suzan-Monti M, La Scola B, and Raoult D (2006) Genomic and evolutionary aspects of mimivirus. *Virus Research* 117: 145–155, with permission from Elsevier.

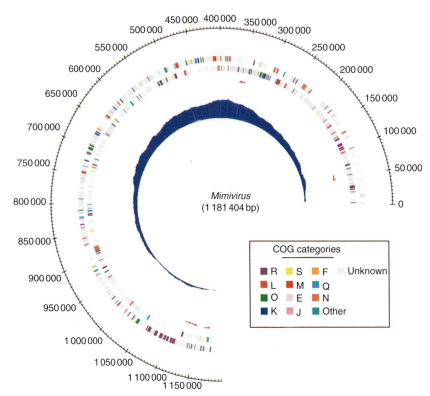

Figure 4 Map of the mimivirus chromosome. The predicted protein-coding sequences are shown on both strands and colored according to the function category of their matching COG. Genes with no COG match are shown in gray. Abbreviations for the COG functional categories are as follows: E, amino acid transport and metabolism; F, nucleotide transport and metabolism; J, translation; K, transcription; L, replication, recombination, and repair; M, cell wall/membrane biogenesis; N, cell motility; O, post-translational modification, protein turnover, and chaperones; Q, secondary metabolites biosynthesis, transport, and catabolism; R, general function prediction only; S, function unknown. Small red arrows indicate the location and orientation of tRNAs. The A + C excess profile is shown on the innermost circle, exhibiting a peak around position 380 000. The leading strand is defined as the one directed outward, from this point. Reproduced from Raoult D, Audic S, Robert C, et al. (2004) The 1.2-megabase genome sequence of mimivirus, *Science* 306: 1344–1350, with permission from American Association for the Advancement of Science.

other strand). Both graphs exhibit a slope reversal near position 380 000 of the genome, as found in bacterial genomes, and associated with the location of the origin of replication. Mimivirus genes are preferentially transcribed in the orientation going away from this putative origin of replication, thus defining a 'leading strand' (with 578 genes) and a 'lagging strand' (with 333 genes). Such a strong asymmetrical distribution of genes is unique among known nucleocytoplasmic large DNA viruses. Despite this local asymmetry, the total number of genes on both strands are very similar (450 'R' vs. 461 'L' genes). The overall amino acid composition of the predicted proteome is strongly biased in favor of residues encoded by codons rich in A + T such as isoleucine, asparagine, and tyrosine. The relative usage of synonymous codons for any given amino acid is also biased by the high A + T percentage. This has made the production of recombinant mimivirus proteins difficult in traditional *Escherichia coli* expression systems. Paradoxically, the preferred codon usage in mimivirus genes is almost the opposite to the one exhibited by its host *Acanthamoeba castellanii* or *A. polyphaga* (the most frequently used codons being the rarest in the amoebal genes). It is also quite different from the more even distribution observed in human and vertebrate genes.

Mimivirus genome contains multiple traces of gene duplications, including a few ancient en bloc duplication events, many dispersed individual gene duplications, and recurrent tandem duplication events leading to large families of co-localized paralogs. Overall, one-third of mimivirus genes have at least one paralog. The analysis of gene collinearity indicates that genome segments [1–110 000] and [120 000–200 000] have been duplicated at positions approximately symmetrical with respect to the chromosome center. The largest paralogous families include an ankyrin-repeat-containing protein (with 66 members), BTB/POZ-domain-containing proteins (26 members), and a 14-paralog family of proteins of unknown function, of which 12 are arranged in a perfect tandem repeat. In contrast to parasitic/endosymbiotic bacteria with genome of comparable size (such as *Rickettsia*), the mimivirus genome exhibits no pseudogene, and thus no sign of an ongoing genome reduction/degradation process. Mimivirus does not appear to be under any evolutionary pressure to decrease the size of its genome. The segmental duplication and massive individual gene duplication explain the origin of a large part of the mimivirus genome, without the need to invoke an exceptional propensity for horizontal gene transfers.

Promoter Structure

A unique feature of the mimivirus genome is the presence of the motif AAAATTGA in the 150 nt upstream region of nearly 50% of the protein-coding genes. From its strong preferential occurrence in the 5′ upstream region (mostly at positions ranging from −50 to −75 from the initiator ATG), it is assumed that this motif corresponds to a specific promoter signal. The presence of the AAAATTGA motif significantly correlates with genes transcribed from the leading strand ($313/578 = 54\%$ vs. $133/313 = 40\%$). It also correlates with predicted gene functions normally associated with the early and late/early phase of the virus replication (nucleotide metabolism, transcription, translation). As the AAAATTGA motif is not prevalent in the genome sequence of amoebal organisms, it is hypothesized that this motif is a core promoter element specifically recognized by the transcription pre-initiation complex encoded by the mimivirus genome (mainly the large and small RNA polymerase subunits and a remote homolog of the TFIID 'TATA box-binding' initiation factor). The reasons behind the lack of variability of this mimivirus promoter element, usually quite degenerate among eukaryotes and eukaryotic viruses, are unknown.

Gene Content

As is usual for viruses, only a third of the 911 protein-coding genes of mimivirus have been associated with functional attributes. About 200 of these genes exhibit significant matches to 108 distinct COG families. The genome also encodes six tRNAs (three $tRNA_{leu}$, one $tRNA_{trp}$, one $tRNA_{cys}$, and one $tRNA_{his}$).

Typical NCLDV Core Genes in Mimivirus

The comparative genomic study of NCLDV suggested the monophyletic origin of four viral families: the *Asfarviridae*, the *Iridoviridae*, the *Phycodnaviridae*, and the *Poxviridae*. These studies identified sets of core genes (shared across these viral families) with various levels of conservation: class I core genes are found in all species from all families, class II core genes are absent in some species from a given family, while class III ones are shared by three families.

A strong argument for classifying mimivirus as a *bona fide* member of the NCLDV is the presence of 9/9 class I core gene homologs in its genome, 6/8 of the class II core genes, and 11/14 of the class III. Remarkably, the two class II core genes missing from mimivirus genomes encode two enzymes central to DNA synthesis: thymidylate kinase (catalyzing the formation of dTDP from dTMP) and the quasi-universal detoxifying enzyme dUTPase (dUTP pyrophosphatase, transforming dUTP into dUMP, to avoid the misincorporation of uracil in DNA).

Phylogenetic analyses using class I core gene sequences let to mimivirus being assigned to a new family, the *Mimiviridae*, distinct from the four previously defined NCLDV families (**Figure 5**).

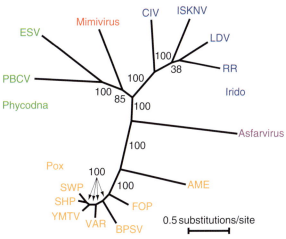

Figure 5 Phylogenetic position of mimivirus among established NCLDV families. Viral species representing the diverse families of NCLDV are included as follows: *Acanthamoeba polyphaga mimivirus*, Phycodnaviridae (*Paramecium bursaria chlorella virus* (PBCV) and *Ectocarpus siliculosus virus* (ESV)), Iridoviridae (*Chilo iridescent virus* (CIV), *Regina ranavirus* (RR), *Lymphocystis disease virus type 1* (LDV), and *Infectious spleen and kidney necrosis virus* (ISKNV)), Asfarviridae (*African swine fever virus*), and Poxviridae (*Amsacta moorei entomopoxvirus* (AME), *Variola virus* (VAR), *Fowlpox virus* (FOP), *Bovine papular stomatitis virus* (BPSV), *Yaba monkey tumor virus* (YMTV), *Sheeppox virus* (SHP), and *Swinepox virus* (SWP)). This tree was built with the use of maximum likelihood and based on the concatenated sequences of eight conserved proteins (NCLDV class I genes): vaccinia virus (VV) D5-type ATPase, DNA polymerase family B, VV A32 virion packaging ATPase, capsid protein, thiol oxidoreductase, VV D6R helicase, serine/threonine protein kinase, and A1L transcription factor. One of the class I genes (VV A18 helicase) is absent in LDV and was not included. The alignment contains 1660 sites without insertions and deletions. A neighbor-joining tree and a maximum-parsimony tree exhibited similar topologies. Bootstrap percentages are shown along the branches. Reproduced from Raoult D, Audic S, Robert C, *et al.* (2004) The 1.2-megabase genome sequence of mimivirus, *Science* 306: 1344–1350, with permission from American Association for the Advancement of Science.

Remarkable Genes Found in Mimivirus

Translation-related genes

The mimivirus genome exhibits numerous genes encoding central components of the protein translation system. This was most unexpected, as viruses are traditionally distinguished from cellular organisms by their inability to perform protein synthesis independently from their host. This incomplete virally encoded translation subsystem includes four amynoacyl-tRNA synthetases (ArgRS, TyrRS, CysRS, and MetRS). These enzymes, responsible for the accurate loading of a given amino acid on its cognate tRNAs, are the true enforcers of the genetic code. The enzymatic activity of the TyrRS has been experimentally validated, as well as its strict specificity for tyrosine and its eukaryotic tRNA$_{Tyr}$.

In addition, mimivirus genome encodes two translation initiation factors: a GTP-binding elongation factor (EF-Tu) and a peptide chain release factor. Finally, mimivirus exhibits the first virus-encoded tRNA (Uracil-5)-methyltransferase. This enzyme is responsible for the U to T modification characterizing the T-loop in all tRNAs, a region involved in ribosome recognition, as well as aminoacylation and the binding of EF-Tu. Besides these unique translation-related enzymes, mimivirus encodes six tRNAs, a common feature in some NCLDV, such as phycodnaviruses.

DNA repair-related genes

The mimivirus genome encodes a comprehensive set of DNA repair enzymes, covering all types of DNA damage: alkylating agents, ultraviolet (UV) light, or ionizing radiations. In particular, mimivirus is the first virus to exhibit a formamidopyrimidine-DNA glycosylase (used to excise oxidized purines), a UV-damage endonuclease, a 6-O-methylguanine-DNA methyltransferase (used to get rid of O^6-alkylguanine), and a homolog to the DNA mismatch repair enzyme MutS. Mimivirus also possesses DNA topoisomerases, the enzymes required for solving entanglement problems associated with DNA replication, transcription, recombination, and chromatin remodeling. There are three mimivirus-encoded topoisomerases: first the type IA topoisomerase identified in a virus, the usual type IIA topoisomerase (such as found in most NCLDVs), and a type IB topoisomerase as found in the family *Poxviridae*.

DNA synthesis

The mimivirus genome exhibits the only known virally encoded nucleoside diphosphate kinase (NDK) protein. This enzyme catalyzes the synthesis of nucleoside triphosphate (other than ATP) from ATP. Its activity has been demonstrated experimentally, and exhibits a specific affinity for deoxypyrimidine nucleotides. This enzyme may help alleviate limited supplies in dTTP and dCTP for DNA synthesis. The genome also encodes a deoxynucleoside kinase (DNK), a thymidylate synthase (dUMP to dTMP), as well as the salvage enzyme thymidine kinase (TMP synthesis from thymidine).

Host signaling interfering pathways

The mimivirus genome encodes a large number of putative proteins exhibiting ankyrin repeats (66), a BTB/POZ domain (26), an F-box domain (10), and a kinase domain (9). Genes harboring these features are frequently involved in protein–protein interactions (ankyrin repeat), ubiquitin-mediated interactions (BTB/POZ, F-Box), and intracellular signaling (kinases). Remarkably, mimivirus exhibits three proteins exhibiting both a cyclin and a cyclin-dependent kinase (CDK) domain. It is likely that these proteins, by interfering with the cell signaling pathways, play a central role in turning the amoebal host into an efficient virus factory. On the other hand, identifying the cellular targets of these viral proteins, and their precise mode of action, might reveal original and valuable new directions for the therapeutic manipulation of eukaryotic cell metabolism.

Miscellaneous metabolic pathways

Previous analyses of NCLDV genomes, in particular from phycodnaviruses, revealed that these viruses possess biosynthetic abilities going well beyond the minimal requirements of viral DNA replication, transcription, and virion-packaging systems.

Mimivirus builds on this established trend by exhibiting a wealth of amino acid-, lipid-, and sugar-modifying enzymes, albeit all of them in the form of apparently incomplete (virus-encoded) pathways. For instance, mimivirus encodes three enzymes related to the metabolism of glutamine: an asparagine synthase (glutamine hydrolyzing), a glutamine synthase, and a glutamine-hydrolyzing guanosine 5′-monophosphate synthase. The reasons behind the special affinity of the virus for glutamine are unknown. Mimivirus also encodes three lipid-modifying enzymes: a cholinesterase, a lanosterol 14α-demethylase, and a 7-dehydrocholesterol reductase. Although these enzymes might be involved in the disruption of the host cell membrane, none of them have been described as participating in the infection process in other viruses. Finally, mimivirus exhibits an impressive array of enzymes normally involved in the synthesis of complex polysaccharides, such as perosamine, a high molecular weight capsular material found in bacteria. Mimivirus particles (see below) retain the Gram stain, a unique phenomenon that might be related to the presence of a densely reticulated (lipo)polysaccharide layer at their surface. In this context, the mimivirus-encoded procollagen-lysine, 2-oxoglutarate 5-dioxygenase, an enzyme that hydroxylates lysine residues in collagen-like peptides, might be central both to the attachment of carbohydrate moieties and to the formation of intermolecular cross-links. It is tempting to speculate that the unique hairy-like appearance of the virion, together with its resistance to chemical and mechanical disruption, is directly linked to the presence of a layer of heavily cross-linked lipo-proteo-polysaccharide material covering the 'regular' capsid.

Intein and introns

Inteins are protein-splicing domains encoded by mobile intervening sequences. They catalyze their own excision from the host protein. Although found in all domains of life (Eukarya, Archaea, and Eubacteria), their distribution is highly sporadic. Mimivirus is one of the few dsDNA viruses exhibiting an intein, inserted within its DNA polymerase B gene. Mimivirus intein is closely related to the one found in the DNA polymerase of heterosigma akashiwo virus (HaV), a large dsDNA virus infecting the single-cell bloom-forming raphidophyte (golden brown alga) *H. akashiwo*. Both appear monophyletic to the archaeal inteins.

Type I introns are self-splicing intervening sequences that are excised at the mRNA level. One type IB intron has been identified in several chlorella viruses species, but they remain rare in eukaryotic viruses. Mimivirus exhibits four self-excising (predicted) introns: one in the largest RNA polymerase subunit gene, the other three in the second-largest subunit. Given that introns are mostly detected when they interrupt the coding sequence of known proteins, some additional instances located within anonymous open reading frames (ORFs) might have escaped detection.

Particle Structure

Morphology

The discovery of the characteristic (i.e., icosahedral) viral morphology exhibited by mimivirus particles (initially mistaken as small Gram-positive intracellular bacteria) under the electron microscope was the turning point in its correct identification as a *bona fide* virus. Despite its unprecedented size, the icosahedral symmetry of the particle is good enough to allow a computer-generated three-dimensional (3-D) reconstruction at a resolution of about 75 Å, from series of cryoelectron microscopy (cryo-EM) images. Mimivirus has a capsid with a diameter of *c.* 0.5 μm, covered by 0.125-μm-long, closely packed fibers (**Figure 6**). The total diameter of a free particle is thus about 0.75 μm, consistent with its visibility in the light microscope following Gram staining (**Figure 1**). The pseudo-triangulation for mimivirus particles is estimated at $T \approx 1180$, predicting that the whole capsid is constituted of *c.* 70 000 individual molecules of the L425 ORF-encoded major capsid protein. Inside a 70-Å-thick protein shell, cryo-EM images suggest the presence of

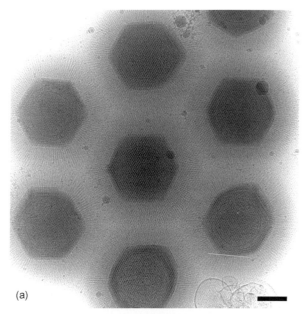

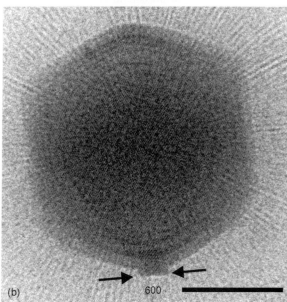

Figure 6 Cryo-EM high-quality images of mimivirus particles. (a) Cluster of mature particles, exhibiting a solid and compact fiber layer. (b) Close-up of one particle (0.75 μm across) exhibiting a densely packed layer of cross-linked fibers, and a unique vertex. Reproduced from Xiao C, Chipman PR, Battisti AJ, *et al.* (2005) Cryo-electron microscopy of the giant mimivirus. *Journal of Molecular Biology* 353: 493–496, with permission from Elsevier.

two 40-Å-thick lipid membranes, a structure also found in other NCLDVs such as the African swine fever virus and some poxviruses.

Virion Proteomics

A detailed study of purified isolated virions using various electrophoresis separation methods followed by mass spectrometry analysis indicated that 114 different mimivirus genes are present in the viral particle. Many of these genes correspond to multiple products, exhibiting a variety of post-translational modifications including glycosylation (such as observed for the major capsid protein L425), proteolysis, and, most likely, phosphorylation.

Besides the expected *bona fide* 'structural' proteins (i.e., capsid proteins, major core protein, A16L-like virion associated protein, lipocalin-like lipoprotein, and a 'spore coat' assembly factor), a large number of proteins with enzymatic functions have been identified. For instance, mimivirus particles appear to possess a complete transcriptional apparatus including all virus-encoded DNA-directed RNA polymerase subunits (five), as well as four transcription factors, its mRNA capping enzyme, and two helicases. Nine gene products all related to oxidative pathways constitute the next largest functional group of particle-associated enzymes. These enzymes are probably important to cope with the oxidative stress generated by the host defense. Five proteins associated to DNA topology and damage repair are also present in the particle. Finally, seven enzymes associated to lipid (lipase and phosphoesterase) or protein modification (kinases and phosphatase) complete the pool of particle enzymes, together with the two tRNA methyltransferases. Yet, the majority of particle-associated proteins are the products of genes (65) of unknown function, of which 45 exhibit no convincing similarity in databases. It is worth noticing that only a small fraction (13/114) of the products associated with the particle are encoded by genes harboring the AAAATTGA promoter signal, confirming its correlation with early and late/early viral functions.

The functions predicted for the particle-associated gene products suggest a possible scenario for the early stage of the virus replicative cycle. First, it is likely that the lipolytic and proteolytic enzymes are part of the process by which the virus penetrates the amoeba cytoplasm from its vacuolar location. The two topoisomerases (IA and IB) might then be involved in the unpacking of the DNA from the particle, and its injection into the cytoplasm. The DNA repair enzymes might then get into action prior to starting transcription. The complete transcription machinery found in the viral particle is very similar to that in poxviruses. This suggests that some transcription could take place in the host cytoplasm immediately upon mimivirus infection.

Particle-Associated mRNAs

In a way reminiscent of members of the family *Herpesviridae*, a number of virus-encoded mRNAs are found associated with the virus particle. They include the messengers for DNA polymerase (R322), the major capsid protein (L425), the TFII-like transcription factor (R339), the tyrosyl-, cysteinyl-, and arginyl-tRNA synthetase (L124, L164, and R663), and four proteins of unknown function. These immediate early gene transcripts may be needed to initiate the first step of the replicative cycle, prior to any viral gene expression.

Evolution: The Position of Mimivirus in the Tree of Life

Of the 63 homologous genes common to all known unicellular organisms from the three domains of life (Eukarya, Eubacteria, and Archaea), seven have been identified in the mimivirus genome: three aminoacyl-tRNA synthetases, the two largest RNA polymerase subunits, the sliding clamp subunit of the DNA polymerase (PCNA), and a 5′-3′ exonuclease. The unrooted phylogenetic tree built from the concatenated sequences of the corresponding proteins indicates that mimivirus branches out near the origin of the Eukaryota domain (**Figure 7**). This position is consistent with several competing hypotheses proposing that large dsDNA viruses either predated, participated in, or closely followed, the emergence of the eukaryotic cell.

Other Possible Members of the Family *Mimiviridae*

Mimivirus was serendipitously discovered within *A. polyphaga*, a free-living ubiquitous amoeba, prevalent in aquatic environments. Many of the mimivirus core genes exhibit a

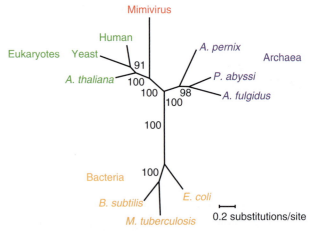

Figure 7 A phylogenetic tree of species from the three domains of life (Eukaryota, Eubacteria, and Archaea) and mimivirus. The tree was inferred with the use of a maximum-likelihood method based on the concatenated sequences of seven universally conserved protein sequences: arginyl-tRNA synthetase, methionylt-RNA synthetase, tyrosyl-tRNA synthetase, RNA polymerase II largest subunit, RNA polymerase II second largest subunit, PCNA, and 5′-3′ exonuclease. Bootstrap percentages are shown along the branches. Reproduced from Raoult D, Audic S, Robert C, *et al.* (2004) The 1.2-megabase genome sequence of mimivirus. *Science* 306: 1344–1350, with permission from American Association for the Advancement of Science.

phylogenetic affinity with members of the family *Phycodnaviridae* (algal and phytoplankton viruses). It can be expected therefore that additional viruses belonging to the family *Mimiviridae* will be found in aquatic/marine environments, for instance in marine protists. The bioinformatic analysis of environmental (metagenomics) DNA sequences strongly suggests that large viruses evolutionarily closer to mimivirus than any known virus species may be found among randomly collected bacteria-sized (passing through 3 μm but retained on 0.3 μm filters) marine microorganisms. Close ecological encounters of mimivirus ancestors with gorgonian octocorals (genus *Leptogorgia*) are also suggested by clear evidence of a lateral transfer of a mismatch repair (MutS) gene. Since very little is known about viruses infecting coral-associated microbial populations, it seems likely that new species of *Mimiviridae* might be hiding within this complex ecological niche.

See also: Origin of Viruses; Nature of Viruses; Virus Evolution: Bacterial Viruses.

Further Reading

Abergel C, Chenivesse S, Byrne D, Suhre K, Arondel V, and Claverie JM (2005) Mimivirus TyrRS: Preliminary structural and functional characterization of the first amino-acyl tRNA synthetase found in a virus. *Acta Crystallographica Section F* 61: 212–215.

Benarroch D, Claverie J-M, Raoult D, and Shuman S (2006) Characterization of mimivirus DNA topoisomerase IB suggests horizontal gene transfer between eukaryal viruses and bacteria. *Journal of Virology* 80: 314–321.

Berger P, Papazian L, Drancourt M, La Scola B, Auffray JP, and Raoult D (2006) Amoeba-associated microorganisms and diagnosis of nosocomial pneumonia. *Emerging Infectious Diseases* 12: 248–255.

Claverie JM, Ogata H, Audic S, Abergel C, Suhre K, and Fournier P-E (2006) Mimivirus and the emerging concept of 'giant' virus. *Virus Research* 117: 133–144.

Ghedin E and Claverié JM (2005) Mimivirus relatives in the Sargasso Sea. *Virology Journal* 2: 62.

La Scola B, Audic S, Robert C, et al. (2003) A giant virus in amoebae. *Science* 299: 2033.

La Scola B, Marrie TJ, Auffray JP, and Raoult D (2005) Mimivirus in pneumonia patients. *Emerging Infectious Diseases* 11: 449–452.

Ogata H, Raoult D, and Claverie JM (2005) A new example of viral intein in mimivirus. *Virology Journal* 2: 8.

Raoult D, Audic S, Robert C, et al. (2004) The 1.2-megabase genome sequence of mimivirus. *Science* 306: 1344–1350.

Raoult D, Renesto P, and Brouqui P (2006) Laboratory infection of a technician by mimivirus. *Annals of Internal Medicine* 144: 702–703.

Renesto P, Abergel C, Decloquement P, et al. (2006) Mimivirus giant particles incorporate a large fraction of anonymous and unique gene products. *Journal of Virology* 80: 11678–11685.

Suhre K (2005) Gene and genome duplication in Acanthamoeba polyphaga mimivirus. *Journal of Virology* 79: 14095–14101.

Suhre K, Audic S, and Claverie JM (2005) Mimivirus gene promoters exhibit an unprecedented conservation among all eukaryotes. *Proceedings of the National Academy of Sciences, USA* 102: 14689–14693.

Suzan-Monti M, La Scola B, and Raoult D (2006) Genomic and evolutionary aspects of mimivirus. *Virus Research* 117: 145–155.

Xiao C, Chipman PR, Battisti AJ, et al. (2005) Cryo-electron microscopy of the giant mimivirus. *Journal of Molecular Biology* 353: 493–496.

Molluscum Contagiosum Virus

J J Bugert, Wales College of Medicine, Heath Park, Cardiff, UK

© 2008 Elsevier Ltd. All rights reserved.

Glossary

Acanthoma A tumor composed of epidermal or squamous cells.
Anamnesis The complete history recalled and recounted by a patient.
Epicrisis A series of events described in discriminating detail and interpreted in retrospect.

History and Classification

The typical molluscum lesion, a smooth, dome-shaped, flesh-colored protrusion of the skin with a typical central indentation, was first described by Edward Jenner (1749–1823) as a 'tubercle of the skin' common in children. Thomas Bateman (1778–1821) first used the term 'molluscum contagiosum' (MC). In 1841, 'molluscum bodies', intracytoplasmic inclusions in the epidermal tissues of MC lesions, were independently observed by Henderson and Paterson. Similarities of molluscum bodies to the 'Borrel' bodies in fowlpox-infected tissues were noted later by Goodpasture, King, and Woodruff. At the beginning of the last century, Juliusberg demonstrated that the etiological agent of MC cannot be removed by filtration through Chamberland filters. Filtrates were used to infect human volunteers, who developed MC between 25 and 50 days after inoculation. Short of fulfilling all the Koch postulates, an animal model for MC has not been established to this day. Smaller elementary bodies inside the molluscum bodies were observed by Lipschütz in 1911. Electron microscopy revealed these elementary bodies to be poxvirus particles with dimensions of 360 nm × 210 nm.

Virion, Genome, and Evolution

Molluscum contagiosum virus (MCV) particles have a typical poxviral morphology (**Figure 1**). The virions are enveloped, pleomorphic, but generally ovoid to brick-shaped, with a dumbbell-shaped central core and lateral bodies similar to those in orthopoxvirus virions. MCV cores show complex structural patterns. Virions are often found to have membrane fragments loosely attached to them, indicating a noncontinuous lipid envelope wrapping the core.

The genome of MCV is a double-stranded DNA molecule of 190 289 bp (GenBank accession U60315: MCV type 1/80) with covalently closed termini (hairpins) and about 4.2 kbp of terminally inverted repeats (**Figure 2**). This excludes 50–100 bp of terminal hairpin sequences that could not be cloned or sequenced because replicative intermediates are not apparent in DNA from MCV biopsy specimens. The genomes of MCV (genus *Molluscipoxvirus*, subfamily *Chordopoxvirinae*), crocodilepox virus, and parapoxviruses stand out in the family *Poxviridae* because they have G+C contents of over 60%.

The MCV genome encodes 182 nonoverlapping open reading frames of more than 45 codons (**Figure 2**), almost half of which have no similarities to known proteins. Hypothetical MCV structural proteins and proteins encoding enzymes of the replication and transcription apparatus share obvious homologies to other poxvirus proteins. Less-obvious homologies exist between MCV and avipoxviruses (MC130, MC133, and MC131 A-type inclusion body-like proteins) and notably between MCV, parapoxviruses, and crocodilepox virus (MC026 modified RING protein and a number of proteins shared between only two of the above poxviruses). Unique MCV nonstructural proteins that are not involved in replication or transcription can be divided into two functional classes: (1) proteins dealing with the host immune system (host-response-evasion factors), such as the MCV chemokine antagonist (MC148) and the interleukin-18 (IL18)-binding protein (MC054), and (2) proteins supporting MCV replication in the host cell or the host tissue (host cell/tissue-modulating factors), such as the antiapoptotic selenoprotein MC066 and the Hrs-binding protein MC162.

An epidermal growth factor (EGF) homolog similar to the ones expressed by other poxviruses was not found in the genome of MCV. The only other poxvirus that does not encode this factor is swinepox virus. However, MCV-infected basal keratinocytes seem to increase EGF receptor and transferrin receptor expression, in comparison to uninfected epidermis. Inducing EGF receptor expression may be an indirect mechanism causing epidermal hyperproliferation. MCV is the only chordopoxvirus that does not encode a J2R-like thymidine kinase.

In a phylogenetic analysis of 26 poxvirus genomes, MCV (representing the molluscipoxviruses) formed a group by itself among the subfamily of chordopoxviruses, separate from avipoxviruses (fowlpox virus), orthopoxviruses (vaccinia and variola viruses), and all other genera.

Four main genetic subtypes of MCV have been identified by DNA fingerprinting. MCV type 1 prototype (p) is the most common genetic type (98%) in immune-competent hosts in Western Europe. MCV type 1 (including variants) is the most common genetic type worldwide. MCV types 2–4 are relatively more commonly seen in immunocompromised individuals. One type has only been described in Japan. MCV genotypes discernible by DNA fingerprinting do not change when the viruses are transmitted between family members or in larger contact groups, indicating a low overall mutation rate.

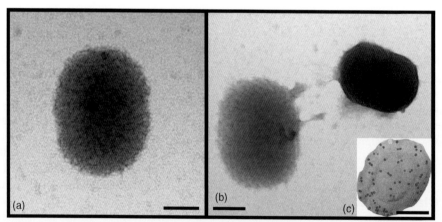

Figure 1 Electron microscope images of MCV particles negatively stained using the ammonium molybdate technique. (a) MCV particle showing the typical core protein pattern. (b) MCV particles: one that has lost the envelope appears larger and shows the typical core protein pattern, while the other is still wrapped in membrane and appears smaller. (c) Five nanometer gold particles bound to mouse antihuman antibodies, detecting human polyclonal patient antibodies binding to the surface of an MCV particle. Scale = 100 nm (a–c). Electron microscopy: Bugert and Hobot, Cardiff University School of Medicine, 2005.

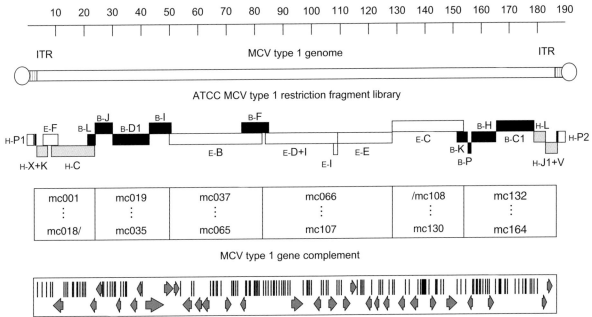

Figure 2 The MCV genome. The scale at the top is in kbp. The inverted terminal repeats (ITRs) are shown as vertical stripes flanking the genome, pointing inward from both covalently closed ends. The ITRs do not contain any complete coding sequences. The area covered by the ATCC MCV genome fragment library is shown as white (EcoRI restriction fragments), gray (HindIII restriction fragments), and black (BamHI restriction fragments) boxes representing the respective size of the viral sequence insert of each clone. The terminal HindIII restriction fragments P1 and P2 do not contain protein-coding regions and were not cloned. MCV genes are given the lowercase prefix 'mc', whereas (in the text) MCV proteins are given the uppercase prefix 'MC'. The MCV genes encoded in different parts of the genome, as defined by the boundaries of defined restriction fragments, are indicated in white boxes. The slashes signify ORFs that straddle a restriction site used for cloning and that are therefore not completely contained in any one plasmid on either side of the restriction site. Two MCV genes (mc036 and mc131) are split between two cloned restriction fragments, and therefore not listed. The numbering of MCV genes ends at mc164, the original number of genes larger than 90 codons. An additional 18 genes have been identified as being larger than 45 codons and are numbered as appendices of preceding genes (e.g., mc004.1). Protein-coding regions of the MCV type 1 gene complement are shown at the bottom as arrows (larger than 1.5 kbp) or small rectangles.

Host Range and Virus Propagation

MCV, like smallpox virus, is considered to be an exclusive pathogen of man. Reports of MCV in a number of animals, including horses, chimpanzees, and kangaroos, have not been supported by DNA sequence confirmation. Many pox-like infections of vertebrates, most of them caused by orthopoxviruses, can be confused with MCV by their clinical appearance. Conventional immune-competent laboratory animals, including mice, rats, guinea pigs, and tree shrews, do not support MCV replication in their skin. MCV-infected human keratinocytes have been transplanted into mice with severe combined immune deficiency (SCID) and typical MCV lesions have subsequently developed in these non-natural hosts. Attempts to passage the virus in SCID mice were unsuccessful. However, despite the absence of molecular evidence, an animal reservoir of MCV cannot be excluded.

MCV has so far not been grown in conventional human cell lines, including immortalized tumorigenic/virus-transformed and nontumorigenic skin keratinocytes (HaCaT, NIKS). Experiments with *ex vivo* cultures of human skin cells (raft cultures) are ongoing. MCV may use a vegetative mechanism for replicating in differentiating keratinocytes.

In the absence of culturable virus, classical virological research on MCV is severely restricted. All progress so far has been made by studying MCV genes in isolation, based on the complete MCV genome sequence gained from an overlapping redundant MCV genome fragment library. This reagent has been made available to the ATCC. The entire MCV gene complement is covered by 18 recombinant bacterial plasmid clones harboring viral sequences from the EcoRI, BamHI, and HindIII restriction fragment libraries of MCV type 1/80 (**Figure 2**).

Further research is being carried out using abortive cell culture systems. MCV induces a remarkable cytopathogenic effect (CPE) in human fibroblasts, both in primary cells (MRC5) and in telomerase-transduced immortal cell lines (hTERT-BJ-1). The CPE starts 4 h post infection (p.i.) and reaches a maximum at 24 h p.i., with the cells looking as if they have been trypsinized, partially detaching from the monolayer, rounding, and clumping. Cells settle down at 48–72 h p.i., but show a morphological transformation from an oblong fibroblast to a more square epithelial-looking cell type. MCV transcribes early mRNA in these cells. The

CPE is not induced by UV-inactivated virions or in the presence of cycloheximide, indicating that expression of viral proteins is required. mRNA transcription can be detected by reverse transcriptase-polymerase chain reaction (RT-PCR) for months in serially passaged infected cells. A productive MCV infection cannot be rescued nongenetically by co-infection with other chordopoxviruses in these cells. MCV can infect human HaCaT keratinocytes and transcribes mRNA in these cells, but does not induce a CPE. MCV cannot infect nonhuman cells. It induces type 1 interferons in mouse and human embryo fibroblasts and IL8 in human lung epithelial cells, which it cannot infect, suggesting involvement of surface pathogen-associated molecular pattern (PAMP) receptors like TLR2 or TLR4. Removal of interferon pathways from cell lines susceptible for MCV infection would be worth investigating, in order to exclude interference causing abortive MCV infections. MCV is currently isolated from human-infected skin biopsies. MCV purified from biopsy material can be used for infection studies, electron microscopy, viral DNA extraction, and analyses of early mRNA synthesized by *in vitro* transcription of permeabilized virions.

Clinical Features and Pathology

The mean incidence of MC in the general population is 0.1–5%. Seroepidemiological studies have shown that antibody prevalence in persons over 50 years of age is 39%. This indicates that MC is a very common viral skin infection, which is supported by its common occurrence in dermatological practice. MCV outbreaks occur in crowded populations with reduced hygienic standards. Outbreaks with more than 100 cases have been described in kindergartens, military barracks, and public swimming pools. MC was found to be very common in the Fiji Islands.

MC is most often seen as a benign wart-like condition (German: *dellwarze*) with light pruritus in preadolescent children, but can occur in immuncompetent adults. MC is more severe in immunocompromised people or individuals with atopic dermatitis, where it can lead to giant molluscum and eczema molluscum. The lesions generally occur on all body surfaces, but not on the palm of the hand or on mucous membranes. They may be associated with hair follicles. MC lesions can grow close to mucous membranes on the lips and eyelids. When situated near the eye, they can lead to conjunctivitis. MC is transmitted by smear-infection with the infectious fluids discharged from lesions and by direct contact with contaminated objects. MC is not very contagious and therefore infection depends on a high inoculating dose. It is a sexually transmitted disease when lesions are located on or in the vicinity of sexual organs.

MC lesions are generally globular, sometimes ovate, 2–5 mm in diameter and sit on a contracted base. Cellular semiliquid debris can be expressed from the central indentation at the top of the lesion. The infection spreads via contact with this fluid. Histologically, the tumors are strictly limited to the epidermal layer of the skin and have a resemblance to hair follicles. They are therefore classified as acanthomas.

If not mechanically disturbed, MC lesions will persist for months and even years in immune-competent hosts, but can disappear spontaneously, probably when virus-infected tissue is exposed to the immune system. To expose the infection by limited (sterile) trauma is a way of treatment.

Scratching or disturbing the lesions leads to a quicker resolution but can complicate the condition through bacterial superinfection. This must be avoided in severely immunocompromised hosts, who develop widespread MC with hundreds of lesions of larger size and can succumb to sepsis following bacterial superinfection. MCV is a marker of late-stage disease in human immunodeficiency virus (HIV)-infected individuals, and in HIV-infected populations the incidence of MC was 30% before the onset of human cytomegalovirus (HCMV) prophylaxis with cidofovir.

MCV probably enters the epidermis through microlesions. The typical MCV lesion contains conglomerates of hyperplastic epithelial cells organized in follicles and lobes, which all develop into a central indentation toward the surface of the skin in a process similar to holocrine secretion. The whole lesion has the appearance of a hair follicle where the hair is replaced by the virus-containing plug. The central indentation is filled with cellular debris and is rich in elementary viral particles in a waxy plug-like structure. This plug becomes mobilized and spreads the infection to other areas of surrounding skin or contaminates objects. The periphery of the MCV lesion is characterized by basaloid epithelial cells with prominent nuclei, large amounts of heterochromatin, slightly basophilic cytoplasm, and increased visibility of membranous structures. These cells are larger than normal basal keratinocytes, they divide faster than normal basal cells, their cytoplasm contains a large number of vacuoles, and they are sitting on top of an intact basal membrane. The lesion is a strictly intraepidermal hyperplastic process (acanthoma). Distinct poxviral factories (molluscum bodies or Henderson–Patterson bodies) appear about four cell layers away from the basal membrane in the stratum spinosum. The inclusion bodies grow and obliterate cellular organelles. Cells with inclusion bodies do not divide further. The cytoplasm of MCV-producing cells shows keratinization, which is not expected at that stage of keratinocyte differentiation and indicates dyskeratinization in the sense of abnormal differentiation.

Immune Response

In undisturbed MC lesions, histological studies have shown a conspicuous absence of effectors of the cellular immune system, in particular skin-specific tissue macrophages (Langerhans cells). This is in contrast to papillomata, where a vigorous cellular immune response, including cytotoxic T lymphocytes (CTLs), is mounted immediately. The absence of macrophages has been attributed to the activity of various MCV genes that are suspected to make the MC lesion immunologically 'invisible'. This includes a biologically inactive IL8 receptor-binding beta-chemokine homolog (MC148), which may suppress the immigration of neutrophils; a major histocompatibility complex (MHC) class I homolog that may upset MHC class I antigen presentation on the surface of infected cells, or natural killer cell recognition; and an IL18-binding protein, which underlines the importance of this cytokine for the local immune response in human skin. As for the humoral response, MCV-specific antibodies have been detected in several studies, showing a seroprevalence of MCV of up to 40% in the general population, much higher than previously expected. However, these antibodies do not seem to confer a neutralizing immunity. MCV genes were expressed in a cowpox virus expression system and two antigenetically prominent MCV proteins identified: mc133L (70 kDa protein: MC133) and mc084L (34 kDa protein: MC084). These proteins are presumably glycosylated and present on the surface of MCV virions, where they allow binding of antibodies and detection by immune electron microscopy (**Figure 1(c)**).

Diagnosis, Treatment, and Prevention

MCV is readily diagnosed by its clinical appearance and by the typical histopathology found in sections of lesion biopsies. After the eradication of smallpox, MCV is the most commonly diagnosed poxviral infection. Ortho- and parapoxviral zoonoses are rare. In the differential diagnosis of MC, smallpox must be considered along with other ortho- and parapoxviruses. The anamnesis must cover contacts with pets, especially gerbils and chipmonks of African origin (monkeypox) as well as local rodents and cats (cowpox). Always looming in the background is the possibility of a smallpox bioterrorist attack. An actual example for the management of such a contingency is the epicrisis of a monkeypox outbreak in Wisconsin, USA, in 2003 published by the Centers for Disease Control and Prevention (CDC). The most likely nonpoxviral differential diagnosis is varicella-zoster virus. Other viral agents can be excluded by -electron microscopy and PCR. Further differential diagnoses include syphilis, papilloma, and skin malignancies such as melanoma.

MC lesions are generally self-limiting, with an average of 6 months to 5 years for lesions to disappear. Patients with immune dysfunction or atopic skin conditions have difficulty clearing lesions. Therapy is recommended for genital MC to avoid sexual transmission, and should include antipruritics to prevent scratching and bacterial superinfection.

Treatment options cover a wide range of invasive and topical treatment strategies (**Table 1**). Topical application of salicylic acid or removal by curettage emerge as the two most successful strategies for large and small numbers

Table 1 Suggested MC treatments, side effects, and success rates

Symptom	Therapy	Side effects	Success rates
Pruritus	Antihistamine ointments	Tiredness	High
Acanthoma	Curettage (surgical) with sharp spoon. Local anesthetics required for children	Scar, pain impractical for large lesion numbers	High
	Lancing of lesion with needle	Pain, infection	High
	Cryosurgery with topical anesthesia	Pain, no scars	Medium
	Topical salicylic acid colloid, for example, occlusal (26% salicylic acid in polyacrylic vehicle)		High
	Immune modulators	Predisposition for bacterial and fungal skin infections	Low
	Tacrolimus 0.1% ointment		
	Pimecrolimus 0.5% ointment		
	Imiquimod 0.5% ointment		
	Podophyllotoxin 0.5% ointment		
	Antivirals (high cost)	Allergy	High
	DNA polymerase inhibitors		
	Acyclic nucleoside phosphonates (e.g., topical cidofovir)		
	Topoisomerase inhibitors		
	Lamellarin		
	Coumermycin		
	Cyclic depsipeptide sansalvamide A		

of lesions, respectively. Needling of lesions has been reported to work and may expose MCV to immune effectors. In immune-suppressed individuals with widespread MC, topical immune modulators like imiquimod have been tried with limited success. Topical antivirals work but are not cost-effective.

MC is best prevented by exposure prophylaxis. The vaccine against smallpox (vaccinia virus) does not protect against MCV, underlining the fundamental antigenic differences between ortho- and molluscipoxviruses.

See also: Capripoxviruses; Cowpox Virus; Electron Microscopy of Viruses; Entomopoxviruses; Fowlpox virus and other avipoxviruses; Leporipoviruses and Suipoxviruses; Mousepox and Rabbitpox Viruses; Parapoxviruses; Poxviruses; Smallpox and Monkeypox Viruses; Vaccinia Virus; Virus Databases; Yatapoxviruses.

Further Reading

Brown T, Butler P, and Postlethwaite R (1973) Non-genetic reactivation studies with the virus of molluscum contagiosum. *Journal of General Virology* 19: 417–421.

Bugert JJ (2006) Molluscipoxviruses. In: Mercer A, Schmidt A, and Webber O (eds.) *Advances in Infectious Diseases*, pp. 89–112. Basel: Birkhauser.

Bugert JJ and Darai G (1991) Stability of molluscum contagiosum virus DNA among 184 patient isolates: Evidence for variability of sequences in the terminal inverted repeats. *Journal of Medical Virology* 33: 211–217.

Bugert JJ and Darai G (2000) Poxvirus homologues of cellular genes. *Virus Genes* 21: 111–133.

Buller RM, Burnett J, Chen W, and Kreider J (1995) Replication of molluscum contagiosum virus. *Virology* 213: 655–659.

Konya J and Thompson CH (1999) Molluscum contagiosum virus: Antibody responses in persons with clinical lesions and seroepidemiology in a representative Australian population. *Journal of Infectious Diseases* 179: 701–704.

McFadden G, Pace WE, Purres J, and Dales S (1979) Biogenesis of poxviruses: Transitory expression of molluscum contagiosum early functions. *Virology* 94: 297–313.

Melquiot NV and Bugert JJ (2004) Preparation and use of molluscum contagiosum virus from human tissue biopsy specimens. *Methods in Molecular Biology* 269: 371–384.

Porter CD and Archard LC (1992) Characterisation by restriction mapping of three subtypes of molluscum contagiosum virus. *Journal of Medical Virology* 38: 1–6.

Postlethwaite R and Lee YS (1970) Sedimentable and non-sedimentable interfering components in mouse embryo cultures treated with molluscum contagiosum virus. *Journal of General Virology* 6: 117–125.

Reed RJ and Parkinson RP (1977) The histogenesis of molluscum contagiosum. *American Journal of Surgical Pathology* 1: 161–166.

Scholz J, Rosen-Wolff A, Bugert J, *et al.* (1998) Molecular epidemiology of molluscum contagiosum. *Journal of Infectious Diseases* 158: 898–900.

Senkevich TG, Bugert JJ, Sisler JR, Koonin EV, Darai G, and Moss B (1996) Genome sequence of a human tumorigenic poxvirus: Prediction of specific host response-evasion genes. *Science* 273: 813–816.

Thompson CH, Yager JA, and Van Rensburg IB (1998) Close relationship between equine and human molluscum contagiosum virus demonstrated by *in situ* hybridisation. *Research in Veterinary Science* 64: 157–161.

Viac J and Chardonnet Y (1990) Immunocompetent cells and epithelial cell modifications in molluscum contagiosum. *Journal of Cutaneous Pathology* 17: 202–205.

Vreeswijk J, Leene W, and Kalsbeek GL (1976) Early interactions of the virus molluscum contagiosum with its host cell. Virus-induced alterations in the basal and suprabasal layers of the epidermis. *Journal of Ultrastructural Research* 54: 37–52.

Mononegavirales

A J Easton and R Ling, University of Warwick, Coventry, UK

© 2008 Elsevier Ltd. All rights reserved.

Glossary

Antigenome An RNA molecule complementary in sequence to the single-stranded genomic nucleic acid of a virus.

Nucleocapsid Internal structure of a virus containing its genetic material and one or more proteins. It may exist in infected cells independently of virus particles.

Introduction

The order *Mononegavirales* comprises four families of viruses, the *Bornaviridae*, *Rhabdoviridae*, *Filoviridae*, and *Paramyxoviridae*. This order of viruses includes the agents of a wide range of diseases affecting human, animals, and plants. The disease may be highly characteristic of the virus (e.g., measles, mumps) or of a more general nature (e.g., respiratory disease caused by various paramyxoviruses). Disease and host range vary to some extent between the viral families; these and other differences between the families are summarized in **Table 1**. The feature which unites these viruses is the presence of a genome consisting of a single RNA molecule which is of opposite polarity to the mRNAs that they encode. The genomic RNA, which has inverted terminal complementary repeats, ranges in size from about 9 to 19 kbp and is usually present in a helical nucleocapsid. Typically, 93–99% of the genomic sequence encodes proteins with the genes encoding the structural proteins common to all members of the order arranged in the same relative position in the genome with respect to each other.

Table 1 Distinguishing features of viruses belonging to the indicated families

	Bornaviridae	Filoviridae	Paramyxoviridae	Rhabdoviridae
Genome size (approximate, nt)	8900	19 000	13 000–18 000	11 000–15 000
Virion morphology	90 nm, spherical	Filamentous, circular, 6-shaped	Pleomorphic, spherical, filamentous	Bullet-shaped or bacilliform
Replication site	Nucleus	Cytoplasm	Cytoplasm	Cytoplasm except nucleorhabdoviruses
Host range	Horses, sheep, cats, ostriches	Primates, (bats?)	Vertebrates	Vertebrates, vertebrates + invertebrates, plants + invertebrates
Growth in cell culture	Noncytopathic	Noncytopathic	Lytic, often with syncytium formation	Lytic for animal viruses, rapid for vesiculoviruses
Pathogenic potential	Behavioral disturbances to severe encephalomyelitis	Hemorrhagic fever	Respiratory or neurological illness	Mild febrile illness to fatal neurological disease

The genome RNA is present in the form of a nucleocapsid complex with at least one structural protein and this complex in turn is enclosed in a lipid membrane which in all cases except the bornaviruses has spikes of about 5–10 nm comprised of the viral glycoproteins. The nucleocapsids are incorporated into virions via interactions mediated by a viral matrix protein which acts as a bridge between the complex and the virus glycoprotein(s). The lipid membrane of the virus is derived from the host cell plasma membrane or internal membranes during a budding process. The shape of the viral particles varies although certain forms predominate in different families of viruses.

Taxonomy

The order *Mononegavirales* is divided into four families on the basis of features such as those listed in **Table 1** and common aspects of the genetic organization, shown in **Figures 1** and **2**. There is little conserved sequence between viruses in different families except for regions of the L (polymerase) protein. A phylogenetic tree based on part of the L protein sequence is shown in **Figure 3** in which the different families can be separated. The taxonomic structure of the order is based on a range of attributes including morphological features of the virion and shape of nucleocapsid rather than the sequence similarities. Consequently, the pneumoviruses are classified as a subfamily of the *Paramyxoviridae* despite the L protein sequence apparently being more similar to that of the filoviruses. A large number of viruses that belong to the order have not been classified as yet in a particular genus. Examples of these are tupaia rhabdovirus and Flanders virus in the *Rhabdoviridae*, various paramyxoviruses isolated from rodents such as tupaia paramyxovirus, Mossman virus, Beilong virus, and J virus, whose L gene sequences cluster near the henipaviruses and morbilliviruses, and fer-de-lance virus with an L gene sequence most similar to that of the respiroviruses.

Virus Structure

All mononegaviruses have an enveloped virus particle which contains glycoprotein spikes projecting from the surface, though Borna disease virus has a less well defined structure with no spikes and an electron-dense center. The shapes of the viruses vary considerably. Viruses in the family *Paramyxoviridae* are highly pleomorphic with the major form usually being roughly spherical, with filamentous and irregular shaped forms also being present. Rhabdoviruses have a more defined shape than the viruses in the other families, being bullet shaped or bacilliform and with a uniform width and length determined by the genome length. Filoviruses such as Ebola virus have a basic bacilliform shape of relatively uniform width (~80 nm) but are often folded into branched, circular, and 6-shaped forms and may form long filaments. Internally, several layers are apparent, the helical outer layer surrounding an electron-dense layer and an axial channel (**Figure 4**).

The nucleocapsid complex structures of the mononegaviruses are all helical which in viruses in the subfamily *Paramyxovirinae* give rise to a herringbone pattern. In contrast, rhabdoviruses and pneumoviruses have a more flexible and less regular structure. Even the more regular structures are nevertheless believed to exist in dynamic states allowing structural changes during RNA synthesis. In the pneumoviruses the width of the nucleocapsid is generally narrower and the helical pitch greater than in other paramyxoviruses, enabling these viruses to be distinguished prior to their molecular characterization.

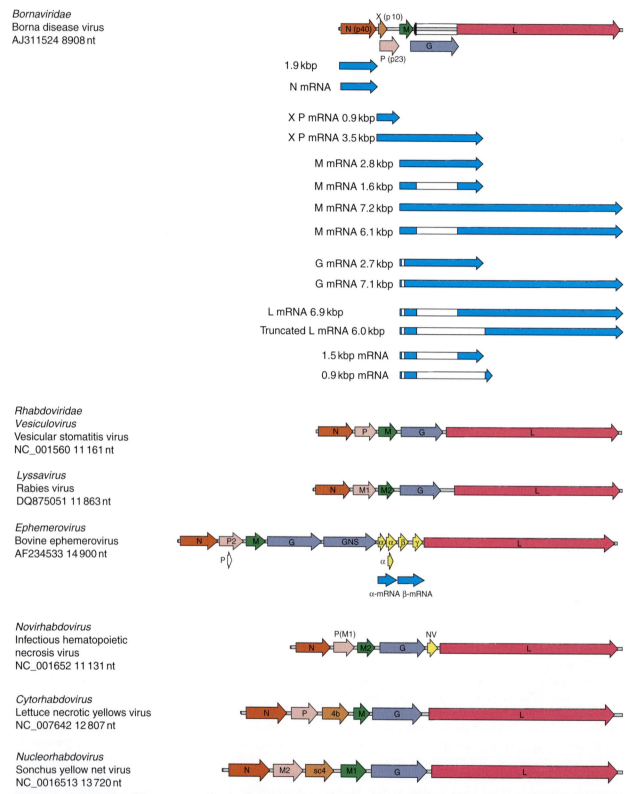

Figure 1 Organization of the genomes of viruses in the families *Bornaviridae* and *Rhadbdoviridae*. Open reading frames are colored to indicate proteins that are believed to have similar functions in the different viruses. Messenger RNAs are only shown where they differ from the monocistronic transcripts normally observed for viruses in the order *Mononegavirales*. Introns in Borna disease virus are indicated by unshaded regions. All the rhabdoviruses have potential second open reading frames in the second (phosphoprotein) gene although this is only shown for bovine ephemerovirus. The P gene was formely known as NS in vesiculoviruses and is sometimes known as M1 in lyssaviruses and novirhabdoviruses and M2 in nucleorhabdoviruses.

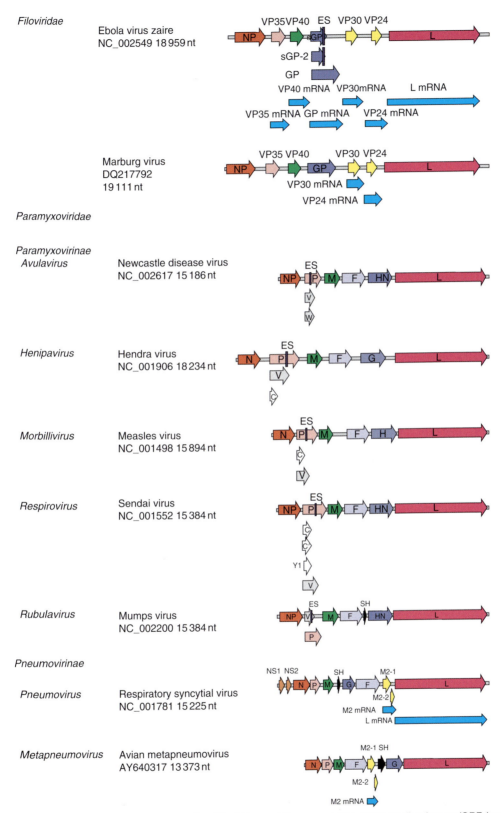

Figure 2 Genomic organization of viruses in the families *Filoviridae* and *Paramyxoviridae*. Open reading frames (ORFs) are colored as in **Figure 1**. Messenger RNAs are again indicated only where they have unusual features such as overlaps between adjacent genes or encoding two ORFs. However, all P genes of viruses in the *Paramyxovirinae* subfamily encode more than one protein, so these mRNAs are not shown. ES indicates an editing site where nontemplated residues are added to generate additional proteins. The ORF in the same line as the ORFs from other genes is the one encoded by the unedited RNA.

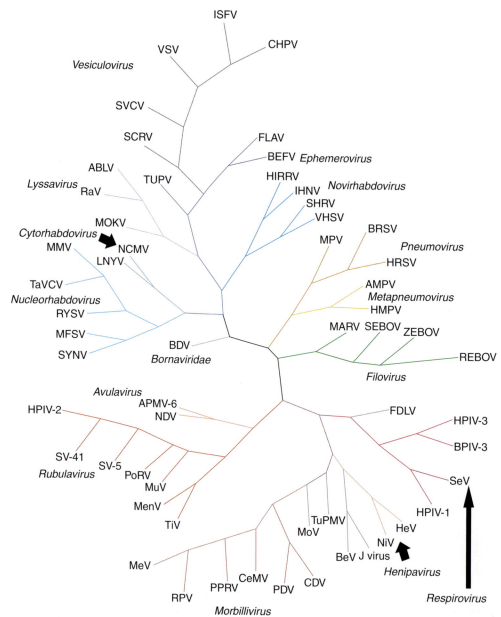

Figure 3 Phylogenetic tree of the largest region of the L protein shared by viruses in each of the mononegavirus genera. Tree branches of members in the same subfamily or family are shaded in similar colors. Virus abbreviations are those in the *Eighth Report of the International Committee on the Taxonomy of Viruses*. Borna viruses: BDV, Borna disease virus. Filoviruses: MARV, Lake Victoria marburgvirus; REBOV, Reston ebolavirus; SEBOV, Sudan ebolavirus; ZEBOV, Zaire ebolavirus Metapneumoviruses: AMPV, avian metapneumovirus; HMPV, human metapneumovirus. Pneumoviruses: BRSV, bovine respiratory syncytial virus; HRSV, human respiratory syncytial virus; MPV, murine pneumonia virus. Vesiculoviruses: CHPV, Chandipura virus; ISFV, Isfahan virus; SVCV, spring viremia of carp virus; SCRV, Siniperca chuatsi virus; VSV, vesicular stomatitis virus (San Juan). Lyssaviruses: ABLV, Australian bat lyssavirus; MOKV, Mokola virus; RaV, rabies virus. Novirhabdoviruses: HIRRV, hirame rhabdovirus; IHNV, infectious hematopoietic necrosis virus; SHRV, snakehead rhabdovirus; VHSV, viral hemorrhagic septicaemia virus. Ephemerovirus: BEFV, bovine ephemeral fever virus. Unassigned animal rhabdoviruses: FLAV, Flanders virus; TUPV, Tupaia rhabdovirus. Cytorhabdoviruses: LNYV, lettuce necrotic yellows virus; NCMV, northern cereal mosaic virus. Nucleorhabdoviruses: MFSV, maize fine streak virus; MMV, maize mosaic virus; RYSV, rice yellow stunt virus; SYNV, Sonchus yellow net virus; TaVCV, taro vein chlorosis virus. Henipaviruses: HeV, Hendra virus; NiV, Nipah virus. Avulaviruses: NDV, Newcastle disease virus; APMV-6, avian paramyxovirus 6. Rubulaviruses: HPIV-2, human parainfluenza virus 2; MenV, menangle virus; MuV, mumps virus; PoRV, porcine rubulavirus; SV-5, simian virus 5; SV-41, simian virus 41; TiV, Tioman virus. Respiroviruses: BPIV-3, bovine parainfluenza virus 3; HPIV-1(3), human parainfluenza virus 1(3); SeV, Sendai virus. Morbilliviruses: CDV, canine distemper virus; CeMV, cetacean morbillivrus; MeV, measles virus; PPRV, peste-des-petit-ruminants virus; PDV, phocine distemper virus; RPV, rinderpest virus. Unclassified paramyxoviruses: BeV, Beilong virus; FDLV, fer-de-lance virus; MoV, Mossman virus; TuPMV, Tupaia paramyxovirus.

Mononegavirales 329

Rhabdovirus nucleocapsids are assembled into a more structured skeleton form in virus particles due to association with a matrix protein.

Features of the Viral Genome RNA

The precise details of the genome RNA vary between viruses with a wide range of lengths possible. An unusual feature of some paramyxoviruses is that the genome length is always a multiple of six nucleotides, referred to as the 'rule of six'. This is due to the nucleoprotein which binds the RNA associating with units of six nucleotides. However, most viruses in the order do not conform to this rule.

Promoter Sequences

The virus RNA contains three functional promoter sequences, the genomic promoters at the 3′ ends of the genome and the antigenome which directs replication of the virion RNA, and the transcriptional promoter which directs transcription from the genome. The genomic and antigenomic promoters share at least a partial sequence similarity due to the inverse complementarity of the virion RNA ends. To date the precise sequence requirements for the genomic and transcriptional promoters have not been defined precisely for most viruses and it is likely that they overlap. In some viruses, a short leader RNA transcript is produced from the genomic promoter. These transcripts, like the full-length antigenome and genome, are neither capped nor, other than for the nucleorhabdoviruses, polyadenylated. Originally it was believed that the transcriptional promoter initiated transcription from the genomic 3′ terminus producing the short leader RNA and then reinitiated transcription from the first gene to produce the first capped mRNA with a termination and reinitiation cycle at each gene junction to produce the remaining mRNAs as described below. More recently this has become less certain and the transcriptional promoter may cause initiation of transcription to occur directly from the first mRNA start signal.

Gene Start and Gene End Signals

Each gene is flanked by a transcription start and termination sequence. The transcription start and transcription termination/polyadenylation signals, though differing

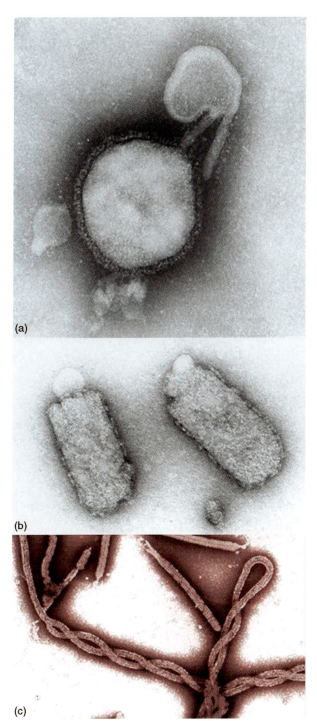

Figure 4 Electron micrographs showing the differing virus particle morphologies seen within the order *Mononegavirales*. (a) Electron micrograph of human parainfluenza virus type 3, a member of the family *Paramyxoviridae*. (b) Electron micrograph of rabies virus, a member of the family *Rhabdoviridae*. (c) Electron micrograph of Ebola virus, a member of the family *Filoviridae*. (a) Reproduced from Henrickson KJ (2003) Parainfluenza viruses. *Clinical Microbiology Reviews* 16: 242–264, with permission from American Society for Microbiology. (b) Reproduced from Mebatsion T, Weiland F, and Conzelmann K-K (1999) Matrix protein of rabies virus is responsible for the assembly and budding of bullet-shaped particles and interacts with the transmembrane spike glycoprotein G. *Journal of Virology* 73: 242–250, with permission from American Society for Microbiology. (c) Reprinted by permission from Macmillan Publishers Ltd: Nature Medicine, vol. 10, pp. S110–S121, Geisbert TW and Jahrling PB, Exotic emerging viral diseases: Progress and challenges, copyright (2004).

between viruses, are conserved between different genes of the same virus, each being *c.* 10 nt in length. The gene end/polyadenylation signals have a series of uridine residues at the end of the gene (the 5′ end in the genomic sense) that are copied by a 'stuttering' mechanism to give a poly(A) tail.

Intergenic Regions

Most viruses of the order *Mononegavirales* have untranscribed regions between the gene end and gene start signals. These occur between all genes in most of the viruses. However, in the *Bornaviridae*, most of the transcription start sites are located upstream of the transcription end/polyadenylation site of the upstream gene so the only intergenic region is between the N and P genes. In the filoviruses and respiratory syncytial viruses, some transcripts also overlap. In the latter case, the overlap is confined to the L gene transcript starting upstream of the M2 gene, in Marburg virus the mRNAs for the VP30 and Vp24 proteins overlap, and in Ebola virus the VP35/VP40, GP/VP30, and VP24/L transcripts overlap. In some viruses, the size of the intergenic regions is conserved between all the genes, for example, two nucleotides in vesicular stomatitis virus (VSV) and three nucleotides in respiroviruses and henipaviruses, whereas in most viruses the length of the intergenic regions is variable.

Transcription

Transcription is initiated from the single promoter near the 3′ end of the genome and requires the N, P, and L proteins in addition to the genome as a minimal requirement. In pneumoviruses, transcription of longer mRNAs is enhanced by the M2-1 protein and in nucleorhabdoviruses, nucleocapsid complexes isolated from nuclei, but not virus particles, are transcriptionally active, thus suggesting a requirement for cellular proteins. The involvement of cellular proteins in virus transcription has been suggested for some other viruses whereas transcription can be achieved *in vitro* with nucleocapsids isolated from VSV and cytorhabdoviruses.

The precise start point of transcription is not known. It is now considered possible that transcription is initiated directly from the start of the first gene, though it is possible for some viruses that transcription begins at the 3′ terminus of the genome with production of a short non-polyadenylated leader RNA. The polymerase initiates mRNA synthesis at the first transcription initiation signal sequence. Having transcribed the gene the polymerase encounters the conserved transcription termination sequence at which point it iteratively adds a polyadenylate tail at the sequence of U residues and ceases transcription.

The polymerase then either detaches from the template and has to reinitiate at the 3′ transcriptional promoter or reinitiates transcription from the next gene start signal. In the case of the L gene of respiratory syncytial virus and most genes in Borna disease virus this gene start signal may be upstream of the gene end signal. This overlapping organization may reduce the level of expression of the downstream gene. More typically an intergenic sequence separates the two genes and the polymerase reinitiates transcription from the downstream gene. In a small proportion of cases the polymerase fails to terminate transcription as normal and continues to the end of the next gene generating a polycistronic transcript. The overall effect of this transcription process is to produce a gradient of transcription with those genes closer to the promoter being expressed at higher levels than those located further away. The steepness of this gradient varies considerably between viruses and between the cell infected (e.g., in measles vs. subacute sclerosing panencephalitis (SSPE)). The attenuation may vary between different genes, so the decline in transcript levels may be in a series of steps rather than a smooth gradient.

Splicing of mRNA

Most viruses in this order replicate in the cytoplasm and produce only unspliced RNA molecules. Nucleorhabdoviruses replicate in the nucleus but splicing has not been observed in these viruses. However, Borna disease virus replicates in the nucleus and three introns have been identified (**Figure 1**). The smallest and most 5′ terminal (mRNA sense) is spliced out to remove the part of the M open reading frame (ORF) and allow translation of the RNA to give the G glycoprotein. A doubly spliced RNA with both the first intron and a second one removing most of the G ORF produces the L mRNA. A third longer intron shares the splice donor site with the second one but has a different splice acceptor site resulting in larger intron that could result in a truncated L protein or proteins in either of the other two ORFs but these have not been observed. The use of spliced RNAs results in a larger number of transcripts beyond that normally observed for a virus in the order *Mononegavirales* with five gene start sites.

Replication of the Viral RNA

Replication of the viral RNA occurs via a positive-sense antigenome intermediate. This RNA, like the genome, is encapsidated as it is synthesized. The factors determining whether a genomic template is used for transcription or replication are not well understood and may vary between viruses. In some cases, the stoichiometry of the N, P, and L proteins constituting the polymerase are known to differ between the two processes. The antigenome RNA is subsequently used as the template for the synthesis of multiple genomic RNA molecules. The inverse complementarity of the terminal sequences is presumably related to their functions in synthesis of full-length antigenomic and genomic RNAs and their encapsidation.

Viral Proteins

A set of five proteins have counterparts in all the viruses of the order whereas others are unique to viruses in specific families or genera.

Nucleocapsid (N) Proteins

All members of the order possess a protein that coats the genomic and antigenomic RNAs to form a nucleocapsid complex along with the phosphoprotein and polymerase protein. In paramyxoviruses conforming to the 'rule of six', the N protein binds six nucleotides of RNA. The N protein is essential for transcription and replication and is usually encoded by the promoter proximal gene transcribed at the highest level (the exception is in the pneumoviruses where there are two short genes encoding nonstructural proteins). In the nucleorhabdoviruses there is a nuclear localization signal near the C-terminus and the N protein is directed to subnuclear structures by association with the P protein. In Borna disease virus, two forms of N protein, p40 and p38, are translated from a single mRNA using an internal translation initiation event. Both are localized in the nucleus, the p38 protein lacking a nuclear localization signal and being translocated in association with other virus proteins.

Phosphoprotein (P)

This protein is encoded by the gene following that encoding the nucleocapsid protein in all the viruses. It is phosphorylated and required for transcription and replication. In some viruses, the ratio of the N and P proteins is believed to influence whether the viral polymerase carries out transcription or replication. The nucleorhabdovirus P protein has a nuclear export signal and is expressed throughout the cell when expressed alone but localized to the nucleus in the presence of N protein. In Borna disease virus, the P protein has a nuclear localization signal and probably has a role in retaining the N and X proteins in the nucleus. The rabies virus P protein inhibits the production of β-interferon by interfering with the function of the TBK kinase that activates IRF3 that in turn activates transcription from the β-interferon promoter. The P protein of Ebola virus (VP35) also inhibits IRF3-mediated activation of the β-interferon promoter but acts upstream of kinases such as TBK.

Other Proteins Encoded by the P mRNA

The P mRNA encodes additional ORFs in the *Bornaviridae* and members of the subfamily *Paramyxovirinae* by use of alternative initiation codons, which are not always the standard AUG. Rhabdoviruses and filoviruses also have additional ORFs that, in the case of some VSV strains, appear to be used to produce C and C' proteins which may enhance transcription. In Borna disease virus, leaky scanning results in the P protein being produced from an ORF different from that of the first which encodes the X or p10 protein. The X protein is localized to the nucleus with other viral proteins and has a nuclear export signal that is blocked by association with P protein. A second AUG codon in the same frame as the P protein results in an N-terminally truncated form of the P protein P' or p16 that associates with the other viral proteins.

In the viruses belonging to the subfamily *Paramyxovirinae* there are two mechanisms for expressing of additional genes from the P mRNA. The C proteins (C, C', Y1, and Y2) are translated in a different ORF from alternative initiation codons that are sometimes not the standard AUG codon. The V proteins (V and W) share a common N-terminus with the P protein but have an editing site where additional G residues are sometimes inserted by the transcriptase (one for V proteins and two for W) into the mRNA giving the protein an alternative C-terminus. In most cases the unedited mRNA transcript encodes the P protein and the edited ones the V or W proteins but in the rubulaviruses the V protein is encoded by the unedited transcript and P by the edited one. These proteins play a role in the inactivation of interferon induction and signaling, although the details of how this occurs vary between viruses.

Matrix Proteins

The matrix protein is typically encoded by the third transcript (except for the pneumoviruses, nucleorhabdoviruses, and cytorhabdoviruses that have additional genes, as described below). It is a major structural component of the viruses and is important for particle morphogenesis. In some cases, virus-like particles can be obtained by expression of M protein alone and the morphology of rabies virus lacking an M gene loses the characteristic bullet shape in addition to there being a greatly reduced yield of virus particles. Different subpopulations of the M protein are believed to be involved in different aspects of morphogenesis (see below). In the case of vesiculoviruses, the M protein inhibits cellular transcription and mRNA export from the nucleus, thus preventing the interferon response. In addition, the M protein is largely responsible for the cytopathic effect seen in cells infected with VSV. In the case of Borna disease virus, there are conflicting reports as to the glycosylation status and subcellular localization of M protein.

Viral Glycoprotein(s)

All the viruses encode between one and three surface glycoproteins that are responsible for attachment of the virus to infected cell membranes and subsequent fusion either with the plasma membrane or in the case of rhabdoviruses with internal membranes. In addition, some rubulaviruses and all pneumoviruses have a less well characterized small hydrophobic protein that is also membrane associated and sometimes glycosylated.

The G proteins of rhabdoviruses and filoviruses and the fusion proteins of the viruses in the family *Paramyxoviridae* form trimers at the surface of infected cells and on virions. These proteins are type I membrane proteins with an N-terminal signal peptide that is cleaved off in the endoplasmic reticulum, a C-terminal membrane anchor, and a number of glycosylation sites on the ectodomain. The Ebola virus G protein is encoded by an edited mRNA with the unedited RNA and a second edited RNA encoding shorter soluble glycoproteins lacking a transmembrane anchor. The fusion proteins of viruses in the families *Paramyxoviridae* and the filovirus G proteins have a protease cleavage site that results in the mature form of each monomer having two disulfide-linked chains comprising a smaller N-terminal chain, F2, and a larger F1 chain. The N-terminus of the F1 chain has the fusion peptide which is located at the end of an extended α-helix following activation by binding to a receptor or exposure to low pH. This peptide inserts into the target membrane and a hairpin-like structure is formed that results in the C-terminal viral membrane anchor being located at the same end of the molecule as the fusion peptide along with their associated heptad repeats.

Viruses in the family *Paramyxoviridae* have a second membrane glycoprotein variously known as H if it has hemagglutination activity, HN if it has hemagglutination and neuraminidase activity, and G if it has neither (although the murine pneumonia virus G protein was named by analogy with that of the other pneumoviruses despite having hemagglutination activity). This protein is a type II membrane protein with an N-terminal, uncleaved signal-anchor and a C-terminal ectodomain that has N-linked glycosylation sites and in pneumoviruses extensive O-linked glycosylation. The proteins form a tetramer in the cases where the structure has been studied and is conventionally regarded as the attachment protein, although in some cases virus lacking this protein can still infect cells. In some viruses (e.g., Newcastle disease virus, human parainfluenza virus type 3, mumps virus, and canine distemper virus), it is also required to obtain full fusion activity, although in others (e.g., simian virus 5, measles virus, and avian metapneumovirus) F protein alone is sufficient.

The rubulaviruses SV5 and mumps along with all the pneumoviruses have a third small type II membrane protein. This is glycosylated in the case of the pneumovirus proteins and in the metapneumoviruses has an additional C-terminal region with conserved cysteine residues. In the pneumoviruses several forms exist, viz. unglycosylated, glycosylated, and a polylactosamine-modified glycosylated form. Little is known about its function although the avian metapneumovirus SH protein suppresses F-protein-mediated cell–cell fusion.

The ephemeroviruses have an additional nonstructural, highly glycosylated protein of unknown function.

The L protein

The L protein is the largest viral protein and is encoded by all viruses in the order *Mononegavirales* at the 5′ end of the genome and is therefore produced from the least abundant mRNA transcript. It contains conserved motifs and is presumed to contain the major enzymatic activities of the RNA polymerase in the nucleocapsid complex in which, together with the N and P proteins (and in pneumoviruses M2-1 protein), it carries out transcription (including capping and polyadenylation) and replication of the viral RNA.

Additional Virus-Encoded Proteins

Several members of the order encode proteins in addition to those described above. Two nonstructural proteins, NS1 and NS2, are encoded by the pneumoviruses and are not found in the metapneumoviruses or any of the other viruses in the order *Mononegavirales*. They are encoded by the most abundant transcripts and appear to have a role in counteracting the host interferon response. Both proteins are required to block the effects of interferon presumably by acting on downstream effectors because they do not interfere with JAK/STAT signaling. They also interfere with IRF3 activation which is required for interferon β-induction.

The pneumoviruses also contain an M2 gene containing two ORFs located immediately upstream of that encoding the L protein. The first ORF encodes the M2-1 protein which shows marked similarity between all these viruses and has been shown to enhance transcription, particularly of longer genes. The second ORF of the M2 gene shows no similarity to other proteins or conservation between viruses. The M2-2 protein is thought to inhibit virus RNA synthesis.

Three plant viruses (the nucleorhabdovirus, rice yellow stunt virus, and the cytorhabdoviruses, northern cereal mosaic virus and strawberry crinkle virus) have poorly characterized ORFs prior to the polymerase gene which may encode additional proteins. There is tentative evidence that these may be related to sequences likely to be involved in RNA replication/transcription. Novirhabdoviruses have an ORF encoding the nonvirion (NV) protein that appears to enhance virus growth.

Cytorhabdoviruses and nucleorhabdoviruses encode between one and four additional proteins at a location between the phosphoprotein and matrix protein genes. These are not well characterized but appear to have similarities to proteins in plants involved in transport between cells via plasmodesmata and are involved in the spread of nucleocapsids between plant cells via the plasmodesmata.

Viral Entry

Mononegaviruses infecting mammalian cells typically attach to receptors on the plasma membrane of cells and fuse either directly with the plasma membrane or, after uptake into endosomal vesicles and exposure to a more acidic environment, with endosomal membranes. These processes are achieved using one or more virus glycoproteins. In viruses of the family *Paramyxoviridae*, these two activities appear to reside on different proteins with the G, H, or HN protein mediating attachment and the F protein mediating fusion. However, in cell culture, viruses lacking the attachment protein gene are often able to grow, suggesting that viral attachment is possible in its absence. The paramyxoviruses generally fuse with the plasma membrane at neutral pH whereas animal rhabdoviruses fuse with endosomal membranes following activation by low pH. Primary infection of plants by the cytorhabdoviruses and nucleorhabdoviruses involves physical injection into plant tissue by invertebrate vectors such as planthoppers, leafhoppers, and aphids, with the latter responsible for viral entry of all but two of these viruses that infect dicotyledenous plants. Fusion of the viral membrane with cellular membranes results in the release of viral nucleocapsids into either the cytoplasm or, in the case of the nucleorhabdoviruses, the nucleus. The animal viruses, with the exception of the bornaviruses, replicate in the cytoplasm where the viral nucleocapsids are released following fusion of the viral membrane with the plasma membrane or endosomal membrane.

Viral Assembly and Budding

The process of viral assembly has been most thoroughly studied in VSV and the general features are thought to be common to all members of the order though the details will differ. The overall process of VSV assembly involves early stages where nucleocapsids are assembled by N protein being released from N–P dimers to bind the genomic RNA as it is being synthesized. At the same time, the G glycoprotein forms microdomains at the plasma membrane. Different populations of M protein are believed to be involved in different steps of assembly. These are, first, binding to regions of plasma membrane enriched in G protein, second, recruitment of the nucleocapsids to these regions, and third, condensation of the nucleocapsids into helical structures which may have M protein at the center as well as between the membrane and the nucleocapsid. While G protein facilitates viral assembly it is not essential, and the M protein is believed to be the key component in assembly with virus-like particles being produced by expression of M protein alone. As the M protein associated with the membrane binds to the condensing nucleocapsid structures, bullet-shaped protrusions occur at the membrane surface. Cellular proteins are believed to associate with the M protein via a so-called late domain (PPPY in the case of VSV) to directly or indirectly cause fusion of the membrane at the base of the protrusions to release virus particles.

The matrix protein is also the main determinant of virus assembly and budding in viruses of the family *Paramyxoviridae* although the precise mechanisms probably vary between viruses. For example, Sendai virus M protein alone can form virus-like particles whereas SV5 M protein cannot. Interaction of viral and cellular proteins probably varies as well since measles and respiratory syncytial virus production are dependent on the presence of cellular actin whereas human parainfluenza virus 3 release is dependent upon the presence of microtubules. These cellular components are required for efficient RNA synthesis as well as virus production. In measles virus, interaction of the M protein and the glycoproteins may occur in lipid rafts to which M protein is recruited independently of the glycoproteins. However recruitment of the hemagglutinin to the rafts may be dependent upon its interaction with the fusion protein. In contrast, the M proteins of Sendai virus and respiratory syncytial virus may require interaction with viral glycoproteins, in particular the fusion protein, for recruitment into lipid rafts. This association is believed to occur during transport to the cell surface.

Budding of most of the animal viruses in the order occurs at the plasma membrane. However, rabies virus buds into internal membrane compartments as do the plant rhabdoviruses. In the case of the nucleorhabdoviruses, there appears to be extensive rearrangement of the intracellular membranes with enlarging of the nuclei due to the insertion of extra membranes through which the virus buds. These membranes are contiguous with the nuclear membrane resulting in the virus ending up in the perinuclear space. Virus is transmitted from here via insect vectors. The matrix protein has an additional role in the polarity of virus release from cells in measles virus where, in contrast to most viruses in the family *Paramyxoviridae*, the HN protein is not transported in a polarized manner and the F protein is directed to the basolateral surface. In the presence of M protein, complexes of M and the glycoproteins are sorted to the apical surface. In other members of this virus family, the glycoproteins are targeted to the apical surface from which the viruses bud, whereas in VSV sorting of G protein and virus budding occur at the basolateral surfaces.

The cytoplasmic tails of the glycoproteins of the viruses in the family *Paramyxoviridae* have roles in enhancing the efficiency and specificity of virus production, recombinant viruses with truncated cytoplasmic tails showing a lack of localization at the membrane surface, lower virus yields, and virus preparations with reduced amounts of viral glycoproteins and increased levels of cellular proteins. VSV G protein also requires a minimal length of cytoplasmic tail to enhance budding efficiency but there does not appear to be any requirement for a specific sequence. Some paramyxovirus fusion proteins can produce virus-like particles.

All the viruses in the order *Mononegavirales* with the possible exception of Borna disease virus therefore seem to have the matrix protein and at least one glycoprotein as major determinants of virus release but the exact details may vary even within members of the same family or subfamily. Budding to release progeny virions of most of the animal viruses occurs at the plasma membrane. However, rabies virus buds into internal membrane compartments as do the plant rhabdoviruses. In the case of the nucleorhabdoviruses, there appears to be extensive rearrangement of the intracellular membranes with enlarging of the nuclei due to the insertion of extra membranes through which the virus buds. These membranes are contiguous with the nuclear membrane resulting in the virus ending up in the perinuclear space, from where they are transmitted via insect vectors.

See also: Animal Rhabdoviruses; Bornaviruses; Bovine Ephemeral Fever Virus; Chandipura Virus; *Ebolavirus*; Filoviruses; Fish Rhabdoviruses; Human Respiratory Syncytial Virus; Marburg Virus; Measles Virus; Mumps Virus; Paramyxoviruses of Animals; Parainfluenza Viruses of Humans; Paramyxoviruses; Plant Rhabdoviruses; Rabies Virus; Rinderpest and Distemper Viruses; Sigma Rhabdoviruses; Vesicular Stomatitis Virus.

Further Reading

Feldmann H, Geisbert TW, Jahrling PB, et al. (2005) Family *Filoviridae*. In: Fauquet CM, Mayo MA, Maniloff J, Desselberger U, and Ball LA (eds.) *Virus Taxonomy: Eighth Report of the International Committee on Taxonomy of Viruses*, pp. 645–653. San Diego, CA: Elsevier Academic Press.

Geisbert TW and Jahrling PB (2004) Exotic emerging viral diseases: Progress and challenges. *Nature Medicine* 10: S110–S121.

Henrickson KJ (2003) Parainfluenza viruses. *Clinical Microbiology Reviews* 16: 242–264.

Jackson AO, Dietzgen RG, Goodin MM, Bragg JN, and Deng M (2005) Biology of plant rhabdoviruses. *Annual Reviews of Phytopathology* 43: 623–660.

Jayakar HR, Jeetendra E, and Whitt MA (2004) Rhabdovirus assembly and budding. *Virus Research* 106: 117–132.

Lamb RA, Collins PL, Kolakofsky D, et al. (2005) Family *Paramyxoviridae*. In: Fauquet CM, Mayo MA, Maniloff J, Desselberger U, and Ball LA (eds.) *Virus Taxonomy: Eighth Report of the International Committee on Taxonomy of Viruses*, pp. 655–668. San Diego, CA: Elsevier Academic Press.

Lamb RA and Parks GD (2006) *Paramyxoviridae*: The viruses and their replication. In: Knipe DM, Howley PM, Griffin DE, et al. (eds.) *Fields Virology*, 5th edn., pp. 1449–1496. Philadelphia: Lippincott Williams and Wilkins.

Lipkin WI and Briese T (2006) *Bornaviridae*. In: Knipe DM, Howley PM, Griffin DE, et al. (eds.) *Fields Virology*, 5th edn., pp. 1829–1851. Philadelphia: Lippincott Williams and Wilkins.

Lyles DS and Rupprecht CE (2006) *Rhabdoviridae*. In: Knipe DM, Howley PM, Griffin DE, et al. (eds.) *Fields Virology*, 5th edn., pp. 1363–1408. Philadelphia: Lippincott Williams and Wilkins.

Mebatsion T, Weiland F, and Conzelmann K-K (1999) Matrix protein of rabies virus is responsible for the assembly and budding of bullet-shaped particles and interacts with the transmembrane spike glycoprotein G. *Journal of Virology* 73: 242–250.

Pringle CR (2005) Order *Mononegavirales*. In: Fauquet CM, Mayo MA, Maniloff J, Desselberger U, and Ball LA (eds.) *Virus Taxonomy: Eighth Report of the International Committee on Taxonomy of Viruses*, pp. 609–614. San Diego, CA: Elsevier Academic Press.

Sanchez A, Geisbert TW, and Feldmann H (2006) *Filoviridae*: Marburg and Ebola viruses. In: Knipe DM, Howley PM, Griffin DE, et al. (eds.) *Fields Virology*, 5th edn., pp. 1409–1448. Philadelphia: Lippincott Williams and Wilkins.

Schwemmle M, Carbone KM, Tomonga K, Nowotny N, and Garten W (2005) Family *Bornaviridae*. In: Fauquet CM, Mayo MA, Maniloff J, Desselberger U, and Ball LA (eds.) *Virus Taxonomy: Eighth Report of the International Committee on Taxonomy of Viruses*, pp. 615–622. San Diego, CA: Elsevier Academic Press.

Tomonga K, Kobayashi T, and Ikuta K (2002) Molecular and cellular biology of Borna disease virus infection. *Microbes and Infection* 4: 491–500.

Tordo N, Benmandour A, Calisher C, et al. (2005) Family *Rhabdoviridae*. In: Fauquet CM, Mayo MA, Maniloff J, Desselberger U, and Ball LA (eds.) *Virus Taxonomy: Eighth Report of the International Committee on Taxonomy of Viruses*, pp. 623–644. San Diego, CA: Elsevier Academic Press.

Mouse Mammary Tumor Virus

J P Dudley, The University of Texas at Austin, Austin, TX, USA

© 2008 Elsevier Ltd. All rights reserved.

Glossary

HRE Hormone response elements found in several viral genomes that bind the activated forms of steroid hormone receptors to increase transcription in the presence of a steroid ligand.

Rem A Rev-like RNA-binding protein that is responsible for efficient nuclear export of unspliced mouse mammary tumor virus RNA.

Superantigen A protein that interacts with entire classes of T cells primarily through the variable region of the beta-chain of the T-cell receptor, leading to signal transduction and the release of cytokines.

History

Mouse mammary tumor virus (MMTV) was first reported in 1933 by the Jackson Memorial Laboratory and by Korteweg in 1934 as an extrachromosomal influence on the incidence of breast cancer in inbred mouse strains. Initial crosses between strains with a high mammary cancer incidence and strains with a low mammary cancer incidence revealed that female progeny invariably had the mammary cancer incidence of the female parent. Subsequently, Bittner showed that an extrachromosomal factor was transmitted through maternal milk, and this factor was later associated with viral particles, called 'B-particles', which caused breast cancer in susceptible mice.

Taxonomy and Classification

MMTV is a prototype species of the genus *Betaretrovirus* in the family *Retroviridae*. These viruses previously were referred to as type B retroviruses based on their appearance by electron microscopy (a characteristic acentric core within particles of *c.* 100 nm). Milk-borne MMTVs are often named for the inbred mouse strain from which they are derived, for example, C3H MMTV or MMTV (C3H). Multiple double-stranded DNA copies are found in the chromosomal DNA of most commonly used laboratory strains of mice (called integrated or endogenous proviruses). These endogenous proviruses presumably represent viral insertions into chromosomal DNA of germline cells and are referred to as *Mtv* followed by an Arabic number, for example, *Mtv8*. Most endogenous *Mtv*s have defects in one or more genes and, therefore, these proviruses often fail to produce infectious virus. Currently, MMTV is classified with other betaretroviruses including Mason–Pfizer monkey virus (MPMV), Jaagsiekte sheep retrovirus (JSRV), and human endogenous retrovirus type-K (HERV-K).

Properties of the Virion

MMTV particles contain a single-stranded positive-sense RNA, which exists as a dimer, and is encapsidated as a helical ribonucleoprotein (RNP) by the nucleocapsid (NC) protein; reverse transcriptase (RT) and integrase (IN) are closely associated with the RNP. The RNP is surrounded by an icosahedral shell (although the exact structure has not been carefully studied) composed of capsid (CA) protein. There is considerable variation in the size of cores, similar to that observed with human immunodeficiency virus 1 (HIV-1). MMTV capsids are bound via the matrix (MA) protein to the viral envelope, a portion of the cellular plasma membrane that has been modified by the insertion of the surface (SU) and transmembrane (TM) proteins.

Properties of the Genome

The viral RNA is bound at either end by a short direct repeat (R) of 15 bp (**Figure 1**). The R regions are adjacent to unique regions of approximately 120 and 1200 bp, respectively, present at the 5′ (U5) or 3′ (U3) ends of the RNA. A cellular tRNA (tRNA3Lys) is bound through 18 bp of complementarity to each copy of the viral RNA

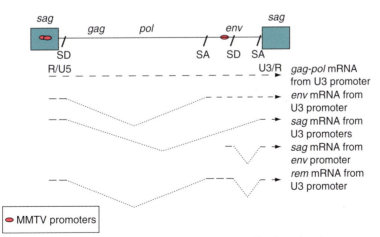

Figure 1 Diagram of the MMTV proviral genome and structure of viral mRNAs. The boxed regions represent the LTRs with two described promoters. The standard promoter gives unspliced transcripts that start with the first base of the R region in the 5′ LTR and end at the last base of the R region of the 3′ LTR. Unspliced transcripts may be directly exported from the nucleus for Gag, Gag-Pro, and Gag-Pro-Pol translation. Alternatively, such transcripts may be spliced to give two singly spliced mRNAs encoding either envelope or superantigen. A fourth transcript from the LTR promoter gives a doubly spliced mRNA that encodes the RNA export protein, Rem. An internal promoter in the envelope gene also allows production of a singly spliced RNA encoding the superantigen. The positions of known splice donors (SD) and acceptors (SA) are also shown.

at the primer-binding site (PBS) located just downstream of U5. Although the RNA packaging site (often referred to as ψ) has not been defined, this site is likely to include the region between the splice donor (SD) and splice acceptor (SA) sites for the envelope (*env*) mRNA. Packaging of viral mRNAs, other than genomic RNA, is prevented by exclusion of the ψ region. The first SD site precedes the group-specific antigen (*gag*) region that encodes a Gag precursor with the nonglycosylated proteins of the virion in the order NH2-MA-p21-CA-NC-COOH. Interestingly, the virus also encodes two other precursor polypeptides, Gag-Pro and Gag-Pro-Pol, from the genomic RNA by ribosomal frameshifting. Gag-Pro encodes the Gag proteins, a dUTPase (DU), and the viral protease (PR), whereas the Gag-Pro-Pol protein also encodes RT, including a ribonuclease H (RNase H) activity, and IN. Both the DU and RT are trans-frame proteins that contain sequences from the preceding protein, NC and PR, respectively. At the 3′ end of the genome, the envelope (Env) proteins are specified in the order NH2-SU-TM-COOH from a singly spliced mRNA. Recent findings indicate that MMTV encodes another protein called regulator of export of MMTV mRNA or Rem from a doubly spliced mRNA. The *rem* mRNA is translated from the open reading frame that also specifies SU and TM. Such data emphasize the efficiency with which MMTV and other viruses use their genetic information. Furthermore, unlike most other retroviruses, which have multiple stop codons within the long terminal repeat (LTR), the MMTV U3 region has yet another gene that encodes a superantigen (Sag) from one or more singly spliced mRNAs.

Virus Replication

Recent evidence indicates that MMTV uses the transferrin receptor 1 (Tfr1) to mediate infection. Tfr1 is ubiquitously expressed in many rodent cells, and this observation may explain infection of numerous rat and mouse cell lines in culture. Differences in mouse Tfr1 and primate Tfr1 appear to be sufficient to prevent MMTV infection of monkey and human cells. Despite numerous reports of MMTV sequences in human tumors, failure of human Tfr1 to allow viral entry is one argument against the validity of these data. However, use of other receptors for MMTV entry, particularly in specific cell types, is possible. Entry through Tfr1 appears to occur through adsorptive endocytosis, and receptors are recycled to the cell surface where they allow entry of additional MMTV particles into previously infected cells. Thus, unlike many retroviruses, MMTV is not susceptible to superinfection resistance, and many chronically infected mouse lines have large numbers of integrated proviruses.

Following entry and partial uncoating in the cytoplasm, the virally encoded RT is activated. Using the cellular lysyl-tRNA primer bound to genomic RNA, RT synthesizes a partial minus-strand DNA bound to plus-stranded RNA. Using the RT-associated RNase H activity and several strand transfers, the RNA template is degraded and a double-stranded provirus is synthesized. However, the product of reverse transcription is different from the starting template so that the U5 and U3 sequences present uniquely in viral RNA are duplicated to give longer repeats at each end of the provirus (LTRs). The LTRs have the structure U3-R-U5 (**Figure 1**). Because nuclear entry of the preintegration complex (PIC) containing the provirus is thought to require nuclear envelope breakdown during mitosis, it is generally believed that MMTV must infect dividing cells. However, MMTV encodes DU, a protein found in many nonprimate lentiviruses that infect nondividing cells. DU prevents misincorporation of uracil and mutation of newly synthesized proviruses in nondividing cells where the ratio of dUTP to TTP is high.

After nuclear entry of the PIC, the provirus integrates using the MMTV-encoded IN protein cleaved from the Gag-Pro-Pol precursor. IN protein introduces an asymmetric cut 2 bp from the linear ends of the provirus as well as an asymmetric break exactly 6 bp apart on opposite DNA strands of host DNA. Proviral integrations are not site specific and may occur at transcriptionally active sites, although this idea has not been tested directly. Following the joining reaction, the repair of virus–cell junctions by cellular enzymes generates a 6 bp direct repeat of cell DNA that flanks the viral LTRs. Such a structure resembles those formed by the transposable elements of bacteria, yeast, and *Drosophila*.

The integrated provirus contains all the signals necessary for recognition by RNA polymerase II, and many of these signals are present in the U3 region of the LTR. Transcription from the standard promoter is initiated in the 5′ LTR starting at the U3/R junction and terminating at the R/U5 junction (**Figure 1**). However, several other promoters have been described, including one approximately 500 bp upstream of the U3/R junction and another within the envelope gene. Termination appears to be reasonably inefficient, and some MMTV transcripts probably terminate in the adjacent cellular DNA. A portion of genome-length MMTV RNA is processed into singly spliced *env* and *sag* mRNAs.

Surprisingly, the *sag* gene appears to be expressed by at least two singly spliced mRNAs from independent promoters (**Figure 1**), possibly allowing different levels of Sag in various cell types. MMTV also produces a doubly spliced mRNA that encodes the RNA export protein, Rem. In most cell types, the levels of *gag-pol* and *env* mRNAs greatly exceed *sag* and *rem* mRNAs.

The unspliced RNA (8.7 kb) is translated into Gag, Gag-Pro, and Gag-Pro-Pol precursor proteins. Since *pro* (the viral protease gene) and *pol* (the polymerase/integrase gene) are out-of-frame with respect to *gag* and each other, one ribosomal frameshift is required for Gag-Pro synthesis, and a second frameshift is necessary to produce Gag-Pro-Pol (**Figure 2**). Because the first frameshift is quite efficient, MMTV RT levels in infected cells are equivalent to those produced by other retroviruses. The *env* mRNA is translated into a precursor protein on membrane-bound polyribosomes. This precursor is modified by glycosylation in both the endoplasmic reticulum and the Golgi, and protein cleavage to SU and TM also occurs in the latter compartment. The *sag* mRNA appears to be translated into a type II transmembrane protein of 36 kDa; this protein is glycosylated and reportedly cleaved to generate a C-terminal fragment of 18 kDa. Sag is associated with major histocompatibility complex (MHC) class II protein at the surface of antigen-presenting cells. Unlike the previously described proteins, Sag is not a known structural component of virions. The Rem protein is translated in the same frame as the envelope protein, including the Env signal peptide (**Figure 2**). The nascent protein escapes signal recognition particle, and the full-length Rem protein (33 kDa) localizes to the nucleolus using motifs found in the signal peptide. Rem synthesis is required for efficient export of unspliced genomic RNA and must precede production of infectious viral particles. The N-terminal one-third of Rem appears to encode all functions necessary for RNA export, and deletion of the C-terminus increases the ability of Rem to function in export assays. These data indicate that unspliced MMTV RNA export and particle production are negatively regulated, at least in some cell types.

The precursors for Gag, Gag-Pro, and Gag-Pro-Pol aggregate within the cell cytoplasm into procapsids called intracytoplasmic A particles. This process is distinct from the maturation of C-type particles that assemble Gag precursors at the cell surface concomitant with the budding process. Presumably, the precursor proteins are folded so that NC and RT proteins are sequestered inside the particle to interact with a dimer of viral RNA and the lysyl tRNA. The viral PR, which is present in a fraction of the Gag precursors, is apparently responsible for the cleavage events that produce the mature virion proteins, MA, p21, p3, p8, CA, and NC. The functions of p21, p3, and p8 are currently unknown. Like HIV, the MMTV Gag precursor contains a P(S/T)AP sequence that likely functions as a late (L) domain; such domains interact with cellular protein complexes (known as ESCRTs) to promote budding. Further cleavages by PR to give functional RT and IN, and mature cores apparently occur after budding.

Transmission and Tissue Tropism

MMTV is transmitted horizontally through maternal milk (called exogenous or milk-borne virus) or vertically through the germline (endogenous viruses). The exogenous viruses are responsible primarily for the high mammary cancer incidence of mouse strains such as RIII and C3H. However, some strains, such as GR, carry an endogenous virus (in this case, *Mtv2*) that is also transmitted through milk. Although most common inbred mouse strains carry endogenous *Mtv*s, some recently inbred strains lack such proviruses. Interestingly, strains that lack *Mtv*s often show resistance to disease induced by exogenous MMTV regardless of the route of transmission.

MMTV particles ingested by newborn mice in maternal milk survive passage through the stomach during the first weeks of life prior to maturation of the intestinal tract (**Figure 3**). These particles enter the small intestine where they cross the epithelium through M cells. Subsequently, MMTV virions encounter and infect dendritic cells and B cells. Following reverse transcription, integration, and transcription, Sag protein is produced and presented at the surface of such cells in association with MHC class II protein. Sag signaling through the T-cell receptor (TCR) leads to the release of cytokines, leading to a pool of dividing lymphoid cells for MMTV infection as well as division of previously infected cells. Both B and T cells are required for MMTV transmission since knockout or transgenic animals lacking either lymphoid subset cannot be infected efficiently by the milk-borne route. This lymphoid cell reservoir is necessary to preserve MMTV infectivity prior to the onset of puberty in mice when a source of susceptible and dividing mammary cells becomes available. Sag-mediated stimulation of lymphoid cells also improves the efficiency of viral transfer within the mammary gland. MMTV production increases during lactation when hormone levels elevate transcription from the hormone-responsive element (HRE) in the LTR U3 region (**Figure 4**). High virion production during lactation ensures that large amounts of particles will be produced at a time when newborn offspring can be infected.

Horizontal transmission of MMTV by seminal or salivary fluid has been reported. The low infectivity of

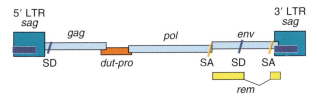

Figure 2 Positions of the open reading frames within MMTV proviral DNA. The large boxes represent the 5′ and 3′ LTRs, whereas the smaller elongated boxes represent the reading frames for the genes shown in italics. The *rem* gene is in the same reading frame as the envelope gene, which is altered by splicing (V shape). The *sag* coding sequence at the 5′ end of the genome is not expressed because it is located upstream of the viral promoters.

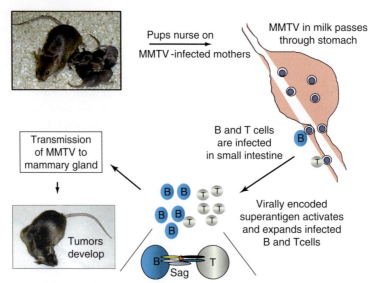

Figure 3 The MMTV life cycle. Infected mothers transmit virus to their pups through milk. In the first few weeks of life, the pups ingest the virus, which passes through the stomach into the small intestine. MMTV then transverses the M cells to deliver virus to gut-associated lymphoid cells. B cells in the gut are infected and express Sag protein at the cell surface in conjunction with MHC class II protein. Recognition of the Sag C-terminal amino acids by the variable portion of the β-chain of the T-cell receptor results in signal transduction and the release of cytokines. Cytokines then stimulate the proliferation of adjacent B and T cells, which act as a reservoir for the virus until transmission to mammary epithelial cells occurs during puberty. Virus expression and release is highest during lactation to ensure milk-borne transmission. Mammary tumors arise after multiple rounds of pregnancy and lactation due to insertional mutagenesis.

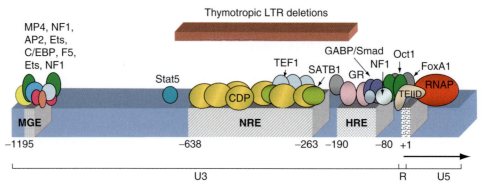

Figure 4 Transcriptional control elements in the MMTV LTR. The major transcriptional start site is located at the U3/R border in the 5′ LTR of integrated proviral DNA. The mammary gland enhancer (MGE) is located near the 5′ end of the LTR. The negative regulatory element (NRE) contains binding sites for Cutl1/CDP, SATB1, and TEF1, but different cell types vary in the level of these factors. The hormone response element (HRE) contains multiple binding sites for the activated forms of several steroid receptors, including glucocorticoid receptor (GR). The GR sites appear to be flanked by binding sites for FoxA1. The region spanning deletions found in thymotropic strains of MMTV is shown above the LTR. TFIID, transcription factor IID; RNAP, RNA polymerase II.

most body fluids (other than milk) may be attributable to virally infected lymphoid cells and the absence of large amounts of infectious MMTV particles.

Genetics and Disease Susceptibility

MMTV causes mammary adenocarcinomas by infection of the mammary epithelium. The genetic factors that influence mammary tumor incidence in mice include (1) the presence or absence of milk-borne virus, (2) the presence of an infectious endogenous MMTV, (3) cellular factors that determine virus entry and replication, and (4) host factors that may influence the immune response (including the Sag response) or hormonal levels in the animals. The presence of milk-borne virus has been demonstrated in mouse strains with a high mammary cancer incidence by foster nursing experiments as originally described by Bittner. Shortly thereafter, it was shown that such high-cancer-incidence strains (e.g., C3H) permanently lost this trait if nursed on mothers with a low mammary cancer incidence (e.g., BALB/c). Such strains, known as C3Hf, have a mammary tumor incidence between 38% and 47% (average latency of 600 days) in multiparous animals, whereas C3H strains expressing milk-borne MMTV have a tumor incidence of 88–95% (average latency of 300 days) in breeding females. Mammary tumors appearing in C3Hf mice are the result of expression of a replication-competent endogenous provirus known as *Mtv1*. Strains, such as BALB/c and C57BL, which have a mammary tumor incidence of 1% or less with long latencies, appear to lack a replication-competent MMTV.

A number of cellular and host factors also influence the appearance of mammary tumors in different mouse strains. Most of these factors have not been defined, but many of them appear to directly or indirectly affect the ability of MMTV to replicate. For example, the resistance of C57BL to MMTV infection appears to be the result of failure to express MHC class II I–E molecules, one of two types of class II antigens expressed in mice. Although many MMTV Sag proteins apparently interact with class II I–A proteins, this interaction is less efficient than the Sag/MHC class II I–E complexes in the stimulation of lymphoid cells. Loss of specific class II molecules is not the only defense against MMTV infection. The presence of endogenous MMTV proviruses that express Sag proteins results in the deletion of specific subsets of T cells. If these endogenous viruses encode Sags with the same TCR reactivity as exogenous MMTVs, milk-borne infection is blocked by preventing viral amplification in lymphoid cells (**Figure 3**). For example, endogenous *Mtv7* expresses a Sag protein that reacts with and causes deletion of Vβ6+ T cells; therefore, the milk-borne MMTV (SW) strain that encodes TCR Vβ6-reactive Sag cannot infect mouse strains that carry *Mtv7* because these mice lack the T-cell subset required for MMTV transmission.

Resistance to exogenous MMTV infection is usually a recessive characteristic, which has been used to determine if various mouse strains have the same or different resistance genes. For example, C57BL and I strain mice are both resistant to C3H MMTV infection, yet F1 hybrids of these strains have a high incidence of mammary tumors when infected by C3H virus. This result suggests that I strain mice (H-2^j) encode a functional MHC class II I–E molecule that complements the defect in C57BL mice and overcomes resistance to MMTV infection. More recently, disease resistance in I strain mice has been attributed to the development of strong neutralizing antibodies to MMTV. Further, development of neutralizing antibodies appears to be dependent on the MMTV strain. MMTVs with weak Sag activities (e.g., C3H MMTV) appear to elicit few neutralizing antibodies, whereas the opposite applies to MMTVs with strong Sag function (e.g., SW MMTV). Another host resistance mechanism is mediated by Toll-like receptor 4 (TLR4) signaling, which normally triggers an innate immune response. Wild-type C3H MMTV is lost during passage in C3H/HeJ animals, which lack functional TLR4, but not in related TLR4+ mice. C3H MMTV interactions with TLR4 trigger the immunosuppressive cytokine IL-10, allowing the virus to subvert the innate immunity by elimination of cytotoxic T cells. Further, YBR/Ei mice exhibit a dominant resistance to MMTV that depends on an adaptive T-cell response to the infection. However, not all MMTV resistance mechanisms may depend on the immune response; NH mice are resistant to the virus, perhaps related to hormonal changes that lead to early reproductive difficulties in this strain.

Traditional types of genetic experiments with MMTV have been difficult for two reasons. First, since MMTV does not form plaques or foci in cultured cells, cloning of viral stocks has not been possible. Second, molecular cloning of an intact MMTV provirus has been difficult for some MMTV strains because of selection against a specific part of the *gag* region during growth of proviral clones in *Escherichia coli*. This difficulty has been overcome by combining the 5′ end of an endogenous provirus (e.g., *Mtv1*) with the 3′ end of an exogenous provirus (e.g., C3H MMTV). Such hybrid proviruses have been used to produce infectious virions that retain oncogenicity for the mammary gland and to show that the MMTV *gag* region contributes to development of mammary tumors.

Pathogenicity

MMTV induces primarily type A and B mammary adenocarcinomas. Insertional mutagenesis is presumed to be the mechanism of MMTV-induced mammary tumors and is consistent with the relatively long latent period for tumor development (6–9 months). However, the MMTV envelope protein may act as a tumor initiator for mammary cells.

Common integration sites can be identified from MMTV-induced tumors. Current data have implicated more than 10 different loci (designated *int* or integration site genes) in MMTV-induced mammary tumors (**Table 1**). Most of these integration sites share the following characteristics:

1. The majority of the MMTV integrations are outside of the gene-coding regions, and the MMTV promoter is rarely used to initiate *int* gene transcription. Thus, an unmodified protein product is produced.
2. The MMTV provirus can activate target gene transcription over considerable distance (in excess of 15 kb). There may be activation of multiple genes in a cluster.
3. Proviruses often are integrated upstream in the opposite transcriptional orientation or downstream in the same orientation as the target gene.

Table 1 Common integration sites found in MMTV-induced mammary tumors

Locus	Gene family/function	Mouse chromosome	Integration frequency
Wnt1 (Int-1)	Wingless/growth factor	15	80% in C3H; 70% in BR6; 30% in GR
Wnt3 (Int-4)	Wingless/growth factor	11	10% in GR
Wnt10b	Wingless/growth factor	15	23% of Fgf3 transgenic
eIF3-p48 (Int-6)	Eukaryotic translation initiation factor	15	6% in Fgf3 transgenic
Rspo2 (Int-7)	R-spondin homolog/ growth factor receptor	15	13% in Czech II
Fgf3 (Int-2)	Fibroblast growth factor	7	65% in BR6; 5% in C3H; 20% in GR
Fgf4 (Fgfk/hst/hst-1/Hstf-1)	Fibroblast growth factor	7	10% in BR6
Fgf8 (Aigf)	Fibroblast growth factor	19	80% in Wnt1 transgenic
Notch1 (Mis6/Tan1)	Notch	2	8% of Erbb2 transgenic
Notch4 (Int-3)	Notch	17	20% in Czech II; 8% in BR6
Cyp19a1 (Cyp19/Int-5/Int-H)	Cytochrome P450/aromatase	9	3 chemically induced BALB/c hyperplasias

4. Target gene transcription is low or undetectable in normal adult mammary glands.
5. Transcription of the *int* genes is regulated developmentally.
6. The *int* genes appear to be conserved evolutionarily, and a number of these genes encode growth factors or truncated growth factor receptors.

Such observations are consistent with the transcriptional activation of conserved genes in the mammary gland by their proximity to MMTV LTR enhancers. Transgenic animals overexpressing the *int* genes often show mammary hyperplasias and sporadic tumors; this observation confirms the involvement of multiple genes during oncogenesis. Mating of animals with different transgenes accelerates this process, indicating cooperation between *int* genes.

MMTV variants that are expressed preferentially in T cells induce T-cell lymphomas. Interestingly, these thymotropic MMTV appear to activate a different set of oncogenes, for example, c-*myc* and *Rorc*, relative to milk-borne MMTVs as a result of LTR alterations.

Transcriptional Regulation

Studies of the MMTV HRE have been a paradigm for hormone-regulated gene expression. MMTV RNA levels increase c. 10–50-fold in the presence of glucocorticoids, progesterone, or androgens at the level of transcriptional initiation. Hormone inducibility is conferred by multiple binding sites to allow cooperative binding with the consensus TGTTCT in the region from −80 to −190 (**Figure 4**). Glucocorticoids are believed to bind to their receptors (GRs) in the cytoplasm and to subsequently translocate to the nucleus where they exchange rapidly with their binding sites. The integrated MMTV LTR is occupied by six nucleosomes (A–F) (not shown) with the A nucleosome located at the transcription initiation site (+1). The standard MMTV promoter contains a TATA element (−30 bp), which binds transcription factor IID (TFIID). Hormone-bound GR is believed to recruit remodeling factors to increase accessibility at nucleosome B and to allow binding of transcription factors Oct1 and NF1 between the TATA box and the HRE.

The tissue distribution of MMTV expression is tightly linked to viral transcriptional control, yet steroid receptors are present in many tissues in which MMTV transcription is repressed. The importance of negative regulation in MMTV transcription and disease specificity is exemplified by the isolation of MMTV variants, such as type B leukemogenic virus (TBLV), which induce T-cell lymphomas rather than mammary tumors. Such variants lack all or part of the negative regulatory element (NRE) present upstream of the HRE and acquire a T-cell specific enhancer. The NRE binds at least two related homeodomain-containing proteins called special AT-rich binding protein 1 (SATB1) and Cut-like protein1/CCAAT displacement protein (Cutl1/CDP), which have different tissue distributions and act as repressors of MMTV transcription. The highest levels of SATB1 are expressed not only in T cells but also in B cells explaining transcriptional repression in most lymphoid tissues. Mutations of the promoter-proximal SATB1-binding site elevate MMTV expression of LTR-reporter constructs in cultured cells and in lymphoid tissues of transgenic mice. Interestingly, CDP expression is high in undifferentiated B cells and mammary cells and decreases during differentiation. In mammary epithelial cells, where SATB1 is also absent, CDP is cleaved during differentiation to yield a dominant-negative protein that interferes with full-length CDP binding to the MMTV LTR. Therefore, the levels of functional CDP repressor are lowest when MMTV particle production is highest at lactation, the period when virus transmission occurs in milk.

MMTV specifies several enhancer elements, including the HRE and a mammary gland enhancer (MGE) near the 5′ end of the LTR. The MGE contains binding sites for multiple nuclear factors, including MP4, NF1, AP2, Ets, and C/EBP. Synthesis of some factors is inducible by prolactin, epidermal growth factor, or tumor necrosis factor α. The absence of NRE-binding repressors and ligand-bound steroid receptors presumably cooperate with the HRE to maximize virus production in milk.

Immune Response

Endogenous MMTVs, like exogenous viruses, express Sag proteins, initially known as minor lymphocyte stimulating (Mls) antigens. Unlike conventional antigens where a peptide associates with the groove formed by the MHC class II α- and β-chains, the central portion of Sag binds to class II protein (**Figure 5**). Since Sag is a type II transmembrane protein, the C-terminal portion is available to interact with the variable region of the β-chain (Vβ) of the TCR on a subset of CD4+ or CD4+CD8+ cells. Therefore, Sag recognizes entire classes of T cells (up to 30% of the entire T-cell repertoire), in contrast to conventional antigens that recognize TCR on less than 1 in 10^4 T cells. Sag stimulation leads to the rapid deletion of reactive cells if Sag is expressed in the thymus, whereas extrathymic deletion appears to occur more slowly.

Sag is relatively conserved among different MMTV strains, except at the C-terminus, which mediates TCR interactions. C-terminal polymorphisms correlate with reactivity with specific TCR β-chains; for example, C3H MMTV reacts with Vβ14 and 15 chains. Molecular switching experiments have shown that the C-terminal half is sufficient to specify TCR reactivity, and virtually any mutation within this region is sufficient to abolish function. Deletion of specific T-cell subsets in various feral and laboratory mouse strains is believed to provide immunity to specific milk-borne MMTVs, and there is evidence that endogenous *Mtv*s shape the immune response against other pathogens.

Future Perspectives

MMTV has been valuable for studies of hormone-regulated gene expression, tissue-specific transcription, immune response, and mechanisms of oncogenesis. The recent discovery that MMTV is a complex retrovirus suggests that this virus will provide a useful model for understanding pathogenesis by human retroviruses, such as HTLV and HIV.

See also: Immune Response to viruses: Antibody-Mediated Immunity; Retroviruses: General Features; Viral Pathogenesis.

Further Reading

Bhadra S, Lozano MM, and Dudley JP (2005) Conversion of mouse mammary tumor virus to a lymphomagenic virus. *Journal of Virology* 79: 12592–12596.

Callahan R and Smith GH (2000) MMTV-induced mammary tumorigenesis: Gene discovery, progression to malignancy and cellular pathways. *Oncogene* 19: 992–1001.

Czarneski J, Rassa JC, and Ross SR (2003) Mouse mammary tumor virus and the immune system. *Immunologic Research* 27: 469–480.

Golovkina TV, Chervonsky A, Dudley JP, and Ross SR (1992) Transgenic mouse mammary tumor virus superantigen expression prevents viral infection. *Cell* 69: 637–645.

Held W, Waanders GA, Shakhov AN, Scarpellino L, Acha-Orbea H, and MacDonald HR (1993) Superantigen-induced immune stimulation amplifies mouse mammary tumor virus infection and allows virus transmission. *Cell* 74: 529–540.

Jude BA, Pobezinskaya Y, Bishop J, et al. (2003) Subversion of the innate immune system by a retrovirus. *Nature Immunology* 4: 573–578.

Kinyamu HK, Chen J, and Archer TK (2005) Linking the ubiquitin-proteasome pathway to chromatin remodeling/modification by nuclear receptors. *Journal of Molecular Endocrinology* 34: 281–297.

Liu J, Bramblett D, Zhu Q, et al. (1997) The matrix attachment region-binding protein SATB1 participates in negative regulation of tissue-specific gene expression. *Molecular and Cellular Biology* 17: 5275–5287.

Mertz JA, Simper MS, Lozano MM, Payne SM, and Dudley JP (2005) Mouse mammary tumor virus encodes a self-regulatory RNA export protein and is a complex retrovirus. *Journal of Virology* 79: 14737–14747.

Ross SR, Schofield JJ, Farr CJ, and Bucan M (2002) Mouse transferrin receptor 1 is the cell entry receptor for mouse mammary tumor virus. *Proceedings of the National Academy of Sciences, USA* 99: 12386–12390.

Zhu Q, Maitra U, Johnston D, Lozano M, and Dudley JP (2004) The homeodomain protein CDP regulates mammary-specific gene transcription and tumorigenesis. *Molecular and Cellular Biology* 24: 4810–4823.

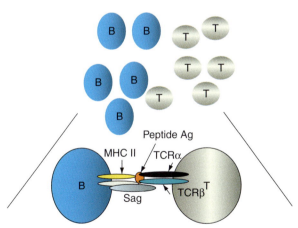

Figure 5 Interaction of MMTV Sag with the TCR. The α and β chains of the TCR on CD4+ cells combine to recognize a peptide (foreign or self) (small circle) residing in the binding pocket of the MHC class II molecule on MMTV-infected antigen-presenting cells. The C-terminal end of Sag interacts with the variable region of the TCR β-chain, although the TCR α-chain has been reported to affect Sag binding. Accessory molecules such as CD4 probably affect the stability of the Sag–TCR interaction.

Mousepox and Rabbitpox Viruses

M Regner, F Fenner, and A Müllbacher, Australian National University, Canberra, ACT, Australia

© 2008 Elsevier Ltd. All rights reserved.

Glossary

Pyknotic Exhibiting a degenerate nucleus by contraction of nuclear contents; a sign of cell death.

Introduction

Ectomelia virus (ECTV), which is the agent of mousepox, and rabbitpox virus (RPXV) share two features; they are both orthopoxviruses and both are known only as infections of laboratory animals, the mouse and rabbit, respectively.

Ectromelia Virus

History

ECTV was discovered in 1930 by J. Marchal, as a spontaneous infection of laboratory mice at the National Institute of Medical Research in London. It was called infectious ectromelia because of the frequent occurrence of amputation of a foot in animals that had recovered from infection. Soon after, J. E. Barnard showed, by ultraviolet (UV) microscopy, that it had oval virions about the same size as those of vaccinia virus (VACV). The only other experiments done with the virus at that time involved studies of experimental epidemics by W. W. C. Topley and his colleagues.

In 1946, F. M. Burnet, in Melbourne, showed that ECTV was serologically related to VACV. During experimental epidemics carried out in Burnet's laboratory, F. Fenner found that in animals that did not die of acute hepatitis there was a rash, and he named the disease mousepox. Subsequent studies led to the development of a classical model explaining the spread of virus around the body in generalized viral infections with rash.

In laboratories in Europe and the USA, the virus was regarded as a major menace to colonies of laboratory mice, and stringent steps were taken to prevent its entry to the USA. Only after extensive outbreaks in several cities of that country in 1979 were studies of the virus undertaken in the USA, in high-security laboratories.

Classification

Ectromelia virus is a species within the genus *Orthopoxvirus* of the family *Poxviridae*, as evidenced by the morphology of the virion, cross-protection tests, and restriction endonuclease mapping. Most strains of *Ectromelia virus* (e.g., Moscow, Hampstead) recovered from naturally infected mouse colonies are highly virulent. However, a substrain of Marchal's original strain was attenuated by serial passage on the chorioallantoic membrane of chicken eggs (Hampstead egg).

Genetics

The ECTV genome is typical for an orthopoxvirus in size and composition. It is $A+T$-rich and comprises 210 kbp. Almost half of the genome consists of a central region of high homology to other orthopoxviruses and this is flanked by significantly more variable terminal regions. The genome comprises 175 potential genes encoding proteins, including numerous predicated host range proteins and host response modifiers. ECTV does not cluster phylogenetically with any other member of the virus family.

Host Range and Virus Propagation

ECTV produces disease in *Mus musculus* and several other species of mice, and is considered a natural mouse pathogen. The rabbit, guinea pig, and rat can be infected by intradermal or intranasal inoculation, with the production of small skin lesions or an inapparent infection. ECTV will grow on the chorioallantoic membrane of the developing chick embryo, and also in cell cultures derived from a variety of species. Quantitation of virus stocks is determined by growth on cell culture monolayers (plaque assay).

Geographic Range and Seasonal Distribution

ECTV has been spread around the world inadvertently by scientists working with laboratory mice, and has been repeatedly reported from laboratories in several countries of Europe and from Japan and China. Mousepox has never been enzootic for prolonged periods in mouse colonies in the USA, but accidental importations sometimes occurred with mice or mouse tissues from European laboratories, with devastating consequences. In contrast, there are extremely limited data regarding the occurrence of ECTV in wild animals.

Epidemiology

In laboratory mice, ECTV is infectious by all routes of inoculation. It always produces a generalized infection but there are local lesions in the lungs after intranasal inoculation and in the peritoneal cavity after intraperitoneal injection. The response of mice is strongly conditioned by mouse genotype (see below).

The usual source of natural infection is via minor abrasions of the skin, which may occur from contaminated bedding or during manipulations by animal handlers. Infection may also occur by the respiratory route, but probably only between mice in close proximity to each other. A primary lesion usually develops at the site of infection. Since mice are readily infected by inoculation, virus-contaminated mouse serum, ascites fluid or mouse cells, tumors or tissues constitute a risk to laboratory colonies previously free of infection.

Enzootic Mousepox

Until the introduction of rigorous screening techniques in the 1960s, mousepox was enzootic in many mouse-breeding establishments in Europe and Japan. A variety of mechanisms probably operated to maintain the virus, without disrupting the mouse-breeding program as to make control mandatory. One important factor was probably the high level of genetic resistance and trivial symptomatology exhibited by many mouse genotypes. Another may have been maternal antibody. Another possible mechanism for maintaining enzootic infection is chronic, clinically inapparent infection, which sometimes occurs after oral administration, when some mice show infection of Peyer's patches, excretion of virus in the feces, and lesions in tail skin.

Susceptibility of Different Strains of Mice

Analysis of spontaneous epizootics and deliberate experiments in the 1980s showed that C57BL/6, 129J, and AKR mice were highly resistant to mousepox, CBA/J and SJL/J intermediate, and BALB/c and A/J mice were highly susceptible. Strain differences are best demonstrated after footpad inoculation or in natural epizootics, since C57BL/6 mice are relatively susceptible by intranasal, intracerebral, or intraperitoneal infection.

Pathogenesis

Fenner's work in the 1930s to 1940s established mousepox as a model system for the study of generalized viral infections. Müllbacher has more recently used the mousepox model for the analysis of immune effector molecules required for resistance to ECTV infections, using specific gene knockout mutant mouse strains. Mousepox is generally considered an excellent mouse model of human smallpox and is now used increasingly by the bioterrorism defense research community.

Mice are usually infected in the footpad and, after an incubation period of c. 7 days (as found in natural infections), a local ('primary') lesion develops at the inoculation site. A few days later some mice die, with no other visible skin lesions but with acute necrosis of the liver and spleen, and in those that survive a rash develops which goes through macular and pustular stages before it scabs.

During the incubation period the virus passes through the mouse body in a stepwise fashion: infection, multiplication and liberation, usually accompanied by cell necrosis, first in the skin and then the regional and possibly the deeper lymph nodes, until it reaches the bloodstream (primary viremia).

C. A. Mims showed in the 1960s that, during the primary viremia, virus is ingested by the phagocytic littoral cells of the liver and spleen. After a day or so, much larger amounts of virus are liberated into the circulation (secondary viremia). Next follows an interval during which the virus multiplies to high titer before visible changes are produced, so that 2 or 3 days usually elapse between the appearance of the primary lesion and the secondary rash. Some animals die before skin lesions appear, but titration experiments and histological examination showed that early skin lesions are present.

Clinical Features of Infection

Early workers described two forms of the disease, a rapidly fatal form in which apparently healthy mice die within a few hours of the first signs of illness and show extensive necrosis of the liver and spleen at autopsy, and a chronic form characterized by ulcerating lesions of the feet, tail, and snout. Fenner found that in natural infections most mice develop a primary lesion, usually on the snout, feet, or belly. Subsequently, virus multiplies to high titer in the liver and spleen. Some mice die at this stage, but if they survive they almost invariably develop a generalized rash (**Figure 1**).

Age affects the response of genetically susceptible mice. Both virulent and attenuated strains produce higher mortalities in suckling mice and in mice about a year old than in 8-week-old mice.

Pathology and histopathology

The pathological changes in naturally occurring mousepox in susceptible mice are quite characteristic. Additional lesions in the peritoneal cavity or lung occur after intraperitoneal or intranasal inoculation, respectively; these are important because mousepox sometimes occurs after unwitting passage of ECTV by these routes.

Intracytoplasmic inclusion bodies

ECTV produces two types of intracytoplasmic inclusion body in infected cells, A-type and B-type. The latter occur in all poxvirus infections and are the sites of viral multiplication; more characteristic of mousepox are the prominent acidophilic inclusion bodies (A-type), which are always found in infected epithelial cells but rarely in liver cells (**Figure 2**).

Skin lesions

The earliest primary lesions that can be recognized macroscopically are the seat of advanced histological changes, for viral multiplication has then been in progress for several days. There is no macroscopic breach of the skin surface, but the dermis and subcutaneous tissue are edematous and there is widespread lymphocytic infiltration of the dermis. Inclusion bodies can be seen in the epidermal cells at the summit of the lesion. Necrosis of these epidermal cells is followed by ulceration of the surface. The exudate forms a scab beneath which healing occurs. Histologically, the changes of the rash are similar to those in the primary lesion.

Lesions of the liver

The liver and spleen are invariably invaded during the incubation period and virus multiplies to high titer here. The liver remains macroscopically normal until within 24 h of death, when it appears enlarged and studded with minute white foci. The necrotic process extends rapidly and at the time of death the liver is enlarged with many large semiconfluent necrotic foci. In animals that survive, the liver usually returns to its normal macroscopic appearance, but occasionally numerous white foci occur.

Histologically, little change is apparent until macroscopic changes have appeared, although immunofluorescence techniques have shown that infection always occurs first in the littoral cells of the hepatic ducts, from which the virus spreads to contiguous parenchymal cells. Numerous scattered foci of necrosis then appear throughout the liver parenchyma and in fatal cases these rapidly extend until they became semiconfluent. The portal tracts show slight infiltration with lymphoid cells. Liver regeneration commences early and is active, especially in nonfatal cases, and fibrosis does not occur.

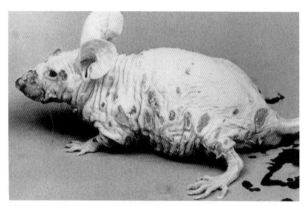

Figure 1 The rash of mousepox as it appears 14 days after infection, in a naturally infected genetically hairless mouse (not athymic). Similar lesions occur beneath the hair of other strains of susceptible mice and can be clearly demonstrated by epilation. Reproduced from Fenner F (1982) Mousepox. In: Foster HL, Small JD, and Fox JG (eds.) *The Mouse in Biomedical Research*, vol II, pp. 209–330. New York: Academic Press, with permission from Elsevier.

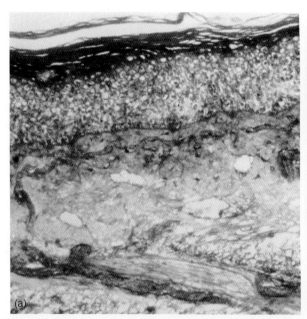 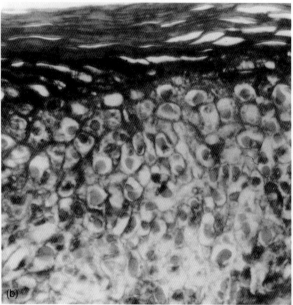

Figure 2 Section of the skin of the foot of a mouse injected with ECTV in the footpad six days earlier. (a) Low power. (b) High power, Mann's stain. Almost every epithelial cell contains an eosinophilic A-type inclusion body. With most strains of ECTV, these A-type inclusion bodies contain large numbers of mature virions. Reproduced from Fenner F (1982) Mousepox. In: Foster HL, Small JD, and Fox JG (eds.) *The Mouse in Biomedical Research*, vol II, pp. 209–330. New York: Academic Press, with permission from Elsevier.

Lesions of the spleen

The spleen shows macroscopic changes at least a day earlier than the liver and higher titers of virus are found in the spleen. Virus reaches the spleen in infected lymphocytes, which initiate infection in the substance of the follicles. While infected follicles are destroyed by the spreading infection, neighboring follicles show the proliferative response characteristic of antibody production.

In surviving mice, lesions of the spleen vary from small raised plaques about a millimeter in diameter to areas of fibrous tissue that, after severe attacks, almost completely replace the normal splenic tissue. These changes constitute reliable autopsy evidence that a mouse has recovered from an attack of mousepox.

Lesions of other organs

The regional lymph nodes draining the site of the primary lesion are enlarged from the time the primary lesions can be detected, and they usually show localized areas of necrosis, with pyknotic nuclear debris in a featureless background. In fatal cases, the gut is often engorged and the lymphoid follicles enlarged. Small necrotic foci with typical inclusion bodies occur in the intestines in most acutely fatal cases of mousepox. Occasionally, especially in very young mice, there are hemorrhagic foci in the kidneys.

Lesions after intraperitoneal inoculation

There is no primary skin lesion, but in acutely fatal cases the necrosis of the liver and spleen resembles that found after natural infection. In addition, there is usually some increase in intraperitoneal fluid and a considerable amount of pleural fluid, and the pancreas is often grossly edematous. In animals that survive the acute infection there is a great excess of peritoneal and pleural fluid, the peritoneal surfaces of the liver and spleen are covered with a white exudate, the walls of the gut are thickened and rigid, and there is often fat necrosis in the intraperitoneal fat. Extensive adhesions between the abdominal viscera develop later.

Lesions after intranasal inoculation

When small doses of virus are inoculated intranasally, there is usually little change in the lungs except patchy congestion; the changes in the liver and spleen are those characteristic of naturally acquired mousepox. With larger doses of virus, congestion of the lungs is more pronounced and consolidation may occur, and when very large doses are given death occurs with patchy or complete consolidation of the lungs and little change in the liver and spleen. The apparent pneumotropism is due to the fact that the local reaction, which occurs after the intranasal inoculation of very large doses of virus, kills the animal before there is time for the characteristic changes in the liver and spleen to occur.

Immune Response

Two weeks after infection mice are solidly immune to reinfection by footpad inoculation of the virus. This immunity declines slowly but even a year after recovery multiplication of the virus after footpad challenge is confined to the local skin lesion.

Humoral immunity

Antibodies generated by a primary infection protect from subsequent challenge. Newborn mice receive maternal antibody via the placenta and in the milk during the first 7 days after birth. Until titers decline to undetectable levels by the seventh week after birth, this maternal antibody confers protection against death, but not against infection, with moderate doses of ECTV. Furthermore, in the absence of functional B cells, clearance of primary infection with ECTV may be deficient, leading to viral persistence and eventual death.

Cell-mediated immunity

Work by Blanden in the early 1970s showed that T cells are critical in recovery from primary infection with ECTV, and mice pre-treated with anti-thymocyte serum die from otherwise sublethal doses of virus due to uncontrolled viral growth in target organs. These mice have impaired cell-mediated responses but normal neutralizing antibody responses, elevated interferon levels in the spleen, and unchanged innate resistance in target organs. The active cells in the immune population are cytotoxic T cells, although natural killer cells probably also play an important role early in the infection.

Virus-specific cytotoxic T cells are detectable 4 days after infection and reach peak levels in the spleen 1–2 days later, while delayed hypersensitivity is detectable by the footpad test 5–6 days after inoculation. In contrast, significant neutralizing antibody is not detectable in the circulation until the eighth day.

Cytotoxic T cells employ two different mechanisms to destroy virus-infected cells before the release of viral progeny: one, the granule exocytosis pathway, is mediated by perforin and granzymes that are stored in cytolytic granules and secreted toward, and enter, the infected cell, inducing apoptosis; the other one via triggering of death receptors (e.g., Fas) on the surface of infected cells, inducing a cascade of caspase activation that also leads to apoptosis. However, poxviruses encode inhibitors of the death receptor pathway of killing (e.g., SPI-2, see below), rendering the granule pathway indispensable for recovery. Consequently, mice genetically deficient in perforin or both principal granzymes (A and B) are highly susceptible to ECTV infection.

T-cell-secreted cytokines are also important factors determining the outcome of infection. Whereas interferon-gamma (IFN-γ) critically contributes to viral clearance, an imbalanced cytokine response of the so-called Th2 type, characterized by lack of IFN-γ and excess of interleukin-4 (IL-4), predisposes to greater susceptibility. A recombinant ECTV expressing IL-4 was found to be highly virulent even in resistant strains of mice and those vaccinated with attenuated virus, prompting concerns about the possible creation of recombinant poxviruses with increased virulence in humans for the purpose of bioterrorism.

Host response modulation

As do other poxviruses, ECTV encodes numerous host-response modifiers in order to evade or suppress the immune response to allow for maximal viral replication. Broadly, these can be divided into inhibitors of the inflammatory response, and anti-apoptotic proteins. Several ECTV-encoded proteins have been shown to neutralize or inhibit key inflammatory cytokine pathways, for example, an IL-18-binding protein and homologs of the IL-1β, tumor necrosis factor (TNF) and IFN receptors. On the other hand, virally encoded inhibitors of components of the caspase cascade inhibit apoptotic pathways. Examples are the ECTV protein p28, a RING finger-domain protein, which is a potent ECTV virulence factor that inhibits UV-induced, but not Fas- or TNF-induced apoptosis; and SPI-2, a serine proteinase inhibitor, which blocks TNF-α-mediated apoptosis via caspase 1/8 inhibition.

Future

Mousepox is now very rare as a natural infection in laboratory colonies of mice but it is likely that, despite the strict controls necessary to protect mouse colonies, mousepox will be investigated more extensively in the future. Workers in Australia, the USA, and the UK are now using it as a model for studies of problems such as immunocontraception and the role of the many homologs of mammalian host response modifier genes that are found in all poxviruses. In addition, the recognition of mousepox as a good mouse model for human smallpox has, in the light of increased interest in poxvirus immunobiology, seen a resurgence of research into its pathologies, immune evasion strategies, and the mechanisms of recovery from this disease.

Rabbitpox Virus

History and Classification

Rabbitpox is a laboratory artifact, due to the infection of laboratory rabbits with VACV, usually with neuro-VACV variants; hence this account of RPXV will omit reference to those aspects that are covered in the article on VACV.

The name rabbitpox was originally given to devastating outbreaks of a generalized disease, likened to smallpox in man, in a colony of laboratory rabbits at the Rockefeller Institute of Medical Research in New York in 1932–34, when other scientists had been working with neuro-VACV in rabbits in an adjacent room prior to the outbreak. The virus recovered from the outbreak was called RPXV and was shown to be very similar to neuro-VACV in its biological properties. Subsequently, the restriction map of the Utrecht strain (see below) was found to be almost identical with that of VACV.

Another outbreak occurred in the Netherlands in 1941. It began among rabbits bought from a dealer a few days after they were introduced into the laboratory colony, and spread among the stock rabbits. The disease was usually lethal, death occurring before there was time for the development of a rash. The virus that caused this outbreak, designated RPXV-Utrecht, caused similar highly lethal epizootics when it 'escaped' in the Institut Pasteur in Paris in 1947; other outbreaks have been described in laboratory rabbits in the USA in the 1960s.

Epidemiology

In all outbreaks, spread appeared to occur by the respiratory route, and experiments confirmed that infection occurs readily by this route. Rabbits infected by contact are not infectious for other rabbits until the second day of illness, which is usually 5 days after infection. Actual contact is not necessary; transmission can occur across the width of a room, and air sampling revealed the presence of RPXV in the air of rooms housing infected rabbits.

Genetics

Since 1960, the Utrecht strain of RPXV virus has been used for genetic studies on poxviruses, since it was found to give rise to white pock mutants on the chorioallantoic membrane of chicken eggs and host range mutants in a pig kidney cell line, both of which entail deletions and transpositions of DNA. Recombination experiments with the white pock mutants were used to construct the first crude 'genetic map' produced for an animal virus.

When the complete coding sequence of the Utrecht strain was reported, it was confirmed that it is most closely related to VACV, with more than 95% sequence similarity. It was also established that RPXV is not a direct evolutionary descendant from VACV, as it contains several genes present in smallpox virus but not VACV.

Pathogenesis

A good deal of experimental work has been carried out on the pathogenesis of rabbitpox as an animal model of smallpox, with results that were largely confirmatory of those obtained with mousepox. In rabbits infected by the intranasal instillation of a small dose of virus, by aerosol, or after intradermal infection or contact infection, there is a stepwise spread of virus through the organs, although the incubation period is shorter than in mousepox and there seems to be little delay at the regional lymph nodes. Viremia is leukocyte associated.

Clinical Features of Infection

RPXV causes an acute generalized disease in which a rash appears in animals that survive long enough, presenting as pocks on the skin and mucous membranes (**Figure 3**). Rabbits dying of hyperacute infection show no obvious skin lesions, the so-called 'pockless' rabbitpox. Such infections are analogous to acutely lethal cases of mousepox and perhaps to early hemorrhagic-type smallpox, in which death occurs before there is time for pustular skin lesions to develop.

Pathology and Histopathology

The most distinctive lesions are the pocks on the skin and mucous membranes and occasionally small areas of focal necrosis are found in the internal organs (liver, spleen, lung, testes, ovaries, uterus, adrenals, and lymph nodes). In the so-called 'pockless' form, a few pocks may occur around the mouth and they may be visible on the shaved skin. The most prominent gross lesions are pleuritis, focal necrosis of the liver, enlarged spleen, and edema and hemorrhage of the testes.

RPXV, being a strain of VACV, produces B-type inclusions (Guarnieri bodies) in infected cells, but not the prominent A-type inclusions found in cells infected with ECTV.

Immune Response

Rabbits that have recovered from rabbitpox are immune to infection with VACV, but in very severe infections rabbits die before there is time for an effective immune response. The importance of enveloped virions in the pathogenesis and immunology of orthopoxvirus infections was fist demonstrated in experiments with RPXV. Passive immunization with sera that did not contain antibody to the viral envelope failed to protect rabbits against challenge infection, even though the neutralization titer of the ineffective antiserum (produced by immunization with inactivated VACV) was much higher, as judged by conventional neutralization tests. This work helped explain the failure of inactivated VACV to provide protection against infection with orthopoxviruses.

Future

Since rabbitpox appears to be a laboratory artifact, due to the introduction of strains of VACV that can spread from one rabbit to another in rabbit colonies, prophylaxis appears to be a matter of preventing such events. These events were rare even when VACV, and especially

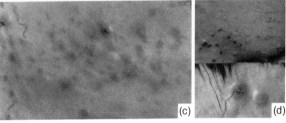

Figure 3 Rabbitpox. (a) and (b) Littermates with different types of disease. The course of infection was mild in the rabbit shown in (a) and external lesions were limited in the skin; the animal in (b) was seriously ill and its posture is a manifestation of acute respiratory distress resulting from extensive mouth lesions. (c) Cutaneous lesions on the trunk. The coat in this area was loose and easily plucked by hand. (d) Skin of a pregnant doe self-plucked for nest fur, showing both dry crusted pustules and others in earlier stages of development. Reproduced from Greene HSN (1934) Rabbit pox: I. Clinical manifestations and course of disease *Journal of Experimental Medicine* 60: 427–440. Copyright 1934 The Rockefeller University Press.

neuro-VACV, were extensively used in animal experiments several decades ago, and it is unlikely that further episodes will occur now that most research with VACV utilizes cultured cells rather than intact animals. Nevertheless, laboratory managers who use rabbits should be aware of the possibility that some strains of VACV can spread naturally from one rabbit to another.

Because of the higher sensitivity of rabbits to low doses with RPXV compared to that of mice to ECTV, and efficient spread via the aerosol and respiratory routes, RPXV has also been recently revisited as a model for human smallpox.

See also: Poxviruses; Smallpox and Monkeypox Viruses; Vaccinia Virus.

Further Reading

Adams MM (2007) Rabbitpox and vaccinia virus infections of rabbits as a model for human smallpox. *Journal of Virology* 81: 11084–11095.

Esteban DJ and Buller RM (2005) Ectromelia virus: The causative agent of mousepox. *Journal of General Virology* 86: 2645–2659.

Fenner F (1982) Mousepox. In: Foster HL, Small JD, and Fox JG (eds.) *The Mouse in Biomedical Research*, vol. II, pp. 209–230. New York: Academic Press.

Fenner F and Buller RML (1996) Mousepox. In: Nathanson N (ed.) *Viral Pathogenesis*, 535pp. Philadelphia: Lippincot-Raven.

Fenner F, Wittek R, and Dumbell KR (1989) *The Orthopoxviruses*. San Diego: Academic Press.

Greene HSN (1934) Rabbit pox. I: Clinical manifestations and course of disease. *Journal of Experimental Medicine* 60: 427–440.

Müllbacher A (2003) Cell-mediated cytotoxicity in recovery from poxvirus infections. *Reviews in Medical Virology* 13: 223–232.

Müllbacher A and Lobigs M (2001) Creation of killer poxvirus could have been predicted. *Journal of Virology* 75: 8353–8355.

Movement of Viruses in Plants

P A Harries and R S Nelson, Samuel Roberts Noble Foundation, Inc., Ardmore, OK, USA

© 2008 Elsevier Ltd. All rights reserved.

Glossary

Ancillary viral proteins Virus-encoded proteins that do not meet the definition of a movement protein, but are required for virus movement.

Intercellular movement Movement between two cells.

Intracellular movement Movement within a single cell.

Microfilaments A component of the cytoskeleton formed from polymerized actin monomers.

Microtubules A component of the cytoskeleton composed of hollow tubes formed from α–β tubulin dimers.

Molecular chaperones A family of cellular proteins that mediate the correct assembly or disassembly of other polypeptides.

Movement protein Virus-encoded proteins that can transport themselves cell to cell, bind RNA, and increase the size exclusion limits of plasmodesmata.

Phloem Vascular tissue that transports dissolved nutrients (e.g., sugars) from the photosynthetically active leaves to the other parts of the plant. In most plants there is only one phloem class, but for some plant families this tissue is divided into two classes: (1) internal phloem (internal or adaxial to xylem) and (2) external phloem (external or abaxial to xylem).

Systemic movement Movement through vascular tissue to all parts of the plant.

Viroid A plant pathogen containing nucleic acid that encodes no proteins.

Xylem Vascular tissue that transports water and minerals through the plant.

Introduction

In order for a plant virus to infect its host systemically, it must be capable of hijacking the host's cellular machinery to replicate and move from the initially infected cell. Plant viruses require wounding, usually by insect or fungal vectors or mechanical abrasion, for an infection to begin. Once inside a cell, the virus initiates transcription (DNA viruses) and translation and replication (DNA and RNA viruses) activities. Some of these viral products are required for virus movement and often interact with host factors (proteins or membranes) to carry out this function. Virus movement in plants can be broken down into three distinct steps: (1) intracellular movement, (2) intercellular movement, and (3) systemic movement. Intracellular movement refers to virus movement to the periphery of a cell and includes all metabolic activities necessary to recycle the host and viral constituents required for the continued transport of the intracellular complex. Intercellular movement refers to virus movement between cells. In order for a plant virus infection to spread between cells, viruses must move through specific channels in the cell wall, called plasmodesmata (PD), that connect neighboring cells. Once intracellular and intercellular movement is established, the virus can invade the vascular cells of the plant and then spread systemically through the open pores of modified PD within the sugar-transporting phloem sieve elements. Upon delivery by the phloem to a tissue distant from the original infection site, virus exits the vasculature and resumes cell-to-cell movement via PD in the new tissue. Although it will not be discussed further in this article, it is important to know that a few viruses utilize the water-transporting xylem vessels for systemic transport.

When contemplating plant virus movement it is critical to understand that each virus movement complex varies in viral and host factor composition over time as it travels within and between cell types. In addition, individual viruses often utilize unique host factors to support their movement. The diverse and dynamic nature of virus movement complexes makes it difficult to summarize plant virus movement in a simple unified model. However, there is evidence that some stages of virus movement, although carried out by apparently unrelated host or virus proteins, do have functional convergence.

Virus movement in plants has been studied with a wide range of virus genera, including, but not limited to, tobamoviruses, potexviruses, hordeiviruses, comoviruses, nepoviruses, potyviruses, tombusviruses, tospoviruses, and geminiviruses. In this article we do not review virus movement by all plant viruses, but rather focus on model viruses within genera that provide the most information on the subject. We review what is currently known about the three steps of virus movement in plants and attempt to convey the complexity of movement mechanisms utilized by members of different virus genera. However, we also highlight recent findings indicating that irrespective of the presence of seemingly unrelated host or viral factors, functional similarities exist for some aspects of movement displayed by viruses from different genera.

Intracellular Movement

Intracellular movement is necessary to deliver the virus genome to PD for cell-to-cell transmission. This has been an understudied area, as researchers have only recently had the ability to label and observe the movement of viral proteins and RNA in near-real-time conditions. Early studies relied on static images of immunolabeled viral proteins from light and transmission electron microscopes to determine their intracellular location. While a few of these studies related the intracellular location of the viral protein to the stage of infection, most did not and thus the importance of the intracellular location for virus movement was not understood. Other early studies of virus movement relied on the mutation of specific viral genes in virus genomic clones and the assessment of the intercellular movement of the resulting mutant virus, through the presence of local (representing intracellular and intercellular movement) or systemic (representing intracellular, intercellular and vascular movement) disease. Although these genetic experiments often determined which viral proteins were important for virus intercellular or systemic movement, they could not determine whether the mutation prevented intracellular or intercellular movement, both outcomes being visually identical. In more recent studies, fusion of viral proteins with fluorescent reporter genes such as the green fluorescent protein (GFP) have given researchers a powerful method to observe both the intracellular movement and final subcellular destination of many viral proteins in near-real-time conditions. However, it is important not to over-interpret movement studies using GFP since GFP maturation for fluorescence emission takes hours and thus the visible movement and position of the GFP or GFP:viral protein fusion may not reflect early movement activity. Additionally, the level of GFP within the movement form of the virus may be too low to detect during critical phases of movement.

Although intracellular virus movement in plants is just beginning to be elucidated, it is clear that specific viral proteins regulate this activity. Chief among these are the virally encoded movement proteins (MPs), named to indicate their genetically determined requirement for intercellular virus movement. MPs are defined based upon three functional characteristics: their (1) association with, or ability to increase, the size exclusion limit (SEL) of PD; (2) ability to bind to single-stranded RNA (ssRNA); and (3) ability to transport themselves or viral RNA cell to cell. Based upon these defining characteristics, a number of proteins have been classified as MPs (**Table 1**). Many viral MPs have similar sequences indicating a shared evolutionary history. However, a considerable number have no obvious sequence similarity between them. The absence of a shared sequence for these MPs suggests convergent evolution for movement function by unrelated predecessor proteins. MPs often interact with host proteins that modify their amino acid backbone (e.g., through phosphorylation) or host proteins associated with intracellular trafficking (e.g., cytoskeletal or vesicle-associated proteins) (**Table 1**). However, the role of MPs in intracellular movement remains largely unknown because technical limitations have prevented visualizing movement of individual viral RNA or DNA associated with MPs in real time. In addition, it is becoming clear that ancillary viral proteins (**Table 1**), which do not fulfill the classical definition of an MP, are essential for virus movement. These proteins are often associated with membranes or cytoskeletal elements and thus likely function primarily for intracellular virus movement. The interaction of MPs with host factors and the impact of the ancillary viral proteins on intracellular virus movement are discussed in detail in the following section. Models for intracellular virus movement of particular genera of viruses are presented based on some of this information (**Figure 1**).

Host factors and intracellular virus movement

Host proteins shown to interact with viral MPs include kinases, chaperones, nuclear-localized proteins (often transcription co-activators), and proteins that are associated with or are core components of the cytoskeletal or vesicle trafficking systems (**Table 1**). In addition, some MPs have been shown to associate with host membranes.

Table 1 Proteins necessary for the cell-to-cell movement of plant viruses

Virus	MP[a]	Ancillary viral proteins[b]	Host protein interactors with MP
Tobacco mosaic virus, Tomato mosaic virus	**30 kDa**	126 kDa	Actin, tubulin, MPB2C, PME, KELP, MBF1, calreticulin
Red clover necrotic mosaic virus	**35 kDa**		
Groundnut rosette virus	ORF4		
Cowpea chlorotic mottle virus	3a		
Brome mosaic virus	**3a**	CP	
Cucumber mosaic virus	**3a**	CP	NtTLP1
Bean dwarf mosaic virus	BC1	BV1	
Tobacco etch virus	CP	CI, HC-Pro, VPg	
Barley stripe mosaic virus	TGBp1	TGBp2, TGBp3	
Potato virus X	**TGBp1**	TGBp2, TGBp3 + CP	TIPs
Cowpea mosaic virus	**48 kDa**	CP	
Cauliflower mosaic virus	**38 kDa**	CP	MPI7, PME
Turnip crinkle virus	p8 + p9		Atp8
Tomato bushy stunt virus	**P22**		HFi22, REF
Potato leaf roll virus	17 kDa		
Tomato spotted wilt virus	**NS$_m$**		DnaJ-like, At4/1
Beet necrotic yellow vein virus	TGBp1	TGBp2, TGBp3, p14	
Grapevine fanleaf virus	2B	CP	Knolle, actin, tubulin
Rice yellow stunt virus	P3		
Rice dwarf virus	Pns6		
Southern bean mosaic virus	ORF1		
Turnip yellow mosaic virus	69 kDa		
Alfalfa mosaic virus	**P3**	CP	
Prunus necrotic ringspot virus	3a		
Tobacco rattle virus	**29 kDa**		
Soil-borne wheat mosaic virus	37 kDa		
Peanut clump virus	P51	P14, P17	
Potato mop top virus	**TGBp1**	TGBp2, TGBp3	
Commelina yellow mottle virus	N-term 216 kDa		
Beet yellows virus	p6, Hsp70h, p64	CPm, CP	
Rice stripe virus	Pc4		
Apple stem grooving virus	36 kDa		
Raspberry bushy dwarf virus	39 kDa		

[a]Regular (i.e., no bold) font indicates marginal classification as MP because the protein either has not been fully tested or has some but not all of the functions classically associated with MP (see text for definition).
[b]Necessary for viral cell-to-cell movement.

For geminiviruses, whose DNA genomes replicate in the nucleus, it is not surprising that nuclear factors may be necessary to transport viral genetic components required for virus replication into or out of the nucleus. For RNA viruses, however, there must be other reasons for an interaction between a nuclear protein and viral MP since these viruses are replicated in the cytoplasm. Some of the nuclear host proteins are non-cell-autonomous factors (e.g., HiF22) and thus it has been suggested that their interaction with MPs may inadvertently aid in virus intracellular and intercellular movement. It is also possible that MP and nuclear protein interactions occur to prevent transcription of host defense proteins or enhance transcription of host proteins necessary for virus movement, either within the infected cell or after transport to uninfected cells at the infection front.

The discovery over 10 years ago that tobacco mosaic virus (TMV) MP associates with microtubules (MTs) and microfilaments (MFs) was the first evidence that the host cytoskeleton might be involved in virus movement in plants. Although results from early studies indicated that disruption of MT arrays or their association with TMV MP could inhibit TMV movement, later studies suggested this was not so. Disruption of MT arrays with pharmacological agents or by tubulin transcript knockdown using virus-induced gene silencing had no effect on TMV movement or MP localization. Other work showed that the association of the MP with MTs happened late in infection, probably after virus movement had occurred. Also, during time-course studies it was determined that the MP disappeared during late stages of infection. This finding, in combination with the discovery that a mutant virus expressing a functionally enhanced MP with limited affinity for MTs moved cell to cell better than the parental virus, led to the idea that the association of MP with MTs is critical for MP degradation rather than to aid virus

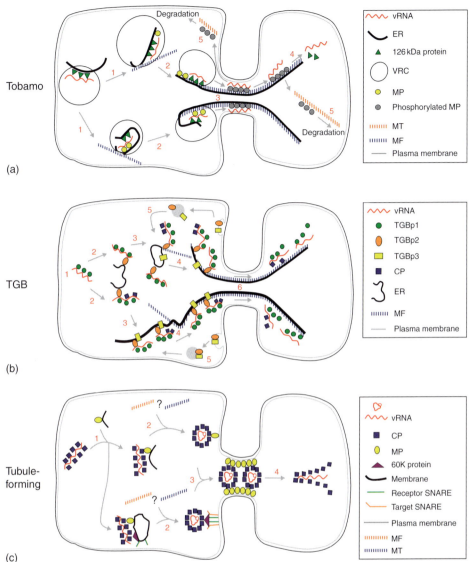

Figure 1 Models for cell-to-cell movement of plant viruses using tobamovirus triple gene blocks (TGBs), or tubule-forming strategies. (a) Viral 126 kDa protein binds both viral RNA (vRNA) and endoplasmic reticulum (ER) forming a cytoplasmic body in the cell termed a VRC. MP associates with the ER and possibly the vRNA within the VRC (step 1). VRCs associated with microfilaments (MFs) traffic toward plasmodesmata (PD; step 2). Here we show an indirect association of the VRC with actin mediated by the ER, but it is also possible that this interaction is mediated directly by the viral 126 kDa protein or MP. At the PD, vRNA is released from its association with the 126 kDa protein and is transported through the PD in association with MP (step 3). Phosphorylation of the MP occurs either within the cytoplasm, the cell wall, or both, and likely regulates transport to and through PD and subsequent translation of the vRNA in the new cell (steps 3 and 4). MP is degraded in the later stages of infection, likely via association with MTs and delivery to specific cellular sites of degradation (step 5). (b) Progeny vRNA binds to TGB protein 1 (TGBp1; step 1). The TGBp1/vRNA complex, either in the presence or absence of coat protein (CP, depending on virus genus), then binds TGBp2 attached to the ER to form a movement-competent ribonucleoprotein complex (RNP, step 2). The RNP then interacts with TGBp3, either directly or indirectly, to be positioned near the PD (steps 3 and 4). RNPs associate with actin MFs likely through an interaction with TGBp2, which may be responsible for transport to the PD. Following delivery of vRNA to the PD, TGBp2 and TGBp3 are likely recycled via an endocytic pathway (step 5). vRNA is actively transported through the PD via an unknown mechanism (step 6), although TGBp1 or CP may be involved, and is released from associated proteins in the next cell to allow replication to initiate. (c) CP-bound vRNA associates with the MP (itself associated with a membrane of unknown origin, step 1). The complex then moves, either as a vesicle directed to the PD through targeting proteins such as those from the SNARE family, or through other unknown targeting signals to the cell periphery (step 2). Interaction between SNAREs and virus may be mediated by a viral 60K protein for cowpea mosaic virus. The requirement for the cytoskeleton in transport of MP–vRNA complex is unclear since the nepovirus, grapevine fanleaf virus, requires cytoskeletal elements for proper delivery of its MP to the cell wall while cowpea mosaic virus does not. At or near the PD, the vesicular or nonvesicular membranes fuse with the plasma membrane and the attached MP directs the CP-associated vRNA through the PD (step 3). The vRNA is then released into the next cell to initiate virus replication and movement (step 4). Reproduced from Nelson RS (2005) Movement of viruses to and through plasmodesmata. In: Oparka KJ (eds.) *Plasmodesmata*, 1st edn., pp. 188–211. Oxford: Blackwell, with permission from Blackwell.

cell-to-cell movement. Further support for this idea came from the finding that the *Nicotiana tabacum* host protein, MPB2C, binds to MP and promotes its accumulation at MTs, yet acts as a negative effector of MP cell-to-cell transport. The role of MTs during TMV movement remains to be fully understood, but at this time it appears that they are more involved with MP degradation or compartmentalization than with virus movement (**Figure 1(a)**).

In contrast to the large body of work focusing on the role of the MT–MP interaction in TMV movement, studies on the role of the MFs in the movement of TMV and other viruses have only recently been published. It was demonstrated that intracellular movement of TMV viral replication complexes (VRCs; large multiprotein complexes comprised of host and viral factors) and cell-to-cell spread of the virus were blocked by MF inhibitors (pharmacological and transcript silencing agents). VRCs were later determined to physically traffic along MFs (**Figure 1(a)**). The interaction of TMV VRCs with MFs may be mediated by the TMV 126 kDa protein (a protein containing helicase, methyltransferase, and RNA silencing suppressor domains), since expression of a 126 kDa protein:GFP fusion in the absence of the virus results in fluorescent protein bodies that, like VRCs, traffic along MFs. VRC association with MFs may be mediated through a direct interaction of the 126 kDa protein with MFs or through an intermediary cell membrane. MFs are known to associate with membranes in plant cells and the 126 kDa protein binds to an integral membrane host protein, TOM1. However, the MP of TMV is long known to bind actin and associate with membranes, so the relative importance of the TMV 126 kDa protein or MP for directing intracellular VRC movement is unclear (**Figure 1(a)**).

Recently, MFs were demonstrated to co-localize with ancillary proteins required for movement of the hordeivirus, potato mop-top virus (PMTV). PMTV encodes a conserved group of proteins termed the triple gene block (TGB) that are required for cell-to-cell virus movement (see **Table 1**). Two of these TGB proteins (TGBp2 and TGBp3) co-localize with motile granules that are dependent upon the endoplasmic reticulum (ER)–actin network for intracellular movement. In addition, the TGBp2 protein from the potexvirus, potato virus X (PVX), localizes to MFs in what are likely ER-derived vesicles (**Figure 1(b)**). The association of TGBp2 from potexviruses and the 126 kDa protein from TMV with MFs and their requirement for successful virus movement provide an elegant example of convergent evolution since TGBp2 and 126 kDa protein have no sequence identity.

The role of the cytoskeleton in the intracellular transport of some plant viruses is unclear. Cowpea mosaic virus (CPMV), for example, does not require the host cytoskeleton for the formation of tubular structures containing MP on the surface of protoplasts. These tubular structures are similar to the tubules formed in modified PD (that likely do not contain cytoskeleton) which are necessary for intercellular movement of this virus (**Figure 1(c)**). The role of tubules in intercellular transport of CPMV is discussed in next section. Work with grapevine fanleaf virus (GFLV), another tubule-forming virus, has revealed the possibility that this virus may be targeted to the PD by membrane vesicle SNARE (v-SNARE)-mediated trafficking. The MP of GFLV co-immunoprecipitates with KNOLLE, a target SNARE (t-SNARE). The 60 kDa protein of CPMV has been shown to bind a SNARE-like protein. t-SNAREs such as KNOLLE act as specific receptors for targeted delivery of Golgi-derived vesicles to sites where fusion with the plasma membrane will occur. Thus, it is possible that the SNARE trafficking machinery delivers viral proteins (and possibly associated viral RNA) to the plasma membrane near PD (**Figure 1(c)**).

There is evidence that, following movement of the viral RNA to the PD, some viral factors involved in this movement may be recycled for further use. The TGBp2 and TGBp3 proteins from PMTV localize to endocytic vesicles as evidenced by labeling with FM4-64 dye, a marker for internalized plasma membrane (**Figure 1(b)**). Additionally, TGBp2 co-localizes with Ara7, a marker for early endosomes. The functional significance of this endocytic association of viral proteins remains to be determined and it is not known whether proteins from other viruses may also traffic in the host cell's endocytic pathway.

Intercellular Movement

Following intracellular movement to the cell periphery, the virus must then move through PD in order to spread into neighboring cells. PDs are plasma membrane-lined aqueous tunnels connecting the cytoplasm of adjacent cells. An inner membrane, termed the desmotubule, is a tubular form of the ER and is an extension of the cortical ER. PDs can be subdivided structurally into simple (containing a single channel) or branched (containing multiple channels) forms. The SELs of PD are increased by the disruption of MFs indicating a role for actin in PD gating and indeed both actin and myosin have been observed in PD. Thus, it is possible that cytoskeletal-mediated transport of viral components results in direct delivery of virus to and passage through PD.

Protein movement through PD is dependent on the developmental stage of the PD. For example, free GFP (27 kDa) moves through simple but not branched PDs. Branched PDs generally have smaller SELs than simple PDs and the presence of more branched PDs in mature photosynthate-exporting (source) versus immature photosynthate-importing (sink) leaves represents a developmental change that limits transport of macromolecules through PD. This developmental change also affects the localization pattern for some viral MPs. For example, both cucumber mosaic virus and TMV MPs are observed predominantly or solely within branched PD in source leaves and not simple PD in sink leaves. The TMV MP expressed in transgenic plants, however, is sufficient to complement the movement of an MP-deficient TMV mutant in both source and sink leaves. Thus, the presence of MP in the central cavity of branched PD in source leaves may not represent a site of function for the MP, but rather the final deposition of inactive MP. Although it is possible that the level of MP binding in simple PD is below the detection limits of the current technology, questions remain about where and how the MP functions in virus movement.

A clue to TMV MP function during virus movement comes from findings showing that a TMV MP–viral RNA complex could not establish an infection in protoplasts, but could do so when introduced into plants. It was suggested that a change in the phosphorylation state of the MP at the cell wall was necessary to weaken the binding between the MP and viral RNA, thereby allowing translation of the viral genome and initiation of infection in the next cell. Indeed, a PD-associated protein kinase has been identified that phosphorylates TMV MP. Thus, the protein kinase in the cell wall may be necessary to end the involvement of MPs in virus movement and release the viral RNA for translation in the new cell (**Figure 1(a)**). Also, considering that there are additional phosphorylation sites on the TMV MP besides those targeted by the PD-associated protein kinase, it is likely that proper sequential phosphorylation of this protein is necessary to allow it to function in both intracellular and intercellular virus movement. For potyviruses, the eukaryotic elongation factor, eIF4E, appears to modulate both virus accumulation, likely by affecting translation, and cell-to-cell movement. Thus, as for TMV, virus accumulation and movement may be linked activities.

Chaperones of host or viral origin may be required for PD translocation of some MPs. Host-encoded calreticulin modulates TMV intercellular movement and co-localizes with TMV MP in PD. The virus-encoded virion-associated protein (VAP) of cauliflower mosaic virus (CaMV) binds MP through coiled-coil domains and co-localizes with MP on CaMV particles within PD. The mechanism by which a molecular chaperone can support intercellular virus movement is illustrated by the virally encoded Hsp70 chaperone homolog (Hsp70h) of beet yellows virus. Hsp70h requires MFs to target it to the PD. The Hsp70h is a component of the filamentous capsid and its ATPase activity is required for virus cell-to-cell movement. These findings led to a model where Hsp70h mediates virion assembly and, once localized to the PD, actively translocates the virion from cell to cell via an ATP-dependent process. The idea that viral proteins may actively participate in plasmodesmal translocation of virus is further supported by the finding that the NTPase activity of the TGBp1 helicase from PMTV is necessary for its translocation to neighboring cells and that the coat protein (CP) of PVX, necessary for virus cell-to-cell movement, has ATPase activity (**Figure 1(b)**). It has also been found that the helicase domain of the TMV 126 kDa protein is required for cell-to-cell movement. In these cases it seems likely that the helicase activity is necessary to remodel viral RNA, thereby easing passage of the virus through PD.

Tubule-forming viruses have adopted another strategy for intercellular movement whereby virus-induced tubules span modified PD that lack a desmotubule in order to transmit capsids from cell to cell (**Figure 1(c)**). Such capsid-containing tubules are known to be composed, at least in part, of MP and have been identified for a number of viruses, including commelina yellow mottle virus, CaMV, CPMV, and tomato ringspot virus.

Although the tubule-forming viruses modify PD differently than those utilizing classical MPs, it was recently determined that the tubule-forming MP from tomato spotted wilt virus can functionally substitute for the non-tubule-forming TMV MP to support TMV movement. This is likely another example where two proteins with no sequence identity and therefore no apparent evolutionary relationship have independently evolved to functionally support movement of viruses.

Systemic Movement

Some viruses are limited to the phloem of plants (i.e., phloem-limited viruses) and require inoculation, often by aphids, directly to vascular cells for infection. Systemic movement of a non-phloem-limited virus through vascular tissue, however, requires that the virus moves from nonvascular cells into veins. Veins are defined as major or minor based on their structure, location, branching pattern, and function (**Figure 2**). Whether major or minor, each vein contains many different cell types with greatly differing structures. Within *N. tabacum*, minor veins include phloem parenchyma, xylem parenchyma, and companion cells, along with sieve elements and xylem vessels (**Figure 2**). All of these cells have distinct structures and locations within the vein which present unique regulatory sites for virus entry. Between plant species, companion cell morphologies vary greatly with an obvious difference being the number of PDs between these cells and other vascular cells. This difference is functionally related to the type of photosynthate transport system exhibited by

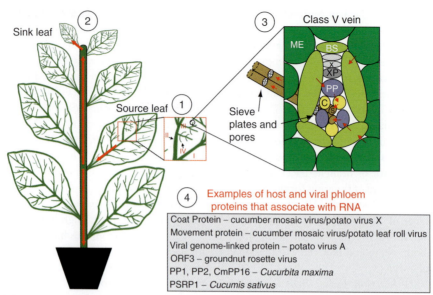

Figure 2 (1) Viral infection of a source leaf occurs by intercellular movement of the virus into the vasculature (class I–V veins indicated). (2) Virus travels through the phloem from the source leaf throughout the plant (red arrows) and exits vascular tissue to resume cell-to-cell movement in sink tissue. (3) In order to enter the phloem of a class V vein, a non-phloem-limited virus must travel through mesophyll cells (ME), bundle sheath cells (BS), and possibly phloem parenchyma cells (PP) before entering the companion cells (C) and finally the sieve elements (S). Movement through SEs requires passage of virus through pores within the sieve plates. A minority of viruses move through the xylem (X). (4) Examples of host and viral proteins that have been identified in phloem and that associate with RNA are indicated.

the plant (i.e., apoplastic versus symplastic). In addition, bundle sheath cells, which have their own unique position and structure surrounding the minor veins, must be considered as potential regulators of virus movement. These complex cell types are difficult to study because it is problematic to directly access or isolate them.

Recently, studies have been conducted that conclusively indicate which veins, minor or major, can serve as entry sites for rapid systemic infection. Using surgical procedures to isolate specific veins and TMV or CPMV modified to express GFP as a reporter, it was determined that either major or minor veins in leaves of *Nicotiana benthamiana* and *Vigna unguiculata* can be invaded directly and serve as inoculum sources for systemic infection. In addition, for TMV, direct infection of cells in transport veins in stems yielded a systemic infection. Considering that major and transport veins do not have terminal endings bounded by nonvascular tissue, it is likely that virus entered these veins by passing through bundle sheath cells and interior vein cells.

Virus transport and accumulation are regulated within vascular tissue. In plants that have internal and external phloem, potyviruses and tobamoviruses accumulate in specific tissue depending on the tissue's position relative to the inoculated leaf. In the inoculated leaf and the stem below, virus accumulates in the external phloem, whereas in the stem and leaf veins above the inoculated leaves, virus accumulates in the internal phloem.

Exit of PVX, TMV, and CPMV from vascular cells in sink tissues only occurs from major veins. For a growing number of viruses, however, exit occurs from both major and minor veins indicating that there is not a uniform exit strategy for all viruses.

The virus and host factors that control systemic virus accumulation are becoming better understood, mostly through genetic studies. Virus factors include CP, some MPs, and some nonstructural proteins such as the 126 kDa protein of TMV. Although CPs are often necessary for systemic infection, it is clear that for some viruses, such as groundnut rosette virus, a CP is not present and the virus still produces a systemic infection in the host. Also, for viruses that normally require the CP for systemic infection, the loss of the CP through mutation or deletion may still allow systemic movement of the virus in specific hosts. Lastly, viroids, which do not encode any proteins, can systemically infect plants. These results indicate that although a capsid may be required to protect viral RNA for systemic transport in some hosts, other viral or host proteins can functionally mimic the CP and allow systemic infection.

MP function during systemic infection has, in one case, been uncoupled from its role during intra- and intercellular transport. Some point mutations in the red clover necrotic mosaic virus MP still allow intercellular movement, but prevent systemic movement. Additional support that MPs function to allow systemic movement comes from studies with the 17 kDa MP of potato leafroll virus, a phloem-limited virus. This MP, when expressed from within an infectious virus sequence in transgenic plants, is uniquely localized to PD connecting the companion cells with sieve elements, even though virus accumulated in both vascular and nonvascular cells. Thus, the PD between companion cells and sieve elements may be uniquely recognized by this MP to allow the virus to only invade vascular tissue. More recently, it has been shown that a host factor, CmPP16, that is thought to function by forming ribonucleoprotein complexes with phloem transcripts has sequence similarity with viral MPs. Thus, some MPs may function to protect RNA while in transit through the phloem.

Other viral proteins such as the 2b protein of CMV, p19 of TBSV, and the 126 kDa protein of TMV have been linked to supporting systemic movement of their respective viruses. Considering that all of these proteins are suppressors of gene silencing, it is possible that this activity is related to their function in supporting systemic movement. It is known that a member of the plant silencing pathway, specifically, the RNA-dependent RNA polymerase, RDR6, functions in sink tissue (e.g., the shoot apex) by responding to incoming signals for RNA silencing. RDR6 has also been shown to control virus accumulation in systemic, but not inoculated, leaves. Thus, it is possible that viral suppressor activity could function to specifically allow systemic accumulation of viruses.

Host factors that modulate virus systemic spread either support or restrict this activity. A protein methylesterase (PME) is involved in both intercellular and systemic movement of TMV. For systemic movement, PME is essential for virus to exit into nonvascular tissue of the uninoculated leaves. A phloem protein from cucumber, p48, was found to interact with CMV capsids and may function to protect the capsid during transport. Host factors that restrict virus systemic movement include the restricted TEV movement (RTM) proteins, which are expressed only in phloem-associated cells and accumulate in sieve elements. RTM1 is related to the lectin, jacalin, while RTM2 has a heat shock protein motif. RTM1 may function in a plant defense pathway within the veins, although the jacalin-like proteins have not been previously linked to virus defense. RTM2 may function as a chaperone to prevent unfolding of a transport form of the virus within the sieve elements. A third protein that serves as a negative regulator is a cadmium-induced glycine-rich protein, cdiGRP. This protein does not act directly to restrict systemic movement. Instead, it induces callose deposits which are thought to restrict intercellular transport of the virus. This could prevent exit of virus from the vascular tissue. Interestingly, cadmium treatment inhibits the systemic spread of RNA silencing, lending support to the idea that spread of specific viruses affected by cadmium treatment (i.e., TMV and turnip vein clearing virus) is functionally similar to that of a host silencing signal.

See also: Brome Mosaic Virus; Bromoviruses; *Carmovirus*; Citrus Tristeza Virus; Cucumber Mosaic Virus; *Furovirus*; *Hordeivirus*; Luteoviruses; *Nepovirus*; Plant Resistance to Viruses: Engineered Resistance; Plant Resistance to Viruses: Geminiviruses; *Potexvirus*; Potyviruses; Tobacco Mosaic Virus; *Tobamovirus*; *Tobravirus*; Tombusviruses; *Tospovirus*; *Umbravirus*; Viral Suppressors of Gene Silencing; Viroids; Virus Induced Gene Silencing (VIGS).

Further Reading

Boevink P and Oparka KJ (2005) Virus–host interactions during movement processes. *Plant Physiology* 138: 1815–1821.

Derrick PM and Nelson RS (1999) Plasmodesmata and long-distance virus movement. In: van Bel AJE and van Kesteren WJP (eds.) *Plasmodesmata: Structure, Function, Role in Cell Communication*, 1st edn., pp. 315–339. Berlin: Springer.

Gilbertson RL, Rojas MR, and Lucas WJ (2005) Plasmodesmata and the phloem: Conduits for local and long-distance signaling. In: Oparka KJ (ed.) *Plasmodesmata*, 1st edn., pp. 162–187. Oxford: Blackwell.

Heinlein M and Epel BL (2004) Macromolecular transport and signaling through plasmodesmata. *International Review of Cytology* 235: 93–164.

Lewandowski DJ and Adkins S (2005) The tubule-forming NSm protein from *Tomato spotted wilt virus* complements cell-to-cell and long-distance movement of *Tobacco mosaic virus* hybrids. *Virology* 342: 26–37.

Lucas WJ (2006) Plant viral movement proteins: Agents for cell-to-cell trafficking of viral genomes. *Virology* 344: 169–184.

Morozov SY and Solovyev AG (2003) Triple gene block: Modular design of a multifunctional machine for plant virus movement. *Journal of General Virology* 84: 1351–1366.

Nelson RS (2005) Movement of viruses to and through plasmodesmata. In: Oparka KJ (ed.) *Plasmodesmata*, 1st edn., pp. 188–211. Oxford: Blackwell.

Nelson RS and Citovsky V (2005) Plant viruses: Invaders of cells and pirates of cellular pathways. *Plant Physiology* 138: 1809–1814.

Oparka KJ (2004) Getting the message across: How do plant cells exchange macromolecular complexes? *Trends in Plant Science* 9: 33–41.

Rakitina DV, Kantidze OL, Leshchiner AD, et al. (2005) Coat proteins of two filamentous plant viruses display NTPase activity *in vitro*. *FEBS Letters* 579: 4955–4960.

Requena A, Simón-Buela L, Salcedo G, and García-Arenal F (2006) Potential involvement of a cucumber homolog of phloem protein 1 in the long-distance movement of cucumber mosaic virus particles. *Molecular Plant Microbe Interactions* 19: 734–746.

Roberts AG (2005) Plasmodesmal structure and development. In: Oparka KJ (ed.) *Plasmodesmata*, 1st edn., pp. 1–32. Oxford: Blackwell.

Scholthof HB (2005) Plant virus transport: Motions of functional equivalence. *Trends in Plant Science* 10: 376–382.

Silva MS, Wellink J, Goldbach RW, and van Lent JWM (2002) Phloem loading and unloading of cowpea mosaic virus in *Vigna unguiculata*. *Journal of General Virology* 83: 1493–1504.

Verchot-Lubicz J (2005) A new cell-to-cell transport model for potexviruses. *Molecular Plant Microbe Interactions* 18: 283–290.

Waigman E, Ueki S, Trutnyeva K, and Citovsky V (2004) The ins and outs of nondestructive cell-to-cell and systemic movement of plant viruses. *Critical Reviews in Plant Sciences* 23: 195–250.

Mumps Virus

B K Rima and **W P Duprex,** The Queen's University of Belfast, Belfast, UK

© 2008 Elsevier Ltd. All rights reserved.

Glossary

Editing Process by which mRNAs are altered during or after transcription of the gene.
Mononegavirales Order of viruses consisting of four families with nonsegmented negative-stranded RNA genomes.
Nucleocapsid The ribonucleoprotein of mumps virus consisting of the RNA genome, the nucleocapsid protein, and associated proteins.
Pleomorphic From the Greek 'having many forms'.
Polyploid Having more than one genome copy per virion.
R_0 Index used in epidemiology, indicating the number of secondary cases that could originate from a single index case of viral infection.
Viremia Presence of virus in blood.
Virion The virus particle.

History

The primary clinical manifestation in mumps is swelling of the salivary glands because of parotitis. This symptom is so characteristic that the disease was recognized very early as different from other childhood illnesses which give rise to skin rashes. Hippocrates described mumps as a separate entity in the fifth century BC. He also noted swelling of the testes (orchitis) as a common complication of mumps. Infection in the central nervous system (CNS) and meninges in some cases of mumps was first noted by Hamilton in 1790. In 1934, Johnson and Goodpasture demonstrated the filterable nature of the causative agent and Koch's postulates were fulfilled by infection of human volunteers with virus propagated in the salivary glands of monkeys.

Taxonomy and Classification

Mumps virus (MuV) is sensitive to ether and other membrane-destroying reagents and has hemagglutinating and neuraminidase (HN) activity. It contains a nonsegmented negative-stranded RNA genome. MuV is thus classified in the family *Paramyxoviridae* and placed in the genus *Rubulavirus*. MuV is the prototype species, and two other human rubulaviruses, human (h) parainfluenza virus 2 (hPIV2) and hPIV4, have been identified. Other rubulaviruses infect a range of vertebrates – for example, Mapuera virus infects bats. Simian virus 5 (SV5) was originally isolated from rhesus and cynomolgus monkey kidney-cell cultures and thus designated as a primate virus. Sometimes SV5 is referred to as canine parainfluenza virus although it has a more extended host range and it has been suggested the virus be renamed PIV5. Porcine rubulavirus (PoRv) caused a fatal encephalomyelitis in neonatal pigs. Newcastle disease virus and a number of avian paramyxoviruses were included in the genus although these have recently been assigned to the genus *Avulavirus*.

Properties of the Virion

Paramyxoviruses consist of an inner ribonucleoprotein (RNP) core surrounded by a lipid bilayer membrane from which spikes protrude (**Figure 1**). The MuV virion appears to be roughly spherical and normally diameters range from 100 to 300 nm when the virus is grown in cell culture or embryonated eggs. Sometimes bizarre rod-shaped and other pleomorphic particles have been observed. Electron microscopy (**Figure 1(a)**) shows that MuV has the typical paramyxovirus structure with a lipid bilayer membrane surrounding an internal RNP complex, the nucleocapsid. The nucleocapsid displays the herring-bone structure characteristic of *Paramyxoviridae* (arrow, **Figure 1(a)**) and is approximately 1 μm in length with a diameter of 17 nm and an internal central core of 5 nm. Some of the pleomorphic particles have been reported to contain more than one RNP structure. The biological significance of such polyploid particles has not been investigated.

The RNP contains the RNA genome covered with nucleocapsid (N) protein as well as a phosphoprotein (P) and the large protein (L). The RNP is surrounded by a lipid bilayer membrane derived from the host cell in which the matrix protein (M) of the virus is embedded (**Figure 1(b)**). This protein interacts with the internal core N protein and the viral glycoproteins. Spikes (10–15 nm in length) protrude from the membrane and these contain the viral glycoproteins, the hemagglutinin–neuraminidase protein (HN) and the fusion (F) protein. The HN protein is probably a homo-tetramer, the fusion protein a homo-trimer. The ratio of HN and F protein molecules in the spike complex has not been elucidated yet. The spikes are also involved in the hemolysis of erythrocytes of different origins. The hemolysis reflects the ability of the virus to fuse with infected cells. Fusion is required for the entry of the RNP cores into cells.

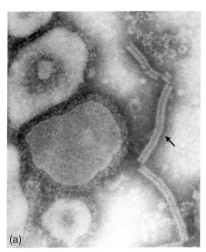 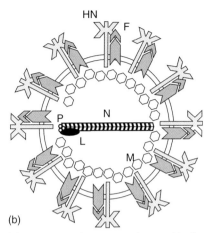

Figure 1 The structure of the mumps virion. In electron micrograph (a) the arrow points to a nucleocapsid released from a virus particle showing the characteristic herringbone structure. Schematic representation (b) indicates the location of the major viral structural proteins (see **Figure 2** for an explanation of the shapes and abbreviations).

Properties of the Genome

MuV has a single nonsegmented negative-stranded RNA genome (**Figure 2(a)**). The nucleotide sequence of the entire genome of several strains is known and all of these are 15 384 nt in length. The order of the genes of MuV and transcription of the genome is similar to that of other paramyxoviruses, especially to that of PIV5. There are seven transcription units which are separated by intergenic (Ig) sequences. The basic unit of infectivity is the negative-stranded, encapsidated RNP (**Figure 2(d)**).

Properties of the Proteins

The properties of the MuV proteins are summarized in **Table 1** and the number of amino acid residues in each viral protein is given in **Figure 2(c)**. It should be emphasized that this assignment is largely based on analogy of the MuV proteins with those of other paramyxoviruses, as gene identifications have not been carried out directly. However, the similarities to other paramyxoviruses are so striking that this assignment is beyond dispute. The presence of six structural proteins, namely, the N, P, M, and L proteins, as well as two glycosylated membrane-spanning proteins, the HN and F proteins, has been demonstrated in MuV virions. Furthermore, the virus induces the synthesis of at least two nonstructural proteins (V and W) from transcripts of the V/W/P gene. The presence of a small hydrophobic (SH) protein has been demonstrated in MuV-infected cells using antisera to peptides derived from the deduced amino acid sequence. It is not clear whether the protein is incorporated into virions. At least one strain (Enders) expresses the SH gene as a tandem readthrough transcript with the F gene in tissue culture. It is unlikely that the SH protein is translated from such an F–SH bicistronic mRNA and growth of this strain in tissue culture may not require the expression of this protein. A recombinant virus lacking the SH protein has been shown to be unaffected in growth in several cell types, formally proving the nonessential nature of the protein.

Physical Properties

The MuV virion is very susceptible to heat and treatment with ultraviolet (UV) light. The UV target size is one genome equivalent of RNA. The virus is inactivated by 0.2% formalin and the presence of the lipid bilayer confers sensitivity to both ether and chloroform. Treatment with 1.5 M guanidine hydrochloride leads to selective inactivation of neuraminidase and not hemagglutination activity of the virus. This indicates that separate domains of the HN molecule are responsible for these two functions.

Replication

MuV is capable of infecting a variety of cells in culture although whether this is the case *in vivo* remains to be determined. The attachment of MuV is mediated primarily through the interaction of the HN protein with a sialic acid, but the exact nature of the molecule(s) to which this moiety is linked remains unknown. Hence, MuV sialoglycoconjugate receptor(s) have not been precisely defined. The cooperative binding of a number of HN

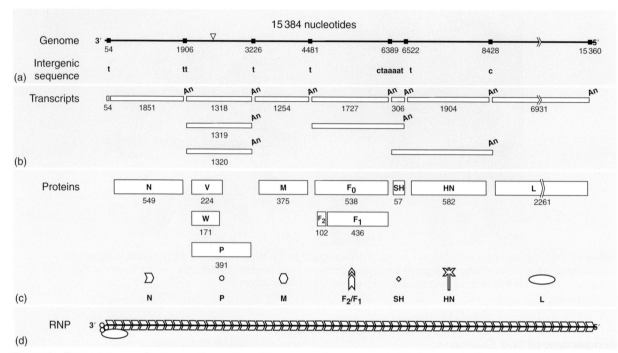

Figure 2 Transcription, replication, and translation of the mumps virus genome. In this figure are indicated: the mumps gene order and size and the position, sequence (positive strand) of the intergenic sequences (a); the major transcripts derived from the MuV genome and their sizes (b); the proteins and their sizes in numbers of amino acid residues and their schematic representations (c), and the replicative intermediates (d).

Table 1 Properties of the proteins of MuV

Protein	Size in SDS-PAGE (kDa)	Function
Nucleocapsid (N)	68–73	Phosphorylated structural protein of RNP; protects genome from RNases, possible role in regulation of transcription and replication (S antigen)
Phosphoprotein (P)	45–47	Phosphorylated protein associated with the RNP; possible role in solubilization of the N protein, role in RNA synthesis
Large (L)	>200	Protein with RdRp polymerase activity associated with the RNP; role in capping, methylation, and polyadenylation
Matrix (M)	39–42	Hydrophobic protein associated with inner side of membrane; role in budding by interactions with the N, HN, and F proteins
Fusion (F)	65–74	Acylated, glycosylated protein F_2–F_1 heterodimer activated by proteolytic cleavage; fusion of virion membrane with the plasma membrane which also involves HN (hemolysis antigen)
Hemagglutinin–neuraminidase (HN)	74–80	Acylated and glycosylated protein with hemagglutinin and neuraminidase activity; accessory role in fusion of virion membrane with the plasma membrane (major V antigen)
Small hydrophobic (SH)	6	Membrane protein with unknown function; present in infected cells but not detected in virions to date
Nonstructural V (earlier NS1)	23–28	Phosphorylated protein with a cysteine-rich domain which may be involved in metal binding; leads to the proteasomal degradation of STAT1 and targets STAT3 for ubiquitination
Nonstructural W (earlier NS2)	17–19	Phosphorylated protein with unknown role, possibly an artifact of misediting

SDS-PAGE, sodium dodecyl sulfate-polyacrylamide gel electrophoresis; kDa, mass of the protein in kilodaltons.

molecules to cell surface molecules probably leads to invagination of the host cell membrane and this may allow the fusion protein in its proteolytically cleaved, activated state to fuse the membrane of the virus and the host cell. Whether viropexis is another mechanism of entry of MuV is not known.

After introduction of the RNP into the cell, primary transcription of the negative-stranded genome occurs.

This is probably mediated by the viral L and P proteins of the virion and leads primarily to the synthesis of positive-strand, monocistronic mRNAs (**Figure 2(b)**). In addition, a number of polycistronic mRNAs have been identified. RNA-dependent RNA polymerase (RdRp) activity has been demonstrated in MuV virions but the role of the various viral proteins in this has not been assessed directly. Therefore, most of what follows is analogous to other paramyxoviruses. The first gene encoding the N protein is preceded by a leader region. No reports have appeared on the presence or absence of encapsidated or unencapsidated leader transcripts in infected cells or on the question of whether the leader region is transcribed by itself, in tandem with the first gene(s), or not at all. The transcription complex recognizes the 3′ end of the genome and transcribes the genes sequentially (**Figure 2(c)**), stopping at each Ig sequence to synthesize the polyadenylate (An) tails of the various mRNAs by repeated transcription on a poly(U) stretch in the genome. This reiterative transcription is sometime referred to as 'polymerase stuttering'. It has not been determined if the mRNAs are capped and methylated at the 5′ end.

The Ig sequences are 1–7-nt-long sequences (**Figure 2(a)**) and they are flanked by highly conserved gene end (GE) and gene start (GS) signal sequences (**Figure 3**). At any intergenic sequence there is a finite chance that the RdRp complex will leave the template. This gives rise to a transcription gradient so that the 3′ proximal N gene is most frequently transcribed and the 5′ proximal L gene is transcribed very infrequently. The gradient has not been quantified in MuV-infected cells. Occasionally, the GE and GS sequences are ignored and tandem readthrough transcripts of two or more genes are generated. Recognition of transcription signals appears to be dependent on host as well as viral factors. Thus, strains which have been adapted to growth in eggs only give rise to monocistronic mRNAs in chicken embryo fibroblasts, whereas only large tandem readthrough transcripts are generated if mammalian cells in culture are infected with these strains.

Co-transcriptional editing is responsible for the generation of mRNAs encoding the P and W proteins of MuV. Insertion of one or up to five extra G residues at a site in the V/W/P gene with a sequence similar to the polyadenylation signal leads to the formation of mRNAs that encode the W and P protein or V, W, or P proteins with an extra glycine residue (**Figures 2(b) and 2(c)**).

The mRNAs are translated by cellular ribosomes and there is no indication for shutoff of host mRNA translation or preferential translation of viral mRNAs in MuV-infected cells. Early reports of temporal translational control and control of transcription and translation of host mRNAs by the presence of viral glycoproteins in the cell membrane have not been confirmed. Some of the viral proteins are modified post-translationally (**Table 1**). The processing pathway for the HN protein differs from that for the F protein. The HN protein forms oligomers during transport through the Golgi apparatus, and carbohydrate side chains are modified slowly in the *trans*-Golgi cisternae, whereas the F protein rapidly matures with respect to its glycosylation. The proteolytic cleavage and consequent activation of the F_0 precursor into the F_1 and F_2 complex occurs in the *trans*-Golgi cisternae. Both the F and the HN proteins appear to play a role in fusion of virus-infected cells and syncytium formation. This is indicated by the observations that first, fusing as well as nonfusing strains contain cleaved activated F protein (i.e., the F_1 and F_2 complex); second, levels of neuraminidase are higher in nonfusing strains; and third, proteolytic cleavage of the HN protein can activate fusion. Co-expression of F and HN from eukaryotic expression vectors leads to the formation of multinucleated syncytia in transiently transfected cells which are indistinguishable from those generated in an MuV infection. These fusion assays have been useful in assessing the effect of site-directed mutagenesis on the biological function of the proteins and in determining where oligomerization takes place in the cell. Co-transfection of an equivalent construct containing the M gene has shown that this does not alter syncytium size.

Replication of the genome starts at the 3′ end of the (−) RNP. In replication mode the RdRp ignores the GE, Ig, and GS signals to generate one single positive-stranded RNA molecule. This RNA is concomitantly encapsidated

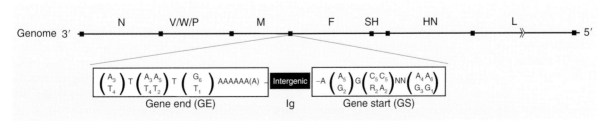

Figure 3 Consensus sequences of the seven gene start (GS) and gene end (GE) sequences. The various consensus sequences are shown as positive strand sequences. The specific intergenic sequences and their positions in the genome are shown in **Figure 2(a)**.

by N protein immediately after the start of its synthesis leading to the formation of a (+)RNP. It is not known for MuV if the transcriptase complex and the replicase complex are identical or whether they involve different host factors, as is the case for vesicular stomatitis virus, another virus in the order *Mononegavirales*. The intracellular concentration of N protein may regulate the balance between transcription and replication. The positive-stranded RNAs in the newly formed RNPs then form the template for further negative-strand synthesis, again with concomitant encapsidation by N protein. The resulting (−)RNPs reach the cell surface by an unknown mechanism where they are incorporated into progeny virions. The budding process itself and its regulation have not been studied in great detail for MuV. Virion budding probably takes place as a consequence of the interaction of the M protein with the RNP (most probably the C-terminal part of the N protein) and the cytoplasmic tails of the viral glycoproteins which accumulate in patches on the plasma membrane. Both (+)RNPs, containing the genome, and (−)RNPs, containing the antigenome, can be incorporated into MuV virions. However, whether the budding process is selective in favoring inclusion of (−)RNPs into virions is not yet clear.

There are a number of outstanding issues with respect to the life cycle of MuV which are not yet clear, some of which are specific to the virus. The role(s) of the V and W (earlier called NS1 and NS2, respectively) nonstructural proteins in the processes of replication, transcription, translation, and assembly has yet to be elucidated. Neither is it clear what role the SH protein plays. A functional role, preventing apoptosis in infected *in vitro* cells, has been suggested (see above). However, whether this is the case *in vivo* remains to be determined. Studies on the assembly of the RNP and the localization of the P and L proteins as well as the transport of the complex to the plasma membrane also remain to be carried out for MuV. Particularly, since the cell cytoskeleton appears to play a role in the replication of other paramyxoviruses, it is important to investigate this aspect of MuV replication.

Geographic and Seasonal Distribution

The virus has a worldwide distribution and requires a minimum population for it to continue to be able to circulate by continuous infection. The minimum population size is assumed to be of a similar order of magnitude to that for measles virus (*c.* 250 000) although no systematic study has been undertaken to ascertain this for MuV. Before successful control of mumps by vaccination had been achieved, outbreaks of mumps were more often observed in the winter and spring than in the summer, at least in the temperate Northern Hemisphere. Such a seasonal pattern was not observed in the tropics.

Host Range and Viral Propagation

Man is the only known host for the virus. Dogs can be naturally infected and show parotid swelling, although they do not pass on the virus. MuV can be used to infect a variety of animals experimentally, including monkeys, cats, dogs, ferrets, and a number of rodent species such as rabbits, suckling rats and mice, hamsters, and guinea pigs. Its adaptation to growth in 8-day-old embryonated eggs allowed the biological activities of the virus to be studied before the advent of tissue culture. The virus infects a wide range of cells in culture and causes a distinct cytopathic effect in most cell cultures with either rounding off or detachment of the cells from the substratum or widespread syncytium formation. Cytopathic effect varies from strain to strain and fusing and nonfusing variants have been described. The virus also readily establishes persistent infection in tissue culture systems, a property shared with many of the other viruses in the family *Paramyxoviridae*. Only low titers of virus are produced in such infections. However, the clinical significance of these observations is questionable as they do not reflect any known pathological outcome of the acute infection.

Serologic Relationships and Variability

MuV is a monotypic virus, and tests with human sera indicate the existence of only a single serotype. Polyclonal sera from infected individuals show a low level of cross-reactivity with hPIV2 and PIV5. Variability between strains has been demonstrated using monoclonal antibodies with the HN and N proteins showing the greatest diversity when single epitopes are examined. Sequence comparisons of the highly variable SH gene of MuV has demonstrated the existence of 12 genotypes. Sequence analysis of the SH gene is routinely used in molecular epidemiological studies (see below).

Evolution and Genetics

Human populations only became dense enough to sustain MuV from about 4000 years ago and, therefore, it has been suggested that the virus must have evolved from an animal pathogen. However, no closely related primate or other animal pathogen has yet been identified. Neither temperature-sensitive nor any other conditional lethal mutants of MuV have been reported nor has recombination been described in any nonsegmented negative-strand RNA virus. Although host range mutants have not been isolated, adaptation of the virus to growth in eggs or in chicken embryo fibroblasts requires a number of blind passages. Strains adapted in this way do not readily grow and fail to generate syncytia when they are used to infect mammalian cells in culture. As MuV is a neuropathogenic virus some clinical isolates have been adapted to grow in the CNS of experimental animals to study this biological property. These viruses are currently being used in reverse genetics approaches to attempt to identify the molecular determinant of neuropathogenesis. At present neutralizing monoclonal antibody escape mutants of the HN protein are the only type of MuV mutants which have been described.

Epidemiology

Outbreaks of mumps show annual periodicities although the length of the cycle may vary from 2 to 7 years as a result of factors that are poorly understood. Historically, the virus caused severe problems when troops were assembled for war and it was observed that male recruits from rural populations were affected in greater numbers. It is, therefore, assumed that children in isolated rural or island populations are exposed to the virus later in life than those in densely populated conurbations. The infection gives rise to lifelong immunity from disease, but it is not clear whether a single dose of live-attenuated vaccine achieves the same. Based on an ongoing outbreak involving a number of vaccinated individuals some have suggested that vaccinees may indeed be susceptible. However, at present this idea should be treated with caution until a systematic retrospective analysis has been performed and the data are considered alongside the known rates of seroconversion after vaccination. The existence of a number of different genotypes has allowed the development of molecular epidemiology for mumps. Specific genotypes appear to dominate in Japan and China, respectively. In some European outbreaks, co-circulation of two genotypes has been described. Changes in the nucleotide sequences of viruses isolated during an epidemic have not been studied although differences between isolates obtained over several years from a given geographical area indicate that the virus gradually accumulates nonexpressed and, to a lesser extent, expressed mutations. However, strains isolated more than 40 years apart can still easily be recognized as belonging to a specific genotype.

Transmission and Tissue Tropism

The virus replicates in the upper respiratory tract and the salivary glands and is transmitted in salivary droplets. Patients are infectious from 3 days before until approximately 4 days after the onset of clinical symptoms. Mumps can also cause viruria but this is not considered important in transmission. Transmission occurs only in the acute phase and from the level of infection observed in naive populations, it can be concluded that the virus is highly contagious ($R_0 = 5-12$) but not as contagious as, for example, measles virus ($R_0 = 16-450$) or chickenpox viruses. The infection is systemic and the virus multiplies in a wide variety of tissues in the human host. The tropism for the pancreas, particularly the β-cells, has been suggested as an explanation for the temporal link between MuV infection and juvenile onset diabetes mellitus. However, direct evidence for such a link is missing (**Table 2**).

Clinical Features of the Infection

In humans the normal MuV incubation period ranges from 14 to 21 days although occasionally this period has been estimated to extend for over 50 days. In infected individuals as many as one-third of MuV infections are subclinical. When symptoms ensue the most common clinical manifestation is parotitis. Other complications, such as orchitis, are not infrequent (**Table 2**). Contrary to popular belief, orchitis has not been linked to an increase in male sterility. Mastitis occurs in females with the same frequency and the incidence of both these clinical features increases with the age at which MuV infection is contracted. Before the development of the currently used live-attenuated vaccine, MuV was the most common cause of viral meningitis and encephalitis in the USA. The encephalitis is usually benign although minor neurological changes, learning, and concentration impairments and sudden deafness are well-documented sequelae in a number of patients. More rare complications are oophoritis, thyroiditis, pancreatitis, otitis, retinitis, conjunctivitis, and keratitis (**Table 2**).

Pathology and Histopathology

Upon infection, the virus replicates primarily in the nasal mucosa and the epithelial layer of the upper respiratory tract. After penetrating the draining lymph nodes, a transient viremia occurs and thereafter various target organs such as the salivary glands, the kidney, pancreas, and the CNS are infected. Infection in the salivary glands produces parotitis which is the most predominant clinical feature of the virus. Viral replication leads to tissue damage and the subsequent immune response leads to inflammation and swelling of the gland. Dissemination into the kidneys can lead to prolonged infection of this organ and viruria. Virus can be isolated from throat swabs, blood, saliva, and urine. Involvement of the CNS may be as high as 50% of cases and parotitis is not required for this to take place. MuV can be readily isolated from the cerebrospinal fluid (CSF) in cases of meningoencephalitis.

Table 2 Clinical features, complications and prognosis of mumps virus infection in various organs

Organ	Clinical features	Frequency	Pathology	Prognosis
Salivary gland	Parotitis (bilateral or unilateral)	95% of symptomatic cases	Blockage of duct of Stensen	Swelling disappears usually in 3–4 days
Submaxillary and sublingual salivary glands	Swelling	Rare		Swelling subsides in 3–4 days
Testes	Orchitis (mostly unilateral; some bilateral)	25% of males (especially adult males)		Good; mostly only transient depression of sperm production
Breast	Mastitis	15% of females		Good; cause of virus in breast milk
Ovaries	Oophoritis	Rare		
Meninges	Meningitis fever/vomiting	Mononuclear pleocytosis in CSF in 40–60% of cases	Ependymal epithelium destroyed by virus	Benign and self-limiting: ataxia (some permanent damage possible)
Brain	Encephalitis	2% of cases (<1% = fatal)	Virus spreads by neuronal pathways	Poor
Kidney				Good/cause of viruria
Pancreas	Pancreatis with nausea and vomiting	50% of cases	Could be due to interferon responses rather than specific tissue infection	Good/no established link to IDDM
Middle ear	Deafness	Very rare (<3%)	Cochlear infection	Permanent deafness is very rare
Heart	Myocarditis	Very rare	Fibroelastosis	Altered ECG; can cause myocardial infection
Blood	Immunosuppression	All cases	Infection in macrophages and lymphocytes	Viremia resolves
Fetus	Abortion in first trimester	Frequent	Virus is widespread in tissues of aborted fetus	In live births CMI response in the absence of humoral response

Pathogenicity

A number of MuV strains give rise to varying degrees of pathogenicity in experimental animal infections, for example, mice and hamsters. However, this does not seem to extend to natural human infections above and beyond some variation in the level of meningitis which tends to be associated with particular MuV strains (see below). The neurovirulence of MuV strains and vaccine batches is assessed using a monkey neurovirulence test (MNVT), but recently an alternative approach was devised based on the level of viral replication and hydrocephalus in neonatal rats. This rat neurovirulence test (RNVT) has apparently better predictive power than the MNVT. Some neutralizing monoclonal antibody escape mutants show alterations in neuropathogenicity in a hamster model.

Immune Response

It is not known whether the humoral or cell-mediated immune (CMI) response is the most important in clearing MuV from an infected host. Both play a role although neither seems to be required exclusively for successful control of the infection. Eleven days after infection, the humoral immune response is well established and the presence of neutralizing viral antibodies probably terminates viremia. Similarly, the appearance of IgA in the salivary fluid stops excretion of infectious MuV in the saliva. The precise time at which MuV reactive cytotoxic T lymphocytes first appear is unclear and these have been demonstrated in both the blood from patients with the natural disease and in vaccinees. The magnitude and effectiveness of this response may be related to the genetic human leukocyte antigen (HLA) background of the host. A complication in the development of the CMI response is the tropism of the virus for T and B cells. Reduction in CMI responses to antigens previously recognized has been observed, although the mechanism is unclear. It is less severe and of shorter duration than the immunosuppression associated with measles virus infection. The virus grows well in activated but not in resting T lymphocytes and infection of the CNS during MuV infection is thought to occur via transfer of infected monocytes into the choroid plexus. Perinatal exposure of the fetus to MuV via the placenta does not appear to lead to infection of fetal tissues. However, this can give rise to anomalous immune responses to MuV in the newborn child. In these children, CMI but no humoral responses are detectable. Neutralizing B-cell epitopes have been defined in the HN gene of MuV but no other B- or T-cell epitopes have been delineated, as yet.

Prevention and Control

Adaptation of MuV to growth in embryonated eggs and chicken embryo fibroblasts allowed the early development of live-attenuated vaccines for mumps. In 1946, Enders observed that adaptation of MuV to chicken cells was associated with the loss of virulence for monkeys. In the past, killed virus preparations were used for human vaccination although these did not lead to lifelong protective immunity and their use has been discontinued. In general, the appearance of atypical cases of measles and respiratory syncytial virus infection after vaccination with killed virus has led to the exclusive use of live-attenuated vaccines for the control of paramyxoviruses. Mumps vaccination has substantially reduced the incidence of the disease worldwide. After successful licensing of the vaccine in 1967 in the USA, the incidence of mumps dropped from 76 to 2 cases per 100 000 population. There are now a number of live-attenuated vaccine strains, for example, the Jeryl Lynn (JL) strain, which are usually administered in a trivalent vaccine containing live-attenuated strains of measles and rubella viruses as well as MuV (MMR). At present most developed countries have chosen to adopt a two-dose schedule for MMR vaccination and the vaccine is administered at 15 months and c. 4 years of age. The JL strain of MuV is most frequently used in the MMR vaccine and interestingly it has been shown to be comprised of two strains, the major (JL5) and the minor (JL2) components. The Urabe strain has been removed from MMR vaccines since reports in several countries indicated a higher incidence of meningitis associated with the use of this strain of MuV.

MuV and Inclusion Body Myositis

In the past MuV has been implicated in inclusion body myositis. However, neither *in situ* hybridization nor the polymerase chain reaction or immunocytochemistry has been able to link MuV to the paramyxovirus-nucleocapsid-like structures observed in muscle cells of patients with this disease. In contrast, myocarditis in patients with fibroelastosis has been associated with the presence of mumps viral RNA with reverse transcription-polymerase chain reaction (RT-PCR). The link to arthritis is also unclear (**Table 2**).

Future Perspectives

Over the last ten years, the development of reverse genetics systems for all members of the *Mononegavirales* has had a significant impact on our understanding of these viruses. Such systems have been generated for PIV5, hPIV2, and MuV, and to date they have been used to begin to define the structural and functional relationships and the roles that various proteins play in attenuating these viruses. The opportunity to examine the contribution of individual proteins from neurovirulent strains, such as Urabe, to neuropathogenesis in animal models should help to identify virulence determinants, something which is important if it proves necessary to develop new MuV vaccines. Expression of individual proteins may also allow a dissection of the humoral and cellular immune response and this should help improve our understanding of the role that these have in virus clearance.

One of the key challenges for the next number of years is to systematically monitor the immune status of vaccines and assess their susceptibility to reinfection by currently circulating MuV genotypes. Furthermore, documenting the occurrence of sequelae after the introduction of large-scale vaccination is vital. This should be underpinned by comprehensive molecular epidemiological approaches to ensure that the currently circulating genotypes are rapidly detected. Such approaches are particularly important given the recent large number of vaccinated individuals who have been reportedly infected in recent and ongoing outbreaks in both the UK and the USA. Linking MuV genotypes to particular phenotypes, such as neurovirulence, is crucial for our understanding of this ubiquitous human pathogen and the key goal in the medium term is to move from unproven association to formal proof. It is in this arena that reverse genetics approaches for MuV should prove most useful.

See also: Central Nervous System Viral Diseases; Epidemiology of Human and Animal Viral Diseases; Measles Virus; Paramyxoviruses of Animals; Parainfluenza Viruses of Humans; Persistent and Latent Viral Infection.

Further Reading

Carbone KM and Rubin S (2006) Mumps virus. In: Knipe DM, Howley PM, Griffin DE, et al. (eds.) *Fields Virology*,, 5th edn, pp. 1527–1550. Philadelphia: Lippincott Williams and Wilkins.

Feldmann HA (1989) Mumps. In: Evans AS (ed.) *Viral Infections of Humans*, ch. 17, p471. New York: Plenum.

Gupta RK, Best J, and MacMahon E (2005) Mumps and the UK epidemic 2005. *British Medical Journal* 330: 1132–1135.

Mungbean Yellow Mosaic Viruses

V G Malathi and **P John,** Indian Agricultural Research Institute, New Delhi, India

© 2008 Elsevier Ltd. All rights reserved.

Glossary

Agroinoculation Delivery of the viral genome through *Agrobacterium* inoculation.

Binary vector In these systems, the T-DNA region containing a gene of interest is contained in one vector and the *vir* region is located in a separate disarmed Ti plasmid. The plasmids co-reside in *Agrobacterium* and remain independent.

Nuclear localization signal Arrangement of basic residues like arginine or lysine in a protein that facilitates entry through nuclear pores.

Replicon Autonomously replicating DNA segments having an independent replication origin.

Rolling circle replication A mechanism for copying single-stranded circular genome by means of double-stranded intermediates.

History

The mungbean yellow mosaic virus (MYMV) and mungbean yellow mosaic India virus (MYMIV) infect a variety of leguminous crop plants and cause devastating yellow mosaic and golden mosaic diseases. A virus disease of mungbean (*Vigna radiata* (L.) Wilczek) exhibiting bright yellow mosaic symptoms was first observed in 1950s in Delhi, India by Nariani. The causal virus was easily transmissible to selected legume hosts through the whitefly vector. Nariani recorded the virus as mungbean yellow mosaic virus in 1960. Yellow mosaic viruses were reported subsequently from several hosts either as strains of MYMV or as separate entities. The annual loss due to yellow mosaic disease caused by these two viruses in blackgram, mungbean, and soybean together is estimated to be $300 million.

Taxonomy and Classification

The two viruses belong to the species *Mungbean yellow mosaic virus* and *Mungbean yellow mosaic India virus* (genus *Begomovirus*, family *Geminiviridae*). The type species *Mungbean yellow mosaic virus* is represented by a mungbean isolate from Thailand. The type species *Mungbean yellow mosaic India virus* is represented by a blackgram isolate from Delhi, India. The members of the family *Geminiviridae* have circular, single-stranded DNA genome of 2.5–2.7 kbp and are distinguished from other viruses by their characteristic geminate particle (about 30×20 nm in size) morphology. In the family *Geminiviridae*, the genus *Begomovirus* is differentiated, among other criteria, from the genera *Mastrevirus*, *Curtovirus*, and *Topocuvirus* by the fact that its members are transmitted by whiteflies (*B. tabaci* Genn) in a persistent manner.

Geographic Distribution

MYMV occurs in Thailand, Pakistan, and in western and southern states of India. MYMIV occurs in northern, eastern, and central states of India, Pakistan, Nepal, and Bangladesh. Identity of yellow mosaic viruses infecting grain legumes in Bhutan and Sri Lanka is not yet elucidated.

Symptoms in Plants

As the name implies, the characteristic symptoms caused by the infection are yellow and golden mosaic in the leaves (**Figure 1**). Symptoms start as scattered small specks or yellow spots in the leaf lamina which enlarge to irregular yellow and green patches alternating with each other on matured leaves. The yellow areas increase, coalesce resulting in complete yellowing of leaves. Plants produce fewer flowers and pods. Pod size and seed size are reduced. In some of the varieties of blackgram, necrotic mottling is seen. Sometimes the green areas are raised and the leaves show puckering and reduction in size. In French bean, affected plants show downward leaf curling and corrugated leaf lamina instead of yellow mosaic.

Virion or Particle Structure

Typical geminate particles measuring $(15-18) \times 30$ nm have been observed in leaf dip preparation of plant extracts of naturally infected plants.

In double antibody sandwich enzyme-linked immunosorbent assay (ELISA) the protein preparations from MYMV- and MYMIV-infected plants reacted to polyclonal and monoclonal antibodies to Indian and African cassava mosaic viruses, although A_{405} values obtained for field-infected plants were very low. The molecular weight of the coat protein (CP) of MYMV and MYMIV is ~28 kDa and consists of ~230 amino acids. There is one α-helix (WRKPRFY) at the N-terminus and the rest of the protein contains potential β-sheets. The recombinant MYMIV CP protein lacking the first N-terminus 22 amino acids was expressed as maltose-binding and glutathione *S*-transferase (GST) fusion protein. The expressed CP bound preferentially to single-stranded (ss) DNA rather than double-stranded (ds) DNA. The blackgram isolate of MYMV-Vig, CP was shown to interact with nuclear import factor, α importin of plants, suggesting nuclear import of CP via α-importin-dependent pathway.

Genome Organization

The MYMV and MYMIV have a genome organization (**Figure 2**) similar to Old World bipartite begomoviruses. The genome consists of two circular ssDNA components. They are designated as DNA-A and DNA-B components which are encapsidated separately. The ~2745 bp DNA-A component codes for the CP and for viral DNA replication and transcription proteins. The ~2616 bp DNA-B encodes for proteins for movement and nuclear localization. DNA-A replicates autonomously and is dependent on DNA-B for movement functions. Replication of DNA-B component is dependent on DNA-A-encoded proteins and both components are essential for viral pathogenicity. DNA-A has two open reading frames (ORFs) in the viral sense, (ORF AV1-CP; AV2-pre-CP) and five in the complementary sense (ORF AC1-replication initiation protein, Rep; ORF AC2-transcription activator protein, TrAP; ORF AC3-replication enhancer protein, REn; ORFs AC4 and AC5). The role of ORFs AV2, AC4, and AC5 is yet to be deciphered. The ORF AV2 is present only in Old World begomoviruses. In DNA-B, there is one ORF in viral sense strand, ORF BV1-nuclear shuttle protein (NSP), and one in complementary sense, ORF BC1-movement protein (MP). The nucleotide sequence identity

Figure 1 Yellow and golden mosaic symptoms in different legumes in the field: (a) mungbean, (b) blackgram, (c) soybean, and (d) cowpea.

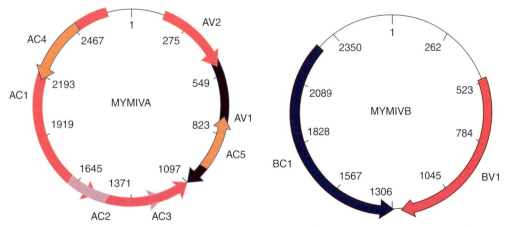

Figure 2 Genome organization of mungbean yellow mosaic India virus (MYMIV). Arrows represent open reading frames of virion and complementry sense. AV1, coat protein; AV2, pre-coat protein; AC1, replication intimation protein; AC2, transcription activator protein; AC3, replication enhancer protein; BC1, movement protein; BV1, nuclear shuttle protein.

between the DNA-A component of the two viruses is only 80%, justifying their recognition as separate species.

One characteristic feature of MYMV occurring in a southern state (Tamil Nadu) of India is the occurrence of multiple DNA-B components along with only one DNA-A component. Between rightward and leftward ORFs, in both the DNA components, there is a noncoding intergenic region. Intergenic region contains a characteristic stem–loop structure which is conserved in all geminiviruses. It lies 29 nt downstream of TATA box. The structure consists of GC-rich stem and an invariant nonanucleotide sequence, TAATATTAC. The Rep protein cleaves this nonanucleotide sequence to initiate replication. In the intergenic region there are 5–8 nt sequences that occur, as repeats and are called as iterons. Iterons represent Rep protein-binding sites, which cleaves the plus DNA strand to initiate replication. The Rep-binding repeat sequence was identified to be ATCGGTGT which occurs as invert and tandem repeat before the TATA box and one copy is also present after the nonanucleotide sequence in MYMIV. In MYMV the iterons are identified as GGTGTAxxGGTGT (x any nucleotide). The arrangement of iterons in MYMV and MYMIV is unique and does not show lineage to other begomoviruses. The term origin of replication (ori) refers to sequences from the tandem repeat of iteron, upstream of TATA box to the invariant nonanucleotide sequence. Intergenic region also contains promoters for rightward and leftward ORFs.

Within the intergenic region, there is a stretch of 180–210 nt sequence that is near identical between DNA-A and DNA-B components of bipartite begomoviruses; and this region called as common region (CR) is specific for a virus. In the case of MYMV and MYMIV, analysis of the CR of DNA-A and DNA-B components of more than 17 isolates revealed that there is considerable divergence in the whole CR in DNA-B compared to DNA-A component. The divergence in the CR is essentially in the origin of replication and ranges from 14% to 23% in the whole CR, 22% to 29% in the origin of replication among different isolates of MYMV and MYMIV.

Replication

The delicate mouth parts of the whitefly vector inject the virion particles in sieve tube cells while sucking the plant juice. From the geminate particles, ssDNA genome is released, whether the intact geminate particle or the genomic DNA enters the nucleus through the nuclear pore is debated. Once the ssDNA enters the nucleus, it is dependent on host DNA polymerase to synthesize a replicative dsDNA. It is the dsDNA which is the template both for transcription of various genes and for replication.

Investigations on geminiviruses have suggested that the viral DNA replicates in a rolling-circle mode (RCR). During RCR, Rep binds to specific iterons present in the CR and hydrolyzes the phosphodiester bond between the seventh and eighth residues of the invariant nonamer sequence 5'TAATATT$\downarrow$AC 3' (arrow indicates site of cleavage). Rep remains bound covalently to the 5' phosphate end and 3' hydroxyl end thus generated becomes available for rolling-circle replication. After a full cycle of replication, the new origin is generated which is again hydrolyzed by Rep. Subsequently, Rep closes the nascent 3' end of the DNA with the previously generated 5' end. In this way one unit genome-length circular, ssDNA molecule, that is, the mature viral genome, is processed.

The Rep protein of MYMIV was overexpressed in *Escherichia coli*. The recombinant and refolded protein bound to CR, in a sequence-specific manner; binding of DNA-A was more efficient than DNA-B. The recombinant protein showed site-specific nicking/closing and type-1 topoisomerase activities. The cleavage function was especially upregulated by ATP, suggestive of ATP-mediated conformational changes required to cleave the nonanucleotides. A large oligomeric complex (approximately 24 mer) of Rep protein was shown to function as helicase. The recombinant Rep protein of MYMIV showed binding with recombinant pea, proliferating cell nuclear antigen (PCNA) protein. The site-specific cleaving and closing activity and ATPase function of Rep were also impaired when bound with PCNA. There was a strong interaction between Rep and CP of MYMIV, the domain of interaction with the CP has been mapped to the central region of Rep. The activities of Rep were downregulated by the CP indicating how geminiviral DNA replication could be regulated by the CP.

Gene Expression and Transcription Regulation

The circular ssDNA genome of geminiviruses replicate through a double-stranded intermediate which is also the template for transcription for polymerase II. Transcription in these viruses is bidirectional and both viral and complementary strand encode different proteins.

Total RNA isolated from young leaves of blackgram seedlings agroinoculated with MYMV-Vig were subjected to transcript analysis by circularized reverse transcriptase-polymerase chain reaction (RT-PCR) method. The study showed that in DNA-A there are two major transcription units, one on viral sense and another one in complementary sense. Both are dicistronic, where rightward transcription unit is used for translation of ORF AV2/AV1 and leftward transcription unit for AC1/AC4. Likewise, there are two major transcripts in DNA-B, one rightward and one in leftward side, ORF BV1 and ORF BC1, are translated from these two transcripts respectively. The two major transcripts in both the DNA are driven by the promoter in the intergenic region and so the promoter is considered to be bidirectional. The promoter present between ORF AC1/AV2 is contained within 252 bp intergenic region in DNA-A. The promoter in DNA-B resides in the larger fragment of the intergenic region (957 bp) between ORFs BC1/BV1. Both the bidirectional promoters have direction-specific core elements located at an optimal distance from the transcription start site. Both rightward and leftward transcription are activated by the transcription factor AC2 encoded by the virus.

A new feature that was seen in MYMV-Vig for the first time is that the Rep protein showed synergistic activity with the AC2 protein and contributed to the activation of rightward promoters.

Besides these four transcripts, a fifth transcript of 1.4 kbp that hybridizes with ORF AC2 probe is predicted to be the transcript from which ORF AC2/AC3 will be translated. This transcript was driven by a strong monodirectional promoter upstream of ORF AC2.

Another unique feature was the transcription unit of the BC1 protein. There is a conserved leader-based intron which is not seen in any other begomoviruses. The 123 nt intron within BC1 transcript appears to have all the features of plant introns. Such an intron with consensus 5′ and 3′ splice sites (AG/GU,CAG/G) and one or more short ORFs (sORFs) appear at the same or nearby location in all the isolates of MYMV and MYMIV.

One characteristic feature observed in all transcripts except that of AC1 and BC1 is multiple transcription initiation sites, closely spaced (3–4 nt) from each other and multiple polyadenylation sites. Translation of ORFs AV1, AC4, and AC3 is predicted to be of leaky scanning type wherein the proximal 5′ AUG is in suboptimal context and is bypassed allowing the second product (AV1, AC4, and AC3) to be translated. Short ORFs present between AV2 and AV1 may also help in reinitiation of transcription of second product.

Infectivity of Cloned Components

Infectivity of blackgram isolate of MYMV, MYMV-[Vig] and blackgram, cowpea, mungbean and soybean isolates of MYMIV, MYMIV-[Bg], MYMIV-[Cp], MYMIV-[Mg], and MYMIV-[Sb] on leguminous hosts have been established by 'agroinoculation'. In this strategy dimers of DNA-A and DNA-B components are cloned in a binary vector, which is mobilized in *Agrobacterium tumefaciens*. The partial or complete tandem repeat constructs are amplified in *A. tumefaciens* in nutrient broth; the bacterial culture is used to deliver the viral inoculum. In all these cases, considerable divergence was seen in the origin of replication in DNA-B component, which did not impair their infectivity. An interesting feature was seen with the cowpea isolate; though it is not whitefly transmissible to hosts other than cowpea and French bean, through agro-inoculation the cowpea isolate was systemically infected and produced yellow mosaic symptoms in blackgram and mungbean. This adaptation to new hosts was maintained by the viral progeny isolated from agroinoculated plants, which could then be easily transmitted by whitefly to other hosts. However, the blackgram isolate of MYMIV did not produce yellow mosaic symptoms in cowpea even by agroinoculation. An improved agroinfection using one strain of *A. tumefaciens* was shown for MYMV-Vig. Partial tandem repeat construct of DNA-A and DNA-B was made in vectors having compatible replicons and introduced into the same *Agrobacterium* cells. When *Agrobacterium* cells having both constructs were inoculated, infectivity rate was 100% in blackgram.

Inoculation with different DNA-B components, KA22 and KA27 with DNA-A of MYMV-Vig, showed differences in symptoms severity. KA22 DNA-B (which is more closely related to MYMIV DNA-B) caused more intense mosaic symptoms with high viral DNA titer in blackgram. In contrast, KA27 DNA-B (closely related to DNA-B of Thailand isolate of MYMV) caused more intense symptoms and high viral DNA titer in mungbean. DNA-B is therefore considered as an important pathogenicity determinant of host range between blackgram and mungbean.

Host Range

The MYMV and MYMIV isolates have very narrow host range infecting only leguminous hosts. Whatever information is available on host range, they have been generated before the demarcation of isolates into two viruses. Whether differentiation into two species based on DNA-A nucleotide sequence will reflect on the biological properties need to be investigated freshly. The isolates from blackgram, mungbean, horsegram, and mothbean are easily transmissible to following leguminous species: mungbean (*V. radiata*), blackgram (*V. mungo*), mothbean (*V. aconitifolia* (Jacq) Marechal), soybean (*Glycine max*), phaemey bean (*Phaseolus lathyroides*), horsegram (*Macrotyloma uniflorum*) and black tapery bean (*P. acutifolius*), pigeonpea (*Cajanus cajan*), and French bean (*P. vulgaris*). These isolates are not transmissible to *Dolichos*. The transmission of these isolates to cowpea is inconsistent and contradictory results are reported. The cowpea isolate of MYMIV is transmissible only to cowpea (*V. unguiculata*), yard long bean (*V. unguiculata* (L.) Walp. f.sp. *sesquipedalis*), and French bean and not to any other host. The pigeonpea and soybean isolates could be transmitted to cowpea, though back transmission from cowpea to soybean or pigeonpea is inconsistent. Thus, it has been difficult to exactly assess the host range of the viruses as they are not sap transmissible and transmission by whitefly is dependent on the biotype of the vector used, its feeding preferences, and susceptibility of the genotypes tested. Weeds like *Brachiaria ramosa*, *Eclipta alba*, and *Xanthium strumarium*, and garden plant *Cosmos bipinnatus* were reported as hosts for MYMIV. Total nucleic acid extracted from above hosts showing symptoms of begomovirus infection did not hybridize with radiolabeled MYMIV DNA-B probe, indicating that the begomoviruses infecting these hosts are different. At present, MYMIV isolates are not transmissible to any other nonleguminous host.

Transmission and Virus–Vector Relationship

All isolates of MYMV and MYMIV are neither sap transmissible (except the Thailand isolate of MYMV) nor are they transmitted through seeds. MYMV and MYMIV are transmitted by whitefly, *B. tabaci* Genn, in a persistent circulatory manner. Young expanding leaves are better sources of inoculum than old leaves. A single whitefly can transmit the virus given an acquisition (AAP) and inoculation access period (IAP) of 24 h. Transmission percentage definitely increases with increased number of whiteflies. Female whiteflies are found to transmit more effectively and retain the virus for a longer period than male whiteflies. Starvation for a duration of 15 min, before acquisition and inoculation access period increases the transmission rate to 50%. Though the vector can acquire and transmit the virus immediately after an AAP and IAP of 10–15 min, the minimum period of 4–6 h is required.

There is a latent period of more than 3 h inside the vector for transmission to occur. The whiteflies can retain the virus for 10 days and transmission rate decreases gradually. The period of retention may differ from one batch of insects to another batch. There is no transovarial transmission. No evidence exists till date for association of any specific biotype of the vector with the epidemic outbreak of the disease.

Ecology and Epidemiology

Bemisia tabaci is a polyphagous vector with wide host range; however, it shows strong host preferences and its feeding behavior is a major factor in deciding the active spread of the virus from one crop species to another. For example, the cowpea isolate of MYMIV does not spread from golden mosaic disease-affected cowpea plants to adjacent blackgram or mungbean fields in northern India.

Bemisia tabaci thrives best under hot and humid conditions that prevail in the tropical and subtropical area in the Indian subcontinent. The population of the vector is influenced by temperature, relative humidity, and rainfall.

Depending on vector population build-up, spread of the virus is gradual, cumulative, and is in the direction of prevalent wind. In northern India with the onset of monsoon rain (June–July), as the population of vector increases, rate of spread of virus increases. Whereas before monsoon

Table 1 Matrix of pairwise identity percentages of complete nucleotide sequences of DNA-A of selected begomoviruses

YMV isolates	MYMIV	MYMIV-Mg[Isl]	MYMV-Vig	MYMV	DoYMV	HgYMV
MYMIV	*	95.8	80.3	80.0	60.5	80.9
MYMIV-Mg[Isl]		*	81.3	81.0	61.7	82.3
MYMV-Vig			*	96.7	62.4	84.1
MYMV				*	62.4	84.0
DoYMV					*	62.0
HgYMV						*

Acronyms are as given in text; '*' indicates homologous comparisons. Matrix was generated using the CLUSTAL algorithms.

Table 2 Matrix of pairwise identity percentages of complete nucleotide sequences of DNA-B of selected begomoviruses

YMV isolates	MYMIV	MYMV-Vig	MYMV-Vig [Mad KA 21]	MYMV-Vig [Mad KA 27]	MYMV-Vig [Mad KA 28]	MYMV-Vig [Mad KA 34]	MYMV	HgYMV
MYMIV	*	91.8	92.3	69.6	92.4	92.0	69.5	66.9
MYMV-Vig		*	96.8	69.7	96.6	98.9	86.3	66.6
MYMV			69.5	95.0	69.8	69.4	*	65.9
MYMV-Vig [Mad KA 21]			*	69.7	98.2	96.9		67.1
MYMV-Vig [Mad KA 27]				*	69.7	69.7		66.6
MYMV-Vig [Mad KA 28]					*	96.7		67.5
MYMV-Vig [Mad KA 34]						*		66.7
HgYMV								*

Acronyms are as given in text; '*' indicates homologous comparisons. Matrix was generated using the CLUSTAL algorithms.

Table 3 List of viruses with their accession numbers from GenBank database used for sequence analysis and phylogenetic comparison

Virus	Acronym	Accession Number
Bean calico mosaic virus	BCaMV	AF110189
Bean dwarf mosaic virus	BDMV	M88179
Bean golden yellow mosaic virus-[Mexico]	BGYMV-[MX]	AF173555
Bhendi yellow vein mosaic virus-[301]	BYVMV-[301]	AJ002453
Bhendi yellow vein mosaic virus-[Madurai]	BYVMV-[Mad]	AF241479
Cabbage leaf curl virus	CaLCuV	U65529
Cotton leaf curl Rajasthan virus	CLCuRV	AF363011
Cowpea golden mosaic virus-[Nigeria]	CPGMV-[NG]	AF029217
Indian cassava mosaic virus-[Maharashtra]	ICMV-[Mah]	AJ314739
Maize streak virus-A[Kom] (RANDOM)	MSV-A[Kom]	AF003952
Mungbean yellow mosaic India virus	MYMIV	AF126406, AF142440
Mungbean yellow mosaic India virus-[Bangladesh]	MYMIV-[BG]	AF314145
Mungbean yellow mosaic India virus-Cowpea	MYMIV-[Cp]	AF481865, AF503580
Mungbean yellow mosaic India virus-Mungbean	MYMIV-[Mg]	AF416742, AF416741
Mungbean yellow mosaic India virus-Mungbean [Akola]	MYMIV-Mg[Akol]	AY271893, AY271894
Mungbean yellow mosaic India virus-[Soybean]	MYMIV-[Sb]	AY049772, AY049771
Mungbean yellow mosaic India virus-Mungbean [Islamabad]	MYMIV-Mg [Isl]	AY269992
Mungbean yellow mosaic India virus-Mungbean [Nepal]	MYMIV-Mg [Nep]	AY271895
Mungbean yellow mosaic India virus-[Soybean TN]	MYMIV-[SbTN]	AJ416349, AJ420331
Mungbean yellow mosaic India virus-Mungbean [Islamabad]	MYMIV-[PK;Cp]	AJ512496
Mungbean yellow mosaic India virus-[MBK-A25]	MYMIV-[MBK-A25]	AY937175, AY937196
Mungbean yellow mosaic virus-Mungbean[Haryana]	MYMV-Mg [Har]	AY271896
Mungbean yellow mosaic virus-Soybean [Islamabad]	MYMV-Sb [Isl]	AY269991
Mungbean yellow mosaic virus-Soybean [Madurai]	MYMV-Sb [Mad]	AJ421642, AJ867554, AJ582267
	MYMV-[Sb]	
Mungbean yellow mosaic virus–Vigna	MYMV-Vig[KA22]	AJ132575, AJ132574
Mungbean yellow mosaic virus-Vigna [KA27]	MYMV-Vig [KA:27]	AF262064
Mungbean yellow mosaic virus-Vigna [Maharashtra]	MYMV-Vig [Mah]	AF314530
Mungbean yellow mosaic virus-Thailand	MYMV-Thailand	D14703, D14704
Hosegram yellow mosaic virus	HgMV-[IN:Coi]	AJ627904, AJ627905
Dolichos yellow mosaic virus	DoYMV-[IN]	AY306241
Mungbean yellow mosaic virus–Vigna [Madurai]	MYMV-[KA:34]	AJ439057
Mungbean yellow mosaic virus–Vigna[Madurai]	MYMV-[KA:28]	AJ439058
Mungbean yellow mosaic virus–Vigna [Madurai]	MYMV-[KA:21]	AJ439059
Potato yellow mosaic virus-[Venezuela]	PYMV-[VE]	D00940
Soybean crinkle leaf virus-[Japan]	SbCLV-[JR]	AB050781
Tomato golden mosaic virus-Yellow Vein	TGMV-YV	K02029
Tomato leaf curl Bangalore virus-[Ban5]	ToLCBV-[Ban5]	AF295401
Tomato leaf curl New Delhi virus-Severe	ToLCNDV-Svr	U15015
Tomato yellow leaf curl Thailand virus-[2]	TYLCTHV-[2]	AF141922

rain B. tabaci populations are nonviruliferous, whiteflies may pick up the inoculum from self-perpetuating weeds harboring viruses. However, the role of any weed or crop plant specifically serving as reservoir for MYMV and MYMIV has not been ascertained. A serious outbreak of yellow mosaic disease caused by MYMV in mungbean occurred in northern Thailand in 1977, resulting in reduction in mungbean cultivation and shift in cropping pattern.

Management

The main components of management of the disease are cultivation of resistant genotypes, cultural practices to ensure weed-free or inoculum-free field, and judicious use of chemicals to reduce the vector population.

The genetics of resistance to MYMIV is understood and it appears to be governed by a single recessive gene or two complementary recessive genes in mungbean. Sources of resistance to both the viruses have been identified by rigorous screening in the field; some of the resistance sources identified are being used for breeding purposes.

Cultural practices like adjustment of sowing dates to avoid the disease, rouging the weeds and plants that get infected early in the season, inter or mixed cropping of mungbean or blackgram with nonhost plants like sorghum, pearl millet, and maize have been adopted and shown effective in bringing down the disease incidence. The vector whitefly is controlled by applying phorate, disulfoton granules @ 2 kg ha^{-1} during sowing. Spray of

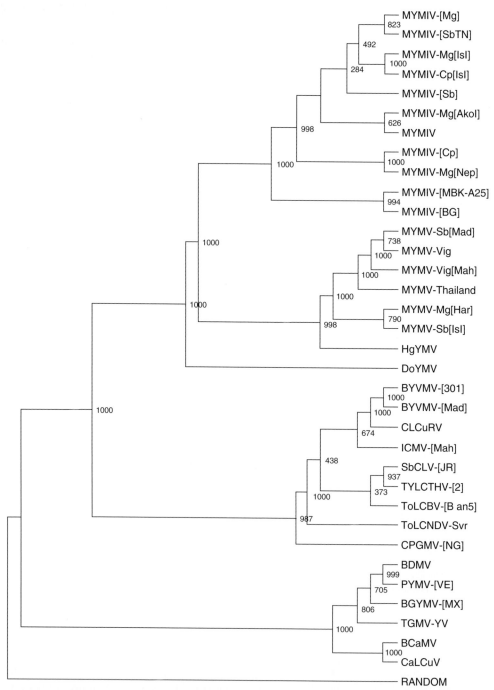

Figure 3 Phylogenetic relationship between selected begomovirus isolates based on complete nucleotide sequence of DNA-A. Dendogram was constructed using the neighbour-joining method with bootstraping (1000 replicates) in CLUSTAL X. Vertical distances are arbitary, horizontal distances are propotional to genetic distances.

metasystox malathion @ 0.1% is known to effectively reduce the virus spread and disease incidence.

Biolistic delivery of RNAi constructs targeting CR sequences of MYMV-Vig isolate resulted in complete recovery from infection (68–77%) in blackgram seedlings, offering an alternative approach for the management of the disease. Possibility of using *Paecilomyces farinosus*, an endophyte of whitefly, as biocontrol agent is being explored.

Phylogenetic Relationship

Complete nucleotide sequence of DNA-A component of 28 MYMV and MYMIV isolates from Indian subcontinent and Thailand is available in the GenBank database. Nucleotide sequence of a limited number of yellow mosaic virus isolates infecting *Dolichos* and horsegram has also been completed. There are currently four different species comprising 'yellow mosaic legume viruses'. They are *Mungbean yellow mosaic virus*, *Mungbean yellow mosaic India virus*, *Horsegram yellow mosaic virus*, and *Dolichos yellow mosaic virus*. Percentage nucleotide identity for DNA-A and DNA-B component between the viruses of the different species is given in **Tables 1** and **2**. Comparison of nucleotide sequence of DNA-A component revealed 81% identity between MYMIV and MYMV isolates; both the viruses share only 60% identity with cowpea golden mosaic virus from Nigeria.

A phylogenetic tree based on complete nucleotide sequence (**Table 3**) of DNA-A component of selected begomoviruses revealed their distinct lineages. MYMV and MYMIV showed a clear dichotomy. Three viruses, MYMV, MYMIV, and DoYMV, occupy three branches separately (**Figure 3**).

The topology of the phylogenetic tree based on nucleotide sequence of DNA-B (**Figure 4**) component is different from DNA-A. There are three major branches: one major branch comprising DNA-B of MYMIV isolates and (four) DNA-B found associated with MYMV-Vig DNA-A. The second branch consisting of one isolate each of soybean (MYMV-Sb [Mad]B1) and blackgram (MYMV-Vig[Mad]KA27) from southern India and one mungbean isolate from Thailand (MYMV); the third branch is highly divergent from others consisting of one soybean isolate of MYMV-Sb[Mad]B2 and HgYMV. Recombination events have been identified in a mungbean isolate of MYMV. Occurrence of multiple DNA-B components with MYMV DNA-A indicates how MYMIV DNA-B could have been captured by MYMV DNA-A during mixed infection of both MYMV and MYMIV.

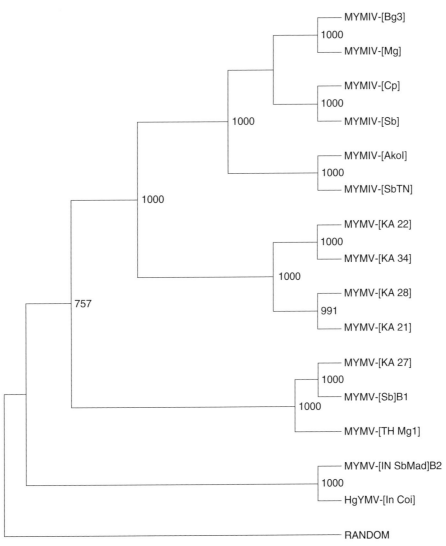

Figure 4 Phylogenetic relationship between yellow mosaic virus isolates based on complete nucleotide sequence of DNA-A dendogram was constructed using the neighbor-joining method with bootstraping (1000 replicates) in CLUSTAL X. Vertical distances are arbitary, horizontal distances are propotional to genetic distances.

A striking feature of DNA-A tree is the unique position occupied by MYMV and MYMIV. They do not cluster with typical begomoviruses of the Indian subcontinent origin, which is unusual. Begomoviruses of a specific geographical origin show high degree of conservation in CP gene, due to selection pressure exerted by the whitefly biotype. On the contrary, MYMV and MYMIV which are transmitted effectively by the same whitefly biotypes in Indian subcontinent like other begomoviruses, yet do not cluster with them. Due to their unique identity, MYMV and MYMIV are separately classified as 'legumoviruses'. They neither show host-related affinity as they are divergent from legume begomoviruses of both Old World (the virus species *Cowpea golden mosaic virus* from Nigeria, *Soybean crinkle leaf virus* from Thailand) and New World (the virus species *Bean golden yellow mosaic virus*, *Bean dwarf mosaic virus*, and *Bean calico mosaic virus*). Probably yellow mosaic begomoviruses of legumes diverged from a common ancestor virus much earlier in evolution and might have gained their unique features while adapting to legume hosts during evolution in the Indian subcontinent.

See also: Bean Golden Mosaic Virus; Tomato Leaf Curl Viruses from India.

Further Reading

Chaudhary NR, Malik PS, Singh DK, Islam MN, Kaliappan K, and Mukherjee SK (2006) The oligomeric Rep protein of Mungbean yellow mosaic India virus (MYMIV) is a likely replicative helicase. *Nucleic Acid Research* 34: 6362–6377.

Girish KG and Usha R (2005) Molecular characterization of two soybean infecting begomoviruses from India and evidence for recombination among legume infecting begomoviruses from Southeast Asia. *Virus Research* 108: 167–176.

Green SK and Kim D (eds.) (1992) *Mungbean Yellow Mosaic Disease. Proceedings of an International Workshop*, Bangkok, Thailand, 79pp. Taipei: AVRDC.

Malathi VG, Surendernath B, Naghma A, and Roy A (2005) Adaptation to new host shown by the cloned components of Mungbean yellow mosaic India virus causing golden mosaic in northern India. *Canadian Journal of Plant Pathology* 27: 439–447.

Malik PS, Kumar V, Bagewadi B, and Mukherjee SK (2005) Interaction between coat protein and replication initiation protein of Mungbean yellow mosaic India virus might lead to control of viral DNA replication. *Virology* 337: 273–281.

Mandal B, Varma A, and Malathi VG (1997) Systemic infection of *Vigna mungo* using the cloned DNAs of the blackgram isolate mungbean yellow mosaic geminivirus through agroinoculation and transmission of the progeny virus by whiteflies. *Journal of Phytopathology* 145: 503–510.

Nariani TK (1960) Yellow mosaic of mung (*Phaseolus aureus* L.). *Indian Phytopathology* 13: 24–29.

Nene YL (1968) A survey of the viral diseases of pulse crops in Uttar Pradesh. In: *First Annual Report, FG-In-358*, Pantnagar, India: Uttar Pradesh Agricultural University,

Shivaprasad PV, Akbergenov R, Trinks D, et al. (2005) Promoters, transcripts and regulatory proteins of Mungbean yellow mosaic geminivirus. *Journal of Virology* 79: 8149–8163.

Stanley J, Bisaro DM, Briddon RW, et al. (2005) Geminiviridae. In: Fauquet CM, Mayo MA, Maniloff J, Desselberger U, and Ball LA (eds.) *Virus Taxonomy: Eighth Report of the International Committee on Taxonomy of Viruses*, San Diego, CA: Elsevier Academic Press.

Trinks D, Rajeswaran R, Shivaprasad PV, et al. (2005) Suppression of RNA silencing by a geminivirus nuclear protein AC2 correlates with transactivation of host gene. *Journal of Virology* 79: 2517–2527.

Usharani KS, Surendranath B, Haq QMR, and Malathi VG (2004) Yellow mosaic virus infecting soybean in northern India is distinct from the species infecting soybean in southern and western India. *Current Science* 86: 845–850.

Varma A, Dhar AK, and Mandal B (1992) MYMV-transmission and its control in India. In: Green SK and Kim D (eds.) *Mungbean Yellow Mosaic Disease. Proceedings of an International Workshop*, pp. 1–25. Bangkok, Thailand, Taipei: AVRDC.

Murine Gammaherpesvirus 68

A A Nash and B M Dutia, University of Edinburgh, Edinburgh, UK

© 2008 Elsevier Ltd. All rights reserved.

Origins and Ecology

Murine gammaherpesvirus 68 (MHV-68) belongs to species *Murid herpesvirus 4*, genus *Rhadinovirus*, subfamily *Gammaherpesvirinae*, family *Herpesviridae*. It was originally isolated from a bank vole (*Clethrionomys glareolus*) in Slovakia. Two other related herpesviruses (MHV-60 and MHV-72) also came from bank voles and two more (MHV-76 and MHV-78) from wood mice (*Apodemus flavicollis*). The five viruses were originally isolated following the inoculation of diluted suspensions of various tissues (lung, spleen, liver, kidney, and heart) into the brains of newborn mice. Different virus isolates were obtained from the brains of mice following either the first, second, or third intracranial passage from mouse to mouse. These Slovakian viruses developed cytopathic effect in epithelial and fibroblast cell lines from a variety of species ranging from chickens to primates. One other isolate (MHV-Brest) was reported in a shrew (*Crocidura russula*). Subsequently, a number of new isolates have arisen primarily from wood mice following a survey of wild rodents in the Wirrell, Liverpool.

Virus Structure and Genetic Content

MHV-68 virion structure is similar to that of other herpesviruses. The capsid, tegument, and glycoprotein genes are homologous to those of other herpesviruses and can be predicted to fulfill the same roles in the MHV-68 virion. The composition of the MHV-68 capsid, tegument, and envelope has been investigated, revealing five capsid proteins (encoded by ORF25, ORF62, ORF26, ORF65, and ORF29), three tegument proteins (ORF75c, ORF45, and ORF11), five envelope proteins (ORF8, ORF51, ORF27, ORF28, and ORF22), and four proteins of undetermined locations (ORF20, ORF24, ORF48, and ORF52). The viral tRNA-like RNAs (vtRNAs), which are unique to MHV-68, are present in the tegument. Other structural proteins are predicted by analogy with other herpesviruses but their presence in the virion is yet to be demonstrated.

The genetic content of MHV-68 is based on the complete DNA sequences of MHV-68 strains g2.4 and WUMS. These are not independent strains, since the latter was derived by limited passage and plaque purification of the former. As a consequence, the two sequences are very similar (g2.4, GenBank accession number AF105037; WUMS, U97553). The proposed gene layout is shown in **Figure 1**.

The linear, double-stranded DNA genome of MHV-68 consists of a single unique region of 118 237 bp flanked by a variable number of 1213 bp terminal repeats. The unique region and terminal repeats have nucleotide compositions of 46 and 77.6% G+C, respectively. Two internal tandem repeats are located within the unique region: a 40 bp repeat located at genome coordinates 26 778–28 191 and a 100 bp repeat located at coordinates 98 981–101 170.

The unique region of the MHV-68 genome is largely collinear with the genomes of other rhadinoviruses, including Kaposi's sarcoma-associated herpesvirus (KSHV) and herpesvirus saimiri. It contains 73 protein-coding open reading frames (ORFs) and also encodes eight vtRNAs and nine predicted microRNAs (miRNAs). As with other rhadinoviruses, MHV-68 possesses a number of cellular homologs, including a complement regulatory protein (CRP, encoded by ORF4), a Bcl-2 homolog (vBcl-2, M11), a cyclin D homolog (vcyclin, ORF72), and a G protein-coupled receptor (vGPCR, ORF74).

MHV-68 also contains a number of unique genes clustered at the left end of the genome. These include four genes encoding proteins (M1, M2, M3, and M4) and eight genes encoding the vtRNAs and miRNAs. In MHV-76, the first 9.5 kbp of the genome, containing M1–M4 and the vtRNAs and miRNAs, is absent;

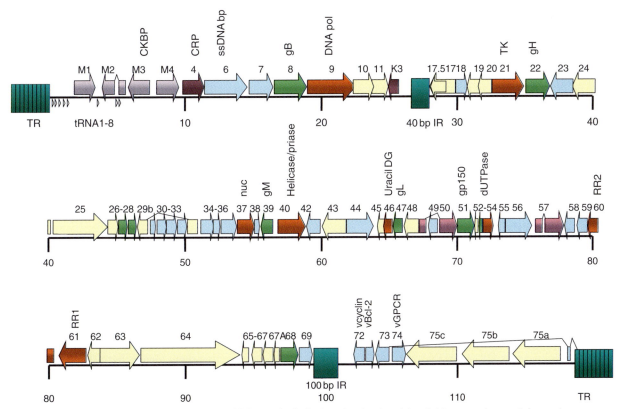

Figure 1 Organization of the MHV-68 genome. Yellow color indicates structural proteins; light green, glycoproteins; red, enzymes; violet, transcriptional transactivators; lavender, unique MHV-68 genes ; plum, immunomodulatory proteins; and dark green, repeats. TR, terminal repeat; IR, internal repeat; CKBP, chemokine-binding protein; CRP, complement regulatory protein; TK, thymidine kinase; nuc; alkaline exonuclease; uracil DG, uracil deglycosylase; RR, ribonucleotide reductase; vGPCR, viral G protein-coupled receptor.

otherwise the genome is identical to that of MHV-68. MHV-72 is deficient in the first 7 kbp of the genome and lacks M1–M3 and the vtRNAs and miRNAs; otherwise the genome is identical to that of MHV-68. It is not known whether MHV-76 and MHV-72 are deletion mutants of MHV-68 that arose during isolation or whether they exist naturally. MHV-Brest is closely related to MHV-68 but appears to represent a different virus species. The sequence of the left end region is particularly highly conserved between MHV-Brest and MHV-68, with greatest divergence in M2.

Infection and Replication

MHV-68 is able to infect and replicate in a range of mammalian epithelial and fibroblast cell lines. A productive infection occurs in the majority of cell lines, but a persistent infection is established in some lymphoid cells. MHV-68 establishes a latent infection in NS0, a myeloma cell line, but not in the thymoma cell line BW5147. The virus is maintained indefinitely in NS0 cells as a latent infection, with approximately 5% of the cells undergoing reactivation and expressing lytic cycle proteins. This scenario is similar to that seen for lymphoblastoid cell lines infected with Epstein–Barr virus (EBV). Other B cell lines can also be infected, including B cell hybridomas and the commonly used A20 cell line. One latently infected B cell line, S11, has been derived from a lymphoma obtained from a MHV-68-infected mouse.

Many MHV-68 genes have been assigned functions based on their orthology to genes of other herpesviruses whose roles are known. The use of signature-tagged transposon mutagenesis has resulted in the identification of a number of MHV-68 genes that are essential for the replication process, as well as several genes that are not essential but significantly enhance viral replication. So far, 41 genes have been shown to be essential for viral replication, of which 17 are essential for replication in all herpesviruses, including MHV-68 ORF6 (encoding the single-stranded DNA-binding protein, ssDNA bp), ORF8 (glycoprotein B, gB), ORF9 (DNA polymerase, DNA pol), ORF22 (glycoprotein H, gH), and ORF64 (large tegument protein). Some essential genes have homologs only within the gammaherpesviruses, including ORF45, which corresponds to the KSHV IRF7-binding protein.

MHV-68 transcription occurs in a temporal fashion and is detected from 3 h post infection (p.i.) in vitro. The replication genes ORF6 (ssDNA bp), ORF9 (DNA pol), ORF60 (ribonucleotide reductase subunit 2, RR2), and ORF61 (RR1) are transcribed with similar kinetics, with a defined early peak of transcription. ORF57, a conserved herpesvirus protein involved in RNA transport and stability that can also act as a transcriptional transactivator, shows similar kinetics. Other genes can be clustered into three general expression patterns. One exhibits a peak in transcript levels at 5 h p.i., followed by a gradual decrease thereafter, and includes ORF50 (Rta), ORF37 (alkaline exonuclease, nuc), and glycoprotein L (gL). The second group has a peak at approximately 8 h p.i., and includes the structural genes ORF25 (major capsid protein), ORF33 (a tegument protein), ORF38 (a tegument-associated membrane protein), and ORF51 (envelope glycoprotein gp150), as well as several genes of unknown function (ORF20 and ORF52). The final group shows a peak of transcription at 12 h p.i., and includes a second group of structural proteins such as ORF19 (a tegument protein), ORF66 (a capsid protein), ORF68 (a glycoprotein), and ORF65 (a capsid protein). From this temporal analysis of viral gene expression, it is possible to build up a profile of the events that take place during MHV-68 replication.

Along with the essential replication proteins, a number of nonessential genes are also expressed during infection in vitro. The M4, K3, and ORF73 genes are expressed with immediate early kinetics, while genes with cellular homologs (encoding vcyclin, vBcl-2, and vGPCR) are expressed with early–late kinetics. The early–late group also includes the abundantly expressed M3 gene, which encodes a chemokine-binding protein.

New virions are assembled in a manner similar to that observed for herpes simplex virus type 1. MHV-68 can exit the cell into the extracellular space or can directly infect adjacent cells. ORF51 (encoding gp150) is necessary for viral egress into the extracellular space, whereas ORF27 (gp48) has been identified in the process of egress directly into adjacent cells. gp48 must be localized to the plasma membrane for efficient movement of virions into new cells, a process that requires the ORF58 protein.

Expression of Unique Viral Genes

The unique MHV-68 genes (M1–M4 and the vtRNA genes) are important determinants of pathogenicity. They are absent from the deletion mutant MHV-76, which replicates with the same efficiency as MHV-68 in vitro but is attenuated in vivo. The functions of these genes have been the subject of intense study but currently they are not fully understood. The M1, M3, and M4 protein sequences are related to each other, but similar proteins have not been detected in sequence databases. Similarly, the M2 protein is not related to known proteins. The M1, M3, and M4 proteins are secreted from infected cells. Studies with mutant viruses have shown that M1 is involved in reactivation from latent infection and in pathogenesis in the lung. M4 is an immediate early gene involved in the establishment of latent infection. M3 encodes an abundant protein that is expressed early in the lytic cycle and found during productive infection and establishment of a latent infection. It is an important determinant of pathogenesis during the acute stages of infection but does not appear to be transcribed during long-term latent infection. It is secreted from cells in large quantities and has a high affinity for specific members of the chemokine family, binding to all classes of chemokines (CC, CXC, C, and CX3X) and functionally inhibiting the ability of chemokines to signal through host G protein-coupled receptors.

M2 is a B-cell-specific gene that is involved in acute latent infection. It encodes a membrane-associated protein required for efficient establishment of latency and control of latent infection in B cells. In its absence, there is a defect in memory B-cell latency and increased long-term latency in germinal center B cells. M2 is likely to play a multifunctional role in MHV-68 infection and potentially interacts with a number of cellular proteins. It binds the guanine nucleotide exchange factor vav, altering the normal lymphocyte signaling process and inhibiting B cell receptor-induced cell-cycle arrest and apoptosis. M2 interacts with the DDB1/COP9/cullin repair complex and the ATM DNA damage transducer, blocking DNA damage-induced apoptosis. There is also evidence that M2 inhibits the cellular response to interferon (IFN) by binding to Stat1/Stat2 proteins.

The vtRNAs are predicted to fold into typical cloverleaf structures and, like cellular tRNAs, are thought to be transcribed by RNA polymerase III since the appropriate promoter elements are present within the genes. Following cleavage of excess nucleotides at both the 5′ and 3′ termini, the transcripts are post-transcriptionally modified by the addition of a 3′ CCA sequence, but are not aminoacylated. Only one vtRNA, vtRNA5, contains all the variant or semivariant bases present in mammalian tRNA sequences. The vtRNAs are expressed during lytic and latent infection and have been used widely as a marker for latent infection *in vivo* as their high level of expression means they are readily detectable by *in situ* hybridization. Their function is unknown. The primary vtRNA transcripts also encode miRNAs. The miRNAs are processed and expressed in infected cells, but no information is available on their specificity and function or on their expression *in vivo*. They are currently the only miRNAs known to be transcribed by polymerase III, and their association with the vtRNAs is intriguing.

Entry and Spread within the Host

The natural route of infection is uncertain, but by analogy with other animal gammaherpesviruses the respiratory tract is likely to be a primary route. Introduction of virus intranasally into inbred mice leads to a productive infection of alveolar epithelial cells, resulting in bronchiolitis. During MHV-68 infection, inflammatory responses in the lung evolve slowly compared to the response seen in MHV-76 infection, where a rapid inflammatory cell infiltrate occurs. This difference could be attributed to the chemokine-binding activity of the M3 protein in MHV-68 delaying the onset of inflammation. During MHV-68 infection, the evolution of the inflammatory response includes an initial wave of macrophages, peaking at day 3, followed by a wave of CD8 T cells peaking at day 7. Inflammation resolves by the second week, although in some mice focal accumulations of mononuclear cells are seen in the lung as late as day 30. In fact, the lung continues to be a site of virus persistence for the life of the infected animal. This indicates a potential role for the lung as a major site for virus transmission.

From the lung, the virus enters the local lymph node (mediastinal lymph node, MLN). Here, dendritic cells and macrophages are initially infected, followed by infection of B cells. Evidence supporting a role for dendritic cells in this process comes from studies using an MHV-68 recombinant expressing green fluorescent protein, in which infected cells are tracked in normal and in B cell-deficient (μMT) mice. In the presence and absence of B cells, virus is detected in CD11c-positive dendritic cells and F4/80-positive macrophages. Infection appears to be transient in the absence of B cells and there is little or no viremic phase. The MLN is considered to be the primary site for B-cell infection and B cells the principal cell population responsible for disseminating virus within the host. Tropism of MHV-68 for B cells may be related to the presence of gp150 on the virion envelope. Experiments using tagged gp150 have demonstrated binding to $CD19^+$ (B cells) and to some $CD19^-$ spleen cells. However, there was no interaction between gp150 and murine epithelial cells. During the first week of infection in the MLN, B cells undergo a rapid expansion accompanied by an increase in the number of latently infected B cells against a background of lymphadenopathy. From the MLN, infected B cells traffic to the spleen and other lymphoid tissues. By the second week of infection, a similar rapid expansion of latently infected B cells is observed in the spleen, concomitant with appearance of splenomegaly. The number of latently infected cells increases from $1/10^7$ to $1/10^4$ spleen cells in the space of a few days, and then the numbers decline to between $1/5 \times 10^5$ and $1/10^6$ by the third or fourth week of infection.

B-cell proliferation, and hence the number of latently infected cells observed during lymphadenopathy and splenomegaly, is controlled by CD4 T cells. This indicates that the virus exploits T–B cell collaboration to its advantage, for example, by utilizing B-cell proliferation as a means to maximize the number of latently infected B cells. However, the virus does not depend completely on this strategy, since in the absence of CD4 T cells a long-term latent infection is still established, suggesting that the virus has the capability directly to manipulate B-cell growth and differentiation in order to establish a latent infection (a possible role for the M2 protein). A remarkable feature of MHV-68 latency is the constant number ($1-2/10^6$ latent cells) found in the spleen for the life of the animal. This number is established whether or not CD4 T cells are present in the host.

The presence of B cells and CD4 T cells is absolutely required for the evolution of splenomegaly, which suggests that cognate interactions occur between these cells similar to those for any other antigenic response. During splenomegaly, there are large increases in both B- and T-cell populations. Germinal centers increase in number and size and act as the principal location for latently infected cells. By the third week of infection there is an increase in the number of circulating lymphocytes, dominated by V$\beta 4^+$ CD8 T cells. This phase of infection is similar to the infectious mononucleosis caused by EBV. The mechanism for this selective increase in V$\beta 4$ usage is not known, but indicates a form of super-antigen-driven proliferation. A summary of the processes involved in entry and spread is shown in **Figure 2**.

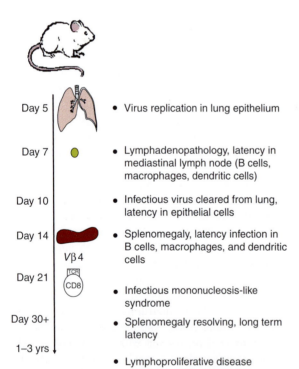

Figure 2 Pathogenesis of MHV-68 infection following intranasal infection. Virus replicates in lung epithelium before spreading to the mediastinal lymph nodes (MLN) which drain the lung. Latent infection is established in the MLN and the virus trafficks throughout the lymphoid system causing transient splenomegaly and expansion of V$\beta 4$+ CD8 T cells. After about 30 days, asymptomatic long-term latency is established. At later times, infection may lead to lymphoproliferative disease.

Host Factors Influencing Infection

The age and immunological status of animals influence the spread of infection and tissue tropism. Infection of young (2–3-week old) or immunocompromised mice results in spread of virus via the bloodstream to a number of tissues, resulting in a productive infection. These include the heart, kidney, liver, adrenal gland, and peripheral nervous system (e.g., trigeminal ganglion). The central nervous system can also become infected when virus is introduced intracranially. Both glial and neuronal cells are observed to undergo a productive infection. These data argue in favor of a promiscuous virus infection, which is supported by observations of the infection of peritoneal exudate cells and epithelial cells in the gut.

Molecular and Cellular Basis of Latency

Viral gene expression changes dramatically from infection in the lung to infection in the spleen. As the virus enters into the latent state, there is a progressive shutdown of gene expression. In MHV-68 infection, this depends to a large extent on the cell type infected. Four cell types have been implicated in maintaining the latent state: B cells, macrophages, dendritic cells, and epithelial cells in the lung. Different populations of B cells are involved in the latent infection. Naive B cells are the initial source of latent virus and, as these cells differentiate following signals from CD4 T cells, they form Germinal Centre B cells (a major site of latency), which may undergo isotype switching and develop into plasma cells and memory B cells, another source of latent infection. Epithelial cells probably represent the main site of persistence, based on the evidence from wood mice where virus is readily isolated from lung tissue, indicating that the respiratory tract may be the principal site of virus shedding.

A number of genes have been linked with MHV-68 latency. These include the vtRNA genes, M2, M3, M4, K3, ORF72 (vcyclin), ORF73, ORF74 (vGPCR), M11 (vBcl-2), and ORF65.

As discussed above, M2 is important for the establishment and maintenance of the initial phases of B cell latency in splenic follicles. However, long-term latency is maintained in the absence of M2 expression. Many of the other genes associated with latency do so by improving the efficiency of the latent infection and/or the reactivation of virus from latency. An exception is ORF73, which is essential for the establishment of MHV-68 latency *in vivo*. The encoded protein is conserved among the gammaherpesviruses and shares with KSHV LANA-1 structural and sequence homology and similar functions such as the ability to tether the viral episome to chromatin to enable the viral genome to be carried to daughter cells during mitosis. Another key function of the ORF73 protein is in inhibiting the activity of Rta, a replication and transcription activator encoded by ORF50, which is a key protein in the reactivation of virus from latency. The balance between the ORF73 protein and Rta is pivotal in defining whether latency or reactivation to a productive infection prevails.

Pathogenesis

Lymphoproliferative disorders are a feature of gammaherpesvirus infection. In MHV-68 infection, a spectrum of disorders of increasing severity occurs, with splenomegaly at one end and lymphomas at the other. The development of lymphoproliferative disease is dependent on mouse strain and host immune status. Lymphomas have been reported in BALB/c mice infected for periods of 9 months or longer (median 14 months). In one study, approximately 10% of infected mice developed tumors in both lymphoid and nonlymphoid tissue (lung, liver, kidney, and heart), of which 50% were classified as high-grade lymphomas. In a separate investigation, the frequency of mice with tumors increased to over 50% following treatment with the immunosuppressive drug, cyclosporin A. The tumors in both experiments were of mixed cell phenotype with CD3+ T cells interspersed among B220+ B cells. The B cells were either kappa or lambda light chain restricted, suggesting a clonal origin of the B-cell population. MHV-68 DNA-positive lymphocytes were found interspersed in the tumor cell mass or on the fringes of lymphomas. In some animals, the number of virus-positive cells was low, whereas in others there were huge numbers of genome-positive cells. The infected cells were not positive for lytic cycle proteins, suggesting that virus reactivation was not occurring in these mice.

BALB/c mice infected with MHV-72 also develop tumors with a frequency similar to MHV-68. The number of tumor-bearing mice increased following immunosuppression with the antifungal agent, FK-506. In 5 of 13 neoplasia-positive mice, virus was isolated directly from the tumors. Lymphoproliferative disease in BALB/c mice has also been reported for MHV-60 and another Slovakian isolate, MHV-Sumava.

BALB/c mice lacking β2-microglobulin (β2m) develop lymphomas and an atypical lymphoid hyperplasia (ALH), with lymphocytosis resembling EBV-associated post-transplant lymphoproliferative disease at a higher rate than wild type BALB/c mice. A total of 67% of the BALB/c β2m mice developed lymphoproliferative disease compared to 22% of mock-infected controls. MHV-68 infected cells were common in the ALH cells but, again, scarce in the lymphomas, suggesting that, while MHV-68 may instigate lymphomagenesis, continued presence of the virus is not necessary in transformed cells.

A number of B cell lines have been established from MHV-68-infected tumor-positive mice, of which S11 is the best characterized. This immunoglobulin M-positive, major histocompatibility complex class II-positive B cell line harbors the virus in a latent form as demonstrated by the presence of a circular genome, a marker of latent infection. As with lymphoblastoid cell lines derived from EBV infection, S11 has around 2–5% of cells expressing lytic antigens. A transcript analysis of S11 revealed that vtRNA and M2, but not M3, were expressed in virtually all latently infected cells. S11 establishes tumors when transferred to nude mice, and has been used to dissect the immunological mechanisms involved in targeting tumor cell growth. In a series of adoptive transfer experiments of MHV-68-specific CD8 and CD4 T cells into S11 tumor-bearing nude mice, regression of tumor cell growth was effectively achieved by CD4 T cells but surprisingly not with CD8 T cells.

The molecular basis for tumor cell induction is not known. Cell lines adapted from lymphomas in mice have multiple chromosome rearrangements, and in the situation where viral DNA was detected in such cells, it is tempting to speculate that virus could initiate tumorigenesis by a hit-and-run mechanism. A number of candidate viral genes could initiate cell transformation, including ORF72 (encoding vcyclin), ORF74 (vGPCR), and M11 (vBcl-2). Transgenic mice expressing vcyclin under the control of the lck promoter, which is active early in thymocyte development, showed increased numbers of immature thymocytes. High-grade lymphoblastic lymphomas developed in the thymuses of 45% of these mice. These mice also show a decrease in the number of mature T cells and an increase in thymic apoptosis, supporting the notion that vcyclin may require the involvement of other factors to promote cell survival and tumor formation. Transfection of MHV-68 ORF74 into 3T3 cell lines leads to the establishment of stable transformed cells. These cells do not, however, develop into tumors in nude mice. vBcl-2 is expressed during the latent phase and is highly efficient at preventing cell death via such immunological mechanisms as tumor necrosis factor α and Fas–Fas ligand interaction. It therefore seems likely that *in vivo* a number of genes act in concert to promote cellular proliferation, survival, and tumor formation.

MHV-68 is also involved with the genesis of other pathologies. In young or immunologically compromised mice (e.g., deficient in the IFN-γ receptor), the virus is associated with arteritis, where persistently infected macrophages colonize the media of the major blood vessels, leading eventually to rupture of the arterial wall and cardiac arrest. Infection of mice deficient in an IFN-γ response results in fibrosis of lymphoid tissue, liver, and lung, from which the animals recover.

See also: Epstein–Barr Virus: General Features; Epstein–Barr Virus: Molecular Biology; Herpesviruses: Discovery; Herpesviruses: General Features; Herpesviruses: Latency; Kaposi's Sarcoma-Associated Herpesvirus: General Features; Kaposi's Sarcoma-Associated Herpesvirus: Molecular Biology; Simian Gammaherpesviruses.

Further Reading

Blaskovic D, Stancekova M, Svobodova J, and Mistrikova J (1980) Isolation of five strains of herpesviruses from two species of free living small rodents [letter]. *Acta Virologica* 24: 468.

Nash AA, Dutia BM, Stewart JP, and Davison A (2001) Natural history of murine gammaherpesvirus infection. *Philosophical Transactions of the Royal Society of London, Series B* 356: 569–579.

Speck SH and Virgin HW (1999) Host and viral genetics of chronic infection: A mouse model of gammaherpesvirus pathogenesis. *Current Opinion in Immunology* 16: 456–462.

Stevenson PG and Efstathiou S (2005) Immune mechanisms in murine gammaherpesvirus-68 infection. *Viral Immunology* 18: 445–456.

Sunil-Chandra NP, Efstathiou S, Arno J, and Nash AA (1992) Virological and pathological features of mice infected with murine gammaherpesvirus 68. *Journal of General Virology* 73: 2347–2356.

Virgin HW, Latreille P, Wamsley P, *et al.* (1997) Complete sequence and genomic analysis of murine gammaherpesvirus 68. *Journal of Virology* 71: 5894–5904.

Mycoreoviruses

B I Hillman, Rutgers University, New Brunswick, NJ, USA

© 2008 Elsevier Ltd. All rights reserved.

The 8th Report of the International Committee for the Taxonomy of Viruses (ICTV) which lists 12 genera of reoviruses that infect mammals, invertebrates, and plants, was the first to also include the fungus-infecting genus *Mycoreovirus*. Members of many of the reovirus genera in the 8th Report replicate in organisms representing more than one kingdom or phylum: for example, all plant reoviruses replicate in and persistently infect their insect vectors, and similarly many of the mammalian reoviruses replicate in their invertebrate vectors. Three fungal virus species have been confirmed to be members of the genus *Mycoreovirus* of the family *Reoviridae*. Two of these viruses were isolated from the filamentous ascomycete fungus *Cryphonectria parasitica*, and the other is from the soil-borne fungus *Rosellinia necatrix*, also an ascomycete but representing a different fungal family.

All of the mycoreoviruses cause disease in their infected hosts, resulting in greatly reduced virulence, reduced growth in culture, reduced laccase accumulation, and reduced sporulation relative to wild-type, virus-free cultures of the same genetic background (**Figure 1**). The mycoreoviruses are most closely related to the genus *Coltivirus* of tick-borne reoviruses (**Figure 2**). Both Colorado tick fever virus (CTFV) and the closely related Eyach virus (EyaV) cause mammalian diseases, and both viruses replicate in their arthropod vectors as well as in their mammalian hosts. The coltiviruses are not well understood at the molecular level and so have shed only a little light on the study of mycoreoviruses. The genomic sequences of both CTFV and EyaV have been published, and they have been found to be more closely related to the genus *Orthoreovirus*, which includes the common human pathogen mammalian reovirus (MRV), than to members of the other two genera of the family *Reoviridae* (*Orbivirus* and *Rotavirus*) that have been well studied at the structural and molecular levels.

Filamentous fungi are valuable subjects for investigation of eukaryotic viruses. This has been especially true for ascomycetes, which usually are haploid throughout their vegetative phases and thus readily amenable to tools of classical genetics and to relatively simple gene knockout and knockdown experiments. *Cryphonectria parasitica* has been exceptional as a host for examination of fungal viruses. The fungus caused the greatest pandemic of a tree species in recorded history, killing 3–4 billion American chestnut trees. The fungus is very stable in culture, showing a consistent morphology upon continued maintenance and subculture, is easily transformed and transfected, and can be examined through classical genetics. Because of its historical importance as a pathogen and the potential for its control using natural or genetically engineered viruses, more viruses that result in stable morphological changes and/or changes in virulence have been identified and characterized in this fungus than in any other.

Structure–Function Relationships

Relatively little is known of the details of mycoreovirus structure, but all indications from electron micrographs of

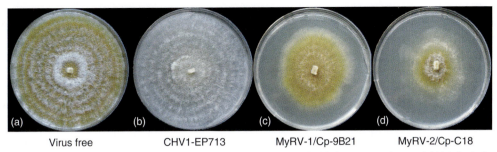

Figure 1 Morphologies of isogenic *Cryphonectria parasitica* colonies infected with three mycoviruses. (a) Uninfected colony, strain EP155; (b) strain EP155 infected with hypovirus CHV-1/EP713; (c) strain EP155 infected with mycoreovirus MyRV-1/Cp9B21; (d) Strain EP155 infected with mycoreovirus MyRV-2/CpC18.

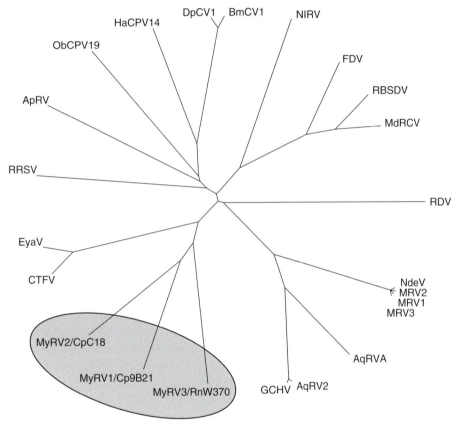

Figure 2 Unrooted neighbor-joining tree based on ClustalW alignments of complete deduced amino acid sequences of RNA-dependent RNA polymerase genes of the reoviruses most closely related to the mycoreoviruses (shaded).

negatively stained virus particles and genome sequence analysis are that it belongs to the orthoreovirus subgroup. The orthoreoviruses are distinct from the orbiviruses and rotaviruses in that they have identifiable pentameric turrets on top of each fivefold axis of the core particle, and these have been demonstrated to be involved in capping of nascent mRNA. The protein that was experimentally demonstrated to be the mycoreovirus guanylyltransferase is homologous to turret proteins of orthoreoviruses that have the same function, supporting the grouping of mycoreoviruses within the orthoreovirus subgroup of the *Reoviridae*.

Relatively little is known about the movement of fungal viruses in general within the mycelium and during horizontal or vertical virus transmission. Viruses that have been investigated are generally found in hyphal tip cells and move with the growing mycelium; however, location of mycoreoviruses within the mycelium has not been investigated. The steps in the infection process of the mammalian orthoreovirus have been investigated in considerable detail. Virus entry into cells via vesicles occurs by receptor-mediated endocytosis, whereupon the μ1 protein, which is myristoylated at its N-terminus and inserted into the inner membrane of the vesicle cleaves autocatalytically, resulting in pore formation in the vesicle membrane and subsequent virus release into the cytoplasm. Both of the *C. parasitica* mycoreoviruses have homologs of the orthoreovirus μ1 protein, with strong myristoylation signals at the amino termini of the corresponding proteins and putative sites for autocatalytic cleavage. This suggests that a similar mechanism of egress from vesicles is a component of the infection cycle of these reoviruses. Interestingly, although the Rosellinia virus, MyRV-3/RnW370, contains a homolog of the μ1 protein, it does not contain a glycine residue at the penultimate position of the N-terminus of this deduced protein, and based on its sequence it has a very low predicted probability of myristoylation. It does, however, contain the N/P putative cleavage motif at a similar position and in a similar environment as those in the other viruses.

It is relatively easy to cure *C. parasitica* and *R. necatrix* from their associated mycoreoviruses. In *C. parasitica*, single asexual spores are usually virus free, and in *R. necatrix*, cultures initiated from excised hyphal tips (the terminal 2–8 cells) are also virus free. These are in contrast to cultures infected with the well-studied hypovirus of *C. parasitica*, cryphonectria hypovirus-1 (CHV-1), in which most or sometimes all conidia contain virus and the virus-infected cultures cannot be cured by hyphal tip isolation. It is likely that details of mycoreovirus movement within the mycelium as well as horizontal and vertical transmission are quite different from those properties of hypoviruses, which have no capsid protein, are more closely related to positive-sense single-stranded RNA viruses such as plant potyviruses and animal picornaviruses, and whose replication is associated with the fungal *trans*-Golgi network.

Taxonomy and Nomenclature

Naming of fungal viruses is often problematic because infectivity studies are very difficult with fungal viruses and have not been done with most. In nature, fungal viruses do not exit completely from an infected isolate and enter an uninfected one exogenously. Instead, they move from one isolate to another only after the hyphae of the two isolates have fused (anastomosed), in which case the contents of one or usually more cells from the infected and uninfected isolate are mixed. Mixed infections are common and symptomless infections are the norm for fungal viruses. For these reasons, fungal virus names usually contain reference not only to the host genus and species, but to the host strain of origin as well. The genus name *Mycoreovirus* was a natural one to use because it is descriptive. For the sake of simplicity and consistency with other viral nomenclature, species are numbered progressively as they are described. The fungal species and isolate in which a particular virus was identified is provided after the species number. In this system, the first mycoreovirus species described was designated *Mycoreovirus-1*. The only isolate of this virus species identified to date was from *C. parasitica* strain 9B21, so the virus is designated mycoreovirus-1/Cp9B21 or MyRV-1/Cp9B21. The reason that MyRV-2 represents a separate virus species even though the two were isolated from the same fungal species and only a few miles away is because it shares much less sequence similarity at both the nucleotide and amino acid levels (<50%) than expected for two viruses in a single species. Surprisingly, both of these species are monotypic: these two virus isolates, each representing a different species, are the only two mycoreoviruses isolated from *C. parasitica*, even though thousands of isolates infected with members of the family *Hypoviridae* have been identified worldwide. In contrast, dsRNA or virus from dozens of isolates of *Rosellinia necatrix* from different parts of Japan have been isolated, but all are closely related to each other, indicating that they represent strains of a single virus species, MyRV-3.

Genome Structures, Organizations, and Relationships

The three fungal reoviruses examined to date appear to have 11 segments of dsRNA that are required for infection. Isolates of one of the viruses, MyRV-3, have been found to have either 12 or 11 segments (see below). There is significant sequence similarity among the larger segments of the three mycoreovirus species and between homologous segments of the mycoreoviruses and their closest relatives, the coltiviruses, but this similarity becomes less apparent in the middle segments and is not apparent at all in the small segments. This feature is common with other members of the family *Reoviridae*, in which the more distant relationships are generally revealed in only the large segments. Each of the 11 required segments of mycoreoviruses appears to contain a single open reading frame. **Figure 3** Segments 1–5 of the mycoreoviruses are homologous to segments 1–5 of the coltiviruses (segments 4 and 5 of MyRV2-CpC18 are transposed relative to the others). Based largely on similarity with other reoviruses, their predicted functions are: S1: RNA-dependent RNA polymerase (core protein); S2: dsRNA-binding (core protein); S3: guanylyltransferase (turret protein); S4: Myristoylated membrane penetration protein (outer capsid); S5: cytoskeleton-interacting (core). Segment 6 of the mycoreoviruses is predicted to encode a nucleic acid binding core protein and is homologous to segment 10 of the two coltiviruses. Segment 7 of MyRV-1/Cp9B21 encodes a proline-rich protein with similarity to several viral and nonviral proline-rich domains. Homologies among the smaller mycoreovirus proteins (from S8–11 or 12) and other reovirus deduced proteins, including those of coltiviruses, are unknown.

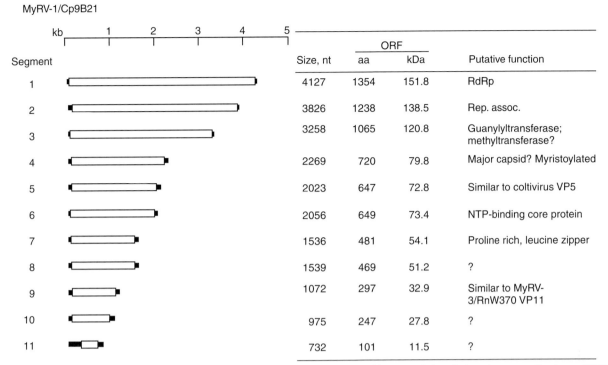

Figure 3 Diagram of the genome segments of mycoreovirus-1/Cp9B21 shown with segment size, deduced open reading frame (ORF) in number of amino acid residues and predicted protein product size, and putative protein function. Only the guanylyltransferase function of segment 3 has been demonstrated experimentally.

One of the indications of close relationship among reoviruses is conserved terminal sequences. This also is a good indication of whether or not pseudorecombinants can be generated by co-infection with two related viruses. In the case of the mycoreoviruses, the three species have different conserved terminal sequences, consistent with their taxonomic separation. Furthermore, although co-infection of a single colony with the two reoviruses has been achieved, pseudorecombinants have not yet been recovered from these doubly infected isolates.

Mycoreovirus-1/Cp9B21

Although MyRV-1/Cp9B21 was not the first of the two *Cryphonectria* reoviruses identified, fungal cultures infected with this virus proved to be more stable and easily studied than those infected with MyRV-2/CpC18 (see below), so early molecular investigations have focused on this virus. When grown on solid media in petri dishes (e.g., a defined complete medium or potato dextrose agar (PDA)), MyRV-1/Cp9B21-infected colonies are deep orange in color and have little aerial hyphae compared to their uninfected counterparts. As with other *C. parasitica* viruses, the phenotype of the infected culture has very little to do with the host isolate but is determined almost entirely by the virus.

Fungal isolates infected with MyRV-1/Cp9B21 are much less virulent than uninfected isolates, and are among the most debilitated of any virus-infected *C. parasitica* cultures studied to date. Although the virus accumulates to reasonably high concentrations in infected colonies, it is transmitted very poorly through conidia, at rates of only 2–5%. This may account in part for the rarity of the virus in nature.

Recently, double infections of the well-characterized hypovirus CHV-1/EP713 and MyRV-1/Cp9B21 were examined. In these analyses, it was found that presence of the hypovirus increased both the concentration and vertical transmission rate through conidia of the reovirus, but that the reovirus did not affect the concentration or transmission rate of the hypovirus. Furthermore, transgenic expression of only the hypovirus protein p29 also resulted in increased accumulation and transmission of MyRV-1/Cp9B21. This is consistent with the prediction that p29 serves as a suppressor of RNA silencing (interference) during hypovirus infection, and that this effect acts *in trans* to support enhanced mycoreovirus replication.

Expression of MyRV-1/Cp9B21 Gene Products

Functional analysis of the MyRV-1/Cp9B21 genome was initiated by cloning the 11 individual segments into a baculovirus expression vector and expressing them in insect cells. All 11 segments were expressed, resulting in 11 identifiable protein products on polyacrylamide gels. Only one of the proteins, the segment 3 product, has been studied functionally. This protein was found to be active in autoguanylylation assays, confirming it as the viral guanylyltransferase. Deletion and site-directed mutational analysis determined that the amino acid sequence EPAGYHPRPSIVVPHYFVFR constituted the catalytically active site of the MyRV-1/Cp9B21 guanylyltransferase. The Hx_8H motif was identified as absolutely conserved in all three members of the genus *Mycoreovirus*, as well as in the structurally related genera *Coltivirus*, *Orthoreovirus*, *Aquareovirus*, *Cypovirus*, *Dinovernavirus*, *Oryzavirus*, and *Fijivirus*. In all of the above genera in which the guanylyltransferase has been identified functionally, the Hx_8H motif has been found within the sequence. The core consensus sequence for the guanylyltransferase within this group of genera was a/vxxHxxxxxxxxHhyf/lvf, with only the H residues being absolutely conserved.

Mycoreovirus-2/CpC18

The first virus that was tentatively identified as a mycoreovirus was MyRV-2/CpC18. Like many of the virus-infected *C. parasitica* cultures, the one that was found to contain this virus was isolated from a canker on an American chestnut tree. The circumstances of the discovery were somewhat unusual in that only one of 36 fungal isolates from that particular canker was virus infected; the rest were virus free. Although it is not unusual to find mixed fungal infections within a canker or to isolate virus-containing and virus-free cultures from one canker, the ratio of 1/36 is extraordinary. The phenotype of the infected culture, designated C-18 (the 18th isolate from canker C on that particular tree) was distinct from the phenotype of MyRV-1/Cp9B21-infected cultures described above in that it was light brown in color and had more aerial mycelium (**Figure 1**). The two viruses have very similar, dramatic negative effects on fungal virulence.

A major difference between the two *C. parasitica* mycoreoviruses is their stability and transmissibility. MyRV-2/CpC18 is easily lost upon subculture of the fungus, while this has never been observed with MyRV-1/Cp9B21. Furthermore, MyRV-2/CpC18 is extremely difficult to transmit from an infected isolate to an isogenic uninfected isolate by hyphal anastomosis, a property that is not seen with other *C. parasitica* viruses. It is presumed that these properties are related, but their reasons have not been elucidated. Both MyRV-1/Cp9B21 and MyRV-2/CpC18 have been transmitted to uninfected fungal isolates by inoculating *C. parasitica* protoplasts with purified virus particle preparations and allowing the protoplast to regenerate cell walls, form a hyphal network, and grow into a single colony. The ability to infect protoplasts efficiently using purified virus particle preparations is an interesting and useful feature of *C. parasitica* reoviruses. This has allowed for infection of different genotypes of *C. parasitica* regardless of vegetative incompatibility group and potential for transmission by hyphal anastomosis.

Mycoreovirus-3/RnW370

White rot is a root disease of fruit trees that can be limiting to production. Control of the fungus that causes the disease, *Rosellinia necatrix*, by chemical means is difficult and not economically feasible. Pathologists in Japan, where the disease is particularly severe, have sought to use virus-infected strains to control the disease, leading to the identification of several viruses. This plant/fungus interaction represents an interesting contrast to the chestnut/*C. parasitica* interaction. As a root disease, there are challenges and opportunities for biological control of a fungal pathogen with viruses that do not apply to aerial diseases. One of the viruses under investigation for biocontrol of *R. necatrix* is the mycoreovirus MyRV-3, which causes reduced virulence of the fungus. Unlike the monophyletic *C. parasitica* reoviruses, different strains of MyRV-3 have been isolated from a variety of *R. necatrix* strains from around Japan.

Of the *R. necatrix* viruses, the virus isolate that has been most thoroughly characterized is MyRV-3/RnW370. In surprising contrast to the *C. parasitica* viruses, MyRV-3/RnW370 was found to contain 12 rather than 11 segments. However, the presence of 12 segments is not a consistent feature of all MyRV-3 isolates. Examination of different virus-infected isolates of *R. necatrix* shows that they may have either 12 or 11 segments. Experiments to investigate virus composition and transmission have been performed on hyphal tip cultures from infected *R. necatrix* isolates, resulting in demonstration that these viruses behave like the *Cryphonectria* mycoreoviruses.

When only 11 segments are present in MyRV-3 isolates, segment 8 is the one that is absent from the full complement. With most reoviruses, sequence conservation among species and genera is evident in the larger segments, but much less so in the smaller segments, and this is true in the mycoreoviruses and related genera. Consistent with this general trend, sequence comparison between the 12 segments of MyRV-3 and the 12 segments of the two coltiviruses has suggested nothing about possible function of the apparently dispensable segment 8.

It is intriguing to think that perhaps segment 8 is vestigial for reoviruses in fungi and is required only in another host, past or present. This would be similar to the leafhopper-transmitted phytoreovirus, wound tumor virus (WTV), in which deletion mutations in any of three segments may be found upon successive serial, insect-free virus passage or long-term virus maintenance in plants, whereupon resulting mutant viruses become defective in their transmission properties and incapable of replicating in their leafhopper vectors. An apparent major difference between MyRV-3 and WTV is that in MyRV-3 there is no evidence for remnants of segment 8; it appears to be either present or entirely absent. In the well-characterized mutants of WTV, the deleted segments are not completely gone, but shorter segments containing the two termini and varying amounts of adjacent sequence remain, ensuring that a total of 12 segments remain. Whether this represents a difference between the dsRNA segment sorting and packaging mechanisms of phytoreoviruses and mycoreoviruses is not known. In this line of inquiry, the close association of fungi with mites in natural settings may be significant to the evolutionary biology of mycoreoviruses: it may be no coincidence that their closest relatives are the tick-borne coltiviruses, with ticks and mites both in the arachnid subclass Acari. Unfortunately, mites are very difficult experimental subjects and cell cultures are not currently available. Furthermore, much less is known about coltivirus gene function than is known about many of the other members of the family *Reoviridae* that are pathogenic to humans, making it more difficult to pursue this line of research from a strictly bioinformatic standpoint.

Effects of Mycoreoviruses on Fungal Gene Expression

Considerable information has been amassed on the impact of CHV-1, a positive-sense RNA virus, on its fungal host. In contrast, studying the mechanisms of mycoreovirus infection of fungi is in its infancy. The first study addressing these questions was done by microarray analysis using mRNA isolated from isogenic fungal isolates of *C. parasitica* infected with either MyRV-1 or MyRV-2, and comparing results to the same strain that was uninfected or infected with either of two different CHV-1 strains, and with fungal mutants defective in virulence characteristics. To date, these experiments have been performed only on EST-based arrays representing ~20% of the total *C. parasitica* gene complement. Overall, there was consistency in the effects of the two *C. parasitica* mycoreoviruses on host gene expression: MyRV-1 infection resulted in differential expression of 6.5% of the genes on the array, whereas MyRV-2 infection affected expression of 5.8% of those genes. As might be expected based on their phenotypes, similar but distinct suites of genes were up- or downregulated in isogenic fungal isolates infected with the two reoviruses. Approximately 60% of the genes whose expression was affected were the same whether infection was by MyRV-1 or MyRV-2, and all but one of those genes were altered in the same direction. Some of these groups of genes are in common with those that are differentially regulated in cultures infected with the unrelated hypoviruses, but there are predictable differences. For example, hypovirus infection of *C. parasitica* results in female infertility, whereas mycoreovirus infection does not, and this is reflected in the expression of two genes predicted to be involved in the *C. parasitica* mating response. Both *mf2-1*, which encodes the fungal pheromone precursor, and *Csp12*, which encodes a homolog of the yeast *Ste12*-like transcription factor, were substantially downregulated in hypovirus-infected fungal isolates, which are female sterile, whereas there was much less effect on expression of these genes in either of the mycoreovirus-infected *C. parasitica* isolates. Virus is transmitted to ascospores at a rate of ~50% or less when the female parent is infected, but there is no virus transmission to ascospore progeny if the male parent in a mating is infected. Sequencing the complete genome of *Cryphonectria parasitica* is now underway. This will allow for the complete set of genes to be represented in an oligo array for more thorough investigation of differential gene expression.

See also: Coltiviruses; Fungal Viruses; Hypoviruses.

Further Reading

Enebak SA, Hillman BI, and MacDonald WL (1994) A hypovirulent *Cryphonectria parasitica* isolate with multiple, genetically unique dsRNA segments. *Molecular Plant-Microbe Interactions* 7: 590–595.

Hillman BI, Supyani S, Kondo H, and Suzuki N (2004) A reovirus of the fungus *Cryphonectria parasitica* that is infectious as particles and related to the *Coltivirus* genus of animal pathogen. *Journal of Virology* 78: 892–898.

Hillman BI and Suzuki N (2004) Viruses of *Cryphonectria parasitica*. *Advances in Virus Research* 63: 423–472.

Kanematsu S, Arakawa M, Oikawa Y, et al. (2004) A reovirus causes hypovirulence of *Rosellinia necatrix*. *Phytopathology* 94: 561–568.

Mertens P and Hillman BI (2005) Genus mycoreovirus. In: Fauquet CM, Mayo MA, Maniloff J, Desselberger U, and Ball LA (eds.) *Virus Taxonomy: Eighth Report of the International Committee on Taxonomy of Viruses*, pp. 556–560. San Diego, CA: Elsevier Academic Press.

Nibert ML and Schiff LA (2001) Reoviruses and their replication. In: Knipe DM and Howley PM (eds.) *Fields Virology*, 2nd ed., vol. 2, pp. 1679–1729. Philadelphia, PA: Lippincott Williams and Wilkins.

Nuss DL (2005) Hypovirulence: Mycoviruses at the fungal-plant interface. *Nature Reviews Microbiology* 3: 632–642.

Osaki H, Wei CZ, Arakawa M, et al. (2002) Nucleotide sequences of double-stranded segments from hypovirulent strain of the white root rot fungus *Rosellinia necatrix*: Possibility of the first member of the *Reoviridae* from fungus. *Virus Genes* 25: 101–107.

Supyani S, Hillman BI, and Suzuki N (2006) Baculovirus expression of the 11 *Mycoreovirus*-1 genome segments and identification of the guanylyltransferase-encoding segment. *Journal of General Virology* 88: 342–350.

Suzuki N, Supyani S, Maruyama K, and Hillman BI (2004) Complete genome sequence of *Mycoreovirus 1*/Cp9B21, a member of a new genus in the family *Reoviridae* isolated from the chestnut blight fungus, *Cryphonectria parasitica*. *Journal of General Virology* 85: 3437–3448.

Wei CZ, Osaki H, Iwanami T, Matsumoto N, and Ohtsu Y (2003) Molecular characterization of dsRNA segments 2 and 5 and electron microscopy of a novel reovirus from hypovirulent isolate, W370, of the plant pathogen *Rosellinia necatrix*. *Journal of General Virology* 84: 2431–2437.

Wei CZ, Osaki H, Iwanami T, Matsumoto N, and Ohtsu Y (2004) Complete nucleotide sequences of genome segments 1 and 3 of Rosellinia anti-rot virus in the family *Reoviridae*. *Archives of Virology* 149: 773–777.

N

Nanoviruses

H J Vetten, Federal Research Centre for Agriculture and Forestry (BBA), Brunswick, Germany

© 2008 Elsevier Ltd. All rights reserved.

Glossary

Rep (replication initiator protein) It initiates replication only of its own DNA molecule and possesses origin-specific DNA cleavage, nucleotidyl transferase activity, and ATPase activity.

Satellite Satellites are subviral agents composed of nucleic acids; they depend for their multiplication on co-infection of a host cell with a helper virus. When a satellite encodes the coat protein in which its nucleic acid is encapsidated it is referred to as a satellite virus.

Introduction

Until the late 1980s, yellowing and dwarfing diseases of legumes and banana whose causal agents were persistently transmitted by aphids, not transmitted by sap and difficult to isolate, were generally thought to be caused by single-stranded RNA (ssRNA) viruses of the family *Luteoviridae*. In attempts to elucidate the etiology of these diseases, unusually small, icosahedral particles measuring only 18–20 nm in diameter were consistently isolated from infected plants. These particles did not contain one type of linear ssRNA but several circular ssDNA molecules, all of which were about 1 kb in size. Because of the disease symptoms (dwarfing) and the small size of the virions and genome components these viruses were referred to as nanoviruses. They differ from geminiviruses, the only other known group of ssDNA viruses of plants, in particle morphology, genome size, number and size of DNA components, genomic organization, mode of transcription, and vector species.

Meanwhile, four of these viruses, namely banana bunchy top virus (BBTV), faba bean necrotic yellows virus (FBNYV), milk vetch dwarf virus (MDV) and subterranean clover stunt virus (SCSV), have been formally described. Because of striking differences to other ssDNA viruses of plants, bacteria, and vertebrates, these viruses have recently been assigned to the family *Nanoviridae*. On the basis of differences in biology (host range, aphid vectors) and both genome size and organization, members of this family ('nanovirids') are subdivided into the genera *Nanovirus* and *Babuvirus*. Moreover, increasing evidence suggests that these viruses are not only quite variable in their biological and molecular properties but also that crops in tropical and subtropical countries of the Old World harbor further nano- and babuviruses, such as abaca bunchy top virus (ABTV) in the Philippines, cardamom bushy dwarf virus (CBDV) in India, and faba bean necrotic stunt virus (FBNSV) in Ethiopia and Morocco. BBTV occurs widely on Pacific Islands (including Hawaii), in Australia and Indochina but has an erratic geographic distribution in South Asia and Africa. SCSV and MDV have been reported only from Australia and Japan, respectively, whereas FBNYV appears to have a much wider geographic distribution (West Asia, Middle East, North and East Africa, and Spain). Nanovirids are not known to occur in the New World.

Particle Properties

Virions of the nanovirids are not enveloped, 17–20 nm in diameter, and presumably of an icosahedral $T = 1$ symmetry structure containing 60 subunits. Capsomeres may be evident, producing an angular or hexagonal outline (**Figure 1**). Virions are stable in Cs_2SO_4 but may not be stable in CsCl. The buoyant density of virions is about $1.24–1.30\ g\ cm^{-3}$ in Cs_2SO_4, and $1.34\ g\ cm^{-3}$ in CsCl. They sediment as a single component in sucrose rate-zonal and Cs_2SO_4 isopycnic density gradients.

Virions have a single capsid protein (CP) of about 19 kDa. No other proteins have been found associated with virions. Up to 12 distinct DNA components each of about 1 kb have been isolated from virion preparations of different species and their isolates. Each ssDNA component appears to be encapsidated in a separate particle.

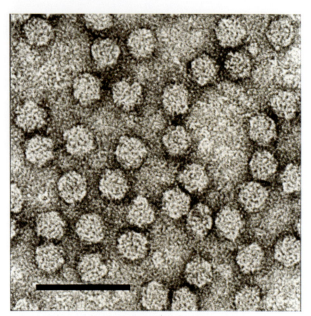

Figure 1 Negative constrast electron micrograph of particles of an isolate of *Faba bean necrotic yellows virus*. Scale = 50 nm. Courtesy of D. E. Lesemann and L. Katul.

Virions are strong immunogens. Most nanovirid species are serologically distinct from one another. However, antisera and some monoclonal antibodies (MAbs) to BBTV cross-react fairly strongly with ABTV and CBDV. On the contrary, antisera to FBNYV and SCSV cross-react weakly with SCSV and FBNYV, respectively, in Western blots and immunoelectron microscopy but not at all in double antibody sandwich enzyme-linked immunosorbent assay (DAS-ELISA). However, MDV antigen reacts strongly not only with an antiserum to FBNYV but also with the majority of MAbs to FBNYV. Therefore, species-specific MAbs are required for the differentiation and specific detection of not only FBNYV and MDV (CP amino acid sequence identity of about 83%), but presumably also BBTV, ABTV, and CBDV.

Biological Properties

Economic Importance

Based on symptomatology and transmission characteristics, all diseases now known to have a nanovirid etiology were earlier suspected to be caused by luteoviruses. Unlike luteoviruses, however, nanovirids generally cause more severe symptoms. In many virus–host combinations, early infections lead to very severe effects and even premature plant death. The disease caused by BBTV is considered the most serious viral disease of banana worldwide. SCSV and FBNYV are also thought to be of great economic importance as they have caused serious diseases of subterranean clover in Australia and of faba bean in Egypt, respectively, leading to repeated crop failures.

Host Range

Individual species have narrow host ranges. FBNYV, MDV, and SCSV naturally infect a range of leguminous species, whereas BBTV has been reported only from *Musa* species and closely related species within the Musaceae, such as abaca (*M. textilis* Née) and *Ensete ventricosum* Cheesem. There are no confirmed non-*Musa* hosts of BBTV. Symptoms of BBTV include plant stunting, foliar yellowing, and most characteristic dark green streaks on the pseudostem, petioles, and leaves. All economically important natural hosts of FBNYV, MDV, and SCSV are legumes, in which these viruses generally cause plant stunting and a range of foliar symptoms, such as leaf deformations and chlorosis or reddening. Although FBNYV infects >50 legume species and only few nonlegume species (*Stellaria media*, *Amaranthus*, and *Malva* spp.) under experimental and natural conditions, major legume crops naturally infected by FBNYV are faba bean, lentil, chickpea, pea, French bean, and cowpea. Likewise, SCSV experimentally infects numerous legume species, but its economically important natural hosts include only subterranean clover, phaseolus bean, faba bean, pea, and medics. MDV is known to cause yellowing and dwarfing in Chinese milk vetch (*Astragalus sinicus* L.), a common green manure crop in Japan, as well as in faba bean, pea, and soybean.

Tissue Tropism and Means of Transmission

SCSV and FBNYV have been shown to replicate in inoculated protoplasts. All members of assigned species are restricted to the phloem tissue of their host plants and are not transmitted mechanically and through seeds. Apart from graft transmission, vector transmission had been the only means of experimentally infecting plants with nanovirids, until infectivity of purified FBNYV virions by biolistic bombardment was demonstrated.

Transmission by Aphids

Under natural conditions, all viruses are transmitted by certain aphid species, in which they can persist for many days or weeks without replicating in their vectors. Whereas only one aphid species (*Pentalonia nigronervosa*) has been reported as vector of BBTV, several aphid species transmit FBNYV, MDV, and SCSV. *Aphis craccivora* appears to be the major natural vector of these viruses as it is the most abundant aphid species on legume crops in the afflicted areas and was among the most efficient vectors under experimental conditions. Other aphid vectors of FBNYV are *Aphis fabae* and *Acyrthosiphon pisum* but in no case were *Myzus persicae* and *Aphis gossypii* able to transmit this virus. SCSV has been reported to be vectored also by *Ap. gossypii*, *M. persicae*, and *Macrosiphum euphorbiae*, but some of these accounts now appear questionable.

Transmission studies showed that aphids are able to transmit FBNYV and SCSV following short acquisition and inoculation access feeding periods of about 30 min each. Although viruliferous aphids often retain transmission ability for life, nanovirids do not multiply in their insect vectors. Together with the observation that longer acquisition and inoculation access feeding periods resulted in higher transmission rates, the strikingly long persistence of nanovirids in their insect vectors indicates that they are transmitted in a circulative persistent manner similar to that of luteoviruses.

For FBNYV it has been demonstrated that purified virions alone are not transmissible by its aphid vector, regardless of whether they are acquired from artificial diets or directly microinjected into the aphid's hemocoel. However, faba bean seedlings biolistically inoculated with intact virions or viral DNA developed symptoms typical of FBNYV infections and were efficient sources for FBNYV transmission by aphids. These observations together with results from complementation experiments suggest that FBNYV (and other nanovirids) require a virus-encoded helper factor for its vector transmission that is either dysfunctional or absent in purified virion preparations.

Genome Organization and Protein Functions

DNA Structure

Up to 12 distinct DNA components each of about 1 kb in size have been isolated from virion preparations of different nanovirid species and their isolates. The majority of them seem to be structurally similar in being positive sense, transcribed in one direction, and containing a conserved stem–loop structure (and other conserved domains) in the noncoding region. Each coding region is preceded by a promoter sequence with a TATA box and followed by a polyadenylation signal (**Figure 2**). By analogy to the BBTV

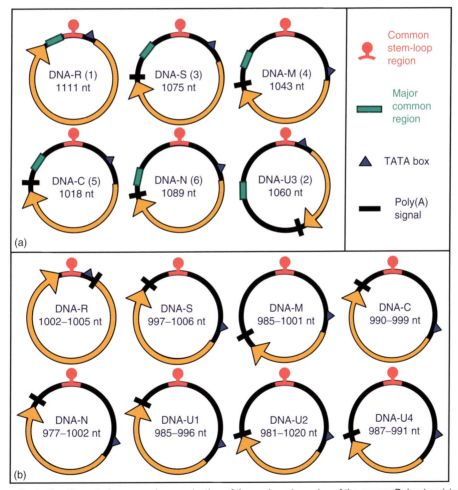

Figure 2 Diagram illustrating the putative genomic organization of the assigned species of the genera *Babuvirus* (a) and *Nanovirus* (b) and depicting the structure of the six and eight identified DNA components identified from the genomes of BBTV (a) and the three nanoviruses (b) FBNYV, MDV, and SCSV (see also **Table 1**). Each DNA circle contains its designated name and its size (range). Arrows refer to the location and approximate size of the ORFs and the direction of transcription. Note that DNA-U2 and -U4 have not been identified for SCSV.

DNA-U3, -S, -C, -M, and -N, each of which has been shown to yield only one mRNA transcript, the majority of the nanovirid DNAs contain only one major gene. Only for BBTV DNA-R two mRNA transcripts were detected, a large one mapping to the major replication initiator protein (Rep)-encoding open reading frame (ORF) and a small one completely internal to the major *rep* ORF.

Viral DNA Types

The understanding of the genomic organization of nanovirids was initially complicated by the fact that several DNAs encoding different replication initiator (Rep) proteins had been found associated with BBTV, FBNYV, MDV, and SCSV isolates. However, it was soon demonstrated that only one of the Rep-encoding DNAs of each nanovirid is an integral part of its genome and required for the replication of the other DNAs that encode other types of viral proteins. In addition to being consistently associated with a nanovirid infection and capable of initiating replication of the other viral DNAs, master Rep-encoding DNAs share with the other DNAs of a nanovirid species, a highly conserved sequence encompassing the stem loop.

The genomic information of each nanovirid is distributed over at least six or eight molecules of circular ssDNA (**Table 1**). The fact that a typical set of six and eight distinct DNAs has been consistently identified from a range of geographical isolates of BBTV and FBNYV (and MDV), respectively, suggests that the babuvirus genome consists of six DNAs and the nanovirus genome of eight DNA components. DNA-R, -S, -C, -M, and -N have been identified from all four assigned species of the family *Nanoviridae* (**Table 1**). DNA-U1 is shared by all three nanovirus species but is absent from the BBTV genome. The apparent absence of DNA-U2 and -U4 in the SCSV genome appears to be due to the fact that these genome components have not been identified yet from this nanovirus. DNA-U1, -U2, and -U4 seem to be absent from the BBTV genome and specific components of the nanovirus genome. In contrast, DNA-U3 appears to be specific of the babuvirus genome and absent from the nanovirus genome. However, the question as to whether DNA-U3 and -U4, which potentially encode similar-sized proteins ($\sim$10 kDa), are functionally distinct, remains to be determined (**Table 1**).

Integral Genome Segments

Despite the aforementioned circumstantial evidence, the number and types of ssDNA components constituting the integral parts of the nanovirid genome are still enigmatic. There has been only one (reported) attempt to use cloned DNAs to reproduce a nanovirid infection. Characteristic symptoms of FBNYV infection were obtained in faba bean, the principal natural host of FBNYV, following biolistic DNA delivery or agroinoculation with full-length clones of all eight DNAs that have been consistently detected in field samples infected with various geographical isolates of FBNYV. However, experimental infection with different combinations of fewer than these eight DNAs also led to typical FBNYV symptoms. Only five genome components, DNA-R, -S, -M, -U1, and -U2, were sufficient for inducing disease symptoms in faba bean upon agroinoculation. Symptomatic plants agroinoculated or bombarded with eight DNAs contained typical FBNYV virions; however, the virus produced after agroinoculation of cloned viral DNAs has not yet been transmitted by *Ap. craccivora* or *Ac. pisum*, two efficient aphid vectors of FBNYV.

Satellite-Like Rep DNAs

In addition to the putative genomic DNAs, a large number of additional DNAs encoding Rep proteins have been described from nanovirid infections. These DNAs are very diverse and phylogenetically distinct from the DNA-R of the nanovirids (**Figure 3**). They are structurally similar and phylogenetically closely related to nanovirid-like *rep* DNAs that have recently been found associated with some begomoviruses (e.g., ageratum yellow vein virus DNA 1 (AJ238493) and DNA 2 (AJ416153); cotton leaf curl Multan virus DNA 1 (AJ132344-5). However, due to the inclusion of an A-rich sequence within the intergenic region, the begomovirus-associated DNAs are larger ($\sim$1300 nt) than the nanovirid-associated DNAs ($\sim$930 to $\sim$1100 nt). In contrast to the genomic DNA-R which encodes the only known Rep protein essential for the replication of the multipartite genome of the nanovirids, these additional *rep* DNAs are only capable of initiating replication of their cognate DNA but not of any heterologous genomic DNA. Since they are, moreover, only erratically associated with nanovirid infections, they are regarded as satellite-like DNAs that depend on their helper viruses for various functions, such as encapsidation, transmission, and movement.

There are no data as to whether these additional *rep* DNAs are of any biological significance to the helper virus. However, recent agroinoculation experiments with eight FBNYV DNAs or with the same eight FBNYV DNAs in combination with a satellite-like *rep* DNA (FBNYV-C11) suggest that FBNYV-C11 can reduce the number of symptomatic (42/77 vs. 21/74) and severely infected faba bean plants (16/42 vs. 2/21) and, thus, interfere with establishment of disease. This may be due to competition between the additional *rep* DNAs and the genomic nanovirid DNAs for factors required for replication, systemic movement, or encapsidation. It is also noteworthy that the protein encoded by another additional *rep* DNA (C1) of FBNYV was about ten times more active in

Table 1 Key properties of assigned and tentative members of the genera *Babuvirus* and *Nanovirus* of the family *Nanoviridae*

	Genus *Babuvirus*			Genus *Nanovirus*			Tentative species
	Assigned species	Tentative species		Assigned species			
	BBTV	ABTV	CBDV	FBNYV	MDV	SCSV	FBNSV
Geographic distribution	Pacific Islands, Australia, southern Asia, Africa	Borneo, Philippines	India	Near East, North Africa, Ethiopia	Japan	Australia, Tasmania	Ethiopia, Morocco
Biological properties							
Major host plants	*Musa* spp.	*Musa* spp. (abaca, banana)	Large cardamom (*Amomum subulatum*)	Legumes	Legumes	Legumes	Legumes
Aphid vectors	*Pentalonia nigronervosa*	*Pentalonia nigronervosa*	*Micromyzus kalimpongensis*	*Aphis craccivora*, *A. fabae*, *Acyrthosiphon pisum*	*Aphis craccivora*, *Acyrthosiphon pisum*	*Aphis craccivora*, *A. gossypii* and other aphid spp.	*Aphis craccivora*
Virion properties							
Morphology	Icosahedral	Icosahedral	Icosahedral	Icosahedral	Icosahedral	Icosahedral	Icosahedral
Particle diameter (nm)	18–20	No data	17–20	18	18	17–19	18
Sedimentation coefficient	46 S	No data	No data	No data	No data	No data	No data
Density (g cm^{-3})	1.28	No data	No data	1.245	No data	1.24	No data
Capsid protein (kDa)	20	No data	No data	20	No data	19	20
Genome properties							
Number of components[a]	6 (9)	6	1	8 (12)	8 (12)	6 (8)	8
Component sizes (nts)	1018–1111	1013–1099	No data	985–1014	977–1022	988–1022	923–1003
Integral genome components[b]							
DNA-R (M-Rep, 33.1–33.6)[c] (U5, 5.0)	+ (1)[d]	+ (1)	+ (1)	+ (2)	+ (11)	+ (8)	+ (2)
DNA-S (CP, 18.7–19.3)	+ (1)	– (1)	– (1)	– (2)	– (11)	– (8)	– (2)
DNA-C (Clink, 19.0–19.8)	+ (3)	+ (3)	No data	+ (5)	+ (9)	+ (5)	+ (5)
DNA-M (MP, 12.7–13.7)	+ (5)	+ (5)	No data	+ (10)	+ (4)	+ (3)	+ (10)
DNA-N (NSP, 17.3–17.7)	+ (4)	+ (4)	No data	+ (4)	+ (8)	+ (1)	+ (4)
DNA-U1 (U1, 16.9–18.0)	+ (6)	+ (6)	No data	+ (8)	+ (6)	+ (4)	+ (8)
DNA-U2 (U2, 14.2–15.4)	–	–	No data	+ (3)	+ (5)	+ (7)	+ (3)
DNA-U3 (U3, 10.3)	+ (2)[e]	+ (2)[e]	No data	+ (6)	+ (7)	–	+ (6)
DNA-U4 (U4, 10 or 12.5)	–	–	No data	–	–	–	–
Additional (satellite-like)							
Rep-encoding DNAs	DNA-S1, -S2, -S3, -Y1, -W2	No data	No data	+ (12)	+ (12)	–	+ (12)
				DNA-C1, -C7, -C9, and -C11	DNA-C1, -C2, -C3, and -C10	DNA-C2 and -C6	No data

[a] Numbers of identified genome components possibly forming the viral genome. Numbers in parentheses give the total number of distinct ssDNA components described from one or various isolates of each virus. For details see text.
[b] DNA(s) encoding proteins that are either functionally equivalent and/or share significant levels of sequence similarities were placed on the same line.
[c] Assigned and tentative functions of the protein encoded by the genome components is given in parentheses: master replication initiator protein (M-Rep), capsid protein (CP), cell-cycle link protein (Clink), movement protein (MP), putative nuclear shuttle protein (NSP), and proteins of unknown functions (U1–U4). The deduced molecular mass (in kDa) of the protein(s) encoded by each genome component is given in parenthesis (behind the protein designation).
[d] A plus (+) and dash (–) indicates as to whether a component (or ORF) encoding a similar protein has been identified from a virus species or not, respectively. Numbers in parentheses give the DNA-component numbering used originally by the research group studying this virus.
[e] A U3-encoding ORF has been identified only from several South Pacific isolates of BBTV, but not from South Asian isolates of BBTV and the two known ABTV strains.

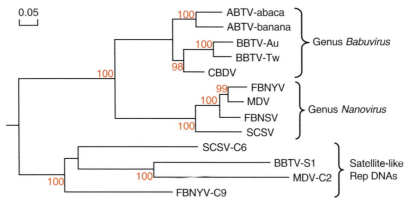

Figure 3 Neighbor-joining dendrogram illustrating the nucleotide sequence relationships in the DNA-R within and between the genera *Babuvirus* and *Nanovirus* of the family *Nanoviridae*. For comparison, four diverse representatives of the numerous satellite-like Rep-encoding DNAs frequently found associated with nanovirid infections were included to demonstrate their phylogenetic distinctness from the DNA-R of the nanovirids. DNA-R sequences used are those of members of the four assigned nanovirid species, Banana bunchy top virus (BBTV), Faba bean necrotic yellows virus (FBNYV), Milk vetch dwarf virus (MDV), and Subterranean clover stunt virus (SCSV), and the tentative species, abaca bunchy top virus (ABTV), cardamom bushy dwarf virus (CBDV), and faba bean necrotic stunt virus (FBNSV). Since some species appear to be particularly diverse, the DNA-R sequences of ABTV isolates from abaca and banana as well as BBTV isolates from Australia (Au) and Taiwan (Tw) representing the South Pacific and Asian groups, respectively, were also included in the comparison. Vertical branch lengths are arbitrary and horizontal distances are proportional to percent sequence differences. Sequence alignments and dendrograms were produced using DNAMAN (version 6, Lynnon Corporation, Quebec, Canada) which uses a CLUSTAL-type algorithm. The dendrograms were bootstrapped 1000 times (scores are shown at nodes).

an *in vitro* origin cleavage and nucleotidyl-transfer reaction than the master Rep protein of FBNYV. Moreover, from many nanovirid infections the additional *rep* DNAs were identified often prior to the master Rep-encoding DNA-R, suggesting that they attain higher concentrations than DNA-R in nanovirid-infected plants.

Proteins

In addition to the CP (about 19 kDa) coded for by the DNA-S transcript, at least 5–7 nonstructural proteins are encoded by the mRNA(s) transcribed from the genomic ssDNAs (**Table 1**, **Figure 2**). The large transcript from DNA-R encodes the master Rep protein (33.1–33.6 kDa). Although a second smaller transcript from the BBTV DNA-R contains a virion-sense ORF completely nested within the master Rep-encoding ORF, there is no experimental evidence that this small ORF potentially encoding a 5 kDa protein (U5) of unknown function is expressed. In addition, a small ORF similar in size and location has not been identified from the DNA-R of both nanoviruses and the tentative babuviruses ABTV and CBDV. DNA-C encodes a 19.0–19.7 kDa protein ('Clink') which contains a conserved LxCxE motif and has been shown to interact with plant proteins involved in cell-cycle regulation. A 12.7–13.7 kDa protein described from all four nanovirid species contains a stretch of 25–30 hydrophobic residues at its N terminus. This together with other experimental evidence obtained for the BBTV DNA-M-encoded protein (MP) indicates that it is involved in cell-to-cell movement of nanovirids. DNA-N encodes a 17.3–17.7 kDa protein which has been identified from all four nanovirids and proposed to act as a nuclear shuttle protein (NSP). Based on significant levels of amino acid sequence identities among some of the other nanovirid proteins, they appear to have similar but unknown functions and are provisionally referred to as U1–U4 proteins (**Table 1**).

The most conserved nanovirid proteins are the master Rep protein (54–97% identity) and NSP (41–91%), followed by the CP (20–84%), Clink (18–72%), and MP (14–76%). Consequently, the babuvirus BBTV shares significant levels of amino acid sequence identity with the nanoviruses only in the M-Rep (54–56%) and NSP (41–45%), whereas the amino acid sequence similarities between the two genera are negligible in the CP (20–27%), the MP (20–23%), and the Clink protein (18–23%). One of the least conserved proteins of the babuviruses and nanoviruses appears to be the protein U3 and U4, respectively.

Nanovirid Replication

Since the nanovirids DNAs and some of the biochemical events determined for nanovirid replication resemble those of the geminiviruses, their replication is also thought to be completely dependent on the host cell's DNA replication enzymes and to occur in the nucleus through transcriptionally and replicationally active double-stranded DNA (dsDNA) intermediates by a rolling-circle type of replication mechanism. Upon decapsidation of viral ssDNA, one of the first events is the synthesis of viral dsDNA with the aid of host DNA polymerase. As the virus DNAs have the ability to self-prime during dsDNA synthesis, it is likely that preexisting primers are used for dsDNA replicative form (RF) synthesis, as has been shown for BBTV. From these dsDNA forms, host RNA polymerase then transcribes mRNAs encoding the M-Rep and other viral proteins required for virus replication. Viral DNA replication is initiated by the M-Rep protein that interacts with common sequence signals on all the genomic DNAs. Nicking and joining within the conserved nonanucleotide sequence TAT/GTATT-AC by the Rep proteins of BBTV and FBNYV has been demonstrated *in vitro*. The nonanucleotide sequence is flanked by inverted repeat sequences with the potential to form a stem–loop structure, a common feature of every nanovirid DNA. Replication of the viral DNAs by the cellular replication machinery is enhanced by the action of Clink, a nanovirid-encoded cell-cycle modulator protein.

Relationships to Other Families of ssDNA Viruses

All Rep proteins of the nanovirid species have most of the amino acid sequence domains characteristic of Rep proteins of ssDNA viruses of other taxa, such as the families *Geminiviridae* and *Circoviridae*. However, the nanovirid Rep proteins differ from those of members of the family *Geminiviridae* in being smaller (about 33 kDa), having a slightly distinct dNTP-binding motif (GPQ/NGGEGKT), and in sharing amino acid sequence identities of only 17–22% with them. A particularly noteworthy feature of the nanovirid M-Rep protein is the lack of the Rb-binding (LxCxE) motif and, thus, the apparent absence of cell-cycle modulation functions from this protein. In contrast to some geminiviruses, whose monopartite genome encodes a Rep protein with a conserved Rb-binding motif, the nanovirids have a separate DNA segment encoding an LxCxE-containing protein involved in cell-cycle regulation. Moreover, nanovirids are clearly distinct from geminiviruses in particle morphology (isometric vs. geminate) and dimensions (18–20 nm vs. 18 × 30 nm), genome size (6.5 or 8.0 vs. 2.6–3.0 or 5.0–5.6) and segments (6 or 8 vs. 1 or 2), mode of transcription (uni- vs. bidirectional) as well as in vector species (aphids vs. whiteflies, leafhoppers, or a treehopper). Although nanovirids and circoviruses share similar particle morphologies, the most notable differences between these two virus taxa are that the circoviruses infect vertebrates (pigs and birds) and have a monopartite genome which is only 1.8–2.0 kb in size and from which the *rep* and *cap* genes are bidirectionally transcribed. All of these viruses have a conserved nonanucleotide motif at the apex of the stem–loop sequence which is consistent with the operation of a rolling-circle model for DNA replication.

See also: Banana Bunchy Top Virus; Circoviruses; Luteoviruses; Maize Streak Virus; Plant Resistance to Viruses: Geminiviruses.

Further Reading

Aronson MN, Meyer AD, Györgyey J, et al. (2000) Clink, a nanovirus encoded protein binds both pRB and SKP1. *Journal of Virology* 74: 2967–2972.

Boevink P, Chu PWG, and Keese P (1995) Sequence of subterranean clover stunt virus DNA: Affinities with the geminiviruses. *Virology* 207: 354–361.

Burns TM, Harding RM, and Dale JL (1995) The genome organization of banana bunchy top virus: Analysis of six ssDNA components. *Journal of General Virology* 76: 1471–1482.

Chu PWG and Helms K (1988) Novel virus-like particles containing circular single-stranded DNA associated with subterranean clover stunt disease. *Virology* 167: 38–49.

Chu PWG and Vetten HJ (2003) *Subterranean clover stunt virus*. AAB Descriptions of Plant Viruses, No. 396. http://www.dpvweb.net/dpv/showadpv.php?dpvno=369 (accessed June 2007).

Franz AW, van der Wilk F, Verbeek M, Dullemans AM, and van den Heuvel JF (1999) Faba bean necrotic yellows virus (genus *Nanovirus*) requires a helper factor for its aphid transmission. *Virology* 262: 210–219.

Karan M, Harding RM, and Dale JL (1994) Evidence for two groups of banana bunchy top virus isolates. *Journal of General Virology* 75: 3541–3546.

Katul L, Timchenko T, Gronenborn B, and Vetten HJ (1998) Ten distinct circular ssDNA components, four of which encode putative replication-associated proteins, are associated with the faba bean necrotic yellows virus genome. *Journal of General Virology* 79: 3101–3109.

Mandal B, Mandal S, Pun KB, and Varma A (2004) First report of the association of a nanovirus with foorkey disease of large cardamom in India. *Plant Disease* 88: 428.

Sano Y, Wada M, Hashimoto T, and Kojima M (1998) Sequences of ten circular ssDNA components associated with the milk vetch dwarf virus genome. *Journal of General Virology* 79: 3111–3118.

Timchenko T, de Kouchkovsky F, Katul L, David C, Vetten HJ, and Gronenborn B (1999) A single rep protein initiates replication of multiple genome components of faba bean necrotic yellows virus, a single-stranded DNA virus of plants. *Journal of Virology* 73: 10173–10182.

Timchenko T, Katul L, Aronson M, et al. (2006) Infectivity of nanovirus DNAs: Induction of disease by cloned genome components of *Faba bean necrotic yellows virus*. *Journal of General Virology* 87: 1735–1743.

Timchenko T, Katul L, Sano Y, de Kouchkovsky F, Vetten HJ, and Gronenborn B (2000) The master rep concept in nanovirus replication: Identification of missing genome components and potential for natural genetic reassortment. *Virology* 274: 189–195.

Vetten HJ, Chu PWG, Dale JL, et al. (2005) Nanoviridae. In: Fauquet CM, Mayo MA, Maniloff J, Desselberger U and Ball LA (eds.) *Virus Taxonomy: Eighth Report of the International Committee on Taxonomy of Viruses*, pp. 343–352. San Diego, CA: Elsevier Academic Press.

Wanitchakorn R, Hafner GJ, Harding RM, and Dale JL (2000) Functional analysis of proteins encoded by banana bunchy top virus DNA-4 to -6. *Journal of General Virology* 81: 299–306.

Narnaviruses

R Esteban and T Fujimura, Instituto de Microbiología Bioquímica CSIC/University de Salamanca, Salamanca, Spain

© 2008 Elsevier Ltd. All rights reserved.

Glossary

Ribozyme RNA with a catalytic activity.

Introduction

The narnaviruses 20S RNA and 23S RNA (ScV20S and ScV23S, respectively) are positive-strand RNA viruses found in the yeast *Saccharomyces cerevisiae*. Currently only these two viruses are ascribed to the genus *Narnavirus* of the family *Narnaviridae*. Like most fungal viruses, they have no extracellular transmission pathway. They are transmitted horizontally by mating, or vertically from mother to daughter cells. It is believed that the high frequency of mating or hyphal fusion that occurs in the host life cycle makes an extracellular route of transmission dispensable for the viruses. The thick cell wall of fungi may also form a formidable barrier. The lack of extracellular transmission may, in turn, explain two prominent features found in narnaviruses. First, they are persistent viruses and do not kill the host cells. If their infection caused damages or disadvantages to the host, then the viruses might have perished during the course of evolution because of the lack of an escape route. Second, because there is no extracellular phase, the viruses do not need to form virions to protect their RNA genomes in the extracellular environment. In addition, they do not need machinery to ensure exit or reentry to a new host. The lack of a virion structure may sound peculiar to those who are familiar with infectious viruses. Considering that viruses are selfish parasites, however, it will be natural for them to shed genes or functions unnecessary for their existence. Consequently, the genomes of narnaviruses are simple and small: they only encode a single protein, the RNA-dependent RNA polymerase (RdRp). This may contribute to their persistence by reducing a number of viral proteins that might interfere with metabolism vital for the host. The simplicity of their RNA genomes encoding a single protein, together with the recent development of 20S and 23S RNA virus launching systems from yeast expression vectors, makes narnaviruses a good model system to investigate replication and the molecular basis for intracellular persistence of RNA viruses.

Historical Background

20S RNA was first described in 1971 as a single-stranded RNA (ssRNA) species accumulated in yeast cells transferred to 1% potassium acetate, a standard procedure to induce sporulation in yeast under nitrogen-starvation conditions. Because of its mobility relative to 25S and 18S rRNAs, the species was named 20S RNA. Later it was found, however, that the accumulation of 20S RNA was not related with the sporulation process because haploid cells that do not sporulate also accumulate 20S RNA under nitrogen-starvation conditions. It was also found that 20S RNA is a cytoplasmic genetic element. The realization of 20S RNA as a viral entity, however, had to wait several years, until the characterization of 20S RNA by cloning and sequencing in 1991. 23S RNA was reported first time in 1992. Both viruses were placed in the genus *Narnavirus* of the new family, *Narnaviridae* (naked RNA virus), a taxonomic group that appeared for the first time in the seventh edition of the International Committee on Taxonomy of Viruses (ICTV). The other genus in the family is *Mitovirus*, whose members are found in mitochondria of fungi, many of them pathogenic to plants. The members of the family have small RNA genomes (2–3 kb) that encode single proteins, their RdRps, and reside either in the cytoplasm (members of the genus *Narnavirus*, narnaviruses) or in the mitochondria (mitoviruses) of the host.

Viral Genomes

Many laboratory strains of *S. cerevisiae* harbor 20S RNA virus and fewer strains contain 23S RNA virus. Both viruses are compatible in the same host. The presence of 20S and 23S RNA viruses does not render phenotypic changes to the host. Under nitrogen-starvation conditions, the amounts of the viral genomes become almost equivalent to those of rRNAs (>100 000 copies/cell; **Figure 1**). In contrast, vegetative growing cells contain much lower amounts of the viral RNAs (5–20 copies/cell). **Figure 2(a)** shows the genome organization of narnaviruses. Both 20S and 23S RNAs are small (2514 and 2891 nt, respectively) and each genome encodes a single protein: a 91 kDa protein (p91) by 20S RNA and a 104 kDa protein (p104) by 23S RNA. The 5′ untranslated regions in both RNAs are extremely short: 12 nt in the case of 20S RNA and only 6 nt in 23S RNA. These RNAs lack poly(A) tails at the 3′ ends and have perhaps no 5′ cap structures. The same RNA can serve as template for translation and also for negative-strand synthesis. The antigenomic (or negative-strand) RNAs have no coding capacity for protein and are present at much lower copy numbers compared to the genomic (or positive-strand) RNAs under the induction conditions. The double-stranded forms of 20S and 23S RNAs are known and called W and T, respectively. These double-stranded RNAs (dsRNAs) accumulate when the cells are grown at 37 °C, a rather high temperature for yeast (the optimal temperature for growth is about 28 °C). These dsRNAs are not intermediates of replication but by-products. Replication proceeds from a positive strand to a negative strand and then to a positive strand.

The proteins encoded in the viral genomes are not processed to produce smaller fragments with distinct functional domains. Both proteins contain amino acid motifs well conserved among RdRps from positive-strand and dsRNA viruses (**Figure 2(a)**) In addition, p91 and p104 share stretches of amino acid sequences (denoted by 1–3 in **Figure 2(a)**) in the same order throughout the molecules, indicating a close evolutionary relationship between these two viruses. Remarkably, their RdRp consensus motifs are most closely related to those of RNA bacteriophages such as Q β.

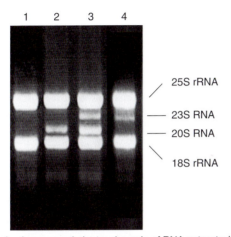

Figure 1 Agarose gel electrophoresis of RNA extracted from virus-free and virus-infected nitrogen-starved yeast cells. RNA from virus-free yeast cells (lane 1) or cells infected with 20S RNA virus alone (lane 2), 23S RNA virus alone (lane 4), or both viruses together (lane 3), was separated in an agarose gel and visualized by ethidium bromide staining. The positions of the 20S and 23S RNAs together with the rRNAs are indicated to the right.

Ribonucleoprotein Complexes as Viral Entities

Yeast is also a natural host for dsRNA totiviruses, called L-A and L-BC. Like narnaviruses, totiviruses have no extracellular transmission pathway. However, these viruses have *gag* and *pol* genes, and their dsRNA genomes are encapsidated into intracellular viral particles. In contrast, 20S and 23S RNA viruses have no capsid genes to form virion structures. Then, how do these viruses exist inside

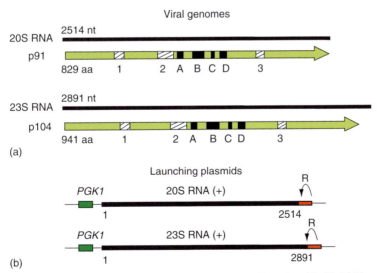

Figure 2 Genomic organization of 20S and 23S RNA viruses (a) and their launching plasmids (b). (a) Diagrams of 20S and 23S RNAs and the proteins encoded by them, p91 and p104, respectively. A–D represents motifs conserved among RdRps from positive strand and dsRNA viruses and 1–3 indicates amino acid stretches conserved between p91 and p104. (b) The complete cDNA of 20S or 23S RNA genome is inserted downstream of the constitutive *PGK1* promoter in a yeast expression vector in such a way that positive strands are transcribed from the promoter. The HDV ribozyme (R) is fused directly to the 3′ end of the viral genome.

the cell and establish a persistent infection without a protective coat? Earlier studies demonstrated that 20S RNA migrated as 'naked RNA' in sucrose gradients. Furthermore, deproteination with phenol had no apparent effect on its mobility. Because protein provides a large part of the molecular mass in virions, these data clearly indicate that narnaviruses lack a virion structure. When specific antibodies against their RdRps became available, however, it was realized that each RdRp is associated with its RNA genome and this interaction is specific; that is, p91 is associated only with 20S RNA and p104 only with 23S RNA. These ribonucleoprotein complexes reside in the cytoplasm and are not associated with the nucleus, mitochondria, or intracellular membranous structures. Further studies indicated that most of the positive strands of 20S and 23S RNA viruses under induction conditions are associated with their own RdRps in a 1:1 stoichiometry. It is not known whether host proteins are present in the complexes. These complexes are called 'resting complexes' to distinguish them from the 'replication complexes' described in the following section. Negative strands are present at much lower amounts compared to positive strands, and available data indicate that they also form complexes with their own RdRps. These findings suggest that the formation of ribonucleoprotein complexes between the viral RNA and its RdRp is important for the life cycles of 20S and 23S RNA viruses.

Replication Intermediates

Lysates prepared from virus-induced cells have an RdRp activity. The activity is insensitive to actinomycin D or α-amanitin, thus independent of a DNA template. The majority of *in vitro* products are positive strands of 20S RNA. Synthesis of negative strands accounts for a small fraction of the RNA products compared to that of positive strands, thus reflecting the high positive/negative-strand ratio in the lysates. There is no, or very little, *de novo* synthesis *in vitro*. Therefore, radioactive nucleotides are unevenly distributed into 20S RNA positive-strand products with more incorporation into the 3′ end region. Replication complexes that synthesize 20S RNA positive strands have a ssRNA backbone and migrate in native agarose gels as a broad band corresponding to an ssRNA in the size ranging from 2.5 to 5 kbp long. These complexes consist of a full-length negative-strand template (2.5 kbp) and a nascent positive strand of less than unit-length, probably held together by the polymerase machinery. Deproteination with phenol converts them to dsRNA. Therefore, W dsRNA is not a replication intermediate but a byproduct. It is likely that the high temperature (37 °C) for growth may destabilize replication complexes, thus resulting in the accumulation of dsRNA. Upon completion of RNA synthesis *in vitro*, the positive-strand products as well as the negative-strand templates are released from replication complexes. It is likely that, in the cell, the released negative strands are immediately recruited to another round of positive-strand synthesis, because the majority of negative strands in lysates are present in replication complexes engaging in the synthesis of positive strands. Interestingly, both positive and negative strands released from replication complexes are associated with protein. Because replication complexes contain at least one p91 molecule per complex, p91 is a good candidate for the protein.

Generation of Narnaviruses *In Vivo*

As mentioned earlier, the presence of narnaviruses does not render phenotypic changes to the host. This has hindered studies on replication or virus/host interactions using yeast genetics. This obstacle has been overcome by recent developments in generating 20S and 23S RNA viruses *in vivo* from a yeast expression vector (**Figure 2(b)**). In either case, the complete viral cDNA was inserted in the vector downstream of a constitutive promoter in such a way that positive strands can be transcribed from the promoter. The 3′ end of the viral sequence was directly fused to the hepatitis delta virus (HDV) antigenomic ribozyme. Therefore, intramolecular cleavage by the ribozyme will create transcripts *in vivo* having the 3′ termini identical to the viral 3′ end. The efficiency of virus launching is high. The 20–70% of the cells transformed with the vector generated the virus. The primary transcripts expressed from the vectors have nonviral sequences (about 40 nt) at the 5′ ends. The generated viruses, however, possessed the authentic viral 5′ ends without the extra sequences. It is likely that the 5′ nonviral extension was eliminated by a 5′ exonuclease. Using these launching systems, it has been demonstrated that each RdRp is essential and specific for replication of its own viral RNA. p91 is essential for 20S RNA replication and does not substitute p104 for replication of 23S RNA virus. Similarly, p104 is essential for 23S RNA replication and does not support 20S RNA replication. Because negative strands cannot be decoded to the RdRps, vectors in which the viral cDNAs were reversed failed to generate the virus. These negative-strand-expressing vectors, however, successfully generated narnaviruses, if active polymerases were provided in *trans* from a second vector. Therefore, both 20S and 23S RNA viruses can be generated from either positive or negative strands expressed from a vector.

cis-Acting Signals for Replication

20S and 23S RNA genomes share the same 5 nt inverted repeats at the 5′ and 3′ termini (5′-GGGGC...GCCCC-OH). Extensive analysis was done modifying each nucleotide at the 3′ ends. It was found that the third and fourth

C's from the 3′ termini are essential for replication in both viruses. While the 3′ terminal and penultimate C's can be eliminated or changed to other nucleotides without affecting virus generation, the generated viruses recovered the wild-type C's at the termini. Therefore, the consecutive four C's at the 3′ terminus are essential for these viruses (**Figure 3**). In contrast, the G at position 5 from the 3′ end is dispensable for replication in both viruses. 23S RNA virus requires an additional 3′ cis-signal for replication. The stem–loop structure proximal to the 3′ end contains a mismatched pair of purines in the stem (**Figure 3**). This mismatched pair is essential for replication but the virus tolerates any combination of purines at this position. On the other hand, changing the purines to pyrimidines or eliminating one of the purines at the mismatched pair blocked virus generation. The distance between the mismatched pair and the 3′ terminal four C's and/or their spatial configuration appears to be critical, because shortening or increasing the length of the stem between the two sites by more than 1 bp abolished virus launching. It is not known whether 20S RNA virus has a similar cis-signal in the stem–loop structure proximal to the 3′ end. The G at position 5 from the 3′ end is located at the bottom of the stem structure. This G, as mentioned earlier, can be changed to another nucleotide without impairing replication, as long as the modified nucleotide is hydrogen-bonded at the bottom of the stem.

The negative strands of 20S and 23S RNA viruses also possess four consecutive C's at the 3′ ends. Using the two-vector system mentioned above, it has been found that the third and fourth C's from the 3′ end are essential for replication. Similar to the positive strands, the 3′ terminal

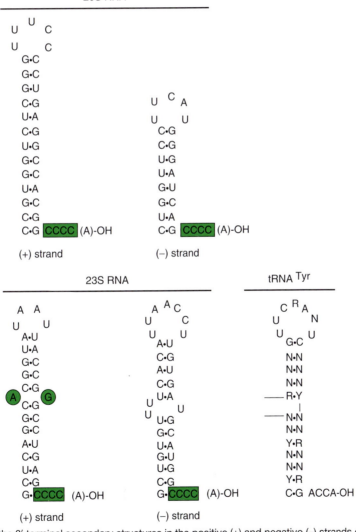

Figure 3 Comparison of the 3′ terminal secondary structures in the positive (+) and negative (−) strands of 20S and 23S RNA viruses, with the top half domain of tRNA^{Tyr}. The nontemplated A residues at the viral 3′ termini are indicated by parenthesis. The consecutive four C's essential for replication are boxed (green). A second cis-signal (the mismatched pair of purines) present in the positive strand of 23S RNA virus is circled (green). Y, R, and N stand for pyrimidine, purine, and any base, respectively.

and penultimate C's can be eliminated or changed to other nucleotides without affecting virus generation and the generated viruses recovered the wild-type four C's at the 3′ ends. Therefore, the consecutive four C's at the 3′ end of the negative strand are again a cis-signal for replication. The 5′ ends of viral positive strands have not been analyzed extensively. Elimination of the 5′ terminal G or changing it to other nucleotide had no effect on virus generation and the generated viruses recovered this G at the 5′ ends.

cis-Signals for Formation of Ribonucleoprotein Complexes

Narnaviruses, as mentioned earlier, exist as ribonucleoprotein complexes in the host cytoplasm. In the absence of the HDV ribozyme, RNA transcribed from the launching vectors failed to generate viruses because of the presence of nonviral extensions at the 3′ end. The transcripts, however, can be decoded to viral polymerases and the polymerases can form complexes in vivo with the transcripts, thus providing an assay system to analyze cis-signals for formation of ribonucleoprotein complexes. By immunoprecipitation with antiserum specific to p104, it has been found that the bipartite 3′ cis-signal for replication (more specifically, the mismatched pair of purines and the third and fourth C's from the 3′ end) is essential for 23S RNA positive strand to form complexes with the polymerase. This highlights the importance of formation of ribonucleoprotein complexes for the life of narnaviruses. The cytoplasm is filled with host RNAs, including a great variety of mRNAs. Formation of complexes between the viral polymerase and its template RNA will therefore facilitate replication and increase its fidelity by discriminating against nonviral RNAs as templates. The importance of the third and fourth C's from the 3′ end for formation of complexes with p104 suggests that these nucleotides are in close contact with p104. Because the 23S RNA virus has no coat protein to protect the viral RNA, it is likely that such interaction protects the 3′ ends of 23S RNA from exonuclease cleavage. In the case of 20S RNA virus, a similar in vivo assay indicates that the 3′ cis-signal for replication (in particular the third and fourth C's from the 3′ end) is also important for formation of ribonucleoprotein complexes with p91. When isolated resting complexes of 20S RNA virus were analyzed in vitro, however, it was found that p91 interacts with 20S RNA not only at the 3′ end but also at the 5′ terminal region of the molecule. The 5′ binding site is located at the second stem–loop structure from the 5′ end. Computer-predicted analysis indicates that the 5′ and 3′ termini of 20S RNA (and 23S RNA) are brought together into close proximity by a long-distance RNA/RNA interaction (**Figure 4**). This may allow a single molecule of p91 to interact simultaneously with both ends of 20S RNA genome in a resting complex. This protein/RNA interaction may provide a clue to understand the molecular basis of narnaviruses persistence (see below).

Narnavirus Persistence in the Host

mRNA degradation in yeast, like in other eukaryotes, is initiated by shortening the 3′ poly(A) tail followed by decapping at the 5′ end. Then the decapped mRNA is degraded by the potent Xrn1p/Ski1p 5′ exonuclease as well as by a 3′ exonuclease complex called exosome. The RNA genomes of narnaviruses, as mentioned earlier, have no 3′ poly(A) tails and perhaps no cap structures at the 5′ ends, thus resembling intermediates of mRNA degradation. This suggests that these RNA genomes are vulnerable to the exonucleases involved in mRNA degradation. In fact, the copy numbers of 20S and 23S RNAs increase several-fold in strains having mutations in SKI genes such as SKI2, SKI6, and SKI8. These mutations were originally identified by their failure in lowering the copy numbers of L-A dsRNA totivirus and its satellite RNA M. It is known that the SKI2, SKI6, and SKI8 gene products are components or modulators of the exosome. These observations suggest that the 3′ end of the viral genome is constantly nibbled by 3′ exonucleases. Therefore, one of the reasons for narnaviruses to form ribonucleoprotein complexes may be to protect their 3′ ends from exonuclease cleavage. The fact that the third and fourth C's from the 3′ end are important to form complexes in both viruses fits this hypothesis because binding of the RdRp to these nucleotides would block progression of the exonuclease and protect the internal region. As described earlier, mutations introduced at the terminal and penultimate positions at the 3′ end had no deleterious effects on virus launching and the generated viruses recovered the wild-type sequences. This suggests that the terminal and penultimate positions at the 3′ ends are not only vulnerable to cleavages but also accessible to the repair machinery. The 3′ ends of these viruses may undergo constant turnover at these positions.

As regards the 5′ end, the first four nucleotides in both 20S and 23S RNAs are consecutive G's (**Figure 4**). It is known that oligo G tracts inhibit progression of the Xrn1/Ski1 5′ exonuclease. Furthermore, these consecutive G's are buried at the bottom of a long stem structure in both viruses. These features thus suggest that 20S and 23S RNAs by themselves are quite resistant to the 5′ exonuclease. The initiation codon of p91 is located in the middle of the long stem structure proximal to the 5′ end. If p91 binds to this stem in the ribonucleoprotein complex, then such a stable binding may interfere with translation of new p91 molecules from the RNA. In this context, it may make sense that the 5′ binding site of p91 in the complexes is located at the second stem–loop structure from the 5′ end. By binding simultaneously to the 3′ end and also

Narnaviruses 397

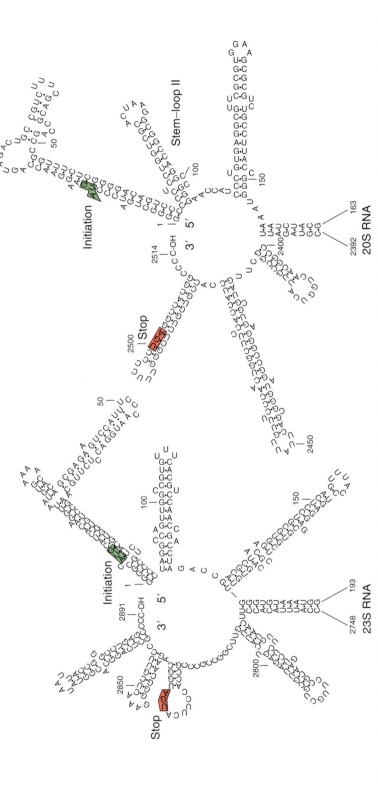

Figure 4 Secondary structures at the 5' and 3' end regions of 23S and 20S RNA positive strands, as predicted by the MFOLD program. The AUG initiation codons (green) and the stop codons (red) for p104 and p91 are boxed. Stem-loop II where the 5' binding site for p91 is located is shown. Note that about 150 nt from the ends in each viral genome there are inverted repeats of 8–12 nt long that bring both 5' and 3' ends to a close proximity.

to the region close to the 5′ end of the same RNA molecule, p91 may stabilize the long-distance RNA–RNA interactions that bring the 5′ and 3′ ends of the RNA into proximity, and thus helps the RNA to form an organized structure in resting complexes.

It is not known whether the 3′ end repair is carried out by the replicase machinery during the replication process, or by host enzymes. However, the following evidence favors the latter case. The 3′ terminal structures of 20S and 23S RNAs resemble a half of tRNA, the so-called 'top-half' domain, consisting of the acceptor stem and T stem (**Figure 3**). The domain provides the determinants necessary for specific interactions with tRNA-related enzymes such as the tRNA nucleotidyltransferase (CCA-adding enzyme). This raises the possibility that 20S and 23S RNAs nibbled at the 3′ ends by 3′ exonucleases are repaired to the wild-type sequences by the CCA-adding enzyme. The fact that 15–30% of both positive and negative strands of 20S and 23S RNAs possess an unpaired A at the 3′ ends supports this possibility. Furthermore, that the 3′ repair is confined to the terminal and penultimate positions is consistent with the catalytic activity expected for the CCA-adding enzyme. Given that narnaviruses are persistent viruses, the underlying mechanism(s) to maintain the integrity of the viral 3′ ends will, therefore, have considerable significance for a long-term infection.

See also: Fungal Viruses; Yeast L-A Virus.

Further Reading

Buck KW, Esteban R, and Hillman BI (2005) *Narnaviridae*. In: Fauquet CM, Mayo MA, Maniloff J, Desselberger U, and Ball LA (eds.) *Virus Taxonomy: Eighth Report of the International Committee on Taxonomy of Viruses*, pp. 751–756. San Diego, CA: Elsevier Academic Press.

Esteban LM, Rodríguez-Cousiño N, and Esteban R (1992) T double-stranded (dsRNA) sequence reveals that T and W dsRNAs form a new RNA family in *Saccharomyces cerevisiae*: Identification of 23S RNA as the single-stranded form of T dsRNA. *Journal of Biological Chemistry* 267: 10874–10881.

Esteban LM, Fujimura T, García-Cuéllar MP, and Esteban R (1994) Association of yeast viral 23 S RNA with its putative RNA-dependent, RNA polymerase. *Journal of Biological Chemistry* 269: 29771–29777.

Esteban R and Fujimura T (2003) Launching the yeast 23S RNA narnavirus shows 5′ and 3′ cis-acting signals for replication. *Proceedings of the National Academy of Sciences, USA* 100: 2568–2573.

Esteban R and Fujimura T (2006) Yeast narnavirus replication. In: Hefferon KL (ed.) *Recent Advances on RNA Virus Replication*, pp. 171–194. Trivandrum, India: Research Ringspot.

Esteban R, Vega L, and Fujimura T (2005) Launching of the yeast 20S RNA narnavirus by expressing the genomic or anti-genomic viral RNA *in vivo*. *Journal of Biological Chemistry* 280: 33725–33734.

Fujimura T and Esteban R (2004) Bipartite 3′ cis-acting signal for replication in yeast 23 S RNA virus and its repair. *Journal of Biological Chemistry* 279: 13215–13223.

Fujimura T, Solórzano A, and Esteban R (2005) Native replication intermediates of the yeast 20S RNA virus have a single-stranded RNA backbone. *Journal of Biological Chemistry* 280: 7398–7406.

Kadowaki K and Halvorson HO (1971) Appearance of a new species of ribonucleic acid synthesized in sporulation cells of *Saccharomyces cerevisiae*. *Journal of Bacteriology* 105: 826–830.

Matsumoto Y and Wickner RB (1991) Yeast circular RNA replicon: Replication intermediates and encoded putative RNA polymerase. *Journal of Biological Chemistry* 266: 12779–12783.

Rodríguez-Cousiño N, Esteban LM, and Esteban R (1991) Molecular cloning and characterization of W double-stranded RNA, a linear molecule present in *Saccharomyces cerevisiae*: Identification of its single-stranded RNA form as 20S RNA. *Journal of Biological Chemistry* 266: 12772–12778.

Solórzano A, Rodríguez-Cousiño N, Esteban R, and Fujimura T (2000) Persistent yeast single-stranded RNA viruses exist *in vivo* as genomic RNA polymerase complexes in 1:1 stoichiometry. *Journal of Biological Chemistry* 275: 26428–26435.

Wejksnora PJ and Haber JE (1978) Ribonucleoprotein particle appearing during sporulation in yeast. *Journal of Bacteriology* 134: 246–260.

Wesolowski M and Wickner RB (1984) Two new double-stranded RNA molecules showing non-Mendelian inheritance and heat inducibility in *Saccharomyces cerevisiae*. *Molecular and Cellular Biology* 4: 181–187.

Widner WR, Matsumoto Y, and Wickner RB (1991) Is 20S RNA naked? *Molecular and Cellular Biology* 11: 2905–2908.

Nature of Viruses

M H V Van Regenmortel, CNRS, Illkirch, France

© 2008 Elsevier Ltd. All rights reserved.

The Discovery of Viruses

By the end of the nineteenth century, it had been established that many infectious diseases of animals and plants were caused by small microorganisms which could be visualized in the light microscope and could be cultivated as pure cultures on synthetic nutritive media. However, in the case of several infectious diseases, it had not been possible to identify the causative agent in spite of many carefully executed experiments. These results suggested that some other type of infectious agent existed which was not identifiable by light microscopy and could not be cultured on conventional bacteriological media.

The first evidence that such infectious agents, later called viruses, were different from pathogenic bacteria was obtained by filtration experiments with porcelain Chamberland filters used for sterilization. These bacteria-retaining filters, which had been used by Louis Pasteur to remove pathogenic microorganisms from water, were used by Dmitri Ivanovsky in St. Petersburg in 1892 in his study of the tobacco mosaic disease. Ivanovky showed that when sap from a diseased tobacco plant was passed through the filter, the filtrate remained infectious and could be used to infect other tobacco plants. Although Ivanovsky was the first person to show that the agent causing the tobacco mosaic disease passed through a sterilizing filter, all his publications show that he did not grasp the significance of his observation and that he remained convinced that he was dealing with a small bacterium rather than with a new type of infectious agent. More than 10 years after his initial experiment he still believed that the filter he used might have had fine cracks which allowed small spores of a microorganism to pass through it.

The same filtration experiment was repeated on 1898 by Martinus Beijerinck in Delft, Holland, who, unaware of Ivanovsky's papers, again showed that the filtered tobacco sap from diseased plants was infectious. However, Beijerinck went further and demonstrated that the infectious agent was able to diffuse through several millimeters of an agar gel. From this, he concluded that the infection was not caused by a microbe but by what he called a *contagium vivum fluidum* or contagious living liquid. He established that the agent could reproduce itself in a tobacco plant and called it a virus. In the same year, a similar type of filtration experiment was done by Friedrich Loeffler and Paul Frosch who were investigating the important foot-and-mouth disease of cattle. These German investigators reported that although the causative agent of the disease passed through a Chamberland-type filter, it did not go through a Kitasato filter which had a finer grain than the Chamberland filter. From this result, they concluded that the causative virus, which was multiplying within the host, was a corpuscular particle and not a soluble agent as claimed by Beijerinck. Within a few years, it became generally accepted that filterable viruses that remained invisible in the light microscope represented a new class of pathogenic agent different from bacteria.

Although all historical accounts of the beginnings of virology mention the work of Ivanovsky, Beijerinck, and Loeffler, there is disagreement among various authors about who should be credited with the discovery that viruses were a new type of infectious agent. This is an interesting debate since it concerns the nature of what is a scientific discovery. It is indeed not sufficient to make a novel observation such as the filterability of an infectious agent but it is also necessary to interpret the observation correctly and to grasp the significance of an unexpected experimental finding. Ivanovsky was the first to observe the filterability of a virus but he did not recognize that he was dealing with a new type of infectious agent. Beijerinck realized that he was dealing with something different from a microbe but he thought that the virus was an infectious liquid rather than a small corpuscular particle. Only Loeffler correctly concluded that the virus causing foot-and-mouth disease was a small particle stopped by a fine-grain Kitasato filter and he therefore came closest to the modern concept of a virus. It was only the work of William Elford with graded collodion membranes, done more than 30 years later, which established that different viruses had particle diameters in the range of 20–200 nm. Only with the advent of the electron microscope was it finally possible to determine the actual morphology of virus particles.

Viruses as Chemical Objects

The perception of what viruses are changed dramatically in 1935 when Wendel Stanley, working at the Rockefeller Institute in Princeton, showed that tobacco mosaic virus (TMV), the agent studied by Ivanovsky and Beijerinck, could be crystallized in the form of two-dimensional paracrystals. This led to the view that viruses were actually chemical objects rather than organisms and it stimulated an intense interest in viruses which now seemed to be entities at the borderline between chemistry and biology. Many scientists became fascinated with viruses and viewed them as 'living molecules' since they seemed to be able to reproduce themselves. The needle-shaped crystals of TMV, visualized by Stanley, suggested that a dead protein could be a living infectious agent and it was believed by some that viruses might hold the key to the origin of life. Stanley had initially reported that TMV was a pure protein, but in 1936 Frederick Bawden and Norman Pirie in Britain showed that TMV also contained phosphorus and carbohydrate and was in fact an RNA-containing nucleoprotein. However, it took nearly another 20 years before it was established that it was the RNA in the virus that was the infectious entity.

Whether or not viruses should be regarded as living organisms has been regarded by some to be only a matter of taste. A definite answer to this question requires that one has a clear understanding of what is meant by 'life'.

What Is Life?

Life is not a material entity, nor a force, nor a property, but a conceptual object made up of the collection of all living systems, past, present, and future. All living systems possess the property of 'being alive' and the concept 'life' corresponds to the abstract, mental representation of this property. Philosophers say that 'life' is the extension of the predicate 'is alive', the extension of a concept being the objects that the concept refers to. Instead of analyzing the concept 'life', it is thus more relevant to ask which characteristics of biological organisms give them the property of being alive. One needs to ascertain to what objects the concept of 'life' refers to and thus to provide an answer to the question: what is a living organism?

What Is an Organism?

A simple answer to this question would be to say that organisms are living agents but this is unsatisfactory for several reasons, one of them being that it would rule out dead organisms. A better answer would be to say that organisms are living agents at some point of their existence. The related claim that all living agents are organisms is equally problematic since organs like hearts or kidneys are clearly not organisms although they are considered to be alive. Equating organisms with living agents is thus not satisfactory and these two concepts need to be differentiated.

A useful approach is to consider the class of living agents as a cluster concept. Such a concept is defined by a cluster of properties, the majority of which have to be present in all members of the class although some properties can be absent in individual members. Many of the properties tend to be present simultaneously because of the existence of underlying relationships between them: this means that if an individual member possesses any one of these properties, it increases the probability that it also possesses some of the others. Such a cluster concept, known as a polythetic class, is useful to group biological entities that show inherent variation and it has also been used to define virus species.

Living agents can be defined by a cluster of properties that include:

1. compositional and structural properties such as the presence of nucleic acids and proteins, and of heterogenous and specialized parts;
2. functional properties such as the capacity to grow and develop, to reproduce, and to repair themselves; and
3. properties such as metabolism and environmental adaptation that arise from the interaction of a living agent with its environment.

Biosystems interact selectively with the environment through a membrane boundary which restricts the types of exchanges that can occur between components of the system and items in its environment. The environment of a system consists of those things that can be influenced by the system or that may act upon the system. The notion of environment is thus limited to the immediate environment of the system and it exists only relative to a given system.

Any agent in order to be living must necessarily possess a sufficient subset of this cluster of properties, although it need not possess them all. For instance, sterile organisms that do not reproduce or plant seeds with a completely dormant metabolism can be included in the class of living agents.

In addition to having many of the properties listed above, living agents in order to qualify as organisms must also belong to a reproductive lineage characterized by a life cycle. Organs, therefore, are not organisms since they do not reproduce themselves as members of a lineage. Organs are replicated when the entire organism reproduces but they lack a life cycle. The various entities that comprise the life cycle of an organism correspond to developmental stages in the life of that organism.

Finally, in order to qualify as organisms, living agents must also possess a functional autonomy that allows them to exercise control over themselves and to be at least partly independent from other organisms and environmental influences. Organs, tissues, or the leaves of a plant, for instance, are living things but they are not organisms since they are not functionally autonomous, their life being dependent on that of the organisms they belong to.

It should be clear from the preceding discussion that things such as DNA molecules and other biochemical constituents of cells, as well as organelles such as ribosomes are neither living agents nor organisms. For the same reasons, viruses, although they are biological systems, are neither living agents nor *a fortiori* microorganisms.

Viruses Should Not Be Confused with Virus Particles or Virions

A virus has both intrinsic properties such as the size of the virus particle and relational properties such as having a host or vector, the second type of property existing only by virtue of a relation with other objects. Relational properties are also called emergent properties because they are possessed only by the viral system as a whole and are not present in its constituent parts. During its multiplication cycle, a virus takes on various forms, for instance, as a replicating nucleic acid in the host cell. One stage of what is metaphorically called the 'life cycle' of a virus corresponds to the virus particle or virion which can be characterized by intrinsic, structural, and chemical properties such as size, mass, chemical composition, and sequence of both coat proteins and nucleic acids.

Compared to a virion, a virus also possesses a number of relational properties that become actualized during transmission and infection, for instance, when the virus becomes integrated into the host cell during the viral replication cycle. A virus cannot be thus reduced to the physical constituents and chemical composition of a virion and it is necessary to include in its description the various biotic interactions and functional activities that make the virus a biological system. Confusing 'virus' with 'virion' is somewhat similar to confusing the entity 'insect', which includes several different life stages, with a single one of these stages such as a pupa, a caterpillar, or a butterfly.

When Eckhard Wimmer of Stony Brook University in New York gives the chemical formula of poliovirus as $C_{332,652} H_{492,288} N_{98,245} O_{131,196} P_{7,501} S_{2,340}$, he provides only the chemical composition of one poliovirus particle. However, this reductionist chemical description of a virion does not amount to a characterization of the entity poliovirus.

The chemical composition or even the sequence of the coat protein of the virion does not give information on the tertiary and quaternary structural elements in the outer capsid that form the three-dimensional assembly of atoms that will be recognized by host cell receptors, allowing the virus to infect its host. The receptor-binding site of the virion is actually a relational, emergent biological entity defined by its ability to be recognized by the complementary receptor molecules present in certain host cells. The viral receptor-binding site is thus not an intrinsic feature of the virion that could be defined independently of an external cellular receptor since its identity as a site depends on the existence of a specific relationship with a host cell. This relationship arose during the process of biological evolution that culminated in the capacity of the virus to infect certain host cells. Such a specific functional relationship with a host is the essential feature that gives viruses their unique status as molecular genetic parasites capable of using cellular systems for their own replication.

The Nature of Viruses

Although viruses are not living organisms, they are considered to be biological entities because they possess some of the properties of living systems such as having a genome and being able to adapt to particular hosts and biotic habitats. However, viruses do not possess many of the essential attributes of living organisms such as the ability to capture and store free energy and they lack the characteristic autonomy that arises from the presence of integrated, metabolic activities. Viruses do not replicate or self-replicate themselves, but are 'being replicated' (i.e., passively rather than actively) through the metabolic activities of the cells they have infected. The replication of viruses involves a process of copying done by certain constituents of host cells. In contrast, cells do not replicate but reproduce themselves by a process of fission.

A virus becomes part of a living system only after its genome has been integrated in an infected host cell and it is not more alive than other cellular constituents such as genes, macromolecules, and organelles. The consensus among biologists is that the simplest system that can be said to be alive is a cell. Viruses are thus nonliving, infectious agents which can be said, at best, to lead a kind of borrowed life. Duncan McGeoch has described viruses metaphorically as "mistletoes on the Tree of Life."

The difference between viruses and various types of organisms becomes obvious when the functional roles of the proteins found in viruses and in organisms are compared. When proteins are divided into three broad functional categories corresponding to energy utilization, information carriers, and communication mediators, the proportion of each protein class found in viruses is markedly different from that found in living organisms.

Viruses have the highest proportion of proteins involved in information processes related to the control and expression of genetic information but have very few proteins of the energy and communication classes (**Figure 1**). This distribution is due to the fact that viruses utilize the metabolic machinery of the host cell and rely entirely on the energy supply systems of the host they infect. In contrast, bacteria have the highest proportion of proteins of the energy class involved in small-molecule transformation, whereas animals have a high proportion of proteins involved in intra- and intercellular communication.

Some authors have argued that viruses are living microorganisms because they share with certain parasitic organisms the property of being obligate parasites. However, the way viruses depend on their cellular hosts for replication is a type of molecular parasitism that is totally different from the metabolic dependency shown, for instance, by rickettsia or by bacteria that colonize the gut of certain animals. Obligate parasitism on its own is clearly not a sufficient criterion for establishing that an entity is a living organism.

Viruses are subcellular infectious agents which at one stage of their replication cycle in the infected cell are reduced to their nucleic acid component. The role of the virion is to allow transmission to new hosts while it also protects the viral genome from degradation by nucleases and from other environmental attack.

It is customary to distinguish acute viral infection which is associated with active replication and production of virions from the asymptomatic type of specific

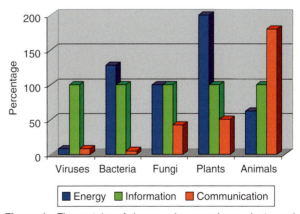

Figure 1 The proteins of viruses, microorganisms, plants, and animals have different functional roles. The vertical bars represent the proportion of proteins in the categories of energy utilization and carrier of information, relative to those in the category of mediator of information. Viruses have the highest proportion of proteins involved in information processes related to the control and expression of genetic information. Reproduced from Patthy L (1999) *Protein Evolution*. Oxford: Blackwell Science, with permission from Blackwell Publishing.

virus–host relationship characterized by latency, persistence, and absence of disease. In the latter case the viral genome is maintained in the host but no virions are produced for long periods of time and no antiviral host immune response is elicited. Small amounts of virions are produced episodically which is sufficient for transmission of the virus to new hosts. If the virus happens to switch to a new type of host, this may involve a changeover from no disease in the latent host to the appearance of disease following reactivation in the new host. This host-switching phenomenon is sometimes responsible for the appearance of new emerging viral diseases.

Fitness and Viral Evolution

Fitness is usually taken to be the property of an organism that ensures its survival and reproductive success in a variable and unpredictable environment. Since viruses are not living organisms, the fitness of a virus is usually measured in terms of virus progeny, that is, the number of virions produced during an acute virus infection. In the case of a latent, persistent viral infection, viral fitness has no equivalent quantitative definition. Whereas for acute virus infections, fitness is measured at the expense of the host (since infected cells die), in the case of latent infections, fitness of the virus merges with fitness of the host and the concept loses its usefulness for describing a differential capacity of the virus.

Even in the case of living organisms, the concept of fitness is of limited value for explaining Darwinian evolution. Natural selection has been defined as the process whereby organisms of a given variety outnumber those of other varieties and prevail in the long run, mainly due to their greater fertility and adaptness to changing environments. In fact, selection amounts only to differential survival and since fitness is defined as anything that promotes the chances of survival, survival of the fittest and natural selection amounts to no more than survival of the survivors.

Because of the small number of phenotypic characters that can be studied with viruses, virus evolution has always been difficult to study. Most studies have concentrated on phylogenetic analyses based on the comparison of genome sequences, with the expectation that genes will diverge in sequence as they evolve from a common ancestor. Like other genomes, viral genomes evolve through such mechanisms as mutation, recombination, and gene reassortment. Molecular analysis of viral evolution has concentrated on measurements of variation in the genotype, without much clarity about what this variation means in terms of survival value for the virus. In many cases, it seems that viruses may in fact have co-evolved with their hosts or vectors, due to the maintenance of host- or vector-restricted molecular constraints on otherwise much higher rates of evolution of the virus itself.

Viral Isolates, Strains, and Serotypes

The term virus isolate refers to any particular virus culture that is being studied and it is thus simply an instance of a given virus.

A viral strain is a biological variant of a virus that is recognizable because it possesses some unique phenotypic properties that remain stable under natural conditions. Characteristics that allow strains to be recognized include (1) biological properties such as a particular disease symptom or a particular host, (2) chemical or antigenic properties, and (3) the genome sequence when it is known to be correlated with a unique phenotypic character. If the only difference between a 'wild type' virus taken as reference and a particular variant is a small difference in genome sequence, such a variant or mutant is not given the status of a separate strain in the absence of a distinct phenotypic characteristic.

Strains that possess unique, stable antigenic properties are called serotypes. Serotypes necessarily also possess unique structural, chemical, and genome sequence properties that are related to the differences in antigenicity. Serotypes constitute stable replicating lineages which allow them to remain distinct over time. The infectivity of individual serotypes of animal viruses can be neutralized only by their own specific antibodies and not by antibodies directed to other serotypes. This inablility of serotype-specific antibodies to cross-neutralize other serotypes is important in the case of animal viruses that are submitted to the immunological pressure of their hosts.

See also: Origin of Viruses; Tobacco Mosaic Virus; Virus Particle Structure: Principles; Virus Species.

Further Reading

Grafe A (1991) *A History of Experimental Virolgy*. Heidelberg: Springer.
Mahner M and Bunge M (1997) *Foundations of Biophilosophy*. Berlin, Springer.
Patthy L (1999) *Protein Evolution*. Oxford: Blackwell Science.
Van Regenmortel MHV (2003) Viruses are real, virus species are man-made, taxonomic constructions. *Archives of Virology* 148: 2481–2488.
Van Regenmortel MHV (2007) The rational design of biological complexity. A deceptive metaphor. *Proteomics* 7: 965–975.
Villarreal LP (2005) *Viruses and the Evolution of Life*. Washington, DC: ASM Press.
Waterson AP and Wilkinson L (1978) *An Introduction to the History of Virology*. Cambridge: Cambridge University Press.
Wilson RA (2005) *Genes and the Agents of Life*. Cambridge: Cambridge University Press.
Witz J (1998) A reappraisal of the contribution of Friedrich Loeffler to the development of the modern concept of virus. *Archives of Virology* 143: 2261–2263.

Necrovirus

L Rubino, Istituto di Virologia Vegetale del CNR, Bari, Italy
G P Martelli, Università degli Studi, Bari, Italy

© 2008 Elsevier Ltd. All rights reserved.

Taxonomy and Classification

As the representative of a monotypic group, *Tobacco necrosis virus A* (TNV-A) was among the 16 groups of plant viruses described in 1971, and became the type species of the genus *Necrovirus* when it was established in 1995. Currently, *Necrovirus* is a genus in the family *Tombusviridae* and comprises seven definitive member species (i.e., *Beet black scorch virus* (BBSV), *Chenopodium necrosis virus* (ChNV), *Leek white stripe virus* (LWSV), *Olive latent virus 1* (OLV-1), *Olive mild mosaic virus* (OMMV), *Tobacco necrosis virus A* (TNV-A), and *Tobacco necrosis virus D* (TNV-D)) and two tentative species (i.e., Carnation yellow stripe virus (CYSN) and Lisianthus necrosis virus (LNV)).

Virion Properties

Necroviruses have very stable particles which resist temperatures in excess of 90 °C. TNV-A virions sediment a single component with a coefficient ($S_{20,w}$) of 118S and have buoyant density of 1.399 g ml^{-1} at equilibrium in CsCl.

Virion Structure and Composition

Virions are approximately 28 nm in diameter, have angular profile and a capsid made up of 60 copies of a trimer consisting of three chemically identical but independent protein subunits (A, B, and C) stabilized by Ca^{2+} ions, arranged in a $T=3$ lattice. Subunit size ranges from 24–27 kDa (BBSV and LWSV, respectively) to 29–30 kDa (other viral species). The capsid has a smooth appearance as protein subunits lack the protruding domain proper of members of the majority of the other genera in the family *Tombusviridae*.

Virions encapsidate a molecule of single-stranded, positive-sense RNA, approximately 3.7 kb in size, constituting *c.* 19% of the particle weight. The genome of five definitive viral species has been fully sequenced: TNV-A, 3684 nt (accession NC001777); TNV-D, 3762 nt (NC003487, U62546); LWSV, 3662 nt (X94660), OLV-1, 3699 nt (X85989); and OMMV, 3683 nt (NC006939).

Genome Organization and Expression

The monopartite genome contains four open reading frames (ORFs) which, in the order from the 5' to the 3' terminus, code for replication-associated proteins (ORF1 and ORF1-RT), movement proteins (ORF2 and ORF3), and the coat protein (CP) (ORF4). Some species possess a fifth ORF either located in the 3' terminal region (TNV-A) or in the middle of the genome (TNV-D^H and BBSV), partially overlapping the C-terminus of ORF1-RT and the N-terminal region of ORF2. The genome is very compact, having noncoding regions of limited size (**Figure 1**).

Genomic RNA acts as messenger for the translation of a protein of 22–24 kDa from ORF1. By translational readthrough of the UAG termination codon of ORF1, a protein of 82–83 kDa is synthesized (**Figure 1**), which contains, in the readthrough portion, the GDD motif of RNA-dependent RNA polymerases (RdRp). This

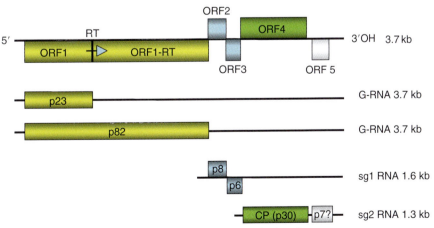

Figure 1 Genome organization and replication strategy of tobacco necrosis virus A.

protein, together with the expression product of ORF1, is indispensable for virus replication.

Genes downstream ORF1-RT are expressed via the synthesis of two subgenomic RNAs, 1.4–1.6 nt and 1.1–1.3 nt in size (**Figure 1**).

ORFs 2 and 3 are two small centrally located ORFs which, depending on the viral species, encode proteins of 7–8 and 6–7 kDa, respectively. Both of these proteins are involved in cell-to-cell transport but are dispensable for viral replication. LWSV ORF2 differs in size from that of other sequenced necroviruses, for it codes for a protein of 11 kDa. The additional central ORFs of TNV-D^H and BBSV code for a 7 kDa and a 5 kDa protein, respectively which, like the products of ORFs 2 and 3, are necessary for cell-to-cell movement.

ORF4 is the CP gene encoding a 24–30 kDa protein, the building block of the capsid, which is required also for efficient systemic virus spreading in the host. No experimental evidence is available of the expression of TNV-A ORF5.

Interspecific Relationships

Necroviruses are serologically distinguishable from one another. Some are unrelated (e.g., LWSV with TNV-A and TNV-D), whereas others share antigenic determinants that result in weak relationships. This is the case of TNV-A and TNV-D that are distantly related with OLV-1 (serological differentiation index from 6 to 9), and of OMMV or BBSV with certain TNV isolates.

The level of molecular similarity varies very much with the viral species. Thus, the RdRp of OMMV, a putative recombinant between TNV-D and OLV-1, shows 91.2% identity at the amino acid level with the comparable protein of OLV-1, whereas the identity level of CP is highest (86.2%) with that of TNV-D. Among the other sequenced species in the genus, the amino acid identity of RdRps and CPs is much lower, ranging from 32% to 34% and 35% to 50%, respectively.

Satellites

TNV-A and TNV-D activate the replication of a satellite virus (TNSV) present in nature as four biologically and serologically distinct strains, supported by different viral isolates. TNSV has isometric particles *c*. 17 nm in diameter, sedimenting as a single component with a coefficient of 50S. Its capsid is constructed with 60 identical subunits *c*. 22 kDa in size, arranged in a $T = 1$ lattice, and contains a single-stranded, positive-sense RNA molecule 1239 nt in size. TNSV RNA accounts for *c*. 20% of the particle weight, and comprises a single ORF encoding the CP.

Whereas satellite viruses are not associated with any other member of the genus, BBSV supports the replication of a small, linear, single-stranded, noncoding satellite RNA 615 nt in size, which is encapsidated in the virions in monomeric or, more rarely, dimeric form. Monomers are thought to be produced from multimeric templates.

TNSV interferes to some extent with helper virus infections for its presence in TNV inocula reduces slightly the virus concentration and the size but not the number of local lesions. By contrast, BBSV satRNA enhances the aggressiveness of the helper virus for more lesions are produced on infected hosts when mixed inocula are used.

Transmission and Host Range

All necroviruses are readily transmitted by mechanical inoculation to experimental herbaceous hosts, which usually react with necrotic local lesions, not followed by systemic infection. Under natural conditions, infection is often restricted to the roots. Some members are transmitted through the soil either by the chitrid fungus *Olpidium brassicae* (TNV-A, TNV-D, BBSV), or without the apparent intervention of a vector (OLV-1). Particles of species transmitted by *O. brassicae* are acquired by the vector from the soil where they are released from roots of infected plants through sloughing off epidermal cell layers and/or following decay of plant debris. Virions are bound tightly to the plasmalemma and the axoneme (flagellum) of the fungal zoospores and are transported inside the zoospore cytoplasm when the flagellum is retracted prior to encystment that precedes penetration into the host. The same mechanism is exploited by TNSV to gain entrance into both vector and host. Since necroviruses and their satellites (TNSV and BBSV satRNA) multiply in the plant cells but not in the fungal plasmodium, zoospores released from infected roots are virus free. Transmission can therefore occur only if they come again in contact with and adsorb virus particles.

Seed transmission of necroviruses has not been reported, except for OLV-1, which was detected in the integuments and internal tissues of 82% olive seeds, and transmitted to 35% of the seedlings.

The natural host range and geographical distribution of necroviruses varies with the species. Thus, TNV-A and TNV-D are ubiquitous and infect a wide range of cultivated and wild plants. OLV-1 was recorded from olive in several Mediterranean countries, citrus in Turkey, and tulip in Japan, OMMV from olive in Portugal, BBSV from beet in China, and LWSV from leek in France.

Virus–Host Relationships

Necroviruses replicate very actively in their hosts, which translates into the production of a large number of virus particles in infected cells of all tissue types, including vessels. Virions are either scattered in the cytoplasm, gathered in bleb-like evaginations of the tonoplast into the vacuole, or arranged in crystalline arrays of various sizes. TNSV particles can also give rise to crystals that can be found in the same cells along with those of the helper virus. Cells infected by TNV and OLV-1 also contain two types of inclusions, that is, clumps of electron-dense amorphous material resembling accumulations of excess CP and fibrous bundles made up of thin filaments with a helical structure. In OLV-1 infected cells, these bundles were identified as accumulations of the 8 kDa movement protein expressed by ORF2. The same protein and the 6 kDa movement protein coded for by ORF3 were detected by immunogold labeling near plasmodesmata. Cytoplasmic clusters of membranous vesicles with fibrillar material, derived from the endoplasmic reticulum or lining the tonoplast, were observed in cells infected by OLV-1 and LWSV.

See also: Carmovirus; Plant Virus Diseases: Ornamental Plants; Tombusviruses.

Further Reading

Castellano MA, Loconsole G, Grieco F, et al. (2005) Subcelluar localization and immunodetection of movement proteins of olive latent virus 1. *Archives of Virology* 150: 1369.

Gua L-H, Cao H-E, Li D-W, et al. (2005) Analysis of nucleotide sequences and multimeric forms of a novel satellite RNA associated with beet black scorch virus. *Journal of Virology* 79: 3664.

Lommel SA, Martelli GP, Rubino L, et al. (2005) Necrovirus. In: Fauquet CM, Mayo MA, Maniloff J, Desselberger U, and Ball LA (eds.) *Virus Taxonomy: Eighth Report of the International Committee on Taxonomy of Viruses*, 926pp. San Diego, CA: Elsevier Academic Press.

Molnar H, Havelda Z, Dalmay T, et al. (1997) Complete nucleotide sequence of tobacco necrosis virus D^H and genes required for RNA replication and virus movement. *Journal of General Virology* 78: 1235.

Oda Y, Saeki K, Takahashu Y, et al. (2000) Crystal structure of tobacco necrosis virus at 2.25 Å resolution. *Journal of Molecular Biology* 300: 153.

Russo M, Burgyan J, and Martelli GP (1994) Molecular biology of Tombusviridae. *Advances in Virus Research* 44: 381.

Uyemoto JK (1981) Tobacco necrosis and satellite viruses. In: Kurstak E (ed.) *Handbook of Plant Virus Infections and Comparative Diagnosis*, 123pp. Amsterdam, The Netherlands: Elsevier/North Holland Biochemical Press.

Nepovirus

H Sanfaçon, Pacific Agri-Food Research Centre, Summerland, BC, Canada

Crown Copyright © 2008 Published by Elsevier Ltd. All rights reserved.

Glossary

Odonstyle Anterior section of the stylet of a longidorid nematode. The odonstyle consists of a needle-like mouth spear used to penetrate plant root cells.

Odontophore Posterior section of the stylet of a longidorid nematode. The odontophore is located at the base of the nematode mouth. Stylet protactor muscles are attached to the ondotophore.

Triradiate lumen Posterior region of the lumen (or food channel) of the esophagus of a longidorid nematode. The triradiate lumen is located in a muscular bulb at the base of the esophagus. The radial muscles attached to the triradiate lumen are used for food ingestion.

Introduction and Historical Perspective

The genus *Nepovirus* was among the first 16 groups of viruses recognized by the International Committee on Taxonomy of Viruses (ICTV). The name stands for nematode-transmitted viruses with polyhedral particles. Although nematode transmission was one of the original defining characteristics of the genus, the primary criteria for inclusion of viruses in the genus are now the structure of the RNA genome and of the particles. As a result, not all viruses belonging to the genus *Nepovirus* are nematode transmitted (**Table 1**), and some nematode-transmitted polyhedral viruses have been reclassified in distinct genera (e.g., strawberry latent ringspot virus is now considered a tentative member of genus *Sadwavirus*).

Taxonomy and Relation to Other Viruses

The genus *Nepovirus*, along with the genera *Comovirus* and *Fabavirus*, belongs to the family *Comoviridae*. Nepoviruses are also related to the unassigned genera *Sadwavirus* and *Cheravirus* which include viruses previously considered as tentative nepoviruses. Common characteristics among members of these five genera include: small polyhedral particles, a bipartite positive-strand RNA genome, and a conserved arrangement of protein domains within the polyproteins encoded by RNA1 and -2. Nepoviruses are distinguished from viruses of the other four genera by their single large coat protein (CP). This classification is supported by phylogenetic comparisons of the RNA

Table 1 Some properties of nepoviruses

Virus name	Abbreviation	Vector[a]	RNA1	RNA2	Satellite RNAs Type B[b]	Type D
Subgroup A (RNA2 M_r 1.3–1.5 × 10^6)						
Arabis mosaic virus	ArMV	*Xiphinema diversicaudatum*	NC006057	NC006056	NC003523	NC001546
Arracacha virus A	AVA					
Artichoke Aegean ringspot virus	AARSV					
Cassava American latent virus	CsALV					
Grapevine fanleaf virus	GFLV	*X. index*; *X. italiae*[c]	NC003615	NC003623	NC003203	
Potato black ringspot virus	PBRSV		AJ616715			
Raspberry ringspot virus	RpRSV	*Longidorus elongatus*; *L. macrosoma*; *Paralongidorus maximus*	NC005266	NC005267		
Tobacco ringspot virus	TRSV	*X. americanum*	NC005097	NC005096		NC003889
Subgroup B (RNA2 M_r 1.4–1.6 × 10^6)						
Artichoke Italian latent virus	AILV	*L. apulus*; *L. fasciatus*		X87254		
Beet ringspot virus[d]	BRSV	*L. elongatus*	NC003693	NC003694		
Cocoa necrosis virus	CNV					
Crimson clover latent virus	CCLV					
Cycas necrotic stunt virus	CNSV		NC003791	NC003792		
Grapevine chrome mosaic virus	GCMV		NC003622	NC003621		
Mulberry ringspot virus	MRSV	*L. martini*			+	
Myrobalan latent ringspot virus	MLRSV					
Olive latent ringspot virus	OLRSV			AJ277435		
Tomato black ring virus	TBRV	*L. attenuatus*	NC004439	NC004440	NC003890	
Subgroup C (RNA2 M_r 1.9–2.2 × 10^6)						
Apricot latent ringspot virus	ALRSV			AJ278875		
Artichoke yellow ringspot virus	AYRSV					
Blackcurrant reversion virus	BRV	*Cecidophyopsis ribis* (mite)	NC003509	NC003502	NC003872	
Blueberry leaf mottle virus	BLMoV		U20622	U20621		
Cassava green mottle virus	CsGMV					
Cherry leaf roll virus	CLRV		Z34265	U24694		
Chicory yellow mottle virus	ChYMV				NC006452	NC006453
Grapevine Bulgarian latent virus	GBLV				+	
Grapevine Tunisian ringspot virus	GTRSV					
Hibiscus latent ringspot virus	HLRSV					
Lucerne Australian latent virus	LALV					
Peach rosette mosaic virus	PRMV	*X. americanum*; *L. diadecturus*[e]	AF016626			
Potato virus U	PVU					
Tomato ringspot virus	ToRSV	*X. americanum*; *X. bricolensis*; *X. californicum*; *X. rivesi*	NC003840	NC003839		

[a]Only cases for which a specific association with a nematode vector has been verified experimentally are reported here. Nonvalidated associations with nematode vectors have been reported for the following viruses: CLRV – *Xiphinema* sp., GCMV – *X. index*, PBRSV – *X. americanum*, PVU – *Longidorus* sp.

[b]The sequence accession number of satellite RNAs is given when known. In the case of MLRSV and GBLV, the + symbol indicates that a type B satellite is known to be associated with the virus but has not been sequenced.

[c]Transmission of GFLV by *X. italiae* has been reported in only one instance. Many other populations of *X. italiae* did not transmit the virus. Thus, *X. index* is considered the main vector of GFLV.

[d]BRSV was previously known as TBRV-S or TBRV-Scottish.

[e]Transmission of PRMV by *L. diadecturus* was reported for only one location in spite of the widespread distribution of the vector in North America. Thus, *X. americanum* is considered the main vector of PRMV.

genomes. Nepoviruses are similar to members of the family *Picornaviridae* in that they share the same modular arrangement of replication proteins on the polyproteins, conserved motifs in these replication proteins, and a similar capsid structure. Nepoviruses have been divided into three subgroups based on the length and packaging of RNA2, sequence similarities, and serological properties (subgroups A, B, and C; see **Table 1**).

Virus Particle Structure

Nepoviruses have isometric particles of 26–30 nm in diameter with sharp hexagonal outlines (**Figure 1(a)**). Equilibrium centrifugation in CsCl of purified virus particles typically reveals the presence of three types of particles. T-particles (top component) sediment at 50S and do not contain an RNA molecule. In electron microscopy, these empty particles are penetrated by a negative stain. B-particles (bottom component) sediment at 115–134S and contain a single molecule of RNA1. In the case of subgroup A nepoviruses, B-particles can also contain two molecules of RNA2. M-particles (middle component) sediment at 86–128S and contain a single molecule of RNA2. M- and B-particles of subgroup C nepoviruses are often difficult to separate due to the larger size of RNA2.

Nepovirus particles contain 60 molecules of a single CP with an M_r of $(53-60) \times 10^3$. Tobacco ringspot virus (TRSV) (see **Table 1** for abbreviations) is the only nepovirus for which the atomic structure of the virus particle has been solved. The pseudo $T=3$ structure was found to be very similar to that of comoviruses and picornaviruses. The single CP contains three functional domains with one β-barrel each. Each functional domain corresponds to one of the three smaller CPs found in picornaviruses. Comoviruses also share a similar structure with one large CP containing two β-barrels and one small CP with a single β-barrel. It was suggested that the CP(s) of nepoviruses, comoviruses, and picornaviruses have evolved from a common ancestor.

Genome Structure

The two RNA molecules of nepoviruses are polyadenylated at the 3' end and are covalently linked to a small viral protein (VPg) at the 5' end. Each RNA codes for one large polyprotein which is cleaved by a viral

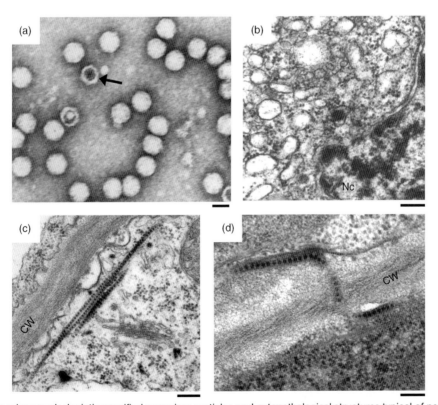

Figure 1 Electron micrograph depicting purified nepovirus particles and cytopathological structures typical of nepovirus-infected cells. (a) Purified ToRSV particles in negative staining. Note the empty particle (T-particle) which is penetrated by the negative stain (arrow). (b) Proliferation of membrane vesicles observed in the vicinity of the nucleus (Nc) in ToRSV-infected cells. (c) Tubular structures containing virus-like particles accumulating near the cell wall (CW) in PRMV-infected cells. (d) Tubular structure traversing the cell wall in ArMV-infected cells. Scale = 25 nm (a), 200 nm (b–d).

proteinase (Pro). RNA1 encodes replication proteins and can replicate independently of RNA2. RNA1 and -2 are both required for cell-to-cell movement of the virus. The identification of conserved sequence motifs and the characterization of cleavage sites recognized by the viral Pro have led to the definition of protein domains within the polyproteins. The genomic organization of a representative virus from each subgroup is shown in **Figure 2**. The RNA2-encoded polyprotein includes the domains for the CP and movement protein (MP) at its C-terminus as well as one (for subgroup A and B nepoviruses) or two (for ToRSV, a subgroup C nepovirus) additional protein domains at its N-terminus. The C-terminal portion of the RNA1-encoded polyprotein contains the domains for the putative helicase (also termed NTB protein because it contains a conserved nucleoside triphosphate-binding sequence motif), VPg, Pro, and RNA-dependent RNA polymerase (Pol). In the case of ToRSV, characterization of cleavage sites recognized *in vitro* by the viral Pro has resulted in the identification of two protein domains upstream of the NTB domain: X1 of unknown function and X2. X2 is a highly hydrophobic protein that shares conserved sequence motifs with the RNA1-encoded 32 kDa protein of comoviruses. An equivalent domain is present in the N-terminal region of the RNA1-encoded polyprotein of subgroup A and B nepoviruses (shown in orange in **Figure 2**). Further characterization of cleavage sites will be necessary to determine whether this region constitutes an independent protein domain in these viruses or whether it is included as part of a larger NTB protein (in the case of GFLV) or 1a protein (in the case of BRSV).

In addition to the coding region, each RNA includes untranslated regions (UTRs) at its 5′ and 3′ ends. The 5′ UTR is 70–300 nt long while the 3′ UTR varies in length from 200–400 nt for subgroup A and B nepoviruses to 1300–1600 nt for subgroup C nepoviruses. A short conserved sequence is present at the immediate 5′ end of the RNAs of many but not all nepoviruses. In addition, conserved structural features have been identified in the 5′ UTRs of subgroup A and B nepovirus RNAs (i.e., a series of stem and loop structures). The presence of conserved sequences and/or structural motifs within the UTRs points to a possible role for these elements in the viral replication cycle. However, this needs to be confirmed experimentally. In addition to these short conserved motifs, RNA1 and -2 often share regions of complete or partial sequence identity in the UTRs (**Figure 2**). In subgroup B and C nepoviruses, the 3′ UTRs are identical between RNA1 and -2 while the 5′ UTRs share 68–100% sequence identity. For example, the 5′ UTRs of ToRSV RNA1 and -2 share a region of 100% sequence identity which extends into the coding region. On the other hand, the 5′ UTRs of the BRV RNAs share only 78% sequence identity and do not extend beyond the UTR. In subgroup A nepoviruses, the 5′ and 3′ UTRs of RNA1 and -2 share homology but are not identical (70–97% sequence identity). It was suggested that the extensive regions of sequence identity detected in the 5′ and 3′ ends of ToRSV RNAs are the result of recombination events occurring during replication of the viral RNAs. Experimental support for this suggestion was provided using pseudorecombinants consisting of GCMV RNA1 and BRSV RNA2. Sequencing of the viral progeny revealed that the 3′ UTR of GCMV RNA1 was transferred to BRSV RNA2 after three passages. In contrast, recombination was not readily observed between the 3′ UTRs of BLMV RNA1 and -2. It was suggested that selection rather than recombination played a role in the conservation of sequence identity in BLMV RNAs.

So far, the production of infectious cDNA clones has only been reported for subgroup A nepoviruses (GFLV, ArMV, RpRSV). Thus, reverse genetics is only possible

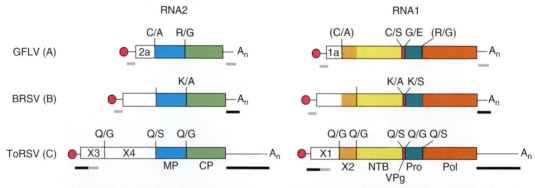

Figure 2 Genomic organization of representative nepoviruses of subgroup A (GFLV), B (BRSV), and C (ToRSV). Each RNA is represented with the covalently attached VPg (pink circle) and the polyA tail (A$_n$). The coding regions are represented by the boxes. Cleavage sites confirmed by *in vitro* processing experiments or by the detection of viral proteins in infected plants are indicated by the continuous vertical lines. Identified or putative (in parentheses) cleavage sites are indicated above each line when known. The function of each protein domain is indicated at the bottom of the figure. Thick bars below each RNA indicate regions with high degree of sequence identity between RNA1 and -2. The black portions indicate regions with 100% sequence identity and the gray portions represent regions with 75–83% sequence identity.

for these viruses. Removal of the covalently linked VPg from purified viral RNAs either decreased (ArMV, RpRSV) or abolished (TRSV, ToRSV, BRSV) their infectivity, although it did not affect the ability of these RNAs to be translated *in vitro*. Possibly, the requirement for an intact VPg is an obstacle for the production of infectious clones for some nepoviruses.

Regulated Polyprotein Processing

Nepoviruses encode a single proteinase which is responsible for processing the two polyproteins. Nepovirus proteinases are related to the 3C-Pro of picornaviruses. The catalytic triad consists of a histidine, aspartic acid, and cysteine. The proteinase also contains a substrate-binding pocket which determines its cleavage site specificity. A conserved histidine is found in the substrate-binding pocket of subgroup C nepovirus, comovirus, and picornavirus proteinases. The cleavage sites recognized by subgroup C nepovirus proteinases contain a glutamine, asparagine, or aspartate at the −1 position (**Table 2**). This is similar to the conserved glutamine or glutamate found at the −1 position of picornavirus cleavage sites. In contrast, the proteinases of subgroup A and B nepoviruses contain a leucine instead of a histidine in their substrate-binding pocket and recognize very different cleavage sites that have a lysine, cysteine, arginine, or glycine at the −1 position.

The RNA1-encoded polyprotein is cleaved predominantly intramolecularly (*cis*-cleavage) although intermolecular processing (*trans*-cleavage) of the N-terminal cleavage site has been reported for GFLV. The RNA2-encoded polyprotein is cleaved *in trans* by the proteinase. The processing cascade results in the release of mature proteins as well as stable processing intermediates containing two or more protein domains. These intermediate polyproteins accumulate in infected plant cells and may have different activities from the mature proteins. In ToRSV-infected cells, several polyprotein intermediates containing the NTB domain are detected in addition to the mature NTB protein. In BRSV-infected cells, an intermediate containing the Pro and Pol domains accumulates rather than the mature Pro and Pol proteins. This suggests preferential recognition of some cleavage sites by the viral Pro. Slow release of mature proteins by processing of stable intermediates at suboptimal cleavage sites may provide a regulatory mechanism to control the accumulation of specific protein species during the replication cycle. In fact, the activity of the proteinase itself is regulated by its release from larger polyprotein precursors. Indeed, the mature proteinase of GFLV and ToRSV cleaves cleavage sites on the RNA2-encoded polyprotein more efficiently than the VPg-Pro precursor.

Viral RNA Replication

Infection of plant cells by nepoviruses results in membrane proliferation and the formation of cytoplasmic inclusion bodies which contain membrane vesicles (**Figure 1(b)**). The use of cerulenin, an inhibitor of *de novo* phospholipid synthesis, has confirmed that membrane proliferation is required for the replication of GFLV RNAs. Brefeldin A also inhibits GFLV replication suggesting a requirement for intact vesicle trafficking between the endoplasmic reticulum (ER) and the Golgi apparatus. The replication complex of two nepoviruses (GFLV and ToRSV) has been shown to co-localize with ER-derived membranes in infected cells. Double-stranded RNA replication intermediates, viral replication proteins, and replication activity are associated with the membrane-bound complexes. GFLV VPg antibodies were used to isolate membrane vesicles that had a rosette-like structure similar to that associated with picornavirus replication complexes.

In the case of ToRSV, the mature NTB protein or a larger intermediate polyprotein containing the NTB domain has been proposed to play a role in anchoring the replication complex to the membranes. The mature NTB and the intermediate NTB-VPg polyprotein are integral membrane proteins that co-fractionate with the replication complex. The NTB protein is targeted to ER membranes when expressed individually and contain two membrane-binding domains: a C-terminal transmembrane domain and a putative N-terminal amphipathic helix. The ToRSV X2 protein is also an ER-targeted integral membrane protein and may play a role in viral replication although its association with the replication complex in infected cells remains to be confirmed. Other replication proteins (Pro and Pol) are soluble when expressed individually but are found in association with

Table 2 Identified cleavage sites in the polyprotein of nepoviruses

Virus	Cleavage site
Subgroup A	
ArMV	R/G
GFLV	R/G, C/A, C/S, G/E
RpRSV	C/A
TRSV	C/A
Subgroup B	
BRSV	K/A, K/S
GCMV	R/A
OLRSV	K/A
CNSV	K/S
Subgroup C	
BRV	D/S
BLMoV	N/S
CLRV	Q/S
ToRSV	Q/G, Q/S

the membrane-bound replication complex in infected cells. Thus, they are probably brought to the replication complex either as part of a larger polyprotein that includes the NTB domain or through protein–protein interaction with the viral membrane anchors. By analogy with picornaviruses, the VPg protein may act as a primer for viral replication, although this has not been demonstrated experimentally for nepoviruses. In the context of the NTB-VPg polyprotein, the VPg domain is translocated in the lumen of the membranes. Because replication presumably takes place on the cytoplasmic side of the membranes, it is unlikely that the luminally oriented VPg present in the NTB-VPg protein plays an active role in replication. Therefore, other intermediate precursors containing the VPg domain (e.g., VPg-Pro or VPg-Pro-Pol) probably act as donors for a replication-active VPg protein. In addition to RNA1-encoded replication proteins, the GFLV RNA2-encoded 2a protein is also associated with the replication complex. The 2a protein is required for the replication of RNA2 but not RNA1 and probably interacts with RNA1-encoded replication proteins.

Cell-to-Cell and Long-Distance Movement in the Plant

Nepovirus-infected cells are characterized by the presence of tubular structures containing virus-like particles in or near the cell wall (**Figures 1(c)** and **1(d)**). These tubules are similar to the ones found in comovirus-infected cells and have been suggested to direct the cell-to-cell movement of intact virus particles. The viral MP is a structural component of the tubular structure. Expression of the GFLV MP alone is sufficient to induce the formation of empty tubular structures in intact plant cells or protoplasts. In the latter case, tubular extensions are found projecting from the surface of the protoplasts, a phenomenon also induced by comovirus MP. The GFLV MP is an integral membrane protein. The secretory pathway and the cytoskeleton are involved in the intracellular targeting of the GFLV MP from its site of synthesis (probably in association with the ER-bound replication complex) to specific foci in the cell wall where it assembles into tubules. The presence of specific sites of tubule formation within the cell wall suggests an interaction between MP and a cellular receptor but this has not been confirmed experimentally. By analogy with como-viruses, it is likely that an interaction between MP and CP is necessary to enable nepovirus cell-to-cell movement. Long-distance movement of TRSV occurs through the phloem resulting in the invasion of most tissues of the plant including meristematic tissues. The virus probably reaches the phloem through cell-to-cell movement from inoculated cells to phloem sieve tubes.

Host Range, Symptomatology, and Interaction of Nepoviruses with the Plant Post-Transcriptional Gene Silencing Pathway

Most nepoviruses have a wide host range that includes woody and herbaceous hosts. In nature, many hosts remain symptomless while others display symptoms such as necrotic or chlorotic rings which can appear as concentric rings (ringspots) or lines. Other symptoms can include leaf flecking and mottling, vein necrosis, plant stunting, and in some cases death. Common experimental hosts used for virus propagation are *Chenopodium quinoa* (in which most nepoviruses induce obvious symptoms), *C. amaranticolor*, *C. murale*, *Cucumis sativus*, *Nicotiana clevelandii*, *N. benthamiana*, *Petunia hybrida*, and *Phaseolus vulgaris*. The intensity of symptoms produced by nepovirus infection depends on the specific virus–host combination and to a large extent on environmental conditions. In many herbaceous hosts and in particular in *Nicotiana* species, symptoms develop on the inoculated leaves and on the first upper systemic leaves. Later in infection, new leaves remain free of symptoms although the virus is present. This phenomenon is termed recovery. Recovered leaves often contain somewhat reduced titer of the virus compared to symptomatic leaves and are resistant to secondary viral infection in a sequence-specific manner. This suggests that induction of the plant RNA silencing machinery plays a role in virus clearance and symptom recovery. It was recently shown that although recovery of *N. benthamiana* from necrotic symptoms induced by ToRSV is accompanied with induction of RNA silencing, the virus titer is not significantly reduced in recovered leaves. Thus, the relationship between symptom recovery and RNA silencing may be more complex than first envisaged. While many plant viruses encode potent suppressors of RNA silencing, analysis of two nepoviruses (TBRV and ToRSV) did not reveal significant silencing suppression activity.

Satellites

Two classes of satellite RNAs (satRNAs) have been found in association with some but not all nepoviruses (**Table 1**). SatRNAs depend on the helper virus for their replication and are encapsidated in virus particles. Type B satRNAs are 1100–1500 nt in length. They are linked to a VPg molecule at their 5′ end, polyadenylated at their 3′ end, and encode a nonstructural protein which is essential for their replication. The exact function of the encoded protein is not known but it has been suggested that it interacts with the viral replication complex. Sequences at the 5′ and 3′ ends of type B satRNAs are also important for their replication. The 5′ ends of type B satRNAs often have short sequence motifs that are identical or nearly identical to the 5' ends of the viral RNAs and are likely recognized by the viral replication complex. One or several copies of type B satRNAs are packaged in the viral particles, either alone or together with one molecule of RNA2. As a general rule, type B satRNAs are replicated specifically by the virus with which they are associated although there are exceptions (e.g., the replication of GFLV satRNA is supported by ArMV). Type B satRNAs are usually found in low concentration in the field or in experimental systems and do not seem to affect significantly the replication of the helper virus or the symptomatology of the disease. For example, type B satRNAs have been found to be associated with only 15–17% of ArMV or GFLV isolates.

Type D satRNAs are less than 500 nt long. They are not linked to a VPg molecule or polyadenylated and do not encode a protein. Type D satRNAs are encapsidated as monomeric or multimeric linear molecules. A circular form of the molecule is present in infected cells and serves as the polymerase template for replication through a rolling-circle mechanism. The multimeric linear forms produced during replication are cleaved in an autocatalytic reaction which allows the release of linear monomers. These monomers are circularized to form new templates of positive or negative polarity. Type D satRNAs have been shown to either attenuate (TRSV) or intensify (ArMV) symptoms associated with the helper virus.

Diseases and Economic Considerations

Nepoviruses cause a wide range of diseases on a variety of crops including: grapevine (ArMV, GBLV, GCMV, GFLV, GTRSV, RpRSV, TBRV, ToRSV, TRSV), soft fruits such as strawberry, raspberry, blueberry, black currant, and red currant (ArMV, BLMV, BRV, CLRV, RpRSV, TBRV, ToRSV, TRSV), fruit trees such as peach, apricot, almond, cherry, plum, walnut, and apple (CLRV, MLRSV, PRMV, RpRSV, ToRSV), and horticultural crops including but not limited to hop, soybean, potato, beet, and tobacco (AILV, ArMV, AVA, BRSV, CGMV, PBRSV, and TRSV). Many nepoviruses also infect and induce diseases in ornamental species. Most nepoviruses are restricted geographically by the natural distribution of their nematode vector. An exception to this is GFLV which has been disseminated worldwide along with its vector. GFLV is the most significant nepovirus at the economic level and can reduce yield in grapevine by as much as 80%. Other nepoviruses can cause significant diseases where they occur.

Transmission

Many nepoviruses are transmitted by soil-inhabiting nematodes belonging to three closely related genera *Xiphinema*, *Longidorus*, or *Paralongidorus* in the order Dorylaimida, family Londigoridae (**Table 1**). The nematodes feed ectoparasitically on the roots using long mouth stylets. There is no evidence that nepoviruses replicate in the nematodes. The acquired viruses remain transmissible for varying periods of time (9 weeks for species *Longidorus* and up to 4 years for species *Xiphinema*), suggesting different modes of retention and release of the virus. Because of the restricted movement of the nematodes through the soil, the spread of nematode-transmitted nepoviruses through an infected field is slow and often occurs in patches. The interaction between nepoviruses and nematodes is usually specific with only one or two species of nematodes transmitting a given nepovirus. A notable exception is RpRSV which is transmitted by nematodes from the genera *Longidorus* and *Paralongidorus*. The viral determinant for the specificity of nematode transmission has been mapped to the CP in the case of GFLV. It is not known whether specific nematode receptors recognize the viral CP. Earlier studies suggested that carbohydrates may be involved in the retention of ArMV particles in its vector. However, further experiments will be required to determine if the viral CP has lectin properties. It has been suggested that pH changes may be involved in the release of viral particles from their site of retention within the nematode. Nepoviruses transmitted by *Longidorus* sp. are usually associated with the odontostyle, while viruses transmitted by *Xiphinema* sp. are found associated with the cuticle lining the lumen of the odontophore and the esophagus. Interestingly, while TRSV and ToRSV are both transmitted by *X. americanum*, immunofluorescence labeling of these viruses in the nematode vector revealed different sites of retention. TRSV is retained predominantly in the lining of the lumen of the stylet extension and the anterior esophagus, while ToRSV is localized only in the triradiate lumen.

Although the predominant vector for transmission of TRSV is a nematode, possible aerial vectors have been suggested including *Thrips tabaci* and *Epitrix hirtipennis* (flea beetle). The importance of these vectors in natural epidemics of TRSV-induced diseases needs to be confirmed. Other nematode-transmitted nepoviruses do not have known aerial vectors. BRV is transmitted by the eriophyid gall mite of black currant (*Cecidophyopsis ribis*) and possibly other *Cecidophyopsis* species but not by nematodes. Plant-to-plant transmission can occur rapidly (in only 4 h). Virus particles have not been found inside mites, suggesting that the transmission may be nonpersistent or semipersistent. Two surface-exposed amino acid triplets are conserved between the CP of BRV and other mite-transmitted viruses from unrelated genera, suggesting that they may play a role in the interaction between the virus and its vector. However, this remains to be determined experimentally.

Seed transmission has been reported for most but not all nepoviruses. For example, BRV is apparently not seed transmitted. Infection can occur through the ovule or the pollen, although nepovirus-infected pollen may not compete effectively with healthy pollen. An exception to this is BLMoV and CLRV which are efficiently transmitted by pollen, even to the mother plant. Many nepoviruses are readily transmitted through grafting and mechanical inoculation. The propagation of infected seed and plant stocks plays an important role in the long-distance movement of nepoviruses. Because many nepoviruses have a wide host range including many common weeds, dormant weed seeds may constitute an important reservoir for the virus in the field.

Population Structures

Examination of nepovirus isolates recovered from different hosts or geographic locations has revealed a degree of sequence diversity of 2–20% at the nucleotide level. Isolates may also differ in their serological properties. Several nepovirus isolates have arisen through recombination implying that mixed infections occur in nature. A detailed analysis of GFLV population structure in an infected vineyard demonstrated the presence of mixed infections and a high degree of recombination between the various isolates. Evidence for mixed infections was also provided from the analysis of an ArMV isolate which revealed the presence of two species of RNA2, each encoding a distinct polyprotein. Analysis of a number of CLRV isolates revealed that the degree of diversity was defined primarily by the host rather than the geographic location. This is an unusual situation among plant viruses and may reflect the fact that this virus is pollen transmitted.

Control

Nepoviruses are mainly controlled through the removal of infected plants and replanting with resistant cultivars (when available) or with virus-free, certified plant material. In the case of nematode-transmitted nepoviruses, soil can be fumigated with broad-range nematicides. However, this method is not always effective because nematode populations can occur at considerable depths in the soil (1 m or more). Further, nematicides are costly and toxic to the environment. Bait plants can be used to test for the presence of viruliferous nematodes in the soil. Careful weed control is recommended to eliminate potential reservoirs for further nepovirus infection. Since few resistant cultivars have been reported for nepoviruses, the usefulness of transgenic approaches has been investigated. Resistance to various nepoviruses has been engineered in herbaceous hosts using the coding region for the viral CP in the sense or anti-sense orientation. Other regions of the viral genome have also been used to engineer resistance to nepoviruses through the induction of RNA silencing. Resistance to GFLV has been reported in transgenic grapevines transformed with the CP coding region. The resistance was effective in a field situation heavily infected with viruliferous nematodes. It is of interest to note that the presence of susceptible and resistant GFLV-CP transgenic grapevine in the field did not increase the occurrence of recombination events in the virus population. It is likely that the next generation of transgenic lines will be aimed at increasing the efficiency of induction of sequence-specific RNA silencing using transgenes that contain only very small portions of the viral genome.

Nepovirus Research in the Future

The next phase of nepovirus research will undoubtedly address several fundamental questions. First, the function and mode of action of nepovirus proteins, in particular protein domains in the N-terminal region of the two polyproteins, require further investigation. Although a putative function has been assigned in some cases (e.g., the GFLV 2a protein was shown to play a role in RNA2 replication), in other cases the role of the protein in the virus replication cycle is unknown (e.g., ToRSV X1, X3, and X4). Second, the role played by host factors in the translation, replication, and cell-to-cell (or systemic) movement of nepoviruses needs to be characterized. Large-scale studies of the interaction of the plant and virus proteomes will be necessary to address this question. In addition, it will be useful to analyze the ability of nepoviruses to infect collections of plants mutated or silenced for the expression of specific genes. Third, further work is necessary to understand the specificity of nepovirus–vector interactions. These questions are not only important to satisfy scientific curiosity, but also for the rational design of alternative approaches for the control of nepoviruses.

See also: Picornaviruses: Molecular Biology; Plant Virus Diseases: Fruit Trees and Grapevine; Plant Virus Diseases: Ornamental Plants; Sadwavirus; Satellite Nucleic Acids and Viruses.

Further Reading

Andret-Link P, Schmitt-Keichinger C, Demangeat G, Komar V, and Fuchs M (2004) The specific transmission of grapevine fanleaf virus by its nematode vector *Xiphinema index* is solely determined by the viral coat protein. *Virology* 320: 12–22.

Brown JF, Trudgill DL, and Robertson WM (1996) Nepoviruses: Transmission by nematodes. In: Harrison BD and Murant AF (eds.) *The Plant Viruses, Vol. 5: Polyhedral Virions and Bipartite RNA Genomes*, pp. 187–209. New York: Plenum.

Chandrasekar V and Johnson JE (1998) The structure of tobacco ringspot virus: A link in the evolution of icosahedral capsids in the picornavirus superfamily. *Current Biology* 6: 157–171.

Gaire F, Schmitt C, Stussi-Garaud C, Pinck L, and Ritzenthaler C (1999) Protein 2 A of grapevine fanleaf nepovirus is implicated in RNA2 replication and colocalizes to the replication site. *Virology* 264: 25–36.

Harrison BD and Murant AF (1996) Nepoviruses: Ecology and control. In: Harrison BD and Murant AF (eds.) *The Plant Viruses, Vol. 5: Polyhedral Virions and Bipartite RNA Genomes*, pp. 211–228. New York: Plenum.

Jovel J, Walker M, and Sanfacon H (2007) Recovery of *Nicotiana benthamiana* plants from a necrotic response induced by a nepovirus is associated with RNA silencing but not with reduced virus titer. *Journal of Virology* 81: 12285–12297.

Laporte C, Bettler G, Loudes AM, et al. (2003) Involvement of the secretory pathway and the cytoskeleton in intracellular targeting and tubule assembly of grapevine fanleaf virus movement protein in tobacco BY-2 cells. *Plant Cell* 15: 2058–2075.

Le Gall O, Iwanami T, Karasev AV, et al. (2005) Family *Comoviridae*. In: Fauquet CM, Mayo MA, Maniloff J, Desselberger U and Ball LA (eds.) *Virus Taxonomy: Eighth Report of the International Committee on Taxonomy of Viruses*, pp. 807–818. San Diego, CA: Elsevier Academic Press.

Mayo MA and Robinson DJ (1996) Nepoviruses: Molecular biology and replication. In: Harrison BD and Murant AF (eds.) *The Plant Viruses Vol. 5: Polyhedral Virions and Bipartite RNA Genomes*, pp. 139–185. New York: Plenum.

Murant AF, Jones AT, Martelli GP, and Stace-Smith R (1996) Nepoviruses: General properties, diseases and virus identification. In: Harrison BD and Murant AF (eds.) *The Plant Viruses, Vol. 5: Polyhedral Virions and Bipartite RNA Genomes*, pp. 99–137. New York: Plenum.

Rebenstorf K, Candresse T, Dulucq MJ, Buttner C, and Obermeier C (2006) Host species-dependent population structure of a pollen-borne plant virus, cherry leaf roll virus. *Journal of Virology* 80: 2453–2463.

Ritzenthaler C, Laporte C, Gaire G, et al. (2002) Grapevine fanleaf virus replication occurs on endoplasmic reticulum-derived membranes. *Journal of Virology* 76: 8808–8819.

Susi P (2004) Black currant reversion virus, a mite-transmitted nepovirus. *Molecular Plant Pathology* 5: 167–173.

Vigne E, Bergdoll M, Guyader S, and Fuchs M (2004) Population structure and genetic variability within isolates of grapevine fanleaf virus from a naturally infected vineyard in France: Evidence for mixed infection and recombination. *Journal of General Virology* 85: 2435–2445.

Zhang SC, Zhang G, Yang L, Chisholm J, and Sanfacon H (2005) Evidence that insertion of tomato ringspot nepovirus NTB-VPg protein in endoplasmic reticulum membranes is directed by two domains: A C-terminal transmembrane helix and an N-terminal amphipathic helix. *Journal of Virology* 79: 11752–11765.

Neutralization of Infectivity

P J Klasse, Cornell University, New York, NY, USA

© 2008 Elsevier Ltd. All rights reserved.

Glossary

Affinity The strength of an intermolecular interaction, for example, of antibody (Ab) binding to antigen (Ag).

Dissociation constant K_d [M], in the law of mass action for a bimolecular association, it corresponds to the concentration of an unbound molecule that yields half-maximal binding to the other molecule. It quantifies affinity: the lower the K_d, the higher the affinity.

Epitope The site on an antigen that makes contact with the paratope of an Ab.

Molecularity The number of molecules involved in the rate-limiting step of a chemical reaction; here it refers to the number of Abs that must bind to a virus particle in order to neutralize it.

Occupancy θ, the degree to which one molecule is ligated by another, for example, the percentage of Ag sites bound by Ab.

Paratope The part of an Ab molecule that makes contact with the epitope of an Ag.

Stoichiometry In the context of neutralization, it is the study of integer ratios of molecules in Ab–virion complexes.

Valency The number of separate, usually identical, areas of contact in bimolecular association.

Introduction

Miscellaneous agents inhibit viral infection. Among them are antibodies (Abs), which have disparate effects on viral infection both *in vivo* and *in vitro*. One such effect is virus neutralization, which can be stringently defined as the reduction in infectivity by interference before the first biosynthetic step in the viral replicative cycle, through the binding of antibodies to epitopes on the surface of the virion. The definition delimits neutralization from other effects of Abs on viral replication and on infected cells. For example, Abs to neuraminidase of influenza virus inhibit viral release from the cell surface, a late step in the replicative cycle. That does not qualify as neutralization. Another example is Abs to viral receptors. Such Abs can block infection, but they do not fulfill the criterion of binding to the virion.

Maybe the definition too restrictively specifies Abs as the neutralizing agents. Obviously, some loosening of this criterion will allow a comparison with the neutralizing effects of fragments of Abs, peptide mimics of paratopes, fragments of receptors and their mimics, or even other ligands, as long as they bind to the viral surface.

What the definition implicitly does include is also noteworthy. Enveloped viruses carry cellular passenger antigens (Ags). If Abs to these inhibit infection, that counts as neutralization. Under certain circumstances, Abs aggregate virions. If that diminishes infectivity, the definition includes it as neutralization – although some researchers have chosen not to. Similarly, the definition can accommodate for effects of complement, in addition to that of the Abs, as enhancement of neutralization. Whether or how such enhancement occurs is a matter for investigation. As a case in point, a report found neutralization through virolysis to be more complement-enhanced when mediated by Abs to a retroviral transmembrane protein than by Abs to the surface unit of the same envelope glycoprotein complex. Notably, the definition stipulates little about mechanism. That is left to hypotheses and their testing by experiment.

A closer look at the defining terms 'reduction', 'first biosynthetic step', and 'binding' raises some questions. First, does any degree of 'reduction' count as neutralization? Does neutralization entail the complete abrogation of the infectivity of at least some virions? Or can it be partial, a mere reduction in the propensity to infect? Second, the 'first biosynthetic step' occurs after viral entry. Why does the definition put the limit at so late a step? How could Abs affect events occurring after the viral genome or core has entered the cytoplasm from the cell surface or from an endosomal vesicle? Controversial cases of postentry neutralization have been promulgated. Therefore, the definition should not exclude them; they must be refuted or corroborated by experiment. Empirically then, which replicative steps does neutralization block? Third, the definition makes Ab 'binding' a necessary condition for neutralization. But is it also sufficient? Or are there Abs that can bind to surface epitopes on virions but still not neutralize? And if so, why?

Experimental Measurement

Passive immunization, that is, transfer of neutralizing antibodies (NAbs), can protect an organism against some viral infections. Many successful vaccines induce high titers of NAbs and neutralization is, quite plausibly, often crucial to immune defenses *in vivo*. But it is measured *in vitro*. Classically, a neutralization assay consists of four steps: First, virus is incubated with test and control Abs. Then these reaction mixes are added to target cells, and the virus is left to adsorb. After free virus is washed off, virus that has entered the cells is allowed to replicate. Finally, the degree of replication is measured.

This basic scheme comes in different forms. A key distinction is between single- and multicycle assays. Early studies measured inhibition of replication in plaque- and focus-forming assays. These quantal assays were adapted for both bacteriophage and animal viruses. Such tools have a theoretical advantage: if the inoculum is dilute, a primary plaque or focus most likely will represent the local spread from a single infectious event. Furthermore, if progeny virus is prevented from forming secondary plaques and foci, the number of infectious units measured will correspond to the first cycle of replication. Then the degree of neutralization can be expressed as the proportion of initial infectious units prevented from infecting.

Today neutralization studies often employ defective viruses: a viral gene has been deleted and replaced with a reporter gene. Envelope glycoprotein genes can be supplied in *trans*. Hence, one can make pseudotype particles with neutralization targets of choice. Like quantal assays, this method limits the infectivity to the first replicative cycle.

Other neutralization assays measure the reduction in viral Ag production during the culture. If multiple replicative cycles are allowed, though, the reduction in Ag may not be proportional to the fraction of virus that was originally inactivated by the binding of Abs.

Antibody–Antigen Binding and Epitope Occupancy on Virions

The binding of Abs to epitopes on virions is the *sine qua non* of neutralization. The concentration of the antibody, together with its affinity for the epitope, determines the occupancy, θ, of Ab on the virion, that is, the proportion of specific epitopes that the Ab ligates. If the occupancy is too low, the virion will not be neutralized. Instead, the infectivity may even be enhanced by subneutralizing occupancies.

The occupancy can be estimated from the law of mass action, provided the molar excess of Ab over Ag is vast. Usually, this is the case since NAbs tend to neutralize in the nM, or higher, concentration range, which viral Ags in inocula rarely reach. This is reflected in the percentage law: the proportion neutralized by an Ab at a certain concentration is approximately constant when the amount of virus is varied.

Suppose we know the number of epitopes on the virion, and the K_d of the Ab for binding to those epitopes. In theory, we can then estimate how many Abs per virion are bound at any given Ab concentration. In practice, however, this is often impossible. Notably, virions do not always have a constant number of epitopes. The virions of some viruses vary in size. Also, enveloped viruses can incorporate varying numbers of envelope glycoprotein molecules; subunits of these can be lost from the virions after budding. Furthermore, the relevant K_d is seldom known. Often the K_d is measured for Ab binding to a form of the neutralization Ag that differs from the functional one on the infectious virion: the binding studies may have used an unprocessed precursor or a truncated, immobilized monomeric subunit of an oligomer. In addition, the oligomeric context of the epitope strongly affects its antigenicity. The spacing of Ag sites on virions, as well as in binding assays, can also affect the avidity, and therefore the functional affinity, of Ab binding. Clearly, the relevant affinity of NAbs is central to an understanding of neutralization.

The IC_{50} of a NAb, that is, the concentration at which it neutralizes half of the inoculum, is also a notable variable. According to the reasoning above, it should stand in a definite relation to the K_d for Ab binding to virions. If the minimum neutralizing occupancy happens to be 50%, the IC_{50} and the K_d will be similar. With higher occupancies required, we get $IC_{50} > K_d$; with lower, $IC_{50} < K_d$ – all according to the formula $\theta = ([Ab]/K_d)/(([Ab]/K_d) + 1)$.

The maximal extent of neutralization is intriguing. As mentioned, some Abs fail entirely to neutralize. Others have only marginally inhibitory effects. Yet others may give a plateau of neutralization at around, say 75% or 90%. We can imagine a whole spectrum. At one extreme end of the spectrum, where we find the most effective NAbs, no residual infectivity is detectable with low-infectivity inocula. But with higher viral doses, even a strong neutralization leaves a detectable residual infectivity. Furthermore, when neutralization is studied over time, it may reach a plateau. Such residual infectivity, called the 'persistent fraction', has been studied extensively. Usually, the persistent and the neutralized fractions do not differ genetically. Nor does spontaneous aggregation of virus explain all persistence. No comprehensive explanation has been found, but some contributing factors deserve mention. Because of both steric hindrance and the stochastic nature of binding, the Ab may not yield a sufficient occupancy for a complete abrogation of the infectivity of all virions. Conformational, and therefore antigenic, heterogeneity among epitopes may leave some unoccupied. Moreover, the Ab binding is in principle reversible, and at the cell surface Abs and receptors may compete for binding to the virion. Finally, NAbs act in parallel with, and may differentially affect, spontaneous decay of the virus. A better understanding of these dynamics may suggest which properties of NAbs render them most effective. Suffice it to say that the maximal extent of neutralization is not predictable from the IC_{50} value of an Ab. And *in vivo*, that maximum may be more decisive than a titration midpoint.

Mechanism

The first replicative step prevented by the Ab is one aspect of the neutralization mechanism. Another is how, at the intermolecular level, the Ab effects this block.

The initial step in the replicative cycle is attachment to a susceptible cell. NAbs to widely different viruses block this step. For example, Abs to distinct epitopes of influenza virus hemagglutinin (HA), when occupying one in four HA trimers, neutralize by blocking viral attachment. Some viruses, however, interact with ancillary attachment molecules on the surface of cells before docking onto their major receptors. In such cases, the NAb may not interfere with the initial attachment but with the subsequent receptor interactions. Moreover, assortments of NAbs to HIV-1, rotavirus, and papilloma virus show widely different capacities to block the attachment of each virus. These differences may stem from how distally on a viral spike or other protrusion the epitope is located, and at what angle to the virus particle the Ab binds; other contributing factors may be the maximum number of epitopes that can simultaneously be occupied, the size of the Ab, and the valency of its binding.

After attachment to the cell, the next obligatory step for many viruses is internalization by endocytosis. Hence, if endocytic uptake is blocked by the Ab, the virus is neutralized. Failing that, neutralization can still occur inside the endosome. By delaying fusion or penetration, the Ab will promote the eventual degradation of the virus in lysosomes.

We can also hypothesize the converse. Some viruses cannot use the endosomal route for productive entry. Therefore shunting the virion–Ab complex onto that route, for example, through Fc–receptor interactions, would instead lead to neutralization.

After docking onto a major receptor, many viruses interact with co-receptors as a necessary entry step. If that interaction is blocked by an Ab, the virus is neutralized. At this stage, however, steric constraints may prevent access to the relevant epitopes, as multimolecular complexes form. Indeed, a Fab binding adjacently to the co-receptor binding site of HIV-1 neutralizes more efficiently than the corresponding immunoglobulin G (IgG).

We can study post-attachment neutralization (PAN) by letting the virus adsorb to cells at a temperature too low to allow entry. After that, the Ab is allowed to bind to the adsorbed virus. As the cells are warmed up, the non-neutralized virus will infect. At which steps, then, can PAN intervene? In spite of the name and experimental setup, PAN does not necessarily affect only pure post-attachment events. In fact, Abs that interfere directly with virus–receptor interactions often mediate PAN. That may be explained by the necessity of recruiting multiple receptors into a fusion or entry complex. The lower temperature prevents collateral diffusion of receptors; the Ab covers remaining sites not engaged by the receptors; then, at the warm-up, the Ab precludes the requisite additional receptor interactions. Or else PAN occurs because the Ab induces the dissociation of the virus particle. Most importantly, we cannot declare that PAN involves major neutralization mechanisms just because we can make it happen. For when Abs are instead first allowed to bind to virions, attachment may be blocked as the complexes later encounter the cells, which pre-empts any further replicative steps. It also pre-empts PAN.

Particularly the neutralization of picornaviruses, such as poliovirus and rhinovirus, was long thought to interfere with a late penetration step and to depend on the induction of conformational changes in their antigens. Many NAbs can change the isoelectric point, pI, of the picornavirus particle. But their capacity to do so correlates poorly with their neutralizing efficacy. Besides, many of these Abs effectively block virus attachment.

Structural studies have shown that sometimes the paratope of a NAb changes more than the epitope upon binding. An extreme example is one of the most potent NAbs against rhinovirus, which remolds its paratope but leaves the Ag conformationally intact. Furthermore, if neutralization depended on conformational changes in the Ag, there ought to be Abs that block neutralization by competing with NAbs, thereby preventing conformational changes. Such antibodies, though, are not readily found.

Of course, viral proteins, while mediating entry, change conformation or even refold. Therefore, hampering such changes may promote neutralization at the postattachment stage. Or blocking interaction with a receptor, the role of which is to trigger conformational changes, will prevent them by default.

The block of receptor interactions is not necessarily direct. On the contrary, because of the relative sizes of surface viral Ags and Abs, indirect steric interference is potentially strong. Furthermore, when an irrelevant epitope is tagged onto an attachment- and entry-mediating viral protein, at least in some cases, Abs to that epitope can neutralize the virus. What seems most crucial, therefore, is simply the Ab occupancy on functional viral proteins.

Kinetics

The dissociation constant, K_d, is equal to the ratio of the two rate constants for the binding reaction, k_{off}/k_{on}. Yet how the kinetics of NAb binding influences the potency and extent of neutralization is largely unexplored. It is known, though, that the higher the density of target cells and receptors, the greater the demands on neutralization. Thus, the kinetic constants for the Ab, as well as for the receptor binding may influence neutralization in the competitive situation at the cell surface.

One kind of kinetic study takes advantage of PAN. When cells with attached virus are warmed up, NAbs differ in the length of time during which they can still neutralize. Several factors may be responsible for such differences: how fast the on-rate is, whether the epitopes are only transiently exposed, and precisely when the Ab interferes in the chronology of molecular entry events.

Kinetic studies of a different sort have influenced the theory of neutralization profoundly. If the reaction between Ab and virus can be stopped, and the mixture rapidly transferred to target cells, the decline in infectivity can be monitored as a function of reaction time. Much debate has focused on the first part of such kinetic plots of declining relative infectivity. The absence of a shoulder on the curve was interpreted as evidence of neutralization through the very first Ab binding events. It has been questioned, though, whether Ab binding is sufficiently slow to allow the recording of the infectivity of virions with single Abs bound. In any case, at low temperatures, with low concentrations of low-affinity Abs, such kinetic neutralization curves do evince shoulders. Even if the first few Abs reduce the infectivity, they may not abrogate the entire infectious potential of a virion. Perhaps they only dent it. Such interpretations depend on whether neutralization and infectivity are all-or-nothing phenomena or more gradual and incremental. Another complication is the need to estimate stochastic, rare multihits when on average only few Abs have bound.

In a seminal study, Dulbecco *et al.* demonstrated that the rate of neutralization of poliovirus and Western equine encephalitis virus, expressed as a function of Ab concentration, follows approximate first-order kinetics. They took this as evidence that each virion is neutralized by a single Ab, so-called 'single-hit neutralization'. The order of a reaction, however, does not have to be an integer; it is not equivalent to the number of Abs required for neutralization of a virion (see section 'Molecularity'). In addition, the discussion above of the molar ratio of Ab to Ag in an ordinary neutralization reaction makes it clear that what Dulbecco and followers have observed are reactions of 'pseudo'-first order. This is still of practical value. For it allows a calculation of the 'neutralization rate constant'. But the kinetic study of reaction order cannot determine whether neutralization requires single or multiple hits.

Molecularity

Molecularity is the minimum number of molecules involved in the rate-determining step of a reaction. In the case of neutralization, the two reactants are Ab and virus. The molecularity is then the minimum number of Abs a virion must bind in order to be neutralized. Stoichiometry describes only relative amounts, but since the goal is to find the number of Abs on each virion, that term is also often used in this context.

The average number of Abs bound to virus particles, at different recorded degrees of neutralization, has been ascertained through radio labeling and electron microscopy. Such data allow us to calculate the minimal number of Abs required for neutralization, if we assume that Ab binding follows a Poisson distribution (**Figure 1**). For example, if a virus were neutralized by the single hit of an Ab, then with an average of 1 Ab per virion, 37% of the initial viral infectivity would remain. That is merely a special case. It therefore makes no sense to identify the number of Ab molecules on the *x*-axis at 37% residual infectivity as the number of hits required for neutralization, except on the single-hit curve. Instead, a proper Poisson analysis involves

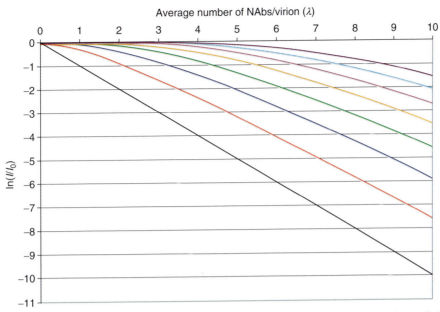

Figure 1 Molecularity of neutralization. The diagram shows the virus infectivities that can be predicted from a Poisson distribution of low numbers of Abs bound to virions. Different absolute thresholds for neutralization are assumed. The natural logarithm of the still infectious fraction of virus, $\ln(I/I_0)$ is plotted on the y-axis as a function of the average number of NAbs per virion, λ, on the x-axis. The latter can be determined, for example, by radio-labeling techniques or electron microscopy. If the minimum number of NAbs per virion required for neutralization is L, then the infectious fraction of virions will be equal to the cumulated fractions with fewer than L NAbs bound to them: $I/I_0 = \Sigma_{r=0}^{L-1}(\lambda^r e^{-\lambda})/r!$. Thus with $L = 1$, the contested single-hit molecularity, and at an average of 1 NAb per virion, $\lambda = 1$, $I/I_0 = e^{-1}$ so that $\ln(I/I_0) = -1$: if the single-hit hypothesis were true, approximately 37% of the infectivity would remain when the virions have on average one NAb bound to them. As can be seen in the diagram only the curve for $L = 1$ (black) goes through the point with the coordinates [1,−1]. The other curves represent other molecularities: $L = 2$ (red), $L = 3$ (blue), $L = 4$ (green), $L = 5$ (orange), $L = 6$ (magenta), $L = 7$ (cyan), and $L = 8$ (purple). These curves do not go through [L,−1]. It has been a common error in the literature to read the x value at $y = -1$, or 37% infectivity on a nonlogarithmic scale, and assume that to be the value of L, the number of hits required for the neutralization. Instead, the diagram shows that the best fitting molecularity can be obtained by comparing the empirical with the theoretical curves.

the plotting of many relative infectivities as a function of the number of bound Abs per virion. Then these empirical data should be compared with the Poisson-derived molecularity curves. The closest one best represents the molecularity.

An additional premise of this Poisson analysis is that a virion is either infectious or not. In other words, there is an absolute threshold of neutralization: with a minimum number of Abs bound, the virion loses all its infectivity at one fell swoop. A radical alternative to this view is that every Ab bound diminishes the infectious propensity of the virion by an equal amount. In fact, there is much evidence of partial reductions in infectivity by NAbs. But a plausible model of neutralization might locate the effects of Abs somewhere in between the two models: there may be a zone of occupancies over which the virion suffers drastic but not total losses in infectivity.

Much direct evidence supports multihit molecularities of neutralization. Four to five Abs are required to give significant neutralization of poliovirus, 36–38 for papillomavirus, 70 for influenza virus, and 225 for rabies virus. Accordingly, Burton and co-workers found an approximately linear relationship between surface area of the virion and the number of Abs required for neutralization. Neutralization could then be seen as the coating of virus particles by Abs, which could interfere with attachment or entry directly or indirectly. The theory can be refined. For example, two enveloped viruses with the same surface area but different numbers of envelope glycoprotein spikes may not require the same number of Abs for neutralization. There is much evidence to support this theory and little to refute it.

Escape and Resistance

Viral genes vary in sequence to widely different extents. The most variable ones encode neutralization antigens; within such antigens surface-exposed elements, often loops containing some of the neutralization epitopes, vary most (**Figure 2**). If a mutation in such a region reduces the affinity of a NAb, then the virus gains a replicative advantage in the presence of the Ab, *in vivo* or *in vitro*. The selection of such mutations constitutes viral escape from neutralization.

Suppose neutralization did not result directly from Ab binding, but indirectly from the induction of conformational changes. Then viral escape from neutralization might sometimes be mediated by mutations blocking those secondary effects of the Abs rather than diminishing the Ab

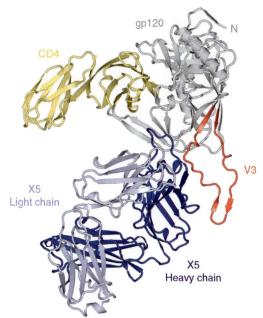

Figure 2 NAbs to HIV-1 are directed to its trimeric envelope glycoprotein, both to the surface unit, gp120, and to the transmembrane protein, gp41. Here, a three-dimensional model of the core of gp120 (gray) is shown. The structure was obtained by high-resolution crystallography of a trimolecular complex. The second molecule in the complex is a two-domain fragment of CD4 (yellow), the major receptor for HIV. Conserved neutralization epitopes overlap the site where CD4 binds. Abs to such epitopes compete with CD4 for binding to gp120 and can thereby directly block a necessary docking event in the viral attachment and entry process. The third molecule in the complex is a neutralizing Fab of the Ab X5. This binds to an epitope generated when CD4 interacts with the full-length gp120; it is located adjacent to the co-receptor binding site. Interestingly, the Fab of X5 neutralizes more efficiently than the whole Ab because its epitope is not induced until the envelope glycoprotein complex docks onto CD4, at which stage access to the epitope is scarce for the larger whole Ab but not for the Fab. Lastly, a variable region, V3, of the gp120 molecule is shown jutting out like a hook (red). This region also contains important neutralization epitopes, but because of the high sequence variability of V3, Abs to these epitopes generally do not cross-neutralize strains. V3 and more conserved adjacent surfaces make contacts with the co-receptor. Reproduced from Huang CC, Tang M, Zhang MY, et al. (2005) Structure of a V3-containing HIV-1 gp120 core. *Science* 310 (5750): 1025–1028. Reprinted with permission from AAAS.

binding itself. Yet, a study of rhinovirus found that all mutations conferring neutralization escape also reduced Ab binding. The routes of escape taken by viruses thus shed light on the mechanisms of neutralization.

Some NAbs bind to more conserved epitopes, which are so intricately involved in attachment or entry functions that the virus can accommodate no great changes there. To produce such Abs has a fitness value for the host; to minimize their elicitation has a fitness value for the virus. A result of these evolutionary forces pitted against each other can be seen in influenza virus. As influenza epidemics rage, the viral HA protein keeps varying residues located on a rim surrounding a receptor-binding pocket. The mutations abrogate the binding of, and hence the neutralization by, Abs that the previous serotype of the virus elicited. But the receptor-binding capacity of the virus, guarded from Ab access in the depth of a pocket, remains intact.

The abrogation of NAb binding does not have to be so direct; mutations outside the epitope can confer resistance; they do not even have to be close in space to the epitope surface. When a neutralization-escape mutant of foot-and-mouth-disease virus was compared with the wild type structurally and serologically, the resistant phenotype was attributed to residues well outside the epitope in the VP1 protein. The mutated residues seem to affect the crucial orientation of the epitope loop from a distance in space. An escape mutant of HIV-1 provides another intriguing example. The subtle substitution of a Thr for an Ala in a highly conserved region of the transmembrane protein confers resistance to NAbs directed to the CD4-binding site on the other, noncovalently linked, moiety of the envelope glycoprotein. Perhaps most intriguing of all, substitutions in the cytoplasmic tail of the HIV-1 transmembrane protein can reduce the sensitivity to NAbs that bind on the other side of the viral membrane, to epitopes on the outer subunit of the envelope glycoprotein complex. This antigenic change must involve intricate conformational effects across the membrane and from subunit to subunit. Even so, the end result points to the most straightforward mechanism of resistance: the affinities of the NAbs are reduced.

Although the molecular basis thus varies for how NAb binding is reduced, such reduction is the route of escape generally taken. One provocative exception has been found. An escape mutant of rabies virus can still infect when coated by NAb; hypothetically, the virus has switched to a new entry mechanism, independent of its highly NAb-occupied envelope glycoprotein.

How does tissue culture affect sensitivity to neutralization? The isolation of a virus removes it from the immune system of the infected host; it entails the lifting of any selective neutralization pressure exerted *in vivo*. Since resistance *in vivo* may have come at a price in basic fitness, when there is no longer any gain from such sacrifices, the replicative functions will evolve anew toward optimum. When propagated *in vitro* the virus may therefore become artificially sensitive to neutralization. That seems to have happened to some strains of HIV.

HIV and many other viruses that survive the onslaught of the host immune system have developed a panoply of phenotypic traits protecting them specifically against neutralization: much of the Ab response is misdirected to epitopes that are exposed only on disassembled oligomers and denatured or degraded protein. Neutralization may require a high occupancy. Glycan shields can render surfaces on envelope glycoproteins nonimmunogenic; the glycan moieties may shift from site to site as a result of escape mutations, providing a malleable shield.

Essential interactions with receptors may be limited to a subset of pocket-lining residues, so that the virus can tolerate variation not just around the receptor-binding site but also within it.

As mentioned, some potent NAbs induce barely any changes in the epitopes. Indeed, less potent Abs to overlapping sites sometimes generate greater changes. When such changes are entropically unfavorable, the adverse effect on affinity and occupancy favor the virus. Thus, entropic masking constitutes yet another viral defense.

Few viruses may exploit the entire repertoire of potential defenses. But these traits attest to the importance of neutralization in natural infection. When combined and richly developed, they also posit a formidable challenge to the induction of NAbs through vaccination.

See also: Antigenic Variation; Immune Response to viruses: Antibody-Mediated Immunity; Persistent and Latent Viral Infection; Vaccine Strategies; Viral Receptors.

Further Reading

Andrewes CH and Elford WJ (1933) Observations on anti-phage sera. I: The percentage law. *British Journal of Experimental Pathology* 14(6): 367–376.

Burnet FM, Keogh EV, and Lush D (1937) The immunological reactions of the filterable viruses. *Australian Journal of Experimental Biology and Medical Science* 15: 227–368.

Burton DR (ed.) (2001) *Antibodies in Viral Infection. Vol. 260. Current Topics in Microbiology and Immunology.* Berlin: Springer.

Burton DR, Desrosiers RC, Doms RW, et al. (2004) HIV vaccine design and the neutralizing antibody problem. *Nature Immunology* 5(3): 233–236.

Della-Porta AJ and Westaway EG (1978) A multi-hit model for the neutralization of animal viruses. *Journal of General Virology* 38(1): 1–19.

Dulbecco R, Vogt M, and Strickland AGR (1956) A study of the basic aspects of neutralization. Two animal viruses. Western equine encephalitis virus and poliomyelitis virus. *Virology* 2: 162–205.

Fazekas de St. GS (1961) Evaluation of quantal neutralization tests. *Nature* 191: 891–893.

Gollins SW and Porterfield JS (1986) A new mechanism for the neutralization of enveloped viruses by antiviral antibody. *Nature* 321 (6067): 244–246.

Huang CC, Tang M, Zhang MY, et al. (2005) Structure of a V3-containing HIV-1 gp120 core. *Science* 310(5750): 1025–1028.

Jerne NK and Avegno P (1956) The development of the phage-inactivating properties of serum during the course of specific immunization of an animal: Reversible and irreversible inactivation. *Journal of Immunology* 76: 200–208.

Klasse PJ and Moore JP (1996) Quantitative model of antibody- and soluble CD4-mediated neutralization of primary isolates and T-cell line-adapted strains of human immunodeficiency virus type 1. *Journal of Virology* 70(6): 3668–3677.

Klasse PJ and Sattentau QJ (2002) Occupancy and mechanism in antibody-mediated neutralization of animal viruses. *Journal of General Virology* 83(Pt 9): 2091–2108.

Mandel B (1978) Neutralization of animal viruses. *Advances in Virus Research* 23: 205–268.

Nybakken GE, Oliphant T, Johnson S, Burke S, Diamond MS, and Fremont DH (2005) Structural basis of West Nile virus neutralization by a therapeutic antibody. *Nature* 437(7059): 764–769.

Skehel JJ and Wiley DC (2000) Receptor binding and membrane fusion in virus entry: The influenza hemagglutinin. *Annual Review of Biochemistry* 69: 531–569.

Nidovirales

L Enjuanes, CNB, CSIC, Madrid, Spain
A E Gorbalenya, Leiden University Medical Center, Leiden, The Netherlands
R J de Groot, Utrecht University, Utrecht, The Netherlands
J A Cowley, CSIRO Livestock Industries, Brisbane, QLD, Australia
J Ziebuhr, The Queen's University of Belfast, Belfast, UK
E J Snijder, Leiden University Medical Center, Leiden, The Netherlands

© 2008 Elsevier Ltd. All rights reserved.

Glossary

3CL^{pro} or M^{pro} 3C-like proteinase, or main proteinase.
ADRP ADP-ribose-1″-phosphatase.
CS TRS Core sequence.
ExoN 3′ to 5′ exoribonuclease.
NendoU Nidovirus endoribonuclease.
O-MT Ribose-2′-O-methyltransferase.
PL^{pro} Papain-like cysteine proteinase.
TRS Transcription-regulating sequence.

Taxonomy and Phylogeny

The order *Nidovirales* includes the families *Coronaviridae*, *Roniviridae*, and *Arteriviridae* (**Figure 1**). The *Coronaviridae* comprises two well-established genera, *Coronavirus* and *Torovirus*, and a tentative new genus, *Bafinivirus*. The *Arteriviridae* and *Roniviridae* include only one genus each, *Arterivirus* and *Okavirus*, respectively. All nidoviruses have single-stranded RNA genomes of positive polarity that, in the case of the *Corona-* and *Roniviridae* (26–32 kbp), are the largest presently known RNA virus genomes. In contrast, members of the *Arteriviridae* have a smaller genome ranging

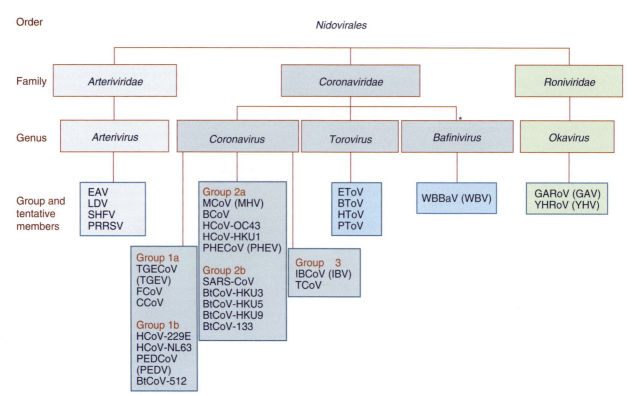

Figure 1 Nidovirus classification and prototype members. The order *Nidovirales* containing the families *Coronaviridae* (including the established genera *Coronavirus* and *Torovirus*, and a new tentative genus *Bafinivirus*), *Arteriviridae*, and *Roniviridae*. Phylogenetic analysis (see **Figure 2**) has confirmed the division of coronaviruses into three groups. In arteriviruses, four comparably distant genetic clusters have been differentiated. To facilitate the taxonomy of the different virus isolates, the types Co, To, Ba, Ro, standing for coronavirus, torovirus, bafinivirus, or ronivirus, respectively, have been included. The following CoVs are shown: human coronaviruses (HCoV) 229E, HKU1, OC43 and NL63, transmissible gastroenteritis virus (TGEV), feline coronavirus (FCoV), porcine epidemic diarrhoea virus (PEDV), mouse hepatitis virus (MHV), bovine coronavirus (BCoV), bat coronaviruses (BtCoV) HKU3, HKU5, HKU9, 133 and 512 (the last two isolated in 2005), porcine hemagglutinating encephalomyelitis virus (PHEV), avian infectious bronchitis virus (IBV), and severe acute respiratory syndrome coronavirus (SARS-CoV); ToV: equine torovirus (EToV), bovine torovirus (BToV), human torovirus (HToV), and porcine torovirus (PToV); BaV: white bream virus (WBV); Arterivirus: equine arteritis virus (EAV), simian haemorrhagic fever virus (SHFV), lactate dehydrogenase-elevating virus (LDV), and three (Euro, HB1, and MLV) porcine reproductive and respiratory syndrome viruses (PRRSV); RoV: gill-associated virus (GAV) and yellow head virus (YHV). Human viruses are highlighted in red. Some nodes are formed by a pair of very closely related viruses (e.g., SARS-CoV and BtCoV-HKU3). Asterisk indicates tentative genus.

from about 13 to 16 kbp. The data available from phylogenetic analysis of the highly conserved RNA-dependent RNA polymerase (RdRp) domain of these viruses, and the collinearity of the array of functional domains in nidovirus replicase polyproteins, were the basis for clustering coronaviruses and toroviruses (**Figure 2**). The more distantly related roniviruses also group with corona- and toroviruses, thus forming a kind of supercluster of nidoviruses with large genomes. By contrast, arteriviruses must have diverged earlier during nidovirus evolution. The current taxonomic position of coronaviruses and toroviruses as two genera of the family *Coronaviridae* is currently being revised by elevating these virus groups to the taxonomic rank of either subfamily or family.

A comparative sequence analysis of coronaviruses reveals three phylogenetically compact clusters: groups 1, 2, and 3. Within group 1, two subsets can be distinguished: subgroup 1a that includes transmissible gastroenteritis virus (TGEV), canine coronavirus (CCoV), and feline coronavirus (FCoV), and subgroup 1b that includes the human coronaviruses (HCoV) 229E and NL63, porcine epidemic diarrhoea virus (PEDV), and bat coronavirus (BtCoV) 512 which was isolated in 2005. Within group 2 coronaviruses, two subsets have been recognized: subgroup 2a, including mouse hepatitis virus (MHV), bovine coronavirus (BCoV), HCoV-OC43, and HCoV-HKU1; and subgroup 2b, including severe acute respiratory syndrome coronavirus (SARS-CoV) and its closest circulating bat coronavirus relative, BtCoV-HKU3. A growing number of other bat viruses has been recently identified in groups 1 and 2. It is currently being debated whether some of these viruses (e.g., BtCoV-HKU5, BtCoV-133 (isolated in 2005), and BtCoV-HKU9) may in fact represent novel subgroups or groups. Avian infectious bronchitis virus (IBV) is the prototype of coronavirus group 3, which also includes several other bird coronaviruses. In arteriviruses, there are four comparably distant genetic clusters, the prototypes of which are equine arteritis

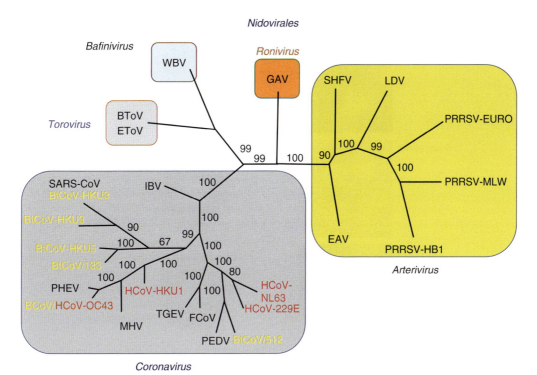

Figure 2 Nidovirus phylogeny. Tree depicting the evolutionary relationships between the five major groups of nidoviruses as shown in **Figure 1** (*Coronavirus, Torovirus, Bafinivirus, Ronivirus,* and *Arterivirus*). This unrooted maximum parsimonious tree was inferred using multiple nucleotide alignments of the RdRp-HEL region of a representative set of nidoviruses with the help of the PAUP*v.4.0b10 software (AEG, unpublished). Support for all bifurcations from 100 bootstraps performed is indicated. The phylogenetic distances shown are approximate. For acronyms, see **Figure 1**.

virus (EAV), lactate dehydrogenase-elevating virus (LDV) of mice, simian hemorrhagic fever virus (SHFV) infecting monkeys, and porcine reproductive and respiratory syndrome virus (PRRSV) which infects pigs and includes European and North American genotypes.

Roniviruses are the only members of the order *Nidovirales* that are known to infect invertebrates. The family *Roniviridae* includes the penaeid shrimp virus, gill-associated virus (GAV), and the closely related yellow head virus (YHV).

More than 100 full-length coronavirus genome sequences and around 30 arterivirus genome sequences have been documented so far, whereas only very few sequences have been reported for toroviruses, bafiniviruses, and roniviruses. Therefore, information on the genetic variability of these nidovirus *taxa* is limited.

Diseases Associated with Nidoviruses

Coronavirus infections are mainly associated with respiratory, enteric, hepatic, and central nervous system diseases. In humans and fowl, coronaviruses primarily cause upper respiratory tract infections, while porcine and bovine coronaviruses establish enteric infections, often resulting in severe economic losses. In 2002, a previously unknown coronavirus that probably has its natural reservoir in bats crossed the species barrier and caused a major outbreak of SARS, which led to more than 800 deaths worldwide.

Toroviruses cause gastroenteritis in mammals, including humans, and possibly also respiratory infections in older cattle. Bafiniviruses have been isolated from white bream fish but there is currently no information on the pathogenesis associated with this virus infection. Roniviruses usually exist as asymptomatic infections but can cause severe disease outbreaks in farmed black tiger shrimp (*Penaeus monodon*) and white pacific shrimp (*Penaeus vannamei*), which in the case of YHV can result in complete crop losses within a few days after the first signs of disease in a pond. Infections by arteriviruses can cause acute or persistent asymptomatic infections, or respiratory disease and abortion (EAV and PRRSV), fatal age-dependent poliomyelitis (LDV), or fatal hemorrhagic fever (SHFV). Arteriviruses, particularly PRRSV in swine populations, cause important economic losses.

Virus Structure

In addition to the significant variations in genome size among the three nidovirus families mentioned above, there are also major differences in virion morphology (**Figure 3**) and host range. Nidoviruses have a lipid envelope which protects the internal nucleocapsid structure and contains

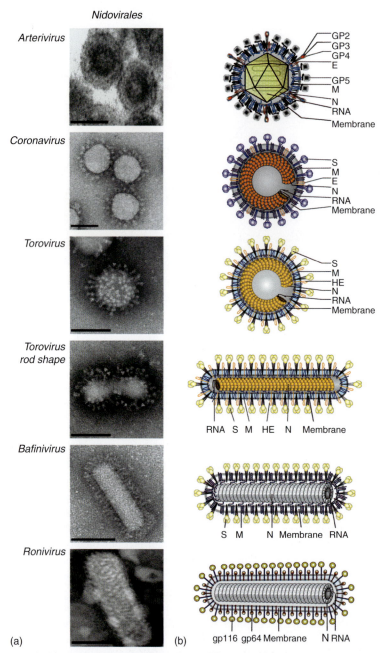

Figure 3 Nidovirus structure. Architecture of particles of members of the order *Nidovirales*: electron micrographs (a) and schematic representations (b) are shown. N, nucleocapsid protein; S, spike protein; M, membrane protein; E, envelope protein; HE, hemagglutinin-esterase. Coronavirus M protein interacts with the N protein. In arterivirus, GP_5 and M are major envelope proteins, while GP_2, GP_3, GP_4, and E are minor envelope proteins. Toro-, bafini-, and roniviruses lack the E protein present in corona- and arteriviruses. Proteins gp116 and gp64, ronivirus envelope proteins. Different images were reproduced with permission from different authors: arterivirus, E. Snijder (Leiden, The Netherlands); ronivirus, P. J. Walker (CSIRO, Australia); bafinivirus, J. Ziebuhr (Queen's University, Belfast) torovirus, D. Rodriguez (CNB, Spain); coronavirus, L. Enjuanes (CNB, Spain).

a number of viral surface proteins (**Figure 3**). Whereas coronaviruses and the significantly smaller arteriviruses have spherical particle structures, elongated rod-shaped structures are observed in toro-, bafini-, and ronivirus-infected cells. The virus particles of the *Corona-* and *Roniviridae* family members carry large surface projections that protrude from the viral envelope (peplomers), whereas arterivirus particles possess only relatively small projections on their surface. Coronaviruses have an internal core shell that is formed by a nucleocapsid featuring a helical symmetry. The nucleocapsid (N) protein interacts with the carboxy-terminus of the envelope membrane (M) protein. The intracellular forms of torovirus, bafinivirus, and ronivirus nucleocapsids have extended rod-shaped (helical) morphology. By contrast, mature (extracellular) toroviruses (but not bafini- and roniviruses) feature a remarkable structural flexibility, which allows them to adopt crescent- and toroid-shaped structures also. Unlike other nidoviruses, arteriviruses have an isometric core shell. In all nidoviruses, the nucleocapsid is formed by only a single N protein that interacts with the genomic RNA.

Both the number and properties of structural proteins vary between viruses of the three families of the *Nidovirales* and may even vary among viruses of the same family. Nidoviruses usually encode at least three structural proteins: a spike (S) or major surface glycoprotein, a trans-membrane (M) or matrix protein, and the N protein (**Figure 3**). Ronivirus particles are unique in that they possess two envelope glycoproteins, gp116 (S1) and gp64 (S2), but no M protein. Coronavirus and arterivirus particles possess another envelope protein called E that is not conserved in toroviruses, bafini-, and roniviruses. Toroviruses and subgroup 2a coronaviruses, such as MHV, have a hemagglutinin esterase (HE) as an additional structural protein, whereas the SARS-CoV has at least four additional proteins that are present in the viral envelope (encoded by ORFs 3a, 6, 7a, and 7b). The proteins may promote virus growth in cell culture or *in vivo*, but they are dispensable for virus replication.

The major envelope proteins are the S and M proteins in coronaviruses and toroviruses, the GP_5 and M proteins in arteriviruses, and the S1 and S2 proteins in roniviruses. Among these, only the corona-, toro-, and bafinivirus S and M proteins share limited sequence similarities, possibly indicating a common origin. Whereas S proteins can differ in size, they share an exposed globular head domain and, with the exception of roniviruses, a stem portion containing heptad repeats organized in a coiled-coil structure. The S proteins of corona- and toroviruses (and most likely those of bafiniviruses) form trimers that bind the cell surface receptor whereas receptor binding in roniviruses is probably mediated by gp116 (S1). The arterivirus envelope proteins form two higher-order complexes: one is a disulfide-linked heterodimer of GP_5 and the M protein; and the other is a heterotrimer of the minor structural glycoproteins GP_2, GP_3, and GP_4. Except for the E and M proteins, all arterivirus structural proteins are glycosylated. By contrast, the M proteins of corona- and toroviruses (and, most likely, bafiniviruses) are glycosylated, and they share a triple-spanning membrane topology with the amino-terminus exposed on the outside of the virions and the carboxy-terminus facing the nucleocapsid. In TGEV, a proportion of the M proteins has a tetra-spanning membrane topology leading to the exposure of both termini on the virion surface.

In the virion, the coronavirus E protein has a low copy number (around 20) and deletion of the E protein gene from the genome of the group 1 coronavirus TGEV blocks virus maturation, preventing virus release and spread. In the group 2 coronaviruses MHV and SARS-CoV, deletion of the E protein results in a dramatic reduction, of up to 100 000-fold, of virus infectivity. The coronavirus E and SARS-CoV 3a proteins are viroporins, that is, they belong to a group of proteins that modify membrane permeability by forming ion channels in the virion envelope.

Genome Organization

Nidovirus genomes contain variable numbers of genes, but in all cases the 5' terminal two-thirds to three-quarters of the genome is dedicated to encoding the key replicative proteins, whereas the 3' proximal genome regions generally encode the structural and, in some cases, accessory (group- and virus-specific) proteins (**Figure 4**). Nidovirus genome expression is controlled at the translational and post-translational levels. Thus, for example, ribosomal frameshifting is required for the expression of ORF1b, and the two replicase polyproteins (pp1a and pp1ab) are proteolytically processed by viral proteases. The proteolytic processing occurs in a coordinated manner and gives rise to the functional subunits of the viral replication–transcription complex. By contrast, the expression of the structural and several accessory proteins is controlled at the level of transcription. It involves the synthesis of a nested set of 3' co-terminal sg mRNAs that are produced in nonequimolar amounts. As in cellular eukaryotic mRNAs, in general only the ORF positioned most closely to the 5' end of the sg mRNA is translated.

The Replicase

The nidovirus replicase gene is comprised of two slightly overlapping ORFs, 1a and 1b. In corona-, toro-, bafini-, and roniviruses, ORF1a encodes a polyprotein (pp1a) of 450–520 kDa, whereas a polyprotein of 760–800 kDa (pp1ab) is synthesized from ORF1ab. Expression of the ORF1b-encoded part of pp1ab involves a ribosomal frameshift mechanism that, in a defined proportion of translation events, directs a controlled shift into the -1 reading frame just upstream of the ORF1a stop codon (**Figure 4**). In arteriviruses, pp1a (190–260 kDa) and pp1ab (345–420 kDa) are considerably smaller in size. Proteolytic processing of coronavirus pp1a and pp1ab generates up to 16 nonstructural proteins (nsps 1–16), while processing of the arterivirus replicase polyproteins generates up to 14 nsps. It is generally accepted that most of the replicase nsps assemble into a large protein complex, called the replication–transcription complex. The complex is anchored to intracellular membranes and likely also includes a number of cellular proteins. Nidoviruses replicase genes include a conserved array of protease, RNA-dependent RNA polymerase (RdRp), helicase (HEL), and endoribonuclease (NendoU) activities. In contrast to other positive-strand RNA viruses,

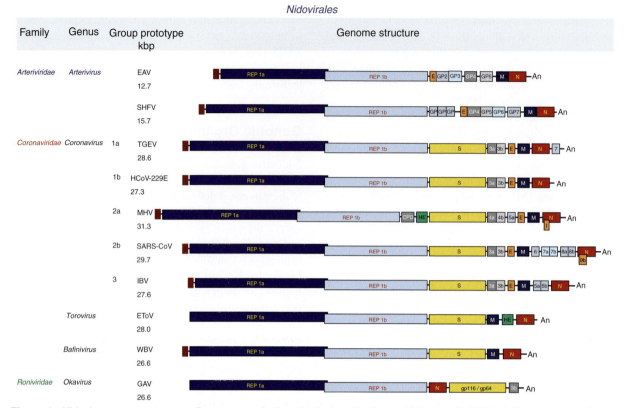

Figure 4 Nidovirus genome structure. Genome organization of selected nidoviruses. The genomic ORFs of viruses representing the major nidovirus lineages are indicated and the names of the replicase and main virion genes are given. References to the nomenclature of accessory genes can be found in the text. Genomes of large and small nidoviruses are drawn to different scales. Red box at the 5′ end refers to the leader sequence. Partially overlapping ORFs have been drawn as united boxes. Spaces between boxes representing different ORFs do not mean noncoding sequences.

they employ an RdRp with a characteristic SDD rather than the usual GDD active site.

The vast majority of proteolytic cleavages in pp1a/pp1ab are mediated by an ORF1a-encoded chymotrypsin-like protease that, because of its similarities to picornavirus 3C proteases, is called the 3C-like protease (3CL^{pro}). Also the term 'main protease' (M^{pro}) is increasingly used for this enzyme, mainly to refer to its key role in nidovirus replicase polyprotein processing (**Figure 5**). Nidovirus–M^{pro} share a three-domain structure. The two N-terminal domains adopt a two-β-barrel fold reminiscent of the structure of chymotrypsin. With respect to the principal catalytic residues, there are major differences between the main proteases from different nidovirus genera. The presence of a third, C-terminal domain is a conserved feature of nidovirus main proteases, even though these domains vary significantly in both size and structure. The C-terminal domain of the coronavirus M^{pro} is involved in protein dimerization that is required for proteolytic activity *in trans*. Over the past years, a large body of structural and functional information has been obtained for corona- and arterivirus main proteases which, in the case of coronaviruses, has also been used to develop selective protease inhibitors that block viral replication, suggesting that nidovirus main proteases may be attractive targets for antiviral drug design.

In arteri-, corona-, and toroviruses, the M^{pro} is assisted by 1–4 papain-like ('accessory') proteases (PL^{pro}) that process the less well-conserved N-proximal region of the replicase polyproteins (**Figure 5**). Nidovirus PL^{pro} domains may include zinc ribbon structures and some of them have deubiquitinating activities, suggesting that these proteases might also have functions other than polyprotein processing. Bafini- and roniviruses have not been studied in great detail and it is not yet clear if these viruses employ papain-like proteases to process their N-terminal pp1a/pp1ab regions (**Figure 5**).

The replicase polyproteins of 'large' nidoviruses with genome sizes of more than 26 kb (i.e., corona-, toro-, bafini-, and roniviruses) include 3′–5′ exoribonuclease (ExoN) and ribose-2′-O-methyltransferase (MT) activities that are essential for coronavirus RNA synthesis but are not conserved in the much smaller arteriviruses (**Figure 5**). The precise biological function of ExoN has not been established for any nidovirus but the relationship with cellular DEDD superfamily exonucleases

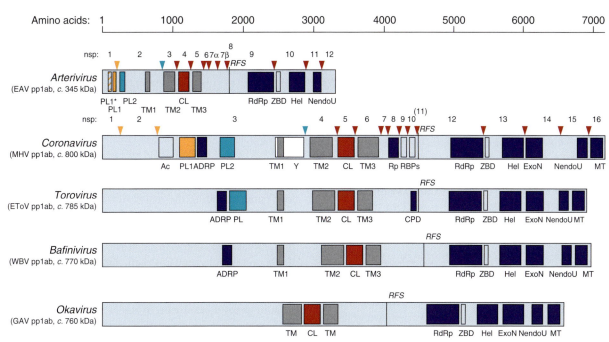

Figure 5 Nidovirus replicase genes. Polyprotein (pp) 1ab domain organizations are shown for representative viruses from the five nidovirus genera. Acronyms as in **Figure 1**. Arterivirus and coronavirus pp1ab processing pathways have been characterized in considerable detail and are illustrated here for EAV and MHV. N-proximal polyprotein regions are cleaved at two or three sites by viral papain-like proteases 1 (PL1) and 2 (PL2), whereas the central and C-terminal polyprotein regions are processed by the main protease, M^{pro}. PL1 domains are indicated by orange boxes and cognate cleavage sites are indicated by orange arrowheads. PL2 domains and PL2-mediated cleavages are shown in green and CL domains and CL-mediated cleavages are shown in red. Note that EAV encodes a second, but proteolytically inactive PL1 domain (PL1*; orange-striped box). For the genera *Torovirus*, *Bafinivirus*, and *Okavirus*, the available information on pp1ab proteolytic processing is limited and not shown here. Other predicted or proven enzymatic activities are shown in blue: ADRP, ADP-ribose 1′-phosphatase; Rp, noncanonical RNA polymerase ('primase') activity; RdRp, RNA-dependent RNA polymerase; HEL, NTPase/RNA helicase and RNA 5′-triphosphatase; ExoN, 3′-to-5′ exoribonuclease; NendoU, nidoviral uridylate-specific endoribonuclease; MT, ribose-2′-O methyltransferase; CPD, cyclic nucleotide phosphodiesterase. Regions with predicted transmembrane (TM) domains are indicated by gray boxes. Other functional domains are shown as white boxes: Ac, acidic domain; Y, Y domain containing putative transmembrane and zinc-binding regions; ZBD, helicase-associated zinc-binding domain; RBPs, RNA-binding proteins. Expression of the C-terminal part of pp1ab requires a ribosomal frameshift, which occurs just upstream of the ORF1a translation stop codon. The ribosomal frameshift site (RFS) is indicated.

and recently published data suggest that ExoN may have functions in the replication cycle of large nidoviruses that, like in the DEDD homologs, are related to proofreading, repair, and recombination mechanisms.

NendoU is a nidovirus-wide conserved domain that has no counterparts in other RNA viruses. It is therefore considered a genetic marker of the *Nidovirales*. The endonuclease has uridylate specificity and forms hexameric structures with six independent catalytic sites. Cellular homologs of NendoU have been implicated in small nucleolar RNA processing whereas the role of NendoU in viral replication is less clear. Reverse genetics data indicate that NendoU has a critical role in the viral replication cycle.

Two other RNA-processing domains, ADP-ribose-1″-phosphatase (ADRP) and nucleotide cyclic phosphodiesterase (CPD), are conserved in overlapping subsets of nidoviruses (**Figure 5**). Except for arteri- and roniviruses, all nidoviruses encode an ADRP domain that is part of a large replicase subunit (nsp3 in the case of coronaviruses).

The coronavirus ADRP homolog has been shown to have ADP-ribose-1′-phosphatase and poly(ADP-ribose)-binding activities. Although the highly specific phosphatase activity is not essential for viral replication *in vitro*, the strict conservation in all genera of the *Coronaviridae* suggests an important (though currently unclear) function of this protein in the viral replication cycle. This may be linked to host cell functions and, particularly, to the activities of cellular homologs called 'macro' domains which are thought to be involved in the metabolism of ADP-ribose and its derivatives.

The CPD domain is only encoded by toroviruses and group 2a coronaviruses. In toroviruses, the CPD domain is encoded by the 3′ end of replicase ORF1a (**Figure 5**), whereas in group 2a coronaviruses, the enzyme is expressed from a separate subgenomic RNA. The enzyme's biological function is not clear. Coronavirus CPD mutants are attenuated in the natural host whereas replication in cell culture is normal, suggesting some function *in vivo*. The available information suggests that nidovirus replicase polyproteins (particularly, those of large nidoviruses) have evolved to include a number of nonessential functions that may provide a selective advantage in the host.

ORF1a of all nidoviruses encodes a number of (putative) transmembrane proteins, like the coronavirus nsps 3, 4, and 6 and the arterivirus nsps 2, 3, and 5. These have been shown or postulated to trigger the modification of cytoplasmic membranes, including the formation of unusual double-membrane vesicles (DMVs). Tethering of the replication–transcription complex to these virus-induced membrane structures might provide a scaffold or subcellular compartment for viral RNA synthesis, possibly allowing it to proceed under conditions that prevent or impair detection by cellular defense mechanisms, which are usually induced by the double-stranded RNA intermediates of viral replication.

Finally, recent structural and biochemical studies have yielded novel insights into the function of a set of small nsps encoded in the 3′-terminal part of the coronavirus ORF1a. For example, nsp7 and nsp8 were shown to form a hexadecameric supercomplex that is capable of encircling dsRNA. The coronavirus nsp8 was also shown to have RNA polymerase (primase) activity that may produce the primers required by the primer-dependent RdRp residing in nsp12. For nsp9 and nsp10, RNA-binding activities have been demonstrated and crystal structures have been reported for both proteins. Nsp10 is a zinc-binding protein that contains two zinc-finger-binding domains and has been implicated in negative-strand RNA synthesis.

Structural and Accessory Protein Genes

In contrast to the large genome of *Coronaviridae*, which can accommodate genes encoding accessory proteins (i.e., proteins called 'nonessential' for being dispensable for replication in cell culture (**Figure 6**)), the smaller genomes of arteriviruses only encode essential proteins (**Figure 4**). Coronaviruses encode a variable number of accessory proteins (2–8), while the torovirus genome contains a single accessory gene encoding a hemagglutinin-esterase (HE). Coronavirus accessory genes may occupy any intergenic position in the conserved array of the four genes encoding the major structural proteins (5′-S-E-M-N-3′), or they may reside upstream or downstream of this gene array. Roniviruses are unique among the presently known nidoviruses in that the gene encoding the N protein is located upstream rather than downstream of the gene encoding the glycoproteins. Several members of the coronavirus group 1a are exceptional in that they contain genes downstream of the N protein gene, which has not been reported for other coronaviruses. The ronivirus glycoprotein gene is also unique in that it encodes a precursor polyprotein with two internal signal peptidase cleavage sites used to generate the envelope glycoproteins S1 and S2 as well as an amino-terminal protein with an unknown function.

The accessory genes are specific for either a single virus species or a few viruses that form a compact phylogenetic cluster. Many proteins encoded by accessory

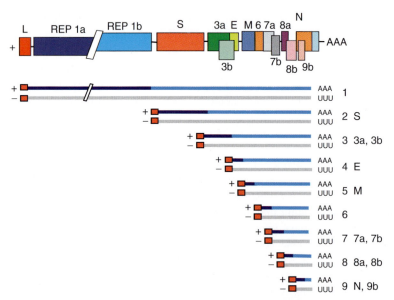

Figure 6 SARS-CoV genome organization and sg mRNA expression. A diagram of CoV structure using SARS-CoV as a prototype. Below the top bar a set of positive- and negative-sense mRNA species synthesized in infected cells is shown. Dark and light blue lines (+), mRNA sequences translated and nontranslated into viral proteins, respectively. Light gray lines (−), RNAs complementary to the different mRNAs. L, leader sequence. Poly(A) and Poly(U) tails are indicated by AAA or UUU, respectively. Rep 1a and Rep 1b, replicase genes. Other acronyms above and below the top bar indicate structural and nonstructural proteins. Numbers and letters to the right of the thin bars indicate the sg mRNAs.

genes may function in infected cells or *in vivo* to counteract host defenses and, when removed, may lead to attenuated virus phenotypes. Group 1 coronaviruses may have 2–3 accessory genes located between the S and E genes and up to two other genes downstream of N gene. Viruses of group 2 form the most diverse coronavirus cluster, and they may have between three and eight accessory genes. In this cluster, MHV, HCoV-OC43, and BCoV form the phylogenetically compact subgroup, 2a, that is characterized by the presence of (1) two accessory genes located between ORF1b and the S gene encoding proteins with CPD and HE functions, (2) two accessory genes located between the S and E protein genes, and (3) an accessory gene, I, that is located within the N protein gene. Of this set of five accessory proteins, only three homologs are encoded by the recently identified HCoV-HKU1, which is the closest known relative of the cluster formed by MHV, HCoV-OC43, and BCoV. In contrast, the most distant group 2 member, SARS-CoV, has seven or eight unique accessory genes, two between the S and E protein genes, four to five between the M and N protein genes, and ORF9b which entirely overlaps with the N protein gene in an alternative reading frame. In group 3 avian coronaviruses, of which IBV is the prototype, several accessory genes, which are expressed from functionally tri- or bicistronic mRNAs, have been identified in the region between the S and E protein genes (gene 3) and between the M and N protein genes (gene 5).

Some functionally dispensable ORF1a-encoded replicase domains may also be considered as accessory protein functions. For instance, MHV and SARS-CoV nsp2 turned out to be nonessential for replication in cell culture.

Replication

Like in all other positive-stranded RNA viruses, nidovirus genome replication is mediated through the synthesis of a full-length, negative-strand RNA which, in turn, is the template for the synthesis of progeny virus genomes. This process is mediated by the viral replication complex that includes all or most of the 14–16 nsps derived from the proteolytic processing of the pp1a and pp1ab replicase polyproteins of arteriviruses and coronaviruses. The replication complex, which is likely to include also cellular proteins, is associated with modified intracellular membranes, which may be important to create a microenvironment suitable for viral RNA synthesis as well as for recruitment of host factors. Electron microscopy studies of cells infected with arteriviruses (EAV) and coronaviruses (MHV and SARS-CoV) have shown that RNA synthesis is associated with virus-induced, DMVs. The origin of DMVs is under debate and different intracellular compartments including the Golgi, late endosomal membranes, autophagosomes, and the endoplasmic reticulum have been implicated in their formation.

Studies of *cis*-acting sequences required for nidovirus replication have mainly relied on coronavirus defective-interfering (DI) RNAs replicated by helper virus. Genome regions harboring minimal *cis*-acting sequences have been mapped to around 1 kb domains of the genomic $5'$ and $3'$ ends. Studies with MHV DI RNAs have indicated that both genome ends are necessary for positive-strand synthesis, whereas only the last 55 nt and the poly (A) tail at the genomic $3'$ end are required for negative-strand synthesis. It has been postulated that the $5'$ and $3'$ ends of the genome may interact directly during RNA replication, as predicted by computer-aided simulations of MHV and TGEV genomic RNA interactions in protein-free media. There is, however, some experimental evidence supporting protein-mediated cross-talk between both genome ends in the form of RNA–protein and protein–protein interactions.

Several experimental approaches have implicated, in addition to the nsps encoded by the replicase gene, the N protein in coronavirus RNA synthesis. Early in infection, the coronavirus N protein colocalizes with the site of viral RNA synthesis. In addition, the N protein can enhance the rescue of various coronaviruses from synthetic full-length RNA, transcribed *in vitro* or from cDNA clones. In contrast, arterivirus RNA synthesis does not require the N protein.

Host factors that may participate in nidovirus RNA synthesis have been identified mainly from studies of coronaviruses and arteriviruses. In coronaviruses (MHV and TGEV), heterogeneous nuclear ribonucleoprotein (hnRNP) A1 has been identified as a major protein binding to genomic RNA sequences complementary to those in the negative-strand RNA that bind another cellular protein, polypyrimidine tract-binding protein (PTB). hnRNP A1 and PTB bind to the complementary strands at the $5'$ end of coronavirus RNA and could mediate the formation of an RNP replication complex involving the $5'$ and $3'$ ends of coronavirus genomic RNA. The functional relevance of hnRNP A1 in coronavirus replication was supported by experiments showing that its overexpression promotes MHV replication, whereas replication was reduced in cells expressing a dominant-negative mutant of hnRNP A1. There is also experimental evidence to suggest that the poly(A)-binding protein (PABP) specifically interacts with the $3'$ poly(A) tail of coronavirus genomes, and that this interaction may affect their replication. Other cellular proteins found to bind to coronavirus genomic RNA, such as aconitase and the heat shock proteins HS40 and HS70, might be involved in modulating coronavirus replication. Similarly, interactions of cellular proteins such as transcription cofactor p100 with the EAV nsp1, or of PTB or fructose bisphosphate aldolase A with SHFV genomic RNA, suggest that, in arterivirus replication also, a number of cellular proteins may be involved.

Transcription

RNA-dependent RNA transcription in some members of the *Nidovirales* (coronaviruses, bafiniviruses, and arteriviruses), but not in others (roniviruses), includes a discontinuous RNA-synthesis step. This process occurs during the production of subgenome-length negative-strand RNAs that serve as templates for transcription and involves the fusion of a copy of the genomic 5′-terminal leader sequence to the 3′ end of each of the nascent RNAs complementary to the coding (body) sequences (**Figure 6**). The resulting chimeric sg RNAs of negative polarity are transcribed to yield sg mRNAs that share both 5′- and 3′- terminal sequences with the genome RNA. Genes expressed through sg mRNAs are preceded by conserved 'transcription-regulatory sequences' (TRSs) that presumably act as attenuation or termination signals during the production of the subgenome-length negative-strand RNAs. In arteriviruses and coronaviruses, the TRSs preceding each ORF are presumed to direct attenuation of negative-strand RNA synthesis, leading to the 'jumping' of the nascent negative-strand RNA to the leader TRS (TRS-L). This process is guided by a base-pairing interaction between complementary sequences (leader TRS and body TRS complement) and it has been proposed that template switching only occurs if the free energy (ΔG) for the formation of this duplex reaches a minimum threshold. This process is named 'discontinuous extension of minus strands' and can be considered a variant of similarity-assisted template-switching that operates during viral RNA recombination. The genome and sg mRNAs share a 5′-leader sequence of 55–92 nt in coronaviruses and 170–210 nt in arteriviruses.

Toroviruses are remarkable in that they employ a mixed transcription strategy to produce their mRNAs. Of their four sg mRNA species, the smaller three (mRNAs 3 through 5) lack a 5′ common leader and are produced via nondiscontinuous RNA synthesis. In contrast, sg mRNA2 has a leader sequence that matches the 5′-terminal 18 nt of the genomic RNA and its production requires a discontinuous RNA-synthesis step reminiscent of, but not identical, to that seen in arteri- and coronaviruses.

Synthesis of torovirus mRNAs 3 through 5, and possibly of the two mRNAs in roniviruses, is thought to require the premature termination of negative-strand RNA synthesis at conserved, intergenic, TRS-like sequences to generate subgenome-length negative-strand RNAs that can be used directly as templates for sg mRNA synthesis. In the case of torovirus mRNA2, a TRS is lacking. Fusion of noncontiguous sequences seems to be controlled by a sequence element consisting of a hairpin structure and 3′ flanking stretch of 23 residues with sequence identity to a region at the 5′ end of the genome. It is thought that during negative-strand synthesis, the hairpin structure may cause the transcriptase complex to detach, prompting a template switch similar to that seen in arteri- and coronaviruses.

In addition to regulatory RNA sequences, viral and host components involved in protein–RNA and protein–protein recognition are likely to be important in transcription. For example, the arterivirus nsp1 protein has been identified as a factor that is dispensable for genome replication but absolutely required for sg RNA synthesis. The identification of host factors participating in nidovirus transcription is a field under development and specific binding assays have recently identified a limited number of cellular proteins that associate with *cis*-acting RNA regulatory sequences. For example, differences in affinity of such factors for body TRSs might regulate transcription in nidoviruses by a mechanism similar to that of the DNA-dependent RNA-polymerase I termination system, in which specific proteins bind to termination sequences.

Origin of Nidoviruses

The complex genetic plan and the replicase gene of nidoviruses must have evolved from simpler ones. Using this natural assumption, a speculative scenario of major events in nidovirus evolution has been proposed. It has been speculated that the most recent common ancestor of the *Nidovirales* had a genome size close to that of the current arteriviruses. This ancestor may have evolved from a smaller RNA virus by acquiring the two nidovirus genetic marker domains represented by the helicase-associated zinc-binding domain (ZBD) and the NendoU function. These two domains may have been used to improve the low fidelity of RdRp-mediated RNA replication, thus generating viruses capable of efficiently replicating genomes of about 14 kbp. The subsequent evolution of much larger nidovirus genomes may have been accompanied by the acquisition of the ExoN domain. This domain may have further improved the fidelity of RNA replication through its 3′–5′ exonuclease activity, which might operate in proofreading mechanisms similar to those employed by DNA-based life forms. It has been suggested that the ORF1b-encoded HEL, ExoN, NendoU, and *O*-MT domains may provide RNA specificity, whereas the relatively abundantly expressed CPD and ADRP might control the pace of a common pathway that could be part of a hypothetical oligonucleotide-directed repair mechanism used in the present coronaviruses and roniviruses. The expansion of the replicase gene may have been associated with an increase in replicase fidelity, thus also supporting the further expansion of the 3′-proximal genome region to encode the structural proteins required to form complex enveloped virions.

Effect of Nidovirus Infection on the Host Cell

Compared to other viruses, the interactions of nidoviruses with their hosts have not been studied in great detail. In many cases, information is based on relatively few studies performed on a limited number of viruses from the families *Coronaviridae* and *Arteriviridae*. Also, most studies have been performed with viruses that have been adapted to cell culture and therefore may have properties that differ from those of field strains. Coronaviruses and arteriviruses are clearly the best-studied members of the *Nidovirales* in terms of their interactions with the host.

Coronavirus infection affects cellular gene expression at the level of both transcription and translation. Upon infection, host cell translation is significantly suppressed but not shut off, as is the case in several other positive-RNA viruses. The underlying mechanisms have not been characterized in detail, but data obtained for MHV and BCoV suggest that they may involve the 5'-leader sequences present on coronavirus mRNAs. The viral N protein was reported to bind to the 5'-common leader sequence and it has been speculated that this might promote translation initiation, leading to a preferential translation of viral mRNAs. Furthermore, host mRNAs were reported to be specifically degraded in MHV- and SARS-CoV-infected cells, further reducing the synthesis of cellular proteins. Another mechanism affecting host cell protein synthesis may be based on specific cleavage of the 28S rRNA subunit, which was observed in MHV-infected cells.

Studies on cellular gene expression following nidovirus infections have mainly focused on the coronaviruses MHV and SARS-CoV. For example, SARS-CoV infection was reported to disrupt cellular transcription to a larger extent than does HCoV-229E. Differences in cellular gene expression have been proposed to be linked to differences in the pathogenesis caused by these two human coronaviruses. Apart from the downregulation of genes involved in translation and cytoskeleton maintenance, genes involved in stress response, proapoptotic, proinflammatory, and procoagulating pathways were significantly upregulated. Both MHV and SARS-CoV induce mitogen-activated phosphate kinases (MAPKs), especially p38 MAPK. In addition, activation of AP-1, nuclear factor kappa B (NF-κB), and a weak induction of Akt signaling pathways occur after SARS-CoV infection and the N and nsp1 proteins were suggested to be directly involved in inducing these signaling pathways.

Nidoviruses have also been reported to interfere with cell cycle control. Infection by the coronaviruses TGEV, MHV, SARS-CoV, and IBV was reported to cause a cell cycle arrest in the G0/G1 phase and a number of cellular proteins (e.g., cyclin D3 and hypophosphorylated restinoblastoma protein) and viral proteins (MHV nsp1, SARS-CoV 3b 7a, and N proteins) have been proposed to be involved in the cell cycle arrest in G0/G1.

Many viruses encode proteins that modulate apoptosis and, more generally, cell death, which allows for highly efficient viral replication or the establishment of persistent infections. Infection by coronaviruses (e.g., TGEV, MHV, and SARS-CoV) and arteriviruses (e.g., PRRSV and EAV) have been reported to induce apoptosis in certain cell types. Apoptosis has also been reported in shrimp infected with the ronivirus YHV and is thought to be involved in pathogenesis. Both apoptotic and antiapoptotic molecules have been found to be upregulated, suggesting that a delicate counterbalance of pro- and antiapoptotic molecules is required to ensure cell survival during the early phase of infection, and rapid virus multiplication before cell lysis occurs. Coronavirus-induced apoptosis appears to occur in a tissue-specific manner, which obviously has important implications for viral pathogenesis. For instance, SARS-CoV was shown to infect epithelial cells of the intestinal tract and induce an antiapoptotic response that may counteract a rapid destruction of infected enterocytes. These findings are consistent with clinical observations of a relatively normal endoscopic and microscopic appearance of the intestine in SARS patients. Furthermore, SARS-CoV causes lymphopenia which involves the depletion of T cells, probably by apoptotic mechanisms that are triggered by direct interactions of the SARS-CoV E protein with the antiapoptotic factor Bcl-xL. Also the MHV E protein has been reported to induce apoptosis. The SARS-CoV 7a protein was found to induce apoptosis in cell lines derived from lung, kidney, and liver, by a caspase-dependent pathway. Apoptosis has also been associated with arterivirus infection but information on underlying mechanisms and functional implications is limited.

Coronavirus and arterivirus infections trigger proinflammatory responses that often are associated with the clinical outcome of the infection. Thus, for example, there seems to be a direct link between the IL-8 plasma levels of SARS patients and disease severity, similar to what has been described for pulmonary infections caused by respiratory syncytial virus. In contrast, despite the upregulation of IL-8 in intestinal epithelial cells, biopsy specimens taken from the colon and terminal ileum of SARS patients failed to demonstrate any inflammatory infiltrates, which may be the consequence of a virus-induced suppression of specific cytokines and chemokines, including IL-18, in the intestinal environment.

Innate immunity is essential to control vertebrate nidovirus infection *in vivo*. The induction of type I IFN (IFN-α/β) varies among different coronaviruses and arteriviruses. Whereas some coronaviruses such as TGEV are potent inducers of type I IFN, other coronaviruses (MHV and SARS-CoV) or arteriviruses (PRRSV) do not stimulate its production, thus facilitating virus escape from innate immune defenses. Type I interferon is a key player in innate immunity and in the activation of effective adaptive immune responses. Upon viral invasion, IFN-α/β is synthesized and secreted. IFN-α/β molecules signal through the type I interferon receptor (IFNR), inducing the transcription of several antiviral mediators, including IFN-γ, PKR, and Mx. IFN-γ is critical in resolving coronavirus (MHV and SARS-CoV), and also arterivirus (EAV, LDV, and PRRSV) infections. Like many other viruses, coronaviruses have developed strategies to escape IFN responses. For example, it has been shown that the SARS-CoV 3b, 6, and N proteins antagonize interferon by different mechanisms, even though all these proteins inhibit the expression of IFN by interfering with the function of IRF-3.

In arteriviruses such as PRRSV, IFN-γ is produced soon after infection to promote Th1 responses. However, PRRSV infections or vaccination with attenuated-live PRRSV vaccines cause only limited IL-1, TNF-α, and IFN-α/β responses. This then leads to IFN-γ and Th1 levels that fail to elicit strong cellular immune responses.

See also: Arteriviruses; Coronaviruses: General Features; Coronaviruses: Molecular Biology; Severe Acute Respiratory Syndrome (SARS); Torovirus; Yellow Head Virus.

Further Reading

de Groot RJ (2007) Molecular biology and evolution of toroviruses. In: Snijder EJ, Gallagher T, and Perlman S (eds.) *The Nidoviruses*, pp. 133–146. Washington, DC: ASM Press.

Enjuanes L (ed.) (2005) *Current Topics in Microbiology and Immunology, Vol. 287: Coronavirus Replication and Reverse Genetics*. Berlin: Springer.

Enjuanes L, Almazan F, Sola I, and Zuniga S (2006) Biochemical aspects of coronavirus replication and virus–host interaction. *Annual Review of Microbiology* 60: 211–230.

Gorbalenya AE, Enjuanes L, Ziebuhr J, and Snijder EJ (2006) Nidovirales: Evolving the largest RNA virus genome. *Virus Research* 117: 17–37.

Masters PS (2006) The molecular biology of coronaviruses. *Advances in Virus Research* 66: 193–292.

Sawicki SG, Sawicki DL, and Siddell SG (2007) A contemporary view of coronavirus transcription. *Journal of Virology* 81: 20–29.

Siddell SG, Ziebuhr J, and Snijder EJ (2005) Coronaviruses, toroviruses, and arteriviruses. In: Mahy BWJ and ter-Meulen V (eds.) *Virology*, 10th edn. vol.1, pp. 823–856. London: Hoddeer-Arnold.

Snijder EJ, Siddell SG, and Gorbalenya AE (2005) The order *Nidovirales*. In: Mahy BWJ and ter-Meulen V (eds.) *Virology*, 10th edn., vol. 1, pp. 390–404. London: Hodder-Arnold.

Snijder EJ and Spaan WJM (2007) *Arteriviruses*. In: Knipe DM, Howley PM, Griffin DE, et al. (eds.) *Fields Virology*, vol. 1, pp. 1205–1220. Philadelphia: Lippincott Williams and Wilkins.

Spaan WJM, Cavanagh D, de Groot RJ, et al. (2005) Nidovirales. In: Fauquet CM, Mayo MA, Maniloff J, Desselberger U, and Ball LA (eds.) *Virus Taxonomy: Eighth Report of the International Committee on Taxonomy of Viruses*, pp. 937–945. San Diego, CA: Elsevier Academic Press.

van Vliet ALW, Smits SL, Rottier PJM, and de Groot RJ (2002) Discontinuous and non-discontinuous subgenomic RNA transcription in a nidovirus. *EMBO Journal* 21(23): 6571–6580.

Walker PJ, Bonami JR, Boonsaeng V, et al. (2005) Roniviridae. In: Fauquet CM, Mayo MA, Maniloff J, Desselberger U, and Ball LA (eds.) *Virus Taxonomy: Eighth Report of the International Committee on Taxonomy of Viruses*, pp. 975–979. San Diego, CA: Elsevier Academic Press.

Ziebuhr J and Snijder EJ (2007) The coronavirus replicase: Special enzymes for special viruses. In: Thiel V (ed.) *Molecular and Cellular Biology: Coronaviruses*, pp. 31–61. Norfolk, UK: Caister Academic Press.

Zuñiga S, Sola I, Alonso S, and Enjuanes L (2004) Sequence motifs involved in the regulation of discontinuous coronavirus subgenomic RNA synthesis. *Journal of Virology* 78: 980–994.

Nodaviruses

P A Venter and A Schneemann, The Scripps Research Institute, La Jolla, CA, USA

© 2008 Elsevier Ltd. All rights reserved.

Introduction

Viruses belonging to the family *Nodaviridae* are small (28–37 nm), nonenveloped, and isometric. These viruses characteristically package bipartite positive-sense RNA genomes that are made up of RNA1 (3.0–3.2 kb) and RNA2 (1.3–1.4 kb). RNA1 encodes protein A, the RNA-dependent RNA polymerase (RdRp), and RNA2 encodes the capsid protein which is required for formation of progeny virions. The family is subdivided into two genera: *Alphanodavirus*, whose members infect insects, and *Betanodavirus*, whose members infect fish. Alphanodaviruses have become model systems for studies on RNA replication, specific genome packaging, virus structure, and assembly, and for studies on virus–host interactions that are required to suppress RNA silencing in animal cells. Betanodaviruses, on the other hand, cause high mortalities in hatchery-reared fish larvae and juveniles, and are therefore economically important pathogens to the marine aquaculture industry.

Taxonomy

Table 1 lists the definitive and tentative nodavirus species, their natural hosts, and geographic origin. A comparison between the capsid protein sequences of alpha- and betanodaviruses shows that these genera are distantly related with an approximate similarity of only 10%. Viruses within each genus share antigenic determinants but also show distinct immunological reactivities. Within the genus *Alphanodavirus*, the species *Flock house virus*, *Black beetle virus*, and *Boolarra virus* are relatively closely related to each other while *Nodamura virus* (the type species of this genus) and *Pariacoto virus* are evolutionarily the most distant alphanodaviruses. Conversely, available sequences for the RdRp and capsid protein of three betanodaviruses show high levels of identity (c. 80% and 90%, respectively). Phylogenetic analysis based on alignments of a variable region within the betanodavirus capsid protein sequence grouped the virus species into four genotypes: *Tiger puffer nervous necrosis virus*, *Striped jack nervous necrosis virus*, the type species of the genus *Betanodavirus*; *Barfin flounder nervous necrosis virus*; and *Redspotted grouper nervous necrosis virus*. Viruses belonging to the *Striped jack nervous necrosis virus* genotype exhibit subtle but distinct serological differences to viruses belonging to the other genotypes.

Host Range and Geographic Distribution

All known alphanodaviruses were originally isolated in Australasia, except for Pariacoto virus (PaV), which was isolated in Peru. Several alphanodaviruses can multiply in a wide range of insect species in addition to their natural insect hosts (Table 1) including bees, beetles, mosquitoes, moths, tsetse flies, and ticks. In the laboratory, larvae of the common wax moth (*Galleria mellonella*) are convenient hosts for most of these viruses. Nodamura virus (NoV) is not only infectious to insects, but also has the unique ability to cause hind-limb paralysis and 100% mortality in suckling mice. Flock house virus (FHV), black beetle virus (BBV), and Boalarra virus (BoV) can readily be propagated to very high yields in cultured *Drosophila melanogaster* cells. PaV is not infectious to *D. melanogaster*

Table 1 Natural hosts and geographic origin of viruses belonging to the family *Nodaviridae*

Virus species	Virus abbreviation	Host	Geographic origin
Genus *Alphanodavirus*			
Black beetle virus	BBV	Scarab beetle (*Heteronychus arator*)	New Zealand
Boolarra virus	BoV	Underground grass grub (*Oncopera intricoides*)	Australia
Flock house virus	FHV	Grass grub (*Costelytra zealandica*)	New Zealand
Gypsy moth virus	GMV	Gypsy moth (*Lymantria ninayi*)	Papua New Guinea
Manawatu virus	MwV	Grass grub (*Costelytra zealandica*)	New Zealand
New Zealand virus	NZV	Unknown	Unknown
Nodamura virus	NoV	Mosquitoes (*Culex tritaeniorhynchus*)	Japan
Pariacoto virus	PaV	Southern armyworm (*Spodoptera eridania*)	Peru
Genus *Betanodavirus*			
Atlantic cod nervous necrosis virus	ACNNV	Atlantic cod (*Gadus morhua*)	Canada
Atlantic halibut nodavirus	AHNV	Atlantic halibut (*Hippoglossus hippoglossus*)	Norway
Barfin flounder nervous necrosis virus	BFNNV	Barfin flounder (*Verasper moseri*)	Japan
Dicentrarchus labrax encephalitis virus	DIEV	Sea bass (*Dicentrarchus labrax*)	France
Dragon grouper nervous necrosis virus	DGNNV	Dragon grouper (*Epinephelus lanceolatus*)	Taiwan
Greasy grouper nervous necrosis virus	GGNNV	Greasy grouper (*Epinephelus tauvina*)	Singapore
Grouper nervous necrosis virus	GNNV	Grouper (*Epinephelus coioides*)	Taiwan
Japanese flounder nervous necrosis virus	JFNNV	Japanese flounder (*Paralichthys olivaceus*)	Japan
Lates calcarifer encephalitis virus	LcEV	Barramundi (*Lates calcarifer*)	Israel
Malabaricus grouper nervous necrosis virus	MGNNV	Grouper (*Epinephelus malabaricus*)	Taiwan
Redspotted grouper nervous necrosis virus	RGNNV	Redspotted grouper (*Epinephelus akaara*)	Japan
Seabass nervous necrosis virus	SBNNV	Sea bass (*Dicentrarchus labrax*)	France
Striped jack nervous necrosis virus	SJNNV	Striped jack (*Pseudocaranx dentex*)	Japan
Tiger puffer nervous necrosis virus	TPNNV	Tiger puffer (*Takifugu rubrides*)	Japan
Umbrina cirrosa nodavirus	UCNNV	Shi drum (*Umbrina cirrosa*)	Italy

cells, but a number of insect cell lines that are susceptible to infection by this virus have been identified, including *Helicoverpa zea* FB33 cells. FHV, NoV, and PaV can also be propagated in baby hamster kidney cells when the genomic RNAs or cDNA clones of these viruses are transfected into these cells. Other heterologous expression systems that have been shown to support FHV replication include plant cells, the yeast *Saccharomyces cerevisiae*, and the worm *Caenorhabditis elegans*.

The incidence of virus nervous necrosis, the disease caused by betanodaviruses, has been reported in Asia, Australia, Europe, Japan, and North America. The natural hosts are predominantly hatchery-reared larvae and juveniles of fish species, but mortalities in adult fish have also been reported. In the laboratory, some betanodaviruses are infectious to a number of fish cell lines, including cultured cells from striped snakehead fish (SNN-1), sea bass larvae (SBL), and orange spotted grouper (GS). Striped jack nervous necrosis virus (SJNNV) also replicates when *in vitro* synthesized transcripts corresponding to its genomic RNAs are introduced into SSN-1 cells.

Virion Properties

The capsids of nodaviruses consist of 180 copies of a single gene product (or protomer) arranged with $T=3$ icosahedral symmetry. The protomers adopt three slightly different conformations based on the three quasi-equivalent positions within the capsid shell. Sixty of these protomers are arranged into 12 pentamers at the icosahedral fivefold axes, while 120 are arranged into 20 hexamers at the threefold axes. In alphanodaviruses, a significant percentage of the packaged genomic RNA is organized as duplex RNA. Specifically, in the case of PaV, X-ray crystallography visualized 25 bp of double-stranded RNA (dsRNA) at each of the 60 icosahedral twofold contacts of the virion (**Figure 1**). Together, these regions of duplex RNA represent 35% of the genome and they give the impression of a dodecahedral cage in the interior of the particle. RNA cage structures are not observed for betanodaviruses. High-resolution X-ray crystal structures and cryoelectron microscopy image reconstructions of several alphanodaviruses as well as a cryoelectron microscopy image reconstruction of the betanodavirus malabaricus grouper nervous necrosis virus (MGNNV) show that capsid structures are not conserved between the two genera (**Figure 2**).

Alphanodavirus Capsids

Alphanodavirus capsids are approximately 32–33 nm in diameter (**Figure 1(a)**). Each protomer in the viral capsid is initially composed of protein alpha (~44 kDa), which spontaneously cleaves into mature capsid proteins beta (~39 kDa) and gamma (~4 kDa) following assembly of virus particles. Beta represents the N-terminal portion of alpha protein, whereas gamma represents a short C-terminal peptide that remains associated with mature virions. Alpha protein (as well as beta protein) contains a central β-barrel motif that forms the spherically closed shell of the capsid. Loops between the β-strands form the exterior surface of the capsid. The N- and C-termini of alpha protein form the interior surface, which harbors the beta–gamma cleavage site and has predominate α-helical secondary structure. Biophysical studies on the dynamic behavior of FHV particles in solution have shown that the termini are transiently exposed at the exterior surface of the capsid.

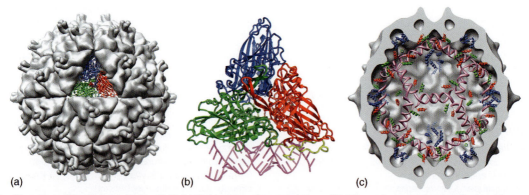

Figure 1 Structure of the alphanodavirus PaV. (a) Molecular surface rendering of PaV based on atomic coordinates. The coat protein subunits or protomers (shown in gray) adopt three different quasi-equivalent positions within the viral capsid. A-subunits are related by fivefold symmetry, while the B- and C-subunits are related by threefold symmetry. The high-resolution structure of one A-, B-, and C-protomer, which together represent one of the 60 icosahedral asymmetric units of the particle, is shown in blue, red, and green, respectively. (b) Top-down view of one icosahedral asymmetric unit showing the interaction between the extended N-terminus (yellow) of the A-subunit (blue) and the duplex RNA (magenta) located at the twofold contacts of the virion. (c) Cut-away view of the PaV virion showing the organization of the internal icosahedrally ordered duplex RNA and the gamma-peptides associated with the A-, B-, and C-subunits. Color coding is as described in panels (a) and (b). The bulk of the coat protein shell (gray) is shown at 22 Å resolution. Panels (a) and (c) were generated with the program Chimera. Courtesy of Dr. P. Natarajan.

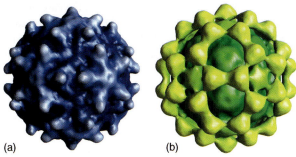

Figure 2 Three-dimensional cryoelectron microscopy reconstruction of (a) native PaV at 22 Å resolution and (b) MGNNV virus-like particle at 23 Å resolution. The inner capsid shell of MGNNV is shown in green, whereas outer protrusions are shown in yellow. Courtesy of Dr. P. Natarajan

The quaternary structure of the termini associated with protomers at the fivefold axes is markedly different from those at the threefold axes:

- Specifically, in PaV, the N-termini of threefold associated protomers are structurally disordered, while the N-termini of fivefold associated protomers are well ordered. Positively charged residues within these termini make extensive neutralizing contacts with the packaged RNA (**Figure 1(b)**). The resultant RNA–protein complexes play an important role in controlling the $T=3$ symmetry of the capsid. This is illustrated by aberrant assembly of FHV capsid proteins with deleted N-termini. These mutants assemble into multiple types of particles, including smaller 'egg'-shaped particles with regions of symmetry that are similar to those of $T=1$ particles.
- The C-terminal α-helices on the gamma-cleavage products of threefold-associated protomers interact with the duplex RNA, while their fivefold-associated counterparts do not contact RNA, but are grouped together into helical bundles along the fivefold axes of the capsid (**Figure 1(c)**). The positioning of these pentameric helical bundles within the capsid and the amphipathic character of each helix makes them ideal candidates for membrane-disruptive agents that are released from the virion during cell entry to facilitate the translocation of genomic RNA into host cells. These helices have in fact been shown to be highly disruptive to artificial membranes *in vitro* as judged by their ability to permeabilize liposomes to hydrophilic solutes.

Betanodavirus Capsids

MGNNV capsids are approximately 37 nm in diameter and thus slightly larger than alphanodavirus capsids. In addition, each of the protomers of the capsid consists of two structural domains as compared to the single domain structures of alphanodavirus protomers. The domains that are more internally located in the capsid form a contiguous shell around the packaged RNA while the outer domains form distinct surface protrusions on MGNNV capsids (**Figure 2**). Each betanodavirus protomer is composed of a single capsid protein (42 kDa) that does not undergo autocatalytic cleavage.

Virion Assembly and Specific Genome Packaging

The assembly pathway of alphanodaviruses is much better characterized than that of betanodaviruses. Alphanodavirus assembly proceeds through a precursor particle, the provirion, while an equivalent assembly intermediate is not evident for betanodaviruses. For alphanodaviruses, 180 copies of newly synthesized capsid precursor protein alpha assemble rapidly in the presence of excess viral RNA into the provirion intermediate, which has identical morphology to that of a mature virion. The assembly of provirions serves as a trigger for a maturation event in which alpha is autocatalytically cleaved into beta and gamma with a halftime of about 4 h. Maturation is required for acquisition of particle infectivity. A dependence on cleavage for infectivity is in agreement with the role of gamma as a membrane disruption agent for RNA translocation during cell entry.

Although mature virions display increased chemical stability compared to provirions, it has been shown that the capsid proteins of mature virions more readily unfold under high pressure. This characteristic combined with the dynamic properties of mature virions could also be critical for their infectivity, because it could accelerate virus uncoating during cell entry.

The two genomic strands of alphanodaviruses are co-packaged into a single virion during assembly. Compelling evidence for co-packaging comes from the discovery that the entire complement of packaged RNA1 and RNA2 forms a stable hetero-complex within FHV virions when these particles are exposed to heat. The mechanism by which FHV co-packages its genome is still unknown, but a number of molecular determinants for specific genome packaging have been elucidated:

- Mutants of capsid protein alpha lacking N-terminal residues 2–31 are unable to package RNA2. The N-terminal residues of this protein are therefore bifunctional for virion assembly, because they control both the specific packaging of RNA2 and the $T=3$ symmetry of capsids.
- The C-terminal residues carried on the gamma-region of alpha protein are not only critical for virus assembly and infectivity, but also for the specific packaging of both viral RNAs.
- A predicted stem–loop structure proximal to the 5' end of RNA2 (nt 186–217) has been proposed to represent a packaging signal for this genomic segment.

Despite the specificity for genomic RNA under native conditions, random cellular RNAs are readily packaged into virus-like particles (VLPs) when alpha is synthesized in a heterologous expression system. This is also true for the betanodaviruses MGNNV and dragon grouper nervous necrosis virus (DGNNV), whose capsid proteins have been synthesized in insect cells and *Escherichia coli*, respectively. The random RNAs packaged within alphanodavirus VLPs are organized into dodecahedral cages that are very similar to those of wild-type virions. Alpha protein is therefore the sole determinant for RNA cage formation, because specific RNA sequences and lengths are not important for this organization of the packaged nucleic acid.

Genome packaging is not only controlled by specific interactions between alpha protein and the viral RNAs, but also by a coupling between RNA replication and virion assembly. Only alpha proteins synthesized from replicating RNAs, as opposed to nonreplicating RNA, can package viral genomic RNA and partake in the assembly of infectious virions. This is exemplified by an ability to synthesize two distinct populations of FHV particles when alpha is co-synthesized from both replicating and nonreplicating RNAs within the same cell: (1) a population of infectious virions derived from the synthesis of alpha from replicating viral RNA, and (2) a population of VLPs containing random cellular RNA derived from the synthesis of alpha from nonreplicating mRNA.

Genome Structure and Coding Potential

The nodaviral genome consists of RNA1 and RNA2, and both RNAs are co-packaged into single virions. The 5′ ends of these RNAs are capped but their 3′ ends are not polyadenylated. Interestingly, the 3′ ends of alphanodavirus RNAs are unreactive with modifying enzymes to the 3′-hydroxyl groups of RNA molecules, while those of betanodavirus RNAs are marginally more reactive. It therefore appears as if these 3′ ends are blocked by an as-yet unidentified moiety or secondary structure.

A schematic representation of the genomic organization and replication strategy of FHV is shown in **Figure 3**. An open reading frame for the 112 kDa RdRp (protein A) nearly spans the entire length of FHV RNA1. RNA3, a 387 nt subgenomic RNA that is not packaged into virions, is synthesized from RNA1. RNA3 corresponds to the 3′ end of RNA1 and carries two open reading frames. One of these is in the +1 reading frame relative to protein A and

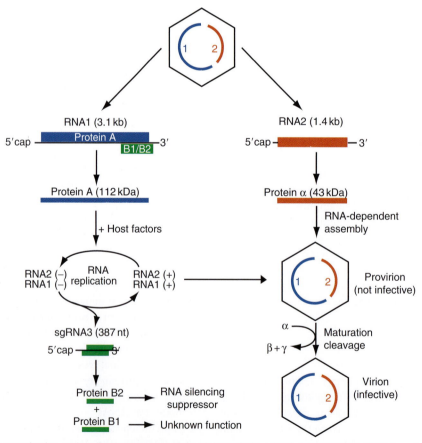

Figure 3 Genomic organization and replication strategy of the alphanodavirus FHV. Adapted from Ball LA and Johnson KL (1998) Nodaviruses of insects. In: Miller LK and Ball LA (eds.) *The Insect Viruses*, pp. 225–267. New York: Plenum.

encodes B2 (106 residues), which is required for the suppression of RNA silencing in infected host cells. The other matches the protein A reading frame and encodes B1 (102 residues) whose function is unknown. The coding potential of RNA1 and RNA3 is conserved within the family *Nodaviridae* with the exception that BoV RNA3 does not encode B1. RNA2 of all the species within this family has a single open reading frame for the viral capsid protein. In the genus *Alphanodavirus*, this protein is in the form of capsid precursor protein alpha, which undergoes cleavage to produce beta and gamma, while in the genus *Betanodavirus* it is in the form of a capsid protein that is not cleaved.

Regulation of Viral Gene Expression

The information on the replication cycle of nodaviruses in this and the next section pertains primarily to alphanodaviruses, which have been studied in much greater detail than betanodaviruses.

Following virion entry, RNA and protein synthesis are temporally regulated with the net effect of dividing the cellular infection cycle of nodaviruses into two phases. In the early phase, replication complexes are established to enable the amplification of viral RNAs, and in the following phase, the synthesis of capsid protein alpha is boosted to high levels to promote virion assembly. As described in the following section, RNA3 fulfills an important role in regulating this process, which is best characterized for FHV among all the nodaviruses.

RNA Synthesis

RNA1 and RNA2 are detectable at 5 h post infection. Rates of synthesis for these RNAs peak at about 16 h, but the RNAs continue to accumulate to very high levels within the cell throughout the remainder of the infection cycle which lasts 24–36 h. RNA3, on the other hand, is co-synthesized with RNA1 in the early hours of infection, but levels of this RNA decrease dramatically after 10 h. Interactions between RNA3 and RNA2 coordinate the synthesis of RNA1 and RNA2 in two ways. First, RNA3 is required as a transactivator for the replication of RNA2, which is a unique function for a subgenomic RNA. Second, RNA2 suppresses RNA3 synthesis, which accounts for the diminished levels of RNA3 at the later stages of the replication cycle.

Protein Synthesis

During the early stages of virus infection, proteins A and B2 accumulate within the cells to establish replication complexes and to suppress the RNA-silencing defenses within the cell, respectively. The rate of protein A synthesis peaks at 5 h post infection and B2 is optimally synthesized between 6 and 10 h. Subsequent to this, the synthesis of these proteins is suppressed, but both proteins are stable and remain detectable at the very late stages of infection. The shutdown of protein A synthesis is partly due to the increased translation efficiency for RNA2 as compared with RNA1. Conversely, the decreased rate of B2 synthesis is due to the RNA2-mediated suppression of RNA3 synthesis. The synthesis of capsid protein alpha lags 2–3 h behind the synthesis of proteins A and B2. Hereafter, alpha is synthesized at a remarkable rate, which peaks at 16 h, and it accumulates to cellular levels that are significantly greater than those of proteins A and B2.

RNA Replication

The replication cycle of nodaviruses occurs exclusively within the cytoplasm of infected cells and leads to the accumulation of RNA1 and RNA2 to levels that approximate those of ribosomal RNAs. The majority of these RNAs are positive stranded, but *c.* 1% are in the form of negative strands. Protein A is the only virus-encoded protein that is required for RNA replication, because it directs the autonomous replication of RNA1 in the absence of RNA2. In addition, the other proteins encoded on FHV RNA1 (B1 and B2) are redundant for replication in cells that do not inactivate FHV via the RNA-silencing pathway. Consistent with its role in RNA replication, protein A of both alpha- and betanodaviruses contains a GlyAsp Asp motif, as well as a number of other conserved motifs for RdRps.

Replication Complexes

The replication of nodaviral RNA occurs within spherules that have a diameter of 40–60 nm and are formed by the outer membranes of mitochondria in infected cells (**Figure 4**). Within each spherule, protein A functions as an integral membrane protein that is exposed on the cytoplasmic face of the outer membrane. The N-terminus of protein A acts both as a transmembrane domain that anchors it into the membrane and as a signal peptide for the targeting of this protein to mitochondria. It is presumed that multiple copies of protein A are associated with each spherule, because several copies thereof are visualized in spherules by immunoelectron microscopy, and this protein was shown to self-interact *in vivo* by co-immunoprecipitation and fluorescence resonance energy transfer.

Several alphanodaviruses are able to replicate their RNAs in diverse cellular environments, including those of insects, yeast, plant, worm, and mammalian cells. This shows that the formation of active replication complexes does not rely on factors that are specific to a particular host. The type of intracellular membrane is also not critical, because protein A mutants with N-terminal targeting sequences to endoplasmic reticulum membranes

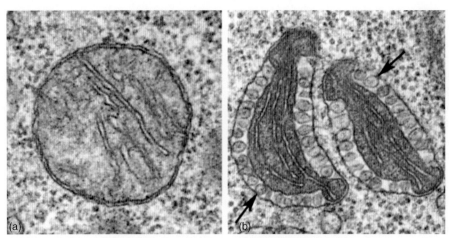

Figure 4 Electron micrographs of mitochondria in (a) uninfected and (b) FHV-infected *D. melanogaster* cells. Arrows point to spherules in which RNA replication occurs. Reproduced from Miller DJ, Schwartz MD, and Ahlquist P (2001) Flock house virus RNA replicates on outer mitochondrial membranes in drosophila cells. *Journal of Virology* 75: 11664–11676.

relocate to these membranes and support robust RNA replication. However, membrane association of protein A is important for complete replication of RNA1, RNA2, and RNA3 as gauged by the *in vitro* activities of partially purified preparations of this protein. Apart from its dependence on intracellular membranes, only one cellular host factor that supports RNA replication has been identified. This factor, the cellular chaperone heat shock protein 90, does not facilitate RNA synthesis *per se*, but plays an important role in the assembly of FHV RNA replication complexes.

cis-Acting Sequences Controlling Replication

During the replication of RNA1 and RNA2, the synthesis of negative strands is governed by approximately 50–58 nt at the 3′ end of positive-strand templates. Conversely, synthesis of positive strands requires fewer than 14 nt at the 3′ ends of negative-strand templates. Internally located *cis*-acting elements on the positive strands have been identified on both FHV RNA1 (nucleotides 2322–2501) and RNA2 (nucleotides 520–720). Furthermore, it is likely that a predicted secondary structure at the 3′ end of positive-strand RNA2 from FHV, BBV, BoV, and NoV acts as a *cis*-acting signal for the synthesis of RNA2 negative strands. This structure consists of two stem loops and is preceded by a conserved C-rich motif. No conserved sequences or secondary structures are discernible at the 3′ termini of RNA1.

A long-distance base-pairing interaction between two *cis*-acting elements on positive strands of RNA1 controls the transcription of the subgenomic RNA3. The base pairing occurs between a short element (approximately 10 nt) located 1.5 kbp upstream of the RNA3 start site and a longer element (approximately 500 nt) that is located proximal to the start site. It is highly likely that this interaction promotes the premature termination of negative-strand RNA1 synthesis, which would result in the synthesis of negative-strand RNA3. Evidence that the synthesis of negative-strand RNA3 precedes the synthesis of positive-strand RNA3 supports this hypothesis. Following its transcription from RNA1, RNA3 is able to replicate in an RNA1-independent manner.

Genetics

The nodavirus RdRp switches templates to generate two types of minor RNA species by RNA recombination: (1) defective interfering RNAs (DI-RNAs) and (2) RNA dimers. DI-RNAs are internally deleted products of the genomic RNAs and often contain sequence rearrangements. RNA dimers, on the other hand, contain head-to-tail junctions between two identical copies of RNAs 1, 2, or 3, or between a copy each of RNA2 and RNA3. Both positive- and negative-strand versions of these RNAs are generated in infected cells and it is unknown whether dimers play a role in RNA replication.

Reassortment between RNA1 and RNA2 could possibly play a role in the evolution of nodaviruses. However, evidence for this is only experimental in nature, as it was shown that the RdRps from FHV, BBV, and BoV can replicate each other's RNAs and that all six of the possible reassortants are infectious.

Suppression of RNA Silencing by Protein B2

B2 plays a critical role in the life cycle of alphanodaviruses as a suppressor of RNA silencing, which acts as an antiviral defense mechanism in the insect hosts of these viruses. RNA silencing is triggered by dsRNA, which may be present at the onset of an alphanodavirus infection as RNA replication intermediates or as structured regions within single-stranded viral RNAs. During the RNA-silencing pathway, this dsRNA is processed into small (21–25 bp) interfering RNAs (siRNAs) by the endoribonuclease Dicer, and the siRNAs, in turn, guide the RNA-induced silencing complex (RISC) to specifically degrade the viral genome.

B2 has been proposed to suppress RNA silencing at two stages of this pathway. First, it binds to the siRNA cleavage products of Dicer and can therefore prevent their incorporation into RISC. Second, it is able to bind to long dsRNAs and may therefore inhibit the formation of siRNAs from full-length viral RNAs. B2 is a homodimer that forms a four-helix bundle (**Figure 5**). The dimer interacts with one face of the RNA duplex and makes several contacts with two minor grooves and the intervening major groove. These contacts are exclusively with the ribose phosphate backbone of the RNA helix, which explains the sequence-independent binding of B2 to dsRNA.

The role of alphanodaviral B2 as a suppressor of RNA silencing is not limited to insects, but extends to the RNA silencing pathways of plants, worms (*C. elegans*), and mammalian cells. Preliminary studies on betanodaviral B2 suggest that this protein functions as an inhibitor of RNA silencing in fish.

Virus–Host Cell Interactions

A general phenomenon of nodavirus infection is the presence of large cytoplasmic paracrystalline arrays of virus particles. In addition, betanodavirus-infected cells exhibit extensive vacuolation, while alphanodavirus-infected cells characteristically show a clustered distribution of mitochondria. The mitochondria are markedly elongated and deformed, and show the presence of numerous spherules on their surface (see **Figure 4**). In cell culture, alphanodavirus infections are accompanied by a progressive shutdown of host protein synthesis and the accumulation of virus particles to very high yields. Despite the high virus yields, most cell types permissive to infection by these viruses remain intact. The exceptions are *D. melanogaster* line 1 cells infected with either FHV or BBV, because these cells undergo cytolysis at 3 days post infection.

Both FHV and BBV are also able to establish persistent infections in *D. melanogaster* cells, rendering these cells immune to superinfection by FHV and BBV. Persistence is not established by mutations within the viral genes and could therefore only be attributed to cellular changes that affect interactions between the virus and its host.

Epidemiology and Pathogenesis

Disease symptoms for the natural insect hosts of alphanodaviruses have not been characterized, but are presumed to be similar to those of experimentally infected *G. mellonella* larvae, which suffer stunted growth, severe paralysis, and 100% mortality. In general, these viruses show a wide histopathological distribution within insects, and are readily detectable in neural and adipose tissues, midgut cells, salivary glands, muscle, trachea, and fat body. NoV causes severe hind limb paralysis in suckling mice, making it the only alphanodavirus pathogenic to a mammalian host. This disease symptom is caused by the replication of this virus in muscle cells and by the degeneration of spinal cord neurons.

Betanodaviruses are the causative agents of virus nervous necrosis in larvae and juveniles of both marine and freshwater fish, and cause significant problems to the marine aquaculture industry. The ability of these viruses to be transmitted both horizontally and vertically is most probably a contributing factor to the epizootic infections that have been caused by these viruses among hatchery-reared fish. Infected fish larvae show abnormal swimming behavior, which is caused by the development of vacuolating encephalopathy and retinopathy. Like alphanodaviruses, betanodaviruses are also able to establish

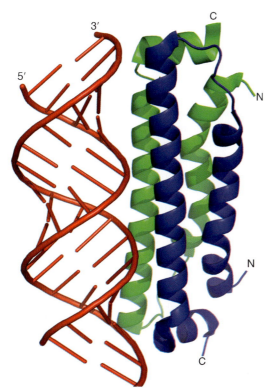

Figure 5 Structure of FHV B2 bound to dsRNA. Blue and green: B2 dimer; red: RNA. B2 forms a homodimer that binds dsRNA in a sequence-independent manner. The interactions occur between the flat face of the B2 molecule and the phosphodiester backbone of the RNA. Courtesy of Dr. P. Natarajan.

persistent infections in their fish hosts. This is evident in the detection of these viruses in large populations of marine fish that show no clinical signs of viral nervous necrosis. The development and instatement of screening methods for the detection of these subclinical infections are therefore of great importance to the marine aquaculture industry.

See also: Fish Viruses; Viral Suppressors of Gene Silencing.

Further Reading

Ball LA and Johnson KL (1998) Nodaviruses of insects. In: Miller LK and Ball LA (eds.) *The Insect Viruses*, pp. 225–267. New York: Plenum.

Chao JA, Lee JH, Chapados BR, *et al.* (2005) Dual modes of RNA-silencing suppression by flock house virus protein B2. *Nature Structural and Molecular Biology* 12: 952–957.

Delsert C, Morin N, and Comps M (1997) Fish nodavirus lytic cycle and semipermissive expression in mammalian and fish cell cultures. *Journal of Virology* 71: 5673–5677.

Li H, Li WX, and Ding SW (2002) Induction and suppression of RNA silencing by an animal virus. *Science* 296: 1319–1321.

Lindenbach BD, Sgro JY, and Ahlquist P (2002) Long-distance base pairing in flock house virus RNA1 regulates subgenomic RNA3 synthesis and RNA2 replication. *Journal of Virology* 76: 3905–3919.

Miller DJ, Schwartz MD, and Ahlquist P (2001) Flock house virus RNA replicates on outer mitochondrial membranes in Drosophila cells. *Journal of Virology* 75: 11664–11676.

Nishizawa T, Furuhashi M, Nagai T, Nakai T, and Muroga K (1997) Genomic classification of fish nodaviruses by molecular phylogenetic analysis of the coat protein gene. *Applied and Environmental Microbiology* 63: 1633–1636.

Schneemann A, Ball LA, Delsert C, Johnson JE, and Nishizawa T (2005) Family *Nodaviridae*. In: Fauquet CM, Mayo MA, Maniloff J, Desselberger U, and Ball LA (eds.) *Virus Taxonomy: Eighth Report of the International Committee on Taxonomy of Viruses*, pp. 865–872. San Diego, CA: Elsevier Academic Press.

Schneemann A, Reddy V, and Johnson JE (1998) The structure and function of nodavirus particles: A paradigm for understanding chemical biology. *Advances in Virus Research* 50: 381–446.

Tang L, Johnson KN, Ball LA, *et al.* (2001) The structure of pariacoto virus reveals a dodecahedral cage of duplex RNA. *Nature Structural Biology* 8: 77–83.

Tang L, Lin CS, Krishna NK, *et al.* (2002) Virus-like particles of a fish nodavirus display a capsid subunit domain organization different from that of insect nodaviruses. *Journal of Virology* 76: 6370–6375.

Venter PA, Krishna NK, and Schneemann A (2005) Capsid protein synthesis from replicating RNA directs specific packaging of the genome of a multipartite, positive-strand RNA virus. *Journal of Virology* 79: 6239–6248.

Noroviruses and Sapoviruses

K Y Green, National Institutes of Health, Bethesda, MD, USA

© 2008 Published by Elsevier Ltd.

Glossary

Fomites An object that in itself is not harmful, but that can harbor a pathogenic organism and therefore serve as an agent of its transmission.

Gastroenteritis An inflammation of the stomach and intestines, often characterized by symptoms of vomiting and diarrhea.

Introduction

Diarrheal illnesses have a major impact on public health. The association of bacteria with such illness was recognized over a century ago with the discovery of bacterial pathogens such as *Vibrio cholerae* and *Shigella dysenteriae*. The role of viruses in diarrheal illness was established much later. Volunteer studies in the 1940s showed that bacteria-free filtrates of feces obtained from individuals with diarrheal disease could induce a similar illness in volunteers who were challenged orally with the inoculum. In 1972, the Norwalk virus was identified in human feces and shown to be the etiologic agent of an outbreak of gastroenteritis that occurred in an elementary school in Norwalk, Ohio in 1968. The Norwalk virus became the prototype strain for a large group of related caliciviruses known now as the noroviruses. A second group of distantly related caliciviruses, the sapoviruses, was discovered soon afterwards.

The family *Caliciviridae* includes several important human and animal pathogens, including the noroviruses and sapoviruses associated with acute gastroenteritis (**Table 1**). The human noroviruses, named for the prototype strain *Norwalk virus*, are now recognized as a major cause of gastroenteritis in all age groups. Norovirus outbreaks often occur in settings such as communities, nursing homes, schools, hospitals, cruise ships, camps, social gatherings, families, and military personnel. Norovirus illnesses can also occur sporadically in the community. A common name for norovirus gastroenteritis is 'stomach flu', but this is a misnomer because the noroviruses are not related to influenza virus. The acute, symptomatic phase of the illness, which often includes either vomiting or diarrhea, or both, generally lasts from 24 to 48 h. In most cases, norovirus gastroenteritis is mild and self-limiting, but it can be

Table 1 Taxonomy of the *Caliciviridae*

Genus	Species	Type strain
Norovirus (NoV)	*Norwalk virus* (NV)	Hu/NoV/GI.1/Norwalk/1968/US
Sapovirus (SaV)	*Sapporo virus* (SV)	Hu/SaV/GI.1/Sapporo/1982/JP
Lagovirus (LaV)	*Rabbit hemorrhagic disease virus* (RHDV)	Ra/LaV/RHDV/GH/1988/DE
	European brown hare syndrome virus (EBHSV)	Ha/LaV/EBHSV/GD/1989/FR
Vesivirus (VeV)	*Vesicular exanthema of swine virus* (VESV)	Sw/VeV/VESV/VESV-A48/1948/US
	Feline calicivirus (FCV)	Fe/VeV/FCV/F9/1958/US

Calicivirus strains are written in a cryptogram format that is organized as follows: Host species from which the virus was obtained/genus/species (or genogroup)/strain name/year of occurrence/country of origin. Abbreviations for the host species are: Fe, feline; Ha, Hare; Hu, Human; Sw, Swine; Ra, Rabbit. Country abbreviations are: DE, Germany; FR, France; JP, Japan; US, United States. GenBank Accession numbers of representative viruses: Norwalk virus, M87661; Sapporo virus, U65427; RHDV, M67473; VESV, AF181082; FCV, M86379.

incapacitating during the symptomatic phase. The illness can sometimes be severe and prolonged in certain individuals such as either the very old or young, or those compromised by pre-existing illness or immunosuppressive therapy.

The sapoviruses, named for the prototype strain, *Sapporo virus*, are characteristically associated with pediatric gastroenteritis, but outbreaks and illness in older individuals can occur. Although illness is characteristically mild, severe disease has been reported. The sapoviruses do not appear to be a major cause of epidemic gastroenteritis, in contrast to the noroviruses. It should also be noted that although sapoviruses and noroviruses have both been associated with pediatric gastroenteritis, and the noroviruses are considered to be the second most important cause of such illnesses, the major cause by far of severe nonbacterial diarrhea in infants and young children is the rotavirus, which belongs to a different virus family, the *Reoviridae*.

Caliciviruses have been detected in a large number of animal species and are associated with a diverse range of clinical syndromes (**Table 2**). Veterinary pathogens of note in the *Caliciviridae* include feline calicivirus (FCV), an important cause of respiratory illness in cats, and rabbit hemorrhagic disease virus (RHDV), an often-fatal acute infection in rabbits.

Taxonomy and Classification

The taxonomy of the noroviruses and sapoviruses is based primarily on the phylogenetic relationships among strains (**Figure 1**). Members of the family *Caliciviridae* apparently share a common ancestor, and the four major phylogenetic groups define the four genera: *Norovirus*, *Sapovirus*, *Lagovirus*, and *Vesivirus* (**Table 1**). Members within each genus share common features in their genomic organization, but host range and disease manifestations may vary among members of each genus. Each genus is further divided into one or more species, with each species represented by a 'type' strain (**Table 1**). The genus *Lagovirus* is comprised of two species, *Rabbit hemorrhagic disease virus* and *European brown hare syndrome virus*, and the genus

Table 2 Natural host range and disease syndromes associated with caliciviruses

Calicivirus Genus	Host	Disease syndrome
Norovirus	Human	Gastroenteritis
	Porcine	Gastroenteritis
	Bovine	Gastroenteritis
	Murine	Asymptomatic enteric infection in wild type mice severe systemic disease in mice lacking STAT1
Sapovirus	Human	Gastroenteritis
	Porcine	Gastroenteritis
	Mink	Gastroenteritis
Vesivirus[a]	Porcine	Vesicular exanthema, abortion
	Sea lion	Vesicular lesions
	Chimpanzee	Vesicular lesions
	Canine	Gastroenteritis
	Walrus	None described
	Mink	None described
	Bovine	None described
	Reptilian	None described
	Skunk	None described
	Amphibian	None described
	Cetacean	None described
	Feline	Upper respiratory, pneumonia, oral ulceration
Lagovirus	Rabbit	Hepatitis, disseminated intravascular coagulation
	Hare	Hepatitis, disseminated intravascular coagulation

[a]There is a report documenting human infection with a vesivirus isolated from a sea lion.

Vesivirus is subdivided into two species, *Vesicular exanthema of swine virus* and *Feline calicivirus*. Presently, the genera *Norovirus* and *Sapovirus* have only one species each, designated as *Norwalk virus* and *Sapporo virus*, respectively.

The official classification system of viruses by the International Committee on Taxonomy of Viruses (ICTV) does not address classification below the species level. However, several genetic typing systems have been developed to facilitate communication among researchers in tracking the spread of epidemic strains. These genetic typing systems are based on sequence relatedness in a selected

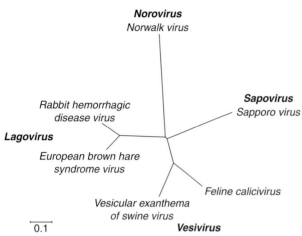

Figure 1 Phylogenetic relationships of the *Caliciviridae*. Caliciviruses in the four genera (*Norovirus*, *Sapovirus*, *Lagovirus*, and *Vesivirus*) cluster into four major groups when the nucleotide sequences of full-length genomes are compared in a phylogenetic analysis (Neighbor-joining, Jukes-Cantor distance parameter). Each genus can be further divided into groups that represent distinct species.

region of the viral genome. One such approach is based on comparisons of the major capsid protein-encoding gene as illustrated for the noroviruses and sapoviruses in **Table 3**. In these genetic typing systems, the noroviruses and sapoviruses are first divided by sequence relatedness into major branches within the genus termed 'genogroups'. For example, noroviruses are presently divided into five distinct genogroups designated I, II, III, IV, and V. Each genogroup is further divided into genetic clusters that are designated as 'genotypes'. For example, genogroup I of the noroviruses is divided into eight genotypes, and genogroup II is divided into 19 genotypes (**Table 3**). The role of genetic diversity in the natural history of the noroviruses and sapoviruses, and the correlation of genotypes with antigenicity is not yet known. However, the classification and genetic typing systems allow the utilization of a cryptogram to describe strains. The cryptogram is organized as follows: Host species from which the virus was obtained/genus/species or genetic type/strain name/year of occurrence/country of origin. For example, Norwalk virus would be designated as: Hu/NoV/GI.1/Norwalk/1968/US.

Virion Structure and Composition

Virus particles are isometric, ranging from 27 to 35 nm in diameter, with icosahedral symmetry. Human caliciviruses such as Norwalk virus (representing the noroviruses) and Parkville virus (representing the sapoviruses) can be observed in stool specimens by negative-stain-electron microscopy (EM) as shown in **Figure 2**. Cryo-EM and computer-generated reconstructions of recombinant virus-like particles produced by expression of the major capsid protein (VP1) of the Norwalk or Parkville virus illustrates the presence of the cup-shaped depressions on the surface, a common feature of caliciviruses that inspired the family *Caliciviridae* (*calix* means 'cup' in Latin) (**Figure 2**). The mature virion contains two structural proteins, VP1 and VP2. VP1 (approximately 60 kDa in most strains) is the predominant capsid protein and is present in 180 copies per virion. The VP1 contains the major antigenic and receptor binding sites of the virus. The VP2 (ranging in size from approximately 8 to 29 kDa) is present in an estimated one or two copies per virion, and may play a role in particle assembly and stability. A third protein found in virions is the VPg, an approximately 15 kDa protein that is covalently linked to the viral RNA genome. The presence of the VPg protein on the viral RNA is required for the efficient initiation of an infection following entry of the viral RNA into cells, most likely because of the proposed role of VPg in mediating interactions between the viral RNA and the host cell translation machinery.

Genome Organization and Expression

The genome of the caliciviruses is a positive-sense RNA molecule ranging in length from approximately 7.4 to 8.3 kbp. The 5′-end of the genome is covalently linked to a protein, VPg, and the 3′-end of the genome contains a poly (A) tract. There is a relatively short nontranslated region (NTR) at both the 5′- and 3′-ends of the genome that flank the open reading frames (ORFs). Genomes of the caliciviruses are organized into either two or three major ORFs, depending on the genus (**Figure 3**). The first ORF (beginning near the 5′-end) of all caliciviruses encodes the nonstructural (NS) proteins of the virus, whereas the terminal ORF encodes the minor structural protein, VP2. The major difference in reading frame usage among the caliciviruses relates to the coding sequence of the major capsid protein, VP1. In the noroviruses and vesiviruses, the VP1 is encoded in a separate reading frame (ORF2), whereas in the sapoviruses and lagoviruses, the VP1 is encoded in the same ORF (ORF1) as the NS proteins. Thus, the norovirus genome is organized into three major ORFs, and the sapovirus genome is organized into two major ORFs.

Caliciviruses encode six to seven mature NS proteins, depending again on the genus. This variation is due to differences among the proteolytic processing strategies used by viruses to cleave the ORF1 polyprotein. In the lagoviruses, the virus-encoded cysteine proteinase cleaves the ORF1 polyprotein at six cleavage sites to release seven mature NS proteins, designated as NS1 through NS7 (**Table 4**). In some genera, cleavage between certain NS proteins has not been detected during virus replication. This includes the NS1 and NS2 precursor of the noroviruses (designated as NS1–2 in this nomenclature system, and known also as the N-terminal protein or p48) and the

Table 3 Norovirus and sapovirus genetic typing systems

Reference virus	Genogroup	Genotype	GenBank accession number
Noroviruses			
Hu/NoV/GI.1/Norwalk/1968/US	I	1	M87661
Hu/NoV/GI.2/Southampton/1991/UK	I	2	L07418
Hu/NoV/GI.3/Desert Shield 395/1990/SA	I	3	U04469
Hu/NoV/GI.4/Chiba 407/1987/JP	I	4	AB042808
Hu/NoV/GI.5/Musgrove/1989/UK	I	5	AJ277614
Hu/NoV/GI.6Hesse 3/1997/DE	I	6	AF093797
Hu/NoV/GI.7/Winchester/1994/UK	I	7	AJ277609
Hu/NoV/GI.8/Boxer/2001/US	I	8	AF538679
Hu/NoV/GII.1/Hawaii/1971/US	II	1	U07611
Hu/NoV/GII.2/Melksham/1994/UK	II	2	X81879
Hu/NoV/GII.3/Toronto 24/1991/CA	II	3	U02030
Hu/NoV/GII.4/Bristol/1993/UK	II	4	X76716
Hu/NoV/GII.5/Hillingdon/1990/UK	II	5	AJ277607
Hu/NoV/GII.6/Seacroft/1990/UK	II	6	AJ277620
Hu/NoV/GII.7/Leeds/1990/UK	II	7	AJ277608
Hu/NoV/GII.8/Amsterdam/1998/NL	II	8	AF195848
Hu/NoV/GII.9/VA97207/1997	II	9	AY038599
Hu/NoV/GII.10/Erfurt546/2000/DE	II	10	AF427118
Sw/NoV/GII.11/Sw918/1997/JP	II	11	AB074893
Hu/NoV/GII.12/Wortley/1990/UK	II	12	AJ277618
Hu/NoV/GII.13/Fayetteville/1998/US	II	13	AY113106
Hu/NoV/GII.14/M7/1999/US	II	14	AY130761
Hu/NoV/GII.15/J23/1999/US	II	15	AY130762
Hu/NoV/GII.16/Tiffin/1999/US	II	16	AY502010
Hu/NoV/GII.17/CS-E1/2002/US	II	17	AY502009
Sw/NoV/GII.18/OH-QW101/2003/US	II	18	AY823304
Sw/NoV/GII.19/OH-QW170/2003/US	II	19	AY823306
Bo/NoV/GIII.1/Jena/1980/DE	III	1	AJ011099
Bo/NoV/CH126/1998/NL	III	2	AF320625
Hu/NoV/GIV.1/Alphatron 98–2/1998/NL	IV	1	AF195847
Mu/NoV/GV.1/MNV-1/2003	V	1	AY228235
Sapoviruses			
Hu/SaV/GI.1/Sapporo/1982/JP	I	1	U65427
Hu/SaV/GI.2/Parkville/1994/US	I	2	U73124
Hu/SaV/GI.3/Stockholm/1997/SE	I	3	AF194782
Hu/SaV/GII.1/London/1992/UK	II	1	U95645
Hu/SaV/GII.2/Mex340/1990/MX	II	2	AF435812
Hu/SaV/GII.3/Cruise ship/2000/US	II	3	AY289804
Sw/SaV/GIII/PEC-Cowden/1980/US	III	1	AF182760
Hu/SaV/GIV/Hou7–1181/1990/US	IV	1	AF435814
Hu/SaV/GV/Argentina39/AR	V	1	AY289803

Note: Norovirus and sapovirus typing systems from Zheng et al., and Farkas et al., respectively.

NS6 and NS7 precursor of the vesiviruses (NS6–7, also known as ProPol). The calicivirus NS proteins and their functions, if known, are summarized in **Table 4**. Key enzymes include an NTPase (NS3), a cysteine proteinase (NS6), and an RNA-dependent RNA polymerase (NS7).

Replication

Cell culture systems are not available for all caliciviruses, but those that grow in cell culture, such as FCV and murine norovirus characteristically grow efficiently and form plaques (**Figure 4**). Like other positive-strand RNA viruses, caliciviruses replicate in the cytoplasm of infected cells. The positive-strand RNA genome functions as a messenger RNA that is translated following entry of the virus particle into the cell. This initial translation event produces NS proteins that are used in the replication of the viral RNA genome and the production of new progeny. Replication of the RNA occurs on intracellular membranes and extensive membrane rearrangements are observed in infected cells (**Figure 4**). The viral capsids can sometimes be seen in infected cells as paracrystalline arrays (**Figure 4**). There are two major positive-strand

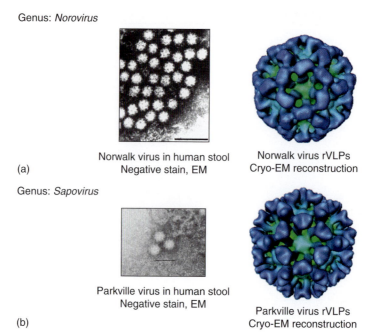

Figure 2 Structural features of noroviruses and sapoviruses. (a) Genus: *Norovirus*. (Left panel) Norwalk virus, a norovirus, as seen in human stool material by negative-staining and immune EM (IEM). The size of Norwalk virus in stool material has been described as 27–32 nm in diameter. Scale = 100 nm. (Right panel) Norwalk virus recombinant virus-like particles (rVLPs) were expressed in the baculovirus system, purified, and analyzed with the technique of cryo-EM. (b) Genus: *Sapovirus*. (Left panel) Parkville virus, a sapovirus, as seen in human stool material by negative staining and EM. Scale = 50 nm. (Right panel) Parkville virus rVLPs were analyzed by cryo-EM. The cryo-EM computer-generated images of both Norwalk virus and Parkville virus rVLPs show that the surfaces of the particles have cup-shaped depressions, a structural hallmark of the caliciviruses. (a, left panel) Image provided by A. Z. Kapikian, National Institutes of Health. (a, b, right panel) Image provided by B. V. Prasad, Baylor College of Medicine. (b, left panel) Image provided by C. D. Humphrey, Centers for Disease Control and Prevention.

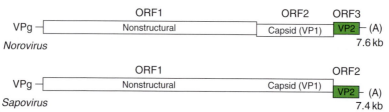

Figure 3 Genome organization of noroviruses and sapoviruses. The positive sense RNA genome of Norwalk virus (representing the noroviruses) is organized into three major ORFs. The genome of Manchester virus (representing the sapoviruses) is organized into two major ORFs, with the VP1 capsid coding sequence in frame with that of the NS proteins. (The Norwalk virus and Manchester virus genome sequences correspond to GenBank accession numbers M87661 and X86560, respectively.)

RNA species detected in infected cells – one corresponds to the full-length genome, and the other corresponds to an abundant subgenomic RNA co-terminal with the 3′-end region of the genome. The NS proteins are translated from the full-length RNA. The subgenomic RNA serves as a messenger RNA for translation of the VP1 and VP2, the two structural proteins.

Evolution

Comparative sequence analysis suggests that members of the *Caliciviridae* have a common ancestor. A distant

Table 4 Calicivirus proteins

Virion proteins (VP)	
VP1	Major structural protein of virion (~60 kDa)
VP2	Minor structural protein, function unknown
VPg	Covalently linked to genomic RNA
Nonstructural proteins (NS)	
NS1	N-terminal protein of ORF1, function unknown
NS2	Function unknown
NS3	NTPase
NS4	Function unknown
NS5	Becomes linked to RNA as VPg, role in translation and replication
NS6	Proteinase (Pro)
NS7	RNA-dependent RNA Polymerase (Pol)

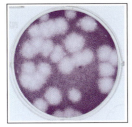
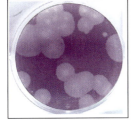

Plaque morphology

Feline calicivirus — Murine norovirus

Intracellular membrane rearrangements

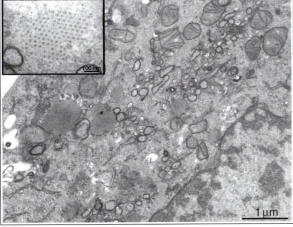

FCV-infected feline kidney cells

Figure 4 (Upper panel) Plaque morphology and membrane rearrangements in an infected cell. Feline calicivirus and murine norovirus induce a rapid cytopathic effect in permissive cell culture. Plaques can be detected in a cell monolayer by 24 h post infection. (Lower panel) Caliciviruses, like other positive strand RNA viruses, induce membrane rearrangements and vesicles in infected cells. The inset shows a paracrystalline array of capsid proteins produced in an FCV-infected cell. Images provided by A. Weisberg.

evolutionary relationship has been proposed between the *Caliciviridae* and members of the family *Picornaviridae*. A striking feature among the caliciviruses is their extensive genetic variation. This variation is due, in part, to the high error rate inherent in positive-strand RNA virus replication, which employs an RNA-dependent RNA polymerase without a known editing function for correcting mistakes. There is compelling evidence also that recombination occurs between related strains, which would then allow the emergence of new viruses. Potential recombination sites have been identified in norovirus genomes at several positions, including the junction between the ORF1 and ORF2, which would allow a virus to emerge with a new subgenomic region that would encode a different VP1 capsid protein.

Transmission and Host Range

Transmission of the noroviruses and sapoviruses occurs by several modes, including direct person-to-person contact, ingestion of contaminated food or water, or exposure to contaminated fomites. Noroviruses are a major cause of foodborne gastroenteritis. The route of transmission is predominantly fecal–oral, and viruses can be shed in feces for days to weeks following infection. The virus has been detected in vomitus, and transmission may occur also via exposure to aerosolized droplets of vomitus. Good personal hygiene and frequent hand washing is important in controlling the spread of these highly infectious viruses.

Noroviruses have been found in several mammalian species, including humans, pigs, cattle, and mice, and it is likely that additional hosts will be identified. Sapoviruses have also been found in several hosts thus far, including humans, pigs, and mink. There is little evidence that zoonotic transmission of animal caliciviruses to humans (or humans to animals) plays a role in the natural history of these viruses, but such transmission might be possible considering that certain human norovirus strains can infect chimpanzees and pigs in experimental challenge studies.

Pathogenicity

Noroviruses and sapoviruses replicate in the enteric tract, and viral antigen-expressing cells (for both groups of viruses) have been identified in the small intestinal epithelial cells of swine. The cell types supporting the replication of the human noroviruses and sapoviruses have not been verified *in vivo*, but intestinal biopsies have shown blunting of the intestinal villi following infection of adult volunteers with Norwalk virus. The mechanisms responsible for diarrhea are not known, but the lesion caused by blunting of the intestinal villi may affect absorption and the osmotic balance. Murine norovirus is shed in feces and can be found in multiple organs of asymptomatic normal mice, including the intestine, liver, spleen, and mesenteric lymph nodes. However, certain strains of murine norovirus can cause a highly lethal infection in immunodeficient mice lacking key components of the innate immune system. In these mice, a lethal, disseminated infection occurs leading to encephalitis, meningitis, vasculitis, hepatitis, and pneumonia.

Clinical Features of Infection

Norovirus gastroenteritis is characterized by a short incubation period and an acute onset of illness. Symptoms include one or more of the following: diarrhea, vomiting, fever, malaise, and abdominal cramps. The illness is generally considered mild and self-limiting, but severe illness can occur. Treatment is aimed toward the prevention of dehydration, and includes either oral rehydration or intravenous administration of fluids. Noroviruses have been identified in several studies as the second most important viral agent (but considerably less than rotaviruses) associated with severe gastroenteritis in infants and young children.

The sapoviruses have been associated predominantly with diarrhea that can range from mild to severe. The illness is usually self-limiting, but the prevention of dehydration is important.

Noroviruses, sapoviruses, and other caliciviruses have been associated with asymptomatic infection, and persistent infection has been documented. Shedding of the virus can occur for prolonged periods (weeks to months) following resolution of symptoms in some patients.

Immune Response

Immunity to the noroviruses is poorly understood, because it has been difficult to study the role of neutralizing antibodies in the absence of a cell culture system. Adult volunteer challenge studies have been the only available approach for the study of resistance to illness in humans. It was noted early in these volunteer studies that the presence of pre-existing serum or local intestinal antibodies to the virus did not correlate with resistance to illness, but rather was associated with susceptibility to infection and illness. In contrast, later studies suggested that local immunity in the intestine played an important role in mediating protection, but immunity to Norwalk virus was short term. Susceptibility to infection may also be associated with host genetic factors relating to the presence or absence of receptors for the virus on intestinal epithelial cells. There is evidence that noroviruses bind to histo-blood group antigens on intestinal epithelial cells, and these antigens might serve as receptors or binding ligands for the virus. Because histo-blood group antigens vary among individuals, and variation among norovirus strains has been detected in the recognition of these antigens, it has been postulated that this might, in part, provide an explanation for varying host susceptibility. The understanding of immunity to the noroviruses is complicated by the marked antigenic diversity among circulating strains. Limited cross-challenge studies have found evidence for at least two distinct serotypes, represented by the Norwalk and Hawaii norovirus strains, but the number of serotypes is presently unknown.

The predominant association of sapovirus gastroenteritis with individuals in younger age groups suggests that immunity to sapoviruses may be acquired early in life.

Prevention and Control

There are presently no vaccines or antiviral therapies for the control of norovirus and sapovirus gastroenteritis. Prevention and control strategies currently consist of standard infection control procedures, such as frequent hand washing, avoidance of exposure, and disinfection of contaminated areas.

Diagnosis

Norovirus outbreaks characteristically have the following epidemiological features, known as the 'Kaplan criteria': (1) a mean (or median) illness duration of 12–60 h, (2) a mean (or median) incubation period of 24–48 h, (3) more than 50% of people with vomiting, and (4) no bacterial agent previously found. Although these features are generally highly specific for the provisional diagnosis of a norovirus outbreak, laboratory tests are recommended for ruling out bacteria and confirming a viral etiology. The most widely used method for detection of noroviruses and sapoviruses is reverse transcriptase-polymerase chain reaction (RT-PCR) analysis of viral RNA in stool specimens. The extensive genetic variation among the noroviruses has been problematic in the design of broadly reactive primer pairs, and many laboratories use more than one primer pair for detection.

Noroviruses and sapoviruses can be observed in stool specimens by EM, but this technique often requires the use of antibodies to facilitate detection (immune EM). Commercial immunoassays are becoming available for clinical use in which broadly reactive antibodies are used to detect norovirus antigen in stool specimens.

Vaccination

Vaccines are not available for the noroviruses, but studies are in progress to identify and evaluate potential vaccine candidates. Most vaccine candidates are based on the production of recombinant norovirus VP1, the major capsid protein. The VP1 will spontaneously self-assemble into virus-like particles (VLPs) that are antigenically similar to native virions and can be produced in high yields in expression systems such as baculovirus. The VLPs are immunogenic when administered orally to humans.

Vaccines have been developed and are in use for caliciviruses such as FCV (for use in cats) and RHDV (for use in rabbits). Because FCV grows efficiently in cell culture, a live attenuated vaccine has been developed, as well as an inactivated vaccine.

Future Perspectives

Noroviruses are the major cause of nonbacterial gastroenteritis outbreaks, and large numbers of illnesses occur each year. The sapoviruses play a lesser role in epidemic gastroenteritis, but further studies are needed. Efforts will continue to focus on the development of improved diagnostic tests and in the establishment of a cell culture system for the human noroviruses and sapoviruses. It will also be important to determine whether vaccines or antiviral drugs will be effective in controlling norovirus disease. A norovirus vaccine candidate based on the expression of recombinant virus-like particles is under investigation in early clinical trials. Promising new tools for study of the noroviruses include the first efficient cell culture system (murine norovirus), a human norovirus infectivity assay using a three-dimensional cell culture system, a cell-based human norovirus replicon system (Norwalk virus), a reverse genetics system for the murine noroviruses, and an animal disease model (pigs) for the human noroviruses. In addition, advances in the elucidation of the structure and function of individual viral proteins may lead to a better understanding of replication in this important group of viruses.

See also: Caliciviruses; Enteric Viruses.

Further Reading

Asanaka M, Atmar RL, Ruvolo V, Crawford SE, Neill FH, and Estes MK (2005) Replication and packaging of Norwalk virus RNA in cultured mammalian cells. *Proceedings of the National Academy of Sciences, USA* 102: 10327–10332.

Atmar RL and Estes MK (2006) The epidemiologic and clinical importance of norovirus infection. *Gastroenterology Clinics of North America* 35: 275–290.

Chang KO, Sosnovtsev SV, Belliot G, King AD, and Green KY (2006) Stable expression of a Norwalk virus RNA replicon in a human hepatoma cell line. *Virology* 353: 463–473.

Cheetham S, Souza M, Meulia T, Grimes S, Han MG, and Saif LJ (2006) Pathogenesis of a genogroup II human norovirus in gnotobiotic pigs. *Journal of Virology* 80: 10372–10381.

Chen R, Neill JD, Noel JS, et al. (2004) Inter- and intragenus structural variations in caliciviruses and their functional implications. *Journal of Virology* 78: 6469–6479.

Estes MK, Prasad BV, and Atmar RL (2006) Noroviruses everywhere: Has something changed? *Current Opinion in Infectious Diseases* 19: 467–474.

Farkas T, Zhong WM, Jing Y, et al. (2004) Genetic diversity among sapoviruses. *Archives of Virology* 149: 1309–1323.

Green KY, Ando T, Balayan MS, et al. (2000) Taxonomy of the caliciviruses. *Journal of Infectious Diseases* 181: S322–S330.

Hansman GS, Natori K, Shirato-Horikoshi H, et al. (2006) Genetic and antigenic diversity among noroviruses. *Journal of General Virology* 87: 909–919.

Hardy ME (2005) Norovirus protein structure and function. *FEMS Microbiology Letters* 253: 1–8.

Jiang X, Graham DY, Wang K, and Estes MK (1990) Norwalk virus genome cloning and characterization. *Science* 250: 1580–1583.

Kapikian AZ (2000) The discovery of the 27-nm Norwalk virus: An historic perspective. *Journal of Infectious Diseases* 181: S295–S302.

Kapikian AZ, Wyatt RG, Dolin R, Thornhill TS, Kalica AR, and Chanock RM (1972) Visualization by immune electron microscopy of a 27-nm particle associated with acute infectious nonbacterial gastroenteritis. *Journal of Virology* 10: 1075–1081.

Le Pendu J, Ruvoen-Clouet N, Kindberg E, and Svensson L (2006) Mendelian resistance to human norovirus infections. *Semin Immunology* 18: 375–386.

Noel JS, Liu BL, Humphrey CD, et al. (1997) Parkville virus: A novel genetic variant of human calicivirus in the Sapporo virus clade, associated with an outbreak of gastroenteritis in adults. *Journal of Medical Virology* 52: 173–178.

Sosnovtsev SV, Belliot G, Chang KO, et al. (2006) Cleavage map and proteolytic processing of the murine norovirus nonstructural polyprotein in infected cells. *Journal of Virology* 80: 7816–7831.

Straub TM, Höner zu Bentrup K, Orosz-Coghlan P, et al. (2007) *In vitro* cell culture infectivity assay for human noroviruses. *Emerging Infectious Diseases* 13(3): 396–403.

Tacket CO, Sztein MB, Losonsky GA, Wasserman SS, and Estes MK (2003) Humoral, mucosal, and cellular immune responses to oral Norwalk virus-like particles in volunteers. *Clinical Immunology* 108: 241–247.

Ward VK, McCormick CJ, Clarke IN, et al. (2007) Recovery of infectious murine norovirus using pol II-driven expression of full-length cDNA. *Proceedings of the National Academy of Sciences, USA* 104: 11050–11055.

Wobus CE, Thackray LB, and Virgin HW (2006) Murine norovirus: A model system to study norovirus biology and pathogenesis. *Journal of Virology* 80: 5104–5112.

Zheng DP, Ando T, Fankhauser RL, Beard RS, Glass RI, and Monroe SS (2006) Norovirus classification and proposed strain nomenclature. *Virology* 346: 312–323.

Ophiovirus

A M Vaira and R G Milne, Istituto di Virologia Vegetale, CNR, Turin, Italy

© 2008 Elsevier Ltd. All rights reserved.

Glossary

Bipartite nuclear targeting sequence Signal involved in nuclear translocation of proteins characterized by (1) two adjacent basic amino acids (Arg or Lys), (2) a spacer region of ten residues, (3) at least three basic residues (Arg or Lys) in the five positions after the spacer region.

Mononegavirales Order of viruses comprising species that have a non-segmented, negative-sense RNA genome. The order includes four families: *Bornaviridae, Rhabdoviridae, Filoviridae,* and *Paramyxoviridae.*

Introduction

The genus *Ophiovirus* does not fit into any existing virus family, and a new family (*Ophioviridae*) has been proposed but is not yet official. The genus comprises a number of viral species, all plant-infecting viruses, relatively recently discovered. In some cases ophioviruses are now known to be the cause of well-known and 'classical' major plant diseases, but in other cases the presence of the virus has not been linked to any specific symptom, because of invariably mixed infections with other viruses. Ophioviruses occur in monocots and dicots, in vegetables, ornamentals, and trees, in the New and Old World, suggesting a well-adapted group of viruses containing more species than the ones already discovered. Where identified, the vectors of ophioviruses have proved to be *Olpidium brassicae,* a soil-inhabiting fungus. Ophioviruses have been slow to emerge mainly because the virions are not easy to see in the electron microscope; indeed, the first ophiovirus particle was observed only in 1988 and its morphology understood in 1994. The generic name derives from the Greek word *ophis,* meaning snake, in reference to the serpentine appearance of the virus particles.

Taxonomy and Classification

There are currently five species in the genus *Ophiovirus,* and at least one tentative species (**Table 1**). The accepted species in the genus are: *Citrus psorosis virus* (CPsV), the type species, *Lettuce ring necrosis virus* (LRNV), *Mirafiori lettuce virus* (MiLV), *Ranunculus white mottle virus* (RWMV), and *Tulip mild mottle mosaic virus* (TMMMV). The family *Ophioviridae,* containing a single genus listing all the species has recently been proposed but is not official. To date, the genomes of three species, CPsV, LRNV, and MiLV, are fully sequenced.

Phylogenetic analysis of the complete RNA-dependent RNA polymerase (RdRp) domain of RWMV and of the RdRp core modules of CPsV, MiLV, and RWMV, provide evidence of a relationship with *Rhabdoviridae* and *Bornaviridae* species, both belonging to the *Mononegavirales.* However, ophioviruses appear to form a monophyletic group, separate from the other negative-stranded RNA viruses; and of course the *Ophiovirus* genome is multipartite, excluding membership in the *Mononegavirales.*

For species demarcation within the genus, different criteria have been considered: the different coat protein (CP) sizes, absence of, or distant serological relationship between the CPs, differences in natural host range and different number, organization, and/or size of genome segments. From available data, there is 94–100% identity among complete CP amino acid sequences of isolates belonging to the same species, in particular for MiLV and CPsV, and alignments between incomplete but considerable parts of CP sequences of two RWMV isolates also show almost 100% identity. The percentage of identity falls to 30–52% for interspecies alignments, with the exception of the identity between MiLV and TMMMV CP sequences, which is about 80%, confirming the closeness of these two viruses. They are currently considered as two different species at least because of different host ranges; more information on TMMMV sequences is needed prior to any revision of their taxonomic position.

The percentage of identity/similarity between CP amino acid sequences may also be proposed as a further

Table 1 Virus members in the genus *Ophiovirus*

Virus (alternative name)	Abbreviations	Accession numbers[a]
Citrus psorosis virus (Citrus ringspot virus)	CPsV (CRsV)	NC006314, NC006315, NC006316
Lettuce ring necrosis virus	LRNV	NC006051, NC006052, NC006053, NC006054
Mirafiori lettuce virus (Mirafiori lettuce big-vein virus)	MiLV (MLBVV)	NC004779, NC004781, NC004782, NC004780
Ranunculus white mottle virus	(RWMV)	AF335429, AF335430, AY542957
Tulip mild mottle mosaic virus	(TMMMV)	AY204673, AY542958
Freesia ophiovirus (freesia sneak virus)	(FOV-FreSV)	AY204676, DQ885455

[a]Complete genome sequences are reported when available.

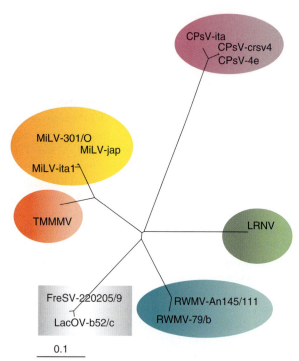

Figure 1 Unrooted phylogenetic tree of established and putative members of genus *Ophiovirus* based on their coat protein amino-acid sequences. The tree was generated by the neighbor-joining method and bootstrap values were estimated using 1000 replicates. All bootstrapping values were above 500. Branch lengths are related to the evolutionary distances. Viral isolates included in the analysis (accession numbers are reported when available): CPsV-4e (AAC41022), CPsV-CRSV4 (AAF00018), CPsV-ita (CAJ43825), MiLV-301/O (AAN60449), MiLV-ita1 (AAU12873), MiLV-jap (AAO49152), TMMMV (AAT08133), LRNV (AAT09112), RWMV-79/b (AAT08132) partial sequence, RWMV-an145/111 partial sequence, FreSV-220205/9 (DQ885455), LacOV-b52/c from *Lachenalia*, partial sequence. Ovals grouping isolates represent established viral species, the rectangle the proposed new species.

tool for species demarcation when complete sequences are available. A phylogenetic tree based on alignment of representative CP amino acid sequences is shown in **Figure 1**.

There has been some further discussion on the position of CPsV, as it is presently the only ophiovirus with a woody natural host, and there is no evidence of soil transmission, contrary to the apparent rule for ophioviruses; Western blotting has shown no serological cross-reaction between CPsV coat protein and other ophiovirus CPs and there are several molecular differences, the main one being the presence of three, not four genome segments. These differences, if confirmed and reinforced, could lead to considering placement of CPsV in a different genus within the proposed family *Ophioviridae*.

Reverse transcriptase-polymerase chain reaction (RT-PCR) amplification, using degenerate primers, of a 136 bp fragment from the RdRp gene, located outside the sequence coding the polymerase domain, is currently the best tool for detecting and identifying species within the genus. Sequencing and phylogenetic analysis of the 45 amino acid string deduced from the amplified fragment of several isolates belonging to different species, fully supports the present species classification and gives an indication of the taxonomic positioning of newly diagnosed isolates. An ophiovirus in freesia provisionally named freesia sneak virus (FreSV), considered a tentative species, was diagnosed and proposed as a new ophiovirus species owing to this procedure, when no other viral genome sequences were known; the complete CP sequence later obtained confirmed the hypothesis. A positive reaction with the genus-specific RT-PCR has recently been obtained for several samples from *Lachenalia*, a monocot in the family *Hyacinthaceae*; in this case the RT-PCR and subsequent analysis of the 45-amino-acid string suggest placing the isolate within the freesia sneak virus tentative species. Also in this case, subsequent CP gene sequencing showed nearly 100% identity with the homologous portion of FreSV at amino acid level, confirming this position.

Properties of Particles

The virions are naked filamentous nucleocapsids about 3 nm in diameter forming circularized structures of different lengths. The virions appear to form internally coiled circles that in some cases can collapse into pseudolinear duplex structures (**Figure 2**). There is no evidence of an

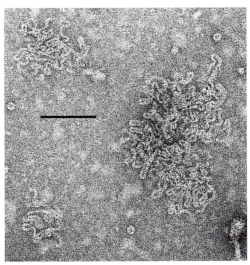

Figure 2 Negative contrast electron micrograph (uranyl acetate) of partially purified virion preparation from field lettuce showing big-vein symptoms. Note large and small ophiovirus particles. The bar represents 100 nm.

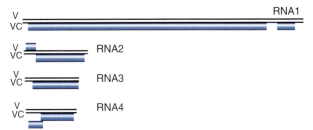

Figure 3 Genomic organization of ophioviruses. The four genomic RNAs are represented. The RNA4 is not described in all species, see text and **Table 2** for details; v-RNA and vc-RNA indicate the viral and the viral complementary RNAs, respectively. Boxes indicate ORFs.

envelope. The 5′ and the 3′ ends of the RNAs have been structurally analyzed in the complete genome sequences available for CPsV and MiLV, to assess the probability of panhandle-like or similar structures. In CPsV, secondary structure predictions do not support significant complementarity between the termini of the viral RNAs to afford panhandle formation. On the contrary, in MiLV, the terminal sequences do carry partial inverted repeats, potentially allowing panhandles to form; the terminal sequences also contain palindromes potentially able to fold into stem-loops, so far not found for CPsV sequences. Thus, the apparent circularization of the particles of all ophioviruses remains incompletely explained.

Genome Organization and Replication

The ophiovirus genome is ssRNA, 11.3–12.5 kbp in size, divided into three or four segments (RNAs 1–4) and mainly negative sense. Positive-sense RNA is also encapsidated, to some extent with CPsV or in nearly equimolar amounts, with MiLV. The 3′ termini of the viral RNAs show sequences of 9–12 nt conserved within the RNAs of each species: A_7GUAUC for CPsV, $A_{4-6}UAAUC$ for MiLV, and $A_7GUAUCA$ and $A_3UA_3GUAUCA$ for LRNV; these may be involved in the recognition of the RNAs by the RdRp.

The genome organization of ophioviruses is as follows (**Figure 3** and **Table 2**). RNA1 is negative sense, shows short 5′ and 3′ untranslated regions, and contains two open reading frames (ORFs) separated by an AU-rich intergenic region. The putative product of ORF1 did not show any significant similarity with others available in databases and its function remains unknown. The product of the large ORF2 contains the core polymerase module with the five conserved motifs proposed to be part of the RdRp active site and shows low but significant amino acid sequence similarities with the L protein of several rhabdoviruses.

RNA2 contains one ORF in the negative strand in all three sequenced viruses, encoding a putative nonstructural protein with no significant sequence similarity to other known proteins and therefore of unknown function. In the amino acid sequence of the CPsV protein, two motifs bear similarity to a putative 'bipartite nuclear targeting sequence' considered to be a nuclear localization signal. An additional minor ORF, in the virus-sense strand, is present in MiLV RNA2, but to date there is no evidence for expression of the predicted 10 kDa protein.

RNA 3 contains one ORF in the vc RNA in all three viruses which has been identified as coding for the nucleocapsid protein.

An RNA4 has been found only in MiLV and LRNV. MiLV RNA4 contains two overlapping ORFs in the negative strand in different reading frames: the first putatively encodes a 37 kDa protein, but the second, overlapping the first by 38 nt, appears to lack an initiation codon. This second ORF has a theoretical coding capacity of 10.6 kDa and is suggested to be expressed by a +1 frameshift. A slippery sequence, GGGAAAU, can be recognized immediately in front of the UGA stop codon of the 37 kDa ORF. This unique feature of ORF positioning has been shown for two different isolates of MiLV, and therefore seems not to be caused by cloning artifacts.

RNAs 1, 2, and 3 of the other partially sequenced species are similar in size to those described. No RNA4 has been found in CPsV, RWMV, and TMMMV, but the possible existence of an RNA4, perhaps in very low concentration or very similar in size to RNA3, cannot be ruled out.

Analysis for the presence of subgenomic RNAs has not extensively been pursued for all species, even though several studies have used minus-strand and plus-strand probes in Northern blot hybridization using RNA extracted from infected tissues as well as virus particles, to obtain information regarding the polarity of encapsidated viral

Table 2 Details of genomic organization of fully sequenced viruses inside the genus *Ophiovirus*

Virus	RNA1	RNA2	RNA3	RNA4
CPsV	8186 nt 2 ORFs in vc RNA ORF1: 24 kDa ORF2: 280 kDa proteins	1645 nt 1 ORF in vc RNA 54 kDa protein	1447 nt 1 ORF in vc RNA 49 kDa	
MiLV	7794 nt 2 ORFs in vc RNA ORF1: 25 kDa ORF2: 263 kDa proteins	1788 nt 1 ORF in vRNA: 10 kDa protein 1 ORF in vc RNA: 55 kDa protein	1515 nt 1 ORF in vc RNA 48.5 kDa	1402 nt 2 ORFs in vc RNA: ORF1: 37 kDa ORF2: 10.6 kDa proteins
LRNV	7651 nt 2 ORFs in vc RNA ORF1: 22 kDa ORF2: 261 kDa proteins	1830 nt 1 ORF in vc RNA 50 kDa protein	1527 nt 1 ORF in vc RNA 48 kDa protein	1417 nt 1 ORF in vc RNA 37 kDa protein

RNA. With CPsV in such studies no subgenomic RNAs were ever found for any of RNA1, 2, or 3.

A putative nuclear localization sequence present in the 280 kDa protein coded by RNA1 and in the 54 kDa protein coded by RNA2 in CPsV, and also observed in available analogous sequences of MiLV and RWMV, together with the polymerase similarities found with isolates from the genera *Nucleorhabdovirus* and *Bornavirus*, may suggest nuclear replication also for ophioviruses; in any case, gold immunolabeling and electron microscopy of RWMV-infected tissues showed accumulation of CP in the cytoplasm.

Viral Proteins

The most studied ophiovirus protein is the CP. There is only one CP, varying somewhat in size according to species. For CPsV, MiLV, and LRNV, this is 48.6, 48.5, and 48 kDa, respectively; Western blotting has given apparent sizes of 43 and 47 kDa for RWMV and TMMMV CPs, respectively. FreSV CP is 48.4 kDa. Ophiovirus CPs are relatively poor antigens. The antisera or, in one case, monoclonal antibodies, have been produced using purified virus preparations or recombinant protein. Western blots show that the CPs of RWMV, TMMMV, MiLV, and LRNV are slightly to moderately related; CPsV CP appears to be unrelated to the others. No clear information is available for FreSV. Genetic variability in the CP genes has been extensively studied in CPsV and MiLV.

Variability of the CP gene of CPsV was assessed serologically and by sequence analysis of two regions located in the 3′ and 5′ halves of the gene. Variability assessed by a panel of monoclonal antibodies to the protein resulted in 14 reaction patterns but no correlation was found between serogroups and specific amino acid sequences, field location, or citrus cultivar. Results from sequence analysis showed limited nucleotide diversity in the CP gene within the population. Diversity was slightly higher in the 5′ region. The ratio between nonsynonymous and synonymous substitutions (d_N/d_S) for the two regions indicated a negative selective pressure for amino acid changes, more intense at the 3′ end. When the entire CP sequences were considered, two clusters were identified, one comprising 19 isolates from Italy and an isolate from Spain and a second one containing only the Florida CPsV isolate.

Phylogenetic analysis of MiLV CP genes has revealed two distinct subgroups; however, this grouping was not correlated with symptom severity on lettuce or the geographic origin of the isolates; whether these two subgroups show different characteristics with regard to virulence in indicator plants or serological relationships remains to be determined; furthermore also in this case, a low value of d_N/d_S ratio was estimated for all MiLV isolates and also between the two subgroups, supporting a negative selection pressure for amino acid changes. In general, under natural conditions, genetic stability seems to be the rule rather than the exception.

The other viral protein, considered mainly for taxonomic purposes, is the putative RdRp, encoded by RNA1 in all species, described earlier.

No information is available regarding the synthesis or the function of other putative ORF products.

Pathogenicity and Geographic Distribution

Formerly a so-called citrus psorosis group of diseases included some of the most widespread graft-transmissible disorders of citrus, in some cases of undemonstrated etiology. Characterization of CPsV has shown that it is the cause of the great majority of psorosis symptoms but that a small but still undefined proportion of trees with 'psorosis-like' bark-scaling shows no evidence of carrying the virus, and the idea of a non-CPsV psorosis-like disease of unknown etiology is emerging. In the past, different kinds of symptoms were described: 'psorosis A', characterized by causing bark-scaling in trunk and limbs of infected field trees; 'psorosis B', causing rampant scaling of thin branches in field trees and chlorotic blotches in old leaves with gummy pustules in the underside; 'ringspot', characterized by presence of chlorotic blotches and rings in the old leaves of inoculated seedlings but apparently no specific symptoms on infected field trees or in other cases, chlorotic flecks and ringspots on leaves, and trunk and fruit symptoms (**Figure 4**). All these appear to be caused by CPsV but in the former 'psorosis group' there are other non-CPsV diseases with symptoms such as chlorotic leaf-flecking and oak-leaf patterns; while graft-transmissible, these diseases have not yet yielded

Figure 4 Bark scaling in citrus, typical of severe psorosis.

to further analysis. Psorosis is an ancient disease; bark-scaling of citrus was first observed in Florida and California in the 1890s. The disease has been brought under control in most advanced citrus-growing countries due to rigorous indexing and quarantine. In Argentina it remains a severe problem, in Mediterranean areas it is also reported, and in citrus-growing parts of Asia the disease may well be widespread, although rigorous testing for the presence or absence of CPsV has generally been lacking.

RWMV has been reported in two species, ranunculus (*Ranunculus asiaticus* hyb.) and anemone (*Anemone coronaria*) in northwest Italy since the 1990s. The pathogenic impact is uncertain as it was almost always found in mixed infection with other viruses commonly infecting the two species. The symptom description 'white mottle' present in its name derives from the bright white mottling symptoms consistently observed on *Nicotiana benthamiana* leaves mechanically infected by RWMV. In some cases the bright mottle has also been observed in naturally infected ranunculus, always in mixed infection. Indexing for the presence of RWMV in ranunculus crops done in 1996 showed an incidence of 2.5% among symptomatic plants; furthermore, the very few plants apparently infected only with RWMV did not show any distinctive symptoms. Infection of ranunculus seedlings by mechanical inoculation results in limited necrosis and deformation of stems and leaves. The pathogenic potential of RWMV is thus still unclear.

Tulip mild mottle mosaic disease, caused by TMMMV, is one of the most serious diseases in some bulb-producing areas of Japan (which produces bulbs on a large scale for the Southeast Asian market) and has been reported since 1979. Symptoms on tulips include color-attenuating mottle on flower buds and color-increasing streak on petals (**Figure 5**); mild chlorotic mottle and mosaic slightly appear along the leaf veins. TMMMV infection is up to now restricted to *Tulipa* species and cultivars, and has not been reported in other geographical areas.

Lettuce big-vein disease was first reported in the United States in 1934 and occurs in all major lettuce-producing areas in the word. The disease becomes serious during cooler periods of the year. The main symptoms are vein-banding in the leaves, due to zones parallel to the vein cleared of chlorophyll (**Figure 6**); there is associated leaf distortion, delayed head formation, and decreased head size.

The causal agent of the disease, long known to be soil-transmitted, was first identified as 'lettuce big-vein virus' (LBVV) (genus *Varicosavirus*); recently, re-evaluation of the etiology has been necessary as a second less easily detected virus, MiLV, was found in lettuces with big-vein symptoms. Following experimental inoculation of the two viruses together and separately, MiLV was shown to be the etiological agent of big-vein, while LBVV, now renamed as lettuce big-vein associated virus (LBVaV),

Figure 5 TMMMV symptoms on tulip (healthy tulip on the left). Courtesy of T. Morikawa, Toyama Agric. Res. Center, Tonami, Toyama, Japan.

Figure 6 Leaf of butterhead lettuce showing big-vein symptoms. Close-up showing vein-banding.

apparently plays no part in the disease, although it is almost always present, and is, like MiLV, transmitted by *Olpidium*. Studies on symptom development in the field have confirmed that both viruses very commonly occur together in lettuce crops. MiLV likely occurs worldwide; it is reported in California (USA), Chile, France, Germany, Italy, Spain, the Netherlands, England, Denmark, Japan, Australia, and New Zealand. To date, natural infection has been reported only in cultivars of *Lactuca sativa*.

LRNV is closely associated with lettuce ring necrosis disease, first described in the Netherlands and in Belgium as 'kring necrosis' and also in France as *maladie des taches orangées*, in the 1980s. It is an increasingly important disease of butterhead lettuce crops in Europe. Definitive proof that LRNV is the cause of ring necrosis is still awaited. In southern France the disease is observed primarily in winter lettuces (September–January) under plastic or glass, when the crop is maturing, the day length is short, and both light intensity and temperature are low. Symptoms depend on the lettuce type and environmental conditions, and mainly consist of necrotic rings and ring-like patterns on leaves, which may render the product unmarketable. LRNV is often found together with MiLV (and, of course, LBVaV) with which it shares host and vector. The virus has also been reported in California.

Although the ophiovirus isolated from freesia has not yet been recognized as an official new ophiovirus species, we would like to say a few words about the freesia disease. Necrotic disorders of freesia (*Freesia refracta* hyb., family Iridaceae), known as 'leaf necrosis' and 'severe leaf necrosis', were first described in freesia crops in the Netherlands before 1970, and a similar disease, named 'freesia streak', was reported in England and Germany in the same years. The 'severe leaf necrosis' appeared to be caused by mixed infection with the potyvirus freesia mosaic virus and a virus with varicosavirus morphology and mode of transmission, that has been tentatively named Freesia leaf necrosis virus. In recent years, in northwest Italy, a necrotic disease of freesia has spread and caused considerable economic losses, even though it was present since 1989. Both in Italy and in the

Netherlands, the disease seems to be linked to the presence of an ophiovirus. Typical symptoms are chlorotic spots and streaks at the leaf tips that expand downwards and turn necrotic, and may vary according to cultivar and climate. Freesia sneak virus has been proposed as the name of this new ophiovirus species.

Experimental Symptoms and Host Ranges

All ophiovirus are, though sometimes not easily, mechanically transmissible to a limited range of test plants, including Chenopodiaceae and Solanaceae, inducing local lesions and in some cases systemic mottle.

Cytopathology

Thin sections of *Nicotiana clevelandii* leaves mechanically infected by RWMV have been examined by electron microscopy but no distinctive inclusions were observed and no virus particles were seen. In classical thin sections, viral nucleic acids stain up well but protein coats are faintly contrasted, so it is not surprising that very thin nucleoprotein threads in random orientations should escape detection. After gold immunolabeling with RWMV-specific polyclonal antibodies, sections of the cytoplasm of parenchyma cells were seen in the EM to be clearly labeled, but nuclei, chloroplasts, mitochondria, and microbodies were unlabeled. No other studies on ophiovirus cytopathology are till now available.

Transmission, Prevention, and Control

Most ophiovirus species are soil-transmitted through the obligately parasitic soil-inhabiting fungus *Olpidium brassicae*; this has been proved for TMMMV, MiLV, and LRNV; FreSV is known to be soil-transmitted, and the freesia leaf necrosis complex (presumably containing FreSV and the varicosavirus freesia leaf necrosis virus) has been transmitted by *O. brassicae*. The virus–vector relationship was recognized as the *in vivo* type, at least for MiLV-*O. brassicae*; virus-free zoospores acquire the virus during the vegetative part of their life cycle in the roots of virus-infected host plants. Zoospores released from infected sporangia apparently carry the virus internally in their protoplasts and transmit it to healthy roots. The virus enters vector resting spores, and can survive for years in the soil in the absence of a growing host plant. Furthermore, disease control is difficult because of the lack of safe and effective treatments against the fungal vector.

As ophioviruses occur in low concentration and are physically labile *in vitro*, it would be interesting to establish whether they multiply in the vector.

No information on natural vectors is available for RWMV. CPsV is commonly transmitted by vegetative propagation and no natural vectors have been identified; in some cases natural spread of psorosis in limited citrus areas has been reported, but the spatial patterns would suggest a hypothetic aerial vector instead of a soil-borne one. The data are however based on symptom observation, not on analysis for the spread of CPsV.

For disease control the use of resistant or tolerant crops may be the best choice. In Japan, the use of resistant tulip cultivars is the most important component of managing TMMMV disease, as it can be highly effective and has no deleterious effect on the environment; resistance assays have allowed researchers to identify highly resistant tulip lines and use them for breeding new resistant cultivars. For lettuce in soil-less cultivation, using ultraviolet (UV) sterilization of nutrients has shown good results in prevention of MiLV and LRNV infection, although for field lettuces the prospect is less good as no classical sources of resistance or tolerance have yet been identified. In the case of CPsV, control of the sanitary status of mother plants for producing propagating material is essential. Shoot-tip grafting *in vitro* associated with thermotherapy or somatic embryogenesis from stigma and style cultures have been successfully used to eliminate CPsV from plant propagating material. Several transgenic citrus lines exist carrying parts of the CPsV genome, and promising resistance may emerge from these.

See also: Plant Rhabdoviruses; Varicosavirus.

Further Reading

Derrick KS, Brlansky RH, da Graça JV, Lee RF, Timmer LW, and Nguyen TK (1988) Partial characterization of a virus associated with citrus ringspot. *Phytopathology* 78: 1298–1301.

García ML, Dal Bo E, Grau O, and Milne RG (1994) The closely related citrus ringspot and citrus psorosis viruses have particles of novel filamentous morphology. *Journal of General Virology* 75: 3585–3590.

Martín S, López C, García ML, et al. (2005) The complete nucleotide sequence of a Spanish isolate of citrus psorosis virus: Comparative analysis with other ophioviruses. *Archives of Virology* 150: 167–176.

Morikawa T, Nomura Y, Yamamoto T, and Natsuaki T (1995) Partial characterization of virus-like particles associated with tulip mild mottle mosaic. *Annals of the Phytopathological Society of Japan* 61: 578–581.

Roggero P, Ciuffo M, Vaira AM, Accotto GP, Masenga V, and Milne RG (2000) An ophiovirus isolated from lettuce with big-vein symptoms. *Archives of Virology* 145: 2629–2642.

Roggero P, Lot H, Souche S, Lenzi R, and Milne RG (2003) Occurrence of mirafiori lettuce virus and lettuce big-vein virus in relation to development of big-vein symptoms in lettuce crops. *European Journal of Plant Pathology* 109: 261–267.

Vaira AM, Accotto GP, Constantini A, and Milne RG (2003) The partial sequence of RNA 1 of the ophiovirus ranunculus white mottle virus indicates its relationship to rhabdoviruses and provides candidate primers for an ophio-specific RT-PCR test. *Archives of Virology* 148: 1037–1050.

Vaira AM, Accotto GP, Gago-Zachert S, *et al.* (2005) Ophiovirus. In: Fauquet CM, Mayo MA, Maniloff J, Desselberger U, and Ball LA (eds.) *Virus Taxonomy: Eighth Report of the International Committee on Taxonomy of Viruses*, pp. 673–679. San Diego, CA: Elsevier Academic Press.

Van der Wilk F, Dullemans AM, Verbeek M, and Van den Heuvel JFJM (2002) Nucleotide sequence and genomic organization of an ophiovirus associated with lettuce big-vein disease. *Journal of General Virology* 83: 2869–2877.

Orbiviruses

P P C Mertens, H Attoui, and P S Mellor, Institute for Animal Health, Pirbright, UK

© 2008 Elsevier Ltd. All rights reserved.

Glossary

Arbovirus Viruses that are transmitted between their vertebrate host species, by insects or other arthropod vectors, replicating in both host and vector species.

Culicoides Blood-feeding dipterous insects also known as biting midges.

Ecchymosis Area of haemorrhage larger than petechiae.

Petechiae Pinpoint- to pinhead-sized red spots under the skin that are the result of small hemorrhages.

Introduction

The reoviruses (a term used here to indicate any member of the family *Reoviridae*) have genomes composed of 9–12 separate segments of linear double-stranded RNA (dsRNA), packaged as exactly one copy of each segment per icosahedral virion. The family *Reoviridae* contains a total of 12 established genera (*Orthoreovirus, Orbivirus, Cypovirus, Aquareovirus, Rotavirus, Coltivirus, Seadornavirus, Fijivirus, Phytoreovirus, Oryzavirus, Idnoreovirus,* and *Mycoreovirus*) as well as three proposed 'new' genera of viruses (*Mimoreovirus, Cardoreovirus,* and *Dinovernavirus*). Closely related reoviruses that infect the same cell can exchange genome segments by a process known as reassortment, generating new progeny virus strains. This ability to 'reassort' is regarded as a primary indication that different virus strains belong to the same virus species, within each of the genera of the *Reoviridae*. The largest genus *Orbivirus* contains 21 distinct virus species and 12 further unassigned viruses (**Table 1**), each of which has a ten-segmented dsRNA genome. The orbiviruses are transmitted between their vertebrate hosts by ticks or hematophagous insects (e.g., mosquitoes or biting midges (*Culicoides* spp.)) in which they also replicate, and they are therefore regarded as 'arthropod-borne viruses' or 'arboviruses'. Some orbiviruses cause severe diseases of domesticated and wild animals, including members of the species: *African horse sickness virus* (AHSV), *Bluetongue virus* (BTV), *Epizootic hemorrhagic disease virus* (EHDV), and *Equine encephalosis virus* (EEV). The prototype *Orbivirus* species, *Bluetongue virus*, has been extensively studied and provides a useful paradigm for other members of the genus.

Historical Overview

Bluetongue (BT) was originally recognized as a disease of sheep and cattle in South Africa in the late eighteenth century and was initially reported in the scientific literature, as 'malarial catarrhal fever'. In 1905, Spreull suggested the name 'bluetongue' to reflect a significant, although infrequent, clinical sign of the disease. He also showed that the agent was filterable and caused an inapparent infection in goats and cattle.

BT was initially regarded as a disease of ruminants that was exclusive to Africa. However, in 1943 an outbreak occurred in Cyprus, killing approximately 2500 sheep (70% mortality in infected animals), and it was suggested that less virulent strains had caused earlier but unrecognized outbreaks on the island. Subsequently outbreaks occurred in 1946, 1951, 1965, and 1977, and have continued sporadically to the present day. BT was recorded in the 1940s in Palestine and Turkey, and by 1950 was present in Israel. In 1948, an apparently new disease known as 'sore muzzle' was recognized in Texas and, in 1952, BTV serotype 10 was isolated from infected sheep in California. BTV serotypes 11, 17, 13, and 2 were subsequently isolated in New Mexico (1955), Wyoming (1962), Idaho/Florida (1967), and Florida (1983), respectively, and these types are now regarded as endemic in North America. More recently, BTV-1 was identified in Louisiana (2004), BTV-3 in Florida and Mississippi (1999–2006), and BTV types 5, 6, 14, 19, and 22 were isolated in Florida (2002–05).

Orbiviruses 455

Table 1 Species in the genus *Orbivirus*

Virus species (virus abbreviation)	Number of serotypes/strains	Vector species	Host species
African horse sickness virus (AHSV)	9 numbered serotypes (AHSV-1 to AHSV-9)	*Culicoides* spp. (biting midges)	Equids, dogs, elephants, camels, cattle, sheep, goats, humans (in special circumstances) predatory carnivores (by eating infected meat)
Bluetongue virus (BTV, *Orbivirus* type species)	24 numbered serotypes (BTV-1 to BTV-24)	*Culicoides* spp. (biting midges)	All ruminants, camelids and predatory carnivores (by eating infected meat)
Changuinola virus (CGLV)	12 named serotypes	Phlebotomines, culicine mosquitoes	Humans, rodents, sloths
Chenuda virus (CNUV)	7 named serotypes	Ticks	Seabirds
Chobar Gorge virus (CGV)	2 named serotypes	Ticks	Bats
Corriparta virus (CORV)	6 named serotypes/strains[a]	Culicine mosquitoes	Humans, rodents
Epizootic hemorrhagic disease virus (EHDV)	10 numbered or named serotypes/strains[a] (EHDV-1 to EHDV-8, EHDV 318, Ibaraki virus (atypical EHDV-2))	*Culicoides* spp. (biting midges)	Cattle, sheep, deer, camels, llamas, wild ruminants, marsupials
Equine encephalosis virus (EEV)	7 numbered serotypes (EEV-1 to EEV-7)	*Culicoides* spp. (biting midges)	Equids
Eubenangee virus (EUBV)	4 named serotypes	*Culicoides* spp., anopheline and culicine mosquitoes	Unknown hosts
Ieri virus (IERIV)	3 named serotypes	Mosquitoes	Birds
Great Island virus (GIV)	36 named serotypes/strains[a]	Argas, Ornithodoros, Ixodes ticks	Seabirds, rodents, humans
Lebombo virus (LEBV)	1 numbered serotype (LEBV-1)	Culicine mosquitoes	Humans, rodents
Orungo virus (ORUV)	4 numbered serotypes (ORUV-1 to ORUV-4)	Culicine mosquitoes	Humans, camels, cattle, goats, sheep, monkeys
Palyam virus (PALV)	13 named serotypes/strains[a]	*Culicoides* spp., culicine mosquitoes	Cattle, sheep
Peruvian horse sickness virus (PHSV)	1 numbered serotype (PHSV-1)	Mosquitoes	Horses
St. Croix River virus (SCRV)	1 numbered serotype (SCRV-1)	Ticks	Hosts unknown
Umatilla virus (UMAV)	4 named serotypes	Culicine mosquitoes	Birds
Wad Medani virus (WMV)	2 named serotypes	Boophilus, Rhipicephalus, Hyalomma, Argas ticks	Domesticated animals
Wallal virus (WALV)	3 serotypes/strains	*Culicoides* spp.	Marsupials
Warrego virus (WARV)	3 serotypes/strains	*Culicoides* spp., anopheline and culicine mosquitoes	Marsupials
Wongorr virus (WGRV)	8 serotypes/strains	*Culicoides* spp., mosquitoes	(Cattle, macropods)
Tentative species			
Andasibe virus (ANDV)		Mosquitoes	Unknown hosts
Codajas virus (COV)		Mosquitoes	Rodents
Ife virus (IFEV)		Mosquitoes	Rodents, birds, ruminants

Continued

Table 1 Continued

Virus species (virus abbreviation)	Number of serotypes/strains	Vector species	Host species
Itupiranga virus (ITUV)		Mosquitoes	Unknown hosts
Japanaut virus (JAPV)		Mosquitoes	Unknown hosts
Kammavanpettai virus (KMPV)		Unknown vectors	Birds
Lake Clarendon virus (LCV)		Ticks	Birds
Matucare virus (MATV)		Ticks	Unknown hosts
Tembe virus (TMEV)		Mosquitoes	Unknown hosts
Tracambe virus (TRV)		Mosquitoes	Unknown hosts
Yunnan orbivirus (YUOV)		*Culex tritaeniorhyncus*	Unknown hosts

[a]In some species the serological relationships between strains has not been fully determined. For more information concerning individual named types, see Mertens PPC, Duncan R, Attoui H, and Dermody TS (2005) *Reoviridae*. In: Fauquet CM, Mayo MA, Maniloff J, Desselberger U, and Ball LA (eds.) *Virus Taxonomy: Eighth Report of the International Committee on Taxonomy of Viruses*, pp. 466–483. San Diego, CA: Elsevier Academic Press.

Several BTV serotypes are also present in Central and South America and, although less well characterized, these include BTV-1, 3, 4, 6, 8, 11, 12, 13, 14, and 17.

BT virus serotype 10 caused a single epizootic in Spain and Portugal in 1956, and was reported in West Pakistan in 1958 and 1960, and in India in 1961. The disease now occurs regularly and is regarded as endemic on the Indian subcontinent, involving many different serotypes. Although Australia was initially considered to be free of BT, in 1978 a virus that was collected in the Northern Territory (during 1976), was identified as BTV. Eight serotypes have subsequently been isolated in Australia (BTV-1, 3, 9, 15, 16, 20, 21, and 23). Initially outbreaks of BT in Europe were infrequent, involving a single virus strain on each occasion, and were generally short lived (4–5 years). However, since 1998, eight distinct BTV strains from six different serotypes (BTV-1, 2, 4, 8, 9, and 16) have invaded Europe, with new introductions almost every year, resulting in the deaths of >1.8 million animals.

African horse sickness (AHS) was recognized as early as 1780, during the early days of the European colonization and importation of horses into southern Africa, resulting in epizootics with high rates of mortality in infected animals. AHSV is only considered to be enzootic in sub-Saharan Africa but has caused occasional major epizootics, with very high levels of associated mortality in infected horses, in the Middle East, the Indian subcontinent, North Africa, and the Iberian Peninsula.

Epizootic hemorrhagic disease (EHD) has occurred as periodic outbreaks in the south-eastern United States since 1890, where it causes a fatal disease of deer known by 'back-woods men' as 'black tongue'. The New Jersey and South Dakota strains were isolated in 1955 and 1956, respectively, and a second serotype was isolated in the Canadian province of Alberta in 1962. EHDV has been isolated from a range of animal species including cattle (a possible 'reservoir' host). Ibaraki virus, which causes an acute febrile disease of cattle, was first recorded in Japan in 1959 and, although it is classified as EHDV serotype 2, it is regarded as atypical. There are also at least 5 serotypes of EHDV in Australia, which are known to infect cattle, buffalo, and deer without causing clinical disease. In 2006, outbreaks of the disease were recorded in cattle in Israel and Morocco.

Members of other orbivirus species are widely distributed around the world and have been isolated in Australia, North and South America, Africa, Asia, and Europe.

Host Range and Transmission

Orbiviruses are transmitted between their vertebrate hosts by a variety of hematophagous arthropods. BTV, AHSV, EHDV, and EEV are all transmitted by adult females of certain *Culicoides* species which bite the mammalian host in order to obtain proteins, prior to laying eggs. BTV is only enzootic in areas where these vectors are present and active for the majority of the year, thus maintaining a continuous infection cycle in the vector and vertebrate host species. Although there are in excess of 1000 species of *Culicoides* worldwide, only 17 have been connected with BTV and 11 are known to be capable of transmitting the virus. These are *C. sonorensis, C. imicola, C. fulvus, C. actoni, C. wadai, C. nubeculosus, C. dewulfi, C. brevitarsis, C. obsoletus, C. pulicaris,* and *C. insignis.* Many of the remaining *Culicoides* species may be refractory to infection, although environmental factors (particularly higher temperatures) can increase vector competence, even in species that are usually refractory. BTV vectors are most active between 18 and 29 °C, and are almost inactive below 10 °C or above 30 °C. The viral polymerase, which is responsible for all viral RNA synthesis, has a temperature optimum between 27 and 35 °C but is almost inactive below 15 °C. Relatively small rises in temperature within the range 15–30 °C can significantly increase both the activity of insect vectors, and the rate of virogenesis within infected individuals, significantly pincreasing their efficiency as vectors.

Infections caused by BTV and EHDV are normally restricted to domesticated and wild ruminants, although there is also evidence for BTV infection of shrews and some rodents. AHS is considered to be a disease primarily of horses and other equids. However, dogs infected with BTV (caused by a contaminated vaccine) developed a fatal viral pneumonia. AHSV infection of dogs (caused by ingestion of infected meat) can be fatal. BTV and AHSV infections have also been reported in large carnivores in Africa, and may significantly affect their numbers. The epidemiological significance of BTV or AHSV infection in dogs and other carnivores, caused by eating infected meat or by other routes, has not been fully assessed. It is unclear if they can develop a high level of viraemia, or can act as a source of infection to adult *Culicoides*. There is serological evidence that AHSV can infect elephants and isolates have occasionally been obtained from camels, cattle, sheep, and goats. Under unusual circumstances (involving inhalation of a freeze dried neurotropic vaccine strain), AHSV has also infected humans, causing encephalitis and retinitis. The potential risk to human health posed by consumption of AHSV infected meat has not yet been fully evaluated.

Orbiviruses can infect marsupials, humans, rodents, bats, monkeys, sloths, and particularly birds (*Great Island virus, Ieri virus,* and *Umatilla virus* species). Under experimental conditions, many orbiviruses can infect mice or embryonated chicken eggs, and these hosts are routinely used for virus isolation from infected tissue samples (e.g., blood) or from insect vectors. Adult mice can show high mortality levels when infected via a nasal route with some strains of AHSV. This may have relevance to the reported cases of AHSV in humans infected via the same route by neurotropic vaccine strains. Experimental animal systems can provide useful models for studies of the immune response to these viruses and for identification of virulence factors, as shown for members of the *Great Island virus* (GIV) and AHSV species.

Epidemiology and Disease

Infection with BTV, EHDV, or AHSV is accompanied by pyrexia and edema/inflammation, particularly around the face, mouth, and nose. BTV replicates in hematopoietic and lymphoid tissues, including the spleen, bone marrow, monocytes, macrophages, neutrophils, and draining lymphoid tissues. Infection of vascular endothelial cells causes vascular thrombosis and hemorrhages that vary from petechiae to ecchymosis and is accompanied by leakage of fluids, into surrounding tissues or the lungs (causing frothing). Clinical signs of BT may be severe, particularly in sheep and can include epithelial lesions or ulceration of the mouth, coronitis, degeneration of skeletal muscle, lameness, and vomiting, leading to pneumonia (which is frequently fatal), hemorrhages of the skin horn junction, torticollis (usually fatal), and occasionally to a cyanotic appearance of the tongue.

BT is effectively restricted to a band around the world between 53° N and 30° S, determined largely by the distribution of vector insects. Temperatures below 0 °C, when maintained for approximately 2 h or more, will kill adult *Culicoides*. Therefore, in those areas at relatively greater latitudes (which experience frosts), adult vectors may not be present or active in significant numbers throughout the entire year. Consequently, the virus may be absent, or maintained at only low levels in line with vector populations, for part of the year (usually in winter). However, in some areas there is evidence for a mechanism that allows the virus to persist through the winter (over-wintering). The exact nature of this mechanism is unknown; it has been suggested that it could involve persistent infection of either the mammalian host or larvae of vector insects, although over-wintering by these mechanisms has not been demonstrated. Seasonal variation in the incidence of disease is illustrated by the outbreaks of AHS in Spain, Portugal, and Morocco (1987–91), where cases were detected only in late summer and autumn, and by the outbreak of BT in northern Europe during the mid to late summer of 2006. The effects of high temperatures, together with relatively low humidity, may also be particularly significant in Africa, where it is evident that the Sahara represents at least a partial barrier to the spread of BT and other orbiviruses to the Mediterranean region.

The outbreaks of BTV in Europe (since 1998) collectively represent one of the clearest examples of disease distribution being affected by climate change. Increases in average temperatures in the region have been reflected by changes in the distribution of *C. imicola* (a major vector species) in southern Europe. This has not only allowed *C. imicola* to transmit BTV in new areas, it also generated a partial overlap with European populations of *C. obsoletus*, *C. pulicaris*, and *C. dewulfi* that are widespread across most of central and northern Europe, allowing the virus to be passed to at least one of these species as a new vector. The more northerly *Culicoides* species can clearly transmit the virus, particularly during the very warm summers that are now affecting much of Europe, as demonstrated by the outbreak of BTV-8 in the Netherlands, Belgium, Germany, France, and Luxemburg during the mid-to late summer of 2006, when record temperatures were experienced in the region. The whole of Europe must now be considered at risk to outbreaks of BT and some other orbivirus diseases that are transmitted by the same vectors (e.g., AHS or EHDV). This conclusion was confirmed by the further spread of BTV-8 and to the UK, Denmark, and Switzerland during 2007.

Although BTV and some of the other orbiviruses are widely distributed, not all serotypes of each virus species are present at each location. Introduction of orbiviruses (e.g., BTV or AHSV) into areas that are usually free of the disease, and which therefore contain immunologically naive populations of susceptible host species, can result in high mortality rates in infected animals. As an example, during the disease outbreak caused by BTV-8 in northern Europe during 2006, it was estimated that up to 50% of those sheep showing clinical signs of infection, eventually died from the disease. Although much smaller numbers of cattle showed clinical signs of infection, approximately 10–15% of these affected animals also died. Even the introduction of a new serotype of BTV into enzootic areas can result in disease in host animals that have neutralizing antibodies against the types already present.

The mortality rate caused by African horsesickness virus (AHSV) in naive horses is frequently cited as above 90%, making it the most lethal and most dangerous of the equine pathogens. The epidemiology of AHSV is discussed elsewhere in this encyclopedia.

BTV can cause a significant reduction in the productivity of domesticated animals in endemic areas. It has been estimated that annual losses in the USA are of the order of $120 million. Infectious BTV has been detected in bull semen and some early experiments (although not reproducible and therefore of uncertain significance) indicated the possibility of long-term persistence of virus and immunotolerance in cattle that were naturally infected *in utero*. Restrictions are imposed on the movement or importation of live animals or germ line materials (semen and ova) from infected areas, although the exact regulations, involving testing, quarantine periods, or even complete import/export bans differ from country to country. These barriers to trade, along with the associated surveillance and testing programs, also represent important causes of financial loss associated with outbreaks of orbivirus diseases.

Viral RNAS, Proteins, Virion Structure, and Properties

The structure, components, properties, and assembly of the BTV particle have been studied extensively by many techniques, including electron microscopy, cryoelectron microcopy, and X-ray crystallography.. The properties of the orbivirus particles and individual viral proteins described here are derived primarily from studies of BTV and AHSV.

Mature orbivirus particles are relatively featureless when viewed by negative-contrast electron microscopy (**Figure 1**). They are icosahedral, nonenveloped, approximately 90 nm in diameter and are composed of three layers of proteins, the subcore (VP3), the core-surface layer (VP7), and the outer capsid (VP2 and VP5) (**Figure 2**). Progeny orbivirus particles can leave infected cells by budding through the outer cell membrane, acquiring an envelope in the process, that can be lost soon after release. This may explain why unpurified virus is usually associated with cellular membranes and cell debris.

The buoyant densities of purified orbivirus particles in CsCl are 1.36 g ml^{-1} (intact virions) and 1.40 g ml^{-1} (cores). Virus infectivity is stable at pH 8–9 but virions exhibit a marked decrease in infectivity outside the pH range 6.5–10.2. This reflects the loss of outer capsid proteins at $\sim$pH 6.5. Virus infectivity is abolished at pH 3.0, reflecting further disruption of the virus core. Viruses held *in vitro* at less than 15 °C in blood samples, serum, or albumin can remain infectious for decades, while virus infectivity is rapidly inactivated at 60 °C. Orbiviruses are relatively resistant to treatment with solvents, and nonionic detergents, or weak anionic detergents (such as sodium *N*-lauroyl sarcosine), although sensitivity varies with virus species. However, strong anionic detergents such as sodium dodecyl sulfate (SDS) will disrupt the particle and destroy infectivity. Freezing reduces virus infectivity by $\sim$90%, possibly due to particle disruption. However, once frozen, virus infectivity remains stable at −70 °C.

The orbivirus capsid contains seven structural proteins, arranged as three concentric capsid layers. The outer capsid layer of BTV and the related orbiviruses is composed of proteins VP2 and VP5 (encoded by genome segment 2 and segment 6, respectively). VP5 is involved in membrane penetration and can cause cell–cell fusion, while VP2 is primarily involved in cell attachment and is the major target for neutralizing antibodies generated by the infected host. VP2 is also the most variable of the virus proteins and the specificity of its interactions with these neutralizing antibodies (as determined by serum neutralization assays) can be used to identify 24 distinct BTV serotypes. The other orbivirus species also contain a variable number of distinct virus 'serotypes' (**Table 1**). The identification of the BTV, AHSV, or EHDV serotype involved in an outbreak of disease is an important aspect of virus diagnosis and would have direct relevance to the selection of an appropriate vaccine 'type'.

The orbivirus outer capsid can be modified by proteolytic enzymes (e.g., cleavage of BTV-VP2 by trypsin or chymotrypsin) and can be completely removed by treatment with divalent metal ions, or reduced pH, to release the virus core. The surface of the orbivirus core has ring-shaped capsomers, composed of the VP7 protein that are readily observed by electron microscopy (**Figure 1**) (hence orbivirus from Latin: *orbis*, meaning 'ring' or 'circle'). The BTV core is also infectious in its own right, for some mammalian cells, and particularly for adult *Culicoides* or *Culicoides* cell cultures, indicating that outer core protein VP7 can also mediate cell attachment and penetration in the absence of either VP2 or VP5. The core structural proteins are more conserved than those of the outer capsid and are serologically cross-reactive (e.g., by enzyme-linked immunosorbent assay (ELISA)) between different serotypes belonging to the same orbivirus species, but not between isolates from different species, providing

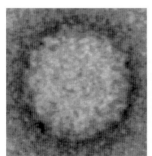
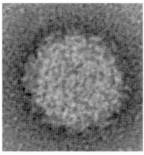
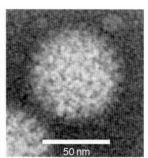

Figure 1 Electron micrographs of bluetongue virus serotype 1. Preparations of purified BTV-1 particles (strain – RSArrrr/01) were stained with 2% uranyl acetate: (left) virus particle, showing the relatively featureless surface structure; (center) infectious subviral particle (ISVP), in which outer capsid protein VP2 has been cleaved by treatment with chymotrypsin, showing some discontinuities in the outer capsid layer; (right) core particle, from which the entire outer capsid has been removed to reveal the structure of the VP7 (T13) core-surface layer and showing characteristic ring-shaped capsomeres.

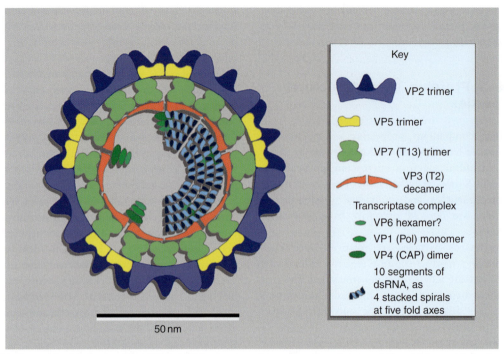

Figure 2 Diagram of the bluetongue virus (BTV) particle structure. A cross-section diagram of the orbivirus particle structure (viewed down a fivefold axis) was constructed using data from biochemical analyses, electron microscopy, cryoelectron microscopy and X-ray crystallography generated for bluetongue virus. Courtesy of P.P.C. Mertens and S. Archibald. Reproduced from Mertens PPC, Maan S, Samuel A, and Attoui H (2005) *Orbivirus, Reoviridae*. In: Fauquet CM, Mayo MA, Maniloff J, Desselberger U, and Ball LA (eds.) *Virus Taxonomy: Eighth Report of the International Committee on Taxonomy of Viruses*, pp. 466–483. San Diego, CA: Elsevier Academic Press, with permission from Elsevier.

a basis for serogroup-(virus-species-) specific assays. VP7 of BTV is particularly immunodominant and represents the major 'serogroup-specific' antigen detected in BTV-specific diagnostic assays, although the other core and nonstructural proteins also show virus-species/serotype-specific cross-reactions.

Orbivirus-infected cells also synthesize three nonstructural proteins (NS1, NS2, and NS3, encoded by BTV genome segments 5, 8, and 10, respectively).

Genome Organization and Replication

The BTV genome represents 12% and 19.5% of the total mass of intact orbivirus particles and cores, respectively. The genome segments range in size from 3954 to 822 bp (total of 19.2 kbp – **Figure 3**) and are identified as 'segments 1 to 10' (Seg-1 to Seg-10) in order of decreasing molecular weight and, hence, increasing electrophoretic mobility in 1% agarose gels. Different isolates belonging to the same orbivirus species usually have genome segments with a uniform size distribution, generating a uniform migration pattern (electropherotype) by agarose gel electrophoresis (AGE) (**Figure 4**), and this can be used to help identify individual virus species. However, variations in primary sequence often cause significant variations in migration patterns during polyacrylamide gel electrophoresis (PAGE) so that in most cases Seg-5 and Seg-6 of BTV migrate in the reverse order. This method can frequently be used to distinguish different virus strains or reassortants within the same virus species.

Orbivirus genome segments usually have a single major open reading frame (ORF) which is always on the same strand of the RNA (see **Table 2**). However, some ORFs can have more than one functional initiation site near to the 5′-end of the RNA, resulting in production of two related proteins (e.g., segment 10 of BTV encoding NS3 and NS3a).

The 5′-untranslated regions (UTRs) of BTV type 10 genome segments range from 8 to 34 bp, while the 3′-UTR are 31–116 bp in length. For other BTV serotypes and other virus species, these lengths can vary. However, in general, the 5′-UTRs are shorter than the 3′-UTRs. The UTRs of almost all of the orbivirus genome segments that have been sequenced (**Table 2**) contain two conserved base pairs at each terminus (5′-GU... AC-3′, in the positive sense). The six terminal base pair sequences at both the 3′- and 5′-UTRs of the ten BTV genome segments are almost invariably conserved in different BTV isolates (**Table 2**). Other orbiviruses have terminal sequences which are comparable to, but not always the same as, those of BTV, although they may not be conserved in all ten genome segments.

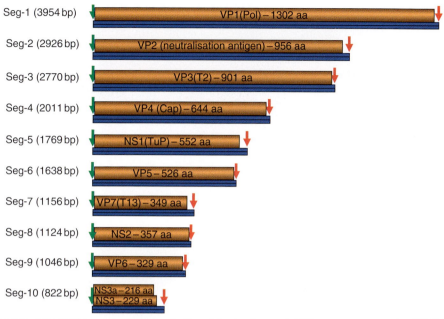

Figure 3 Organization of the BTV genome. The organization of the ten linear, dsRNA, genome segments of BTV-10. With the exception of Seg-10, each genome segment encodes a single viral protein. Seg-10 has two in-frame and functional initiation codons near to the upstream end of the segment (see http://www.iah.bbsrc.ac.uk/dsRNA_virus_proteins/BTV.htm). Like other members of the family, each AHSV genome segment contains conserved terminal sequences immediately adjacent to the upstream and downstream termini ((green arrow), 5′-GUUAAA..............ACUUAC-3′ (red arrow) in the positive-sense strand) (see **Table 2**).

Orbiviruses replicate in a variety of mammalian and insect cell lines, including BHK21 (baby hamster kidney), Vero (African green monkey kidney), KC (*Culicoides sonorensis*), and C6/36 (*Aedes albopictus*) cells. Intact orbivirus particles bind to the host cell surface via their outer capsid proteins, leading to endocytosis and cell penetration. However, BTV core particles are also infectious, indicating that the core surface protein (VP7 of BTV) can also mediate cell attachment and entry. Cell-surface receptors for BTV have not yet been identified, although core particles can bind to glycosaminoglycans. VP2 is the BTV hemagglutinin (binds sialic acid residues) and (along with VP5, NS1, and NS3) appears to determine virulence of AHSV, which may reflect its role in cell entry and initiation of infection. VP5 can induce cell fusion, suggesting an involvement in cell membrane penetration (i.e., release from endosomes).

The details of cell infection and the intracellular replication cycle of BTV represent a paradigm for other orbiviruses and are discussed elsewhere in this encyclopedia.

Evolutionary Relationships among the Orbiviruses

Structural similarities, serological cross-reactions, and significant levels of nucleotide or amino acid sequence identities in specific genes or proteins clearly demonstrate that the different orbiviruses have a common ancestry. However, the members of distinct orbivirus species can be distinguished by a failure to cross-react in 'serogroup specific' serological assays, which target the more conserved core or nonstructural proteins (e.g., VP7). Different orbivirus species also show relatively large sequence differences even in their most conserved RNAs/proteins. These can be used to distinguish them and have been used to design species-specific primers for diagnostic reverse transcriptase-polymerase chain reaction (RT-PCR) assays. The nonstructural proteins and structural proteins of the orbivirus core are usually the most conserved components of each virus species (serogroup), while showing significantly higher levels of variation between different species. For example, the sub-core shell protein (VP3 of BTV) is very highly conserved between members of the same orbivirus species (showing >73% amino acid identity between BTV strains) but shows significantly lower levels of conservation between virus species (serogroups 21.4% to 72%), reflecting more distant evolutionary relationships (**Figure 5**).

The protein components of the outermost capsid layer of the virus interact with the host defenses (including antibodies and cellular components of the mammalian immune system) and are therefore subjected to selective pressure, to change and avoid recognition. The outer surface components of different orbivirus species have also evolved to mediate transmission, cell attachment, and penetration in different host and vector species. Consequently, these outer capsid proteins (and the genome segments from which they are translated) usually show a greater degree of diversity

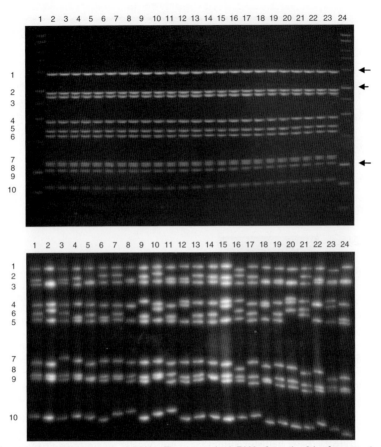

Figure 4 Electrophoretic analysis of BTV genomic dsRNAs. The genomic dsRNAs from the 24 reference strains of bluetongue viruses (BTV-1 to BTV-24: lanes 1–24) were analyzed by electrophoresis in a 1% agarose gel (top panel). Genome segments are referred to by numbers, in order of increasing electrophoretic mobility as indicated at the sides of gel. A consistent electropherotype was observed, which (despite similarities to that of some EHDV isolates) is different from the majority of other orbiviruses. The same set of dsRNAs were also analyzed by 11% polyacrylamide gel electrophoresis (PAGE) (bottom panel). Genome segments 5 and 6 migrate in a reverse order during PAGE, for most BTV isolates. Some of the more intense bands contain two genome segments which co-migrate. Differences were detected in the RNA migration pattern (PAGE-electropherotype) of each of the BTV isolates, which reflect variations in their primary nucleotide sequences.

Table 2 Conserved terminal sequences of orbivirus genome segments

Virus isolate	Conserved RNA termini (positive strand)
Bluetongue virus (BTV)	5'-GUUAAA..................................UUAC-3'
African horse sickness virus (AHSV)	5'-GUU $^A/_U$ A$^A/_U$....................AC$^A/_U$UAC-3'
Epizootic hemorrhagic disease virus (EHDV)	5'-GUUAAA...........................$^A/_G$CUUAC-3'
Great Island virus (BRDV)	5'-GUAAAA.......................A$^A/_G$GAUAC-3'
Palyam virus (CHUV)	5'-GU $^A/_U$ AAA.....................$^A/_G$CUUAC-3'
Equine encephalosis virus (EEV)[a]	5'-GUUAAG...........................UGUUAC-3'
St. Croix River virus (SCRV)	5'-$^A/_G$UAAU$^G/_{A/U}$..........$^G/_{A/U}$$^C/_U$$^C/_A$UAC-3'
Peruvian horse sickness virus (PHRV)	5'-GUUAAAA..................$^A/_G$$^C/_G$$^A/_G$UAC-3'
Yunan orbivirus (YUOV)	5'-GUUAAAA............................$^A/_G$UAC-3'

[a]Based on genome segment 10 (only) of the seven different serotypes.

within each species, than the protein components of the core or the nonstructural proteins.

In BTV, AHSV, or EHDV, the larger of the two outer capsid proteins (VP2 – encoded by genome segment 2) mediates cell attachment and is the major target for neutralizing antibodies. VP2 shows up to 27% amino acid sequence variation within a single serotype, while different BTV types can show as much as 73% variation in the sequence of this protein. Variations in the amino acid sequence of VP2 also reflect the serological relationships (cross-reactions) between different serotypes (**Figure 6**). The other orbivirus outer coat protein (VP5) can also

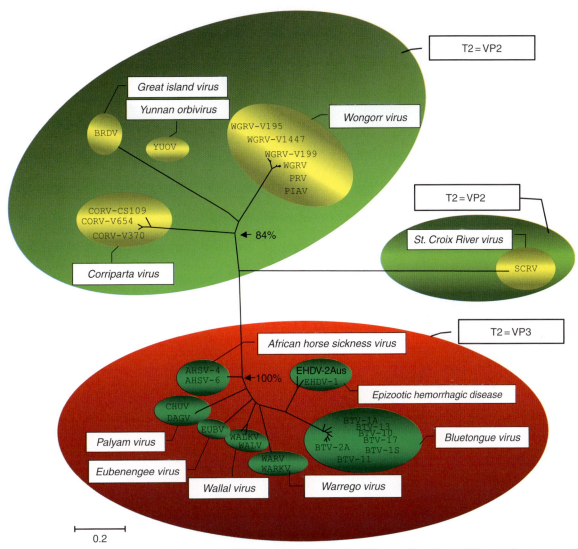

Figure 5 Phylogenetic comparison of sub-core shell (T2) proteins of different orbiviruses. Unrooted neighbor-joining tree showing relationships between the deduced amino acid sequences of the sub-core shell protein (T2) of different orbivirus species. This NJ tree was constructed using MEGA program version 3.1 and the p-distance algorithm based on partial sequences for VP3/VP2, as indicated (amino acids 393–548 relative to BTV-10 sequence). Similar trees were obtained with the Poisson correction and the gamma distance. Two major clusters of virus species were detected, one group with T2 proteins (VP2) encoded by genome segment 2 (supported by bootstrap values of >84%), and a second group with T2 protein (VP3) encoded by genome segment 3 (e.g., BTV and AHSV). These two groups appear to represent distinct evolutionary lineages. SCRV was the most divergent virus from either insect-borne or the other tick-borne orbiviruses and forms a separate small cluster.

influence serotype, although variations in the amino acid sequence of BTV VP5 show only a partial correlation with virus serotype. In some other orbiviruses (e.g., the Great Island viruses), it is the smaller outer capsid protein that exerts a greater influence over serotype. Despite its biological significance and at least partial control of cross-protection between different virus strains, the serotype of the virus currently has no formal taxonomic significance.

Within a single virus species, many of the genome segments also display sequence variations that reflect the geographic origin of the virus isolate (topotypes). For example, BTV and EHDV isolates can be divided into eastern and western groups, based on sequence comparisons of several genome segments/proteins (i.e., 'eastern' viruses from Australia, India, and Asia, and 'western' viruses from Africa and America). Since 1998, both eastern and western strains of BTV have invaded Europe, providing a potentially unique opportunity for these viruses to co-infect the same host and exchange (reassort) genome segments. Regional variations in VP2/Seg-2 and VP5/Seg-6 indicate that different BTV serotypes evolved before they became geographically dispersed, and then acquired further point mutations that distinguish these different topotypes.

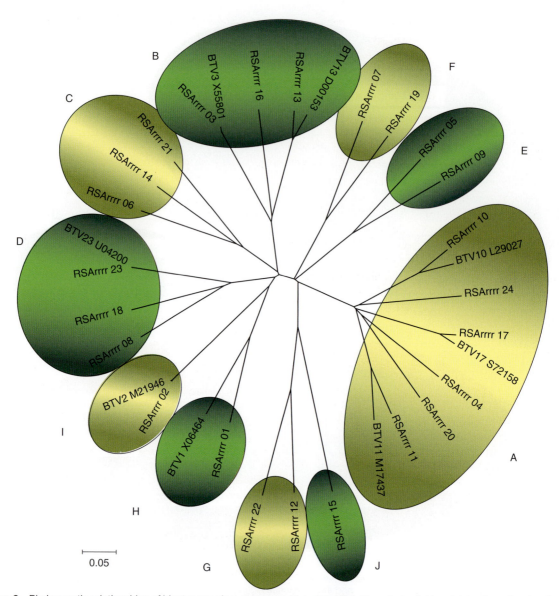

Figure 6 Phylogenetic relationships of bluetongue virus outer capsid protein VP2. Unrooted neighbor-joining tree showing the relationships between deduced amino acid sequences of VP2 (encoded by genome segment 2) from the 24 BTV serotypes. This NJ tree was constructed using MEGA program version 3.1 using the p-distance algorithm and the full-length VP2 sequences of the 24 BTV types. The different serotypes are distinct but show some relationships (green bubbles) that mirror the serological relatedness (cross-reactions) that are known to exist between different serotypes. These groupings are reflected in the nucleotide sequences of genome segment 2 and the ten distinct groups have previously been identified as nucleotypes A–J.

Sequence analyses of BTV Seg-2/VP2 provide rapid and reliable diagnostic methods to determine both the serotype and origin of individual isolates (molecular epidemiology).

However, VP7/genome segment 7 and NS3/genome segment 10 of BTV, EHDV, and AHSV show significant variations that do not appear to reflect virus serotype, or the geographic origins of the isolate. Indeed NS3 of AHSV is the second most variable of the viral proteins (after VP2), showing greater diversity than the smaller outer capsid protein VP5. As VP7 and NS3 can play roles in infection and virus release/dissemination that may be essential within the insect vector, it has been suggested that these proteins and genome segments might show variations that relate to the species or population of vector insects by which they are transmitted.

Orbiviruses of Humans

Orbiviruses and antibodies indicative of orbivirus infection have been isolated from and/or detected in humans. Changuinola virus (one of twelve 'named' serotypes within the *Changinola virus* species) was isolated in Panama from a human with a brief febrile illness. The virus has also been isolated from phlebotomine flies, and antibodies have been detected in rodents. Changuinola virus replicates in mosquito cells (C6/36) without producing CPE and is pathogenic for newborn mice or hamsters following intracerebral inoculation.

Kemerovo, Lipovnik, and Tribec viruses are among 36 tick-borne virus serotypes from the *Great island virus* species. These viruses were implicated as causes of a nonspecific fever, or neurological infection in the former USSR (Kemerovo) and Central Europe (Lipovnik and Tribec). More than 20 strains of Kemerovo virus were isolated in 1962 from patients with meningoencephalitis, and from *Ixodes persulcatus* ticks in the Kemerovo region of Russia. The virus was also isolated from birds and infects Vero or BHK-21 cells. Lipovnik and Tribec viruses may be involved in Central European encephalitis (CEE) and >50% of CEE patients had antibodies to Lipovnik virus. The virus is also suspected in the etiology of some chronic neurological diseases including polyradiculoneuritis and multiple sclerosis. Antibodies against a Kemerovo-related virus have been detected in Oklahoma and Texas, in patients with Oklahoma tick fever. Sixgun City virus is one of seven tick-borne serotypes of the *Chenuda virus* species isolated from birds. Several Oklahoma tick fever patients also had antibodies to Sixgun City virus, although no virus was isolated.

Lebombo virus type 1 (the only serotype of the *Lebombo virus* species) was isolated in Ibadan, Nigeria, in 1968, from a child with fever. The virus replicates in C6/36 cells without CPE and lyses Vero and LLC-MK2 (Rhesus monkey kidney) cells. It is pathogenic for suckling mice and has also been isolated from rodents and mosquitoes (*Mansonia* and *Aedes* species) in Africa.

The *Orungo virus* species (ORV) contains four distinct serotypes (ORV-1 to ORV-4), transmitted by *Anopheles*, *Aedes*, and *Culex* mosquitoes. ORV is widely distributed in tropical Africa where it has been isolated from humans, camels, cattle, goats, sheep, monkeys, and mosquitoes. ORV was first isolated in Uganda during 1959 from the blood of a human patient with fever and diarrhoea, who developed weakness of the legs and generalized convulsion. The weakness progressed to flaccid paralysis. Only a few clinical cases were reported (involving fever, headache, myalgia, nausea, and vomiting), despite a high prevalence of virus infection and three deaths. ORV causes lethal encephalitis in suckling mice and hamsters. It also causes CPE and plaques in Vero and BHK-21 cells, and it replicates when inoculated into the thorax of adult *Aedes aegypti* mosquitoes. High rates of co-infection with yellow fever and Orungo viruses have been reported, reflecting their similar geographic distribution and transmission by Aedes mosquitoes, as the principal vectors.

See also: African Horse Sickness Viruses; Bluetongue Viruses; Reoviruses: General Features; Reoviruses: Molecular Biology.

Further Reading

Anthony S, Jones H, Darpel KE, et al. (2007) A duplex RT-PCR assay for detection of genome segment 7 (VP7 gene) from 24 BTV serotypes. *Journal of Virological Methods* 141: 188–197.

Grimes JM, Burroughs JN, Gouet P, et al. (1998) The atomic structure of the bluetongue virus core. *Nature* 395: 470–478.

Maan S, Maan NS, Samuel AR, Rao S, Attoui H, and Mertens PPC (2006) Analysis and phylogenetic comparisons of full-length vp2 genes of the twenty-four bluetongue virus serotypes. *Journal of General Virology* 88: 621–630.

Mertens PPC and Attoui H (eds.) ReoID: Orbivirus reference collection. http://www.iah.bbsrc.ac.uk/dsRNA_virus_proteins/ReoID/viruses-at-iah.htm (accessed July 2007).

Mertens PPC, Attoui H, and Bamford DH (eds.) The RNAs and proteins of dsRNA viruses. http://www.iah.bbsrc.ac.uk/dsRNA_virus_proteins/ (accessed July 2007).

Mertens PPC and Diprose J (2004) The Bluetongue virus core: A nano-scale transcription machine. *Virus Research* 101: 29–43.

Mertens PPC, Duncan R, Attoui H, and Dermody TS (2005) *Reoviridae*. In: Fauquet CM, Mayo MA, Maniloff J, Desselberger U, and Ball LA (eds.) *Virus Taxonomy: Eighth Report of the International Committee on Taxonomy of Viruses*, pp. 447–454. San Diego, CA: Elsevier Academic Press.

Mertens PPC, Maan NS, Prasad G, et al. (2007) The design of primers and use of RT-PCR assays for typing European BTV isolates: Differentiation of field and vaccine strains. *Journal of General Virology* 88: 621–630.

Mertens PPC, Maan S, Samuel A, and Attoui H (2005) Orbivirus, Reoviridae. In: Fauquet CM, Mayo MA, Maniloff J, Desselberger U, and Ball LA (eds.) *Virus Taxonomy: Eighth Report of the International Committee on Taxonomy of Viruses*, pp. 466–483. San Diego, CA: Elsevier Academic Press.

Mertens PPC and Mellor PS (2003) Bluetongue. *State Veterinary Journal* 13: 18–25.

Mo CL, Thompson LH, Homan EJ, et al. (1994) Bluetongue virus isolations from vectors and ruminants in Central America and the Caribbean. Interamerican Bluetongue Team. *American Journal of Veterinary Research* 55: 211–215.

Owens RJ, Limn C, and Roy P (2004) Role of an arbovirus nonstructural protein in cellular pathogenesis and virus release. *Journal of Virology* 78: 6649–6656.

Purse BV, Mellor PS, Rogers DJ, Samuel AR, Mertens PPC, and Baylis M (2005) Climate change and the recent emergence of bluetongue in Europe. *Nature Reviews Microbiology* 3: 171–181.

Sellers RF (1980) Weather, host and vectors: Their interplay in the spread of insect-borne animal virus diseases. *Journal of Hygiene (Cambridge)* 85: 65–102.

Shaw AE, Monaghan P, Alpar HO, et al. (2007) Development and validation of a real-time RT-PCR assay to detect genome bluetongue virus segment 1. *Journal of Virological Methods*. 145: 115–126.

Relevant Websites

http://www.rcsb.org/pdb – Information concerning protein structures, RCSB Protein Data Bank.

http://viperdb.scripps.edu – Information concerning virus structure, virus particle explorer database(viperdb).

http://www.oie.int – OIE data on AHSV outbreaks, OIE homepage.

Organ Transplantation, Risks

C N Kotton, Massachusetts General Hospital, Boston, MA, USA
M J Kuehnert, Centers for Disease Control and Prevention, Atlanta, GA, USA
J A Fishman, Massachusetts General Hospital, Boston, MA, USA

© 2008 Elsevier Ltd. All rights reserved.

Donor-Derived Infections

Donor-derived viral infections represent a significant risk to recipients in the solid organ and hematopoietic stem cell transplantation (SOT and HSCT) setting. Viruses can be transmitted due to latent or active infection in the donor at the time of donation; rarely, donor infection is very acute and transmission occurs during the incubation period of infection, before symptoms or other evidence of infection can arise, for example, when the donor is transfused with infected blood near the time of death. Clinical consequences of infection in the transplant recipient depend on the virulence of the infecting pathogen, immune status of the recipient to the pathogen in question, and level of host immunosuppression, among other factors. Immunocompromised patients may have unusual presentations of infection, and even organisms of low virulence may cause high morbidity and mortality in these recipients.

Routine donor screening is in place for cytomegalovirus (CMV), Epstein–Barr virus (EBV), herpes simplex virus (HSV), varicella zoster virus (VZV), hepatitis B (HBV), hepatitis C (HCV), and human immunodeficiency virus (HIV), although positive donor testing does not preclude organ transplantation, with the exception of HIV. Transplantation of an organ from an infected donor is at the discretion of the transplant center and transplanting surgeon. The most commonly recognized transplant-transmitted pathogens are the herpesviruses (e.g., HSV, VZV, CMV, EBV, human herpes virus 8 (HHV-8)), hepatitis viruses (e.g., HBV, HCV), and retroviruses (e.g., human T-cell leukemia virus (HTLV)), although transmission of HIV is now unusual given current screening using nucleic acid testing (NAT). Most notable are infections that are either emerging in donor populations, or are otherwise unusual in causing illness in transplant patients. Many of these pathogens are zoonotic in nature. Some examples include West Nile virus (WNV), rabies, and lymphocytic choriomeningitis virus (LCMV) (**Table 1**).

WNV transplant-associated transmission was recognized during the first large epidemic seasonal activity in the US in 2002, when four organ recipients developed infection from a common donor (three complicated by neuroinvasive disease, the other by febrile illness), and was associated with a donor who contracted the infection from blood transfusion. The risk of infection through transfusion has been greatly reduced by blood donor screening initiated by NAT in 2003, although rarely, breakthrough infections can still occur. Subsequently, there has been an additional report of transplant transmission from a donor who likely acquired infection by mosquito in 2005; in that investigation, the donor was found to be IgM positive, indicating evidence of recent acquisition of infection, but viremia could not be detected by NAT. WNV infection has also been suspected to be transmitted through HSCT. Meningitis and encephalitis of unknown etiology in a patient recently post-transplant should cause clinicians to consider transplant transmission of WNV. Although WNV neuroinvasive disease occurs in less than 1% of WNV infections overall, transplant patients who acquire infections have an estimated 40-fold risk for developing neuroinvasive disease compared with the general population.

In 2004, four recipients of kidneys, a liver, and an arterial segment from a common organ donor died of encephalitis later diagnosed as rabies. Encephalitis developed in all four recipients within 30 days of transplantation and was associated with delirium, seizures, respiratory failure, and coma. Antibodies against rabies virus were present in three of the four recipients along with the donor, who had reportedly been bitten by a bat. This investigation also outlines the importance of careful accounting and management of tissue and organ-associated conduits in hospitals; during the investigation, the source of rabies infection in the arterial segment recipient was not clear.

Two separate occurrences of transplant transmission of LCMV have also been reported in 2003 and 2005. The transplant recipients had abdominal pain, altered mental

Table 1 Viral pathogens in transplantation recipients

Herpes simplex virus
Varicella zoster virus
Epstein–Barr virus
Cytomegalovirus
HHV-6
HHV-7
HHV-8/KSHV
Parvovirus B19
West Nile virus
Rabies
Hepatitis B and C virus
Papillomavirus
Polyomavirus BK/JC
Adenovirus, RSV, influenza, parainfluenza viruses
Lymphocytic choriomeningitis virus (LCMV)
Metapneumovirus
HIV
SARS-associated coronavirus

status, thrombocytopenia, elevated transaminases, coagulopathy, graft dysfunction, and either fever or leukocytosis within 3 weeks after transplantation. Seven of the eight recipients died. In both investigations, LCMV could not be detected in the organ donor, requiring further epidemiologic investigation to try to confirm the source. No source of LCMV infection was found in the 2003 cluster; in the 2005 cluster, the donor had had contact in her home with a pet hamster infected with a viral strain identical to the LCMV detected in the transplant recipients.

Transplant-transmitted infection is rare and might be difficult to recognize, but physicians should consider the possibility, particularly when unexplained neurologic complications occur. These investigations underscore the challenge in detecting and diagnosing infections that occur in recipients of organs or tissues from a common donor. The potential for disease transmission from donor source may not be considered in recipient evaluation. In these investigations, the ability to connect illnesses to a common organ donor was facilitated by the fact that multiple recipients were hospitalized at the same facility. As organ and tissue transplantation becomes more common, the potential risks of disease transmission may also increase.

Because of improved diagnostic assays, donor-transmitted infections are increasingly recognized, although often times, the impact of infection is not well understood. The advent of polymerase chain reaction (PCR) and other genetic material-based tests have allowed for detection of active viremia, in contrast to serologic tests, which reflect past acquisition of infection. This is important not only for recipient diagnosis, but also for recognition of donor infection retrospectively; donor diagnosis is only possible if appropriate specimens are stored postmortem.

Recognition of viral infection transmitted through transplantation is increasing, likely both due to increased recognition of unexpected symptoms consistent with transmission, and due to increasingly sophisticated diagnostic testing in both the recipient and the donor. Investigation of potential donor-transmitted infection requires rapid communication among physicians in transplant centers, organ procurement organizations, and public health authorities. An immediate system for tracking and disseminating pertinent patient data to evaluate donor-derived infection and associated adverse event outcomes is needed. Until such a system can be established, clinicians should report unexpected outcomes or unexplained illness in transplant recipients to their local organ and tissue procurement organizations.

Reactivation of Latent Infections

Given the frequency of latent viral infection, notably among herpesviruses, reactivation of latent infection provides a major source of infection after transplantation.

The specific virus, the tissue infected, stimuli for activation, and the nature of the host immune response impact the nature of viral latency. Some viruses are metabolically inactive when latent, while others continue to replicate at low levels that may be determined by the effectiveness of the host's immune response. Multiple factors contribute to viral reactivation after transplantation, including graft rejection and therapy, immune suppression (especially reduction of T-cell mediated, cytotoxic immunity), inflammation, and tissue injury. Numerous cellular pathways are involved in the control of viral replication and are activated after transplantation, such as nuclear factor κB, IκB, and JAK-STAT (the Janus family of protein tyrosine kinases (JAKS) and signal transducers and activators of transcription (STAT) proteins). Antirejection therapy can also result in a significant release of pro-inflammatory cytokines which may increase viral replication. In general, reactivation of viruses, especially late after transplantation, should suggest new immune defects (e.g., cancer) or relative over-immunosuppression.

Latency and reactivation has been best studied in the herpesviruses, which establish lifelong, latent infection after primary infection. In general, latency is considered to be the absence of viral replication, with viral genomes present in the cell without replication or spread. Studies of other viruses, such as Friend virus in mice, suggest that protective antiviral immunity is an active process mediated by 'leaky' (low-level) viral replication. The existence of true latency, as opposed to low-level replication, remains controversial. Herpesviruses make 'latency' proteins that both control viral persistence within the target cell and influence other cellular processes. The latent state is characterized by low levels or the absence of detectable viral antigens, minimal transcription of productive or lytic cycle genes, and expression of the latency-associated viral transcripts. Viral latency may be occasionally interrupted, leading to reactivation and spread of infectious virus with or without recurrent disease. EBV establishes latency in B lymphocytes in association with expression of a limited set of viral genes. Immune control of HSV infection and replication occurs at the level of skin or mucosa during initial or recurrent infection and in the dorsal root ganglion, where latency and reactivation are controlled by immune mechanisms mediated by interferons, myeloid and plasmacytoid dendritic cells, CD4(+) and CD8(+) T cells, and other cytokines. Despite similarities, the molecular details and mechanisms of latency and reactivation vary considerably among the herpesviruses. Mechanisms responsible for maintenance of latency are unclear.

Reactivation of CMV has been extensively studied. CMV viral genomes can be found in CD14+ monocytes and CD34+ progenitor cells, although the primary reservoir for latent CMV and the mechanisms by which latency is maintained are unknown. Allogeneic immune responses and fever (via tumor necrosis factor-α (TNF-α)) have been shown *in vitro* to increase both CMV promoter activity and viral replication. Immune suppression is not essential for the reactivation of latent CMV, but serves to perpetuate such infections once activated. Subclinical activation of CMV is common and increasing diagnosed by sensitive molecular assays.

For other viruses such as BK polyomavirus, specific types of tissue damage such as warm ischemia and reperfusion injury may precipitate viral activation; they have been linked to an inflammatory state in grafts (via activation of TNF-α, nuclear factor kappa B (NF-κB), neutrophil infiltration, and nitric oxide synthesis), tubular-cell injury, and enhanced expression of cell-surface molecules, all of which may contribute to viral activation. Thus, immune injury, inflammatory cytokines, and ischemia-reperfusion injury stimulate viral replication and change expression of virus-specific cell-surface receptors. The hosts' direct pathway antiviral cellular immune response within allografts is less effective due to mismatched major histocompatibility antigens between the organ donor and host with dependence on indirect pathways of antigen presentation. These factors may render the allograft more susceptible to viral infection.

Common reactivation infections after transplantation include CMV, HBV, HCV, HIV, HSV-1 and HSV-2, HPV, and VZV (as zoster). Other less clinically common viral infections related to reactivation include the polyoma viruses BK and JC, human herpes virus 6 (HHV-6), human herpes virus 7 (HHV-7), and HHV-8. Reactivation of one virus may lead to reactivation of others; multiple studies have shown that infection with HHV-6 and/or HHV-7 are risk factors for CMV disease and CMV infection may trigger HHV-6 and HHV-7 reactivation. While some reactivation infections routinely cause significant clinical disease, such as CMV, HSV, and VZV, others may cause more variable illness. HHV-6, for example, commonly reactivates with immunosuppression, especially after HSCT, often with clinically significant infection. By contrast, the role of both HHV-6 and HHV-7 in SOT recipients is less well defined; while reactivation is common, clinical disease is generally not evident.

New Infections

Based on epidemiologic exposures, new infections from the environment are commonly acquired after transplantation. The respiratory viruses are the most common new infections after transplantation, including RSV, influenza, parainfluenza, and adenovirus. New respiratory pathogens (metapneumovirus and SARS coronavirus) also cause major infections in immunocompromised hosts. Gastrointestinal viruses such as rotavirus or Norwalk virus are common and can cause significant diarrhea and dehydration; diarrheal syndromes may alter absorbance of calcineurin inhibitors (e.g., cyclosporine and tacrolimus), with unexpectedly elevated levels of tacrolimus. Nonimmune patients can acquire primary EBV, CMV, VZV, parvovirus B19, and other infections in the post-transplantation period. In the absence of previous immunity and with the attenuation of immunity due to the immunosuppressive regimen, new infections are often more severe and prolonged than in the general population. For example, parvovirus B19 infection is often more persistent and relapsing in transplantation patients, occasionally complicated by the unusual findings of hepatitis, myocarditis, pneumonitis, glomerulopathy, arthritis, or transplantation graft dysfunction.

'Direct Effects' and 'Indirect Effects' of Viral Infection

The effects of viral infection are conceptualized as 'direct' and 'indirect' (see **Table 2**). This classification serves to separate the tissue-invasive viral infection (cellular and tissue injury) from effects mediated by inflammatory responses (e.g., cytokines) or by alterations in host immune responses. Syndromes such as fever and neutropenia (e.g., with CMV infection) or invasive disease resulting in pneumonia, enteritis, meningitis, and encephalitis are considered direct effects. Indirect effects of viral infections are generally thought to be immunomodulatory responses to viral infections mediated by cytokines, chemokines, and/or growth factors. The impact of these

Table 2 Direct and indirect effects of CMV infection

Direct effects of CMV infection	Indirect effects of CMV infection
Fever and neutropenia syndrome (leukopenia, fever, myalgia, fatigue, thrombocytopenia, hepatitis, nephritis)	Increases risk of secondary infection by bacteria, fungi, and viruses
Myelosuppression	Increases risk of graft rejection
Pneumonia	Increases risk of PTLD
Gastrointestinal invasion with colitis, gastritis, ulcers, bleeding, or perforation	May increase risk of HHV-6 and HHV-7 infection
Hepatitis	
Pancreatitis	
Chorioretinitis	

effects is diverse and includes systemic immune suppression predisposing to other opportunistic infections (notably with CMV or HCV infections). In addition, viral infection may alter the expression of cell-surface antigens (e.g., major histocompatibility antigens) provoking graft rejection and/or cause disregulated cellular proliferation (contributing to atherogenesis in cardiac allografts, obliterative bronchiolitis in lung transplantation, or to oncogenesis). Increased viral replication and persistence may contribute to allograft injury (fibrosis) or chronic rejection. Infection with one virus may stimulate replication of other viruses in a form of viral 'cross talk'. As was noted above, infection with HHV-6 and/or HHV-7 serve as risk factors for CMV disease and vice versa. The direct and indirect effects of HHV-6 reactivation can be significant: HHV-6 infection is associated with high levels of IL-6 and TNF-α, and in pediatric renal or bone marrow transplantation, HHV-6 reactivation is strongly associated with acute rejection. Co-infection with HCV and CMV predicts an accelerated course for hepatitis. Co-infection with CMV and EBV increases the risk for post-transplantation lymphoproliferative disorders (PTLD) by 12–20-fold. A more theoretical concern is that T-cell responses against viral infections are thought to produce cross-reacting immune responses against graft antigens, possibly via 'alternative recognition' within the T-cell receptor. This cross-reactivity is termed 'heterologous immunity' and may provoke abrogation of graft tolerance.

Virus Specific Syndromes

The HHV family has eight human members, all of which can cause significant disease in transplantation recipients. The risk of many of these infections is reduced by the use of valganciclovir or acyclovir after transplantation. HSV-1 and -2 (HHV-1 and -2) usually cause oral and genital ulceration, although it may occur in more unexpected areas as well. Recurrent disease can be problematic for some patients and warrants consideration of secondary prophylaxis with antiviral therapy. VZV (HHV-3) is common with an incidence of herpes zoster among 869 patients after SOT of 8.6% (liver 5.7%, renal 7.4%, lung 15.1%, and heart 16.8%), with a median time of onset 9.0 months, as reported by Razonable et al. in 2005. After allogeneic HSCT, in one study by Koc et al. 41 of 100 (41%) developed VZV reactivation a median of 227 days (range 45–346 days) post-transplantation. Both primary and disseminated VZV infection can be lethal in these populations. Nonimmune transplantation patients should be monitored carefully after exposure to clinical varicella and the use of antiviral therapy or varicella immunoglobulin for prophylaxis should be considered. EBV (HHV-4) can cause febrile systemic illness, lymphocytosis, leukopenia, hepatitis, and mediates post-transplantation lymphoproliferative disease (PTLD). PTLD constitutes a spectrum of disease, which is often responsive to reduced immunosuppression in previously immune hosts. EBV-seronegative individuals with primary infection after transplantation are at increased risk for EBV-mediated PTLD. The clinical presentation of CMV (HHV-5) can range from a 'CMV syndrome' including fever, malaise, leukopenia, to a 'flu-like' illness with myalgias and fatigue, to a more significant end-organ disease with pneumonitis, colitis, encephalitis, hepatitis, or chorioretinitis. CMV is the single most important pathogen in transplantation recipients due to direct and indirect effects (see above).

HHV-6 commonly reactivates after transplantation, especially after HSCT where it is associated with hepatitis, pneumonitis, CMV reactivation, bone marrow suppression, and encephalitis. HHV6 causes less symptomatic clinical infection after SOT, although the indirect effects of reactivation have not been studied. HHV-7 commonly reactivates after transplantation. The clinical symptoms caused by HHV-7 are uncertain, although neurological symptoms seem to be significant, especially in children. HHV-8 causes Kaposi's sarcoma and is seen in SOT recipients at a rate 500–1000 higher than the general population, with a prevalence of 0.5–5% depending on the patient's (and donor's) country of origin.

Hepatitis B and C are among the most common indications for liver transplantation, and can complicate other transplantations as well. Hepatitis C is currently the most common indication for liver transplantation, accounting for 40–45% of cases in recent times. Recurrent post-transplantation hepatitis C infection poses a conundrum between treating the hepatitis C and reducing immunosuppression without precipitating rejection. Given the risk of precipitating graft dysfunction, hepatitis C treatment with interferon and ribavirin is often deferred in extrahepatic transplant recipients. For hepatitis B, the goal is complete viral suppression before and after transplantation, using hepatitis B immunoglobulin as well as antiviral agents with lower-dose immunosuppression. At this time, it is unclear which antivirals and immunosuppressive regimens are optimal for this population. Liver transplantation for HBV with combination viral prophylaxis and hepatitis B immunoglobulin results in survival rates equivalent to other indications for liver transplantation.

Respiratory viruses are the most common community-acquired infections in transplantation recipients. Given the increased rates of pneumonia and bacterial and fungal superinfection, prevention (vaccination, avoidance of sick individuals) is essential. Diagnosis of respiratory viruses within a few hours via enzyme-linked immunosorbent assay (ELISA) or immunofluorescent staining is available in most medical centers. Viral cultures are time consuming and expensive. Respiratory syncytial virus and parainfluenza are the most common community-acquired respiratory viruses, followed by influenza and adenovirus. Antiviral medications (rimantidine, amantidine, or oseltamivir) may prevent or reduce the severity of illness. The use of ribavirin or RSV immune globulin in adults to prevent RSV infection is unproven. Ribavirin is commonly used for documented RSV infections of the lower respiratory tract.

Metapneumovirus and severe acute respiratory syndrome-associated coronavirus (SARS-CoV) are emerging pathogens in transplantation patients. The clinical spectrum of disease from metapneumovirus ranges from symptomatic (even fatal) to asymptomatic cases. Some groups have suggested a possible correlation with graft rejection in lung transplant recipients. Diagnosis is often made using molecular assays; the full impact of this infection in transplantation is yet to be realized. SARS, caused by a zoonotic coronaviruses, is a highly contagious and rapidly progressive form of viral pneumonia, which spread from Asia to many parts of the world in early 2003. A number of transplantation patients were infected, some of whom died. The impact of SARS and resulting infection control issues was significant for both active organ transplantation (i.e., concerns about transmitting donor-derived infections) as well as routine follow-up care for transplantation patients, some of who deferred healthcare visits.

Gastrointestinal viruses such as rotavirus or Norwalk virus may cause significant diarrhea and dehydration. Enteroviral infections in the summer months in the northern hemispheres are common and can have a more complicated and prolonged course in renal transplantation recipients.

BK virus is associated with a range of clinical syndromes in immunocompromised hosts: viruria and viremia, ureteral ulceration and stenosis, and hemorrhagic cystitis. The majority of patients with BK virus infections are asymptomatic. Infection by JC polyomavirus has been observed in renal allograft recipients as both nephropathy (in association with BK virus or alone) and/or progressive multifocal encephalopathy (PML). JCV establishes renal latency but receptors are present in multiple tissues including the brain. Infection of the central nervous system generally presents with focal neurologic deficits or seizures and may progress to death following extensive demyelination.

Human papilloma virus (HPV) infections can cause significant disease in renal transplantation recipients, including oral, skin, genital, and rectal lesions ranging from warts and dysplasia to malignancy (especially squamous cell carcinoma). The recent arrival of a vaccine for genital HPV infections may help reduce these infections.

In transplantation recipients, parvovirus B19 infection can cause erythropoietin-resistant anemia, pancytopenia, myocarditis, or pneumonitis. Direct renal involvement with glomerulopathy and allograft dysfunction has been reported in renal transplantation recipients. Clinical and virologic responses to treatment with intravenous immunoglobulin are usually excellent.

Several zoonotic viruses have caused major illness and death in the transplantation setting, including WNV, rabies, and lymphocytic choriomeningitis virus (LCMV). All have been recently reported as donor-derived infections related to SOT, with clinically subtle infections in the donor and often deadly infections in the recipients. WNV is more morbid after SOT; the risk of meningoencephalitis in a SOT patient infected with WNV has been estimated to be 40%, compared with <1% in normal hosts. Donors in endemic areas should be screened for WNV, as the prevalence can be high. Aside from donor-derived infections, rabies and LCMV have not been reported.

Although HIV-infected patients are living longer and dying less often from complications related to acquired immunodeficiency syndrome (AIDS), they are experiencing significant morbidity and mortality related to end-stage liver and renal disease. Preliminary studies suggest that both patient and graft survival are similar in HIV-negative and HIV-positive kidney and liver transplantation recipients. However, HCV infection appears to be accelerated even in controlled HIV infection. Ongoing multicenter trials of transplantation in HIV infections are continuing. Drug interactions between the immunosuppressive regimen and antiretroviral drugs necessitate careful monitoring. The profound and long-lasting suppression of the CD4+ T-cell count in patients who receive thymoglobulin induction therapy has been associated with an increased risk of infections requiring hospitalization.

Diagnostic Assays

Rapid and sensitive molecular biology-based assays for many of the common viruses after transplantation have replaced, for the most part, serologic testing and *in vitro* cultures for the diagnosis of infection. Serologic assays are generally less sensitive in transplant patients, as humoral immune responses may be delayed or absent. In one series, 29% of patients with parvovirus B19 infection as shown by PCR assay had a negative IgM assay. Quantitative molecular tests allow the optimization and individualization of antiviral therapies for prevention and treatment of infection. This advance is most significant in the management of CMV, EBV, hepatitis B, and hepatitis C viruses, where quantitative assays (such as viral loads or antigenemia tests) guide antiviral therapy. Nonquantitative (i.e., qualitative) assays are less useful in management as they do not assess responses to therapy and cannot differentiate primary infection, from reactivation or reinfection. For example, chromosomal integration of latent HHV-6 DNA (which happens in 1–3% of immunocompetent subjects) leads to high levels of viral DNA, whether or not the infection is active; one group concludes that any diagnosis of HHV-6 encephalitis should not be made without first excluding chromosomal HHV-6 integration by measuring

DNA load in CSF, serum and/or whole blood. Latent infections due to EBV and CMV may be qualitatively positive by PCR, confirming the need for quantitative assays.

Blood tests may not always accurately reflect the level of end-organ diseases; thus, it may be useful to test specific affected tissues as well the blood. In this regard, histologic evaluation of tissues using pathogen-specific immunohistochemistry may augment systemic assays. Patients with CMV colitis, for example, may have negative molecular or antigenemia blood assays for CMV. In addition, patients may shed CMV in secretions without true infection, limiting the diagnostic capacity of a positive culture. BK polyomavirus may be detectable in the urine (either by cytology, looking for the classic decoy cells, or by PCR) before it is detectable in the blood, providing a window of opportunity for reducing immunosuppression possibly in advance of invasive disease. Adenovirus may be detectable in local infections, such as cystitis, with negative blood assays.

Therapy

The treatment of viral infections in the renal transplantation recipient includes: the reduction of immunosuppression, antiviral therapy, diagnosis and treatment of co-infections (such as CMV, EBV, HHV-6, or -7), and use of adjunctive therapies such as immunoglobulins or colony stimulating factors. The overall level of immunosuppression has a major impact on both the risk of reactivation of latent infection and the ability to clear such an infection. Reducing the immunosuppressive regimen during active viral infection can be a major contribution toward clearing infection, although it presents a risk of graft rejection. As protective cytotoxic immunity to viruses is generally T-cell (CD8+) mediated, an initial reduction of antimetabolites (if neutropenic) and calcineurin inhibitors merits consideration. In contrast, a reduction of the steroid dose during the acute phase of a febrile illness may cause acute adrenal insufficiency.

When available, antiviral therapies (such as acyclovir, ganciclovir, ribavirin, lamivudine, and oseltamivir) are often essential. The toxicity of some agents (such as cidofovir, foscarnet, and ribavirin) may complicate management, notably in the face of reduced renal function in patients receiving calcineurin inhibitors as part of the immunosuppressive regimen. The duration of therapy is often longer in transplant patients than in normal hosts, and often reflects the ability to reduce the overall level of immunosuppression, that is. less immunosuppression may result in a shorter treatment time. The increased information provided by molecular diagnostics may allow for more directed treatment regimens.

Antiviral therapy is often used for prophylaxis in the post-transplantation period, especially for individuals at risk for primary infection. In general, acyclovir and related agents are used to prevent HSV and VZV and ganciclovir or valganciclovir to prevent CMV as well as HSV and VZV. The relative advantages of universal prophylaxis (i.e., the use of antiviral medication in all susceptible patients for a period after transplantation) or monitoring with preemptive (early) therapy remain to be established. Meta-analyses suggest that universal prophylaxis with antiviral medications in SOT recipients reduces CMV disease and CMV-associated opportunistic infections, graft rejection, and mortality; their use is recommended for high risk individuals (CMV-positive recipients and in CMV-negative recipients of CMV-positive organs). Some studies indicate that universal prophylaxis and preemptive therapy are effective in reducing the incidence of CMV disease. In the HSCT setting, where the risks of bone marrow toxicity from valganciclovir are higher, many programs choose to use preemptive monitoring and therapy for CMV.

Various adjunctive therapies have been helpful in treating and preventing viral infections. Immunoglobulins (i.e., intravenous immune globulins (IVIG), CMV, and HBV hyper-immune globulins (both prepared from plasma preselected for high titer antibodies to CMV and HBV, respectively), as well as monoclonal antibodies such as those for RSV) have been helpful in preventing and treating viral infections, likely due to both direct and indirect immunomodulatory properties. A significant percent of patients have post-transplantation hypogammaglobulinemia, which has been linked to increased mortality and may benefit from globulin repletion. This may be most apparent in the setting of active infections, as well as prophylaxis. Immunostimulatory agents, such as interferon-alpha used to treat hepatitis C, can be helpful at treating the viral infection but may perturb the relationship between the graft and the host, precipitating rejection. Reversal of neutropenia can be done using colony stimulating factors such as G-CSF, which is generally well tolerated in sold organ transplantation patients.

Vaccination

Vaccines should be given to patients as early as possible in the course of organ failure and well in advance of transplantation to optimize immune responses. In general, a response to vaccination is more likely to occur when given pre-transplantation rather than after transplantation. Nonlive viral vaccines such as hepatitis B, hepatitis A, injectable influenza, rabies, HPV, injectable polio may be given both pre- and post-transplantation (see **Table 3** for classifications). Live viral vaccines such as attenuated influenza (delivered by nasal spray), measles, mumps, rubella, varicella (both Varivax, for protection against varicella in nonimmune subjects, as well as Zostavax, for protection against zoster), yellow fever, oral polio, and

Table 3 Viral vaccines, classified by type of vaccine

Virus	Live attenuated[a]	Killed	Subunit/protein
Measles	x		
Mumps	x		
Rubella	x		
Varicella-zoster (both for varicella and zoster)	x		
Yellow fever	x		
Smallpox	x		
Influenza	x (intranasal)	x (injectable)	x (injectable)
Japanese encephalitis	x	x	
Poliovirus	x (oral)	x (injectable)	
Hepatitis A virus		x	
Rabies virus		x	
Hepatitis B			x
Human papilloma virus			x

[a]Live-attenuated vaccines are generally contraindicated in immunocompromised hosts.

vaccinia (smallpox) should generally be avoided after SOT and in patients who are chronically immunosuppressed (such as those with GVHD); some may be given after HSCT. Consideration of future travel or trips home should also be considered in the pre-transplantation period. Vaccines for CMV, EBV, RSV, and other viral pathogens are under further investigation.

See also: Transmissible Spongiform Encephalopathies.

Further Reading

Anonymous (2004) Guidelines for vaccination of solid organ transplant candidates and recipients. *American Journal of Transplantation* 4(supplement 10): 160–163.

Cervera C, Marcos MA, Linares L, et al. (2006) A prospective survey of human herpesvirus-6 primary infection in solid organ transplant recipients. *Transplantation* 82: 979–982.

Eid AJ, Brown RA, Patel R, and Razonable RR (2006) Parvovirus B19 infection after transplantation: A review of 98 cases. *Clinical Infectious Disease* 43: 40–48.

Fischer SA, Graham MB, Kuehnert MJ, et al. (2006) Transmission of lymphocytic choriomeningitis virus by organ transplantation. *New England Journal of Medicine* 354: 2235–2249.

Fishman JA and Rubin RH (1998) Infection in organ-transplant recipients. *New England Journal of Medicine* 338: 1741–1751.

Hirsch HH, Knowles W, Dickenmann M, et al. (2002) Prospective study of polyomavirus type BK replication and nephropathy in renal-transplant recipients. *New England Journal of Medicine* 347: 488–496.

Hodson EM, Barclay PG, Craig JC, et al. (2005) Antiviral medications for preventing cytomegalovirus disease in solid organ transplant recipients. *Cochrane Database of Systematic Reviews* CD003774.

Kotton CN (2007) Zoonoses in solid-organ and hematopoietic stem cell transplant recipients. *Clinical Infectious Disease* 44: 857–866.

Kotton CN and Fishman JA (2005) Viral infection in the renal transplant recipient. *Journal of the American Society of Nephrology* 16: 1758–1774.

Kotton CN, Ryan ET, and Fishman JA (2005) Prevention of infection in adult travelers after solid organ transplantation. *American Journal of Transplantation* 5: 8–14.

Kumar D, Drebot MA, Wong SJ, et al. (2004) A seroprevalence study of West Nile virus infection in solid organ transplant recipients. *American Journal of Transplantation* 4: 1883–1888.

Ljungman P, Engelhard D, de la Camara R, et al. (2005) Vaccination of stem cell transplant recipients: Recommendations of the Infectious Diseases Working Party of the EBMT. *Bone Marrow Transplantation* 35: 737–746.

Mylonakis E, Goes N, Rubin RH, Cosimi AB, Colvin RB, and Fishman JA (2001) BK virus in solid organ transplant recipients: An emerging syndrome. *Transplantation* 72: 1587–1592.

Small LN, Lau J, and Snydman DR (2006) Preventing post-organ transplantation cytomegalovirus disease with ganciclovir: A meta-analysis comparing prophylactic and preemptive therapies. *Clinical Infectious Disease* 43: 869–880.

Ward KN, Leong HN, Thiruchelvam AD, Atkinson CE, and Clark DA (2007) HHV-6 DNA level in CSF due to primary infection differs from that in chromosomal viral integration and has implications for the diagnosis of encephalitis. *Journal of Clinical Microbiology* 45: 1298–1304.

Origin of Viruses

P Forterre, Institut Pasteur, Paris, France

© 2008 Elsevier Ltd. All rights reserved.

Glossary

Archaea A domain of prokaryotic microorganisms whose informational mechanisms (DNA replication, transcription, and translation) are closely related to those of eukaryotes.

Homology Two biological structures are homologs if they originated from a common ancestral structure.

> The homology cannot be quantified (a protein is or is not homologous to another one). Many biologists without an evolutionary background still confuse homology and similarity (only the latter can be quantified).
> **LUCA** (Last Universal Common Ancestor) The most recent common ancestor shared by all modern cellular organisms. The modern genetic code was already established in LUCA. Other features of LUCA (cellular vs. acellular, RNA vs. DNA genome) are still highly controversial.
> **RNA world** The period of life evolution before the appearance of DNA. Depending on the authors, the RNA world is viewed as either cellular (a world of RNA cells) or acellular (a world of free-living macromolecules).
> **Universal tree of life** The tree based on 16S/18S rRNA comparison, in which the cellular living world is divided into three domains: Archaea, Bacteria, and Eukarya. The evolutionary relationships between these domains are controversial.

The Classical View of Virus Origin and Its Consequences

The origin of viruses and their evolutionary relationships with cellular organisms are still enigmatic, but recent advances from comparative genomics and structural biology have produced a new framework to discuss these issues on firmer grounds. Historically, three hypotheses have been proposed to explain the origin of viruses: (1) they originated in a precellular world ('the virus-first hypothesis'); (2) they originated by reductive evolution from parasitic cells ('the reduction hypothesis'); and (3) they originated from fragments of cellular genetic material that escaped from cell control ('the escape hypothesis'). All these hypotheses had specific drawbacks. The virus-first hypothesis was usually rejected firsthand, since all known viruses require a cellular host. The reduction hypothesis was difficult to reconcile with the observation that the most reduced cellular parasites in the three domains of life, such as Mycoplasma in Bacteria, Microsporidia in Eukarya, or Nanoarchaea in Archaea, do not look like intermediate forms between viruses and cells. Finally, the escape hypothesis failed to explain how such elaborate structures as complex capsids and nucleic acid injection mechanisms evolved from cellular structures, since we do not know any cellular homologs of these crucial viral components.

Because of these drawbacks, the problem of virus origin was for a long time considered untractable and not worth of serious consideration (similarly, the study of bacterial evolution was considered a hopeless and futile task prior to the pioneering work of Carl Woese). However, since the problem of the origin is so entrenched in the human mind, it was never completely ignored. Much like the concept of prokaryotes became the paradigm on how to think about bacterial evolution, the escape hypothesis became the paradigm favored by most virologists to solve the problem of virus origin. This scenario was chosen mainly because it was apparently supported by the observation that modern viruses can pick up genes from their hosts. In its classical version, the escape theory suggested that bacteriophages originated from bacterial genomes and eukaryotic viruses from eukaryotic genomes (**Figure 1(a)**). This led to a damaging division of the virologist community into those studying bacteriophages and those studying eukaryotic viruses, 'phages' and viruses being somehow considered to be completely different entities. The artificial division of the viral world between 'viruses' and bacteriophages also led to much confusion on the nature of archaeal viruses. Indeed, although most of them are completely unrelated to bacterial viruses, they are often called 'bacteriophages', since archaea (formerly archaebacteria) are still considered by some biologists as 'strange bacteria'. For instance, archaeal viruses are grouped with bacteriophages in the drawing that illustrates viral diversity in the last edition of the *Virus Taxonomy Handbook*. Hopefully, these outdated visions will finally succumb to the accumulating evidence from molecular analyses.

Viruses Are Not Derived from Modern Cells

Abundant data are now already available to discredit the escape hypothesis in its classical adaptation of the prokaryote/eukaryote paradigm. This hypothesis indeed predicts that proteins encoded by bacterial viruses (avoiding the term bacteriophage here) should be evolutionarily related to bacterial proteins, whereas proteins encoded by viruses infecting eukaryotes should be related to eukaryotic proteins. This turned out to be wrong since, with a few exceptions (that can be identified as recent transfers from their hosts), most viral encoded proteins have either no homologs in any cell or only distantly related homologs. In the latter cases, the most closely related cellular homolog is rarely from the host and can even be from cells of a domain different from the host. More and more biologists are thus now fully aware that viruses form a world of their own, and that it is futile to speculate on their origin in the framework of the old prokaryote/eukaryote dichotomy. As for all other aspects of microbiology, the problem of the nature and origin of viruses made a great leap forward when this dichotomy was successfully challenged by the trinity concept introduced

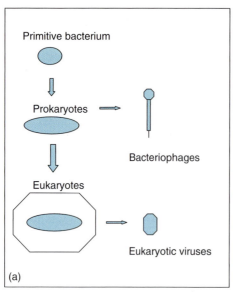
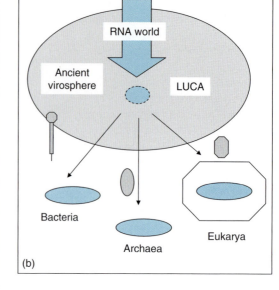

Figure 1 Two conflicting views of virus origin and evolution.

by Carl Woese, that is, the division of the living cellular world into three domains, Archaea, Bacteria, and Eukarya.

The building of a universal tree of life based on rRNA sequence comparisons, and the idea that all living organisms did not diverge from a 'primitive bacterium', as previously assumed in most textbooks, but from a less-defined last universal common ancestor (LUCA) opened the way to think about the origin and evolution of viruses in a new context (**Figure 1(b)**). Indeed, the main problem with the initial formulation of the three major hypotheses for the origin of viruses was that all of them were based on our knowledge of modern cells (themselves viewed as either prokaryotes or eukaryotes). Hence, modern viruses need modern cells to replicate, modern cells cannot regress to viral forms, and free RNA or DNA does not recruit today's proteins from modern cells to form capsids and other elaborated viral structures. The major innovation introduced by the work of Carl Woese was to open a window on the possibility of ancient worlds populated by cells most likely very different from modern ones. As early as 1980, Woese and co-workers coined the term urkaryote to name the cellular lineage that gave rise to modern eukaryotic cells (prior to the mitochondrial endosymbiosis and possibly before the origin of the nucleus itself). Later on, Woese discussed in detail the fact that the organisms that populated the basal branches of the universal tree were probably not yet modern cells, and suggested that some proteins with puzzling phylogenetic patterns could have originated in ancestral cell lineages that have now disappeared. These theoretical considerations laid the ground to the interpretation of observations that could not have been understood in the classical linear and dichotomic view of cellular evolution, such as the discovery that viruses infecting either prokaryotic or eukaryotic hosts could share homologous features (see below). In the new framework introduced by Woese's universal tree of life, these observations led to the idea that viruses might be relics of lost domains, or else that viruses predated LUCA. As a result, viruses now appear in the literature as additional branches in the universal tree or the universal tree itself is immersed in a 'viral ocean'.

The metaphor of the viral ocean, proposed by Dennis Bamford, illustrates both the concept of virus antiquity and their predominance in the modern biosphere. Another major breakthrough in recent viral research was indeed the realization that viruses are much more abundant than cells and much more diverse than previously suspected. It is thus currently assumed that viral genomes represent the major throve of genetic diversity on Earth. All these trends make it irrelevant to ask whether viruses are alive. As recently pointed out by Jean-Michel Claverie, the question of the nature of viruses has been for a long time obscured by the confusion between virus and virus particle, whereas the major components of the virus life cycle correspond to the intracellular viral factory. All this concurs to put again the question of virus origin on the agenda. Thus a brief summary of the main data that presently point to an ancient origin of viruses and the discussion of how the three major hypotheses explaining the origin of viruses have been rejuvenated in this new framework are presented here.

Viruses Are Ancient

For a long time, virologists thought that the various virus families were evolutionary unrelated, indicating a polyphyletic origin. In recent years, this view has progressively changed with the identification of more and more relationships (sometimes totally unexpected) between different viral lineages, making it possible to define a limited number of large viral groups whose hosts encompass the three cellular domains. For example, it has been clearly established that some double-stranded RNA viruses infecting bacteria are homologous to those infecting eukarya, suggesting that they predated the divergence between bacteria and eukaryotes. Since the RNA replicases/transcriptases of these double-stranded RNA viruses are homologous to those of single-stranded RNA viruses, all RNA viruses presently known seem to be evolutionary related (at least in term of their replication apparatus). In the case of DNA viruses, structural analyses of their capsid proteins have revealed an unexpected close evolutionary relationship between some bacterial spherical viruses, eukaryotic viruses of the nucleocytoplasmic-large-DNA viruses (NCLDV) superfamily, and an archaeal virus isolated at Yellowstone, USA. Structural comparative analyses also indicate that head and tailed bacterial viruses (*Caudavirales*) share homologous features with herpes viruses. These data clearly indicate that some viral specific proteins originated before the divergence between the three domains (hence before LUCA). It is tempting to suggest that a similar ancient origin also explains why most viral proteins either have no cellular homologs (except for plasmid versions or viral remnants in cellular genomes) or are only very distantly related to their cellular homologs. All these considerations thus push back the origin of viruses before the emergence of modern cells. We will now discuss how the three classical hypotheses for the origin of viruses have been revisited in this new context.

The Virus-First Hypothesis

The virus-first hypothesis was for a long time politically incorrect. It clashes with the cellular theory of life and the traditional assumption that viruses are nonliving entities. This hypothesis was first revived in the 1980s by Wolfram Zillig, who suggested that viruses originated in a prebiotic word, using the 'primitive soup' as a host. Such hypothesis has gained strength in recent years, in parallel with the suggestion that cellular organisms originated only at a late stage of life evolution. The idea that 'life' first evolved in an acellular context can be traced back to the first version of the RNA world theory. More recently, it was boosted by the discovery that archaeal lipids are dramatically different from the bacterial ones (with an opposite stereochemistry of the glycerol backbone linkages and a different type of carbon chains). To explain this dichotomy, several authors have proposed that LUCA was not a cellular entity and that cellular membranes originated independently after the divergence of Archaea and Bacteria. A more elaborate version of this scenario has been proposed by William Martin and Eugene Koonin, who suggested that life originated and evolved in the cell-like mineral compartments of a warm hydrothermal chimney. In that model, viruses emerged from the assemblage of self-replicating elements using these inorganic compartments as the first hosts. The formation of true cells occurred twice independently only at the end of the process (and at the top of the chimney), producing the first archaea and bacteria. The latter escaped from the same chimney system as already fully elaborated modern cells. In the model, viruses first co-evolved with acellular machineries producing nucleotide precursors and proteins (**Figure 2(a)**). This acellular 'life' evolved by competition between different machineries and associated viruses to infect more and more compartments of the hydrothermal system.

Cellular versus Acellular Evolution of Early Life

The acellular model of early life evolution (up to LUCA) raises several problems. First, comparative genomics analyses indicate that some membrane proteins (ATP synthetases, signal recognition particle receptors) are homologous and ubiquitous in the three domains of life, hence were probably already present in LUCA. Some authors have further stressed that the emergence of the RNA world involves at least the existence of complex mechanisms to produce ATP, RNA, and proteins. This means an elaborated metabolism to produce ribonucleotide triphosphate (rNTP) and amino acids, RNA polymerases and ribosomes, as well as an ATP-generating system. If such a complex metabolism was present, it appears unlikely that it was unable to produce lipid precursors, hence membranes. If this is correct, then 'modern' viruses did not predate cells, but originated in a world populated by primitive cells. The proponents of this scenario consider that it fits better with the contention that Darwinian selection requires competition between well-defined individual entities. It has been often assumed that RNA viruses are relics of the RNA world. In that case, viruses might have originated in a world of primitive cells with RNA genomes. In that context, it is even possible that cellularization occurred before the emergence of the modern protein synthesizing machinery and that RNA cells existed that contained no proteins (at least no proteins produced by an RNA machinery related to modern ribosomes). Modern viroids may be relics of this stage, whereas true viruses might have only appeared after the establishment of the ribosome-based mechanism for protein synthesis. In such a cellular scenario, now one has to explain how RNA viruses originated from RNA cells. Interestingly, this has led to a revival of the reduction and escape hypothesis, but in a new context.

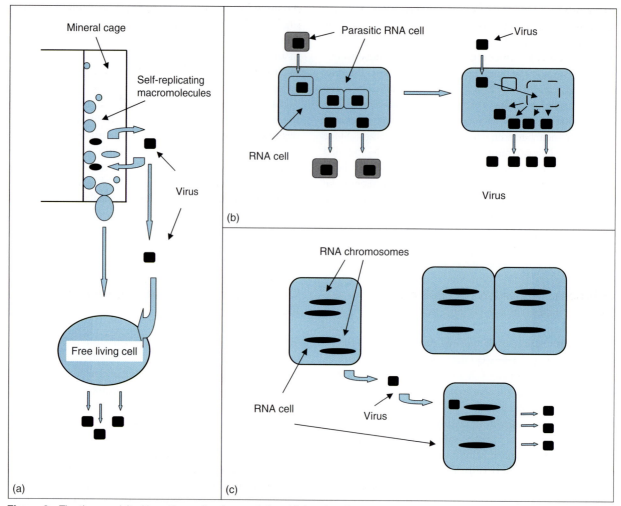

Figure 2 The three revisited hypotheses for virus evolution: (a) the virus-first hypothesis; (b) the reduction hypothesis; (c) the escape hypothesis.

The Reduction Hypothesis

The reduction hypothesis revisited in the context of pre-LUCA cells posits that RNA viruses originated by reduction from parasitic RNA cells, by losing progressively their own machinery for protein synthesis and for energy production (**Figure 2(b)**). An analogy for the possible mechanism of reduction can be seen in the reductive evolution that led to modern *Chlamydia*. Indeed, an interesting parallel can be drawn between the viral particles (the virion) and the infectious form of this bacterium (the elementary body) that is small and metabolically inactive. The main difference between viruses and *Chlamydia* is that the intracellular form of the latter, called the 'reticulate body', is a fully developed intracellular bacterium that uses its own ribosomes for protein synthesis, whereas the intracellular viral factories (although often physically separated from the host cytoplasm) have somehow a direct access to the host ribosomes. A first step in the evolution of a parasitic RNA cell toward a viral state might thus have been the division of its cell cycle between an inert extracellular stage (the protovirion) and an intracellular stage. The second step would have been the dissolution of the membrane of the intracellular parasitic cell to gain access to the protein machinery of the host. One can argue that the transformation of an intracellular parasitic cell into a viral-type factory became impossible with modern cells (such as *Chlamydia*) because, as stated by Carl Woese, the latter are too complex and too integrated to be 'deconstructed' into free-living subentities. In contrast, small parasitic RNA cells reproducing inside larger RNA cells were probably much simpler and could have been easily reduced into a viral factory by loss of their own membrane and translation machinery.

The Escape Hypothesis

The escape hypothesis is also easier to defend in the context of an ancestral world of RNA cells. It has been often argued that the genomes of ancestral RNA cells may have been fragmented and composed of semiautonomous chromosomes that were replicated independently and transferred randomly from cell to cell. The coupling between the segregation of the cellular genome and of the cellular machinery for protein synthesis (the genotype and the phenotype) was probably not so efficient in these RNA cells than it is in modern cells. The reproduction of such primitive RNA cells could have produced a mixture of progenies, some of them containing both systems (chromosomes plus ribosomes) but others containing only either RNA chromosomes or ribosomes. The latter two types of progenies would have died, except if a cell containing only chromosomes turned out to be able to infect a complete cell (or a cell containing only ribosomes). The RNA chromosomes carrying genes facilitating specifically their infectious ability and/or protecting their integrity during their resting stages would have been selected in such a situation (**Figure 2(c)**).

Relationships between RNA and DNA Viruses

If the first viruses were RNA viruses infecting RNA cells, one is left with a major question; did DNA viruses originate independently from RNA viruses or did they evolve from RNA viruses? One possibility is that DNA viruses originated either by escape or reduction from primitive DNA cells, much like RNA viruses could have originated from RNA cells. Such hypothesis supposes the existence of primitive DNA cells (less integrated than modern ones) that lived either before LUCA (if the latter was already a cellular organism) or shortly after LUCA (corresponding to early branches of the universal tree which have disappeared without descendants). The possibility that some large DNA viruses originated by reduction from extinct DNA cells evolutionarily related to early eukaryotic cells was recently boosted by the discovery of the giant mimivirus whose genome size (1.2 Mb) is three times larger than the smallest genomes of parasitic archaea or bacteria. This virus encodes a few components of the translation system that could be relics of ancient cellular lineages now extinct. On the other hand, the huge diversity of DNA viruses suggests that different lineages of DNA viruses could have originated at different periods and by different mechanisms. Indeed, there are arguments to suggest that at least some DNA viruses originated from RNA viruses. In particular, the RNA replicases/transcriptases of RNA viruses are homologous to the reverse transcriptase of retroviruses and to DNA polymerases of the A family encoded by many DNA viruses. Similarly, RNA and DNA viruses encode homologous RNA/DNA helicases. The hypothesis of an evolutionary transition from RNA to DNA viruses could explain the existence of intermediate forms such as retroviruses (with an RNA genome and an RNA–DNA–RNA cycle) and hepadnaviruses (with a DNA genome and a DNA–RNA–DNA cycle). Interestingly, retroviruses and hepadnaviruses are evolutionarily related, suggesting that the transition from RNA to DNA occurred in the virosphere.

Viruses and the Origin of DNA

Considering the possibility that at least some DNA viruses originated from RNA viruses, it has been suggested that DNA itself could have appeared in the course of virus evolution (in the context of competition between viruses and their cellular hosts). Indeed, DNA is a modified form of RNA, and both viruses and cells often chemically modify their genomes to protect themselves from nucleases produced by their competitor. It is usually considered that DNA replaced RNA in the course of evolution simply because it is more stable (thanks to the removal of the reactive oxygen in position $2'$ of the ribose) and because cytosine deamination (producing uracil) can be corrected in DNA (where uracil is recognized as an alien base) but not in RNA. The replacement of RNA by DNA as cellular genetic material would have thus allowed genome size to increase, with a concomitant increase in cellular complexity (and efficiency) leading to the complete elimination of RNA cells by the ancestors of modern DNA cells. This traditional textbook explanation has been recently criticized as incompatible with Darwinian evolution, since it does not explain what immediate selective advantage allowed the first organism with a DNA genome to predominate over former organisms with RNA genomes. Indeed, the newly emerging DNA cell could not have immediately enlarged its genome and could not have benefited straight away from a DNA repair mechanism to remove uracil from DNA. Instead, if the replacement of RNA by DNA occurred in the framework of the competition between cells and viruses, either in an RNA virus or in an RNA cell, modification of the RNA genome into a DNA genome would have immediately produced a benefit for the virus or the cell. It has been argued that the transformation of RNA genomes into DNA genomes occurred preferentially in viruses because it was simpler to change in one step the chemical composition of the viral genome than that of the cellular genomes (the latter interacting with many more proteins). Furthermore, modern viruses exhibit very different types of genomes (RNA, DNA, single-stranded, double-stranded), including highly modified DNA, whereas all modern cellular organisms have double-stranded DNA genomes. This suggests a higher degree of plasticity for viral genomes compared to cellular ones. The idea that DNA originated first in viruses could also explain why many DNA viruses encode their own enzymes for deoxynucleotide triphosphate (dNTP) production, ribonucleotide reductases (the enzymes that produce deoxyribonucleotides from ribonucleotides), and thymidylate synthases (the enzymes that produce deoxythymidine monophosphate (dTMP) from deoxyuridine monophosphate (dUMP). Because, in modern cells, dTMP is produced from dUMP, the transition from RNA to DNA occurred likely in two steps, first with the appearance of ribonucleotide reductase and production of U-DNA (DNA containing uracil), followed by the appearance of thymidylate synthases and formation of T-DNA (DNA containing thymine). The existence of a few bacterial viruses with U-DNA genomes has ben taken as evidence that they could be relics of this period of evolution.

If DNA first appeared in the ancestral virosphere, one has also to explain how it was later on transferred to cells. One scenario posits the co-existence for some time of an RNA cellular chromosome and a DNA viral genome (episome) in the same cell, with the progressive transfer of the information originally carried by the RNA chromosome to the DNA 'plasmid' via retro-transposition.

New Hypotheses about the Role of Viruses in the Origin of Modern Cells

The idea that viruses 'invented' DNA implies that they have been major players in the origin of modern cells. Indeed, several provocative hypotheses have been proposed in recent years that put viruses as central players of various evolutionary scenarios In the context of an ancient DNA virosphere, it has been argued that different lineages of DNA viruses would have 'invented' different enzymatic activities to replicate, repair, and recombine their DNA, explaining why the proteins dealing with DNA are now so diverse, often belonging to several nonhomologous protein families. It has thus been proposed that many (possibly all) cellular enzymes involved today in cellular DNA replication, repair, and/or recombination first originated in viruses before being transferred to cells. More specifically, several authors suggested that either the bacterial DNA replication mechanism, the eukaryotic/archaeal ones, or both, are of viral origin, in order to explain why the major proteins of the DNA replication machineries in eukaryotes and archaea (DNA polymerase, helicase, and primase) are not homologous to their functional analogs in bacteria (suggesting that LUCA had still an RNA genome). In order to explain why the archaeal and eukaryotic DNA replication machineries also exhibit some crucial differences (besides a core of homologous proteins), it was even suggested that the three cellular domains (Archaea, Bacteria, and Eukarya) originated from the independent fusions of three RNA cells and three large DNA viruses. In the latter scenario, the replacement of an RNA genome by a DNA genome at the onset of domain formation would have produced a drastic reduction in the rate of protein and rRNA evolution, explaining why proteins evolved apparently much less rapidly after the formation of the three domains than during the period between LUCA and the last common ancestor of each domain. The formation of each domain from three different types of RNA cells could have also selected three groups of RNA and DNA viruses specific for each domain, those that were able to infect these three ancestral RNA cells and their immediate descendants. This could explain a paradox in the modern biosphere, that each domain is characterized by its own set of viruses (for instance, NCLDVs are specific for eukaryotes) despite the fact that these viruses probably originated from a virosphere that predated LUCA. It must be remembered that NCLDVs share homologous capsid proteins with some bacterial and archaeal viruses.

Another area in which evolutionists have now recruited viruses for help is the problem of eukaryote origin. The 'viral eukaryogenesis' hypothesis posits that the eukaryotic nucleus originated from a large DNA virus, possibly related to NCLDV. This hypothesis was inspired by the analogies between the life cycle of the nucleus and those of poxviruses. In particular, both the eukaryotic nucleus and poxviruses build their membrane from the endoplasmic reticulum. Again, the discovery of the mimivirus (a member of the NCLDV family) led credence to such hypothesis. The relationships between the eukaryotic nucleus and giant viruses might have been even more complex. Hence, it was also recently proposed that a bidirectional evolutionary pathway was operating early on, with both large DNA viruses producing nuclei by infecting ancestral proto-eukaryotic cells, and also infectious nuclei producing new large DNA viruses. Finally, it was suggested that several different viruses might have been involved in eukaryogenesis to explain the presence of multiple RNA and DNA polymerases in eukaryotic cells.

Although most hypotheses previously discussed will probably always lack definitive proof, comparative genomics analyses have recently revealed a clear case of viral intervention in the formation of modern eukaryotic cells, that is, the viral origin of the DNA transcription and replication apparatus of mitochondria. This was inferred from the discovery that the RNA polymerase, DNA polymerase, and DNA helicase operating in mitochondria are of viral origin. These enzymes probably originated from a provirus that was integrated into the genome of the α-proteobacterium at the origin of mitochondria, since proviruses encoding homologs of these enzymes have been detected in the genome of several proteobacteria.

Conclusions

The idea that modern viruses are not simple extensions of prokaryotic or eukaryotic cells but derived from an ancient virosphere whose evolution encompassed the RNA world and the period of the RNA-to-DNA transition has far-reaching consequences. One of the most important in terms of practical consequences for all biologists is that modern viruses (and plasmids, which most likely originated from them) would have inherited from this ancient virosphere many molecular mechanisms that have disappeared from modern DNA cells. This would explain why the molecular biology of the viral world for transcription, replication repair, and recombination is more diverse than that of the cellular world (despite the fact that we have only explored a tiny fraction of the modern virosphere). If this view is correct, many still unknown molecular mechanisms (and their associated proteins) remain to be discovered in viruses. The exploration of viral diversity will be for sure one of the major challenges of biology in this new century.

See also: Evolution of Viruses; Nature of Viruses; Virus Evolution: Bacterial Viruses.

Further Reading

Bamford DH, Grimes JM, and Stuart DI (2006) What does structure tell us about virus evolution? *Current Opinion in Structural Biology* 15: 655–663.

Bell PJL (2001) Viral eukaryogenesis: Was the ancestor of the nucleus a complex DNA virus? *Journal of Molecular Evolution* 53: 251–256.

Claverie JM (2006) Virus takes center stage in cellular evolution. *Genome Biology* 7: 1–10.

Fauquet CM, Mayo MA, Maniloff J, Desselberger U, and Ball LA (eds.) (2005) *Virus Taxonomy: Eighth Report of the International Committee on Taxonomy of Viruses*. San Diego, CA: Elsevier Academic Press.

Filée J and Forterre P (2005) Viral proteins functioning in organelles: A cryptic origin? *Trends in Microbiology* 13: 510–513.

Forterre P (2005) The two ages of the RNA world, and the transition to the DNA world, a story of viruses and cells. *Biochimie* 87: 793–803.

Forterre P (2006) Three RNA cells for ribosomal lineages and three DNA viruses to replicate their genomes: A hypothesis for the origin of cellular domain. *Proceedings of the National Academy of Sciences, USA* 103: 3669–3674.

Forterre P and Krish H (2003) Viruses: Origin, evolution and biodiversity. *Research in Microbiology (Special Issue)* 154: 223–311.

Hamilton G (2006) Virology: The gene weavers. *Nature* 441: 683–685.

Koonin EV and Martin W (2005) On the origin of genomes and cells within inorganic compartments. *Trends in Genetics* 21: 647–654.

Ortmann AC, Wiedenheft B, Douglas T, and Young M (2006) Hot crenarchaeal viruses reveal deep evolutionary connections. *Nature Reviews Microbiology* 4: 520–528.

Prangishvili D, Forterre P, and Garrett RA (2006) Viruses of the Archaea: A unifying view. *Nature Reviews Microbiology* 4: 837–848.

Raoult D, Audic S, Robert C, *et al.* (2004) The 1.2-megabase genome sequence of mimivirus. *Science* 306: 1344–1350.

Villarreal LP (2005) *Viruses and the Evolution of Life*. Washington: ASM Press.

Woese CR, Kandler O, and Wheelis ML (1990) Towards a natural system of organisms: Proposal for the domains Archaea, Bacteria, and Eucarya. *Proceedings of the National Academy of Sciences, USA* 12: 4576–4579.

Orthobunyaviruses

C H Calisher, Colorado State University, Fort Collins, CO, USA

© 2008 Elsevier Ltd. All rights reserved.

Glossary

Arbovirus A virus transmitted to vertebrates by hematophagous (blood-feeding) insects.
Orthobunyavirus A virus in the genus *Orthobunyavirus*.
Reassortant A virus having genomic RNAs of two different viruses.
Transovarial transmission Vertical transmission, from mother to offspring.
Sympatrically Occupying the same or overlapping geographic areas without interbreeding.
Teratogenic Causing malformations of an embryo or fetus.

Introduction

From the late 1950s and onward, Jordi Casals, Robert Shope and co-workers at the World Health Organization's Collaborating Centre for Arbovirus Reference and Research collected and studied the antigenic relationships among the many viruses that had been and were being collected by Rockefeller Institute workers investigating epidemics of yellow fever and other virus diseases. Through their meticulous studies, they were able to show that some of these viruses were related to each other serologically (antigenically) and formed 'groups' of viruses. However, at least one virus in each of certain groups reacted with antibody to at least one virus in another group or groups. This was confusing because a group logically comprises certain viruses and not others, else the 'others' would be considered in the group. Casals then proposed that these tenuously interrelated viruses formed a 'supergroup', the Bunyamwera supergroup, and suggested that there might be other supergroups to be found. Subsequently, many viruses were shown to have similarities antigenically, by size, morphogenetics, and morphology, and by molecular means, yet also could be distinguished by these methods, such that groups, supergroups, and, eventually, genera, were included in a family of viruses, the *Bunyaviridae* (named after *Bunyamwera virus*).

This family comprises five genera: *Orthobunyavirus*, *Nairovirus*, *Phlebovirus*, *Hantavirus*, and *Tospovirus*. Viruses of the first three are transmitted by hematophagous arthropods and infect vertebrates; hantaviruses are not known to be transmitted by arthropods but infect vertebrates; and tospoviruses are transmitted by plant-feeding thrips and do not infect vertebrates. Any virus in the family is 'a bunyavirus' so, to avoid confusion, the original genus name *Bunyavirus* was changed to *Orthobunyavirus*.

According to the International Committee on Taxonomy of Viruses, 48 species are recognized within the genus and these 48 species include 160 viruses, including various strains, plus three viruses that are, at this time, considered 'tentative species' (**Table 1**). As the current ICTV list is somewhat confusing (it lists strains and synonyms), **Table 1** provides a slightly modified list of species and viruses placed in genus *Orthobunyavirus*.

Many studies of the orthobunyavirus-type species (prototype), *Bunyamwera virus*, have yielded a commensurate amount of information. Much that is known about the orthobunyaviruses is known by extrapolation from studies of one or more viruses of the genus or of viruses in a particular group of viruses. However, whereas all orthobunyaviruses share some characteristics, they are distinct. The differences, which are relatively or seemingly trivial, may be biologically significant. For example, La Crosse and Snowshoe hare viruses share considerable RNA sequence similarities and are considered 'subtypes' or 'varieties' of the same virus. Their geographic ranges overlap and their natural cycles include being transmitted between small mammals by mosquitoes and they cause disease in humans, usually young humans. However, the small mammals they employ as principal vertebrate hosts for amplification differ (chipmunks [*Tamias* spp.] and squirrels [*Sciurus* spp.] for La Crosse virus, hares [*Lepus americanus*] for Snowshoe hare virus); their principal mosquito vectors differ (*Aedes triseriatus* for La Crosse virus, other *Aedes* spp. for Snowshoe hare virus); La Crosse virus is the primary cause of pediatric arboviral encephalitis in the US, Snowshoe hare virus is a rare cause of such infections; and La Crosse virus has been isolated as far south as Texas and Louisiana, Snowshoe hare virus has been found in Alaska, and in Canada in the Yukon and Northwest Territories, British Columbia, Alberta, Saskatchewan, Manitoba, Ontario and Quebec in Canada, but only as far south as Montana, Minnesota, Wisconsin, Ohio, Pennsylvania, New York and Massachusetts in the US. Clearly, differences between 'species' (a taxonomic term) and 'virus' (a nomenclatural term) may be confusing but they are epidemiologically and diagnostically relevant.

Most orthobunyaviruses were discovered during routine or epidemic surveillance efforts. Many have been isolated only once or, at most, a few times; some have been isolated many times in a single location or a few times in many locations; some are isolated with considerable frequency; orthobunyaviruses are found worldwide. Most have not been associated with disease in humans, livestock, or wildlife but those that have been cause uncomplicated illnesses (fever, headache). However, certain orthobunyaviruses are recognized as the etiologic agents of severe disease, for example, the aforementioned La Crosse virus.

La Crosse Virus

In 1964, this virus was isolated from brain tissue of an encephalitic child from Minnesota who had died in 1960 in a hospital in nearby La Crosse, Wisconsin. Till then, the only recognized human pathogenic California serogroup virus in North America was California encephalitis virus, but La Crosse virus was soon shown to differ from that virus. Most infections are subclinical or cause mild illnesses, but the more severe infections can lead to illnesses characterized by frank encephalitis progressing to seizures with coma. The case–fatality rate is <1%, with neurological sequelae often requiring several years to resolve, if they do resolve and there are individual and social costs from the adverse effects on IQ and school performance. Short-term to long-term hospitalization costs can exceed $450 000. Since its recognition, La Crosse virus has been shown to cause childhood encephalitides each year. In the period 1964–2003, 3190 (a mean of 80 per year) human California group (mostly La Crosse virus infections) have been diagnosed in the US. In comparison, 4632 infections with St. Louis encephalitis (a flavivirus), 640 infections with Western equine encephalitis, and 215 infections with Eastern equine encephalitis (both togaviruses) occurred in the same period. Inapparent:apparent infection rates as high as 26:1 have been determined. Other California group viruses occasionally cause febrile illnesses with infrequent central nervous system involvement in Europe and North America.

Akabane Virus

Epizootics of congenital defects and abortion 'storms' had been observed in cattle, sheep, and goats in Japan and Australia since the 1930s, but an etiologic agent was not identified until 1959, when the orthobunyavirus Akabane virus was isolated from *Aedes vexans nipponii* and *Culex tritaeniorhynchus* mosquitoes in Japan; in Australia, the principal arthropod vector is the biting midge *Culicoides brevitarsis*. In other places, other vectors have been identified.

This virus now is known to occur widely in Africa and the Middle East and, wherever it occurs, has been responsible for malformations of the fetus. The range and

Table 1 Species and viruses[a] placed in the genus *Orthobunyavirus*

Species	Viruses
Acara virus	Acara, Moriche
Akabane virus	Akabane, Sabo, Tinaroo, Yaba-7
Alajuela virus	Alajuela, San Juan
Anopheles A virus	Anopheles A, Las Maloyas, Lukuni, Trombetas
Anopheles B virus	Anopheles B, Boraceia
Bakau virus	Bakau, Ketapang, Nola, Tanjong Rabok, Telok Forest
Batama virus	Batama
Benevides virus	Benevides
Bertioga virus	Bertioga, Cananeia, Guaratuba, Itimirim, Mirim
Bimiti virus	Bimiti
Botambi virus	Botambi
Bunyamwera virus	Batai, Birao, Bozo, Bunyamwera, Cache Valley, Fort Sherman, Germiston, Iaco, Ilesha, Lokern, Maguari, Mboke, Ngari, Northway, Playas, Potosi, Santa Rosa, Shokwe, Tensaw, Tlacotalpan, Tucunduba, Xingu
Bushbush virus	Benfica, Bushbush, Juan Diaz
Bwamba virus	Bwamba, Pongola
California encephalitis virus	California encephalitis, Inkoo, Jamestown Canyon, Keystone, La Crosse, Lumbo, Melao, San Angelo, Serra do Navio, Snowshoe hare, Tahyna, Trivitattus
Capim virus	Capim
Caraparu virus	Apeu, Bruconha, Caraparu, Ossa, Vinces
Catu virus	Catu
Estero Real virus	Estero Real
Gamboa virus	Gamboa, Pueblo Viejo
Guajara virus	Guajara
Guama virus	Ananindeua, Guama, Mahogany Hammock, Moju
Guaroa virus	Guaroa
Kairi virus	Kairi
Kaeng Khoi virus	Kaeng Khoi
Koongol virus	Koongol, Wongal
Madrid virus	Madrid
Main Drain virus	Main Drain
Manzanilla virus	Buttonwillow, Ingwavuma, Inini, Manzanilla, Mermet
Maritoba virus	Gumbo Limbo, Marituba, Murutucu, Nepuyo, Restan
Minatitlan virus	Minatitlan, Palestina
M'Poko virus	M'Poko, Yaba-1
Nyando virus	Nyando, Eret-147
Olifantsvlei virus	Bobia, Dabakala, Olifantsvlei, Oubi
Oriboca virus	Itaqui, Oriboca
Oropouche virus	Facey's Paddock, Oropouche, Utinga, Utive
Patois virus	Abras, Babahoya, Pahayokee, Patois, Shark River
Sathuperi virus	Douglas, Sathuperi
Simbu virus	Simbu
Shamonda virus	Peaton, Sango, Shamonda
Shuni virus	Aino, Kaikalur, Shuni
Tacaiuma virus	Tacaiuma, Virgin River
Tete virus	Bahig, Matruh, Tete, Tsuruse, Weldona
Thimiri virus	Thimiri
Timboteua virus	Timboteua
Turlock virus	Lednice, Turlock, Umbre
Wyeomyia virus	Anhembi, BeAr-328208 (unnamed), Macaua, Sororoca, Taiassui, Wyeomyia
Zegla virus	Zegla
Tentative species in the genus	
Leanyer, Mojui dos Campos, Termeil	

[a] The International Committee for Taxonomy of Viruses lists additional strains of certain of these viruses. Strain designations have been omitted from this list.

severity of its effects are related to the stage of gestation at infection of the dam. In adult animals, infection is subclinical and preexisting immunity in the dam provides protection to the fetus, such that the presence of Akabane virus may go undetected in areas enzootic for the virus. When it does cause congenital malformations, they may present as increased incidence of abortions and premature births occurring during seasons when arthropods are most abundant. The principal malformations are hydranencephaly (congenital absence of the cerebral hemispheres in which the space in the cranium that they normally occupy is filled with fluid) and arthrogryposis (permanent fixation of a joint in a contracted position), but dystocia (slow or difficult labor or delivery) may occur, necessitating cesarian section to save the dam. Surviving offspring manifest various developmental deficits and do not thrive. Other orthobunyaviruses have been shown to be teratogenic and Cache Valley virus has been associated with congenital defects in ruminants and, perhaps, in humans.

Oropouche Virus

Another example of the variations among orthobunyaviruses is Oropouche virus, isolated in 1955 from a febrile forest worker in Trinidad. Since then, a great deal has been learned as a result of data accrued during numerous epidemics. Oropouche disease is not fatal, but patients suffer from a mélange of signs and symptoms, including abrupt onset, fever, headache, myalgias, arthralgias, anorexia, dizziness, chills, photophobia, and an assortment of other symptoms. The acute phase of illness lasts 2–5 days but recurrence of symptoms can occur in patients who resume strenuous activities prior to complete resolution of their illness. Repeated epidemics of thousands of cases of Oropouche disease have occurred in urban population centers throughout the Brazilian states of Para, Amapa, Amazonas, Tocantins, Maranhao, Rondonia, and Acre. Epidemics of Oropouche fever were also reported in Panama in 1989 and in the Amazon region of Peru in 1992 and 1994.

The principal vector is thought to be the biting midge *Culicoides paraensis*, although isolations of Oropouche virus have also been made from mosquitoes. During epidemics, transmission of Oropouche virus can be maintained in a vector-human cycle. Oropouche disease transmission does not occur year-round and investigators have shown that monkeys, sloths, and mosquitoes comprise a sylvatic maintenance cycle.

Reassortment

Among the many orthobunyaviruses causing uncomplicated febrile illness, the group C viruses are notable. These viruses (Apeu, Bruconha, Caraparu, Ossa, Vinces, Madrid, Gumbo Limbo, Marituba, Murutucu, Nepuyo, Restan, Itaqui, and Oriboca) are related one to another in various ways. They have been detected only in the Americas, from Florida to South America and have been isolated from mosquitoes, bats, rodents, marsupials, and humans. When the first group C viruses were isolated in Brazil in the 1950s antigenic analyses demonstrated complex patterns of relationships. That is, whereas pairs of group C viruses might be closely related by complement-fixation (CF) tests, they were not related, much less closely related, by hemagglutination-inhibition (HI) and neutralization (N) tests. Using CF tests, Apeu and Marituba, Caraparu and Itaqui, and Oriboca and Murutucu were shown to be related but Apeu and Itaqui, Oriboca and Caraparu, and Marituba and Murutucu were shown to be related by HI and N tests. Because these viruses occur sympatrically and are transmitted between vertebrate hosts by mosquitoes of the same species, it was thought that these results were an indication of a series of natural reassortments and that these viruses might serve as a model of viral evolution.

As detailed elsewhere in this volume, viruses of the family *Bunyaviridae* have tripartite genomes consisting of three RNA segments: designated small (S), medium (M) and large (L). The S segment encodes the nucleocapsid (N) protein and a nonstructural protein (NSs); the M segment encodes a polypeptide that is post-translationally cleaved to produce surface glycoproteins Gn and Gc, as well as a nonstructural protein (NSm); the L segment encodes a large protein containing the RNA-dependent RNA polymerase for replication and transcription of the genomic RNA segments. We now know that the group C viral CF antigen is the N protein and that the group C hemagglutinin is a surface glycoprotein, critical for virus attachment and, therefore, for neutralization. Recent molecular genetic studies of these viruses have corroborated earlier antigenic, ecologic, and genetic studies, and have shown that many of these viruses are genetic reassortants. That is, that they have various combinations of S, M, and L RNAs and that the various combinations lead to the various antigenic characteristics, patterns of cross-protection, and enzootic persistence in nature.

Naturally occurring reassortants of La Crosse virus and Patois group viruses have been documented; a Simbu group virus, Jatobal virus has been shown to contain the S RNA of Oropouche virus; Tinaroo and Akabane viruses are naturally occurring reassortants; and Ngari virus, a Bunyamwera group virus, is a reassortant comprising the S and L segments of Bunyamwera virus and the M segment from Batai virus. The Ngari virus strains (named Garissa virus at the time) had been isolated from human hemorrhagic fever patients, one in Kenya and one in Somalia, during a large outbreak of Rift Valley fever (Rift Valley fever virus is a phlebovirus) in East Africa. Thus, reassortment of orthobunyaviral RNAs produced a virus with characteristics atypical of Bunyamwera virus, which has not been associated with severe disease, and which mimic the clinical characteristics of Rift Valley fever. It is likely that other orthobunyaviruses will be shown to be reassortants, that reassortment is an on-going evolutionary process, and that reassortment will provide us with emerging diseases into the future.

Indeed, laboratory studies of experimental reassortants of La Crosse, Snowshoe hare, and other California group orthobunyaviruses have been useful in providing us with insights to understand the structure–function relationships of the various RNAs (neurovirulence or neuroinvasiveness and neuroattenuation map to the L RNA (which encodes the viral polymerase)). Lifelong infection of the arthropod vector, as well as transovarial transmission and venereal transmission, provide substantial opportunity for the virus to evolve by genetic drift or, under suitable circumstances of mixed infections, by segment reassortment. Orthobunyavirus reassortment, however, is limited to closely related viruses, reducing the possibility of unrestricted orthobunyaviral evolution.

Diagnosis and Treatment

Diagnosis of orthobunyavirus infections is by detection of viral RNA by PCR, as well as serologic assays including HI, CF, N, enzyme-linked immunosorbent assays, and immunofluoresence. Identification of orthobunyaviruses is by PCR, and the serologic assays mentioned above, the most specific of which is the N test. Treatment of illnesses caused by orthobunyaviruses is not needed for the more mild illnesses but when called for is merely symptomatic.

See also: Bunyaviruses: General Features; Tomato Spotted Wilt Virus.

Further Reading

Anderson CR, Spence L, Downs WG, and Aitken THG (1961) Oropouche virus: A new human disease agent from Trinidad, West Indies. *American Journal of Tropical Medicine and Hygiene* 10: 574–578.

Casals J (1963) New developments in the classification of arthropod-borne animal viruses. *Anais de Microbiologia* 11: 13–34.

Edwards JF, Karabatsos N, Collisson EW, and de la Concha Bermejillo A (1997) Ovine fetal malformations induced by *in utero* inoculation with Main Drain, San Angelo, and La Crosse viruses. *American Journal of Tropical Medicine and Hygiene* 56: 171–176.

Endres MJ, Griot C, Gonzalez-Scarano F, and Nathanson N (1991) Neuroattenuation of an avirulent bunyavirus variant maps to the L RNA segment. *Journal of Viology* 65: 5465–5470.

Fauquet CM, Mayo MA, Maniloff J, Desselberger U, and Ball LA (eds.) (2005) *Virus Taxonomy: Eighth Report of the International Committee on Taxonomy of Viruses*, pp. 699–716. San Diego, CA: Elsevier Academic Press.

Gerrard SR, Li L, Barrett AD, and Nichol ST (2004) Ngari virus is a Bunyamwera virus reassortant that can be associated with large outbreaks of hemorrhagic fever in Africa. *Journal of Virology* 78: 8922–8926.

Hammon WMcD and Reeves WC (1952) California encephalitis virus, a newly described agent. Part I. Evidence of natural infection in man and other animals. *California Medicine* 77: 303–309.

Iroegbu CU and Pringle CR (1981) Genetic interactions among viruses of the Bunyamwera complex. *Journal of Virology* 37: 383–394.

Nunes MRT, Travassos da Rosa APA, Weaver SC, Tesh RB, and Vasconcelos PFC (2005) Molecular epidemiology of group C viruses (Bunyaviridae, Orthobunyavirus) isolated in the Americas. *Journal of Virology* 79: 10561–10570.

Saeed MF, Wang H, Suderman M, *et al.* (2001) Jatobal virus is a reassortant containing the small RNA of Oropouche virus. *Virus Research* 77: 25–30.

Shope RE and Causey OR (1962) Further studies on the serological relationships of group C arthropod-borne viruses and the association of these relationships to rapid identification of types. *American Journal of Tropical Medicine and Hygiene* 11: 283–290.

Thompson WH, Kalfayan B, and Anslow RO (1965) Isolation of California encephalitis group virus from a fatal human illness. *American Journal of Epidemiology* 81: 245–253.

Ushijima H, Clerx-van Haaster M, and Bishop DH (1981) Analyses of the Patois group bunyaviruses: Evidence for naturally occurring recombinant bunyaviruses and existence of viral coded nonstructural proteins induced in bunyavirus-infected cells. *Virology* 110: 318–332.

Watts DM, Pantuwatana S, DeFoliart GR, Yuill TM, and Thompson WH (1973) Transovarial transmission of La Crosse virus (California encephalitis group) in the mosquito *Aedes triseriatus*. *Science* 182: 1140–1141.

Relevant Website

http://www.cdc.gov – Centers for Disease Control and Prevention (accessed on 11 December 2006).

Orthomyxoviruses: Molecular Biology

M L Shaw and P Palese, Mount Sinai School of Medicine, New York, NY, USA

© 2008 Elsevier Ltd. All rights reserved.

Glossary

Antigenic drift The gradual accumulation of amino acid changes in the surface glycoproteins of influenza viruses.

Antigenic shift The sudden appearance of antigenically novel surface glycoproteins in influenza A viruses.

Cap snatching The process whereby the influenza virus polymerase acquires capped primers from host mRNAs to initiate viral transcription.

Recombinant virus A virus generated through reverse genetics techniques.

Reverse genetics Techniques that allow the introduction of specific mutations into the genome of an RNA virus.

Subtype Classification of influenza A virus strains according to the antigenicity of their HA and NA genes.

Classification and Nomenclature

Influenza viruses are members of the family *Orthomyxoviridae* which is divided into five genera: *Influenzavirus A, Influenzavirus B, Influenzavirus C, Thogotovirus,* and *Isavirus*. These viruses share many common features but are defined by having a segmented, single-stranded, negative-sense RNA genome that replicates in the nucleus of infected cells. Influenza A and B viruses both have eight genome segments which can encode 11 proteins while influenza C viruses have seven genome segments that encode nine proteins. Division of influenza viruses into the *Influenzavirus A, B,* or *C* genera is based on cross-reactivity of sera to the internal viral antigens, whereas different strains within each genus are distinguished by the antigenic characteristics of their surface glycoproteins. On this basis, 16 different hemagglutinin (HA) subtypes (H1–H16) and nine different neuraminidase (NA) subtypes (N1–N9) have been described for influenza A viruses. In theory, any combination of H and N subtypes is possible but the only subtypes known to have circulated in the human population are the H1N1, H2N2, and H3N2 subtypes. All remaining H and N subtypes have been identified in viruses isolated from animals, particularly avian species. Individual influenza virus strains are named according to their genus (i.e., A, B, or C), the host from which the virus was isolated (omitted if human), the location of the isolate, the isolate number, the year of isolation, and, for influenza A viruses, the H and N subtypes. As an example, the first isolate of a subtype H3N8 influenza A virus isolated from a duck in Ukraine in 1963 is named A/duck/Ukraine/1/63 (H3N8). Influenza A viruses can undergo antigenic shift through reassortment of their glycoprotein genes. In contrast, influenza B and C viruses have only one antigenic subtype and do not undergo antigenic shift; however, all influenza viruses are characterized by minor antigenic differences, known as antigenic drift.

Virion Structure and Composition

Influenza virions are pleomorphic, enveloped particles (**Figure 1**). Spherical influenza virus particles have a diameter of approximately 100 nm but long filamentous particles (from 300 nm to several micrometers in length) have also been observed, particularly in fresh clinical isolates. The viral glycoproteins are embedded in the host-derived lipid envelope and are visible as spikes that radiate from the exterior surface of the virus when particles are viewed under the electron microscope. For influenza A and B viruses, the glycoproteins consist of the HA protein, which is the major surface protein and the NA protein. Influenza C viruses have only one surface glycoprotein, the HEF (hemagglutinin/esterase/fusion) protein. Additional minor components of the viral envelope are the M2 protein (influenza A viruses), NB and BM2 proteins (influenza B viruses), and the CM2 protein (influenza C viruses). The most abundant virion protein, M1, makes up the matrix which lies beneath the lipid membrane and surrounds the ribonucleoprotein (RNP) complexes. RNPs consist of the viral RNAs which are coated with the nucleoprotein (NP) and associated with the heterotrimeric polymerase complex (PB1, PB2, and PA proteins). Finally, small amounts of the nuclear export protein (NEP/NS2) have also been found within influenza A and B virus particles.

Physical Properties

Because influenza viruses contain a lipid membrane, they are highly sensitive to delipidating and other denaturing agents. Differences in pH, ionic strength, and ionic composition of the surrounding medium influence the viral resistance to physical and chemical agents. The infectivity can be preserved in saline-balanced fluid of neutral pH at low temperatures. Influenza viruses are also relatively thermolabile, and are rapidly inactivated at temperatures higher than 50 °C. Agents affecting the stability of membranes, proteins, or nucleic acids, such as ionizing radiation, detergents, organic solvents, etc., reduce or completely destroy the infectivity of the virus.

Properties of the Viral Genome and Proteins

The genomes of influenza A and B viruses consist of eight separate negative-sense, single-stranded RNA segments known as vRNA. The influenza C virus genome has seven such segments. Each viral RNA segment exists as an RNP complex in which the RNA is coated with nucleocapsid protein (NP) and forms a helical hairpin that is bound on one end by the heterotrimeric polymerase complex. Noncoding sequences are present at both 5′ and 3′ ends of the viral gene segments; at the extreme termini are partially complementary sequences that are highly conserved between all segments in all influenza viruses (5′ end: AGUAGAAACAAGG and 3′ end: UCG(U/C)UUUCGUCC). When base-paired, these ends function as the viral promoter which is required for replication and transcription. The additional noncoding sequences contain the polyadenylation signal for mRNA synthesis as well as parts of the packaging signals required during virus assembly.

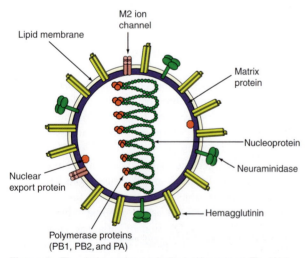

Figure 1 The viral proteins are indicated by arrows. The HA is a trimer while the NA and M2 proteins are both homotetramers. The matrix protein lies beneath the host-derived lipid envelope. Eight ribonucleoproteins form the core of the particle. These consist of the RNA genome segments coated with the nucleoprotein and bound by the polymerase complex. Small amounts of the nuclear export protein are also found within the virus particle.

The eight segments of the influenza A virus genome code for the viral proteins (**Table 1**). The three largest segments each encode one of the viral polymerase subunits, PB2, PB1, and PA. The second segment also encodes an accessory protein, PB1-F2, from an alternate open reading frame within the PB1 gene. PB1-F2, which is unique to influenza A viruses, localizes to mitochondria and has pro-apoptotic activity. Segment 4 codes for the HA protein. The mature HA protein is a trimeric type I integral membrane glycoprotein which is found in the lipid envelope of virions and on the surface of infected cells. HA undergoes several post-translational modifications including glycosylation, palmitoylation, proteolytic cleavage, disulfide bond formation, and conformational changes. Cleavage of the precursor HA0 molecule into its HA1 and HA2 subunits (which are linked by a disulfide bridge) is mediated by host cell proteases and is essential for the fusion activity of HA. HA is also responsible for binding to host cell-surface receptors (sialic acid) and is the major target of neutralizing antibodies. Segment 4 of the influenza B virus genome also encodes HA but in influenza C virus, this segment codes for the HEF protein. HEF has attachment, fusion, and receptor-destroying activity and therefore incorporates the functions of HA and NA (see below) into one protein, explaining why influenza C viruses possess one less genome segment. HEF also recognizes a different cellular receptor, namely 9-O-acetylneuraminic acid.

The NP is encoded by segment 5. This is a highly basic protein whose main function is encapsidation of the viral RNA (an NP monomer binds approximately 24 nt of RNA) which is necessary for recognition by the polymerase. NP also plays a crucial role in transporting the viral RNPs into the nucleus which it achieves through interaction with the host nuclear import machinery. Segment 6 of influenza A and B viruses encodes the NA protein which is the second major viral glycoprotein (**Table 1**). This type II integral membrane protein is a tetramer and its sialidase (neuraminidase) activity is required for efficient release of viral particles from the infected cell. In influenza B viruses, the NA gene also encodes the NB protein from a second open reading frame. Little is known about this protein apart from the fact that it is a membrane protein and is required for efficient virus growth *in vivo*.

Segment 7 of influenza A viruses encodes two proteins, the matrix protein, M1, and the M2 protein. M1 is expressed from a collinear transcript, while M2 is derived from an alternatively spliced mRNA. M1 associates with lipid membranes and plays an essential role in viral budding. It also regulates the movement of RNPs out of the nucleus and inhibits viral RNA synthesis at late stages of viral replication. M2 is a tetrameric type III membrane protein that has ion channel activity. It functions primarily during virus entry where it is responsible for acidifying the core of the virus particle which triggers dissociation of M1 from the viral RNPs (uncoating). The BM2 protein of influenza B virus plays a similar role but it is expressed from the M1 transcript via a 'stop/start' translation mechanism. The role of CM2 in uncoating of influenza C viruses is less well established. This protein is encoded on segment 6 and is produced by signal peptide cleavage of a precursor protein (p42). The matrix protein of influenza C virus, CM1, is expressed from a spliced transcript.

The shortest RNA (segment 8 in influenza A and B viruses and segment 7 in influenza C viruses) encodes the NS1 protein from a collinear transcript and the NEP/NS2 protein from an alternatively spliced transcript. NS1 is an RNA-binding protein that is expressed

Table 1 Influenza A virus[a] genes and their encoded proteins

Genome segment	Length in nucleotides	Encoded proteins	Protein size in amino acids	Function
1	2341	PB2	759	Polymerase subunit, mRNA cap recognition
2	2341	PB1	757	Polymerase subunit, endonuclease activity, RNA elongation
		PB1-F2[b]	87	Pro-apoptotic activity
3	2233	PA	716	Polymerase subunit, protease activity
4	1778	HA	550	Surface glycoprotein, receptor binding, fusion activity, major viral antigen
5	1565	NP	498	RNA binding activity, required for replication, regulates RNA nuclear import
6	1413	NA	454	Surface glycoprotein with neuraminidase activity, virus release
7	1027	M1	252	Matrix protein, interacts with vRNPs and glycoproteins, regulates RNA nuclear export, viral budding
		M2[c]	97	Integral membrane protein, ion channel activity, virus assembly
8	890	NS1	230	Interferon antagonist activity, regulates host gene expression
		NEP/NS2[c]	121	Nuclear export of RNA

[a]Influenza A/PR/8/34 virus.
[b]Encoded by an alternate open reading frame.
[c]Translated from an alternatively spliced transcript.

at high levels in infected cells. Gene knockout studies have shown that NS1 is required for virus growth in interferon-competent hosts but not interferon-deficient hosts, indicating that NS1 acts to inhibit the host antiviral response. Influenza A virus NS1 has also been shown to interfere with host mRNA processing. Influenza B virus NS1 also has interferon antagonist activity and inhibits the conjugation of ISG15 to target proteins. The NEP/NS2 protein mediates the nuclear export of newly synthesized RNPs, corresponding with its expression at late times during viral infection.

Viral Replication Cycle

Virus Attachment and Entry

Influenza viruses bind to neuraminic acids (sialic acids) on the surface of cells via their HA proteins. The receptors are ubiquitous on cells of many species and this is one of the reasons why influenza viruses are able to infect many different animals. There appears to be some specificity associated with certain HAs, which recognize sialic acid bound to galactose in either an α-2,3- or an α-2,6-linkage. The latter linkage is more common in human cells while the α-2,3-linkage is more common in avian cells. Consequently, the HAs of human influenza viruses have a preference for the α-2,6-linkage and the HAs of avian influenza viruses are more likely to bind to α-2,3-linked sialic acids (**Figure 2**).

Entry of influenza viruses into cells involves internalization by endocytic compartments. This process is thought to occur via clathrin-coated pits, but non-clathrin-, non-caveolae-mediated internalization of influenza viruses has also been described. The actual entry process requires a low-pH-mediated fusion of the viral membrane with the endosomal membrane. The fusion activity can only occur after a pH-mediated structural change in the HA of the virus has occurred and if the HA was previously cleaved into an HA1 and an HA2 subunit. The fusion peptide at the N-terminus of the HA2 subunit interacts with the endosomal membrane, and through the acid pH-mediated conformational change the viral and endosomal membranes are brought together and fuse, opening up a pore that releases the viral RNPs into the cytoplasm. This uncoating step also depends on the presence of the viral M2 protein, an ion channel protein. The uncoating can be inhibited by amantadine (and rimantadine), which are FDA-approved anti-influenza drugs targeting the M2 protein. These compounds interfere with the influx of H^+ ions from the endosome into the virus particle, which

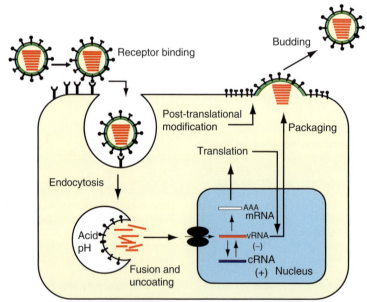

Figure 2 Virus attachment to the host cell is mediated by interaction of the HA protein with sialic acid-containing receptors. During endocytosis the interior of the endocytic vesicle becomes acidified. This induces conformational changes in HA which triggers its fusion activity and leads to fusion between the viral and endosomal membranes. The interior of the virus also becomes acidified due the ion channel activity of the M2 protein. The low pH dissociates the M1 protein from the RNP complexes, which are then transported into the nucleus. Viral RNAs are transcribed into mRNA and are replicated through a cRNA intermediate. Following export into the cytoplasm, the mRNAs are translated into viral proteins. The HA, NA, and M2 proteins are transported to the surface via the endoplasmic reticulum and Golgi (where they undergo post-translational modification). The remaining viral proteins are transported into the nucleus where they are required for either replication or nuclear export of newly synthesized vRNA. Virus assembly takes place at the apical plasma membrane which involves packaging of the eight RNPs into budding virus particles. Efficient release of budding particles requires the activity of the viral neuraminidase.

seems to be required for the disruption of protein–protein interactions, resulting in the release of RNPs into the cytoplasm, free of the other viral proteins.

An important characteristic of the influenza virus life cycle is the dependence on the nuclear functions of the cell. Once the RNP has been released from the endosome into the cytoplasm, it is then transported into the nucleus, where RNA transcription and replication take place (**Figure 2**). This cytoplasmic nuclear transport is an energy-driven process involving the presence of nuclear localization signals (NLSs) on the viral proteins and an intact cellular nuclear import machinery.

Viral RNA Synthesis

All influenza virus RNA synthesis occurs in the nucleus of the infected cell. The viral polymerase uses the incoming negative-sense vRNA as a template for direct synthesis of two positive-sense RNA species, mRNA and complementary RNA (cRNA). The viral mRNAs are incomplete copies of the vRNA and are capped and polyadenylated. cRNAs are full-length copies of the vRNA and in turn serve as a template for the generation of more negative-sense vRNA. Both vRNA and cRNA promoters are formed by a duplex of the partially complementary 5′ and 3′ ends, which increases the affinity for polymerase binding. Several studies have also indicated the importance of secondary structure in this region leading to the proposal of a corkscrew configuration. The stem–loop structures in this model have been shown to be critical for transcription.

Transcription

The viral transcription reaction is dependent on cellular RNA polymerase II activity because initiation requires a 5′ capped primer which is cleaved from host cell pre-mRNA molecules. This unique mechanism is known as cap snatching. Initiation of transcription begins with the binding of the vRNA promoter to the PB1 subunit. PB2 then recognizes and binds the cap structure on host pre-mRNAs which stimulates the endonuclease activity of PB1. The host mRNA transcripts are cleaved 10–13 nt from their 5′ caps, and the addition of a G residue which is complementary to the C residue at position 2 of the vRNA 3′ end serves to initiate transcription. Elongation is catalyzed by the PB1 protein and proceeds until the polyadenylation signal is encountered. This signal is located approximately 16 nt from the 5′ end of the vRNA and consists of five to seven uridine residues. On reaching this position the polymerase stutters, thereby adding a poly(A) tail to the viral mRNA transcripts.

Certain influenza virus proteins are expressed from alternatively spliced transcripts. The virus uses the cellular splicing machinery for this process but unlike host mRNAs which undergo complete splicing, the splicing of viral transcripts is relatively inefficient (1:10 ratio of spliced to unspliced transcripts). Of course this is necessary to ensure the production of proteins from both spliced and unspliced mRNAs and consequently several regulatory mechanisms are in place. These include the rate of nuclear export, the activity of host cell splicing factors, as well as *cis*-acting sequences within the NS1 transcript.

Replication

The replication step involves the synthesis of positive-sense cRNA from the incoming vRNA and subsequent generation of new negative-sense vRNA for packaging into progeny virions. In contrast to viral mRNAs, cRNAs are not prematurely terminated (and not polyadenylated) and do not possess a 5′ cap. Therefore the initiation and termination reactions for these two positive-sense RNAs are very different. Initiation of cRNA (and vRNA) synthesis is thought to be primer independent but it is not well understood and we also do not know how the polymerase manages to bypass the polyadenylation signal when it is in replication mode. In contrast to mRNAs, newly synthesized cRNAs and vRNAs are encapsidated and it has been proposed that the availability of soluble NP (i.e., not associated with RNPs) controls the switch between mRNA and cRNA synthesis. Other theories are that there are conformational differences between the transcription- and replication-competent polymerases and also that cellular proteins may be involved.

Packaging, Assembly, and Release of Virus

Influenza viruses assemble their components at (and in) the cytoplasmic membrane. The M1 protein has been shown to play an important role in recruiting the viral RNPs to the site of assembly at the cytoplasmic membrane, and together with the HA, NA, and M2 proteins it is the driving force in particle formation. Several influenza virus proteins, including the HA and NA, have apical sorting signals, which partially explains the observation that influenza viruses bud from the apical side of polarized cells. This asymmetric process may also provide the mechanism by which influenza viruses are largely restricted to the surface side of the respiratory tract rather than reaching the internal side, which would result in a systemic disease.

Assembly of an infectious particle requires the packaging of a full complement of the RNA genome. The precise mechanism by which this is achieved remains unclear. The first model, the random incorporation model, suggests that influenza virus RNAs, by virtue of their 5′ and 3′ termini, are recognized by viral proteins at the cytoplasmic membrane and that the correct eight RNAs are randomly packaged. If, for example, an average of 12 RNAs gets packaged, a reasonable percentage of particles would then possess a full set of the eight influenza virus RNAs. The other model, the selective incorporation model, suggests that the newly made particle has only eight RNA segments because specific signals on each RNA result in the incorporation of one segment each of the eight different RNAs. There is increasing evidence to support the validity of the latter model: there appear to be unique packaging signals on each gene segment which include both coding and noncoding sequences near the 5′ and 3′ ends of the eight different RNAs. However, it is still not clear how (or even whether) such sequences define structural elements required for the packaging of each different RNA.

Infectious particles are only formed after budding is complete. Infectious influenza virus is not found inside of cells, as is the case for poliovirus, adenovirus, and others. As influenza virus particles bud off the cytoplasmic membrane, the viral NA is needed to remove sialic acid from the cell surface as well as from the carbohydrate-carrying viral glycoproteins themselves. In the absence of a functional NA or in the presence of NA inhibitors, virus particles are not released but instead form aggregates on the surface of infected cells. The NA is thus a releasing factor which is needed in order for the newly formed virus to break free and move on to infect neighboring cells. Oseltamivir and zanamivir are two FDA-approved neuraminidase inhibitors which have strong prophylactic and therapeutic activity against influenza A and B viruses.

Reverse Genetics of Influenza Viruses

Influenza viruses are negative-strand RNA viruses, and thus their genomic RNA, when transfected into cells, does not result in the formation of infectious virus. (In contrast, the RNA of poliovirus, a positive-strand RNA virus, after transfection into cells gives rise to infectious poliovirus particles.) Initial experiments involving the transfection of a synthetic influenza virus RNA mixed with purified viral polymerase and superinfection with helper virus resulted in the formation of infectious virus which had at least one cDNA-derived gene (into which mutations could be introduced). More recently, transfection of plasmids expressing the viral RNAs and viral proteins led to a helper-free virus rescue system (**Figure 3**). The term reverse genetics refers to the fact that viral RNA is first reverse-transcribed into DNA, which can be modified. This mutated DNA is then used in the plasmid-only transfection system to create infectious influenza virus. The advances of reverse genetics have been invaluable for the study of the structure/function relationships of the different influenza virus genes and proteins. In many cases, the definitive role of a protein (or domain of a protein) can only be explored by making the appropriate knockout mutant or by changing the genes in a virus and then studying its effect on virus replication.

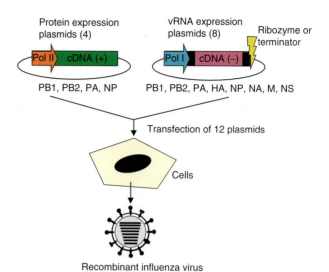

Figure 3 The negative-sense cDNA for each viral segment is cloned between a polymerase I promoter and the hepatitis delta virus ribozyme or polymerase I terminator. These plasmids will give rise to exact copies of the eight viral RNA segments. These eight plasmids are transfected into mammalian cells along with four expression plasmids for the polymerase proteins and NP, which are required for replication of the viral RNAs. The viral RNAs are replicated and transcribed by the reconstituted viral polymerase and result in the formation of infectious recombinant influenza virus.

Reverse genetics has also been used to reconstruct the extinct 1918 influenza virus using sequence data from tissue samples of victims of the 1918 pandemic. The study of this virus showed that the 1918 virus was intrinsically more virulent than other human influenza viruses and that this fact, more than any other, was responsible for the high morbidity and mortality observed in the 1918–19 pandemic.

Another application of reverse genetics techniques relates to making improved influenza virus vaccines. In the case of the avian H5N1 viruses, most strains are highly virulent by virtue of their having a basic cleavage peptide between the HA1 and HA2 subunits of the HA. The presence of this sequence of basic amino acids in the cleavage peptide allows for rapid cleavage of the HA in a variety of cells and has been found to be associated with rapid growth and high virulence. For vaccine-manufacturing purposes, it is advisable to remove this sequence signature of virulence by reverse genetics techniques. Also, the master seed strains for use in killed and live influenza virus vaccines could be constructed by reverse genetics, which would greatly accelerate the time-consuming process of selecting the right strains during interpandemic years.

Reverse genetics also allows the construction of influenza viruses which can express foreign genes. Such constructs are not only helpful in the laboratory for studies of the effects of the foreign genes in the context of an infectious virus, but they may also be useful for the development of combination vaccines. Reverse genetically modified influenza viruses represent a potential platform to express foreign antigens which can induce protective immunity against several disease agents. Genetically modified influenza viruses might also be used in the fight against cancer, as influenza viruses have been shown to be oncolytic and to induce a vigorous T-cell response. The latter could be taken advantage of by expressing tumor antigens from an influenza virus vector, which then would result in the induction of a vigorous cytolytic T-cell response against cancer cells expressing such antigens.

See also: Antiviral Agents; Human Respiratory Viruses; Orthomyxoviruses: Structure of antigens.

Further Reading

Cros JF and Palese P (2003) Trafficking of viral genomic RNA into and out of the nucleus: Influenza, Thogoto and Borna disease viruses. *Virus Research* 95: 3–12.

Deng T, Vreede FT, and Brownlee GG (2006) Different *de novo* initiation strategies are used by influenza virus RNA polymerase on its cRNA and viral RNA promoters during viral RNA replication. *Journal of Virology* 80: 2337–2348.

Fodor E, Devenish L, Engelhardt OG, et al. (1999) Rescue of influenza A virus from recombinant DNA. *Journal of Virology* 73: 9679–9682.

Gamblin SJ, Haire LF, Russell RJ, et al. (2004) The structure and receptor binding properties of the 1918 influenza hemagglutinin. *Science* 303: 1838–1842.

Garcia-Sastre A and Biron CA (2006) Type 1 interferons and the virus–host relationship: A lesson in *détente*. *Science* 312: 879–882.

Hayden FG and Palese P (2002) Influenza virus. In: Richman DD, Whitley RJ, and Hayden FG (eds.) *Clinical Virology,* 2nd edn., pp. 891–920. Washington, DC: ASM Press.

Krug RM, Yuan W, Noah DL, and Latham AG (2003) Intracellular warfare between human influenza viruses and human cells: The roles of the viral NS1 protein. *Virology* 309: 181–189.

Neumann G, Watanabe T, Ito H, et al. (1999) Generation of influenza A viruses entirely from cloned cDNAs. *Proceedings of the National Academy of Sciences, USA* 96: 9345–9350.

Noda T, Sagara H, Yen A, et al. (2006) Architecture of ribonucleoprotein complexes in influenza A virus particles. *Nature* 439: 490–492.

Palese P (2007) Influenza and its viruses. In: Engleberg NC, DiRita V, and Dermody TS (eds.) *Schaechter's Mechanisms of Microbial Disease,* 4th edn., pp. 363–369. Philadelphia, PA: Lippincott Williams and Wilkins.

Palese P and Shaw ML (2007) *Orthomyxoviridae*: The viruses and their replication. In: Knipe DM, Howley PM, Griffin DE, Lamb RA, and Martin MA (eds.) *Fields Virology,* 5th edn., pp. 1647–1689. Philadelphia, PA: Lippincott Williams and Wilkins.

Pinto LH and Lamb RA (2006) Influenza virus proton channels. *Photochemical and Photobiological Sciences* 5: 629–632.

Schmitt AP and Lamb RA (2005) Influenza virus assembly and budding at the viral budozone. *Advances in Virus Research* 64: 383–416.

Stevens J, Blixt O, Tumpey TM, et al. (2006) Structure and receptor specificity of the hemagglutinin from an H5N1 influenza virus. *Science* 312: 404–410.

Tumpey TM, Basler CF, Aguilar PV, et al. (2005) Characterization of the reconstructed 1918 Spanish influenza pandemic virus. *Science* 310: 77–80.

Orthomyxoviruses: Structure of Antigens

R J Russell, University of St. Andrews, St. Andrews, UK

© 2008 Elsevier Ltd. All rights reserved.

Glossary

Hemagglutinin The surface glycoprotein responsible for receptor binding and cell fusion.
Influenza The prototypical member of the family *Orthomyxoviridae* of viruses.
Neuraminidase The surface glycoprotein necessary for viral release.

Introduction

The family *Orthomyxoviridae* is defined by viruses that have a negative-sense, single-stranded, and segmented RNA genome. There are five different genera in the family: *Influenzavirus A, B,* and *C, Thogotovirus,* and *Isavirus.*

Influenza viruses are classified into three types: A, B, and C. Type A influenza viruses are further classified into subtypes based on the antigenic properties of their surface antigens, hemagglutinin (HA) and neuraminidase (NA). There are 16 different HA subtypes (H1–H16) and nine different NA subtypes (N1–N9) of influenza A. Viruses containing HAs from each of the 16 HA subtypes and NAs of the nine NA subtypes, in different combinations, have been isolated from avian species. Human viruses that caused pandemics in the last century were of the combination H1N1 in 1918, H2N2 in 1957, and H3N2 in 1968. Viruses that are currently circulating are of the H1N1, H3N2 combination and B. Influenza viruses also infect a range of other animals including swine and equine, with H1N2 and H3N2 currently infecting the former and H3N8 the latter.

Different influenza virus strains are named according to their type, the species from which the virus was isolated (omitted if human), place of virus isolation, the number of the isolate and the year of isolation, and in the case of the influenza A viruses, the HA and NA subtypes. For example, the 282nd isolate of an H5N1 subtype virus from chickens in Hong Kong in 2006 is designated: influenza A virus (A/chicken/Hong Kong/282/2006(H5N1)).

The genus *Thogotovirus* contains two different species, *Dhori virus* and *Thogoto virus*. Viruses from both these species were isolated from ticks, and therefore are different from influenza viruses with respect to their host range. The genus *Isavirus*, with the type species *Infectious salmon anemia virus*, is also distinct from influenza viruses A, B, and C, although numerous studies identify these isolates as members of the family *Orthomyxoviridae*.

The majority of the research into the family *Orthomyxoviridae* of viruses has been concerned with influenza viruses and therefore this article will focus exclusively on the structures of the influenza viral antigens.

Influenza Virus Surface Glycoproteins

The surface of type A and B influenza viruses has two major surface glycoproteins, HA and NA, whereas influenza C viruses have a single glycoprotein, hemagglutinin–esterase-fusion (HEF). Antibodies are raised to both HA and NA, although HA elicits the dominant antigenic response, and it is the natural variation in amino acid sequence (antigenic drift) and subsequent structural changes that limits the long-term effectiveness of an anti-influenza vaccine. Consequently, annual updates to the vaccine composition are required to maintain a match to the viruses that are currently circulating.

Influenza A and B Hemagglutinin

In type A and B influenza viruses, HA mediates attachment of the virus to the host cell via binding to sialic acid residues on the termini of cell surface glycoproteins. HA is also responsible for the fusion of viral and host cell membranes.

HA is synthesized as a single-chain polypeptide (HA0) of approximately 550 amino acids, which is subsequently proteolytically cleaved into two chains, HA1 and HA2. An interchain disulfide bond exists between cys-14 (HA1) and cys-137 (HA2). The monomers then associate noncovalently to form the functional trimer. The highly conserved and hydrophobic N-terminal residues of HA2, known as the fusion peptides (see below), are buried from solvent upon trimerization. It is known that these residues insert into the host membrane to facilitate virus membrane–cell membrane fusion. The C-terminal end of HA2 has a transmembrane anchor that tethers the molecule to the viral membrane.

HA is a highly N-glycosylated protein. For example, the H1 HA from the virus that caused the pandemic of 1918 has five glycosylation sites in HA1 and one in HA2. The number of glycosylation sites varies between virus species and has a marked impact on the antigenicity of the molecule (see below). H1 HA isolated from viruses circulating in 2002 had accumulated an additional six potential glycosylation sites with respect to the 1918 H1 HA. In addition to their influence on the antigenicity of HA, glycosylation sites that are conserved in all 16 HAs play a role in the co-translational folding of HA via binding to host cell chaperones.

To facilitate structural studies of HA, the protein is proteolytically cleaved from the surface of the virus, which removes the hydrophobic transmembrane anchor from the molecule. To date, no structure has been elucidated of a full-length HA.

The first crystal structure of HA to be determined was from the A/Aichi/2/68 (H3N2) virus, and showed the protein to be composed of a membrane-proximal, triple-stranded α-helical stem-like structure, that is composed of residues predominantly from HA2, that supports a membrane-distal globular multidomain structure, that is composed of residues solely from HA1 (**Figure 1(a)**). The membrane-distal part of each monomer can be subdivided into a vestigial esterase domain and a receptor-binding domain, the latter located at the very tip of each monomer. The receptor-binding domain consists of three secondary structure elements – the 190 helix (residues 190–198), the 130 loop (residues 135–138), and the 220 loop (residues 221–228) – that form the sides of each site, with the base made up of the conserved residues Tyr-98, Trp-153, His-183, and Tyr-195.

HA mediates the first stage of virus infection via binding to terminal sialic acid sugars on host glycoproteins and glycolipids. Sialic acids are usually found in either α-2,3- or α-2,6-linkages to galactose, the predominant penultimate sugar of N-linked carbohydrate side chains. The binding preference of an HA for one of these linkage types correlates with the species that the virus infects. The avian enteric tract has predominantly sialic acid in the α-2,3-linkage and all HAs found in the 16 antigenic subtypes found in avian influenza viruses bind preferentially to this linkage. In cells of the human upper respiratory tract, however, sialic acid in the α-2,6-linkage predominates. Indeed, HAs from human influenza viruses display a binding preference for sialic acid in the α-2,6-linkage. An avian origin has been proposed for human influenza viruses and therefore a change in binding specificity is required for cross-species transfer. The precise details of the amino acid changes underlying this switch in specificity seem to be HA subtype dependent but it has been shown that for both H1 and H3 HAs, only a small number of mutations are necessary. For example, a single mutation in H3 HA of Gln-226 to Leu-226 causes a switch in specificity from

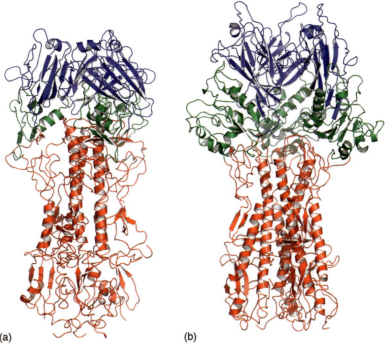

Figure 1 (a) Cartoon representation of influenza A HA. The receptor-binding domain of HA1 is colored blue and the vestigial esterase domain green. HA2, which forms the central α-helical stem, is colored red. (b) Cartoon representation of influenza C HEF. The receptor-binding domain of HEF1 is colored blue and the esterase domain green. HEF2, which forms the central α-helical stem, is colored red.

α-2,3- to α-2,6-binding preference. Insights into the amino acids involved in avian and human receptor binding have also been gained from elucidating the structure of HA in complex with receptor analogs.

The structure of HA0 has also been elucidated via crystallization of a protein in which Arg-329 had been mutated to a glutamine to prevent proteolytic cleavage of the polypeptide. The structure revealed that the cleavage site is located in a prominent surface loop which lies adjacent to a cavity which is not present in the cleaved-HA structure. There are three ionizable residues in this cavity which are buried from solvent after cleavage due to the burial of the nearly formed N-terminus of HA2 (fusion peptide). Thus it has been proposed that cleavage of HA0 results in a metastable form of the protein in which a low-pH trigger has been set due to the burial of ionizable residues. A characteristic of highly pathogenic influenza viruses such as the currently circulating H5N1 viruses is an insertion of a series of basic residues adjacent to the cleavage site. This polybasic insertion would create a much more extended surface loop which would facilitate enhanced intracellular cleavage.

In addition to sialic acid receptor binding, HA mediates virus membrane–cell membrane fusion. Subsequent to receptor binding, the influenza virus is internalized via endocytosis, exposing the virus to a low-pH environment which triggers a dramatic and irreversible conformational change in HA. Exposure of HA to low pH causes an alteration in the protonation state of residues that are buried by the fusion peptide, resulting in its removal from its buried site. Subsequently, the middle of the original long α-helix of HA2 unfolds to form a reverse turn, jack-knifing the C-terminal half backward toward the N-terminus, and results in HA2 adopting a rod-like coiled-coil structure. HA1, which was missing from the crystal structure, de-trimerizes and swings away from the HA2 but the interchain disulfide bond is maintained. As a consequence of these molecular rearrangements, the N-terminal fusion peptide, which inserts into the host membrane, and the C-terminal transmembrane anchor of HA2 are placed at the same end of the HA molecule, thereby facilitating membrane fusion (**Figure 2**).

The structure adopted by HA2 at the pH of membrane fusion shares a number of features with equivalent parts of the membrane fusion proteins of human immunodeficiency virus (HIV) and Ebola virus. In each case, the molecules contain central triple-stranded coiled coils that are surrounded by three α-helices that pack antiparallel to the core helices. As a consequence, the fusion peptides and membrane anchor regions of all of the proteins are at the same end of a rod-shaped molecule.

At present, no crystal structure of influenza B HA has been elucidated but it is expected to adopt the same conformation as influenza A HAs.

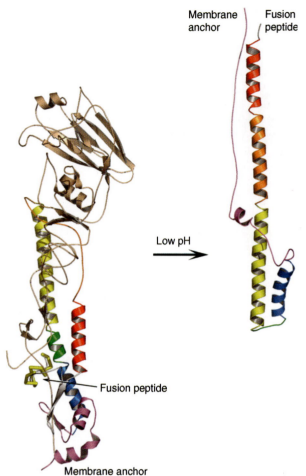

Figure 2 Schematic representation of the molecular rearrangements of HA upon exposure to low pH. The location of the fusion peptide and membrane anchor is shown in both structures. HA1 is colored in gold and HA2 is colored to highlight equivalent regions of the polypeptide. Note that only the yellow part of HA2 adopts the same position and conformation in both forms.

Influenza C HEF

In contrast to the HA of influenza A and B viruses, the major glycoprotein, HEF, of the C viruses has a receptor-destroying activity, as well as receptor-binding and fusion activities. As a consequence, influenza C viruses lack the NA glycoprotein. Despite sharing only 12% sequence identity with influenza A HAs, both HEF and HA are structurally similar (**Figure 1**). HEF has a membrane-proximal, α-helical stem-like structure and a membrane-distal globular multidomain structure. The receptor-binding domain of both HEF and HA are structurally similar despite HEF binding 9-O-acetyl sialic acid rather than sialic acid. The receptor-destroying domain shows structural homology to bacterial esterases, in keeping with its activity as a 9-O-acetylesterase. The stem region of HEF is similar to that observed in HA except that the triple-stranded, α-helical bundle diverges at both its ends and that the fusion peptide is partially exposed to solvent. The receptor-binding domain is inserted into a surface loop of the esterase domain which itself is inserted into a surface loop of the stem. Thus all three functions of the HEF are segregated into structurally distinct domains.

Influenza A and B Neuraminidase

After virus replication, NA removes sialic acid from virus and cellular glycoproteins to facilitate virus release and the spread of infection to new cells, otherwise viral aggregation would occur. Thus NA is commonly known as a receptor-destroying enzyme.

NA is synthesized as a single polypeptide chain but unlike HA no post-translational cleavage occurs. NAs have a highly conserved short cytoplasmic tail and a hydrophobic transmembrane region that provides the anchor for the stalk and the head domains. Like HA, NA is a highly N-glycosylated protein and serves to influence the antigenicity of NA.

NA suitable for structural studies is made via proteolytic cleavage from the virus surface which removes the membrane anchor and stalk domain. A deletion in the stalk domain is characteristic of highly pathogenic influenza viruses but as yet no structural information is available for this domain of NA.

The first crystal structure of the head domain of NA was of a N2 subtype, and showed that the molecule was homotetrameric with circular fourfold symmetry. Each monomer is composed of six topologically identical four-stranded antiparallel β-sheets that are themselves arranged like the blades of a propeller (**Figure 3**).

Sialic acid, the product of catalysis, binds in a deep pocket on the surface of the molecule, roughly in the middle of each monomer. No conformational changes occur upon binding of sialic acid. The amino acids in the active site that interact with sialic acid are highly conserved across all NA subtypes. The active sites of all influenza NAs contain three arginine residues, Arg 118, Arg 292, and Arg 371, that bind the carboxylate of the substrate sialic acid; Arg 152 that interacts with the acetamido substituent of the substrate; and Glu 276 that forms hydrogen bonds with the 8- and 9-hydroxyl groups of the substrate.

These properties of the active site made NA an attractive target for structure-based drug design programs, and resulted in the synthesis of oseltamivir (Tamiflu) and zanamivr (Relenza), two clinically licensed anti-influenza drugs.

The crystal structure of influenza B NA has also been elucidated and is essentially identical to influenza A NA,

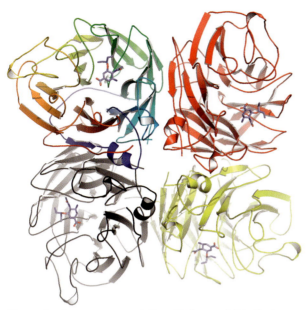

Figure 3 Cartoon representation of influenza A NA. Each monomer of the tetramer is colored differently, with the monomer at the top-left colored to emphasize the six-bladed β-propeller structure. Oseltamivir is bound to the active site and shown in stick representation and colored blue.

with the structure having the same homotetrameric arrangement and a high degree of conservation of the active site.

Influenza A Subtype Variation in Antigen Structure

To date, the crystal structures of H1, H3, H5, H7, and H9 HAs have been elucidated, which represent the four phylogenetic clades of influenza A HAs. Sequence identity between clades is between 40% and 60%. All of the HAs have the same molecular fold but HAs representative of each clade are distinguished by differences in the orientation of their membrane distal globular domains relative to the central trimeric coiled coil. For example, the receptor-binding domain of H3 HA is rotated clockwise by 24° relative to H1 HA. Also, the conformation of the interhelical loop of HA2 is different in HA from each clade. The clade-specific differences between HAs are clustered in regions that undergo conformational changes in membrane fusion, for example near to the fusion peptide, and it has been suggested that differences in the stability of HAs to pH and temperature may represent selection pressures in the evolution of influenza A HAs.

Influenza A NAs form two distinct phylogenetic groups. Crystal structures of N1, N4, and N8 have been elucidated as representatives of group 1 and N2 and N9 as representatives of group 2. All NAs have the same homotetrameric conformation but group-specific differences occur in the active site. Relative to group 2 NAs, a large cavity exists in group 1 NAs adjacent to the 4-amino group of oseltamivir. Upon binding of inhibitors to group 1 NAs, a conformational change of the 150-loop occurs which results in the active site of the two groups of NA being essentially identical. Chemical exploitation of this cavity is currently being undertaken in the development of a new generation of anti-influenza drugs to address the problem of mutations in NA that cause resistance to oseltamivir and zanamivir.

Antigenic Variation and Antibody Binding

Influenza viruses undergo a process known as 'antigenic drift' whereby natural variants of HA and NA occur, due to the error-prone nature of the viral RNA polymerase. HA is the major molecule recognized by the adaptive immune system of the host, although NA does elicit an immune response. Upon infection of a cell, the virus causes an immune response that commonly results in the production of neutralizing antibodies, in the case of HA, which acts to prevent the virus from binding to cells. Antibodies against NA, while they do not inhibit virus entry, prevent spread of the virus and afford some protection against challenge with the same or similar virus.

In the last century, three viruses were introduced into the human population from the avian reservoir of influenza viruses, and the subsequent pandemics caused millions of deaths. Following them, immune pressure causes antigenic variation that has the potential to continue until the fitness of the virus is compromised via deleterious mutations in HA which affects its receptor-binding properties.

Examination of the positions of amino acid substitutions of natural variants of HAs show that they are scattered throughout the molecule. Of the changes that are retained by circulating viruses, so-called 'fixed' substitutions, the majority are located on the surface of the molecule. Whereas, in the case of H3 HA subtype virus isolated between 1968 and 2005, two-thirds of the residues that are not retained are buried. This suggests that the 'fixed' substitutions have been selected because they alter the local structure of HA and prevent antibody binding.

This is supported by the fact that the location of amino acid substitutions in antigenic variants of HAs that have been selected by growing virus in the presence of anti-HA monoclonal antibodies (mAbs) map to the same location as the 'fixed' substitutions. Thus sites of mAb-selected mutations indicate the sites at which selecting antibodies bind. These mutants escape neutralization by the selecting antibody. The antibodies nevertheless still bind but with a much reduced affinity. There appears to be no

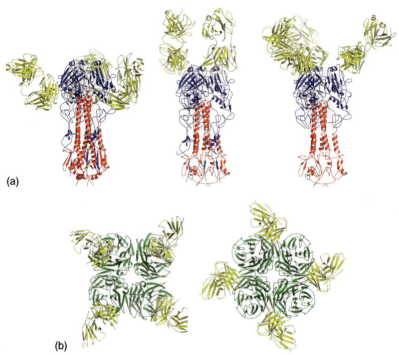

Figure 4 Antibody binding to influenza antigens. (a) Three different crystal structures of influenza HA bound to Fab fragments. (b) Two different crystal structures of influenza NA bound to Fab fragments.

preference in the type of amino acid change that occurs in selected variants.

Prevention of antibody recognition frequently occurs through the introduction of new glycosylation sites in HA. Oligosaccharide attachment at these sites prevents recognition first by covering of the protein surface and thus no antibodies being selected and second, because the sugars in the oligosaccharides are made by cellular enzymes, they are deemed by the immune system to be antigenically 'self'.

Sites of amino acid substitutions that occur in natural and antibody-selected variants are predominantly on the surface of the membrane distal part of HA, and commonly surround the receptor-binding site. This suggests that there is a link between the neutralization of viral infectivity and the prevention of virus binding to cells.

Crystal structures have been elucidated for HA and NA in complex with antibodies (**Figure 4**), and has shown that fragment antigen binding (Fab) antibody fragments do indeed bind to different parts of the molecules, supporting the observation that natural variants are scattered throughout the molecule. It is interesting to note that the stoichiometry of Fab binding is not always three molecules of Fab to one trimeric spike of HA. Of the three anti-HA antibodies that have been studied structurally, two of the binding sites overlap. All three of the antibodies have been shown to neutralize infectivity by the prevention of virus binding to cells. One of the antibodies, however, also blocks the conformational change of HA that is necessary for fusion.

Conclusion

Knowledge of the three-dimensional structures of the surface antigens of influenza viruses have had a major impact on the understanding of their role in the life cycle of influenza viruses. Insights into receptor binding, fusion mechanisms, and drug discovery have been gained and the study of these antigens from a range of influenza viruses has given insights into their structural evolution in relation to immune recognition.

See also: Orthomyxoviruses: Molecular Biology.

Further Reading

Lamb RA and Krug RM (2001) *Orthomyxoviridae*: The viruses and their replication. In: Knipe DM, Howley PM, Lamb RA, *et al.* (eds.) *Fields Virology*, 4th edn., p. 1487. Philadelphia: Lippincott Williams and Wilkins.

Palese P and Shaw ML (2006) *Orthomyxoviridae*: The viruses and their replication. In: Knipe DM, Howley PM, Griffin DE, *et al.* (eds.) *Fields Virology*, 5th edn., pp. 1647–1689. Philadelphia: Lippincott Williams and Wilkins.

Skehel JJ and Wiley DC (2000) Receptor binding and membrane fusion in virus entry: The influenza hemagglutinin. *Annual Review of Biochemistry* 69: 531.

Wright PF, Neumann G, and Kawaoka Y (2006) Orthomyxoviruses. In: Knipe DM, Howley PM, Griffin DE, *et al.* (eds.) *Fields Virology*, 5th edn., pp. 1691–1740. Philadelphia: Lippincott Williams and Wilkins.

Oryctes Rhinoceros Virus

J M Vlak, Wageningen University, Wageningen, The Netherlands
A M Huger, Institute for Biological Control, Darmstadt, Germany
J A Jehle, DLR Rheinpfalz, Neustadt, Germany
R G Kleespies, Institute for Biological Control, Darmstadt, Germany

© 2008 Elsevier Ltd. All rights reserved.

Glossary

GC content Guanosine + cytosine content.
Gonad Male reproductive organ in insects.
Nuclear hypertrophy Enlargement and structural alterations of cell nuclei.
Oryctes rhinoceros A beetle insect that is an economic pest of coconut and palm plantations.

History

Oryctes rhinoceros virus (OrV) was first identified in 1963 in Malaysia through a field study in Southeast Asia to develop a strategy to control the rhinoceros beetle, *Oryctes rhinoceros*. This scarab beetle is native in Southeast Asia, but was accidentally introduced into some South Pacific Islands and became a major pest in coconut and palm plantations greatly affecting local economies. Since its initial discovery, OrV is found in rhinoceros and related beetles throughout Southeast Asia and used as an effective biocontrol agent of beetles in many countries, especially in Southeast Asia and the Pacific. Due to its structural similarity to baculoviruses, its replication in the nucleus, and the facultative occurrence of polyhedra-shaped occlusion bodies, OrV was originally considered a nonoccluded baculovirus. OrV is known in the literature under various names, such as rhabdionvirus, Oryctes virus (OrV), Oryctes baculovirus, Oryctes nonoccluded baculovirus, Oryctes nudibaculovirus, and abbreviations, such as OrV, Or-V1, and OrNOB.

Pathology

Larvae and adult beetles are infected by the oral route taking up OrV released from either disintegrated deceased larvae or from defecated infected midgut cells of chronically infected beetles. The first signs of OrV infection of larvae are extreme lethargy and lack of appetite. The larvae become translucent and their fat bodies in state of disintegration become visible through the integument. The midgut epithelial cells are the primary site of infection of both larvae and adult beetles. In the latter, the infected midgut is strikingly swollen and whitish in appearance. After replication of the virus in the larval midgut epithelial cells (**Figure 1**), the virions then transfer by an unknown mechanism into the interior where they infect other tissues such as fat body to establish a systemic infection. In the end, the larvae become translucent, flaccid, and finally disintegrate, often under septicemic conditions after 1–4 weeks. In beetles, the infection remains initially restricted to the midgut, where the epithelial cells are sloughed off regularly upon the infection into the gut lumen. This stimulates the crowds of regenerative crypts sitting on the midgut to start an enduring vigorous cell proliferation, by which the midgut lumen is completely filled up with myriads of cells that are packed with virus in their hypertrophied nuclei (**Figure 2**). This process can take weeks, before the infection becomes systemic and the beetles die. Strong nuclear hypertrophy of midgut and fat body cells and a ring zone in the nucleus recognize infection microscopically, which is characteristic for virus-harboring regions (**Figure 2**).

Replication and Assembly

OrV replicates in a cell line derived from the beetle *Heteronychus arator* (DSIR-HA-1179) and this allows the dissection of the replication of OrV both morphologically and biochemically. Virions derived from midgut tissue of infected beetles are used for infection. OrV enters cells most likely by pinocytosis and enters the nucleus as nucleocapsid through nuclear pores. The first signs of infection are the rounding up of the fibroblast-like *H. arator* cells followed by chromatin rearrangement and nuclear hypertrophy and finally, 3–4 days after infection, cell lysis. As evidenced by electron microscopy, virus assembly starts with the accumulation of so-called nucleocapsid shells in chromatin-free areas of the nucleus and their enwrapping by virion envelopes. Later electron-dense material, presumably viral nucleoprotein core with DNA, condenses into these preformed enveloped 'shells'. Later in infection, mature rod-shaped virions accumulate in large arrays at the nuclear periphery (**Figure 1**). The virions then enter the cytoplasm through the nuclear pores and ultimately leave the cell by budding from the cell membrane. The assembly process is a salient feature of OrV infection which also occurs during white spot syndrome virus (WSSV) morphogenesis.

Virion Properties

OrV in solution has a rod- to ovoid-shaped appearance and its nucleocapsid shows a short tail-like appendage at one end (**Figure 3**). As such, OrV looks very similar to several other helicoverpa zea viruses, HzV-1, Hz-2V, and GSV (gonad-specific virus), and WSSV of shrimp, except that the appendage is associated with the envelope rather than in the latter case the nucleocapsid as is the case with OrV. The OrV virions are about 220 nm × 120 nm in size and contain a single rod-shaped nucleocapsid, which is wrapped in an envelope consisting of two unit membranes that are sensitive to nonionic detergents (**Figure 5**). The nucleocapsid contains a circular double-stranded DNA molecule in a supercoiled form. The flexuous proteinacious tail-like appendage measures about 270 nm in length and 10 nm in width. Its function is unclear.

Virions isolated from defecations of *Oryctes* adults showed at least 12 major structural proteins ranging in size from 10 to 76 kDa by sodium dodecyl sulfate polyacrylamide gel electrophoresis (SDS-PAGE). The smaller two, 10 and 13 kDa, are the most abundant and make up 50% of the total virion protein. At least 27 proteins have been found associated with OrV virions obtained from *H. arator* cell culture (cell line DSIR-HA-1179), ranging in size from 9.5 to 215 kDa; 14 of these are thought to be envelope proteins. The reason for this discrepancy in protein composition between OrV produced in beetles and in beetle cells is not known, but it could be the result of the production method employed. There is evidence to suggest that OrV and some baculoviruses share common antigens in immunological tests.

In OrV-infected *H. arator* cells only eight out of the 27 virion proteins were detected by pulse-labeling cells with radioactive amino acids ($[^{35}S]$-methionine). Two of these structural virion proteins, p4.6 and p10, were found as early as 4 h post infection (hpi), and the remaining six, p46.5, p40, p27, p25, p22, and p13, at 6 hpi. Since host protein synthesis is not shut off as a consequence of infection, the other OrV virion proteins may have been made and labeled if they contain methionine, but could then be submerged in the large amounts of labeled host-cell proteins. Immunoprecipitation of infected cell

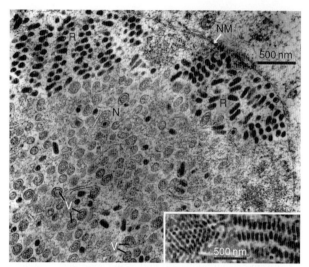

Figure 1 Electron micrograph of a thin section showing part of the nucleus (N) of a fat body cell of a third instar *Oryctes* beetle larva with heavy OrV infection. Note the accumulation of virus rods (R) at the nuclear periphery and the single- and double-membraned vesicles (V) in the nuclear centre. NM, nuclear membrane. Insert: Virus rods in cross (left) and longitudinal (right) section arranged in a pseudocrystalline pattern. Reprinted from *Journal of Invertebrate Pathology*, 89, Huger AM, The Oryctes virus: Its detection, identification, and implementation in biological control of the coconut palm rhinoceros beetle *Oryctes rhinoceros* (Coleoptera: Scarabaeidae), pp. 78–84, Copyright (2005), with permission from Elsevier.

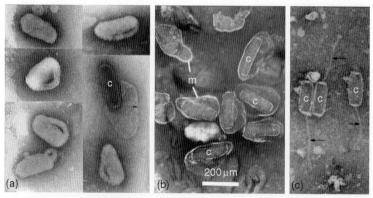

Figure 2 Electron micrographs with structural details of Oryctes rhinoceros virus rods, negatively stained with phosphotungstic acid. (A) Virions unpenetrated by stain, often being artificially mug-shaped; middle right: the virus membrane (arrow) is shedded from the nucleocapsid (c). (B) Virions with longer penetration by stain, thus displaying the nucleocapsids (c) and the surrounding viral membrane (m). (C) Three nucleocapsids (c) showing the typical thread-like appendix (arrows). Reprinted from *Journal of Invertebrate Pathology*, 89, Huger AM, The oryctes virus: Its detection, identification, and implementation in biological control of the coconut palm rhinoceros beetle *Oryctes rhinoceros* (Coleoptera: Scarabaeidae), pp. 78–84, Copyright (2005), with permission from Elsevier.

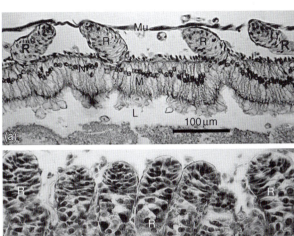

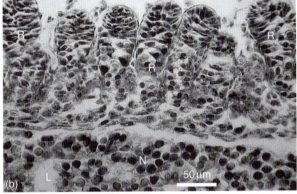

Figure 3 Light micrographs showing the midgut epithelium of rhinoceros beetle adults in longitudinal sections stained with Heidenhain's hematoxylin. (a) Midgut epithelium (M) of a healthy adult with a black-stained row of cell nuclei (N); the epithelium is outside occupied by regenerative crypts (R) and bounded by the muscularis (Mu); L, midgut lumen. (b) Midgut epithelium of an OrV-diseased adult with greatly enlarged regenerative crypts (R); the latter are massively proliferating cells from their apical region into the midgut lumen, where the hypertrophied cell nuclei (N) produce large amounts of virus. Reprinted from *Journal of Invertebrate Pathology*, 89, Huger AM, The Oryctes rhinoceros virus: Its detection, identification, and implementation in biological control of the coconut palm rhinoceros beetle *Oryctes rhinoceros* (Coleoptera: Scarabaeidae), pp. 78–84, Copyright (2005), with permission from Elsevier.

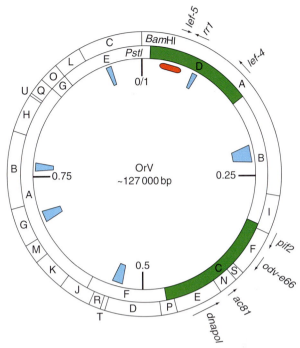

Figure 4 Genetic map of the OrV genome. The genome is drawn as a circular molecule and the *Bam*HI and *Pst*I restriction sites are indicated in the outer and inner circle, respectively. The site between *Bam*HI A and C is arbitrarily chosen as the zero point of the map. The approximate location of repeat regions is indicated (quadrangulars in blue). The genes conserved with baculoviruses and their direction of transcription are indicated (arrows) as well as the location of the primer sets for PCR identification (red bar).

proteins with OrV virion-specific antibodies is needed to reveal the remaining OrV virion proteins. OrV-specific nonstructural proteins would have remained undetected.

Genome Structure

The circular OrV genome is about 127 kbp in size with a GC content of 43–44%. Digestion of the viral DNA with restriction enzymes *Bam*HI, *Eco*RI, *Hin*dIII, and *Pst*I produced 21, 43, 23, and 7 DNA fragments, respectively, of different size. A physical map of the OrV genome has been constructed for these four restriction enzymes and six regions with reiterated sequences have been identified and mapped on the viral genome (**Figure 4**). Many of these restriction fragments have been cloned in bacterial plasmids and two of these (*Pst*I-C and *Pst*I-D) have been analyzed in detail.

Genetic variation between OrV field isolates exists as evidenced by DNA restriction fragment length polymorphism (RFLP). These variations are often small insertions and deletions in the reiterated repeat sequences. On the basis of genomic differences, at least three strains can be identified, which also differ in virulence. An OrV-specific DNA detection test based on a polymerase chain reaction (PCR) has been developed (**Figure 4**).

Portions of the OrV genome, more specifically OrV fragments *Pst*I-C (19 805 bp) and *Pst*I-D (17 146 bp), have been sequenced. Forty open reading frames of 50 or more amino acids were identified. Only 15 showed homology with other DNA viruses and these include DNA polymerase (*dnapol*), subunits of DNA-dependent RNA polymerase, ribonucleotide reductase (*rr1*), and thymidylate synthase, all enzymes involved in DNA replication and transcription. Ten open reading frames had homologs with HzV-1.

Phylogeny and Taxonomy

The OrV particle is rod-shaped and the virus replicates in the nucleus, hence it was originally considered to be a member of the family *Baculoviridae*. Since the virus particles are mostly not occluded into occlusion bodies in infected hosts, it was assigned as a baculovirus of the nonoccluded type (subgroup C baculoviruses: nonoccluded baculoviruses). Later OrV was taken out of the family *Baculoviridae* due to the lack of genetic information and orphaned. OrV remained unassigned. Suggestions to establish a new Oryctes virus family were made, but genetic evidence was not available. Recent sequence and phylogenetic analysis provides compelling evidence that OrV deserves species status and that it can be a member of a new genus (*Nudivirus*) to be established. OrV is distantly related to baculoviruses, but not to whispoviruses despite its morphological similarities.

Phylogeny reconstructions using DNA polymerase gene (dnapol) sequences of OrV and a number of other viral DNA polymerases indicated that OrV formed a monophyletic group with HzV-1 and this might suggest that they have a common ancestor (**Figure 5**). On this basis OrV may be accommodated together with HzV-1 in a new genus, for which the name *Nudivirus* has been proposed. On the basis of the limited amount of sequence information of OrV, already ten open reading frames are closely related to HzV-1 and more are expected upon completion of the OrV genome sequence analysis. Furthermore, OrV and HzV-1 share structural and replication properties warranting inclusion in the same genus. The major difference between OrV on the one hand and the *H. zea* viruses on the other is their genome size; the HzV-1 genome is about twice the size (228 kbp) and the virus has never been isolated from insect larvae. There is a distant relationship of OrV and HzV-1 with baculoviruses at the level of the 29 conserved baculovirus genes, such as *dnapol*,

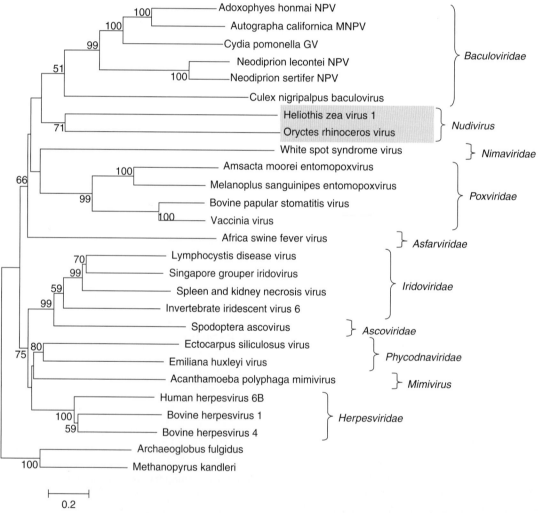

Figure 5 Phylogenetic affiliations of *Oryctes rhinoceros virus* (OrV) to other dsDNA viruses using DNA polymerase. Bootstrap values indicate the robustness of the minimal evolution tree (1000 replicates) and the values above 50% are given next to the nodes. Neighbor-joining and maximum parsimony analyses revealed a similar tree topology (not shown). Both OrV and HzV-1 are indicated on a gray background. The tree is rooted using the DNA polymerases of two archaeal species: *Archaeoglobus fulgidus* and *Methanopyrus kandleri*. The scale bar represents a distance of 20%. Adapted from Wang Y, van Oers MM, Crawford AM, Vlak JM, and Jehle JA (2006) Genomic analysis of Oryctes rhinoceros virus suggests genetic relatedness to Heliothis zea virus. *Archives of Virology* 152: 519–531.

lef4, and *lef5*. These viruses share common early (CAGT) and late (DTAAG) transcription promoter motifs. Therefore an inclusion of the proposed genus *Nudivirus* in the family *Baculoviridae* may be a future possibility. Genetically, OrV is not related to WSSV (*Nimaviridae*) despite its morphological similarities.

Transmission

OrV is taken up orally by larvae of *Oryctes* beetles at the breeding sites, which are usually located in rotting logs and stumps of palm trees and also in any decomposing organic matter. The infection in these larvae becomes systemic and at the ultimate stage they disintegrate. The virus is released at the breeding site, where it can be taken up by both new uninfected larvae or adult beetles. Horizontal disease transmission between beetles most frequently occurs by oral contact with fecal virus during mating or by co-occupation of the same habitat. In adult beetles the infection becomes chronic and develops differently; the virus initially only replicates in the midgut but not beyond. After infected midgut cells have been sloughed off into the gut lumen, large amounts of virus are produced in masses of cells proliferating from the regenerative crypts (**Figure 3**), and the virus is released through the feces. The infection of the beetles is usually without symptoms, although they stop feeding and egg laying. Such chronically infected beetles act as 'flying virus-spraying machines', flying around for weeks and defecating virus into their natural habitats and breeding sites, thus providing for an effective autodissemination and colonization of the virus disease over large areas. This horizontal transmission of OrV by autodissemination takes advantage of the ecological behavior of beetle populations and this is a key element in the success of OrV as a biocontrol agent.

Application as Biocontrol Agent

The first field trials to control the rhinoceros beetle were undertaken in 1967 in Samoa (then known as Western Samoa). Breeding sites were artificially contaminated with OrV and the disease became endemic over time. The virus spread autonomously in the beetle population and this autodissemination became the method of *Oryctes* control in coconut and oil palm until the present day. The adult beetles themselves are the natural vectors of OrV and responsible for the spread of the virus through fecal deliveries. The method of contamination of breeding sides was superseded by the release of trapped and subsequently infected beetles in affected sites. The virus became endemic and a permanent component of the *Oryctes* beetle ecology. *Oryctes* populations were effectively controlled in the Pacific Islands and other countries, and OrV provided long-term suppression. The use of OrV is a classical but single example of successful inoculative control of insects by virus. After the initial success of this strategy in the 1970s and 1980s, the interest in OrV faded away, in part due to the absence of severe beetle attacks. Recent outbreaks of the beetle in the Pacific as a consequence of careless policies for land clearance and replanting of palm trees have initiated a revival of *Oryctes* control by OrV.

See also: Baculoviruses: Molecular Biology of Granuloviruses; Baculoviruses: Molecular Biology of Nucleopolyhedroviruses; Insect Viruses: Nonoccluded; White Spot Syndrome Virus.

Further Reading

Bedford GO (1980) Biology, ecology and control of palm rhinoceros beetles. *Annual Review of Entomology* 25: 309–339.

Burand JP (1998) Nudiviruses. In: Miller LK and Ball LA (eds.) *The Insect Viruses*, pp. 69–90. New York: Plenum.

Crawford AM, Ashbridge K, Sheehan C, and Faulkner P (1985) A physical map of the Oryctes baculovirus genome. *Journal of General Virology* 66: 2649–2658.

Crawford AM and Sheehan C (1985) Replication of Oryctes baculovirus in cell culture: Viral morphogenesis, infectivity and protein synthesis. *Journal of General Virology* 66: 529–539.

Crawford AM, Zelazny B, and Alfiler AR (1986) Genotypic variation in geographical isolates of Oryctes baculovirus. *Journal of General Virology* 67: 949–952.

Hajek AE, Delalibera I, and McManus ML (2000) Introduction of exotic pathogens and documentation of their establishment and impact. In: Lacey LA and Kaya HK (eds.) *Field Manual of Techniques in Invertebrate Pathology*, pp. 339–369. Dordrecht, The Netherlands: Kluwer.

Huger AM (2005) The Oryctes virus: Its detection, identification, and implementation in biological control of the coconut palm rhinoceros beetle *Oryctes rhinoceros* (Coleoptera: Scarabaeidae). *Journal of Invertebrate Pathology* 89: 78–84.

Huger AM and Krieg A (1991) *Baculoviridae*: Nonoccluded baculoviruses. In: Adams JR and Bonami JR (eds.) *Atlas of Invertebrate Viruses*, pp. 287–319. Boca Raton, FL: CRC Press.

Jackson TA, Crawford AM, and Glare TR (2005) Oryctes virus – time for a new look at a useful biocontrol agent. *Journal of Invertebrate Pathology* 89: 91–94.

Mayo MA (1995) Unassigned viruses. In: Murphy FA, Fauquet CM, Bishop DHL, et al. (eds.) *Virus Taxonomy: The Sixth Report of the International Committee on Taxonomy of Viruses*, pp. 504–507. Vienna: Springer.

Mayo MA, Christian PD, Hillman BI, Brunt AA, and Desselberger U (2005) The unassigned viruses. In: Fauquet CM, Mayo MA, Maniloff J, Desselberger U, and Ball LA (eds.) *Virus Taxonomy: Eighth Report of the International Committee on Taxonomy of Viruses*, pp. 1129–1144. San Diego, CA: Elsevier Academic Press.

Mohan KS and Gopinathan KP (1991) Physical mapping of the genomic DNA of Oryctes rhinoceros baculovirus, Kl. *Gene* 107: 343–344.

Ramle M, Wahid MB, Norman K, Glare TR, and Jackson TA (2005) The incidence and use of Oryctes virus for control of rhinoceros beetle in oil palm plantations in Malaysia. *Journal of Invertebrate Pathology* 89: 85–90.

Richards NK, Glare TR, Aloali'i I, and Jackson TA (1999) Primers for the detection of Oryctes virus from Scarabaeidae (Coleoptera). *Molecular Ecology* 8: 1552–1553.

Wang Y, van Oers MM, Crawford AM, Vlak JM, and Jehle JA (2006) Genomic analysis of Oryctes rhinoceros virus suggests genetic relatedness to Heliothis zea virus. *Archives of Virology* 152: 519–531.

Ourmiavirus

G P Accotto and **R G Milne,** Istituto di Virologia Vegetale CNR, Torino, Italy

© 2008 Elsevier Ltd. All rights reserved.

Introduction

This is a fascinating group of viruses with a unique virion structure but very little is known about them. Cassava virus C (CsVC) has been reported only once, in cassava from Malawi, in a mixed infection with African cassava mosaic virus. The damage it may cause is as unresearched as its molecular biology. Epirus cherry virus (EpCV) was found in a single cherry tree (since dead) showing rasp-leaf symptoms, with leaf distortion and development of enations, but this may not have been a single infection, and the virus was not isolated and returned to cherry. In a collaboration between Iranian researchers and the Turin plant virus laboratory, Ourmia melon virus (OuMV) has been detected on several occasions in melons from northwestern Iran, but generally in mixed infections with other cucurbit viruses, so the extent of economic losses ascribable to this virus is unknown.

Potentially the viruses are damaging since they have been experimentally transmitted to a rather wide range of dicot species including important crop plants, producing symptoms of mosaic, mottle, and necrosis. Data on the genome are too limited to assign the genus to any family, but if one had to guess, that family might be the *Bromoviridae*. Unless otherwise indicated, the data given below refer to OuMV.

Taxonomy

OuMV, CsVC, and EpCV have been placed within the genus *Ourmiavirus* firstly because all the virions have the same morphology, which is distinct and unique to the group. In addition, the cytopathology of cells experimentally infected with OuMV and CsVC is similar but distinct from that of other viruses (the cytopathology of EpCV has not been reported). The experimental host range of the three viruses, tested by mechanical inoculation, is similar, and the symptoms are generally systemic ringspots, mottle, mosaic, and necrosis. The coat proteins of OuMV and EpCV are of similar but not identical size and are distantly serologically related. The three positive-strand single-stranded RNAs (ssRNAs) comprising the genomes of OuMV and EpCV are of similar size (there is no information for CsVC).

The rather poor data available have suggested placement of the three viruses in separate species, for the following reasons. The viruses have nonoverlapping natural host ranges (cucurbits, cherry, cassava). OuMV and EpCV coat proteins are only distantly related serologically and are of distinct sizes. CsVC is reportedly not serologically related to OuMV. The three RNAs of EpCV are slightly but consistently smaller than those of OuMV, and the respective RNA-2's and RNA-3's show no evident sequence homologies (no data for CsVC).

Similarity with Other Taxa

The architecture of the virions differs clearly from that of other bacilliform or quasi-bacilliform particles of similar size such as alfalfa mosaic virus, olive latent virus 2, or ilarviruses, but this morphology, together with the presence of an ssRNA positive-strand genome in three segments, may indicate affinities with members of the *Bromoviridae*.

Properties of Particles

The particles (**Figure 1**) are unenveloped. They have cylindrical bodies 18 nm diameter capped by sharply conical (probably hemi-icosahedral) ends and are found in discrete lengths. The bodies of the particles are composed of a series of double disks, the commonest particles having two disks (particle length 30 nm), a second common particle having three disks (particle length 37 nm), with rarer particles having four disks (particle length 45.5 nm) and six disks (particle length 62 nm). A particle with five disks (hypothetical length 54 nm) has not been detected. The buoyant density in CsCl of particles of all sizes is $1.375 \, \text{g cm}^{-3}$.

The particles are stable near pH 7 (infective in suspension for at least 35 days at 20–25 °C), and thermally they are relatively stable; infectivity survives in crude sap after heating for 10 min at 70 °C but not 80 °C. The particles survive treatment with chloroform but not *n*-butanol and survive treatment with 1% Triton X100 detergent.

Particle preparations contain ssRNA of three sizes, estimated as 0.9, 0.35, and 0.32×10^6 (M_r). There is one coat protein (estimated size 25.2 kDa).

Genome Organization

The ssRNA is positive stranded. RNA-1 presumably encodes the polymerase. The coat protein is encoded by either RNA-2 or RNA-3.

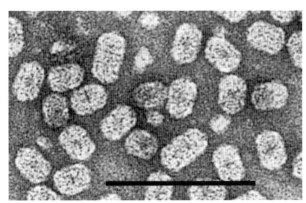

Figure 1 Purified preparation of OuMV particles negatively stained in uranyl acetate. Scale = 100 nm.

Viral Proteins

There is one coat protein. One nonstructural protein (possibly with a transport function) forms tubules in the cytoplasm of infected cells; virions are frequently found within the tubules.

Pathogenicity

The only natural host of OuMV so far detected is melon (*Cucumis melo*), which develops a mosaic-ringspot disease. The virus can easily be mechanically inoculated to many different hosts (34 species in 14 families). Local lesions appear in 3–4 days on *Chenopodium quinoa* and *Gomphrena globosa*, and these are followed, on *Nicotiana clevelandii*, *Nicotiana benthamiana*, and *Nicotiana megalosiphon*, by systemic necrotic mosaic. The virus particles accumulate to a relatively high titer. Among experimentally inoculated plants of commercial value tested, systemic disease developed in *Capsicum annuum* (two of three cultivars tested), tomato (one of two cultivars tested), *Cucumis melo* (melon; three out of three cultivars tested), *Cucumis sativus* (cucumber; four out of four cultivars tested), *Ocimum basilicum* (sweet basil), *Petunia hybrida*, *Spinacia oleracea* (spinach), *Viola tricolor* (pansy), and *Zinnia elegans*.

Geographic Distribution

OuMV have been found repeatedly in northwestern Iran; CsVC was reported from Malawi; EpCV was reported from Greece.

Cytopathology

All tissues are invaded. Membranes proliferate in the cytoplasm, with damage to chloroplasts. Many long tubules composed of protein, about 25 nm in diameter and packed with virions, develop in infected cytoplasm, giving a highly characteristic appearance to ultrathin sections viewed in the electron microscope.

Transmission, Prevention, and Control

No vector has yet been identified, and the method of field spread is unknown, although experimental seed-transmission rates of 1–2% in *N. benthamiana* and *N. megalosiphon* have been reported. Several aphid species, the whitefly *Trialeurodes vaporariorum* and the mite *Tetranychus urticae*, failed to transmit the virus experimentally.

See also: Bromoviruses.

Further Reading

Accotto GP, Riccioni L, Barba M, and Boccardo G (1997) Comparison of some molecular properties of Ourmia melon and Epirus cherry viruses, two representatives of a proposed new virus group. *Journal of Plant Pathology* 78: 87–91.

Aiton MM, Lennon AM, Roberts IM, and Harrison BD (1988) Two new cassava viruses from Africa. Abstracts, 5th International Congress of Plant Pathology, Kyoto, Japan, 20–27 August 1988, p. 43.

Avgelis A, Barba M, and Rumbos I (1989) Epirus cherry virus, an unusual virus isolated from cherry with rasp-leaf symptoms in Greece. *Journal of Phytopathology* 126: 51–58.

Lisa V, Milne RG, Accotto GP, Boccardo G, Caciagli P, and Parvizy R (1988) Ourmia melon virus, a virus from Iran with novel properties. *Annals of Applied Biology* 112: 291–302.

Milne RG (2005) Genus *Ourmiavirus*. In: Fauquet CM, Mayo MA, Maniloff J, Desselberger U, and Ball LA (eds.) *Virus Taxonomy: Eighth Report of the International Committee on Taxonomy of Viruses*, pp. 1059–1061. San Diego, CA: Elsevier Academic Press.

Set
978-0-12-373935-3

Volume 3 of 5
978-0-12-373938-4